# Handbook of Die Design

## Other McGraw-Hill Handbooks of Interest

AVALLONE & BAUMEISTER • *Marks' Standard Handbook for Mechanical Engineers*

BOTHE • *Measuring Process Capability*

BRATER • *Handbook of Hydraulics*

BRINK • *Handbook of Fluid Sealing*

CORBITT • *Standard Handbook of Environmental Engineering*

CZERNIK • *Gasket Handbook*

GIECK • *Engineering Formulas*

HARRIS • *Shock and Vibration Handbook*

HARRISON • *Environmental, Health, and Safety Auditing Handbook*

HARRISON • *Supplement to the Environmental, Health, and Safety Auditing Handbook*

HICKS • *Standard Handbook of Engineering Calculations*

JURAN & GRYNA • *Juran's Quality Control Handbook*

LINGAIAH • *Machine Design Data Handbook*

NAYAR • *The Metal Databook*

PARMLEY • *Standard Handbook of Fastening and Joining*

ROTHBART • *Mechanical Design Handbook*

SHIGLEY & MISCHKE • *Standard Handbook of Machine Design*

WALSH • *Electromechanical Design Handbook*

WALSH • *McGraw-Hill Machining and Metalworking Handbook*

WATLINGTON • *Contract Engineering, Start and Build a New Career*

YOUNG • *Roark's Formulas for Stress and Strain*

# Handbook of Die Design

Ivana Suchy

McGraw-Hill

New York San Francisco Washington, D.C. Auckland Bogotá
Caracas Lisbon London Madrid Mexico City Milan
Montreal New Delhi San Juan Singapore
Sydney Tokyo Toronto

**Library of Congress Cataloging-in-Publication Data**

Suchy, Ivana.
Handbook of die design / Ivana Suchy.
p. cm.
Includes index.
ISBN 0-07-066671-7
1. Dies (Metal-working—Design—Handbooks, manuals, etc.
I. Title.
TS253.S79 1997
621.9′84—dc21 97-54307
CIP

*McGraw-Hill*

*A Division of The* ***McGraw-Hill*** *Companies*



1 2 3 4 5 6 7 8 9 0 DOC/DOC 9 0 2 1 0 9 8 7

ISBN 0-07-066671-7

*The sponsoring editors for this book were Robert Esposito and Harold B. Crawford, the editing supervisor was Peggy Lamb, and the production supervisor was Pamela A. Pelton. It was set in Century Schoolbook by Donald A. Feldman of McGraw-Hill's Professional Book Group composition unit.*

*Printed and bound by R. R. Donnelley & Sons Company.*

*This book is printed on recycled acid-free paper containing a minimum of 50% recycled de-inked fiber.*

*This book is dedicated to all enthusiasts in this technical field*

# Contents

## Chapter 14. Materials and Surface Finish 14-1

# Preface

The value of knowledge is enormous. It is an asset surpassing in importance many other personal qualities and traits. It's a dynamic and supportive capacity which enables us to attain our goals, avoid difficulties, be creative, be assertive, be wise. If used properly, knowledge is the main tool of creation in all areas of undertaking, be it manufacturing, business application, agriculture, or any other venture. Our personal lives are immensely affected by the amount of knowledge we have acquired and the same way are influenced our communities, companies, and whole nations.

But the cost of not knowing is still greater. It may run into hundreds, thousands, even ten-thousands of dollars, marks, pounds, or whatever currency, and sometimes even higher. It is an expense none of us can afford. The phantom of ignorance may silently creep about our companies, unrecognized, unchecked, until it bankrupts the very institution. Many firms may go belly up without ever knowing the real culprit behind their problems. And many firms do go belly up without ever knowing such culprits.

The lack of knowledge is a dangerous asset. It corrupts ideas, it cripples work methods, wrecks inventiveness, and institutes a job security syndrome in many. In the manufacturing domain, an alibi is usually produced in the form of unreasonably tightened dimensions, excessive production meetings, purchase of expensive equipment and gadgets, or an outright search for scapegoats. And this extravaganza is followed by a stream of complaints, complaints, and complaints from everywhere.

Knowledge is the best available weapon to curb such excesses. It doesn't matter if it is imposed from the top toward the bottom or from any other direction, as long as it brings about the necessary comprehension and enlightenment. The method of administration may differ, since some prefer learning from their mentors while others learn from their mistakes.

Naturally, learning from books is much easier, more merciful, and rewarding than learning from mistakes. In this particular book a student of die design and die manufacturing may find all the material needed to master the subject without ever resorting to guessing or taking chances. He or she may be a mechanical engineer or a shop worker, an estimator, a toolmaker, an owner of the company or a machinist, a production supervisor or a buyer—they all may find an abundance of material pertaining to their field of expertise, most of it not published for more than the past twenty years.

The die cost estimating section can help the engineer or estimator to determine not only the cost of designing and building a die, but the cost of running it in production.

The worldwide chart of major steel and alloy materials may assist every buyer and purchaser in any manufacturing field.

Keywords throughout the text may help anyone to acquire not only an air of a person thoroughly familiar with the subject, but the actual knowledge behind them as well.

A descriptive and detailed treatment of die design is layered for different levels of application. The basic portrayal presented at the beginning is exploded into a detailed and elaborate treatment of the subject. Sometimes more than one approach to a given problem is presented, to allow for a cross-comparison so often needed in this field.

At the beginning, the student/reader's perception is trained to distinguish between different metal altering processes and ensuing differences in their die construction. An attempt at the maturation of the "feel" for the sheet metal behavior during fabricating processes follows. It is aimed at foreseeing the pitfalls of research and development stage and choosing the proper manufacturing technique for a given task.

Finally, on learning how to come up with the suitable design approach, when providing the reader with all designing tools in the form of formulas, charts, graphs, and illustrations, the book is topped off by a section on cost estimating.

I hope all readers of this book will benefit from the material it contains. It was collected in the form of notes, test reports, examples of manufacturing methods, sketches, and actual work samples over the span of many years. Additionally, I would like to express my gratitude to many die and sheet metal design and manufacturing experts for allowing me to benefit from their experiences be it in the form of discussions or by reading their published materials.

*Ivana Suchy*

Chapter

# 1

# Basic Die Design and Die-Work Influencing Factors

## 1-1. Sheet-Metal Stamping in Comparison with Other Metal Fabricating Processes

In today's practical and cost-conscious world, sheet-metal parts have already replaced many expensive cast, forged, and machined products. The reason is obviously the relative cheapness of stamped, or otherwise mass-produced, parts, as well as greater control of their technical and esthetic parameters. That the world slowly turned away from heavy, ornate, and complicated shapes and replaced them with functional, simple, and logical forms only enhanced this tendency. Remember bathtubs? They used to be cast and had ornamental legs. Today they're mostly made of coated sheet metal. Manufacturing methods for picture frames, chandeliers, door and wall hardware, kitchen sinks, pots and pans, window frames, and doors were gradually replaced by more practical and less costly techniques.

But, sheet-metal stampings can also be used to imitate handmade ornamental designs of previous centuries. Such three-dimensional decorations can be stamped in a fraction of the time the repoussé artist of yesterday needed.

Metal extrusions, stampings, and forgings, frequently quite complex and elaborate, are used to replace handmade architectural elements. Metal tubing, metal spun products, formings, and drawn parts are often but cheaper substitutes of other more expensive merchandise.

Metal stampings, probably the most versatile products of modern technology, are used to replace parts previously welded together from

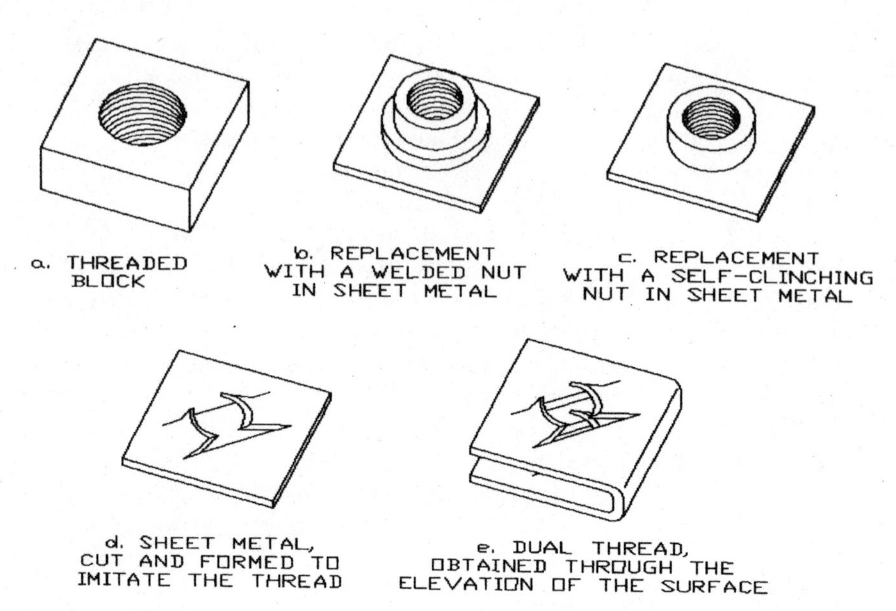


**Figure 1-1** Threaded part, replaced by other, less expensive means.

several components. A well-designed sheet-metal stamping can sometimes eliminate the need for riveting or other fastening processes (Fig. 1-1). Stampings can be used to improve existing designs that often are costly and labor-intensive. Even products already improved upon, with their production expenses cut to the bone, can often be further improved, further innovated, further decreased in cost.

The **metal stamping die** (Fig. 1-2) is an ideal tool that can produce large quantities of parts that are consistent in appearance, quality, and dimensional accuracy. It is a press tool capable of cutting the

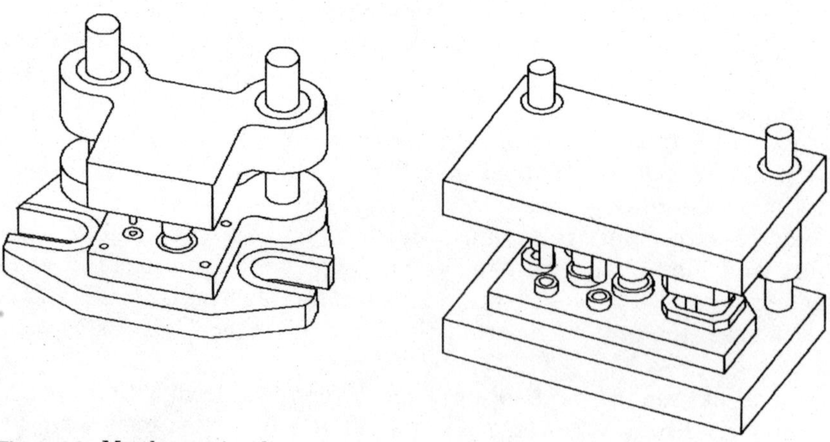

**Figure 1-2** Metal stamping dies.

metal, bending it, drawing its shape into considerable depths, embossing, coining, finishing the edges, curling, and otherwise altering the shape and outline of the metal part to suit the wildest imaginable design concepts. Figure 1-3 shows samples of these products.

The word "die" in itself means the complete press tool in its entirety, with all the punches, die buttons, ejectors, strippers, pads, and blocks, simply with all its components assembled together.

When commenting on these little ingenuities, it is important to stress the role of designers of such products, both artistic and technical. Their thorough knowledge of the manufacturing field will definitely enhance not only the appearance but the functionality, overall manufacturability, and cost of these parts.

Metal stamping die production output can be enormous, with huge quantities of high-quality merchandise pouring forth from the machine (Fig. 1-4). For that reason technical ignorance is not readily excusable, as the equal quantities of rejects can be generated just the same way.

### 1-1-1. Grain of material

Often, parts produced by various manufacturing methods can be redesigned to suit the sheet-metal mass production (Fig. 1-5).

When designing such replacements, there are several aspects to be evaluated. The first and probably the most important is the grain of the material (Fig. 1-6).

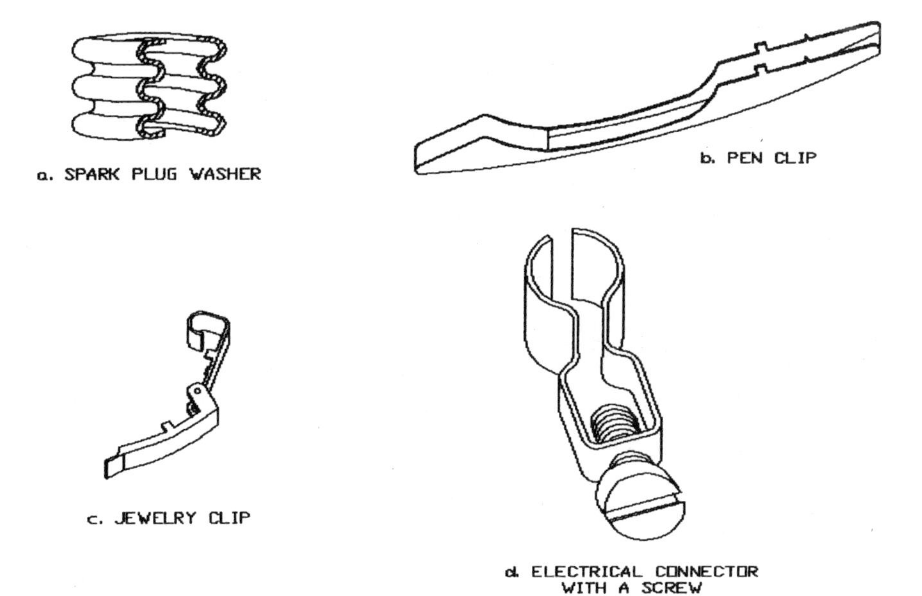


**Figure 1-3** Various sheet-metal products.

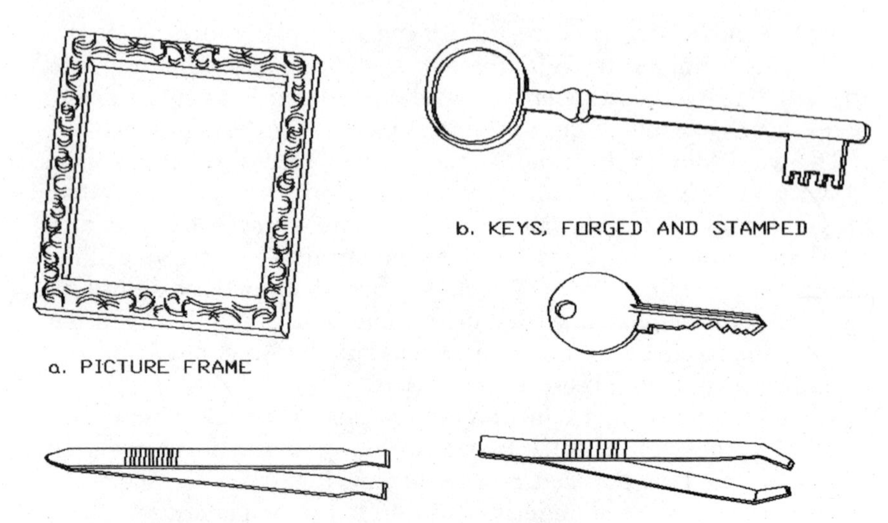


**Figure 1-4** Metal-stamped replacements.

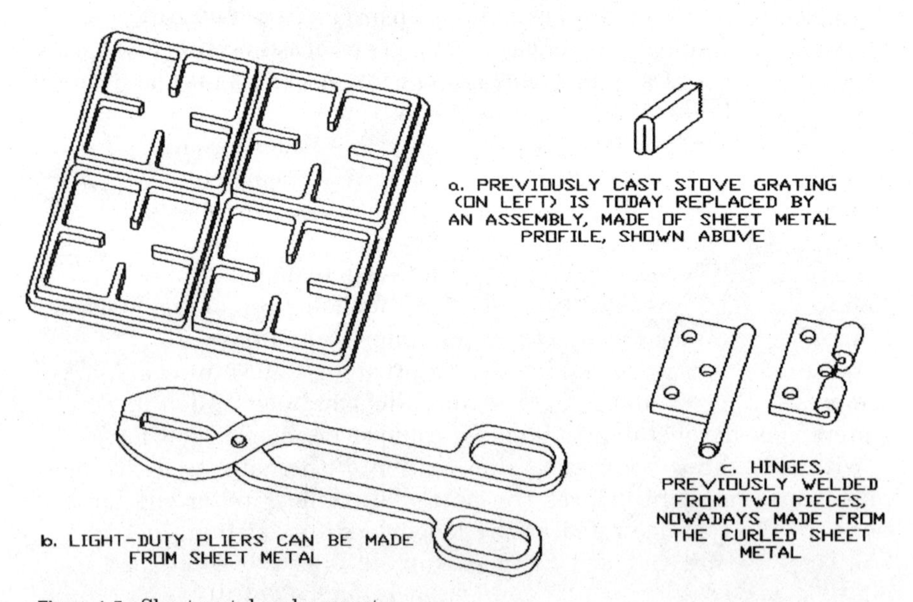


**Figure 1-5** Sheet-metal replacements.

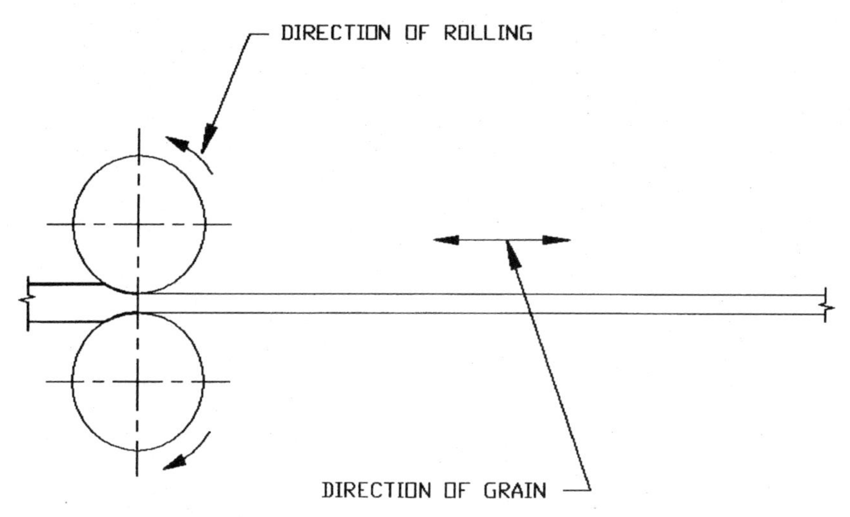


**Figure 1-6** Grain of material in sheet-metal strip.

Sheet metal of every form, be it a strip or a sheet, displays a definite grain line. It is the direction along which the material was produced in the mill-rolling process. In strips, the grain line usually runs lengthwise along its longest edge. The grain direction in sheets may vary, and designers must always make themselves familiar with it prior to planning a production run of any kind.

In contrast, cast or forged parts display a different grain direction, and in sintered powder metal parts the grain is completely gone. For this reason, each of these manufacturing methods can be used to produce items for different applications.

For example, a part, shown in Fig. 1-7, will display a different reaction to various forces and stresses when made by the forging method than when obtained through other manufacturing processes.

Where the forging would possess a great resistance to tensile and compressive forces along the *A-A* line, the same part, when made of sintered powder metal, may break or collapse under the same force.

With this shape being cast, the location of the gate is of extreme importance, as it influences the part's sturdiness in various directions. In the casting, gated at the longer end (as pictured in Fig. 1-7*b*), the opposite end will be more susceptible to breakage, as the molten metal will reach that portion later, when already cooling down. The existence of an opening in that area will divide the flow of material and thus create the so-called knit line, along which a separation, resulting in defects and possible breakage, may occur.

The same casting, when gated in the middle (Fig. 1-7*c*), will have an equal breakage proneness at both ends. However, these ends will

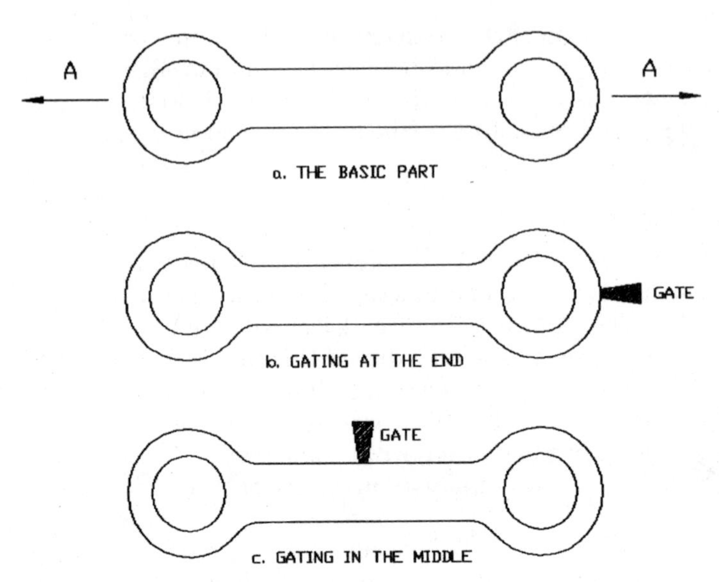


**Figure 1-7** Forces applied to a casting.

be somewhat sturdier, as the molten metal will reach them sooner than in the case of Fig. 1-7*b*. Of course, the existence of openings may have the same detrimental effect described earlier.

A similar product, made of sheet metal, as pictured in Fig. 1-8, will also display a grain-dependent behavior; the part with the lengthwise grain will be considerably sturdier along the *A-A* line of force than the same shape positioned across the grain line.

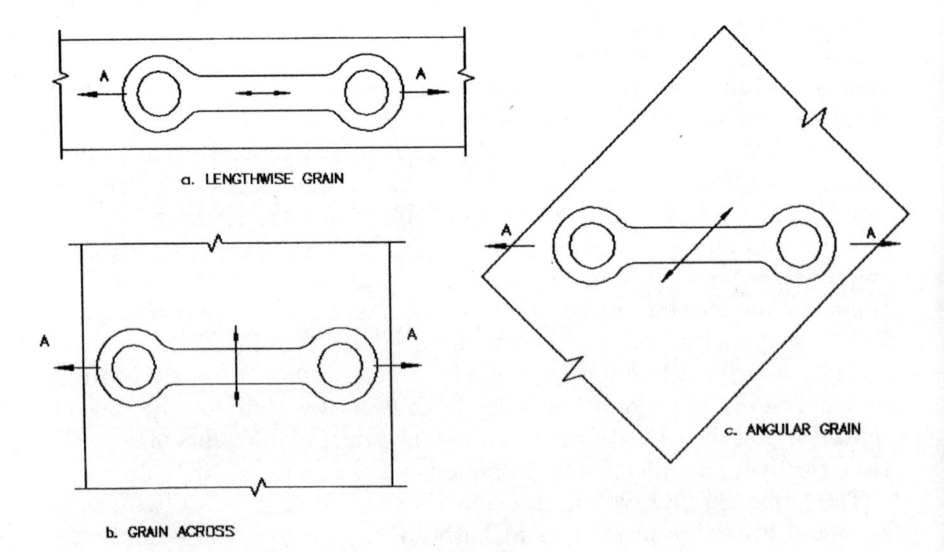


**Figure 1-8** Grain variation in sheet-metal stamping.

Sometimes the economy of the strip or sheet, the shape of the product, the location of its bends, and other criteria do not allow for the proper lengthwise positioning of the grain. In such instances, an angularly located grain may often be the answer (see Fig. 1-8*c*).

### 1-1-2. Edge formation

Another important aspect to be considered when designing sheet-metal replacements for parts manufactured by other methods is the formation of the edge. A cast part (Fig. 1-9*a*) will always exhibit a parting line to some degree. The visibility of this line is dependent on tool quality; with well-manufactured and well-maintained tooling, the line can be almost invisible, but with worn-out dies, that area may bulge out and perhaps even show a burr at some places. The existence of draft angle in cast parts is another necessity the designer has to take into account.

The same part when forged will have edge characteristics similar to those of its cast counterpart. But sheet-metal products' edges will be completely different; these will be straight if first punched, drawn, and then blanked (Fig. 1-10).

Usually there is a slight burr on stamped parts, and its location is always at the end of the cutting stroke, as shown in Fig. 1-11. The height and amount of this distortion are tool-dependent and with sharper tools and tighter tolerances can be minimized.

The word "die" describes the part, which receives the punch and the slug. Sometimes the term "die button" may be used as well.

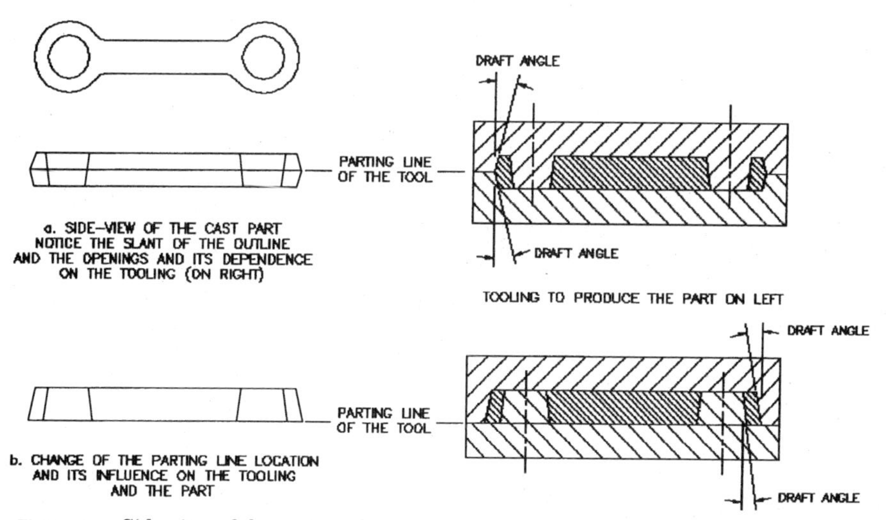


**Figure 1-9** Side view of the cast product.

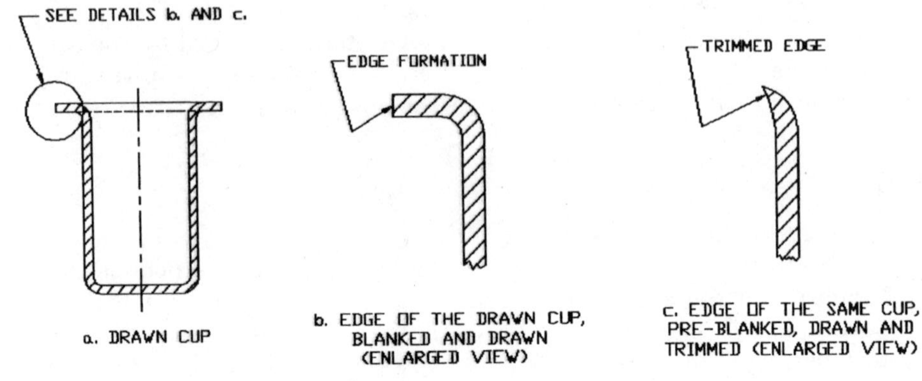


**Figure 1-10** Edge formation in drawn parts.

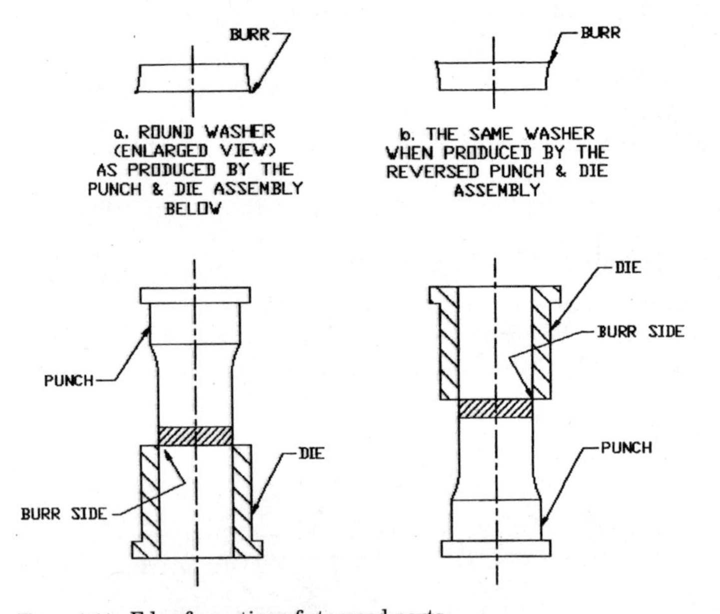


**Figure 1-11** Edge formation of stamped parts.

The burr on metal-stamped products is a great aid in evaluation of the sequence of the manufacturing process, as it clearly indicates the direction of punching (or blanking) of each opening.

In drawn parts, the formation of the cross section of the drawn portion further influences the product's characteristics. There is often some thinning of the wall due to the drawing process, and the deeper the draw, the thinner the wall becomes (Fig. 1-12).

The reason for this is obvious: The material needed for the expanded length of the drawn portion has to be taken from somewhere, and practically (and mathematically) the volumnar content of that section

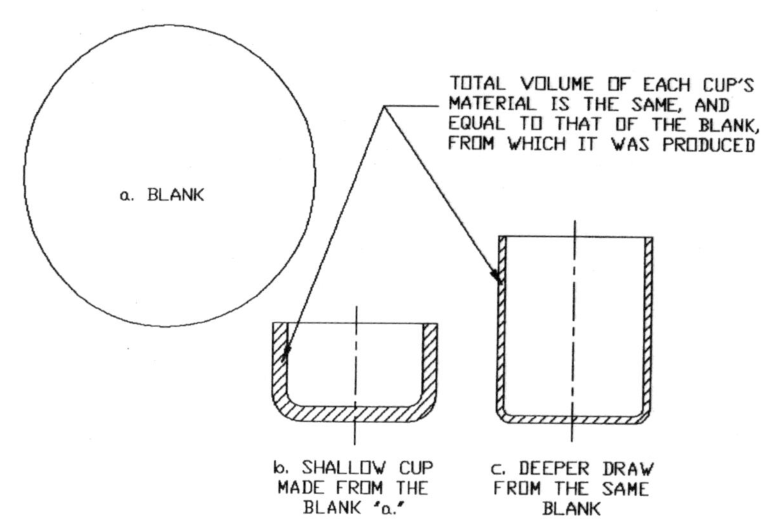


**Figure 1-12** Volumnar changes in drawing operation.

must be equal to that portion of the flat piece from which it is produced.

## 1-2. What Constitutes Suitability for Die Production?

When evaluating a part for die production, the most restrictive aspect to be considered is the cost of the tooling. To build a metal stamping die is a costly process, involving many people, many machines, and several technologies. For that reason, the demand for tooling must first be economically justified.

The quantitative demands per given time span should be evaluated first, because a scenario of 50,000 washers to be delivered each month requires a different treatment from 50,000 washers to be delivered each week.

The actual time needed to punch out these washers must be established on the basis of:

- Availability of the appropriate press
- The equipment's running speed
- The length of production shifts
- Scheduling for the needed time interval

For a small run with few repetitions, a single line of tooling may be chosen. However, if the quantities are large and the time constraint exists, a multiple-part-producing tool must be built. Such a die, generating at least two or more complete parts with each hit, will speed up production admirably. But increasing its size necessitates the use of a larger and more powerful press and may even require a nonstandard width of a strip which may be presently out of stock.

With parts other than simple washers, the shut height of the press versus the height of the part (and subsequently the height of the die) is another production-influencing factor. The width of the opening in the press plus the width of the proposed die must be in congruence.

The possibility of reorders should be considered at this point, as they may result in an extended production run, greater material demands, and longer occupancy of the press. Such longer runs are usually beneficial from the economical standpoint, as they save on die-mounting procedures and press adjustments, while creating a continuity of inspection processes.

On the other hand, a problem of storage of these extra parts may arise, along with the existence of (temporarily) unrewarded financial investments into the purchase of material, work-force demands, and overhead.

To properly evaluate the situation, all applicable expenditures should be added up as follows:

1. Cost of the storage space (prorated rent or property taxes, cost of the building and improvements, cost of extra storage containers)
2. Cost of all material and other production-related necessities
3. Overhead, such as electricity, cost of heating, cooling, water, and fuel
4. Spoilage of possible storage-sensitive material and the scrap rate
5. Cost of labor, including possible overtime
6. Proportional cost of paperwork handling and storage
7. Interest rate which the monies allocated to the above activities could have generated when invested otherwise

The combined expenses 1 through 7, when added up, should be equal to or less than the combined:

1. Cost of the removal of a die from the press
2. Cost of the installation of a die in the press (for the subsequent run)
3. Cost of the machine's downtime during the die removal and installation
4. Cost of the press operator's standby, if applicable
5. Cost of the press adjustments and trial runs

6. Cost of the first piece inspection and the cost of further adjustments and approvals, if applicable
7. Cost of the material and supplies, which must be purchased ahead of time even if not immediately utilized
8. Overhead, such as cost of electricity, heating, cooling, water, and fuel
9. Cost of subsequent billing and paperwork
10. Combined interest (per going rate) the finances allocated to the above causes would have generated when invested otherwise

The length of each run and its influence on the need for sharpening and maintenance of tooling must be evaluated for the entire production run. Should a maintenance-related interruption be necessary, a possible split of the previously planned combined run should be considered.

*A definite advantage of the die production* is its unrivaled consistency in the product's quality and dimensional stability. In absence of design and construction mistakes, the die, once built, needs minimal alteration aside from regular sharpening.

Some dies, true, are more sensitive than others, which is mostly attributable to excessive demands on close tolerances of parts and the material's thickness. With some bending and drawing operations, the consistency in hardness of stock can be essential as well. But a regular die, well designed and well built, can deliver a great load of products before its punches begin to wear and a need for sharpening arises.

Generally, it may be claimed that if the conditions of the running process are kept the same, the parts from the die will emerge consistent with previous loads.

## 1-3. Design Criteria for Die-Manufacturable Products

Today's world places greater and greater demands on products and materials from which they are made. Years ago, nobody figured out stress and strain, elasticity, fatigue, or similar values. If it broke, then you just made it 2 inches thicker (or 3 inches, or 5, whatever you preferred).

But that's not how current manufacturing is governed. Resources are getting scarcer, perhaps even limited in some cases, and designers are forced to economize. After all, why should a car body be thick and heavy, when a thinner-gauge galvanized steel will bring about the same, if not better, results.

Demands for special alloys are continuously expanding, and they are in equal competition with all the new and increasingly better alloys that are being produced. Ferrous and nonferrous alloys, titanium and its alloys, alloys with traces of rare metals added for additional qualities, all are available to fill that specific gap where they're needed.

Manufacturing methods are next on the list of economizing designers. Avoiding secondary operations whenever possible, designers apply cost-conscious strategies and planning not only in small shops but in medium and large plants as well.

This certainly is a good approach to any given problem, since every product has its price. If manufacturing costs become greater than the value of a product's services, such an item becomes unsalable.

For these reasons, manufacturability of products is extremely important. Almost anything can be manufactured somehow, if people put their minds to it. But at what cost? And who will be willing to pay for it?

Out of this ever-present regard for price versus actual value, new methods are being devised daily, new approaches to old problems sought for. Crowds of engineers, designers, toolmakers, model makers, and representatives of other professions are nitpicking new, almost new, or old problems, in an attempt to come up with a simple, straightforward, and cost-effective answer.

Sometimes, however, shortcuts are taken, where cheaper materials, thinner coatings, less durable tools, or less experienced labor are used. These steps are just what they present themselves as: shortcuts. They usually produce more returns, more repairs, more problems around their drawbacks, and even more expenses. There is a time and a place for everything, but these remedies are not always helpful. You pay for them later.

A good, sound design and overall manufacturability cannot be replaced by trinkets. The old saying "if it isn't good, fix it" should perhaps be replaced by "if it isn't good, redesign it!"

### 1-3-1. Manufacturability aspects

The manufacturability of products depends on many factors. Sometimes a lack of space may prevent a mechanic from reaching the area of concern, and long hours may be lost before this obstacle is overcome. Or a wrong sequence of operations will cause the final product to become distorted. Sometimes an adhesive may not hold because the part was not degreased enough, or a screw may fall out because someone forgot to add that second nut or loctite.

In die work, the manufacturability of parts is dependent on a much narrower range of influences. The main areas of concern are

1. Grain direction of the material
2. Openings, their shape and location
3. Bends and other three-dimensional alterations, their shape and location
4. Outline of the part and its size

**1-3-1-1. Grain direction of the material.** The ever-present grain of material must be taken into consideration first. Unless absolutely necessary, it should not appear alongside a bend, a joggle, or any other deflection and elevation in the part's surface.

Every sheet-metal material behaves differently alongside the grain line and across it. Forming, drawing, even simple punching, may sometimes show differences in the size and shape of the hole when evaluated for the grain influence. An extruded opening, shown in Fig. 1-13, illustrates this claim: By cutting across the grain line, it behaves almost as if constantly in tension, which—when forcibly removed (by the cutting process)—causes the material to back off.

If a bracket such as the one shown in Fig. 1-14*a* will be rotated 90° and positioned on the strip with its bends along the grain line, these flanges may sometimes crack in forming or even in service afterward. For that reason, wherever the problem of multiple bends occurs and there is no chance of avoiding their placement alongside the grain line, an angular positioning on the strip or sheet, shown in Fig. 1-14*b*, should be considered.

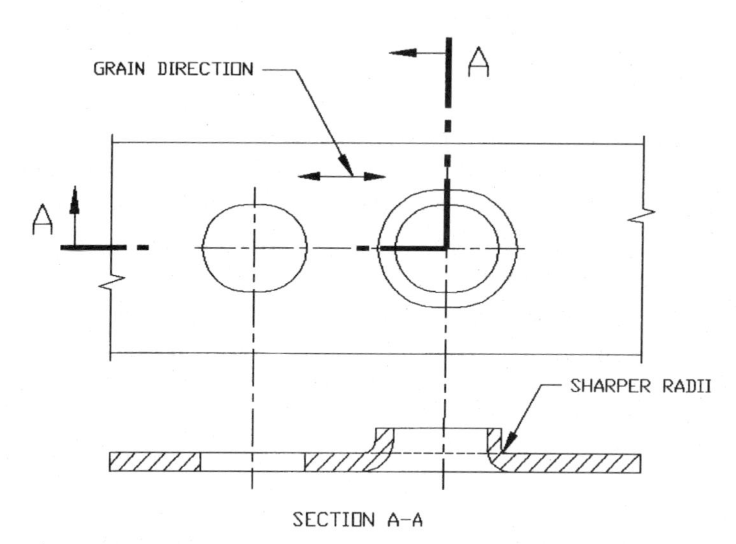


**Figure 1-13** Elongation of openings (exaggerated) caused by the direction of grain.

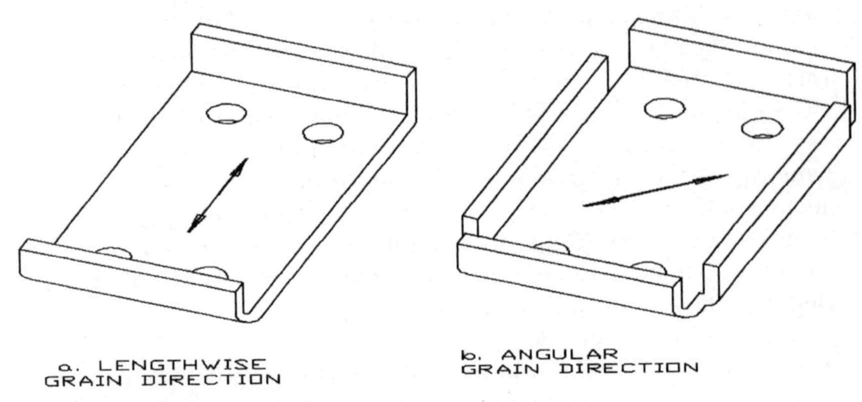


**Figure 1-14** Grain direction in formed sheet-metal parts.

Such a grain-line pattern should be used quite habitually with materials of the 6061-T4 (T6) aluminum group, as they are prone to cracking. Especially if—for some reason—parts are belt-sanded in flat prior to bending, their proneness to cracking will be enhanced. A greater bend radius as well as vibratory sanding may help to alleviate the problem.

In parts with various formed sections, the shear strength and resistance to columnar stress of their flanges will vary with their variation from the material's grain, as shown in Fig. 1-15.

Should a force *A* be applied to the bend in a bracket parallel with the grain line, the greatest shear strength will be encountered.

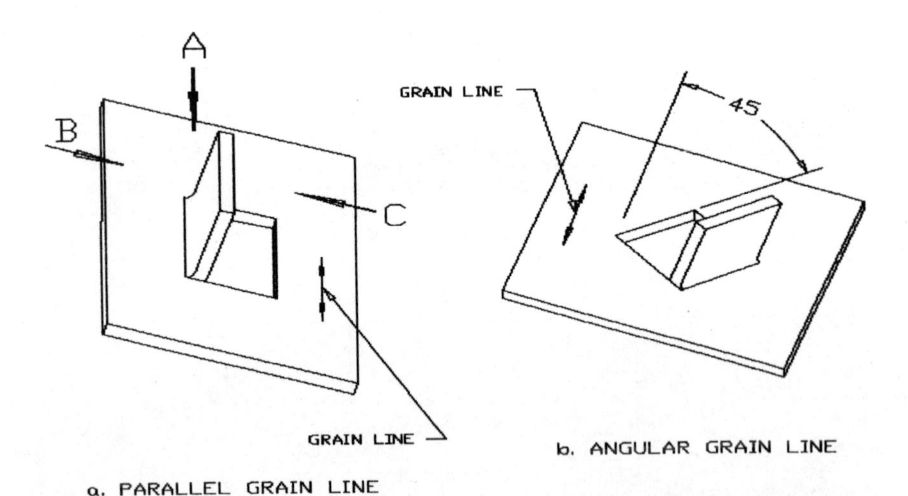


**Figure 1-15** Stresses and their relation to the direction of grain. (*From Frank W. Wilson,* "Die Design Handbook," *New York, 1965. Reprinted with permission from The McGraw-Hill Companies.*)

However, we already know that bends running parallel with the bend line are prone to cracking in forming and are not recommended.

Intermediate shear strength will be encountered in the direction of the *C* force line in Fig. 1-15*a,* whereas the *B* force line will display the least shear strength, as the flange will tend to crack under it. Whenever a bent-up flange is acted upon by a bending force, it has a tendency to follow that force's direction only if consistent with the initial movement of the flange in forming. A force applied against the direction of bending will not flatten the material but will break it.

Bending shown in Fig. 1-15*b,* with flanges at 45° off the grain line, is considered a fair practice.

The value of the bend radius is another factor influencing the part's behavior in forming: The smaller the radius, the greater the material's proneness to cracking. There is a certain minimal bend radius for various materials and thicknesses which is discussed in Chap. 8.

**1-3-1-2. Openings, their shape and location.** Openings in the part should not be located too close to each other and certainly not too close to the edge of the sheet or strip (Fig. 1-16).

At this point, it should suffice to compare the sheet-metal cutting operation to that of slicing a block of Swiss cheese: The closer to each other the cuts are placed, the more distorted the cut will be.

The shape of openings other than round has a considerable effect on the part's behavior in further manufacturing as well as in service. Sharp edges in cutouts become the points of accumulated stresses and

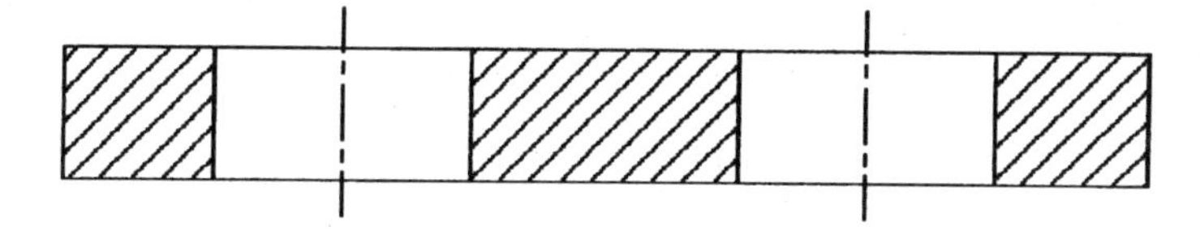

a. PROPER DISTANCE BETWEEN OPENINGS

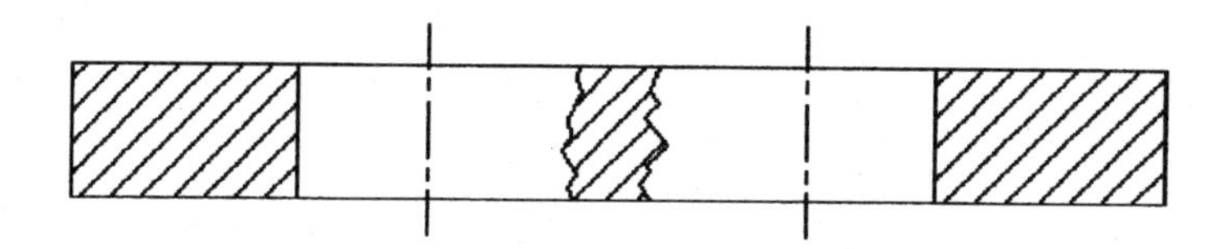

b. IMPROPER DISTANCE - OPENINGS TOO CLOSE
PUNCHING OPERATION CAUSES THE WALL IN-BETWEEN
TO COLLAPSE

**Figure 1-16** Distances between openings.

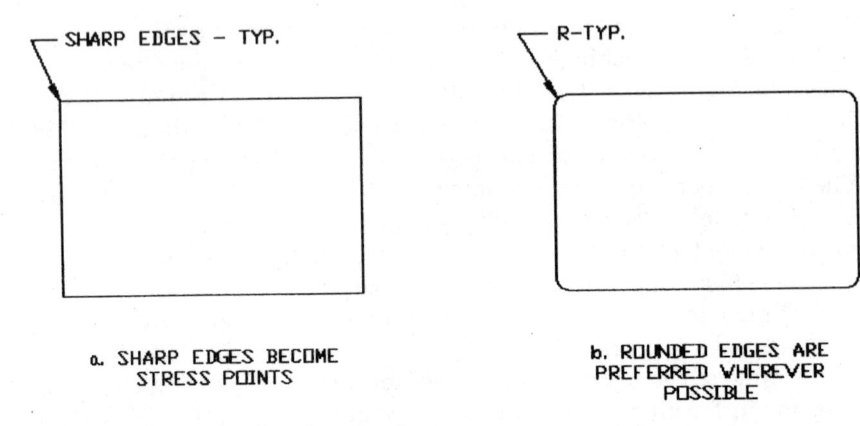


**Figure 1-17** Openings other than round.

may become points of failure (Fig. 1-17). Sharp edges are also difficult to protect from rust and corrosion, which may seep into the part through these areas. For that reason, rounded edges are preferable wherever possible.

Some minimal dimensions for punched parts are shown in Fig. 1-18. Should an opening be located too close to a bend, the recommended practice would be to first produce the bend and only subsequently to pierce the opening. By following this procedure the dimensional stability can be achieved. If such an opening is pierced first and the bend produced afterward, distortion of the opening will occur (Fig. 1-19).

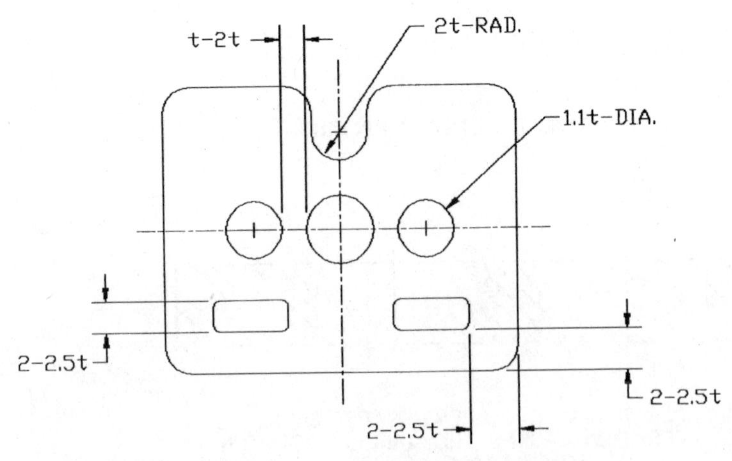


**Figure 1-18** Minimal practical punching and blanking dimensions.

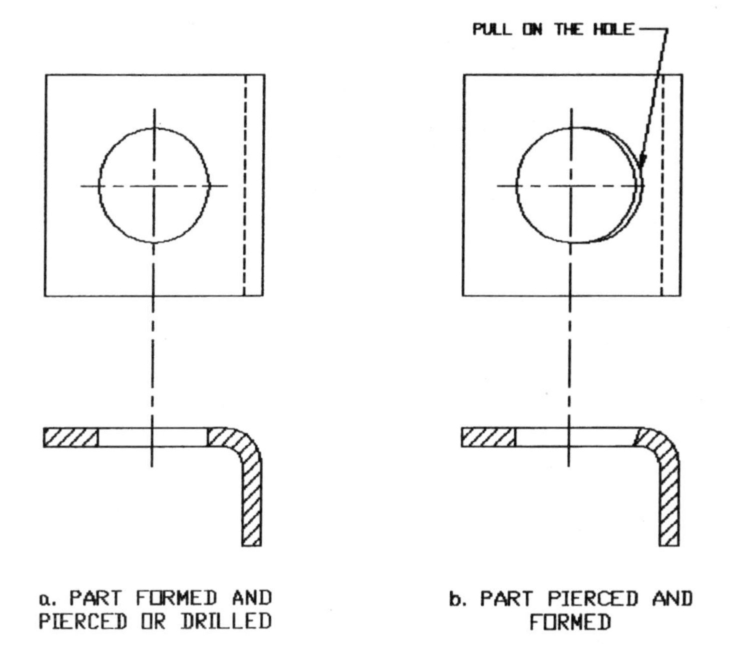


**Figure 1-19** Influence of bending and piercing sequence of operations.

**1-3-1-3. Bends and other three-dimensional alterations to the flat part, their shape and location.** The location of formed portions and their dependence on the direction of grain was addressed in Sec. 1-3-1-1.

With formed enclosures, there are many methods to their bending, but only three basic solutions to joining of their flanges together at corners, as shown in Fig. 1-20.

The seams between the flanges may be further improved if necessary or required. For face plates, the gaps are usually weld-filled, sanded smooth, and painted. For unexposed areas, or where another plate will be used as a cover, sanding will suffice.

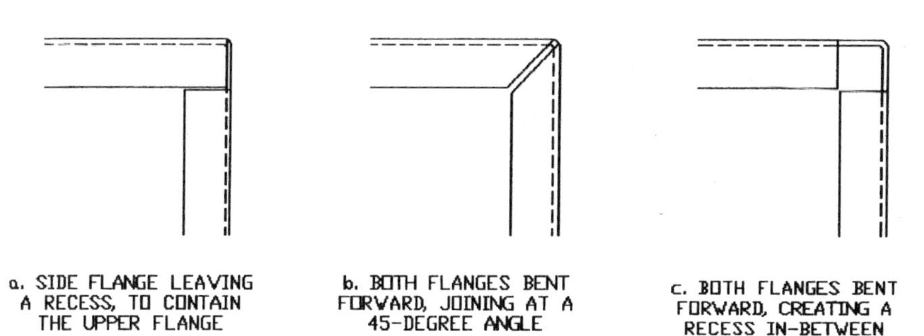


**Figure 1-20** Three ways of forming the corner flange.

Gaps may be large, small, or almost nonexistent. Their size and quality depend on the bend calculations, condition of tooling, and experience of the operator (in manual bending operations).

All bent-up portions should be provided with proper bend reliefs (Fig. 1-21). These not only ease the bending operation but also prevent the material from being pulled in the wrong direction, wrinkled, or torn. Figure 1-22 shows flange-bending examples.

**1-3-1-4. Outline of the part and its size.** Razor-sharp edges (or featheredges) must be avoided, especially if the parts are to be further handled by hand. These types of cuts are detrimental to the tooling as well, for if a punch does not engage the majority of its surface area in cutting, it leans toward one side, breaking afterward. Figure 1-23 shows examples of edge trimming.

In metal stamping, featheredges may result in formation of chips and small break-offs, which tend to remain on the die surface and impair further work. These little pieces of metal may scratch the advancing strip, may become embedded in finished parts or forced into their surface by the die operation, or may even be randomly flung around and endanger shop personnel.

Designers should also beware of phantom bends (Fig. 1-24), and for that reason a flat layout of every bent-up part should be produced prior to any design work.

Phantom bends are those which appear to be correct on the bent-up drawing but actually cannot be produced for various reasons. Most

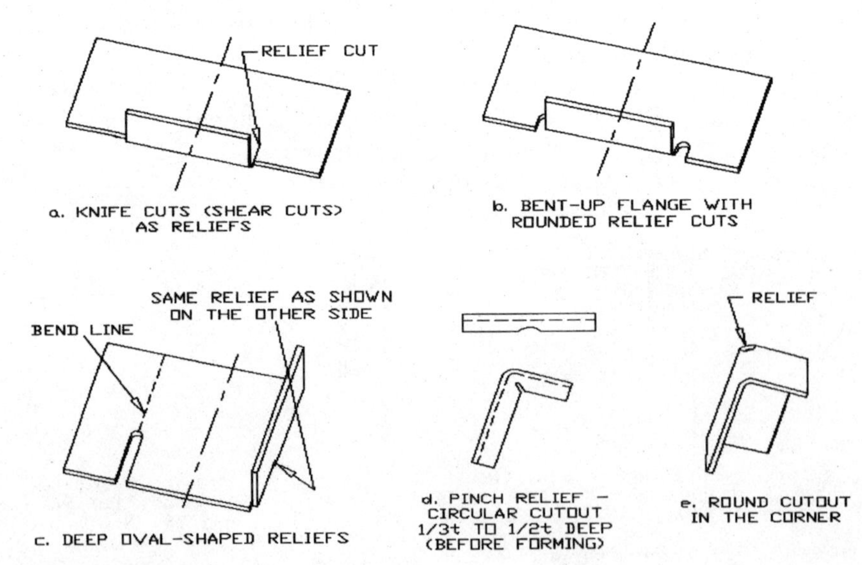


**Figure 1-21** Samples of bend reliefs.

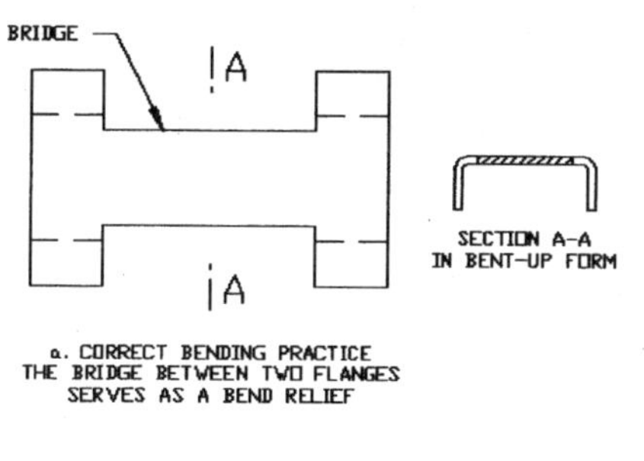


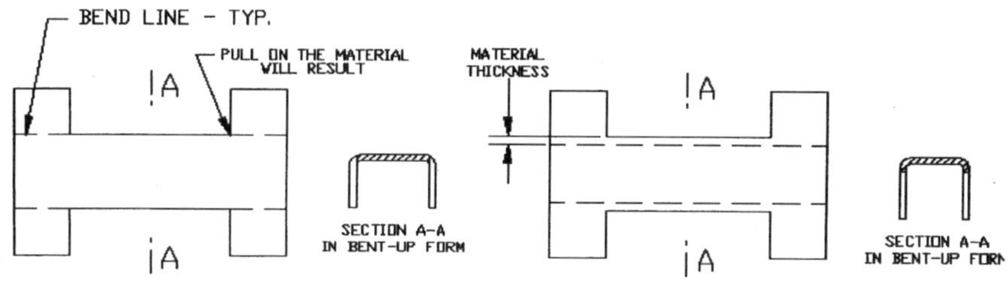


**Figure 1-22** Flange-bending examples.

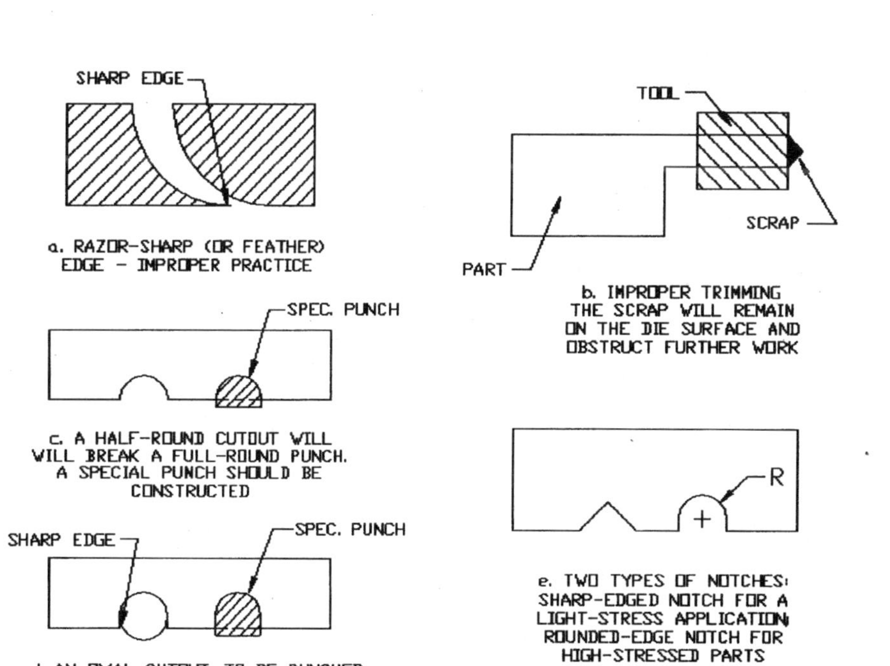


**Figure 1-23** Samples of edge trimming.

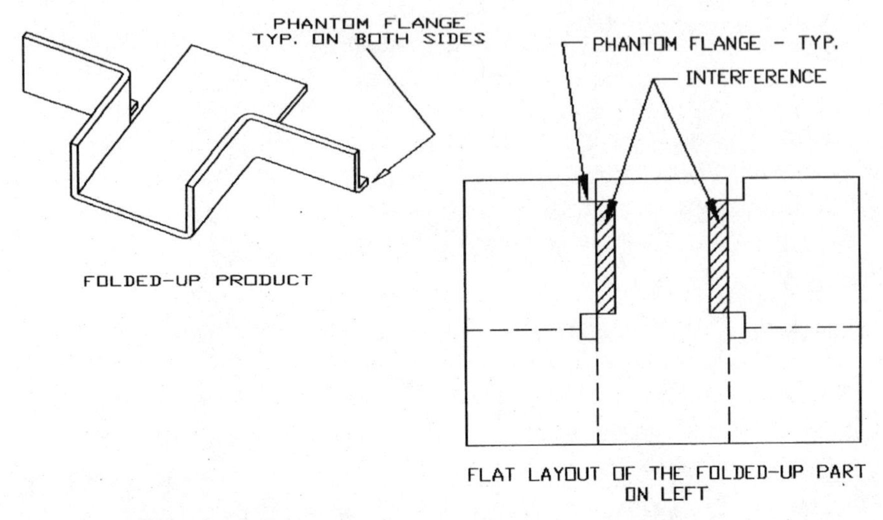


**Figure 1-24** Example of phantom bends.

often there is not enough material to form the bent-up portions or one section of the part interferes with another. These flaws are not always obvious from the part's drawing, especially where the product is complex in shape. An accurate flat layout not only provides for spotting these problems beforehand, it also displays the extent of it and a possible solution.

Additionally, flat layouts are important for a proper assessment of the size of the blank, as shown in Fig. 1-25. Where a part itself may

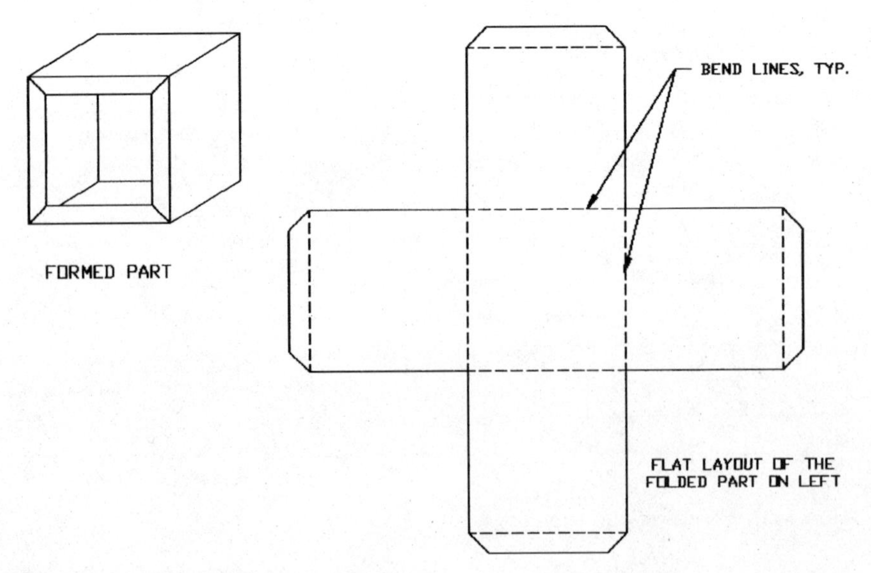


**Figure 1-25** Formed part and its flat layout.

seem small, its blank may be considerably larger than anticipated. This is most often caused by the size and location of scrap areas, attributable either to the part's shape or to the method of bending. If the part-forming procedure is not specified on the drawing, manufacturers may feel free to combine bends and seams to suit production practices. These alterations allow for a manipulation of the shape of the blank, shown in Fig. 1-26. By changing the part's outline, while still producing the same formed product, more economical shapes may be arrived at.

However, in sheet-metal stamping, even seemingly wasted areas of scrap may be decreased, if not minimized, just by rearranging parts on the strip (see Fig. 1-27). Naturally, the size of the resulting strip, and consequently the size of the die, must be kept in mind in the course of such evaluation.

### 1-3-2. Functionability aspects

Another method of evaluating a product is its functionability. To be functional, a part must sustain the anticipated amount of work cycles, while performing all its intended duties without any unusual wear, without excessive need for repairs, without succumbing to rust or corrosion, without significant changes in its outward characteristics, and without causing damage to any other part of the assembly or manufacturing system.

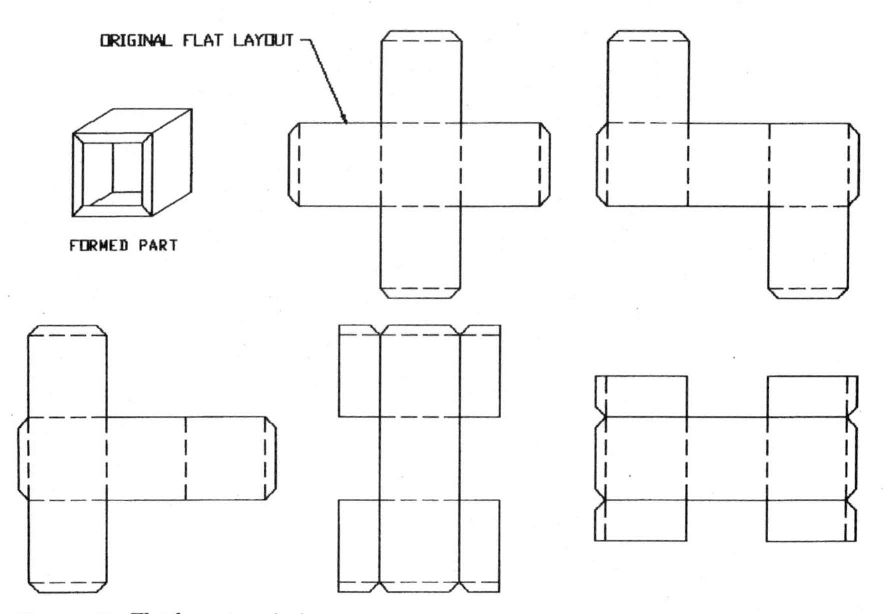


**Figure 1-26** Flat layout variations.

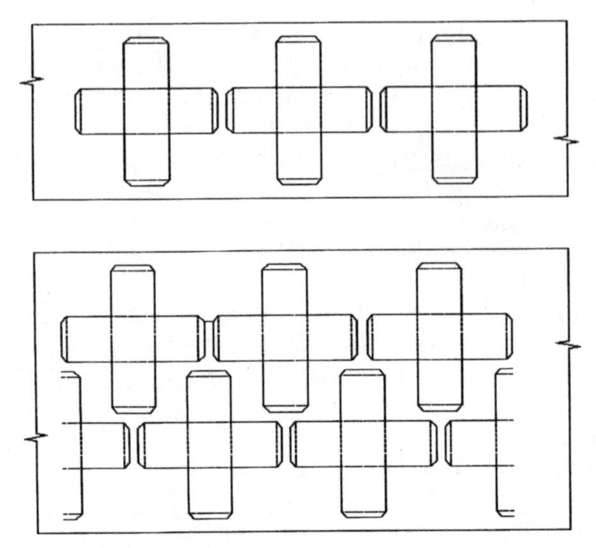

**Figure 1-27** Two variations of strip layout.

A well-designed, well-manufactured, and well-functioning part must be sturdy enough but not exaggerated in size or weight. It must use the supportive function of its grain structure in places where expected or necessary. It must not become detrimental to the function of surrounding parts or mechanisms and must not mar the surfaces of adjoining elements (including the hands of the operating personnel) even in the absence of protective means.

If the design calls for a part which may be considered aggressive to its surroundings, be it for its shape, sharp edges, or unfinished corners, proper barriers or protective devices should be used in manufacturing, transport, and storage.

Where possible, parts should be designed to allow for stacking. Their size and shape must fit the packaging material freely, without any constraints, yet with no excessive free space left to allow for their movement.

The amount of parts in a shipping container must be well proportioned to their weight, so that the load of the cargo will not cause any damage to the bottom layers of the batch.

*Sturdiness of a sheet-metal part* is often aided by the inclusion of

1. Beads and ribs
2. Bosses or buttons
3. Flanges
4. Lightening holes

These indentations protect the part's surface from deformation, buckling, or so-called oilcan effect, while strengthening the material structure by the cold work during the forming operation.

**1-3-2-1. Beads and ribs.** There are two types of these formations: Internal beads and external beads.

**Internal beads.** Internal beads can be produced either by rubber pad forming or by a set of matching dies.

In rubber pad forming (Fig. 1-28*a*), the entire surface of sheet-metal strip comes into contact with the rubber pad at the same moment. As the pressure increases and the metal is forced into the die recess, the surrounding material is already restrained from movement by the pressure of the pad. Therefore, the only deforming portion is that of the bead itself, while the surrounding material is not influenced by the metal movement.

With die forming of internal beads (Fig. 1-28*a*), the outward-shaped die reaches the material first and starts to form the bead without establishing firm retaining contact with the remaining material. The bead area gets stretched and begins to pull on the surrounding metal, distorting it in the process. Where such distortion may be objectionable, a pressure pad securing the metal in a fixed position and a double-action press should be used.

The maximum possible internal bead depth $a$ (shown later in Fig. 1-30) depends primarily on the width of bead $A$, standard beads commonly having a ratio of width to depth between 4 and 6, or

$$\frac{A}{a} = 4 \text{ to } 6$$

Beads should be spaced as closely together as possible to give maximum strength and to avoid wide flat areas, which are inherently weak. Between parallel beads the minimum spacing is about $8a$ to allow full bead formation without fracturing the metal. Between a bead and flange at right angles to the bead, allow $2a$; between beads at right angles to each other, allow $3a$; and between a bead and a flange parallel to the bead, allow $5a$.

**External beads.** With external beads, the pressure of the rubber pad is first applied to the top of the bead (see Fig. 1-28*b*). Metal is locked at this point, and with increasing pressure the area between bead strips is stretched until it bottoms on the form block. Deformation being thus spread progressively over a large area, an external bead can be formed considerably deeper than an internal bead of the same curvature. Somewhat disadvantageous is the necessity of a rather

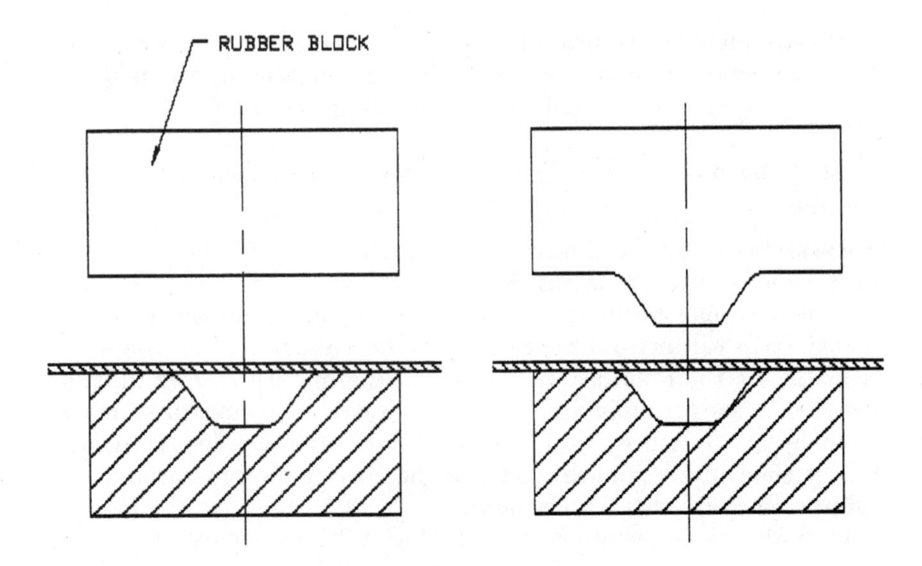


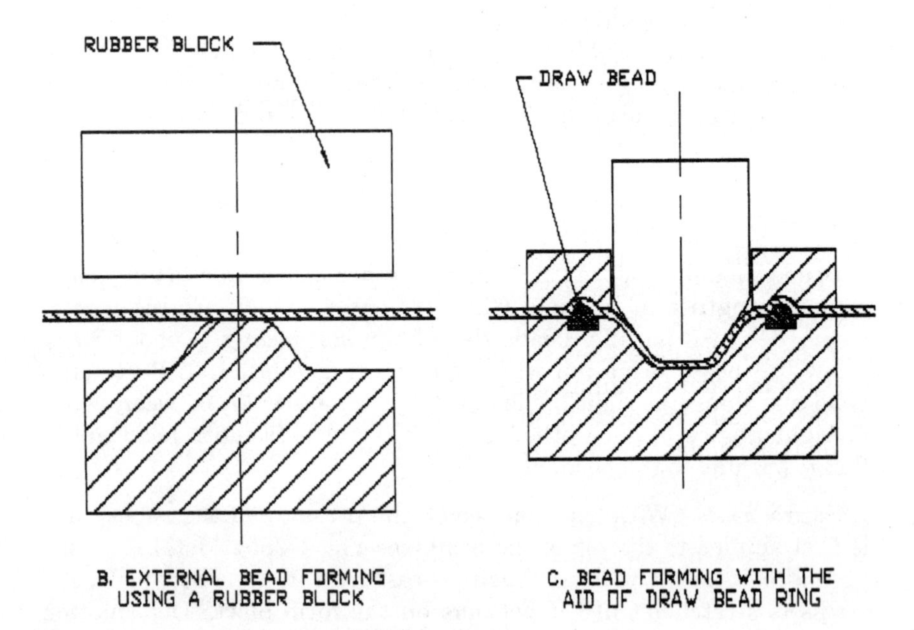


**Figure 1-28** Internal and external bead forming.

large edge radius. But because of the sharper resulting contours, external beads are more efficient stiffeners than internal beads.

**Draw beads.** A material-restraining action can be provided by draw beads (Fig. 1-28*c*). These not only secure the material in a given position, they further prevent its wrinkling during forming action.

The disadvantage of this application is the size of the draw radii: The draw radius of the punch should be four times the material thickness and the draw radius of the die still greater. If a smaller set of radii is used, the material will tear. However, using greater than necessary radii will not aid the manufacturing process either. In such a case, the strip will not be restricted in its movement, and it may flow along with the forming or drawing action, resulting in the formation of wrinkles.

The only way to adapt the final radius to the requirements of the drawing or those of the practicality is to restrike that area of the part afterward with properly sized tooling.

**Shapes.** Shapes of beads or ribs can be various. There are corner stiffening ribs, reinforcing beads, and hole reinforcing beads. Figure 1-29 shows corner bead designs. Locked-in beads are those that end sooner than the edge of the part. These should always be connected with the remaining flat surface by liberal radii.

A circumferential bulge, a thread, a curl can all be considered beads, for they all provide the part with reinforcing action. A drawn box can be reinforced by making its sides slightly convex; a container can have a convex bottom (or a recessed concave bottom) not only to strengthen its construction but to flatten its base circumferential area as well.

The bead design, useful for flat, curved, and angular surfaces, is shown in Fig. 1-30, and the design data are given in Table 1-1.

Reinforcing circumferential ribs (Fig. 1-31) are usually formed around openings. Since the ribs are the last to be formed, with the hole already in place, they should be as far away from that opening as possible in order to minimize its distortion.

These beads are actually the size of their radius deep. The radius value is dependent on the stock thickness, type of material, and forming pressure as follows:*

$$R = \frac{2(TS)}{P} \qquad \text{for circular ribs} \qquad (1\text{-}1a)$$

*Formulas in this chapter are based on those given in Frank W. Wilson, *Die Design Handbook,* New York, 1965. Reprinted with permission from The McGraw-Hill Companies.

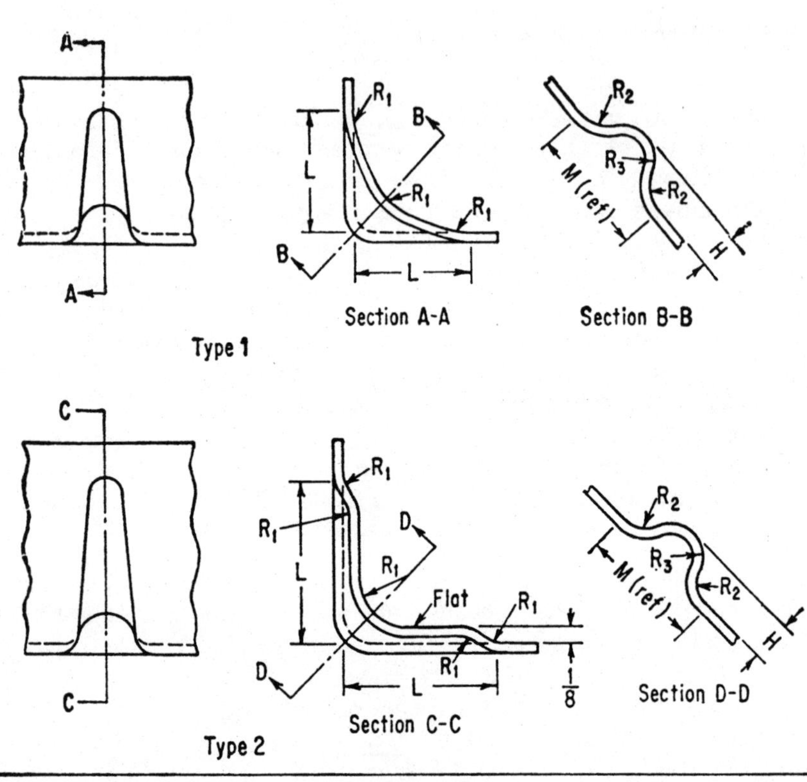


DESIGN DATA

| Size L | Type | $R_1$ | $R_2$ | $R_3$ | H | M (ref) | Spacing between beads |
|---|---|---|---|---|---|---|---|
| 1/2 | 1 | 1/4 | 23/64 | 3/16 | 1/8 | 23/32 | 2-1/2 |
| 3/4 | 1 | 5/16 | 41/64 | 17/64 | 13/64 | 1-5/32 | 3 |
| 1-1/4 | 2 | 11/32 | 55/64 | 21/64 | 17/64 | 1-1/2 | 3-1/2 |

Dimensions are in inches

**Figure 1-29** Corner bead design. (*Reprinted with permission from* "Product Engineering Magazine.")

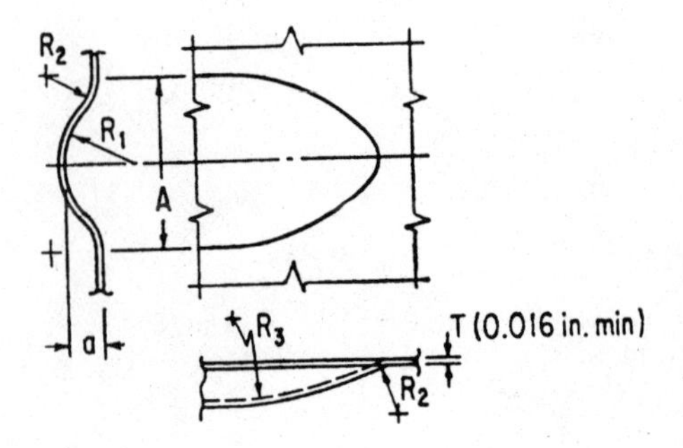


**Figure 1-30** Bead design. (*Reprinted with permission from* "Product Engineering Magazine.")

**TABLE 1-1 Bead Design Data**

| $a$ | $A$ | $R_1$ | $R_2$ | $R_3$ | $T$* | $T$† |
|---|---|---|---|---|---|---|
| Low-carbon Steel, Aluminum, Magnesium | | | | | | |
| 0.050 | 1/4 | 7/64 | 0.070 | 1/2 | 0.016 | 0.032 |
| 0.100 | 1/2 | 7/32 | 0.140 | 1 | 0.025 | 0.040 |
| 0.150 | 3/4 | 21/64 | 0.210 | 1 1/2 | 0.032 | 0.064 |
| 0.200 | 1 | 7/16 | 0.280 | 2 | 0.040 | 0.064 |
| 0.250 | 1 1/4 | 35/64 | 0.350 | 2 1/2 | 0.051 | 0.064 |
| 0.300 | 1 1/2 | 21/32 | 0.420 | 3 | 0.064 | 0.064 |
| 0.350 | 1 3/4 | 49/64 | 0.490 | 3 1/2 | 0.064 | 0.064 |
| 0.400 | 2 | 7/8 | 0.560 | 4 | 0.064 | 0.064 |
| 0.600 | 3 | 1 5/16 | 0.840 | 6 | 0.064 | 0.064 |
| 0.800 | 4 | 1 3/4 | 1.120 | 8 | 0.064 | 0.064 |
| 1.000 | 5 | 2 3/16 | 1.400 | 10 | 0.064 | 0.064 |
| 1.200 | 6 | 2 5/8 | 1.680 | 12 | 0.064 | 0.064 |
| Titanium | | | | | | |
| 0.38‡ | 1.33 | 0.46 | 0.250 | 1.12 | | 3.88¶ |
| 0.38 | 1.90 | 0.94 | 0.250 | 1.12 | | |
| 0.50‡ | 1.75 | 0.72 | 0.250 | 1.50 | | 2.50¶ |
| 0.50 | 2.50 | 1.56 | 0.250 | 1.50 | | |
| 0.70 | 3.50 | 2.32 | 0.250 | 2.00 | | |
| 0.88‡ | 3.08 | 1.46 | 0.250 | 2.52 | | 1.76¶ |
| 0.88 | 4.40 | 3.00 | 0.250 | 3.52 | | |
| 1.12‡ | 3.92 | 2.00 | 0.250 | 3.00 | | 1.26¶ |
| 1.12 | 5.60 | 3.82 | 0.250 | 3.00 | | |

*Maximum thickness for rubber-pad-formed beads on a hydraulic press.

†Maximum thickness for beads formed by punch and die on a mechanical press.

‡Use when edge of bead to edge of sheet does not exceed the dimension shown.

¶Maximum distance between edge of bead and edge of sheet. See Fig. 1-30.

NOTE: Beads of these proportions may be formed in titanium pure AMS 4901, hot or cold. Further in titanium alloy RE-T-41 and in cold only RE-T-32. In the latter case, the radius $R_2$ should be 0.312″ for alloy of 0.063″ thickness.

SOURCE: Reprinted with permission from *Product Engineering Magazine.*

and

$$R = \frac{TS}{P} \quad \text{for elongated ribs} \tag{1-1b}$$

where $R$ = bottom radius, in
$T$ = material thickness, in
$S$ = tensile strength, lb/in$^2$
$P$ = forming pressure, lb/in$^2$

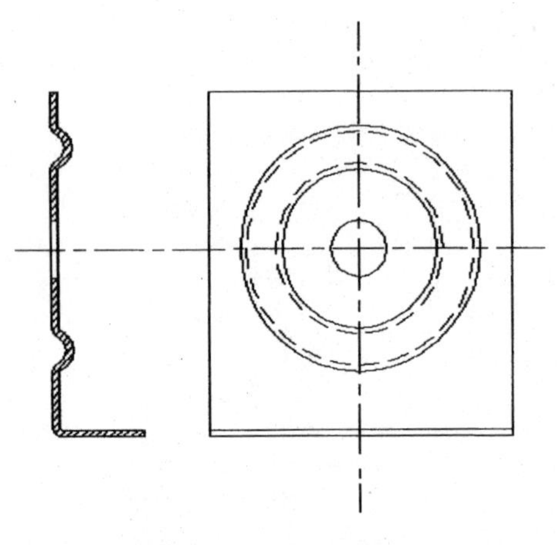


**Figure 1-31** Circumferential ribs in stampings. (*From Frank W. Wilson,* "Die Design Handbook," *New York, 1965. Reprinted with permission from The McGraw-Hill Companies.*)

**1-3-2-2. Bosses or buttons.** These are flat-bottomed circular depressions or elevations in sheet (see Fig. 1-32). They are most often used for offsetting purposes, be it for hardware or for other applications.

**1-3-2-3. Flanges.** These can be either straight or curved. Straight flanges are made by simple bending of a portion of sheet-metal part, with no material flow involved. Curved flanges also use simple bending, as well as additional stretching or compressing action on the material. The material flow is similar to that in drawing or other cold working.

With curved flanges, there is always a certain amount of deformation involved. In convex, or shrink, flanges (Fig. 1-33*b*) the material is compressed in order to produce the required shape. In concave, or stretched, flanges (Fig. 1-33*c*) the material is elongated. The amount of deformation, when calculated, can be used to determine the exact type of the flange:

$$\% \text{ of deformation} = 100\left(\frac{R_2}{R_1} - 1\right)$$

where $R_1$ = edge radius before forming, in flat, in
$R_2$ = edge radius after forming, in

If the deformation percentage comes out as a positive number, an elongation of material (stretch) is involved. With a negative number,

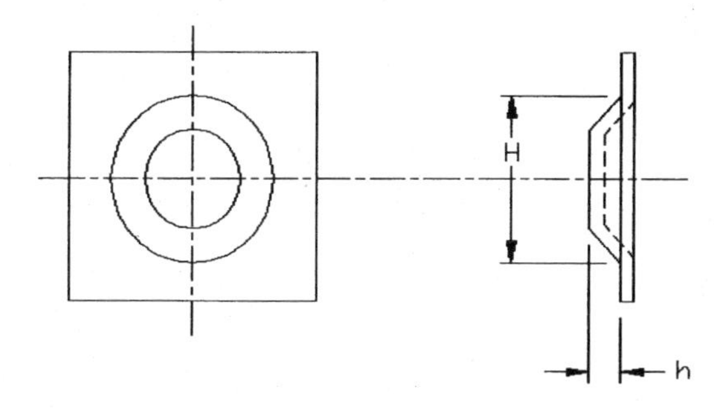


| HEIGHT "h" | MATERIAL THICKNESS | SUGGESTED RATIO, H/h |
|---|---|---|
| 1/16" | .020"<br>.031"<br>.048" | 12 - 20<br>14 - 24<br>16 - 26 |
| 3/32" | .020"<br>.031"<br>.048" | 12<br>14<br>16 |
| 1/8" | .020"<br>.031"<br>.048" | 11<br>12<br>13 |

**Figure 1-32** Round beads or bosses.

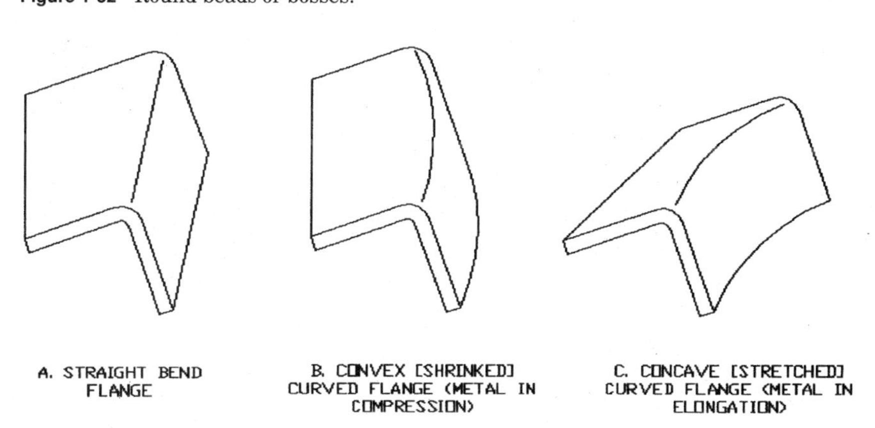


**Figure 1-33** Three types of sheet-metal flanges.

the compression (shrink) is indicated. Table 1-2 gives maximum forming limits.

The amount of setback for all flanges can be determined from Fig. 1-34 by connecting the radius scale at the value $R$ to the thickness scale at the value $T$ with a straight line. The setback value $J$ is read at the point where this line intersects the horizontal line representing the bevel of the bend.

Flat pattern flange width $Y$ can be calculated by using the following formula:

**TABLE 1-2 Maximum Forming Limits for Flanges**

| Material type | Stretch flanges (elongation) | | Shrink flanges (compression) | |
|---|---|---|---|---|
| | Rubber tooling, % | Solid die, % | Rubber tooling, % | Solid die, % |
| Aluminum 3003-O | 20 | 30 | 5 | 40 |
| Aluminum 3003-T | 5 | 8 | 3 | 10 |
| Aluminum 5052-O | 20 | 25 | 5 | 35 |
| Aluminum 5052-T | 5 | 8 | 3 | 12 |
| Aluminum 6061-O | 21 | 22 | 8 | 35 |
| Aluminum 6061-T | 5 | 10 | 2 | 10 |
| Aluminum 7075-O | 10 | 18 | 3 | 30 |
| Aluminum 7075-T | 0 | 0 | 0 | 0 |
| Steel 1010 | | 38 | | 10 |
| Steel 1020 | | 22 | | 10 |

$$Y = W - J \tag{1-2}$$

where $Y$ = flange width, in flat, in
$W$ = formed flange width, in
$J$ = value of setback, from Fig. 1-34

Dimensioning of stretch, shrink, and special flanges is given in Fig. 1-35. Dimensions for 90° flanges can be determined from Fig. 1-36, as can be the percentage of elongation (stretch) or compression (shrink) in the metal of a given flange.

Connect the flange width value $Y$ to the amount of compression (shrink) with a straight line. Where this line crosses the mold line radius graph, that value is applicable to the given problem.

Dimensions for open or closed flanges can be determined from Fig. 1-37; the method for the chart's use is similar to that described earlier.

The flange width $W$ or the projected flange width $H$ can be determined from the lower scale. The approximate deformation of the free edge of curved flanges, percentagewise, is determined on the upper scale.

*Permissible strain in stretched flanges* depends on the edge condition of the metal, flange width (from Fig. 1-35), and method of forming. For 90° flanges, this value may be approximated by using the following formula:

$$e = \frac{W}{R_2} \tag{1-3}$$

where $e$ = elongation (strain) factor at free edge of flange
$W$ = flange width, in
$R_2$ = contour radius of bent-up flange, in

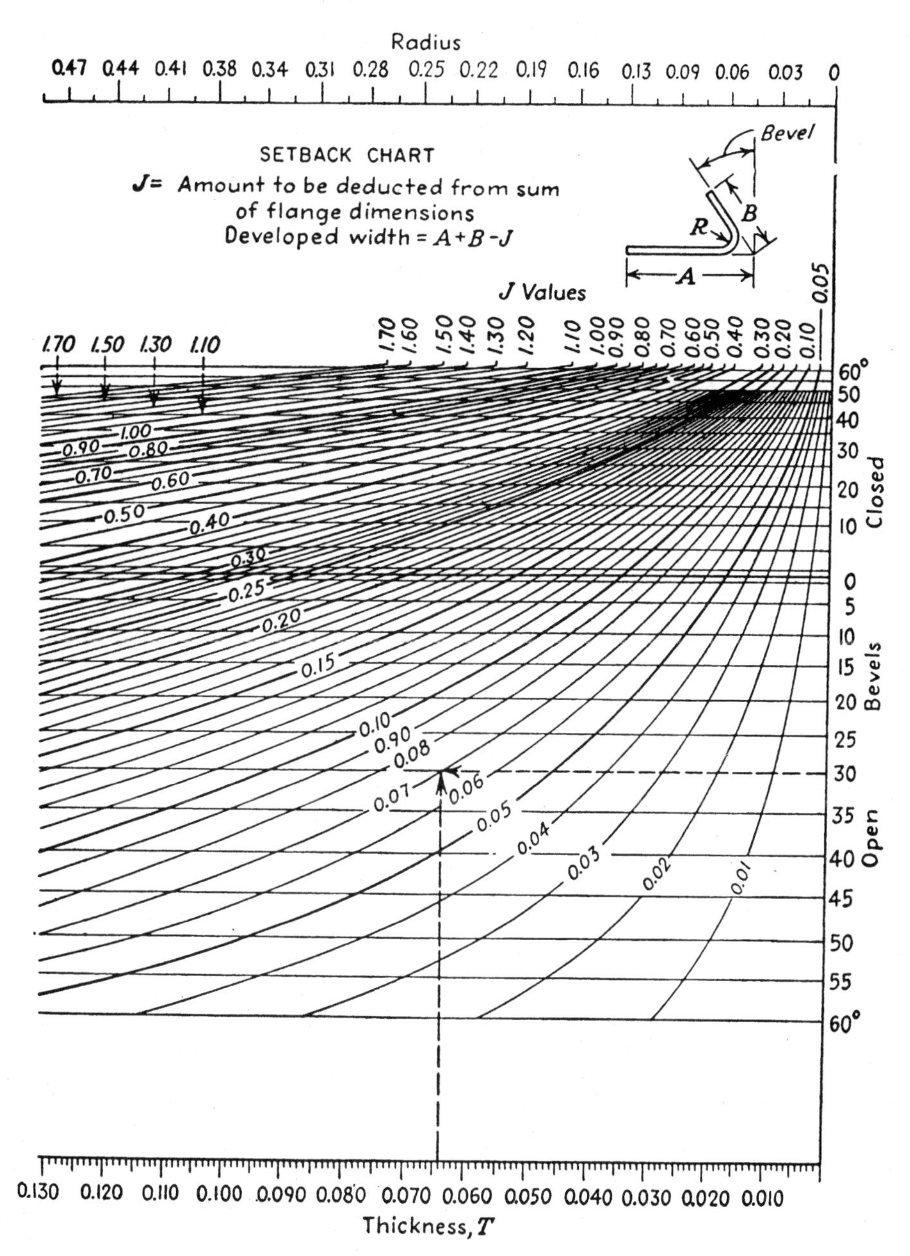


**Figure 1-34** Setback chart. (*From Frank W. Wilson,* "Die Design Handbook," *New York, 1965. Reprinted with permission from The McGraw-Hill Companies.*)

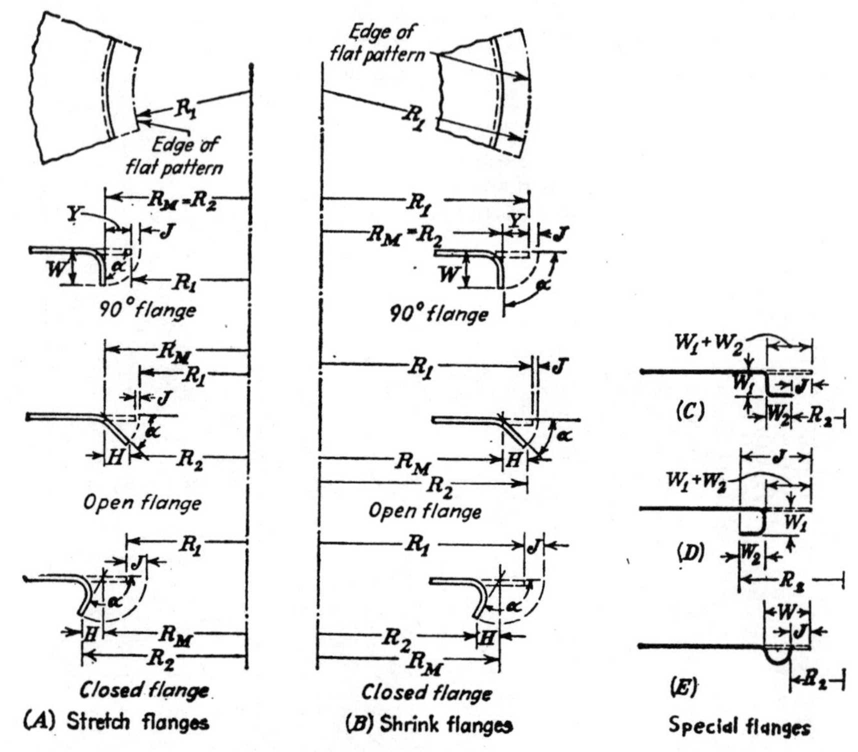


**Figure 1-35** Dimensioning of flanges: (*a*) for stretch flanges; (*b*) for shrink flanges; (*c*), (*d*), and (*e*) for special flanges. For 90° flanges use Fig. 1-36. For other flanges, use Fig. 1-37. (*From Frank W. Wilson,* "Die Design Handbook," *New York, 1965. Reprinted with permission from The McGraw-Hill Companies.*)

For 2024-0, -T3, and -T4 aluminum 90° flanges, 0.10 is a safe value for $e$ where edges are smooth; 0.06 is a safe value for sheared edges. A larger degree of stretch occurs where contour radius $R_2$ is small or where the stretch flange is adjacent to a shrink flange.

Equation (1-3) for 90° stretch flanges also applies to 90° shrink flanges. Here, however, the metal is in compression, and the sheet must be supported against buckling or wrinkling. With rubber forming, there is practically no support against buckling, and only slight shrinking can be accomplished, so that rubber forming is limited to very large flange radii or vary narrow widths. For 2024-0 aluminum, without subsequent rework, shrink is limited to not over 2 or 3 percent; shrink is limited to 0.5 percent for 2024-T3 and -T4.

U.S. Air Force specifications indicate that there is danger of cracking when elongation exceeds 12 percent in 2 in. Therefore, for safety, $e$ = 0.12, and

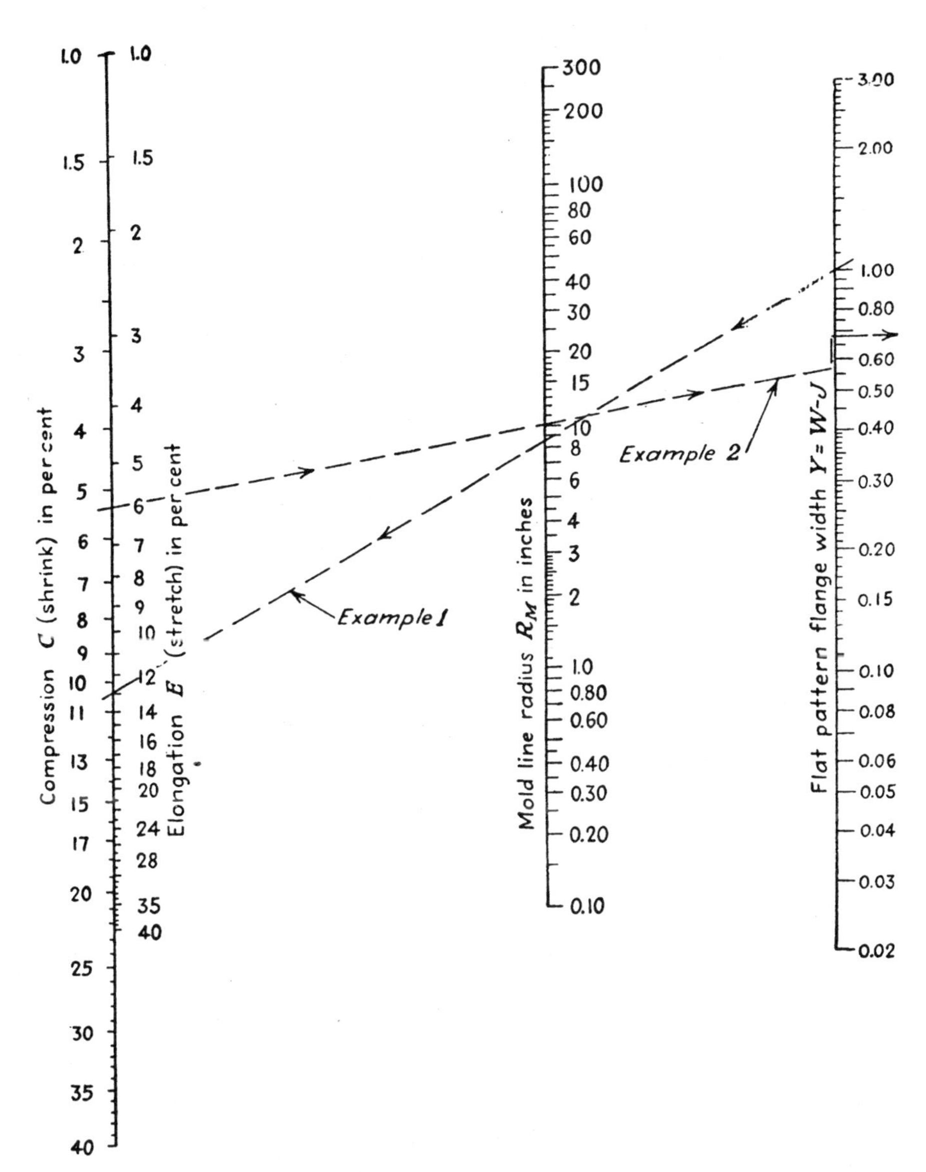


**Figure 1-36** Chart for calculating 90° flange width and percentage of deformation. (*From Frank W. Wilson,* "Die Design Handbook," *New York, 1965. Reprinted with permission from The McGraw-Hill Companies.*)

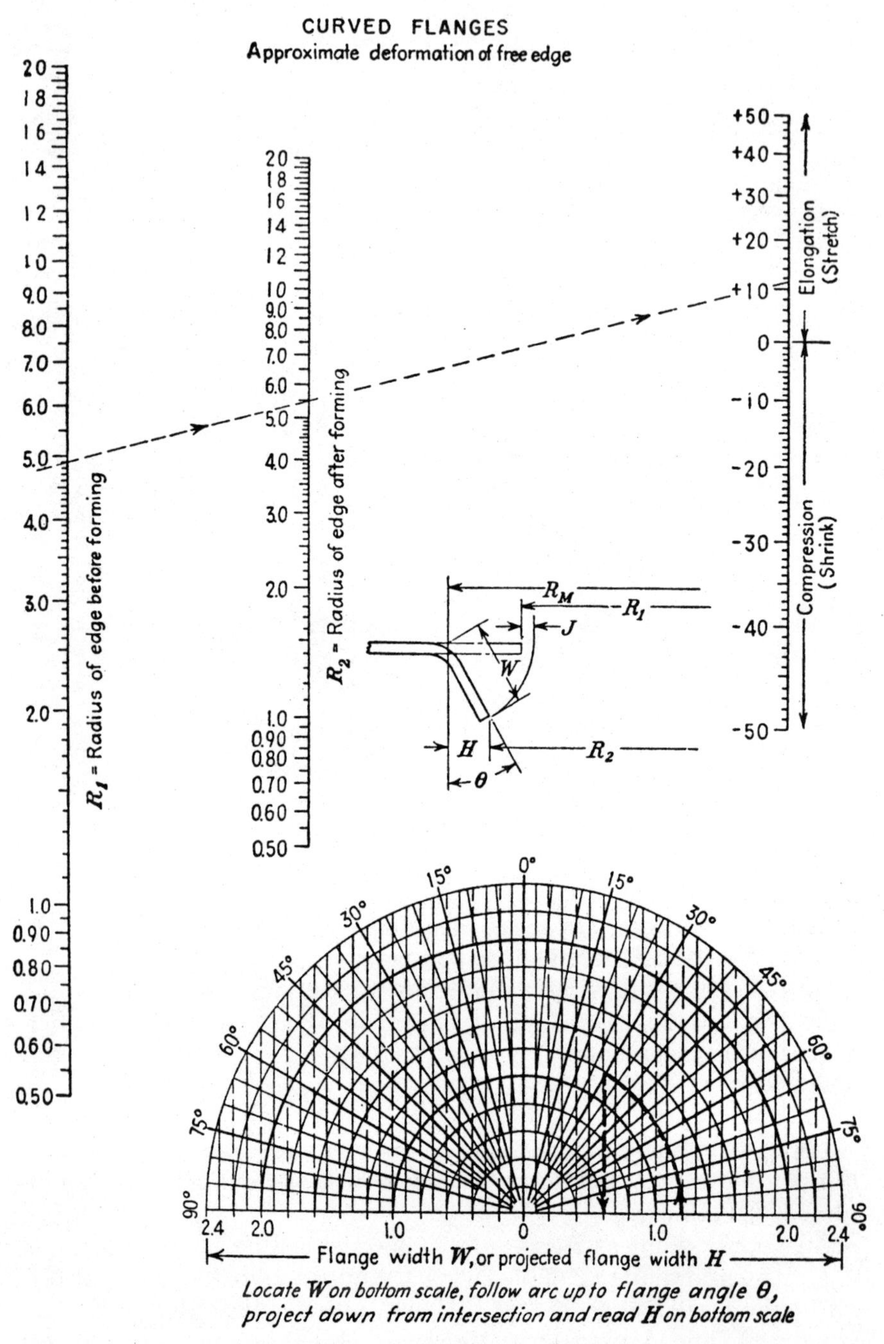


**Figure 1-37** Chart for calculating widths of open or closed flanges. (*From Frank W. Wilson,* "Die Design Handbook," *New York, 1965. Reprinted with permission from The McGraw-Hill Companies.*)

$$\frac{R_2 - W}{R_2} = 0.88 \qquad (1\text{-}4a)$$

For open flanges (angles smaller than 90°, see Fig. 1-35) the formula is

$$e = \frac{W(1 - \cos \alpha)}{R_2} \qquad (1\text{-}4b)$$

Values of $e$ for some other shaped flanges are as follows: For flanges in Fig. 1-35*c*,

$$e = \frac{W_1}{R_2} \qquad (1\text{-}4c)$$

For flanges in Fig. 1-35*d*,

$$e = \frac{W_1 + 2W_2}{R_2} \qquad (1\text{-}4d)$$

For flanges in Fig. 1-35*e*,

$$e = \frac{J}{R_2} \qquad (1\text{-}4e)$$

Cold forming changes the mechanical properties of carbon steel strip and produces certain useful combinations of hardness, strength, stiffness, ductility, and other characteristics. Temper numbers indicate degrees of strength, hardness, and ductility produced in cold-rolled carbon steel strip. These temper numbers are associated with the ability of each temper to withstand certain degrees of cold forming.

The No. 1 temper is not suited to cold forming; temper Nos. 4 and 5 are used in producing parts involving difficult forming or drawing operations.

**1-3-2-4. Lightening holes.** These can be flanged in annealed aluminum alloys up to 0.125″ thick and in heat-treated 2024 aluminum up to 0.064″ thick.

Austenitic stainless steel up to 0.060″ can be formed with external hole flanges and up to 0.050″ thick with internal flanges. The quarter-hard stainless steel up to 0.040″ thick can be externally flanged and internally flanged up to 0.030″ thick.

Rubber-sheared lightening hole minimum diameters, in relation to aluminum alloy thicknesses, are shown in Table 1-3.

**TABLE 1-3 Lightening Holes for 35° Flange***

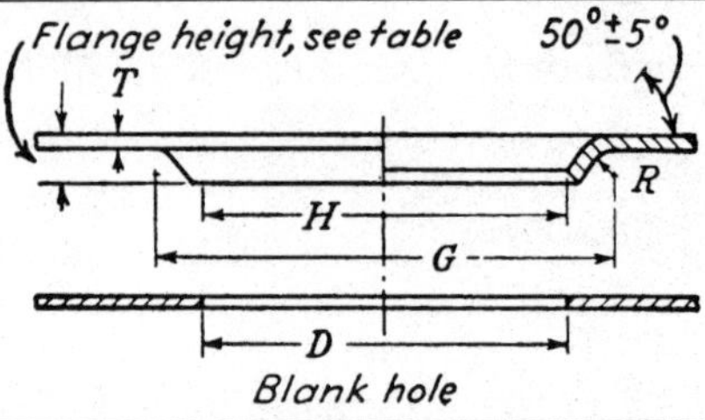

Flange height, see table
T
50° ± 5°
R
H
G
D
Blank hole

| *D*, in | *H*, in | *G*, in | *T*, in | *R*, in |
|---|---|---|---|---|
| | | 1/8-in Flange Height | | |
| 0.445 | 0.812 | 1.223 | 0.020–0.040 | 3/16 |
| 0.400 | | 1.212 | 0.051–0.072 | 1/4 |
| 0.570 | 0.938 | 1.348 | 0.020–0.040 | 3/16 |
| 0.525 | | 1.337 | 0.051–0.072 | 1/4 |
| 0.695 | 1.062 | 1.473 | 0.020–0.040 | 3/16 |
| 0.650 | | 1.462 | 0.051–0.072 | 1/4 |
| 0.820 | 1.188 | 1.598 | 0.020–0.040 | 3/16 |
| 0.775 | | 1.587 | 0.051–0.072 | 1/4 |
| | | 5/32-in Flange Height | | |
| 0.900 | 1.312 | 1.800 | 0.020–0.040 | 3/16 |
| 0.852 | | 1.791 | 0.051–0.072 | 1/4 |
| 1.150 | 1.562 | 2.050 | 0.020–0.040 | 3/16 |
| 1.082 | | 2.041 | 0.051–0.072 | 1/4 |
| 1.276 | 1.688 | 2.175 | 0.020–0.040 | 3/16 |
| 1.208 | | 2.166 | 0.051–0.072 | 1/4 |
| 1.400 | 1.812 | 2.300 | 0.020–0.040 | 3/16 |
| 1.332 | | 2.294 | 0.051–0.072 | 1/4 |
| | | 3/16-in Flange Height | | |
| 1.606 | 2.062 | 2.625 | 0.020–0.040 | 3/16 |
| 1.543 | | 2.617 | 0.051–0.072 | 1/4 |
| 1.490 | | 2.611 | 0.081–0.102 | 5/16 |
| 1.856 | | 2.875 | 0.020–0.040 | 3/16 |
| 1.793 | 2.312 | 2.867 | 0.051–0.072 | 1/4 |
| 1.740 | | 2.861 | 0.081–0.102 | 5/16 |
| 1.982 | | 3.000 | 0.020–0.040 | 3/16 |
| 1.919 | 2.438 | 2.992 | 0.051–0.072 | 1/4 |
| 1.866 | | 2.987 | 0.081–0.102 | 5/16 |
| 2.106 | | 3.125 | 0.020–0.040 | 3/16 |
| 2.043 | 2.562 | 3.117 | 0.051–0.072 | 1/4 |
| 1.990 | | 3.111 | 0.081–0.102 | 5/16 |
| 2.356 | | 3.375 | 0.020–0.040 | 3/16 |
| 2.293 | 2.812 | 3.367 | 0.051–0.072 | 1/4 |
| 2.240 | | 3.361 | 0.081–0.102 | 5/16 |
| 2.606 | | 3.625 | 0.020–0.040 | 3/16 |
| 2.543 | 3.062 | 3.617 | 0.051–0.072 | 1/4 |
| 2.490 | | 3.611 | 0.081–0.102 | 5/16 |
| 2.856 | | 3.875 | 0.020–0.040 | 3/16 |

**TABLE 1-3 Lightening Holes for 35° Flange* (*Continued*)**

| *D*, in | *H*, in | *G*, in | *T*, in | *R*, in |
|---|---|---|---|---|
| | | 3⁄16-in Flange Height | | |
| 2.793 | 3.312 | 3.867 | 0.051–0.072 | 1⁄4 |
| 2.740 | | 3.861 | 0.081–0.102 | 5⁄16 |
| 3.106 | | 4.125 | 0.020–0.040 | 3⁄16 |
| 3.043 | 3.562 | 4.117 | 0.051–0.072 | 1⁄4 |
| 2.990 | | 4.111 | 0.081–0.102 | 5⁄16 |
| 3.356 | | 4.375 | 0.020–0.040 | 3⁄16 |
| 3.293 | 3.812 | 4.367 | 0.051–0.072 | 1⁄4 |
| 3.240 | | 4.361 | 0.081–0.102 | 5⁄16 |
| 3.606 | | 4.625 | 0.020–0.040 | 3⁄16 |
| 3.542 | 4.062 | 4.617 | 0.051–0.072 | 1⁄4 |
| 3.490 | | 4.611 | 0.081–0.102 | 5⁄16 |
| 3.856 | | 4.875 | 0.020–0.040 | 3⁄16 |
| 3.793 | 4.312 | 4.867 | 0.051–0.072 | 1⁄4 |
| 3.740 | | 4.861 | 0.081–0.102 | 5⁄16 |

*These lightening holes may be formed in 5052-H32, 6061-T, 2024-T, 2014-T, R-301-T, and 7075-T aluminum alloy and AMC52SO and FS-Ia magnesium alloy or any softer condition of any of these alloys.

SOURCE: Frank W. Wilson, *Die Design Handbook,* New York, 1965. Reprinted with permission from The McGraw-Hill Companies.

### 1-3-3. Control of close-toleranced dimensions

The control of precise dimensions in die work is still unequaled by other manufacturing processes. The method of sheet progression itself is often used to restrict variations of sensitive dimensions to a high degree (Fig. 1-38).

Figure 1-38 shows a group of closely spaced *b* openings and off their location a rather accurate dimensioning being done. The spacing between the groups would be jeopardized if a wrong sequence of die work were used. This is because such closely spaced piercing produces a distortion of the material structure, which results in its expansion in that particular area, often to the point of bulging above the remaining surface. This condition can be controlled to a degree with

- Specially shaped punches
- Special cutting conditions
- Staggered cutting, which is the least effective method

All these techniques are discussed in detail later. At this point, it suffices to plan the appropriate sequences of operations for the given part. Method No. 1 utilizes the following scenario:

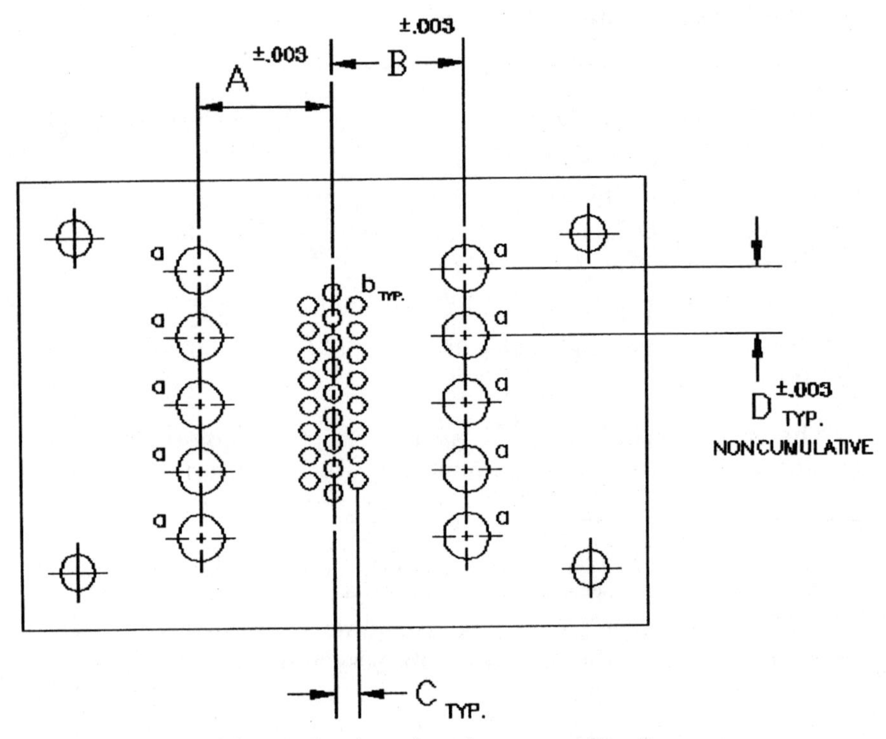


**Figure 1-38** Accurate dimensioning through work-sequence alteration.

1. Pierce the cluster of small *b* openings
2. Flatten the part in the die
3. Pilot on two opposite holes for further punching
4. Pierce two rows of *a* holes, positioned at distances $A \pm 0.003$ and $B \pm 0.003$ off center

This sequence of operations will place the *a* holes a reasonably exact distance off the center, wherever it may be. The center itself is not being firmly established here; therefore, it can be shifted slightly off because of the distortion of material in punching.

To obtain yet more accurate results, the Method No. 1 sequence can be slightly altered, and it becomes Method No. 2:

1. Pierce one row of *a* holes, located at either $A \pm 0.003$ or $B \pm 0.003$ dimension off the center
2. Pilot on two openings
3. Pierce the cluster of small *b* holes in the middle
4. Flatten the part

5. Pilot on two openings
6. Pierce the remaining row of *a* holes

This way the linear distance between the *a* holes can be controlled slightly more. But the ultimate results will be obtained with Method No. 3:

1. Pierce all openings slightly smaller than they should be. The sequence of operations is negligible.
2. Pierce the central cluster to the accurate hole size.
3. Pierce the remaining *a* holes at $A \pm 0.003$ and $B \pm 0.003$ distances off the center.

What will happen during Method No. 3 piercing is that the first cutting creates stresses in the material, which will certainly make it bulge. But the second cutting cuts off these stressed portions, located around the holes. After these small areas are removed, the material flattens itself almost miraculously, and a perfectly flat part emerges from the die.

Unfortunately, the last method cannot always be used in die work, as such double punching may prove too cost-intensive.

The sequence of operations to produce a part in Fig. 1-39 should be planned with the overall picture of the part in mind. In Fig. 1-39*a* we have a bent-up flange the location of which is not controlled, while in Fig. 1-39*b* it has a definite dimensional importance.

The part's sequence of operations (Fig. 1-39*a*) should be as follows:

1. Pierce the holes *b* and *a*
2. Cut the flange relief
3. Bend the flange

The sequence of the piercing may be reversed (first the flange relief, then *a* and *b* holes), but the forming should be last.

The sample's sequence in Fig. 1-39*b* must be oriented about the flange, which must be produced first:

1. Cut the flange relief
2. Bend the flange
3. Bank off the edge of flange and pierce holes *a* and *b*

The reason for the sample's altered sequence (Fig. 1-39*b*) is the bending operation, which is hardly as accurate as piercing. To locate the openings off the flange in flat is hazardous, since the flange may or may not be exactly where it is expected to appear. The grain and

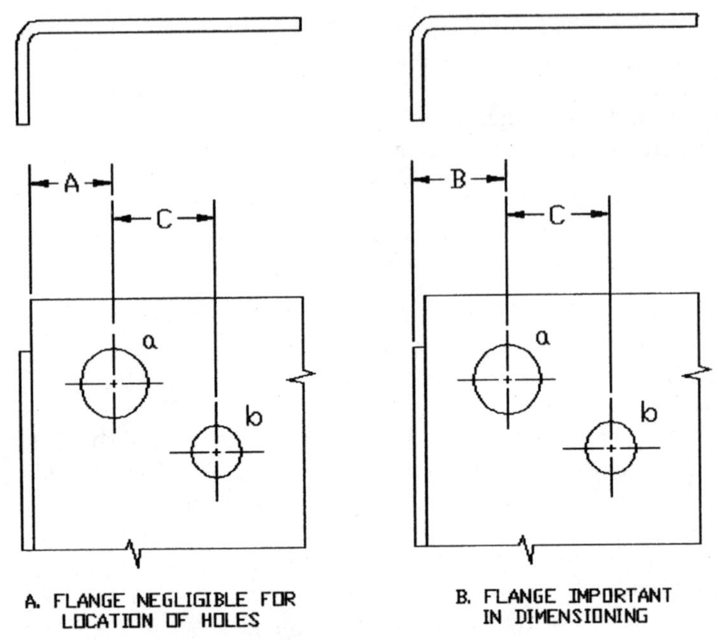


**Figure 1-39** Dimensions and their control.

hardness of material, the condition of tooling, and the variation in material thickness all have a tremendous influence on the distance *B* and subsequently distance *C* in the figure. Therefore, to play it safe, always first produce the flange, bank off its edge (outer or inner, whatever is appropriate), and produce the remaining work on the part.

Drawn parts are another challenge for an eager die designer. The changes that can be produced in a part by the drawing process are often mind-boggling.

In Fig. 1-40, there are four *a* openings, located on the flange of the drawn part. From the previous text and illustrations, it can be assessed that these holes would be impossible to pierce before the drawing sequence is done. Such an assumption is correct, and the progression of work should be adjusted accordingly:

1. Preblank the part, leaving it attached to the strip by small bridges.
2. Draw the cup shape.
3. Pierce the four small holes *a*.
4. Blank the outline.

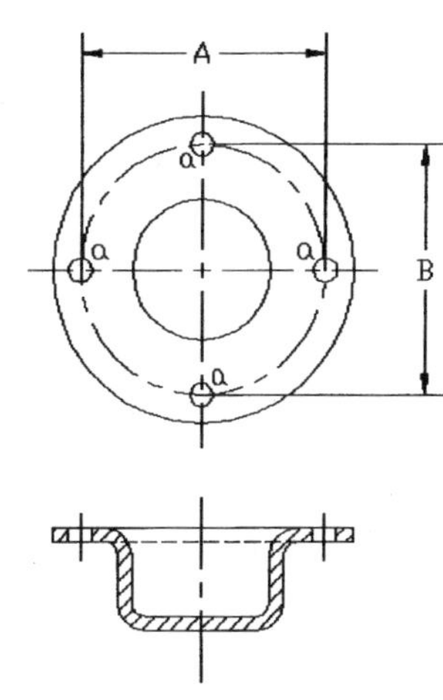


**Figure 1-40** Dimension control in drawn parts.

Such logical evaluation of each given problem must be based on an accurate flat layout of the part, and that's usually all the help die designers need. Slowly, sequence after sequence, they should evaluate the information on the drawing, look for hidden connections, read every little line, and follow up on all the cuts and cross sections, before they resort to sketching the actual strip layout. If the part is too complex, they should break it into small areas and evaluate one after another.

Often it may prove helpful to cut the flat layout out of a piece of paper and demonstrate its progression through the die, adding all bending operations by folding it where appropriate and observing the influence on the part's shape and size.

An accurate flat layout is a gem of information. To produce it, the designer must carefully follow the drawing of the part and omit no information contained on it. The drawing-reading process is complex and involved, and generally the word "read" should be replaced by "study." All drawings should be carefully studied, screened for information they contain, with follow-up on all the coded data, such as specification numbers, coded finish markings, and other procedures that product designers include on them. Often a code, consisting of three letters and two alphabetic symbols, may be used in lieu of specifications several pages long.

Serious students of die design should never neglect the importance of actual observation of running dies. They should be out there, in the shop, whenever possible and watch the tooling in operation or in sharpening, observe the strip, check for the progression sequence, evaluate the direction of punching, and scrutinize the toolmakers as they put the dies together.

Theory is great and highly relevant. But without practical experience, it may often prove inadequate.

## 1-4. Elimination of Secondary Operations through Better Part Design

Secondary operations are always undesirable. They can be slow and laborious and tie up the factory floor for days. They can also be tedious and damage-prone. But above all, they can drive the cost of production to unplanned heights. Therefore, secondary operations should be avoided whenever possible.

The first place to look for their elimination is the design of the part. Often a problematic operation may not actually be needed, or perhaps can be replaced by a much less troublesome process.

Tighter than customary tolerances are sure indicators of a problem within the part's design or function. There the designers are trying to protect themselves, so that in the case of emergency they can point an accusing finger at the shop, which didn't hold a particular dimension that often would be impossible to hold at all.

Often the product's application and function should be scrutinized when looking for the operations worth eliminating. Yet, in die design, the most obvious problems to alleviate are replacements of welded-on segments and replacements of other costly and work-intensive additions.

### 1-4-1. Welding of added-on segments as replaced by bent-up portions of the original part

Welding is a very expensive process, where every inch-long portion of a seam may cost in the vicinity of $10 and more. Welding creates high stresses in the welded part, and it causes distortions and breakages. Often a nondetected flaw may be the reason for a failure of the weld. Additionally, aluminum is very difficult to weld at all.

For these reasons, if welding can be replaced by some other method of joining, that method should be used by all means. But which methods are available to choose from?

True, parts can be joined with adhesives, or rivets, or screws and nuts, to name but a few methods. However, these methods cannot be applied in all circumstances, and certainly few of them can be compared to the sturdiness of the properly welded section.

It seems obvious that when possible, welding should be replaced by such a design change, that still allows for its complete elimination. Some ideas appear in Fig. 1-41.

In Fig. 1-41, two grooved metal blocks are attached to a sheet-metal bracket by two screws each. The groove is intended to retain a shaft, and the bracket is welded to the supporting block. The whole assembly is to fit four openings in a table (not shown).

The points of importance are:

- Groove in the two blocks
- Height of the groove off the block's surface
- Four mounting openings in the block
- Height of the block

If we sketch these areas on a piece of paper and evaluate how they can be made of sheet metal, it only remains to fill in the spaces and the replacement part is found. No assembly work is to be done, no additional hardware is to be used, no costly steel blocks. The part is simple and light, and if made from the proper grade of steel, it will surely be sturdy enough to replace the complicated assembly on the left.

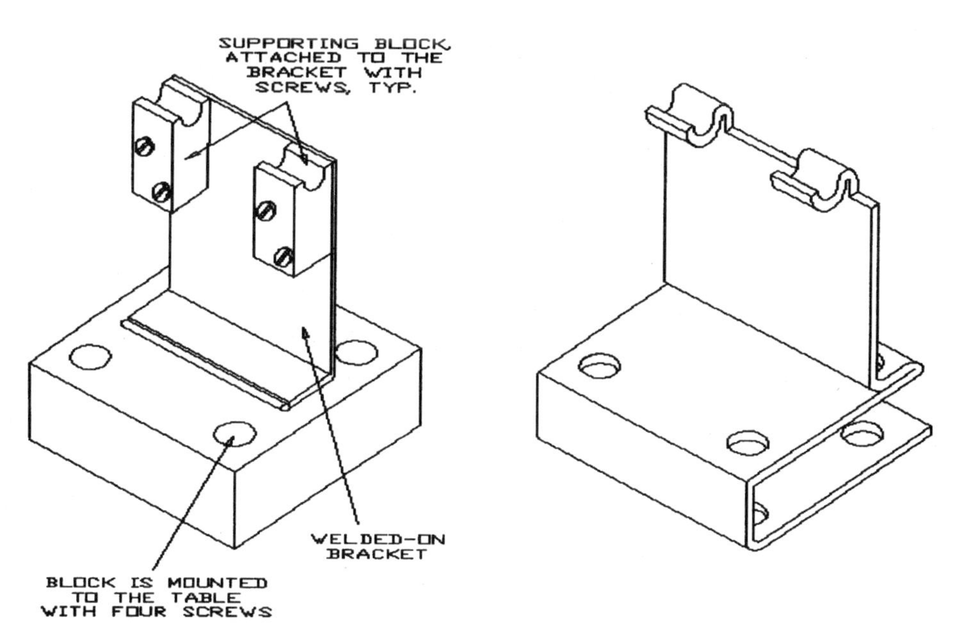


**Figure 1-41** Welded assembly and its sheet-metal replacement.

Shown in Fig. 1-42 are three brackets, attached to the basic sheet-metal part by spot welding.

The spot-welding operation is costly, it may drive the scrap rate much higher than anticipated, and it requires some elaborate fixtures, since the two openings (indicated as bearing surfaces) must line up.

Further, there are four parts to this assembly, all die products, and all costly to make.

Again, the portions most important for the proper evaluation are:

- Two bearing holes
- Their height off the bottom of the bracket
- Two sets of mounting holes on each flange
- Bottom bracket with additional three openings

If we place these in correct proportions on a piece of paper, we may immediately realize that all these points of interest have enough metal sheet space in between to form all these brackets, with no harm done to the main part. Not only will such a redesigned part be cheaper to produce, but the location and accuracy of the two bearing surfaces will be much more controllable and precise.

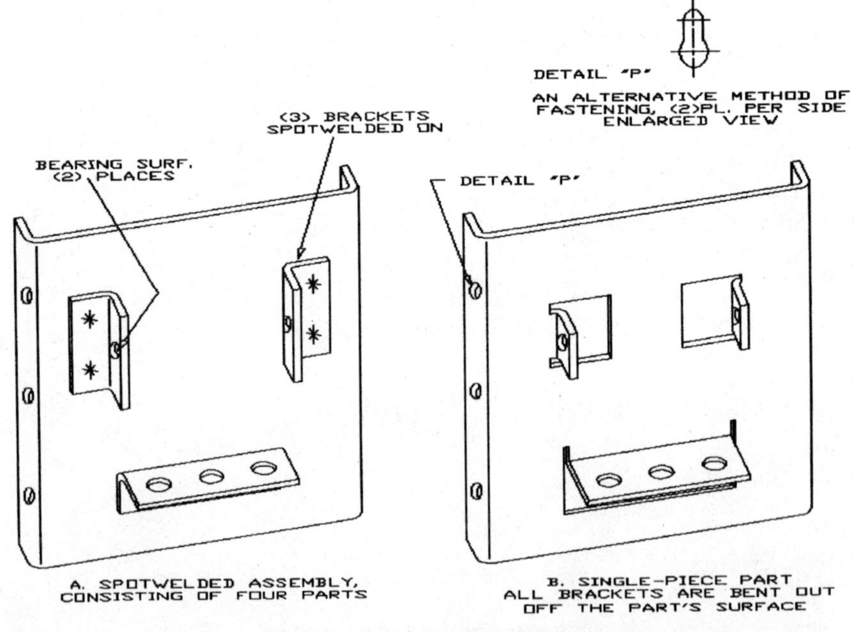


**Figure 1-42** Spot-welded assembly, as replaced by a single sheet-metal part.

If we scrutinize the part further, we may begin to question the necessity of three mounting holes on each flange. A single screw can carry quite a load before breaking off. It also takes much more time to assemble six screws with washers and nuts than two. If the part were to be used for equipment produced a few pieces at a time, the mounting holes would be better left alone. But if these products run into hundreds or thousands, a change in design would be more than justified.

The best way to replace a group of assembly screws is to create posts on the adjoining members of the assembly, slide the bracket over them (using perhaps an alternative type of opening, shown in detail P of Fig. 1-42), and secure in place with only one screw per side. The posts may be bent-up pieces of sheet metal of the connecting part, formed to resemble hooks, and they surely will carry the load quite nicely. Screws would be there mainly for security against shifting or otherwise displacing.

### 1-4-2. Use of sheet-metal hardware in replacement of drilled and tapped holes, offsets, bosses, etc.

Sheet-metal hardware is quite commonly used in replacement of previously drawn and threaded bosses, in lieu of additional blocks, often attached by quite labor-intensive means, or as a replacement of various brackets, supports, or standoffs.

For heavier work, welded nuts, screws, and pins can be used. They will suffice for materials of 0.025 to 0.187″ sheet thickness, and where spot welding is not objectionable for design or function requirements.

Three projections are located on the side to be attached to the connecting surface (Fig. 1-43), and they allow for easy fusing of the hardware to the other component (Fig. 1-44). The method of assembly is either spot welding or projection welding.

For lighter-gauge applications, up to 0.090″ sheet thickness, clinching fasteners will prove quite useful. They come in a wide range of designs and applications:

**Figure 1-43** Projection weld screw. (*Reprinted with permission from The Ohio Nut and Bolt Co.*)

**Figure 1-44** Flanged weld nut. (*Reprinted with permission from The Ohio Nut and Bolt Co.*)

- Self-clinching fasteners are easily squeeze-installed into a punched or drilled hole, where they lock themselves in place, flush with the opposite side of the sheet.
- Floating fasteners allow for ± 0.015″ mating hole misalignment.
- Blind and threaded standoffs are commonly used for printed circuits application.
- Self-clinching studs (Fig. 1-45) and nuts (Fig. 1-46), pins, and standoffs with a snap top adjustment are a continuation of a wide assortment of this type of hardware.

### 1-4-3. Die-produced parts to replace other products, such as castings, forgings, or plastics

Many die-stamped parts are used to replace castings, forgings, and parts made by various other methods. Die production is faster and more precise and consistent. Dimensions are easily controlled and no subsequent grinding, drilling, metal removal, and other extensive finishing is required.

Lately, however, the trend is to replace some sheet-metal parts with plastics, often to the disadvantage of the buyer. Everything has its time and place, and plastic materials, even though displaying many admirable properties, sometimes just don't suffice. Sheet metal is defi-

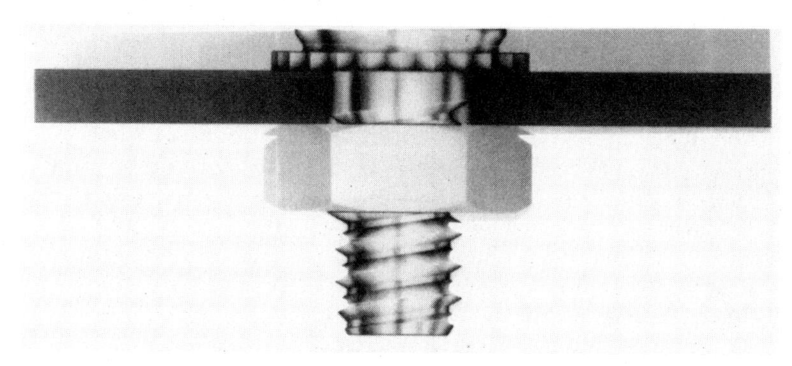

**Figure 1-45** Concealed-head, self-clinching threaded stud. (*Reprinted with permission from Penn Engineering & Mfg. Corp., Danboro, PA.*)

**Figure 1-46** Self-clinching nut. (*Reprinted with permission from Penn Engineering & Mfg. Corp., Danboro, PA.*)

nitely more sturdy and flexible, the effect of its aging is more negligible, and the weathering effect has no influence on it if properly coated.

Plastics are much less sturdy, they change colors with time, they suffer from weather influence, and the aging process may cause their total disintegration. If they are soft and yielding, their life cycle is often threatened. If they're hard and inflexible, they break. In some cases, where plastic parts replace products previously made of metal, a wire frame or other metal framework must be used for support. Extensive ribbing and wall support is common with plastics as well, and it makes the products harder to service and maintain.

Plastic parts used in high-heat areas such as car front panels often disintegrate with time and literally "dust away." A threaded plastic part takes the screw only once, the first time. With every additional removal and attachment the functionality of the thread vanishes.

It is obvious that there are many opportunities for the eager die designer who is willing to devote the time and brain capacities to the task of innovating the innovated. The sky is the limit.

## 1-5. Advantage of Sheet-Metal Application

A definite advantage of sheet-metal parts is the uniform thickness of the material, which can be controlled up to a quite close range of tolerances. Industry standards on flat sheets are as shown in Table 1-4. Anything tighter is more costly.

Tolerances for die-punched holes can be considered quite impressive. Some basic numbers are given in Table 1-5.

Perhaps it should be pointed out that the tightest linear tolerance range of current automatic punch presses is ±0.005″ for short distances only.

**TABLE 1-4 Thickness Tolerances of Strip Stock**

Thickness tolerance plus or minus. All values are in inches. Tolerance is measured across the width, but not less than 3/8″ off the edge.

Aluminum Sheet

| Stock thickness, in | Width: Up to 18″ | 18–36″ | 36–48″ | 48–54″ |
|---|---|---|---|---|
| 0.011–0.017 | 0.0015 | 0.0015 | 0.0025 | 0.0035 |
| 0.018–0.028 | 0.0015 | 0.002 | 0.0025 | 0.0035 |
| 0.029–0.036 | 0.002 | 0.002 | 0.0025 | 0.004 |
| 0.037–0.045 | 0.002 | 0.0025 | 0.003 | 0.004 |
| 0.046–0.068 | 0.0025 | 0.003 | 0.004 | 0.005 |
| 0.069–0.076 | 0.003 | 0.003 | 0.004 | 0.005 |
| 0.077–0.096 | 0.0035 | 0.0035 | 0.004 | 0.005 |
| 0.097–0.108 | 0.004 | 0.004 | 0.005 | 0.005 |
| 0.109–0.140 | 0.0045 | 0.0045 | 0.005 | 0.005 |
| 0.141–0.172 | 0.006 | 0.006 | 0.008 | 0.008 |
| 0.173–0.203 | 0.007 | 0.007 | 0.010 | 0.010 |
| 0.204–0.249 | 0.009 | 0.009 | 0.011 | 0.011 |

Carbon Steel, Cold Rolled

| Stock thickness, in | Width: Up to 3″ | 3–6″ | 6–9″ | 9–12″ |
|---|---|---|---|---|
| 0.011–0.017 | 0.0075 | 0.0075 | 0.001 | 0.001 |
| 0.020–0.023 | 0.001 | 0.001 | 0.0015 | 0.0015 |
| 0.020–0.026 | 0.001 | 0.0015 | 0.0015 | 0.0015 |

**TABLE 1-4 Thickness Tolerances of Strip Stock (*Continued*)**

Thickness tolerance plus or minus. All values are in inches. Tolerance is measured across the width, but not less than 3/8″ off the edge.

Carbon Steel, Cold Rolled (*Continued*)

| | Width | | | |
|---|---|---|---|---|
| Stock thickness, in | Up to 3″ | 3–6″ | 6–9″ | 9–12″ |
| 0.026–0.032 | 0.0015 | 0.0015 | 0.002 | 0.002 |
| 0.032–0.035 | 0.0015 | 0.002 | 0.002 | 0.002 |
| 0.035–0.040 | 0.002 | 0.002 | 0.002 | 0.002 |
| 0.040–0.069 | 0.002 | 0.0025 | 0.0025 | 0.0025 |
| 0.069–0.100 | 0.002 | 0.0025 | 0.003 | 0.003 |
| 0.100–0.161 | 0.002 | 0.003 | 0.003 | 0.003 |
| 0.161–0.200 | 0.0035 | 0.004 | 0.004 | 0.0045 |
| 0.200–0.250 | 0.004 | 0.0045 | 0.0045 | 0.005 |

Carbon Steel, Hot Rolled

| | Width | | | |
|---|---|---|---|---|
| Stock thickness, in | Up to 3″ | 3–6″ | 6–9″ | 12–15″ |
| 0.025–0.034 | 0.003 | | | |
| 0.034–0.044 | 0.003 | 0.003 | | |
| 0.044–0.056 | 0.003 | 0.003 | 0.004 | 0.005 |
| 0.056–0.097 | 0.004 | 0.005 | 0.005 | 0.006 |
| 0.098–0.118 | 0.004 | 0.005 | 0.005 | 0.007 |
| 0.118–0.187 | 0.005 | 0.005 | 0.005 | 0.007 |
| 0.187–0.203 | 0.006 | 0.006 | 0.006 | 0.007 |
| 0.203–0.230 | | | 0.006 | 0.007 |

Stainless Steel

| | Width | | | |
|---|---|---|---|---|
| Stock thickness, in | Up to 3″ | 3–6″ | 6–9″ | 9–12″ |
| 0.011–0.012 | 0.001 | 0.001 | 0.001 | 0.001 |
| 0.012–0.013 | 0.001 | 0.001 | 0.001 | 0.0015 |
| 0.013–0.019 | 0.001 | 0.001 | 0.0015 | 0.0015 |
| 0.020–0.025 | 0.001 | 0.0015 | 0.0015 | 0.0015 |
| 0.026–0.029 | 0.0015 | 0.0015 | 0.0015 | 0.0015 |
| 0.0291–0.034 | 0.0015 | 0.002 | 0.002 | 0.002 |
| 0.035–0.040 | 0.002 | 0.002 | 0.002 | 0.002 |
| 0.041–0.049 | 0.002 | 0.0025 | 0.003 | 0.003 |
| 0.050–0.099 | 0.002 | 0.003 | 0.003 | 0.003 |
| 0.100–0.160 | 0.002 | 0.003 | 0.004 | 0.004 |
| 0.161–0.187 | 0.003 | 0.004 | 0.004 | 0.004 |

Sheet-metal parts are often of utmost importance as covers and liners in electrical products, where they provide for radio-frequency (rf) shielding. With covers made of plastic, the particles of rf shielding material can be added into the plastic melt or later spray-applied to the inner surfaces of finished products. However, this can become impractical or too costly.

**TABLE 1-5 Hole Size Tolerances**

| Hole diameter, in | Material thickness | | | | |
|---|---|---|---|---|---|
| | 0.020–0.040″ | 0.041–0.070″ | 0.071–0.093″ | 0.094–0.156″ | 0.157–0.250″ |
| 0.125 | 0.002 | 0.005 | 0.008 | 0.010 | |
| 0.141 | 0.003 | 0.006 | 0.008 | 0.010 | |
| 0.250 | 0.004 | 0.008 | 0.010 | 0.011 | 0.020 |
| 0.437 | 0.006 | 0.008 | 0.010 | 0.011 | 0.020 |
| 0.687 | 0.009 | 0.010 | 0.010 | 0.011 | 0.020 |

The magnetic or nonmagnetic properties are other factors in sheet-metal use. The material's resistance to heat and fire, along with its capability of heat dissipation, are exceeded only by ceramics.

Chapter

# 2

# Theory of Sheet-Metal Behavior

## 2-1. Sheet Metal and Its Behavior in the Metal Stamping Process

The metal stamping process can alter sheet-metal material in many ways. Parts may be blanked, pierced, drawn, formed, or embossed, just to name a few basic operations. Each of these processes exerts its influence upon the structure of the material: that of the part and that of the scrap.

Often, a congested piercing can cause stresses that will produce an increase in area measurements of that particular section, which is called bulging, or oilcan effect (from the oilcan's snapping back and forth). Forming or drawing, on the other hand, can produce wrinkling, tearing, ironing, or undesirable folding of metal.

## 2-2. Plasticity Theories

In an attempt to control the material-related defects in sheet metal, several theories on its plasticity developed. Plasticity of metal is the capacity to withstand the application of force, which—when excessive—may produce its deformation. When this force is small enough to fit within that material's *yield strength* limit, the deformation is only temporary, and after the release of the load, the material returns to its initial state. However, when the force exceeds the *yield strength* of that particular material, the resulting deformation of its crystal lattice remains permanent, and the part is permanently deformed, or perhaps, formed.

Stress force applied to the material can be categorized to fit into one of the three following groups:

- **Linear influence of stress,** during which the stress is applied along a single axis only, with the two other axes remaining stress-free
- **Plane-type influence of stress,** with stresses applied along any two axes, while the third axis remains stress-free
- **Volumetric influence of stress,** where stresses act along all three axes

Needless to say, the vast majority of metal-forming processes belong in the third, volumetric, group.

*Plasticity theory* is a mathematically oriented approach to evaluation of metal-altering (forming) processes. However, its mathematical equations, when used to solve actual problems, were often found quite inadequate and invalid. For that reason, a method called an *elementary plasticity theory* was developed.

According to the elementary plasticity theory, various forces can be calculated without immediately considering the metallurgic properties of the material, and this way their behavior and their influence during metal's shape-altering (forming) process can be predicted. This theory, based on actual records of macroscopic observations of the deformation stage, deals with concrete data pertaining to particular qualities of the material in question, such as the stress-strain rate and yield criteria in tension and compression. A material's background data, such as metallurgical processes chosen for its manufacture, are mostly omitted.

During the metal-forming (or deforming) process, a displacement of metal material occurs, and its direction coincides with the direction of forces acting upon it. If we take a small particle of material and make it a representative of the whole mass, the displacement looks similar to that shown in Fig. 2-1.

The amount of time such a displacement needs to occur is called $\Delta t$. During that period points $A$, $B$, $C$ will change their location to $A'$, $B'$, and $C'$, in congruence with the direction of forces acting upon them, and this movement will permanently alter the shape of that particle. We may generalize that if (at least) two points in any material change their relative location within the time interval $\Delta t$, that particle is exposed to the influences of deformation, or strain.

The type of deformation most often considered in metal-forming processes is permanent deformation. The nonpermanent deformation, or elasticity, is usually neglected unless it falls into the category called *spring-back,* which will be discussed later.

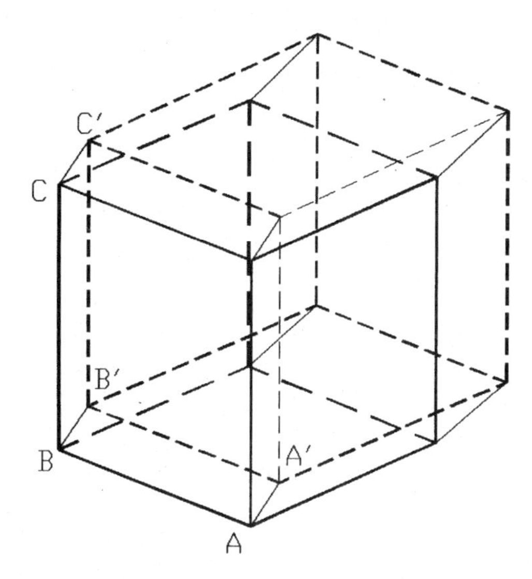


**Figure 2-1** Displacement of a particle in time $\Delta t$.

Two definitions of the material's condition apply to the process of deformation: *homogeneous* and *anisotropic*. A material is homogeneous when each of its particles has the same properties. The properties of an anisotropic material differ with the coordinate system. For example, wood may be a homogeneous material, since its characteristics are constant throughout its entire volume. But when in deformation, its reaction to the force applied *along* its fibers is different from its reaction to the force coming *across* them. For that reason, wood is an anisotropic yet a homogeneous material.

Metal, prior to forming, is usually homogeneous and isotropic. But after being permanently deformed, it may become *anisotropic,* and if an uneven deformation has been achieved, it would be *inhomogeneous* as well.

*Anisotropy* is the opposite from isotropy. An *isotropic* polycrystalline metallic material is one whose properties are the same in all directions. In forming and drawing operations, anisotropy affects the material flow. Even the deep drawing limits are based on the effects of anisotropy.

Forming abilities of all metal material are known to be temperature-dependent: With higher temperatures, the forming limit of metal is expanded. This is mainly due to the recrystallization processes within the matter, which is temperature-dependent as well and which allows for considerably more elastic behavior when warmed up. (Note: Recrystallizing temperatures may be those below 200°C or ~400°F, held for an hour, as the exposure to these does not alter the material's strength and yet it alters its lattice.)

If we cut a body, upon which various stresses are acting, in half, we disrupt the forces holding that body together. We then add a resultant to these stresses and attach it to the plane of the cut to balance their influence out. According to the direction of this resultant, we recognize normal stress δ (compressive or tensile) or tangential stress $\tau$.

Several theories describe the flow of material during various forming processes. All these approaches may be grouped within two basic categories:

1. **Analytical methods,** consisting of: 2-2-1. Strip or slab theory; 2-2-2. Slip-line theory; 2-2-3. Bound theorems; 2-2-4. Finite element analysis; and 2-2-5. Theory of weighted residuals.
2. **Experimental-analytic methods,** such as: 2-2-6. Axisymmetric forming processes; 2-2-7. Methods utilizing the visibility of plastic flow; 2-2-8. Hardness method; 2-2-9. Macroscopic methods; and 2-2-10. Miscellaneous theories.

### 2-2-1. Strip or slab theory (the equilibrium method, see Fig. 2-2a)

Strip (or slab) theory is used for evaluation of processes that fall into the plane deformation group, such as rolling. This method isolates a representative volumnar element in the material, $d_x$, and on the basis of its behavior, it assumes that:

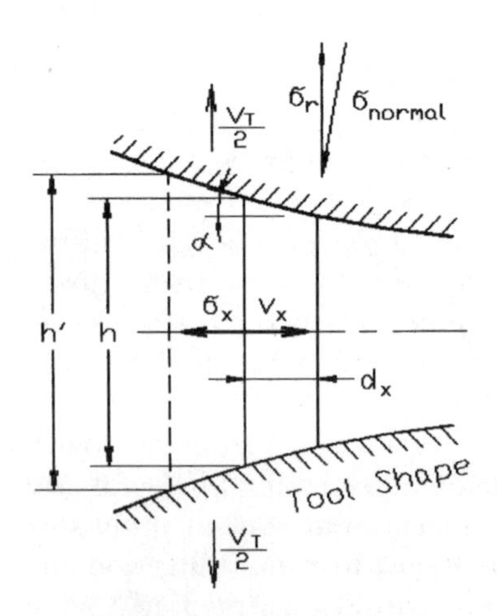


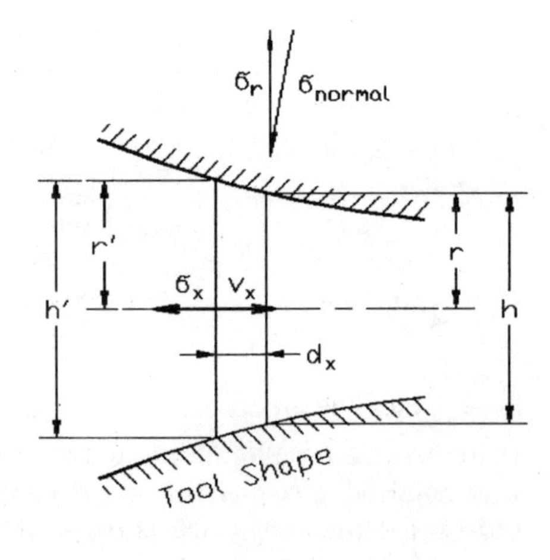


**Figure 2-2** Plasticity theories.

- The velocity of material $v_x$ is constant at every point of a cross section of the flow (see Fig. 2-2*a*).
- The volume of material is constant as well.
- Friction between the material and the tooling exists. Such friction is constant, and the tooling is symmetrical.
- The rate of change in height of the strip $h_R$ depends on the slope of the tool in point of contact, on the velocity $v_T$ at which the opposite boundaries approach each other, and on the velocity of the strip $v_x$.
- The weight and inertia forces are negligible.

A formula which calculates the rate of change in height of a moving tool is

$$h_R = \frac{\delta h}{\delta t} + v_T = \frac{\delta h}{\delta x} v_x + v_T \tag{2-1}$$

where $h_R$ is the rate of change between $h$ and $h'$. This formula, when applied to a stationary tool, becomes

$$h_R = \frac{\delta h}{\delta t} = \frac{\delta h}{\delta x} v_x \tag{2-2}$$

This type of motion in metal results in a homogeneous matter after forming. The natural strain along $y$ axis $\varphi_y$ can be calculated as follows:

$$\varphi_y = \ln \frac{h'}{h} \tag{2-3}$$

and the strain rate $\varphi_R$ becomes

$$\varphi_R = \frac{\delta \varphi_y}{\delta t} = \frac{h_R}{h} \tag{2-4}$$

### 2-2-2. Slip-line theory

Prior to the development of three-dimensional models, this theory was used for assessment of the plane deformation, since it indicates the direction of the maximum shear stress and maximum shear strain rates. These values are used for a numerical evaluation of stress distribution throughout the deformed portion and its boundaries, upon which the loads exerted by tools are acting. Later, this theory was extended to some cases of axial symmetry, even though it could not be reasonably applied to the majority of axisymmetric problems. Its name is derived from the physical observation of plastic, the flow of which seems to be a result of microscopic slip on an atomic scale, along its crystallographic plane.

The application of this method consists of constructing a network of lines, also called Lüders' lines (shown in Fig. 2-3). These lines, when applied to the surface of the part, are used for evaluation of those forming processes during which there is an absence of strain hardening. The technique of assessment is not time-dependent and the forming pressure is considered static and isothermic.

It is given that the slip lines α and ß are orthogonal and that they cut through main vectors of tension under an angle $\pi/4$. Differential equations of these two types of lines will come out as tangents:

$$\alpha\text{-calculation:} \quad \frac{dy}{dx} = \tan\varphi \tag{2-5}$$

$$\beta\text{-calculation:} \quad \frac{dy}{dx} = \tan\left(\varphi + \frac{\pi}{2}\right) = -\cot\varphi \tag{2-6}$$

### 2-2-3. Bound theorems

This method of evaluation divides the surface into triangular segments (see Fig. 2-4) sliding across each other. Inside these segments is a homogeneous velocity field, with constant velocity values at any point within the given area. It is speculated that the rate of work done by such surficial segments is:

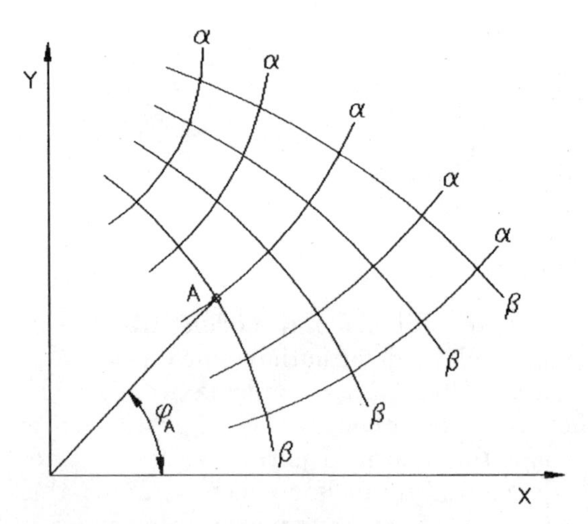


**Figure 2-3** Network of slip lines.

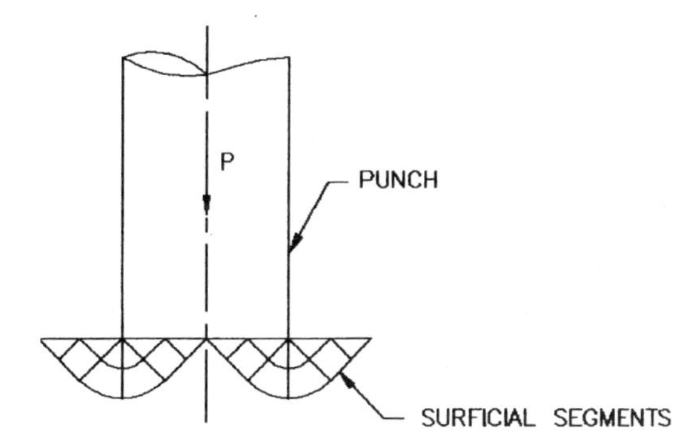


**Figure 2-4** Bound theorem's representation.

- With *lower bound theorem* greater than or equal to the rate of work performed by any other segments, that can satisfy the same equilibrium conditions and stress boundary descriptions. A sum of instantaneous outputs of inner forces is considered equal to the forming pressure needed to alter the material. However, it is not always easy to locate statistically similar stress areas, for which reason upper bound theorem is often preferred as a means of evaluation.
- With *upper bound theorem* the rate of work done by surficial segments is smaller than or equal to the rate of work attributable to any other segments, displaying the same velocity arrangements, whose material content can be considered incompressible. Calculation of such output will provide an approximate assessment of the forming force on the basis of which tooling and machinery may be chosen.

### 2-2-4. Finite element method

Lately, this is the most often used method for its complex calculations, which can now be so readily solved by computers. The material in question is divided into minute particles, joined at their corners. The stress throughout each particle is considered uniform, and the distortion is computed by using any conventional mathematical theory. The final behavior of the material is obtained through evaluation of the influences these particles exert on each other.

Naturally, since all actual processes within the material are very complex, their mathematical and graphical representation is simplified. Even elements constituting the representation of the part's construction are but simple geometrical shapes. For example, an analysis of axisymmetrical tension uses triangular or rectangular cross-sectional cuts, as shown in Fig. 2-5.

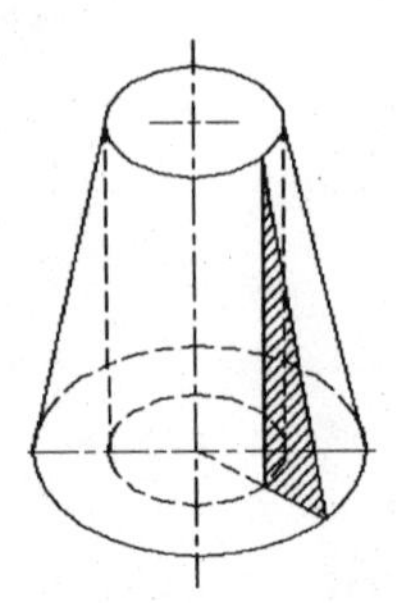
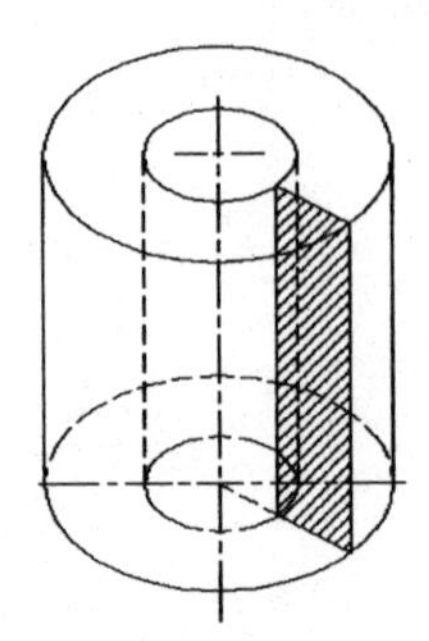

**Figure 2-5** Axisymmetric geometric shapes used in finite element analysis.

### 2-2-5. Theory of weighted residuals

This method can be applied to simple deformation processes, while delivering acceptable results through approximation of stress and strain cases of complex problems. It is based on the law of energy conservation, with the volume of flowing material considered constant, and equilibrium conditions satisfied. Each stress is represented by a series of functions, such as polynomial expressions, outfitted with arbitrary parameters.

It maintains that for small virtual movements ($\delta_u$, $\delta_v$, $\delta_w$), the work of all inner and outer forces is minimal, which is attributable to its equilibrium condition. Therefore the following equation applies:

$$\delta W_{\text{in}} - \delta W_{\text{out}} - \delta W_F = 0 \tag{2-7}$$

Considering the constancy of volume and ignoring elastic changes within the material, the formula may be rewritten as

$$W_{\text{in}} - W_{\text{out}} - W_F = 0 \tag{2-8}$$

This equation actually claims that the work of inner forces ($W_{\text{in}}$), used up for plastic deformation of the material, is in equilibrium with the work of outer active forces ($W_{\text{out}}$) and the value of surficial friction ($W_F$).

### 2-2-6. Axisymmetric forming processes, or disk (tube) theory

This theory assumes that:

- Tooling has no velocity, the only velocity in the process being that of $v_x$.
- The material flow is symmetric along one axis.

The natural strain $\varphi$ can be calculated by using the formula:

$$\varphi = 2 \ln \frac{r'}{r} \tag{2-9}$$

and the strain rate $\varphi_R$ becomes

$$\varphi_R = 2 \frac{r_R}{r} = - \frac{\delta v_x}{\delta x} \tag{2-10}$$

stresses along the material flow $\delta_f$ (see Fig. 2-2*b*) will be

$$\pm \delta_f = \delta_x - \delta_r \tag{2-11}$$

### 2-2-7. Methods utilizing the visibility of plastic flow

Since exact mathematical representation of changes within the formed part is practically impossible, allowing for visibility of changes using the elementary surface representation may be utilized instead. There are three methods of such depiction.

**2-2-7-1. Visioplasticity.** Visioplasticity is based on an assessment of vectors of velocity, using a grid structure, which is assigned to the surface of material, often by photographic means. This pattern, showing a distortion in areas of deformation, creates a dependency of directional tensors of tension and those of deformation.

The velocity of the flow can be figured out from the distances that points under deformation have moved. These velocities are considered constant, and stresses are represented by grid lines, not by arbitrary points. The equilibrium is no longer applicable with this method. Such a technique cannot be used to predict further behavior of the material, but it is valued for the detailed analysis of the distribution of stresses, strain, and strain rate in any portion of the deformed area.

**2-2-7-2. Photoelasticity.** Photoelasticity uses models made of linear elastic materials (such as organic glass), which are exposed to outside influences within their elastic areas only. In comparison to this method, photoplasticity uses models made of linear plastic materials.

Both these methods are based on the principle of optical refraction, where the force-affected areas display a difference in light intensity. Dark areas contain similarly oriented tension, while colored stripes depict differences between tension and deformation.

**2-2-7-3. Moiré method.** This method is based on the principle of geometric interferences. A network applied to the product's surface changes its linear arrangement under deformation. By attaching a reference network of the same lines, unaltered by any outside influences, a comparison of the two patterns is possible. Dark and light abnormalities, or moiré stripes, may be observed where changes of the material structure occur.

### 2-2-8. Hardness method

As a method of assessment, the hardness comparison method considers a dependency between the hardness of deformed material and that which is unaltered in any way. By the distribution of variation in hardness, the area of deformation and the amount of tension within the material may be established.

### 2-2-9. Macroscopic methods

Macroscopic methods allow for observation of material changes, depicted by alteration of patterns, assigned to the surface of parts. The network of lines may be printed upon the part or attached by any other means. A sample of patterns is shown in Fig. 2-6.

### 2-2-10. Miscellaneous theories

During the process of development and refinement of plasticity theories, various researchers added their opinions in the form of calcula-

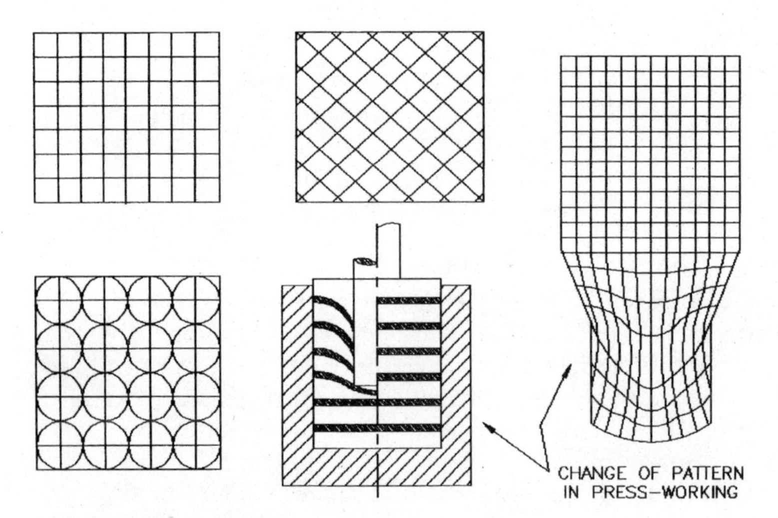


**Figure 2-6** Sample patterns.

tions or specifications. The slip-line theory is based on the research of Hencky and Prandtl, with von Mises adding his flow rule to it. Geiringer supplied the equations for the velocity assessment. The weighted residuals method gained through the work of Lagrange, while Thompson established the dependency of the speed of change and tension of material in visioplasticity. A considerable input was supplied in the form of Prandtl-Reuss and von Mises research.

**2-2-10-1. Prandtl-Reuss plasticity theory.** This theory assumes that with application of force on a material, elastic strain prevails until yield strength has been arrived at. From there, the force does not have to be increased for the permanent deformation to occur (an elastic perfectly plastic material).

**2-2-10-2. Von Mises plasticity theory.** This theory is based on a simplified assumption that the material is under no strain until the yield strength has been attained. The material therefore behaves like a rigid body (a rigid, perfectly plastic material). This theory is not so erroneous as it may seem at the first glance, for permanent strains are much greater than their smaller and consequently negligible elastic counterparts.

However, the exact solution of the plasticity of material is not easily obtainable because of the great complexity of the problem. Therefore, only simple results must be sought, with complex problems simplified and approximated.

## 2-3. Tribology—Surface Friction in Forming and Drawing

Friction in metal stamping can have many beneficial as well as detrimental effects on the tooling and quality of produced parts. It increases the surficial pressure between the tool and sheet-metal material, which results in deformation of both, with subsequent degradation of surface quality and wear of tooling. The demand for press force is increased, sometimes considerably.

Since the area of contact between the part and its tooling constantly changes, the distortion and degradation influences widespread portions of it. The roughing effect on the surface of tooling causes the actual contact areas to diminish in size, or localize, which subsequently increases the frictional influences in each of them, and a faster deterioration of the tooling and parts follows.

The heat, along with the damaging effect of surficial pressure, tears out small portions of sheet-metal material, attaching it permanently to the tooling or elsewhere within the place of contact. Such small pieces are as if welded; they're difficult to remove and their presence further affects the quality of parts, their dimensional accuracy, and the condition of the tooling. For example, the force needed to overcome friction during the backward extrusion of a cup was found to amount to approximately 40 percent of the total force exerted by the punch.

The problem of friction is quite complex and cannot be readily solved. On the other hand, some processes, such as metal forming, depend on a certain amount of friction, the removal of which may not be beneficial to the forming process at all. In the absence of this friction, grave problems with material retention may emerge, which may result in parts that are perhaps impossible to form at all. Additionally, such a condition may generate a completely different set of forces acting against the tooling, which may produce such an inner strain within its material structure that an internal distortion and collapse are unavoidable.

The only means of controlling friction are lubricants. These separate adjoining surfaces by providing an isolated layer between them with completely different physical and mechanical properties. Application of various different lubricants may alter their effect upon the particular frictional forces to almost perfection.

There are lubricants that are immune to higher temperatures, lubricants that tolerate extreme pressures, high-viscosity lubricants, low-viscosity lubricants, and other variations.

### 2-3-1. Types of friction

Various materials, in combination with various types of lubricants, will produce different types of friction:

- **Static, or dry friction,** created between two metallic surfaces, with no lubricant added. The friction mechanism depends on the physical properties of the two materials in contact. A metallic lubricant (e.g., lead, zinc, tin, or copper) may improve this condition.
- **Boundary friction,** where two surfaces are separated by a layer of some nonmetallic lubricant a few molecules thin. The shear strength of the lubricating material is low, resulting in low friction.
- **Hydrodynamic friction,** where two surfaces are totally separated by a viscous lubricant of hydrodynamic qualities. In such a case, friction depends strictly on the properties of the lubricant.
- **Mixed friction,** the mixture of the above conditions. This type is the most frequently encountered situation in metal-forming processes.

Out of all metal-forming processes, only a few do not require any surface treatment or coating when it comes to friction. These are open die forming, spreading, some bending operations, and extrusion of easily deformable materials. All other metal forming depends on the use of proper lubricants. Even die forging requires a surface treatment of raw material, in this case for the protection of the die itself.

### 2-3-2. Lubricants

The lubricant's main duty is to diminish the influence of friction between the tooling and the material. Ideally, lubricants should also act as a coolant and thermal insulator, while not being responsible for any detrimental action against the tooling or the material, the press equipment, or the operator. The lubricant should not cause rusting of metal parts and should be easily removable by some accessible means.

Lubricants are of utmost importance in the forming and drawing processes, which are divisible into two categories, based on the type of lubricant used:

- *Wet drawing,* using mineral oils, vegetable oils, fat, fatty acids, soap, and water
- *Dry drawing,* using metallic coatings (Cu, Zn, brass) with graphite or emulsions, Ca-Na stearate on lime, borax or oxalate, chlorinated wax, or soap phosphate

In metal forming, the danger of entrapping the lubricant with the fast action of the tooling presents additional possibilities of surface deformation. Usually, areas affected by a restrained lubricant display a sudden roughness, often resembling a matte finish.

The actual process of lubrication is provided by only a few basic ingredients added to the composition of lubricating materials. These are:

- *Mineral oils,* which are petroleum derivatives such as motor oil, transmission fluid, and SAE oils.
- *Water-soluble oils,* which are a combination of mineral oils, adjusted by an addition of other elements to become emulsifiable with water.
- *Fats and fatty oils,* most often of vegetable or animal origin, such as lard oil, fish oil, tallow, all vegetable oils, and beeswax.
- *Fatty acids,* such as oleic and stearic acids, generated from fatty oils.
- *Chlorinated oils,* a combination of fatty oils and chlorine.
- *Soaps,* which are basically water-soluble portions of fatty acids, combined with the alkali metals.

- *Metallic soaps,* which are insoluble in water, such as aluminum stearate and zinc stearate.
- *Sulfurized oils,* or hydrocarbons, treated with sulfur.
- *Pigments,* such as graphite, talc, or lead. These are actually minute particles of solids, not soluble in water, fats, or oil. They are often supplied in a mixture of oils or fats, where these ingredients provide for their retention and spreading.

## 2-4. Strain Hardening

The action of material forming has a distinct cold-working effect on its structure, producing a *strain hardening* of its mass. Once strain-hardened, a part requires increased forces for additional forming. The influence of strain hardening can be partially alleviated by heat working of the part, which may not always be applicable, as it causes distortion of the surface, coupled with a diminished accuracy.

Another method of neutralization of the strain-hardening effect is the application of heat treatment to the affected parts between forming processes.

Strain hardening is sometimes considered beneficial to the product because of its effect on the part's hardness, with subsequent increase in tensile strength. Often such an influence may justify the use of materials of inferior strength, and cold working them to the required or expected level.

## 2-5. Shear of Metal in Cutting Operation

During any metal-cutting operation, the material is compressed between the punch and die until parted by the act of shearing. The development of compressive and tensile stresses, accompanied by an alteration of the part's edges, causes the material to separate, as shown in Fig. 2-7.

During the metal-cutting process, there are several stages during which the transformation of material takes place (Fig. 2-8). An explanation of the process is necessary in order to understand the behavior of sheet under the punch:

1. **Figure 2-8A:** Clearance between the punch and die is clearly visible, and its amount is crucial to the success of the metal-cutting process. Clearance is the space between the two cutting edges, those of the punch and those of the die. Clearance not only allows for the body of a punch to be contained in the cavity of a die; it also provides for the development of fractures during the cutting process (see Fig. 2-8*c*).

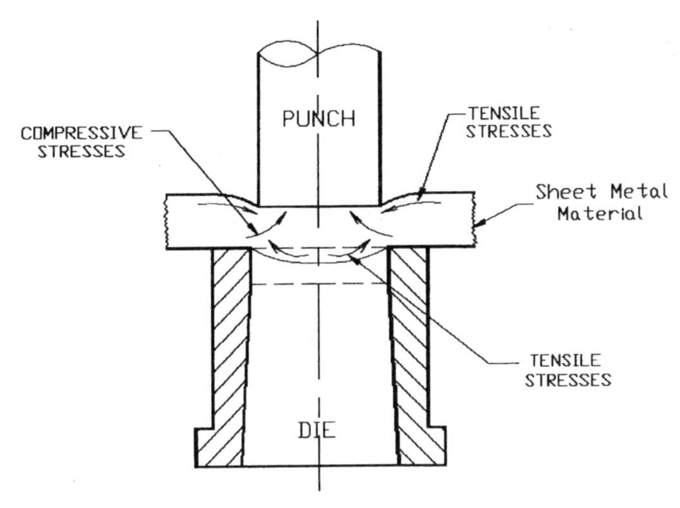


**Figure 2-7** Stresses in shear operation.

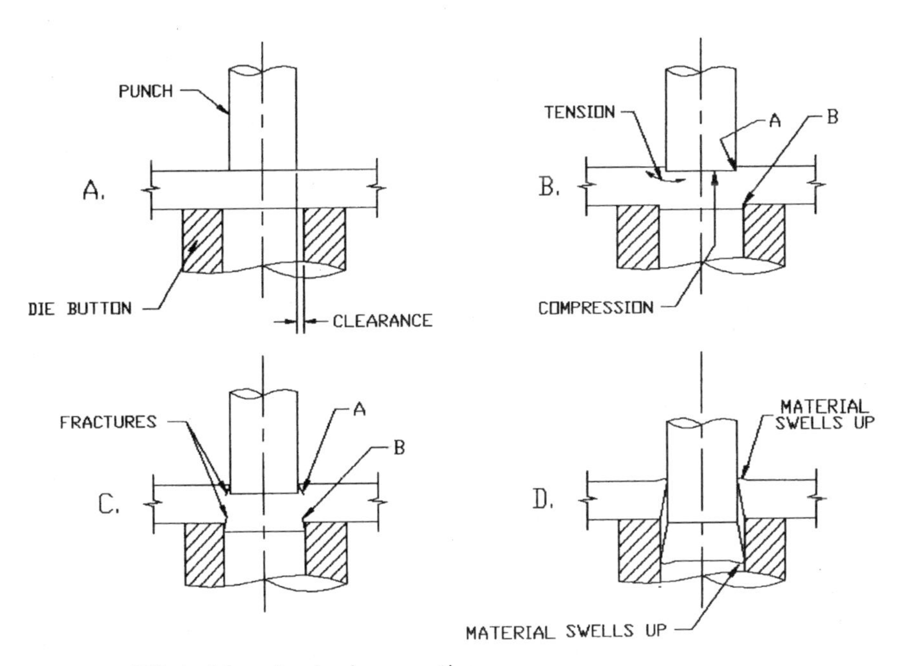


**Figure 2-8** Effect of shear in piercing operation.

2. **Figure 2-8B:** The punch moves down and presses into the material. Stretching occurs at points *A* and *B*, where the stock is in tension; the remaining material under the punch is compressed. However, the material's elastic limit has not been exceeded yet.

3. **Figure 2-8C:** The material is pushed farther down, and fractures begin to form around the corners of both punch and die as the elastic limit of the material is being exceeded. The angle of these fractures depends on the die clearance: If the clearance should be either excessive or too small, this angle may not allow for a smooth connection of the upper and lower fractions, and a rough, jagged cut may result.

4. **Figure 2-8D:** With further descent of the punch, fractures deepen and finally meet. The cutout is separated from the strip and pushed into the die. There, owing to inner stresses thus created, it swells up; the strip also tightens around the punch, prompted by forces from within.

The fracture, angular in its cross section, actually spans through the area of tolerance between punch and die (see Fig. 2-9). Notice the smooth, flat, circumferential band (*A*), usually about one-third of the total material thickness (*t*) in size, with well-sharpened tooling. The remaining two-thirds of the stock thickness are called the breakoff. The upper surface is called the *burnishing side,* and the bottom is the *burr side.* In every punching, piercing, or blanking operation, the burr side is always opposite the punch.

The proper identification of the burr side is of great importance in some secondary operations such as shaving, blanking, and burnishing. Also the visual appeal and the functionability of the part may be ruined should the burr appear at the wrong side.

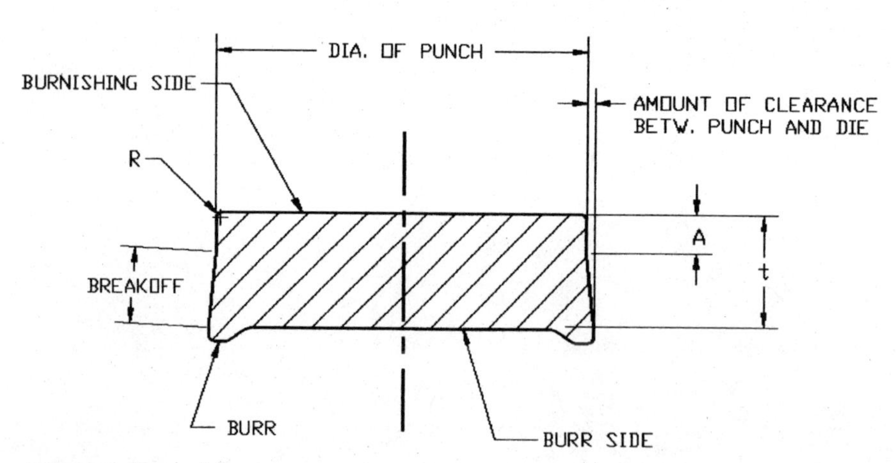


**Figure 2-9** Sheared slug.

## 2-6. Bending and Forming of Sheet-Metal Material

During simple bending, sheet-metal material remains homogeneous and isotropic. No stress residues remain within its mass on termination of a simple bending process.

According to the type of tooling used, bending can be divided into various processes. The three most commonly used are:

- V-die bending
- U-shaped bending
- Rotary bending

Other bending methods include stretch forming (or wrap forming), air forming, and roll forming. However, none of the latter methods are applicable to die work.

In forming, as in bending, there is always one boundary of metal stretched and the opposite one shrunk. In between, somewhere around the middle of the stock thickness, there's an imaginary axis, which is considered neutral. Some believe it to be exactly in the middle, others place it in one third, and the rest use a host of additional ratios.

Similarly, various calculations differ in approach to the location of the neutral axis, as well as in results. Many times the condition of tooling, or the prevailing methods used within the particular shop, material variables, etc., render all such formulas unsuitable. Therefore, with sensitive parts, where the blank dimension is difficult to assess, or when working with an unknown material, it is advisable to construct few temporary punches and dies and run actual tests, recording the results and comparing them to previously performed calculations.

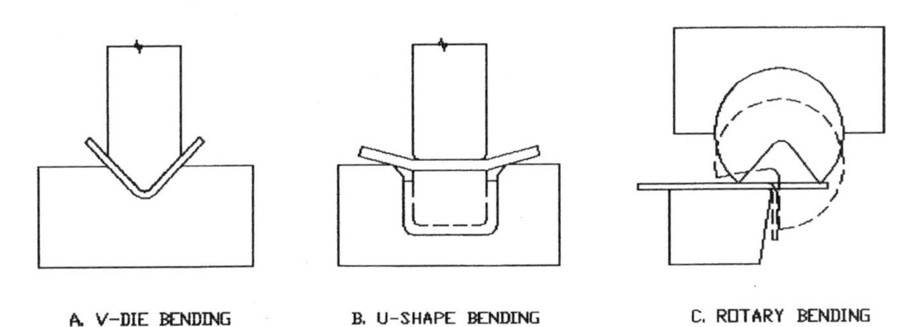


**Figure 2-10** Three basic types of bending.

In bending, as in forming, the size of the bend radius is of great importance. Often a drawing may call for a sharp-corner bend, which someone put down without realizing that such bends are virtually impossible to obtain. After all, if sheet metal were forced into such a bending extreme, it would be cut. The existence of some corner radius is absolutely necessary, and the greater in size, the easier the bending process is. The smallest radii per different stock thicknesses are discussed in Chap. 8.

Forming, even though similar to bending, differs in that it adds some drawing action to the process. Forming utilizes the plastic capacities of the material in a wide range of applications. Mill rolling, extruding, heading, drop forging, and even drawing, swaging, spinning, and bulging can all be considered metal-forming operations.

According to the condition of the formed material, forming processes can be divided into two basic groups: cold forming and hot forming. Both processes may produce parts with their mechanical properties:

- Unaltered
- Temporarily altered
- Permanently altered

This classification is based on elastic limits of various materials. Further division can be obtained by sorting all forming processes with regard to the distribution of stresses in the material as:

- **Tensile forming,** where the deformation is achieved by application of various singular or multitudinal tensile stresses. Examples of such forming are stretch forming, stretch drawing, bulging, expanding, and embossing.
- **Compressive forming,** where the alteration of the part is achieved with the aid of various compressive forces acting upon it. This type of forming is represented by coining, forging, rolling, heading, plunging, and swaging.
- **Tensile and compressive forming combined** include metal spinning, deep drawing, ironing, some types of bulging, flange forming, and pultrusion.

## 2-7. Variations of Stock Thickness in Bending and Forming Operations

In any type of metal-altering processes, the variation in cross section of the sheet-metal material is in direct proportion to the following influences of the:

- Condition and construction of tooling
- Friction between the tooling and the strip
- Compressing forces against the surface of the material
- Influence of the material's own mechanical properties

In bending, should the die surface be rough or should the clearance between the punch and die be inadequate, drawing of the metal will be produced in the bend or in its section. Such alteration of the process is mostly undesirable, as it changes the material's cross section, which in turn alters the size of the finished part. The accumulation of material in certain areas may also be promoted by such a change, along with an excessive burr formation, bulging, buckling, and many other defects.

The material is predisposed to differences in the outcome of various operations because of its grain structure. An additional distortion in thickness may only add to problems and discrepancies.

As mentioned earlier, in simple bending, the material is shrunk on one side of the bend and stretched on the opposite side. This variation is not consistent with all types of bends and materials. Thinner stock and smaller radii will bring about different-sized parts than thicker stocks with larger radii.

Therefore, we may generalize that a bent-up part's final dimensions depend excessively on the radius of the bend with regard to stock thickness. For example, material 0.031″ thick with (inner) bend radius of 0.062″ decreases in size after bending some −0.007″ per bend; the same material with 0.125″ bend radius will decrease lengthwise −0.034″ per bend. (For bend radii allowances, see Chaps. 7 and 8.) But not all material thicknesses and radii sizes decrease the linear length of the part. For example, material 0.062″ thick with an inner bend radius of 0.062″ will increase in length after bending approximately +0.016″; stock 0.125″ thick at a 0.125″ bend radius will increase +0.025″.

It seems obvious that the amount of compression or elongation of the bent-up material varies and therefore the neutral line (refer to Fig. 2-11) cannot be positioned in the middle of the stock. Rather its location will vary along with the thickness of the material and bend radius.

In a drawing operation, where the sheet metal's flat shape is deformed into a cuplike profile, all its available thickness is used up during such a transformation. Depending on the depth of the draw, the metal logically must get thinner and thinner, up to a complete fracture, tearing, and distortion should the process continue. The opposite of metal thinning is its increase in thickness, which can be

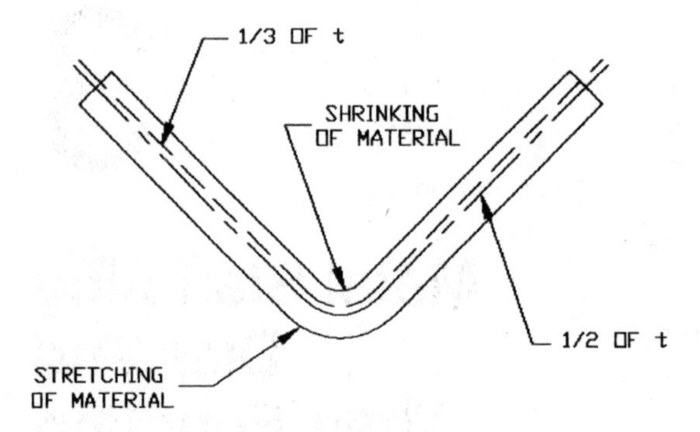


**Figure 2-11** Neutral axis in bending operation.

observed in some drawing operations, where wrinkles and folds are formed.

With coining, necking, forging, and similar work processes, a portion of the part may get thinner, while its other portion will expand. However, such processes, where the material is restricted from free movement by the shape of a die, display a more or less controlled form of thinning and thickening of stock.

Chapter

# 3

# Metal Stamping Dies and Their Function

## 3-1. Description of a Die

A die set is the fundamental portion of every die. It consists of a *lower shoe* (or a *die shoe*) and an *upper shoe,* both machined to be parallel within a few thousandths of an inch. The upper die shoe is (sometimes) provided with a punch shank, by which the whole tool is clamped to the ram of the press. Because of their much greater weight, larger dies are not mounted this way. They are secured to the ram by clamps or bolts. However, sometimes even large die sets may contain the shank, which in such a case is used for centering of the tool in the press. Figures 3-1 and 3-2 show these parts in a compound and a progressive die.

Both die shoes, upper and lower, are connected by *guide pins,* or *guideposts.* These provide for a precise alignment of the two halves during the die operation. The guide pins are made of ground, carburized, and hardened tool steel, and they are firmly embedded in the lower shoe. The upper shoe is equipped with bushings, into which these pins slip-fit.

The *die block,* containing all die buttons, nests, and some spring pads, is firmly attached to the lower die shoe. It is made of tool steel, hardened after machining. The die block is usually a block of steel (solid or sectioned) in which the openings are machined to the contours of punches, plus the amount of tolerance. The opening must have a straight, precise portion at the top, where the punch meets the metal, and an angular clearance hole afterward.

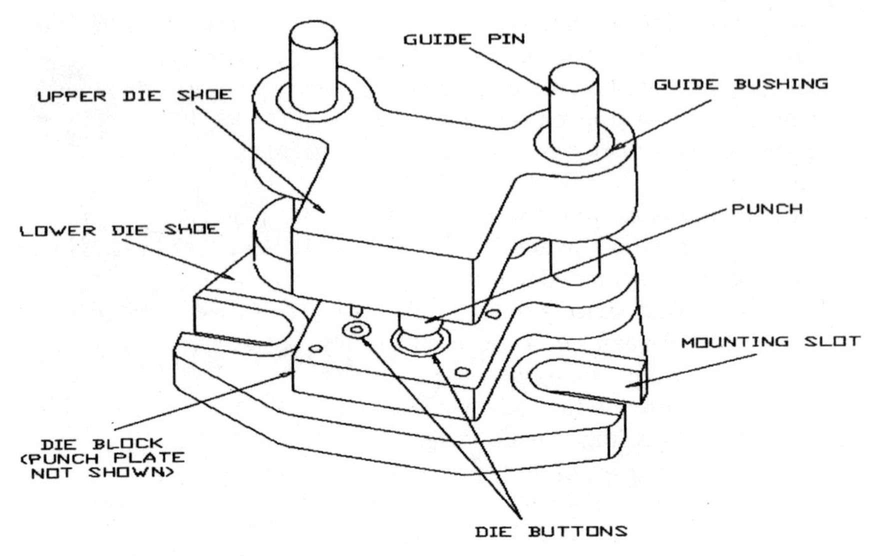


**Figure 3-1** Compound die.

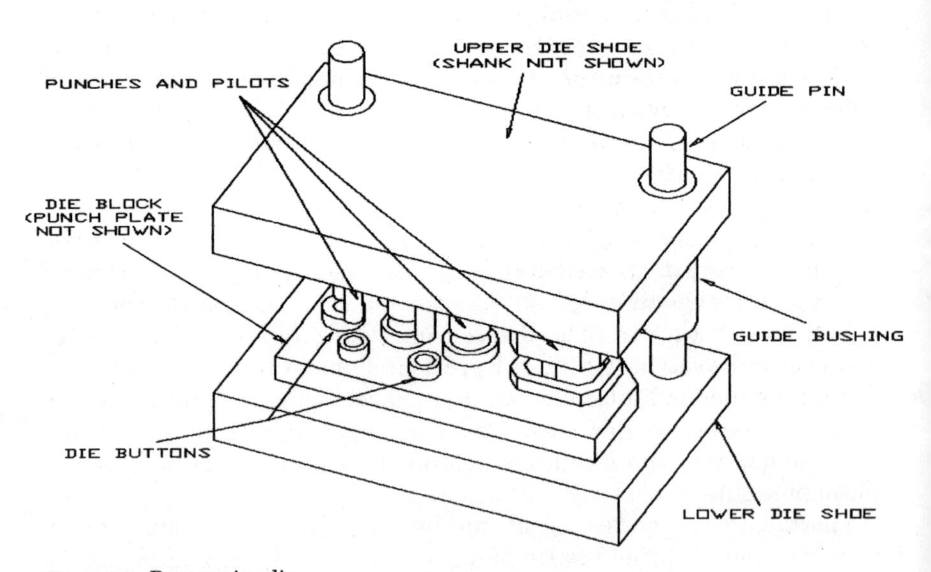


**Figure 3-2** Progressive die.

The *punch plate* is mounted to the upper shoe in much the same manner as the die block. Again, it is made of a hardened tool steel, and it may consist of a single piece of steel or be sectioned. It holds all punches, pilots, spring pads, and other parts of the die. Their sizes and shapes conform with those of the die, minus the tolerance amount.

Both the die block and the punch plate are separated from the die shoe by *backup plates,* whose function is to prevent the punches and dies from becoming embedded in the softer die shoe.

The sheet-metal strip is fed over the die block's upper surface, and it is secured between *guide rails,* or *gauges.* There are two types of gauges: *side gauges,* for guiding the sheet through the die, and *end gauges,* which provide for the positioning of stock under the blanking punch at the beginning of each strip.

The strip is covered up (either whole, or its portions) by the *stripper,* which provides for stripping of the pierced material off the punch. The stripper is often made of cold-rolled steel, and its openings are made to the size and shape of punches, plus some clearance. Sometimes these shapes are cut through the block itself; often they are bushed.

The *stationary stripper* is mounted firmly to the face of the die block, and a channel running its entire bottom length receives the steel strip. The *spring-loaded stripper* is held in its location by the force of springs, and in such a case it is attached to the punch plate.

With reversed punching, where the punch is mounted in the die block and the die is up in the punch plate, the stripping arrangement is reversed to conform.

The cross section of a typical die set is shown in Fig. 3-3. There the knockout pins are going through the head of the punch, their stripping pressure being provided by a spring. The pins push the stripping insert against the material so that the blank is held down when the punch moves upward. The die contains a similar set of pins, here called push pins. These lift up the cup off its face after forming.

The stripper is stationary, and it prevents the remainder of the strip from moving along with the forming-blanking punch action. This punch cuts the blank out of the strip with its circumference, forming it afterward with its face area and bottoming on a forming support.

### 3-1-1. Die shoe types

The upper and lower die shoe, along with guideposts, can be purchased in various sizes. The two basic types of these die sets are:

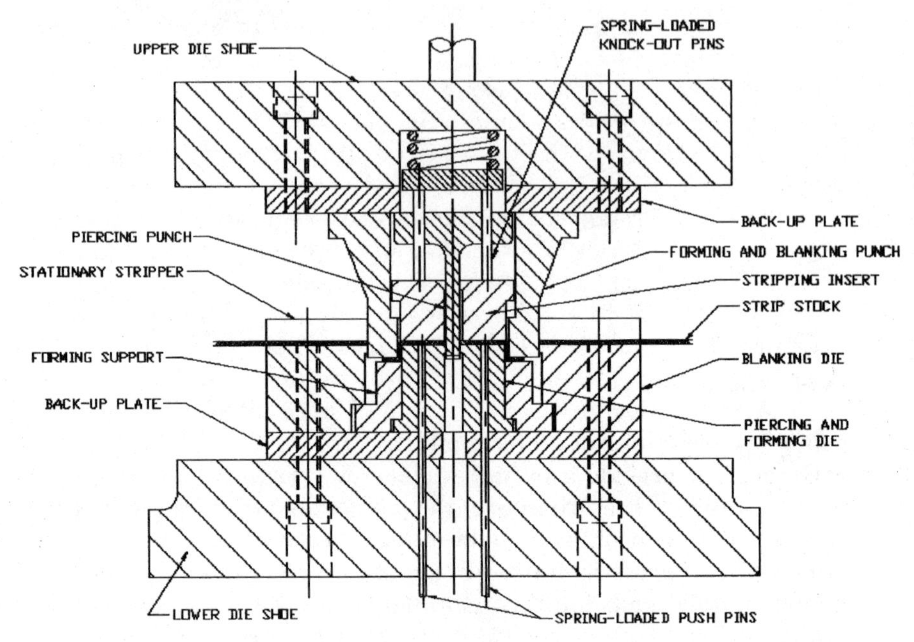


**Figure 3-3** Compound die, producing a pierced cup.

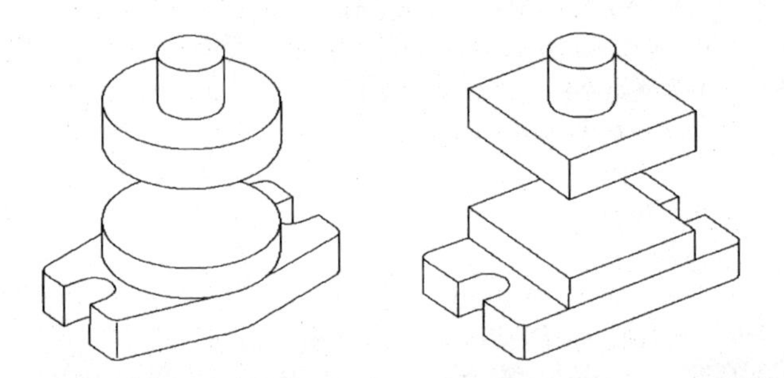

**Figure 3-4** Open die sets.

- **Open die set** (Fig. 3-4) which is used for manufacture of simple parts in small quantities or where no close tolerances are required. It is the most inexpensive die set, but since the guideposts are not there to secure the alignment of the two halves, setting up of these tools in press is often problematic.
- **Pillar die set** (Fig. 3-5) comes in a wide range of shapes, sizes, and combinations. The pillars, or guideposts, can be located in various places. *Back post die sets* have two guideposts located in the back, *two post die sets* have the posts placed either diagonally or opposite each other and along the centerline of the shoe. *Four post die sets* contain one guidepost in each corner.

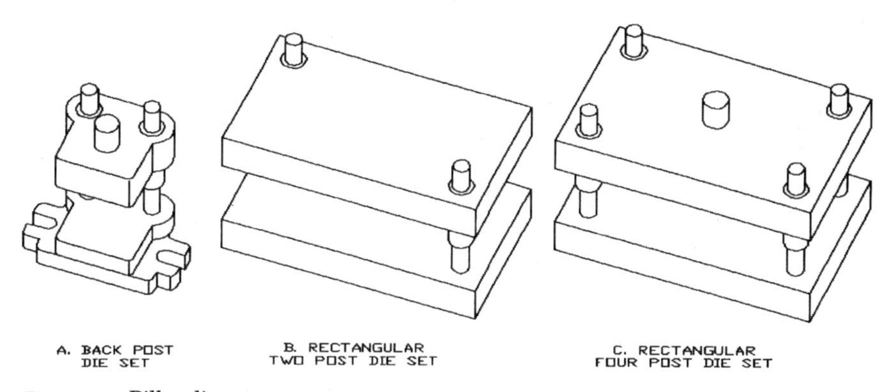


**Figure 3-5** Pillar die sets.

Guideposts provide a perfect alignment between the two halves of the die. They keep the punches and die buttons in a fixed location against each other, which protects their cutting edges from damage. The press-mounting demands are decreased, as the die alignment is already built in. The storage and transportation of the die places no strain on its elements, thus guarding their working surfaces and extending the die life.

The vast majority of die work is done with die sets that have two guideposts. But where greater accuracy is required, for heavy-gauge strips, four post die sets are a better choice.

### 3-1-2. Die set selection guidelines

Die sets are manufactured in three accuracy groups:

1. *Commercial die sets,* with tolerances between guideposts and bushings from 0.0004 to 0.0008″. Commercial die sets should be used for dies where no piercing, blanking, or any other cutting is performed, such as forming and bending dies.
2. *Precision die sets,* where the alignment between guideposts and bushings is further perfected by precision grinding of the bushing's inner opening, as well as its outer diameter, which is press-fit assembled into the die shoe. The alignment of these dies is excellent, and they should be specified for cutting, piercing, blanking, and perforating dies.
3. *Ball-bearing die sets,* with ball-bearing arrangement in place of plain sleeve bushings. These die sets are very tight-fitting, and they completely eliminate the possible development of thrust stresses, or so-called side play. Die sets with ball bearings are rec-

ommended for materials over 0.015″ thick; pin sets may be used for all sheet stock under 0.015″ in thickness.

Die shoes are manufactured from various types of material, the choice of which depends on the demands for strength. The three choices of die shoe materials are:

1. *Semisteel die sets,* actually made of cast iron, with some 7 percent of steel added. Semisteel sets cannot be used where large openings in the lower shoe are required, since they crack under the press-induced operational stresses on the die.
2. *All-steel die sets* are used where large openings (for blank removal, etc.) are to be cut into the shoe or where milling of pockets is involved.

   However, since all die shoes come from their respective manufacturers stress-relieved, no milling or large cutting should be introduced afterward. If such operations are required, drawings of such alterations should accompany the die set order, and the manufacturer should implement these changes prior to the delivery because the die set must be stress-relieved after each cutting or milling operation.

   Should a die set not be stress-relieved, all stresses remaining within the material would be slowly released over time, which will ruin the consistency of the die material and eventually ruin the die with all its components.
3. *Combination die sets,* with an all-steel lower shoe (die holder) and semisteel upper shoe (punch holder).

### 3-1-3. Die shoe accessories

The *punch shank* is either welded to the upper die shoe or can be purchased separately and screwed on. With semisteel die sets, the punch shank is cast along with the shoe and machined to size afterward. The size of the shank depends on the mounting dimensions of the press the die is intended for.

With some die sets of greater weight, an additional holding provision is furnished in the form of socket cap screws, attached through the upper die shoe to the underside of the ram.

*Guideposts* are precision ground pins, made of hardened, centerless ground steel for commercial die sets, and of hardened, centered ground steel for precision die sets. To reduce friction and to increase guideposts' resistance to wear, the posts used in precision die sets are hard chromium plated.

Usually guideposts are embedded in the lower die shoe, but sometimes removable guideposts can be used. This is in situations where they must be removed for sharpening of the die, especially where more than two back posts are used (see Fig. 3-6*e*).

*Guideposts' length* should be sufficient so that they never come out of their bushings during the press operation. This requirement is essential for the safety of work and alignment as well. They should be ordered ¼″ shorter than the shut height of the die. The *shut height of the die* is the distance between the outer surfaces of the upper and lower die shoe, with the die in its lowest position. This dimension does not include the length of the die shank.

The ¼″ distance off the die shut height, which is the minimal working height of the die, is an adequate grinding allowance. It also provides for clearance between the two halves of the die during its operation.

Some manufacturers supply their die sets with one guidepost longer than the other one, the usual difference being ½″. It is expected that the upper die shoe first enters the longer guidepost, aligns itself around it, and only then engages in the second, shorter guidepost.

*Removable guideposts* are usually located on a taper pin, which is attached to the lower die shoe with a screw.

Two types of bushings are used with die sets: headless (a plain sleeve) bushings and shoulder bushings. The shoulder bushing type is recommended for all cutting, piercing, and blanking dies. Like the guideposts, bushings are press-fitted into the die shoe.

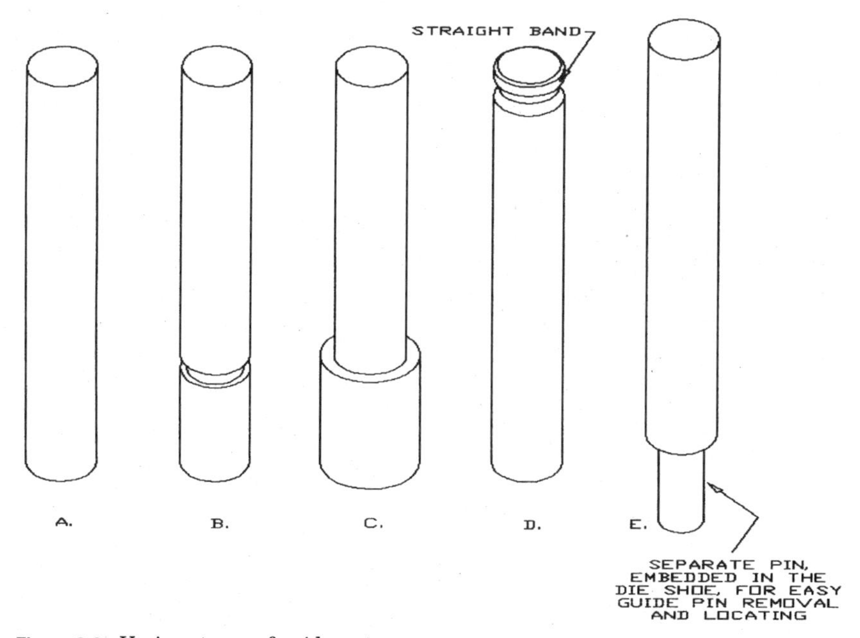


**Figure 3-6** Various types of guideposts.

Helical grooves, crisscrossing the inner surface of bushings, allow for proper lubrication during the die operation. Some bushings are self-lubricating: Made of porous powdered alloy steel, the lubricant is entrapped in their pores. Such lubrication usually lasts the entire life of the bushing.

The contact surfaces between the guidepost and the bushing are machined into such a fine finish that they tend to stick together. This problem occurs especially at the beginning, when the die is assembled together, just before the bushing is fully engaged by the guidepost. To alleviate this problem, the ends of these guideposts have been altered, as shown in Fig. 3-6*d*.

The narrow band enters the bushing first, and because its width allows for rocking of the part, no sticking may occur. The slanted surface guides the post farther into the bushing.

## 3-2. Dies, According to Their Construction

There are considerable differences in the way dies are built to function. In some, the metal strip is fed through the die, which produces the desired part in stages. Another die makes a complete part with each stroke of the press.

According to their construction and functionability, all dies can be separated into the following four groups:

- Compound dies
- Progressive dies
- Steel-rule dies
- Miscellaneous dies

### 3-2-1. Compound dies

Compound dies (shown earlier in Figs. 3-1 and 3-3) produce very accurate parts, but their production rate is quite slow and they are quite expensive to build. These dies consist of a single station, where the part is most often blanked out and either formed, embossed, pierced, or otherwise altered in a single stroke of the press. No progression of the strip is involved, as each stroke of the press produces a single, complete part.

Compound dies, as their name indicates, perform two different operations during each stroke of the press. *Combination dies* combine at least two operations during each stroke of the press. Other than that, these two types of dies are so similar in their construction that their applications can be considered interchangeable.

Some shops, however, make a distinction between the two types, calling any cutting and forming die a combination die, while the compound die is considered only a cutting die.

### 3-2-2. Progressive dies

Progressive dies (shown earlier in Fig. 3-2) are a mixture of various single dies operating as different stations and grouped into the same die shoe. These stations are positioned to follow a proper sequence of operations needed to produce the required part.

Usually, the die sequence is arranged side by side, or horizontally. The vertical arrangement of operations is shown in Fig. 3-7. Such dies are called *tandem dies* and are used mostly for drawing of shell types of products.

*Gang dies* (Fig. 3-8) or *multiple dies* are used where a large amount of simple blanks is required. The die consists of duplicate punches and dies, which cut as many blanks as there are tools during each stroke of the press.

*Lamination dies* require very precise and accurate work to be done on very thin and very hard material (Fig. 3-9). Most often, silicon steel is used, which is extremely tough. Its thickness runs between 0.014 and 0.017″.

The tooling to produce this type of work is a problem. Laminations must be manufactured with practically no burrs, and for that reason, the clearance between punches and dies is almost none. Further, these tools are usually made in sections whenever possible, to allow for quick and easy replacement.

Perforating such a hard material will soon render inadequate all common carbon steel punches and dies. Therefore, high-chrome, high-carbon steel must be used on all laminating work.

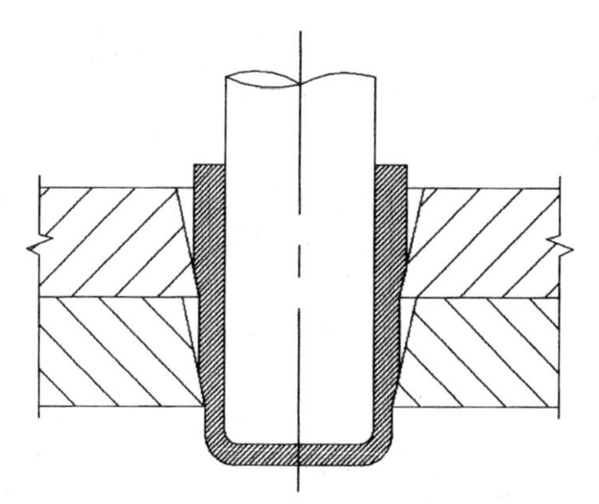

**Figure 3-7** Tandem die.

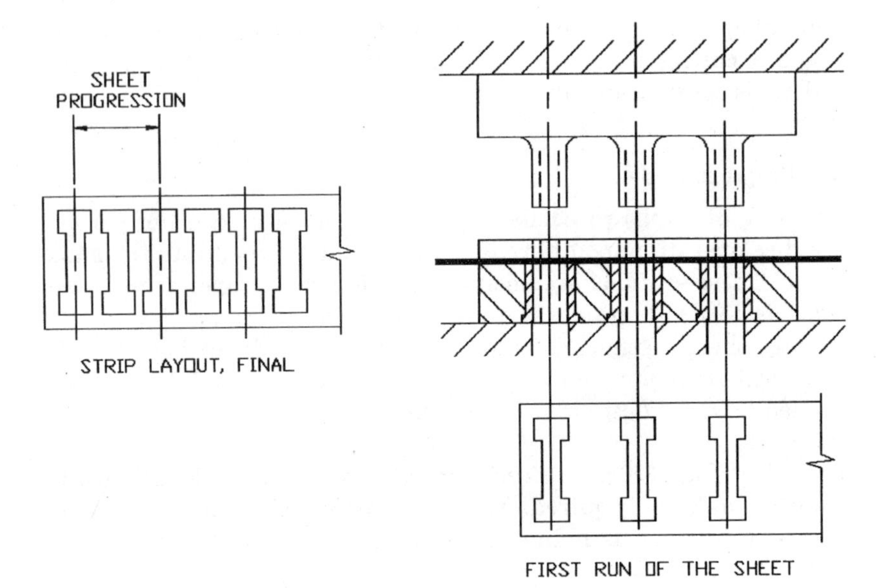


**Figure 3-8** Gang die.

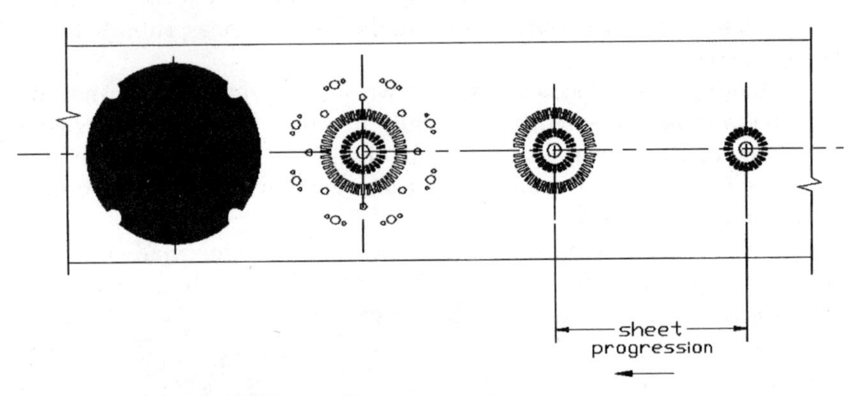


**Figure 3-9** Cutting and blanking of laminating strip.

Even though for the required level of precision the compound die would be the appropriate production tool in laminating work, progressive dies are most often used, since they run faster and allow for stacking of parts. Often, a compound blanking station may be utilized within the progressive die. In such a case, the blanked part is forced back into the strip and carried farther to another station, where it is pushed out and stacked. Such blanking is called *return blanking* (Fig. 3-10).

During the operation of such a die, the strip is fed over the bottom punch, surrounded by two spring-loaded stripping inserts. When the

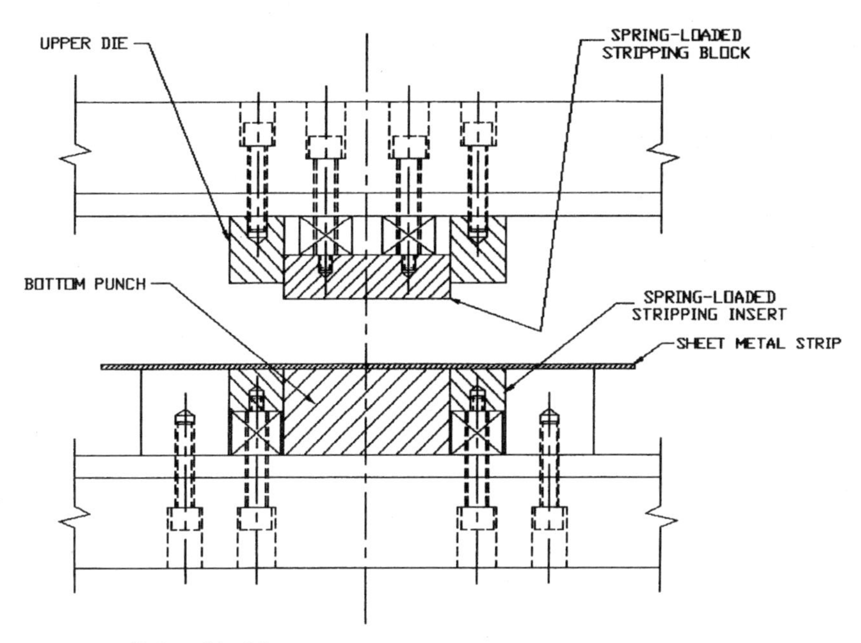


**Figure 3-10** Return blanking.

upper portion of the die slides down, the spring-loaded stripping inserts are pushed down against the force of their springs, while cutting is performed by the two upper die segments. During the cutting process, the two spring-loaded stripping inserts yield to the pressure of the die, and as soon as the upper portion begins to recede, they push the cut pieces up, back into the strip. Owing to the force exerted by its springs, the upper stripping block holds the pressure against the sheet longer than it takes for the punch to withdraw, thus holding down the section into which the cut pieces are being forced.

### 3-2-3. Steel-rule dies

Also called metalform dies, steel-rule dies consist of a heavy strip steel mounted in a standard die set, which serves as the cutting edge. Plastic and paper cutting steel-rule dies may be mounted on a heavy plywood plate.

The way the strip steel is attached to the block (at a 90° angle), its function is the same as that of a cutting knife. But with its thin profile, it will display a tendency to swelling and buckling. For that reason, the strip steel's compressive force should be twice the amount of the cut material shear coefficient.

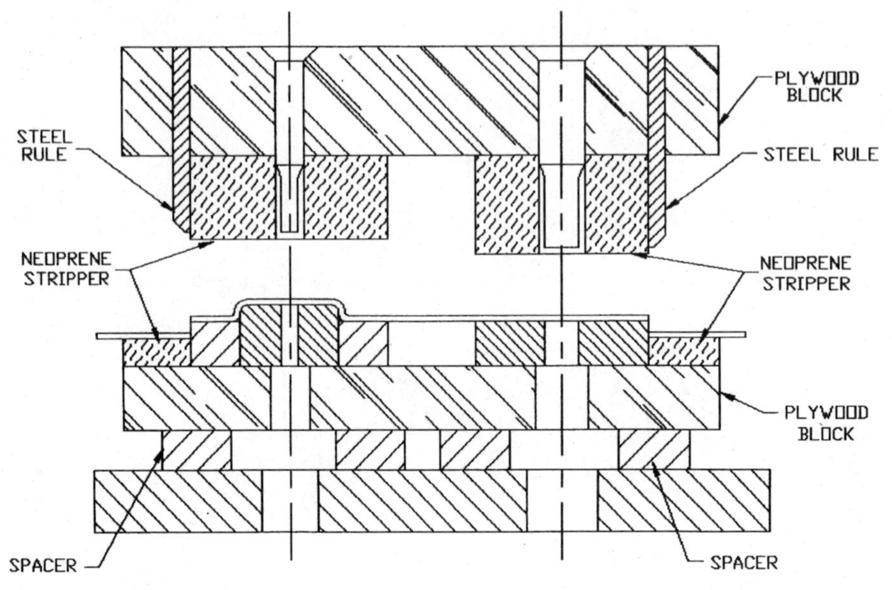


**Figure 3-11** Steel-rule die.

Steel-rule dies (Fig. 3-11) are usually equipped with neoprene or rubber strippers, slightly exceeding the height of the blades. Punches and dies are standard; they may be used to pierce openings while the steel rule (Fig. 3-12) is cutting the outline of the part. Tolerances for steel blades and their mating die sections are the same as those for regular punches and dies (Table 3-1).

Steel-rule dies may perform blanking, trimming, piercing, forming, embossing, and extruding operations. They are often utilized for cutting papers, fibers, card stock, rubber, felt, leather, and similar soft materials. Major areas of interest for their applications in metalworking are: limited quantities of parts, short lead times, large shapes, and tolerances above ±0.005″.

### 3-2-4. Miscellaneous dies

*Sectional dies* are built where some experimental work has to be carried out. Often a difficult operation or a combination of various tasks may require the tooling to be groomed to the correct size on the basis of actual tests.

Two samples of sectional dies are illustrated in Fig. 3-13. Notice the uneven location of screws and dowels. In die work, this is a common practice to prevent any variation from the proper installation of the block.

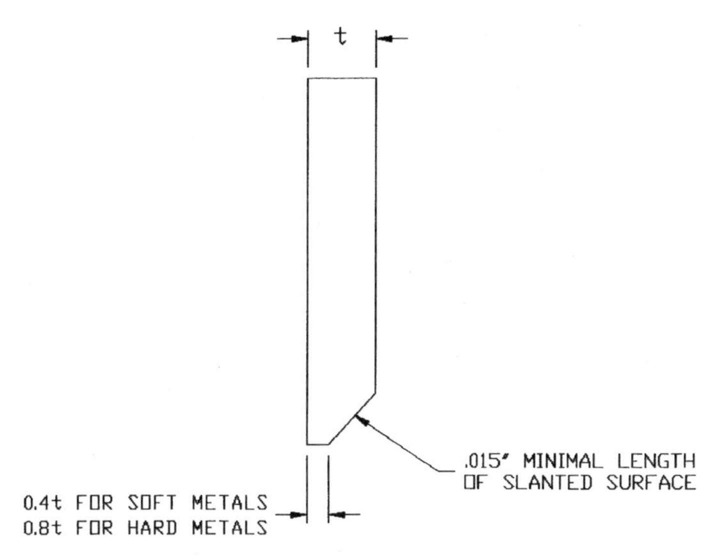


**Figure 3-12** Steel rule.

**TABLE 3-1 Recommended Steel Rule Thicknesses by Points for Various Materials***

| | Material | | | | | | |
|---|---|---|---|---|---|---|---|
| Material thickness | Alum, soft | Brass ½ H | Alum 2024-T | Alum 7075-T | Steel, mild | Steel 4130 | Stainless 302 |
| 0.010 | 3 | 4 | 4 | 4 | 4 | 4 | 4 |
| 0.031 | 4 | 4 | 4 | 4 | 4 | 4 | 4 |
| 0.062 | 4 | 4 | 4 | 4 | 4 | 4 | 6 |
| 0.078 | 4 | 4 | 4 | 4 | 4 | 6 | 6 |
| 0.093 | 4 | 4 | 6 | 6 | 6 | 6 | 6 |
| 0.125 | 6 | 6 | 6 | 6 | 6 | 8 | 8 |
| 0.150 | 6 | 6 | 8 | 8 | 8 | 8 | |

*One point equals approximately 0.014".

SOURCE: Frank W. Wilson, *Die Design Handbook,* New York, 1965. Reprinted with permission from The McGraw-Hill Companies.

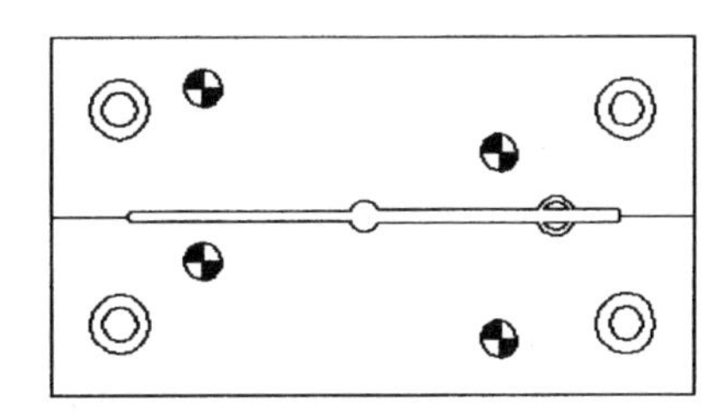


**Figure 3-13** Die blocks made of sections.

The use of temporary tooling may save a lot on material, additional work, or costly operations. If a punch and die has to be tried out and discarded and a new set has to be manufactured for another trial, the cost of such proceedings can be tremendous. With sectional dies, as with other temporary tooling, the parts already tried out are altered, used again, and altered further if necessary, until the produced part meets the predetermined requirements. The same experimental tooling may later be implemented in the production die.

*Dies with interchangeable parts* are those that can produce different parts in a single die. Usually this is achieved by using standard openings for punches and dies, and interchanging these or their sequence of operations with requirements. Often, by just pulling a punch or two, a completely different part can be produced.

*Indexing and transfer dies.* Indexing dies (Fig. 3-14) are useful for repeated notching or perforating, where the operation is copied around a circumference of the part. An indexing die usually produces several cuts at a time, rotates (transfers) the blank, and pierces again.

This procedure can be used for parts of limited quantities, where complicated tooling would be cost-prohibitive.

Transfer dies are suited for long runs of parts, since the die shoe, equipped with a transfer mechanism (Fig. 3-15), often cannot be adapted for other work.

This type of die transfers single pieces of work from station to station, shuttling them in a linear rather than circular fashion. The die construction itself is that of a progressive die, with no strip feeding provision. Sometimes the transfer device is a built-in feature of the press, in which case the press is called a *transfer press.*

## 3-3. Dies, According to Their Effect on the Structure of Material

When evaluated for the influence a die operation exerts on the structure of sheet metal, dies can be grouped into several categories,

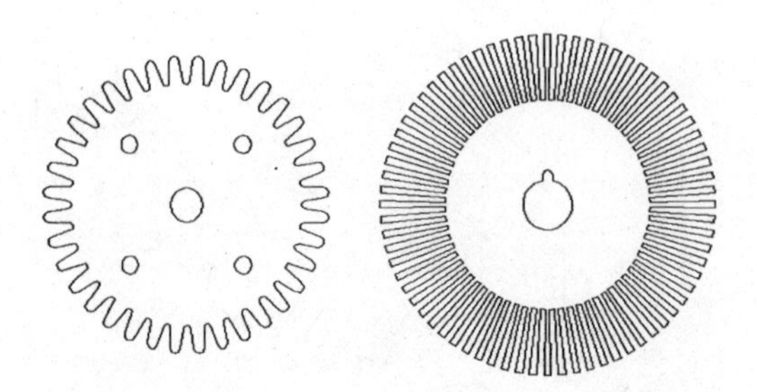

**Figure 3-14** Indexing die samples.

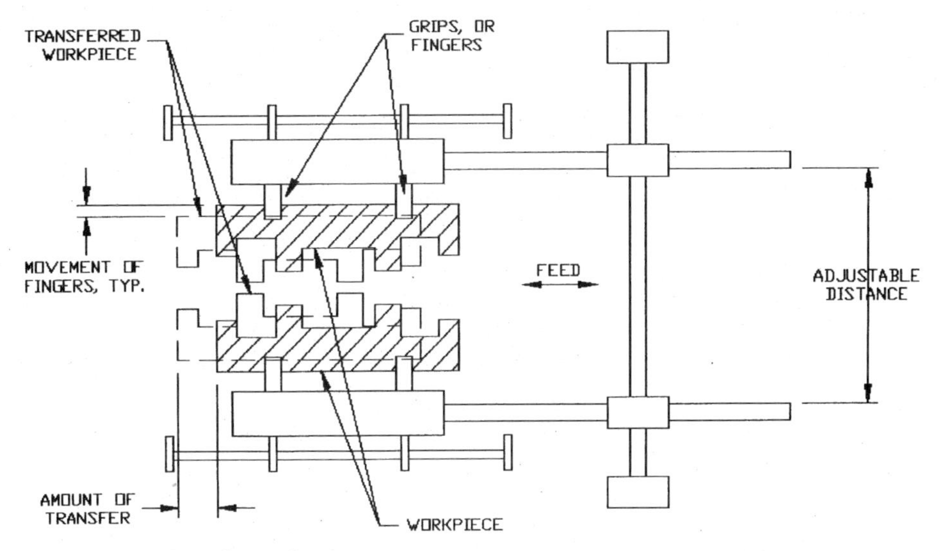


**Figure 3-15** Transfer mechanism.

named after the operations they perform. There are blanking dies and piercing and shearing dies, which all cut the material, separating the slug from the part. Other tools are those that force the metal to flow into predetermined locations, tools that form the metal, stretch it, expand it, or compress it.

All the various types of dies fit loosely into five categories, where they are grouped according to the type of work they produce. These are:

- Cutting dies
- Bending and forming dies
- Drawing dies
- Compressive dies
- Miscellaneous dies

### 3-3-1. Cutting dies

These dies separate pieces of metal from the main blank by the cutting process. They include blanking, piercing, perforating, notching, lancing, slitting (or cutoff), plunging, trimming, shaving, and pinch trim tools.

*Blanking dies* cut out the complete outline of the part in a single operation. *Piercing dies* pierce singular holes, either for pilots' engagement or where piercing is required after bending, drawing, and other shape altering processes (Fig. 3-16).

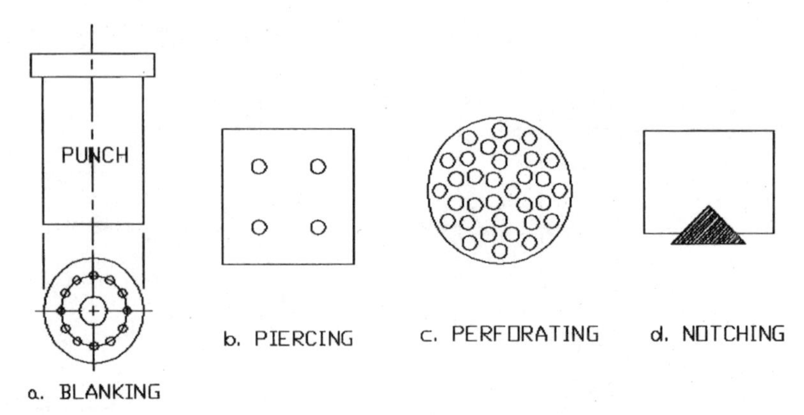


**Figure 3-16** Cutting operations.

*Perforating dies* produce a multitude of openings called perforations. This type of work is used in production of strainers, sifting devices, for objects of shielding or for their decorative appeal.

*Cutoff dies* chop off the pieces of flat material, tubing or wire, from the continuous supply. Often the cutoff operation is the last of the sequence of a progressive die, in which case it cuts actual products off the strip.

*Notching* is basically the same operation as blanking, the only difference being in the imbalance of the cut. This is because notches usually appear along the edges of parts, and if the tooling exceeds the width of the strip, that side of its shape would be unsupported. To prevent this type of complication, notches are usually punched out first. For example, a corner fillet should be removed before the sides of the part are cut off.

Heeled punches are also found supportive in notching operations, along with special-shaped tooling.

*Lancing* (Fig. 3-17), as found in sheet-metal work, is commonly used to produce short tabs, where a single punch cuts the metal and forms it at the same time.

The forming portion of the lancing punch must have the same radius as any other bending tool. Its face area must be ground at an angle, beginning at the point where the cutting of metal first occurs. No nesting of the part is necessary in this operation.

*Slitting,* or *cutoff* (Fig. 3-18), is another cutting operation during which the cut is only partial and the cutout remains attached to the strip for further processing. This type of die work is used for cutting off certain areas, which will be further processed by the progressive die, or to produce louver shapes, in which case a special-shaped tooling is needed.

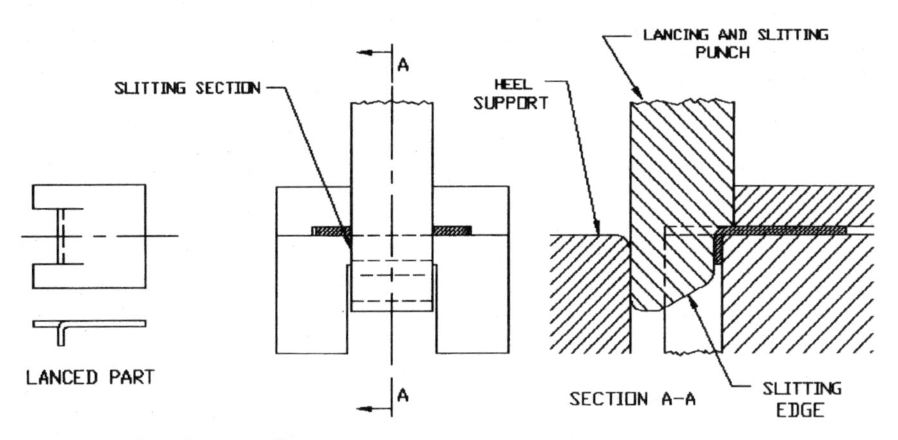


**Figure 3-17** Lancing operation.

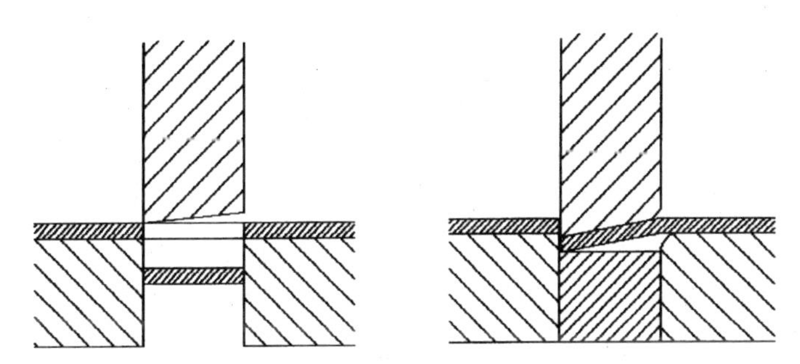

**Figure 3-18** Cutoff and slitting operations.

Instead of piercing of the material, sometimes a method called *plunging* (Fig. 3-19) is used. The punch in this operation is called the plunger, and it does not cut through the strip but rather draws the metal, extending its shape and breaking through at the end of the stroke.

*Extruding* is a similar operation, only the metal is prepierced and the extruding punch only pushes the opening into an extended length. With plunging, the edges are rough and extrusions have smooth and even edges.

*Trimming dies* (Fig. 3-20) are used for removal of portions of material distorted during previous operations, such as forming or drawing.

*Shaving dies* are actually very accurate trimming tools, yet they operate on a different principle. They are used to remove minute amounts of material, surrounding previously performed cuts. A shaving operation is used for removal of burrs and to flatten the edges of precision parts. The shaved part has a straight, clean, smooth edge, which can be held to accurate tolerances (Tables 3-2 and 3-3).

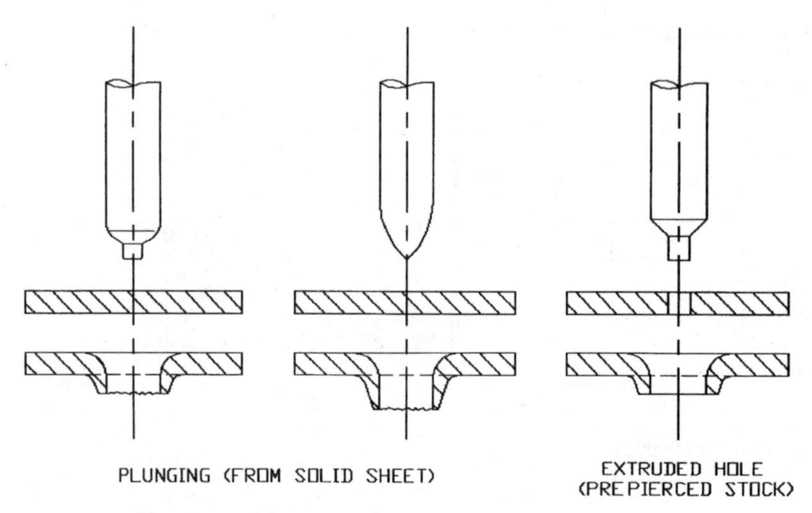


**Figure 3-19** Plunging and hole extruding.

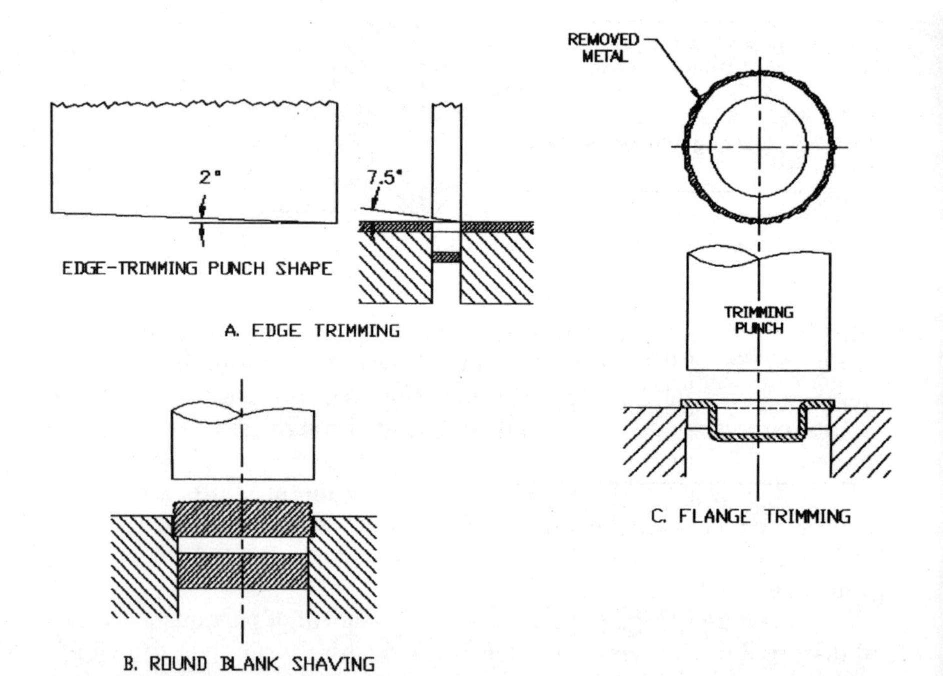


**Figure 3-20** Trimming and shaving die.

**TABLE 3-2 Shaving Allowances per Side (Inches) for Steel, Brass, and German Silver Stock**

| | Steel | | | |
|---|---|---|---|---|
| Thickness of blank, in | Hardness, 50–66 $R_B$ | Hardness, 75–90 $R_B$ | Hardness, 90–105 $R_B$ | Brass and German silver |
| Where One Shave Is Necessary | | | | |
| 3/64 (0.0468) | 0.0025 | 0.003 | 0.004 | 0.005 |
| 1/16 (0.0625) | 0.003 | 0.004 | 0.005 | 0.006 |
| 5/64 (0.078) | 0.0035 | 0.005 | 0.006–0.007 | 0.007 |
| 3/32 (0.0938) | 0.004 | 0.006 | 0.007–0.008 | 0.008 |
| 7/64 (0.1094) | 0.005 | 0.007 | 0.009–0.011 | 0.010 |
| 1/8 (0.125) | 0.007 | 0.009 | 0.012–0.014 | 0.014 |
| Where a Second Shaving Operation Is Necessary | | | | |
| 3/64 (0.0468) | 0.00125 | 0.0015 | 0.002 | 0.0025 |
| 1/16 (0.0625) | 0.0015 | 0.002 | 0.0025 | 0.003 |
| 5/64 (0.078) | 0.00175 | 0.0025 | 0.003–0.0035 | 0.0035 |
| 3/32 (0.0938) | 0.002 | 0.003 | 0.0035–0.004 | 0.004 |
| 7/64 (0.1094) | 0.0025 | 0.0035 | 0.0045–0.0055 | 0.005 |
| 1/8 (0.125) | 0.0035 | 0.0045 | 0.006–0.007 | 0.007 |

SOURCE: Frank W. Wilson, *Die Design Handbook,* New York, 1965. Reprinted with permission from The McGraw-Hill Companies.

**TABLE 3-3 Shaving Allowances for Aluminum**

| Thickness, in | First shave allowance, in | Final shave allowance, in |
|---|---|---|
| 0.03 | | 0.004 |
| 0.05 | | 0.006 |
| 0.06 | | 0.007 |
| 0.08 | 0.007 | 0.003 |
| 0.100 | 0.008 | 0.004 |
| 0.125 | 0.010 | 0.005 |
| 0.175 | 0.013 | 0.007 |
| 0.250 | 0.020 | 0.010 |

SOURCE: Frank W. Wilson, *Die Design Handbook,* New York, 1965. Reprinted with permission from The McGraw-Hill Companies.

Proper nesting is a necessity in shaving operations. The shaving punch then pushes the part through a slightly smaller opening, forcing the material to conform with that opening's surface.

Another process used for trimming of rough edges of drawn cups is *pinch trimming* (Fig. 3-21).

Despite all good intentions, a pinch trim cut usually comes out slightly squeezed, as shown earlier in Fig. 1-10*c*.

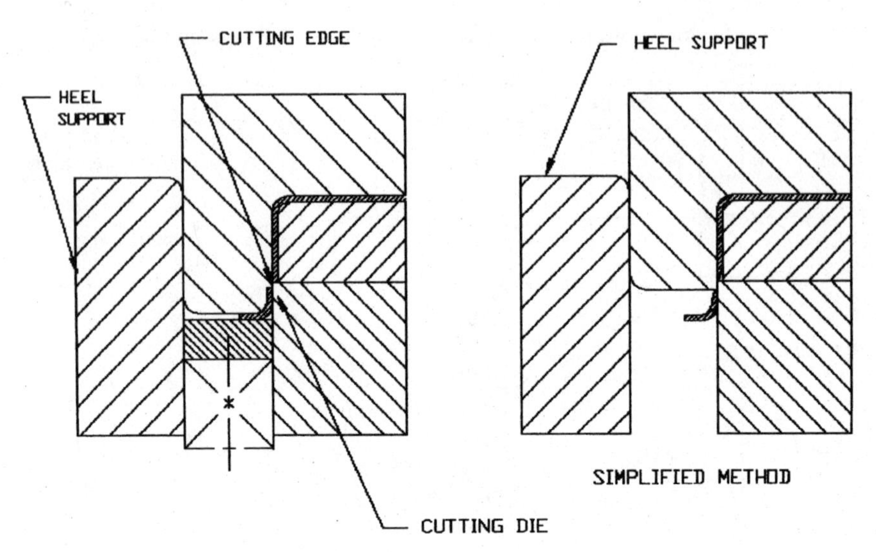


**Figure 3-21** Pinch trim.

### 3-3-2. Bending and forming dies

Bending dies can be used to form, fold, or offset parts without subjecting their material structure to flow and plastic deformation. Bending-operation dies include curling dies, twisting dies, burnishing or sizing dies, and straightening or flattening dies. Forming dies, however, belong in this category only marginally, as they fit in with the drawing dies as well.

*Bending dies* deform a flat part into an angular shape. The bend line is straight, with no plastic deformation present. *Forming dies* deform a flat part in much the same manner, but the line of a bend may be curved, with plastic deformation evident around the vicinity of the bend. (For examples, see Fig. 1-33.)

*Curling dies* form the edge of a part into a circular, hollow ring (Fig. 3-22). Sometimes a wire can be installed in such a shape. Common representatives of curled parts are hinges.

If more than one material is to be curled in the same die, two preforming operations are necessary. The first preform should bend the flanges to the 45° angle with the horizontal, creating a V-shaped cross section at each side of the part. The second preform should bring these V arms down to obtain a 90° bend with the horizontal. The final stage is the curl.

The reason behind this requirement is the variation in the material composition: With a different stock, a different length of the curl will be obtained. Preforming of metal as described will eliminate the pos-

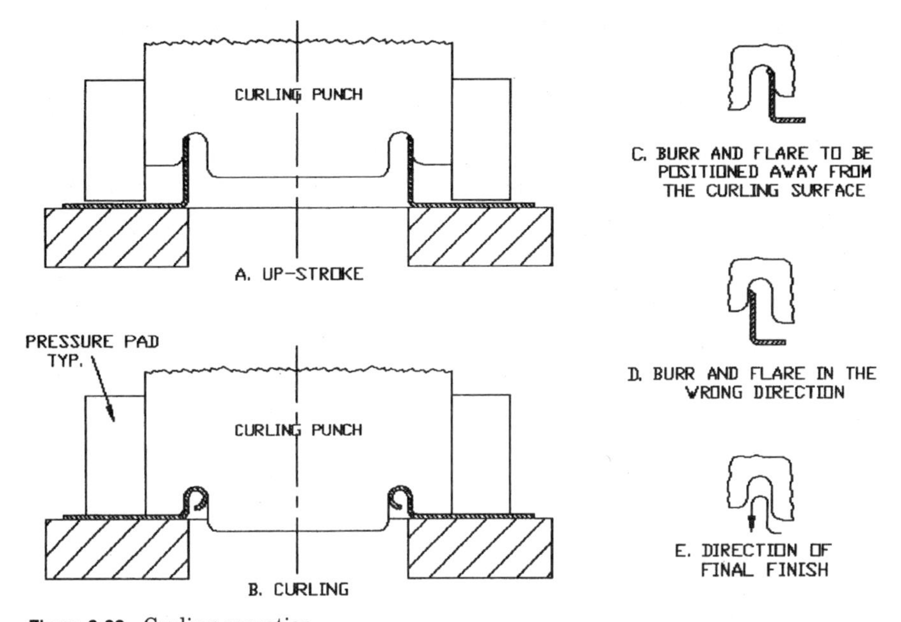


**Figure 3-22** Curling operation.

sibility of metal stretching, along with eliminating its spring-back as well, and the parts, even though made from various materials, will emerge from the die the same in size.

However, even with a single type of material being used in a curling die, a flare in the part's edge must sometimes be provided to ease the curling process. The location of the flare or burr is important, as it always should be turned away from the curling surface (see Fig. 3-22*c* and 3-22*d*). If reversed, the burr or the flare would scratch the curling die's surface, perhaps even dig into it, and obstruct the curling operation or even ruin the tooling.

The curling groove should be well proportioned, its size being dependent on the thickness of material. Some basic groove sizes are shown:

| Stock thickness, in | Recommended curling dia., in | Stock thickness, mm | Recommended curling dia., mm |
|---|---|---|---|
| 0.010 | 0.062–0.078 | 0.25 | 1.50–2.00 |
| 0.018 | 0.125–0.156 | 0.50 | 3.25–4.00 |
| 0.030 | 0.203–0.250 | 0.75 | 5.00–6.00 |
| 0.042 | 0.281–0.343 | 1.00 | 6.75–8.00 |

If a larger than required diameter of the curling die is used, the material will ignore its guidance and form a curl of its own, smaller in size.

Curling dies should be made of hardened tool steel, since they suffer from a great amount of wear. The grooves must be very smooth, well polished, preferably lapped, to aid a uniform sliding movement of the material. Even though the grooves will be produced in conventional machinery, the final polishing must be done in the direction of curling (see Fig. 3-22*e*). Otherwise the curled metal may get obstructed by the grooves, remaining in the surface after the lathe work or other metalworking machinery.

Normally, curling dies run in presses with long strokes, for the length of the curl must be in congruence with the travel of the ram.

*Twisting dies* can twist the strip material. A slight plastic deformation may be evident in this operation.

### 3-3-3. Drawing dies

Drawing dies force the material to flow in conjunction with the movement of the punch, creating plastic deformation of its structure. During this process, the volumnar amount of flat blank is transformed into a drawn, shell-like shape. Often, a marked thinning of the part's cross section may occur. These dies include drawing and redrawing dies and ironing, reducing, and bulging dies (Fig. 3-23).

*Ironing dies* are the same as drawing dies, the only difference being the tolerance between the drawing punch and the die, which in ironing dies is smaller. This change results in drawn parts becoming forcibly thinner, and the shell's wall surface uniform and smooth.

*Bulging dies* (Fig. 3-24) expand a drawn shell to conform to the shape of the die. There are two kinds of bulging dies: those utilizing rubber as an expanding material and those using water or other fluid. The latter are also called *fluid dies.*

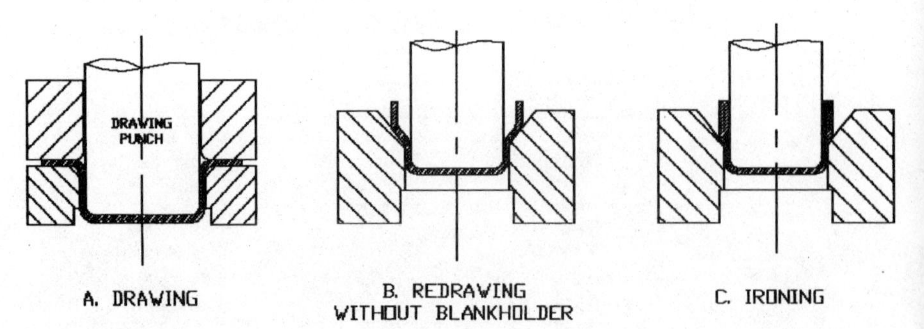


**Figure 3-23** Drawing, redrawing, and ironing.

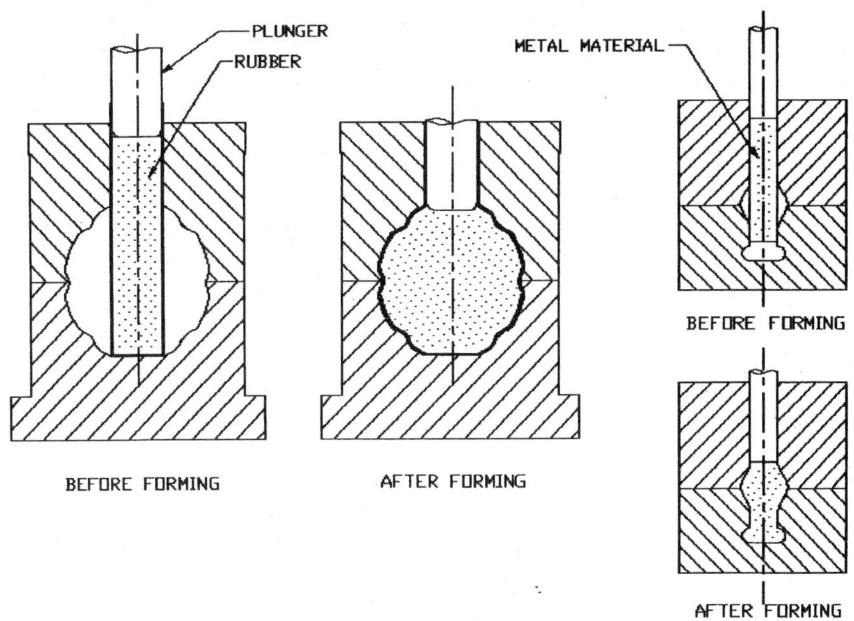


**Figure 3-24** Bulging process.

Parts bulged with rubber display a smooth surface, but the disadvantage of this medium is that it wears out quickly. Probably because of the continuous expansion and contraction, aided by the corrosive effect of lubricants, the rubber material tires out and easily deteriorates.

Where fluid is used as the expanding material, the parts are free of tool marks, and their walls are evenly thick, with no thinning even in radiused areas. The metal necessary for the increase of shape is taken from the height of the shell, which at the end of the bulging operation is found lowered.

Bulging with a material-retaining flange added on top of the shell produces bulges that do not decrease in height. Naturally, in such a case, the walls of the bulged part will come out thinner.

The liquid is forced into the bulging die under a pressure the amount of which is dependent on the stock thickness. It is advisable to begin forming with a lower pressure and increase it gradually, for if it should become excessive, the shell may burst open.

The bulging die is made of two halves, which are taken apart for the finished part removal.

The expected circumferential bulging increase should be about 30 percent in a single operation. Any bulging greater than that has to be performed in stages, with annealing of the bulged material in between.

*Rubber and fluid forming.* In rubber forming, the rubber pad is attached to the ram of the press, and on coming down it forces the flat sheet to conform to the shape of the form block underneath it. The rubber piece being flat can take on any shape, and a single rubber pad can therefore be used to form parts of various shapes.

The pressure which the rubber pad exerts on the metal is uniform, so that the forming process creates no thinning of the walls or radii. The radii are, however, more shallow than those produced in conventional dies.

The disadvantage of this method of forming is that rubber easily tears. The continuity of expansion and contraction also places a great strain on this material and subjects it to greater than usual wear.

Several processes utilize the rubber pad forming techniques, described as follows:

1. **Guerin process** (Fig. 3-25), in which the pad is a fairly soft rubber block, either solid or put together from laminated slabs. The height of the block is usually three times the depth of the formed part. It is held in a sturdy cast-iron or steel retainer, as the forming pressures may be as high as 20,000 lb/in$^2$ in some applications.
2. **Verson-Wheelon process** (Fig. 3-26) is based on the previously described Guerin process, with higher forming pressure, supplied by a flexible hydraulic device. The pressure is applied toward the rubber pad, which may serve as either a punch or a die, uniformly

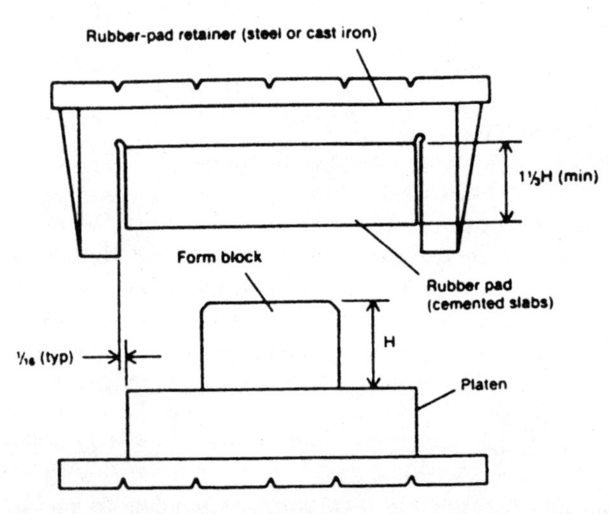


Figure 3-25 Guerin forming process, tooling and setup. (*From O. D. Lascoe, "Handbook of Fabrication Processes," ASM International. Reprinted with permission from ASM International, Materials Park, OH.*)

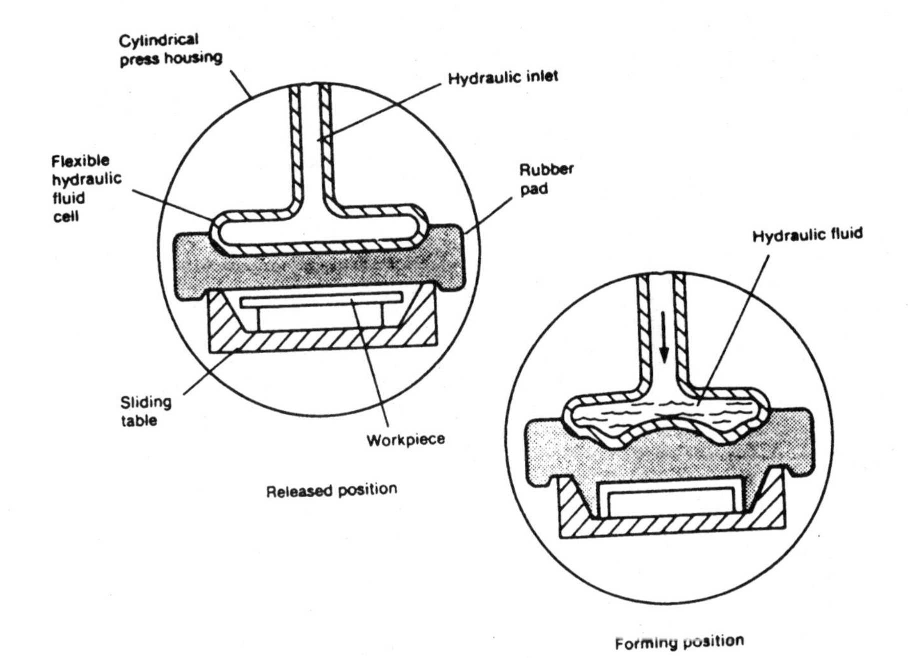


**Figure 3-26** Verson-Wheelon forming process principle. (*From O. D. Lascoe, "Handbook of Fabrication Processes," ASM International. Reprinted with permission from ASM International, Materials Park, OH.*)

distributed all over its surface. Such well-distributed pressure allows for formation of wider flanges, shrink flanges, beads, ribs, joggles, etc. These formations display rather a sharp detail, be they made of aluminum, low-carbon steel, or even titanium steel. The parts produced by this process are limited in depth.

3. **Marform process** (Fig. 3-27) is another utilization of the cheap tooling used in the Guerin and Verson-Wheelon processes. However, here it is applied to deep drawing and forming of wrinkle-free shrink flanges. The rubber pad is thick, coupled with a hydraulic cylinder, whose function is controlled by a pressure-regulating valve. The blank is held firmly between the blankholder and the rubber pad, when the pressure is applied to it, drawing it over the form block.

Aside from the rubber block, the main part in this type of forming is the form block. Form blocks can be made of wood, masonite, or aluminum, kirksite, or similar materials. The blank is positioned on the block with at least two locating pins, whose height should not obstruct the action of the rubber pad. Several rubber forming blocks and auxiliary forming tools are shown in Fig. 3-28.

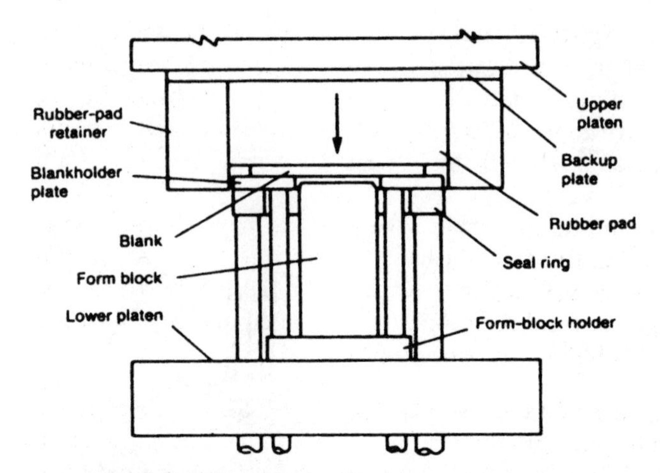


**Figure 3-27** Marform forming process, tooling and setup. (*From O. D. Lascoe,* "Handbook of Fabrication Processes," *ASM International. Reprinted with permission from ASM International, Materials Park, OH.*)

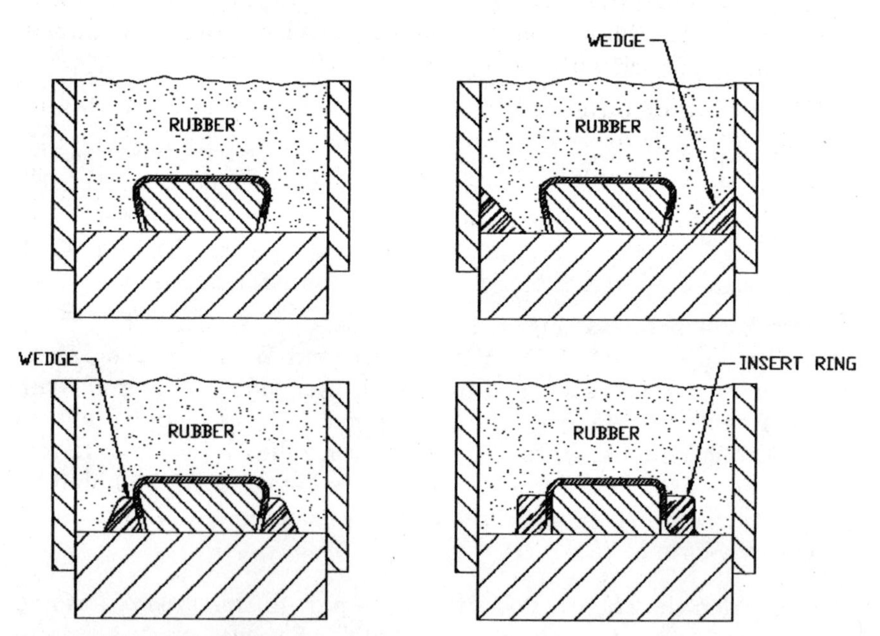


**Figure 3-28** Forming with rubber.

Stretch flanges (Fig. 3-29) can be easily and quite accurately formed with the rubber forming process. Such a production technique is more economical than that utilizing regular hard tooling. Shrink flanges (Fig. 3-29) are more difficult to obtain, and without various mechanical forming aids, shown in Fig. 3-28, their production would be quite impaired.

*Fluid forming* is similar to rubber forming, but here the rubber is replaced by the forming fluid, in this case oil. Such forming is a useful method of producing complex parts fast and economically. Basically, there are three types of fluid forming: forming in cavity dies, forming over a rigid punch, and expanding or bulging.

1. **Fluid forming in cavity dies** (Fig. 3-30). The form unit consists of a rubber diaphragm, filled with oil, acting under hydraulic pressure on the flat piece of metal positioned over the die cavity. At the end of the cycle, the formed part is blown out by compressed air coming in through the bottom of the die.
2. **Fluid forming over a rigid punch** (Fig. 3-31) utilizes a principle similar to the previous example. Here the system of valves allows for a variation of the blankholder pressure during the working cycle. The blank, positioned on the draw ring, is wrapped around the punch with the descent of the ram.
3. **Expanding or bulging** (Fig. 3-32) replaces the wear pad with an oil-filled rubber sack. During the forming cycle, the fluid pressure is transmitted equally in all directions. There is virtually no spring-back on the part, because the fluid acts simultaneously as a sizing element.

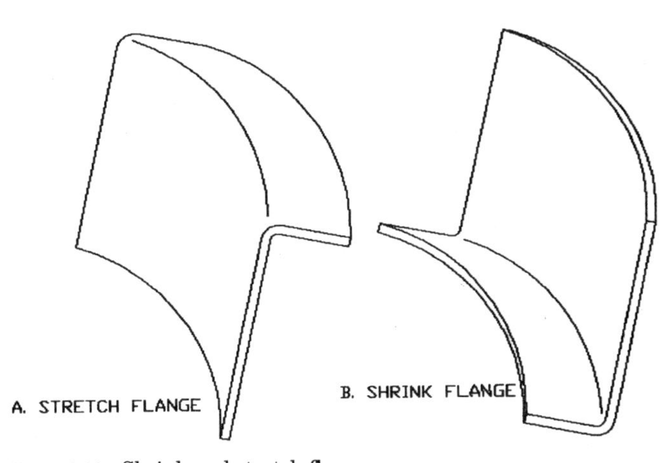


**Figure 3-29** Shrink and stretch flanges.

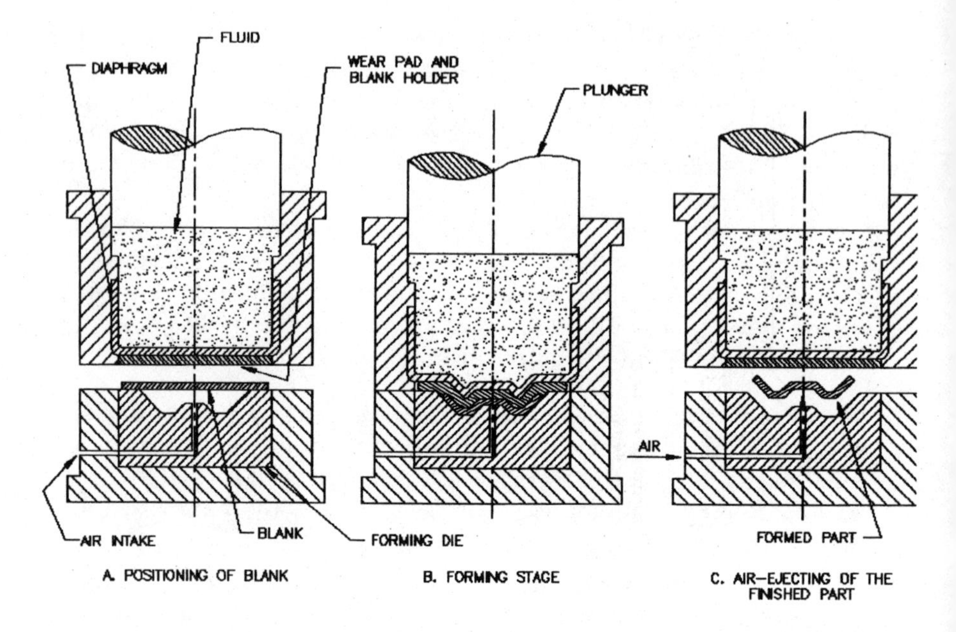


**Figure 3-30** Fluid forming in a cavity die.

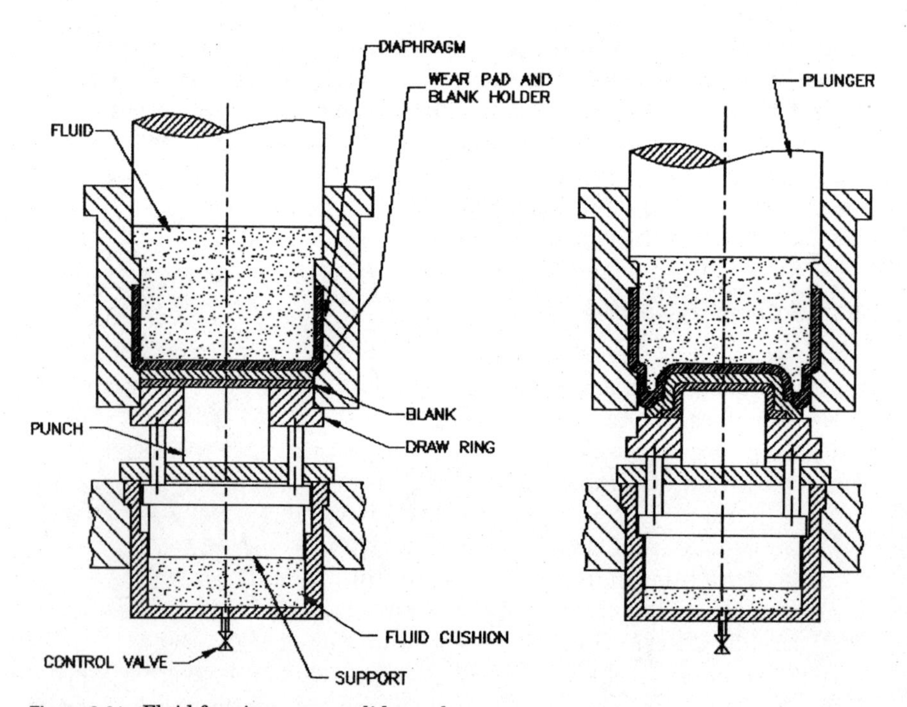


**Figure 3-31** Fluid forming over a solid punch.

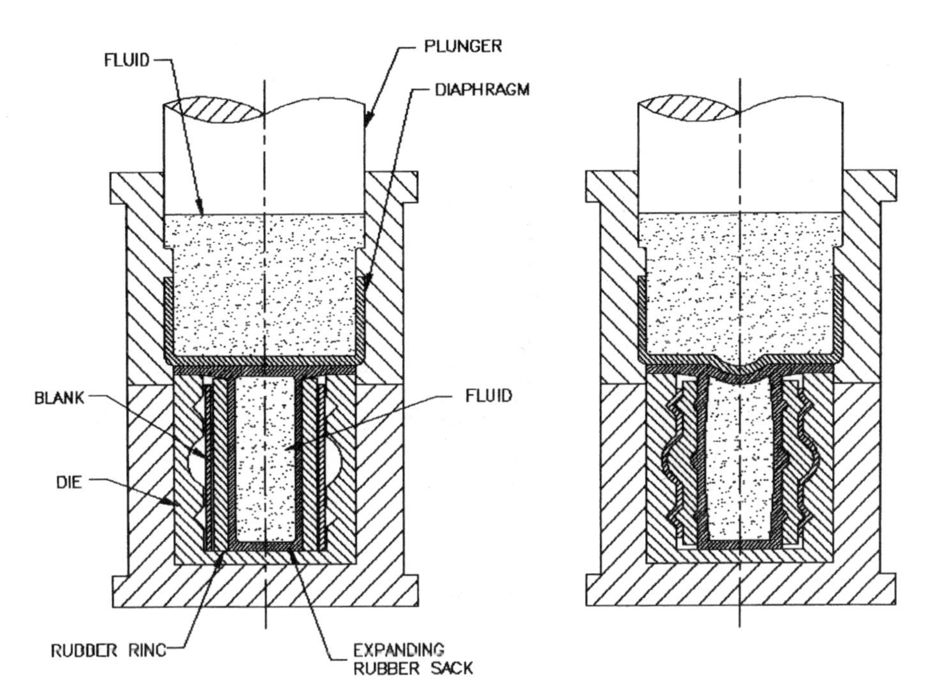


**Figure 3-32** Bulging with fluids.

*Reducing dies* are also called *swaging* or *necking dies,* and they are the exact opposite of bulging dies. Here the shell is reduced in diameter, with subsequent elongation of its length (Fig. 3-33).

The advantage of this operation lies in the fact that the shape is altered without any need for machine-induced removal of material, and therefore many additional finishing operations are eliminated as well.

### 3-3-4. Compressive dies

Compressive dies force the material to flow into a cavity and fill all its crevices. These dies are called coining, embossing, extruding, impact extruding, forging, heading, riveting, upsetting, and staking.

*Coining and embossing dies* are cold-forming tools which force the material into a structural cavity by exerting a considerable pressure on it. The metal is squeezed, with resulting displacement of its portions, until it fills the whole cavity. The closer the cavity to be filled is located to the filler material, the easier the flow process is.

*Embossing* (Fig. 3-34) is a metal stretching and compressing operation, already described in Sec. 1-3-2, "Functionability Aspects."

If an embossing operation is to be included in a progressive die sequence, it should be placed at one of the beginning stages, for the

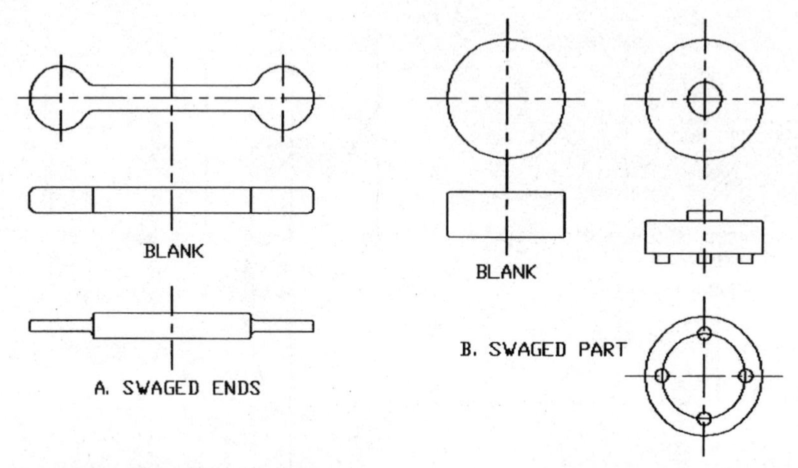


Figure 3-33 Examples of swaging process.

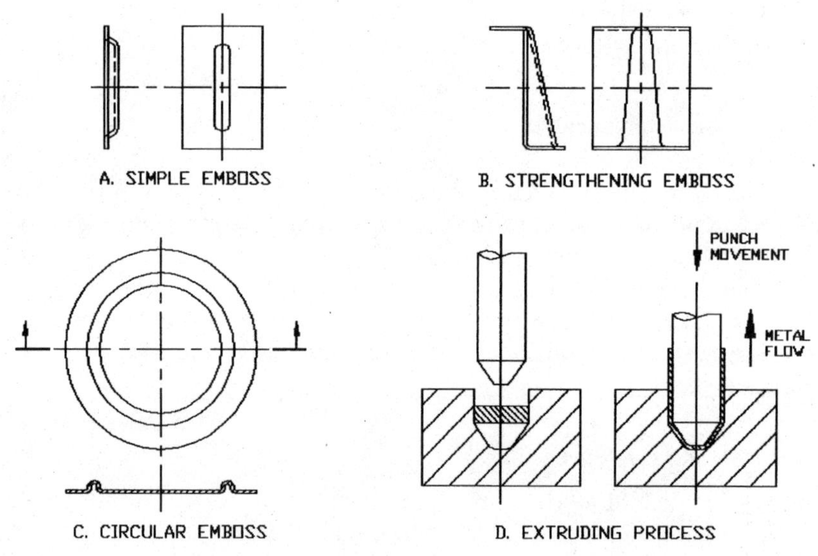


Figure 3-34 Embossing and extruding.

emboss will draw the metal from its immediate surroundings for its shape, which may affect the final outline of the part.

*Extruding dies* can form a flat piece of metal into a tubelike shape by first forcing it into the cavity and then shooting it up by the pressure of the extruding punch (Fig. 3-34*d*).

*Impact extrusion* (Figs. 3-36 and 3-37) is used for the manufacture of hollow, thin-walled, deep-recessed parts. Depending on the type of

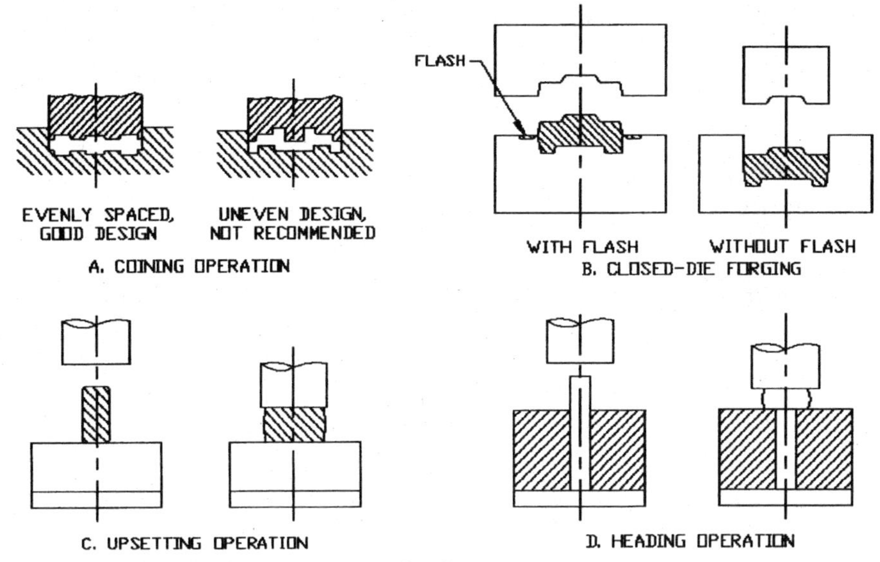


**Figure 3-35** Coining, forging, upsetting, heading.

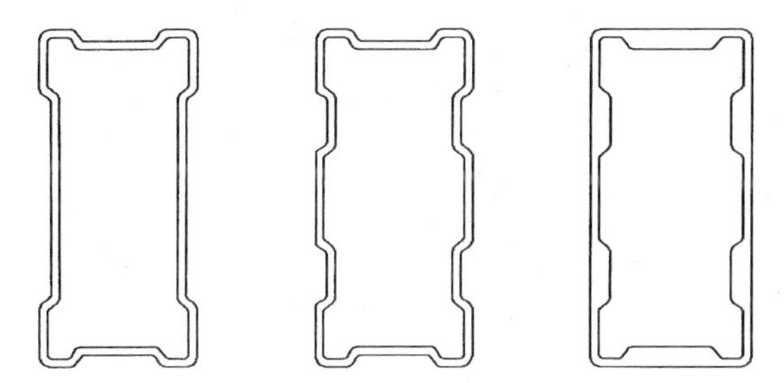

**Figure 3-36** Impact extrusion. Cross section of parts.

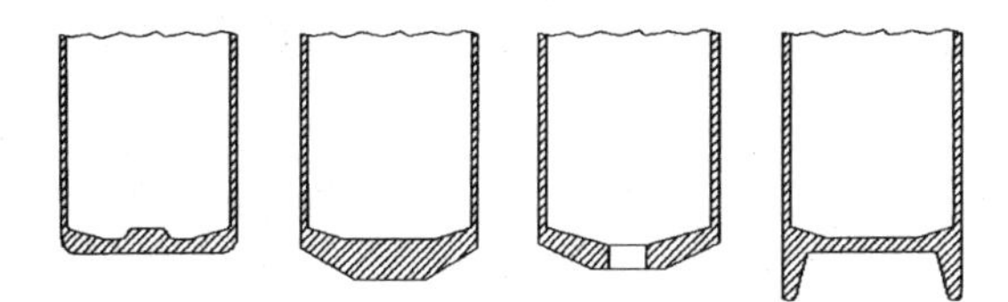

**Figure 3-37** Shapes of bottoms in impact extrusion.

metal, it can be performed with cold slugs as well as at elevated temperatures. Aluminum alloys, tin, and lead are formed cold, while zinc uses elevated temperatures around 300°F or 150°C.

*Forging dies* (see Fig. 3-35) are similar in obtained results to the impact extrusion process, with the only difference being the source of the pressure on the part, which in this case is mostly a hammer which is dropped on the workpiece.

There are two kinds of drop-hammer forging: gravity-hammer forging and double-action hammer forging. Gravity-hammer forging depends on the weight of the hammer, which is lifted to a certain height and allowed to fall on the workpiece. The double-action hammer is accelerated in velocity during its fall by an addition of steam or air pressure.

Forging dies can run *hot* or *cold*. Hot forging achieves the deformation of the blank with a single stroke of the hammer. Cold forging, even though called "cold," is actually exposed to certain thermal influences as well, these being induced by the forming action on the metal.

In forging, the entire volume of the flat blank is forced into a die impression. Closed forging dies with no flash can actually be called coining dies. The exact volume of metal blank must be well calculated for this operation, as there is no provision for the excess to flow out of the die and turn into a flash.

*Heading, riveting, upsetting, and staking dies* are cold-forming tools which force the material to take the desired form. (See Fig. 3-35.)

These operations are similar in their outcome, even though they produce parts for different applications. Heading is used to form screw heads and similar parts. According to its functional aspects, it may actually be called open die forming. Upsetting, on the other hand, is quite close to the forging process and may be defined as free forming.

The upset ratio in a part must be proportioned in order to limit buckling, as

$$0.5 = \frac{l_0}{d_0} = 2$$

where $l_0$ = initial length
$d_0$ = initial diameter

### 3-3-5. Miscellaneous dies

Miscellaneous dies include marking and numbering dies, straightening or flattening dies, burnishing or sizing dies, horn dies, cam dies, hemming dies, crimping dies, assembly dies, and subpress dies.

*Marking and numbering dies* are used for stamping numbers, characters, and symbols on metal parts. *Straightening or flattening dies* will bring a part to size, or within the drawing dimensions by striking it between two surfaces, without allowing for its walls to become thinner.

*Burnishing and sizing (or calibrating) dies* force the blank through the die, where there is no cutting clearance. The edges of the part forcibly become even, smooth, and accurate. There are two types of this process: Either the punch presses the part through the die with rounded corners or the die has sharp corners, in which case the punch is larger than the die opening. (See Fig. 3-38.)

Material to be added for the burnishing process is about 0.006 to 0.016 in or 0.15 to 0.40 mm. With complicated shapes, these values should be doubled.

The radius of the burnishing die (Fig. 3-38*a*) is dependent on metal thickness usually being 0.020 to 0.062 in. The spring-back runs in the vicinity of 0.02 to 0.06 percent of the blank's size.

For aluminum alloys, brass, and bronze, the second method should be used (Fig. 3-38*b*), where the punch is larger than the die opening. The difference between punch and die should be 0.004 to 0.020 in or 0.1 to 0.5 mm.

*Horn dies* are equipped with a horn, which is a sort of protruding stake, or a mandrel, the shape of which conforms to the inner configuration of the part. Drawn or formed parts are positioned over the horn for the application of secondary operations.

Finished products may be stripped off by the action of springs, cams, levers, or air-blowing devices. Sometimes the stripping arrangement may be dependent on the movement of the ram.

*Cam dies* transform the vertical motion of the die movement into a horizontal (or angular) movement. With its aid, many side-piercing operations can be performed.

The cam and horn die in Fig. 3-39 utilizes the spring-loaded movement of the forming tool, or slide (one for each side of the part), while the spring-dependent horn supports the formed part. On lowering of the upper die member, the horn engages the part, and soon afterward the cam driver (one on each side) pushes the sliding form tool toward

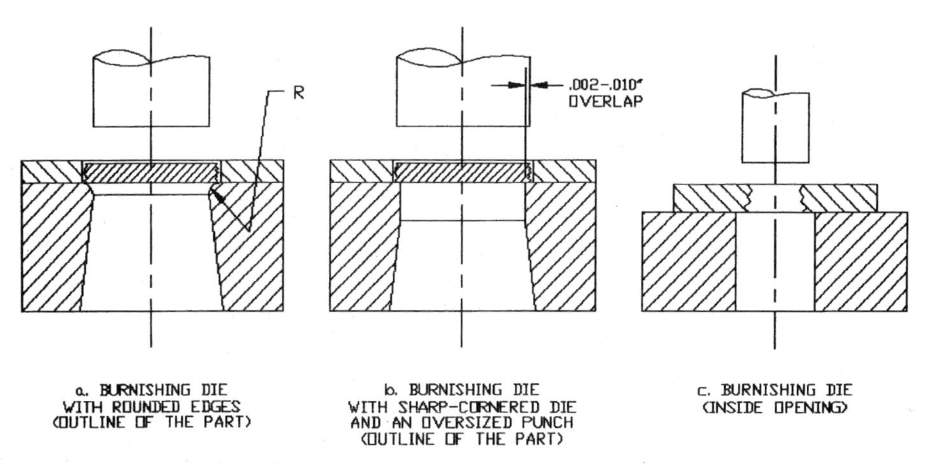


**Figure 3-38** Burnishing dies.

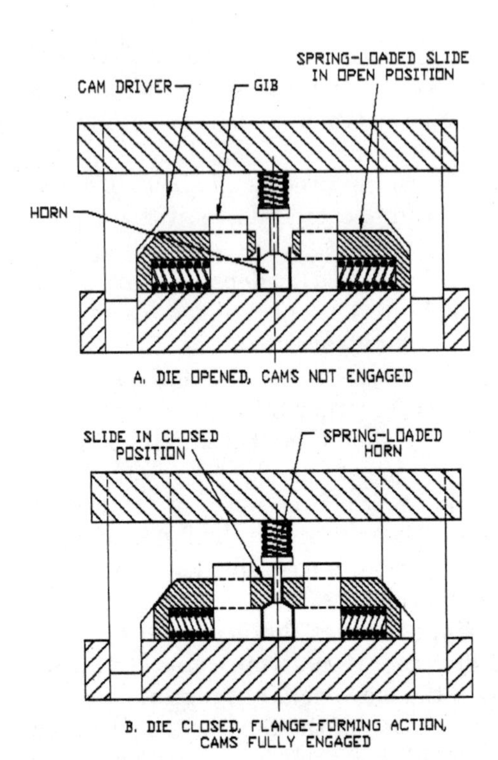


**Figure 3-39** Cam and horn die.

the part to be formed. The slide is restricted from other than intended movements by the gib. The gib assembly is usually made of hardened and ground tool steel. Sometimes the slide mechanism may contain an addition in the form of a hardened wear plate, located underneath its body.

The cam driver may be constructed of tool steel, cold-rolled steel, or just iron, and it may be welded together from pieces. Its work angle must be between 20 and 40° off the vertical (see Fig. 3-40), which should not be exceeded either way.

Generally, the closer the driving angle is to the vertical, the better the mechanical function of the cam mechanism will be.

*Hemming dies* can fold an edge of a sheet-metal piece back onto itself, which is often used as an edge reinforcement (Fig. 3-41).

*Crimping dies* are used to create additional surfaces, used for retention of other parts of the assembly. These dies operate by bending, denting, louvering, or otherwise forming the retainer's shape. The crimping operation (Fig. 3-42) forces the metal to form a flange around a cup, to be retained.

*Assembly dies* are built to assemble various parts, and they utilize riveting, staking, forming, press fitting, and similar operations.

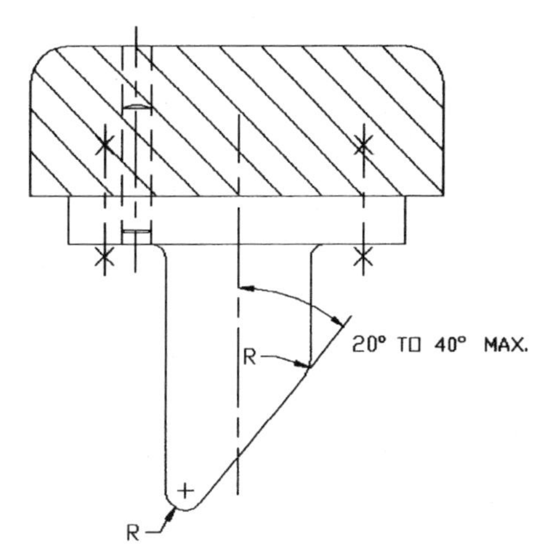


**Figure 3-40** Cam driver detail.

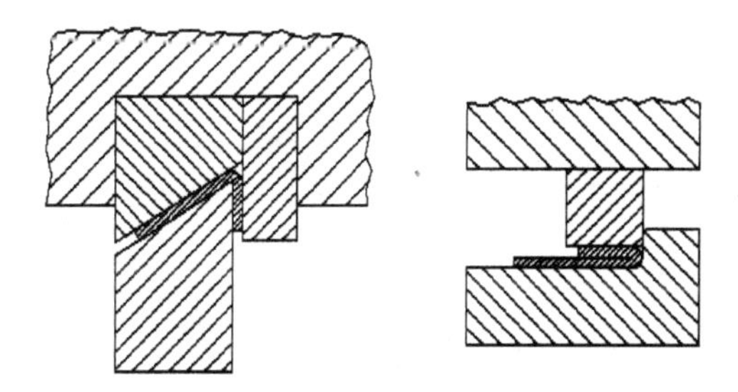

A. PRE-FORMING AT AN ANGLE AND
FINAL FORMING OF A FLANGE

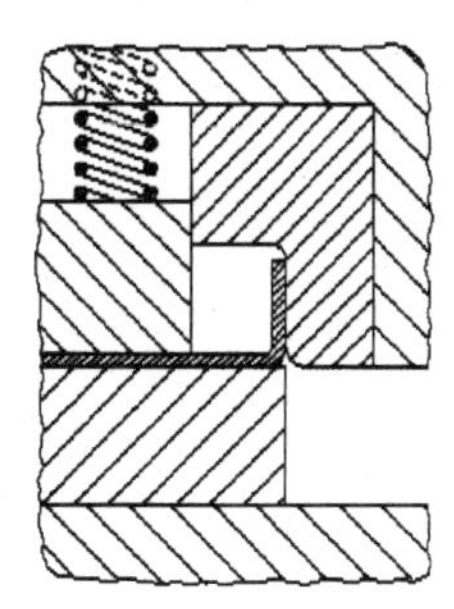

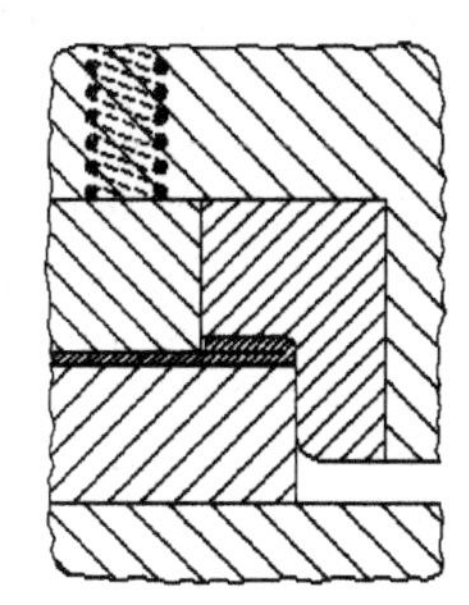

B. PRE-FORMED FLANGE AT THE
BEGINNING AND THE END OF
HEMMING OPERATION

**Figure 3-41** Hemming operation.

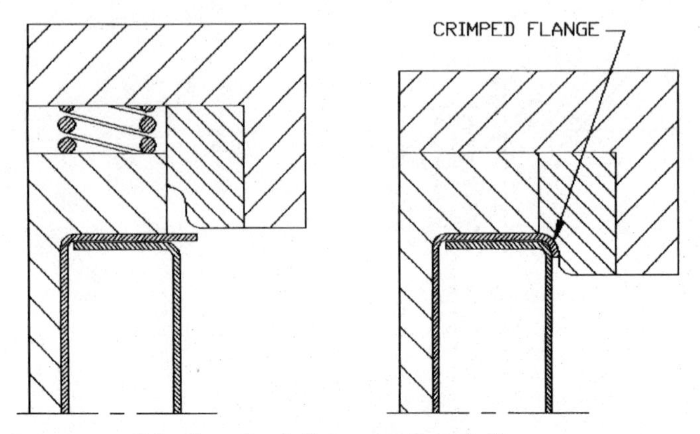


**Figure 3-42** Crimping of a shell over another shell.

*Subpress dies* (Figs. 3-43 and 3-44), or self-guiding dies, are valued for their great accuracy in punching out minute, precise parts, such as watch hands, geared wheels, and watch dials. There are two types of these dies: the cylindrical subpress and the pillar-type subpress.

In the cylindrical subpress the up-and-down moving plunger is guided by a bearing surface, which is firmly attached to the sturdy casting of its body. The pillar type uses pillars for the support of its movement.

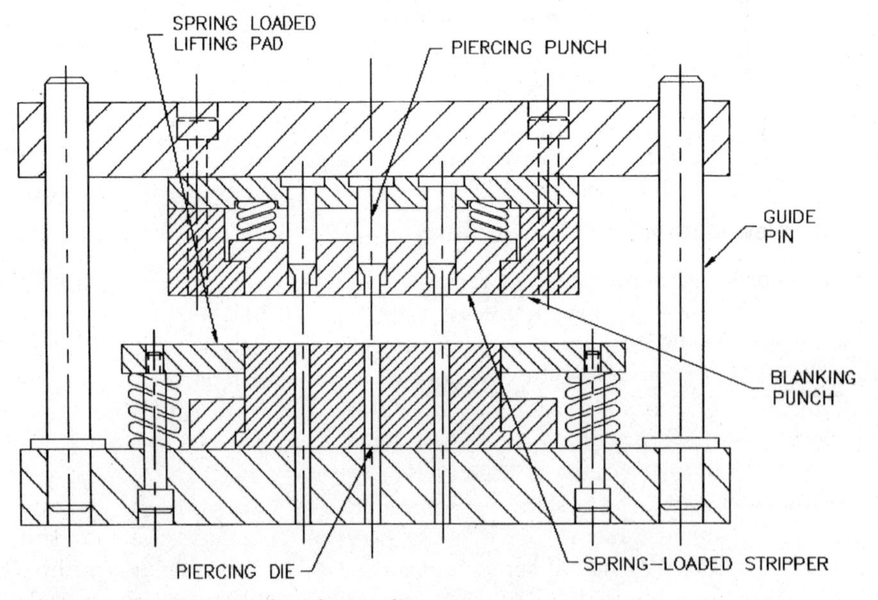


**Figure 3-43** Section view of a subpress die.

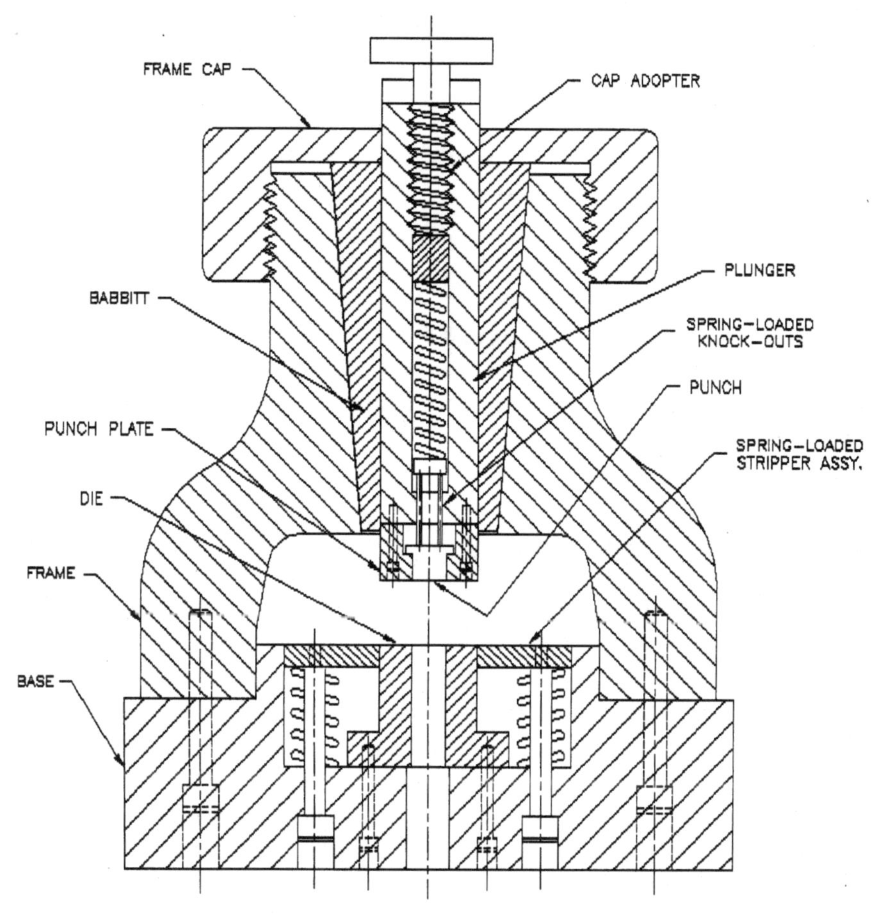


**Figure 3-44** Cylindrical subpress die.

## 3-4. New Methods in Metalworking

Metal-forming techniques, as recognized today, are often scrutinized for any applicable improvement with the addition of other, new processes. These new means of manufacturing brought quite surprising results: for example, in the field of materials' superplasticity, an elongation of over 4000 percent was achieved, with no thinning or necking of the material. Forming with superimposed vibrations cut the needed press tonnage to unbelievable ranges.

However, not all these new processes are yet adaptable to our means. Many may seem to be promising in their outcome, but the conditions for arriving at such methods are not always acceptable. Vibrations may lower the press force required for bending the metal, but its influence would reach far beyond that steel, affecting the manufacturing personnel, perhaps surrounding machinery, maybe even the structure of buildings—who knows?

Many of these techniques are too new for their long-term effects to be known. Only time can evaluate these processes and choose the most appropriate combination of manufacturing ease and human tolerance.

### 3-4-1. Electromagnetic forming

During this process, the energy stored in condenser batteries is released in the form of electric impulses. These are guided through the coil, which is placed within the part to be formed. With smaller parts, the part is placed inside the coil.

The pulsating impact of the current creates a primary magnetic field around the coil, turning it into the forming (or cutting) tool. The induced turbulent current forms a secondary magnetic field around the part to be formed. A reaction between these electric fields brings about the part's alteration.

Such a forming process is carried out without any physical contact between the tool and the workpiece. Therefore, no tool marks are left on the formed part and there is no friction or surface contamination.

The magnetic field affects only electrically conductive materials, which may be considered either an advantage or a disadvantage.

The forming pressure is equal throughout the range of the field, but it is quite difficult to apply it uniformly to a part with openings, notches, or embosses.

Because of the required forming frequency of above 15 kHz for steel materials, only objects larger than 12″ can be formed.

### 3-4-2. Electrohydraulic forming

In this process, the energy, stored in capacitors, is discharged over a spark gap located in a tank with water. This creates a sudden release of steam, which along with ionization causes the development of a high-pressure shock wave within the liquid. The die, containing the part to be formed, is immersed in the tank as well. When exposed to the shock wave, the part is forced to take on the shape of its die.

This process may be found useful for all tube-altering processes, such as bulging and expanding, that alter the tube's profile and shape.

### 3-4-3. Forming with explosives

Explosive forming is not a new process. It has been around for years, with differing results. Some consider it a superb method of manufacturing; others have lost their buildings to it in an explosion. It is a process in which safety cannot be overemphasized.

The energy derived from explosives can be of tremendous intensity, and the use of such force for forming processes is certainly tempting.

During the forming process, the explosive material, either in pieces or encapsulated, is placed in a water-filled tank alongside (or within) a die with the material to be formed. The charge, when detonated, prompts release of a great amount of steam and gas during a relatively short time interval. Such an action creates a strong shock wave in the liquid medium, which affects the part to be formed, forcing it to take on the shape of its die.

Objects suitable for utilization of such a manufacturing process are tubes, which may be bulged, expanded, or squeezed to tight tolerances and uneven shapes. Metal plates may be drawn to various shapes, many of them unattainable otherwise.

### 3-4-4. Superimposed vibrations

Ultrasonic waves, when applied to the molten metal, promote the development of additional currents within its mass, which in turn produce a more effective mixing, resulting in an improved homogeneity of the metal. When applied to the metal as it begins to solidify, ultrasound dissolves microfractures, removes gaseous entrapments, and drives out impurities.

In solidified metals, high-intensity ultrasound alters their structural defects by bringing the material into the stage of plastic deformation and rearranging its structure. Application of ultrasound reduces friction between metal particles, which in turn causes a free movement of metal layers with respect to each other, aiding the forming process and improving homogeneity of the outcome. The speed of the forming process is increased as well, with lessened friction between the material and its tooling, which subsequently decreases the wear of the latter.

Ultrasound enhances mechanical properties of materials, increasing their hardness, preventing structural changes due to deformation, and lowering stresses caused by manufacturing processes, while improving the quality of the product's surface. Many brittle materials, such as bismuth, were possible to form only after ultrasound was applied to the process. This is explained by the effect of vibrations on a metal crystal, which under their influence develops a series of linear defects which lower its yield stress range.

When applied to the forming process, ultrasonic vibrations greatly reduce the amount of force necessary for the metal alteration.

However, this type of manufacturing is not widely practiced yet. Its possible negative effects on the equipment, on the manufacturing personnel, and perhaps on the fabricated part have not yet been assessed.

### 3-4-5. Lasers and their application

Lasers operate on the basis of a concentration of their output to a small area of operation, approximately 0.002 to 0.010″ in diameter. One of their advantages is the lack of contact between the tool (laser) and the workpiece.

The laser cutting process is very fast, achieving high-quality burrless sharp edges, achieved through the high temperature of the laser ray, which causes the metal material to evaporate on contact. Such high temperature, attained in a very short period of time, allows for no surface distortion, as the surrounding material has no time to heat up to the temperature of the cutting process.

## 3-5. Fineblanking

Fineblanking is a special form of blanking which not only produces finished edges on a cut part but also works to close tolerances, attaining a superb consistency over high volumes of production. Fineblanking is performed in a die, yet it is a process quite similar to cold extrusion.

In fineblanking (Fig. 3-45) the material is firmly retained around the punch by the pressure ring, which—on descending—partially enters the material with its grips. The punch follows, piercing an opening in the sheet. The pressure of the retaining ring is not released. Instead a counterpressure to a die pad is applied. This pressure drives the blanked part up, along with the punch. At the die bushing level, the pressure ring releases its grip on the metal, while the blank is ejected from the die by the rising bottom pressure pad.

This process uses very tight clearances between the punch and the die, which amount to some 0.5 percent of the material thickness. Blanks, while being taken down and up through the die, have their cut edges forced into conformity with the surface of the opening. Possible disadvantages of fineblanking operation are

- A tapered edge of blanked parts, which is due to a friction between the blank and the die opening. This taper is greater with thicker materials or with those of higher carbon content.
- Rounding of the bottom edge, as shown in Fig. 3-46. In fineblanked parts, the burr appears on the punch side, as shown.

The advantage of this process is the high precision of the work. Openings of 0.125″ diameter can be punched even in 0.187″ thick sheet, with the hole tolerance range of ±0.0004″.

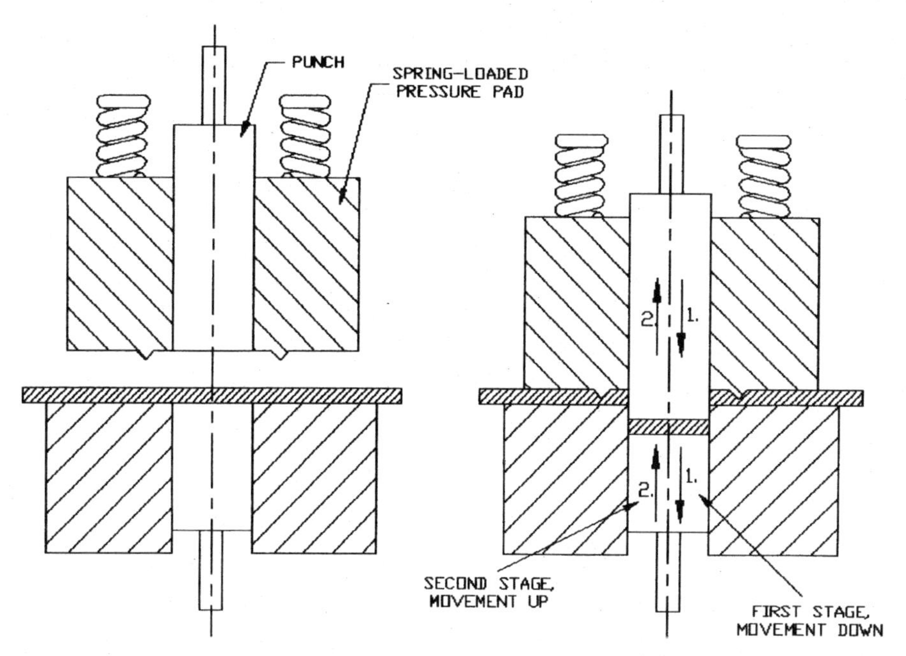


**Figure 3-45** Fine-blanking principle.

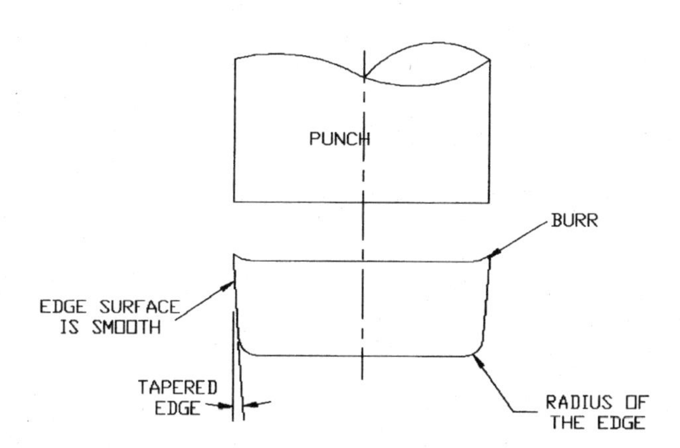


**Figure 3-46** Fine-blanked part.

The work-retaining efficiency of the grips is quite relevant to the quality of the cut. Their shape digs into the material before the punch descends to cut it. They not only provide for positioning of the sheet under the punch; their function is also that of stretching the sheet material in all directions.

The grips are most often located on the face of a pressure pad, bordering the punch along its entire shape. With materials thicker than

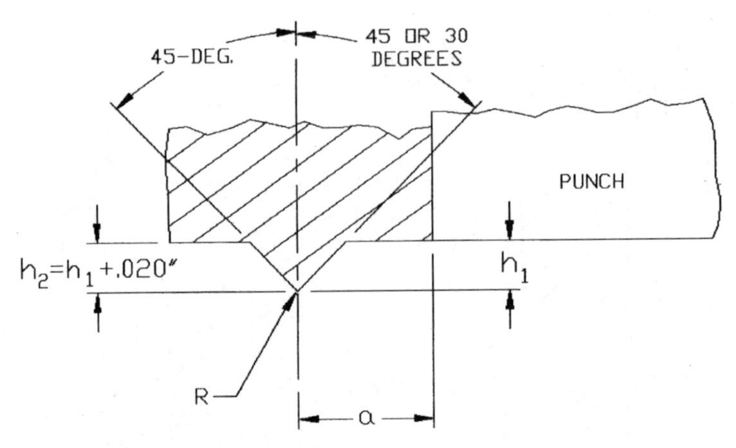


**Figure 3-47** Shape of the grip.

0.156″, or where rounding of cut edges is to be kept to a minimum, additional grips may be located on the upper surface of the die.

The shape of the grip, as shown in Fig. 3-47, has two variations: either 45°-45° angles on both sides or a 45°-30° angle combination. The height $h_1$ depends on the material thickness and its quality. It may vary along these sizes:

$$h_1 = 0.167t \text{ for hard materials}$$

$$h_1 = 0.333t \text{ for softer materials}$$

The distance off the edge of the punch $a$ depends on the height of the grip, and its percentile value should be

$$a = (0.6 \text{ to } 1.2)h$$

The height of the pressure pad behind the grip's edge is usually relieved, or

$$h_2 = h_1 + 0.020 \text{ in}$$

Chapter

# 4

# Metal Stamping Dies and Their Construction

## 4-1. Parts of a Die

A die, as mounted in the press, is a complex-action mechanism, producing parts in predetermined sequence. The lower half of the die, mounted on the lower die shoe, is firmly attached to the press bed, while the upper portion is bolted to the ram, sliding up and down along with it.

Aside from a die shoe, every die consists of several other blocks, which hold or support the punches, dies, bushings, inserts, and other elements.

The die in Fig. 4-1 has several piercing punches, with the last punch being the blanking station. All punches are assembled into a block called a punch plate, which is separated from the upper die shoe by a backup plate. Backup plates protect the die shoe from the effect of forces generated during the operation of the die. These plates are made of hardened steel, usually $\frac{3}{8}''$ thick, with a $\frac{1}{2}''$ exception for heavy work.

The stripper, shown in Fig. 4-1, is stationary, meaning that it does not ride along with the upper half of the die, being firmly attached to the die block. Usually a milled channel for guidance of the metal strip is added to its bottom surface.

A stripper prevents the piercings from sticking to punches. It also restrains the rest of the strip from moving along with the upper half of the die by keeping it positioned on the face of a die block.

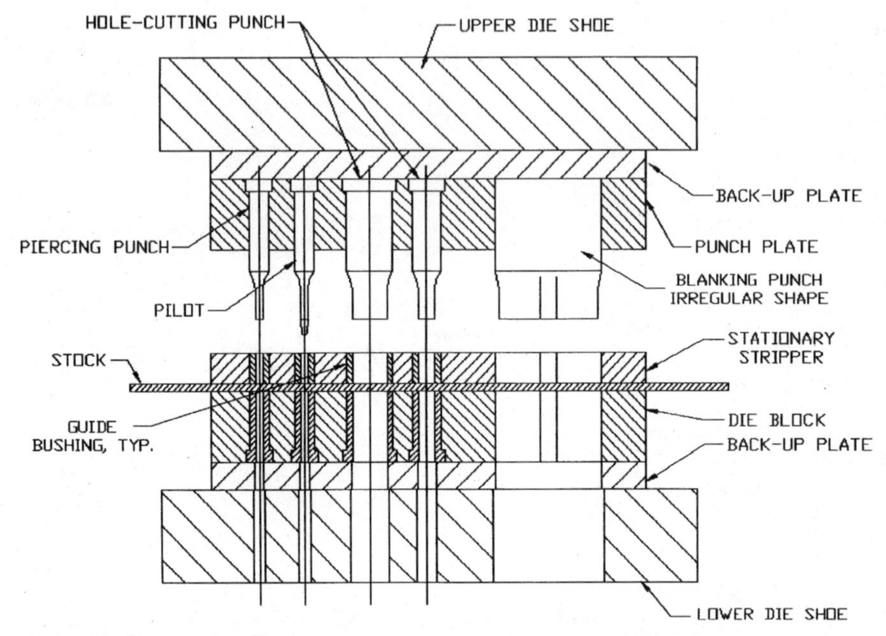


**Figure 4-1** Progressive die.

The die block contains all bushings, forming dies, or cutting inserts. It is supported by another backup plate positioned between this block and the lower die shoe.

All cutting, forming, and other material-altering punches and dies are assembled into their respective blocks using two methods of attachment: Either their body diameter $D$ is press-fit within the block, with their heads remaining loose, or their head diameter is press-fitted, while the body remains loose.

### 4-1-1. Punches and dies

A sample of a typical punch, its dimensioning and tolerances, is shown in Fig. 4-2. Notice the diameter of the cutting portion $P$ is quite precise. *This dimension is always that of the opening to be pierced.* The cutting tolerance is added to the die opening.

Mounting of this punch is evaluated in Fig. 4-3 with respect to the two mounting techniques already described: Either the shank is press-fit within the block while the head is loosely contained in the counterbore (Fig. 4-3*A*), or the head is press fit and the shank is loose (Fig. 4-3*B*). The second method of mounting is reserved for special instances, whereas the first method is commonly used for mounting of majority punches.

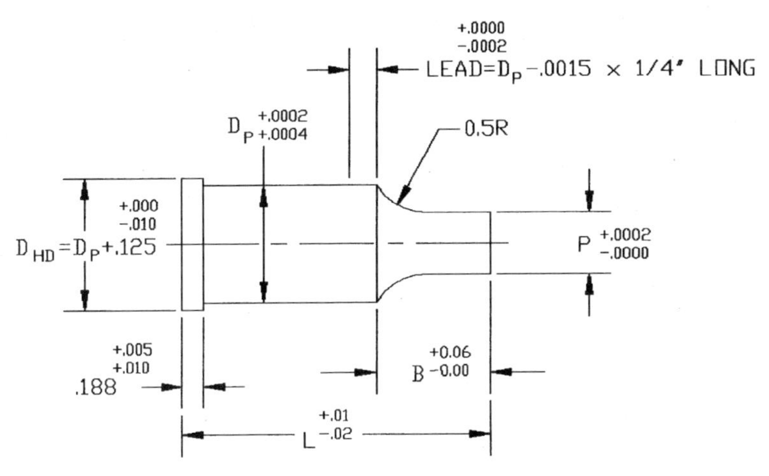


Figure 4-2 Punch detail.

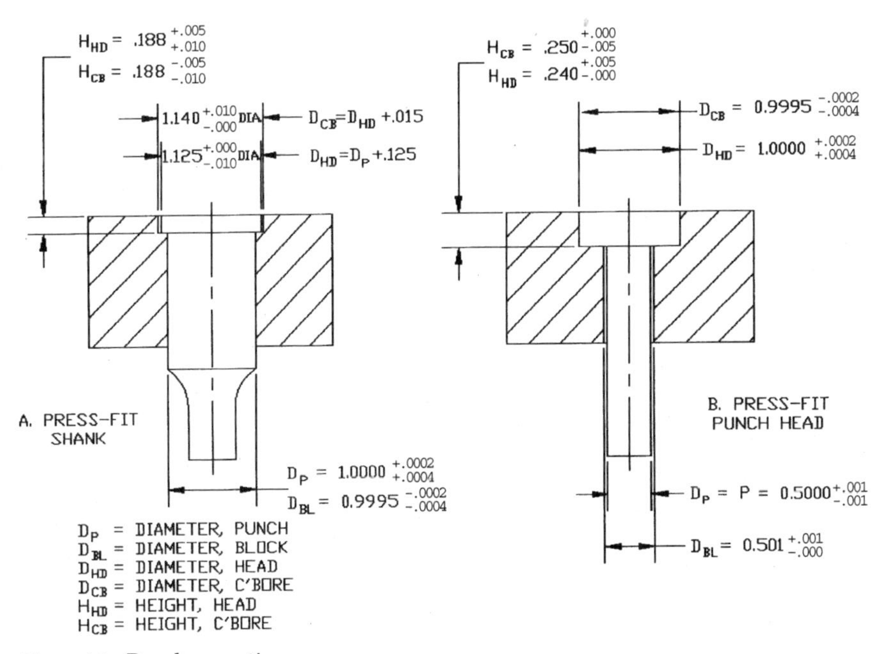


Figure 4-3 Punch mounting.

The proper length of a punch has a considerable effect on the overall performance of the die. With too long punches, the compressive stress on them may be excessive, resulting in frequent breakages. The maximum length of a punch may be calculated with the aid of the formula

$$L = \frac{\pi d}{8} \sqrt{\frac{Ed}{St}} \tag{4-1}$$

where $L$ = maximum length of the punch
$d$ = punch diameter
$t$ = thickness of punched material
$E$ = modulus of elasticity
$S$ = shear stress of the material, lb/in$^2$

The ratio of the punch diameter to the stock thickness must satisfy the condition

$$\frac{d}{t} = 1.10 \text{ minimum}$$

Smaller ratios are generally not recommended in sheet-metal practice, not that smaller holes cannot be produced, but their method of manufacture is much more complicated. Better punch materials with higher compressive strengths may be required, and additional stripping and tool-guiding arrangements may be needed, along with greater clearances between the punch and die.

Other restrictive conditions, dealing with similar situations, may be followed up in Sec. 6-8, "Scrap and Hole Size Recommendations: Minimum Punch Diameter."

**Types of fits in assembly of parts.** The inch-based measuring system has one great advantage, that it may establish several layers of dimensions for easy application of tolerances. In die design, we most often have:

- Three decimals, or x.xxx, ranging in tolerance ±0.005″
- Two decimals, or x.xx, with tolerance ±0.010″
- Fractions, which are ±0.015″

These tolerance ranges are rather common in this type of manufacturing. Other areas of production may use different sets of numbers. For example, with longer or larger parts, the fractions may run up to ±0.031″ and perhaps even more.

Quite precise tolerances for glass cutting are:

x.xxx ±0.015″    x.xx ±0.031″    fractions ±0.062″

Minimal tolerances in woodworking, where NC equipment is utilized, are approximately:

x.xxx ±0.062″ x.xx ±0.125″ fractions ±0.25″ and more

Every manufacturing field alters the tolerance ranges to suit its needs. The fits, however, are a rather different story, as they always involve two parts, assembled together. Here the tolerance range must be somewhat standardized and the amounts of tolerance tightened. After all, we're not fitting wooden shafts into openings drilled through glass.

The general range of tolerances, as published by the American Standards Association in 1925 (ASA Standard B4a-1925), runs as follows:

| Class of fit | Clearance | Interference | Hole tolerance | Shaft tolerance |
|---|---|---|---|---|
| 1. Loose fit | $0.0025\sqrt[3]{D^2}$ | | $+0.0025\sqrt[3]{D}$ | $-0.0025\sqrt[3]{D}$ |
| 2. Free fit | $0.0014\sqrt[3]{D^2}$ | | $+0.0013\sqrt[3]{D}$ | $-0.0013\sqrt[3]{D}$ |
| 3. Medium fit | $0.0009\sqrt[3]{D^2}$ | | $+0.0008\sqrt[3]{D}$ | $-0.0008\sqrt[3]{D}$ |
| 4. Snug fit | 0.0000 | | $+0.0006\sqrt[3]{D}$ | $-0.0004\sqrt[3]{D}$ |
| 5. Wringing | | 0.0000 | $+0.0006\sqrt[3]{D}$ | $+0.0004\sqrt[3]{D}$ |
| 6. Tight | | $0.00025D$ | $+0.0006\sqrt[3]{D}$ | $+0.0006\sqrt[3]{D}$ |
| 7. Medium force | | $0.0005D$ | $+0.0006\sqrt[3]{D}$ | $+0.0006\sqrt[3]{D}$ |
| 8. Heavy force | | $0.001D$ | $+0.0006\sqrt[3]{D}$ | $+0.0006\sqrt[3]{D}$ |

The use of this table is based on the hole dimension being the nominal size, toleranced on the plus side, with negative tolerance range equal to zero.

The shaft is handled in the opposite way, its tolerance ranges being negative, with plus tolerance equal to zero.

**Press-fit in assembly of parts.** Press-fitted assembly depends on very accurate hole-and-part dimensioning and surface finish. Often the block has to be heated and the punch frozen for assembly, since the punch is always slightly larger than the opening. The maximum amount of interference between any two press-fit objects is 0.0014″. Any amount greater than that will not allow for the assembly of such manufactured parts.

Some may argue that the tolerance range of parts should be made more precise, more tight, almost none, but that's not an easy solution either. The more precise the required dimensions are, the more expensive operations the part has to go through, and the higher its cost is going to be.

In order to find the happy medium, the tolerances are given within a certain range, and their buildup is evaluated so that it does not exceed the limits of interference or 0.014″ TIR (total indicator reading). In our case, the punch body diameter, including its highest amount of tolerance (see Fig. 4-3A), will be

$$1.0000 + 0.0004 = 1.0004''$$

The punch plate opening into which the punch will be assembled (here we subtract the higher of the two tolerance amounts) is

$$0.9995 - 0.0004 = 0.9991''$$

The total variation between these two numbers is

$$1.0004 - 0.9991 = 0.0013''$$

This interference is acceptable, and yet the ranges of tolerance for the two vital dimensions are not out of the ordinary. Through such evaluation we may assess that if the two parts were to be made to the fullest extent of their tolerance ranges, they still could be press-fit together.

Now the lower amount of press-fit scenario has to be assessed, where the minimum amount of interference between two press-fit objects is 0.0003″. This lower end of the tolerance range should be evaluated as well, to provide for a situation where both parts may be made to their smallest possible press-fit dimensions.

The punch size, including the lower value of the two tolerance ranges, will be

$$1.000 + 0.0002 = 1.0002''$$

The punch plate opening minus the smaller amount of tolerance is

$$0.9995 - 0.0002 = 0.9993''$$

Subsequently,

$$1.0002 - 0.9993 = 0.0009''$$

0.0009″ is quite a tight fit, which clearly indicates that the assembly will be adequately stable. But the absolute minimum of interference, such as that where no tolerance buildup of any kind will be generated on either of the two parts, must be judged as well. This condition is obtained by comparing the two basic dimensions as follows:

$$1.0000 - 0.9995 = 0.0005''$$

With 0.0005″ being the lowest possible interference and 0.0013″ being the highest, we have an acceptable level of press fit for the two metal parts. However, dimensions of actual products are rarely found on either extreme side of their tolerance range but are rather somewhere in between. Therefore, it is not important in which section of the tolerance range these parts are made. As long as they are made within the drawing's dimensional requirements, they will fit.

The second method of assembly, shown in Fig. 4-3*B*, should be evaluated the same way. Here we control the tolerance buildup of the press-fitted punch head. Its size and tolerances were purposely made the same so that the equality of both methods can be easily demonstrated.

The height of the punch head in this type of assembly is usually greater, since the increased length of the press-fit area is vital to the stability of the tool.

In assembly, the process of inserting punches and dies into their openings in blocks is aided by the presence of a lead. The lead is a ¼″ wide band on the circumference of the punch shank (or die), which is slightly smaller in diameter, for an easy entry of the large part into the smaller, press-fit opening.

**Depth of the counterbore versus the height of the punch head.** When comparing the thickness of the punch head to the depth of the counterbored opening, it is obvious that the metal of the block is purposely being allowed to exceed the height of the head. This is because after all punches are assembled, the whole block is placed into the surface grinder, where it is leveled down so that it will be flush with all punch heads.

Tolerance ranges for the height of punch head lean toward the plus side, while tolerances for the depth of counterbore go in the minus direction.

The height of the punch head in Fig. 4-3*A*, is dimensioned as

$$0.188 \begin{array}{l} +0.005 \\ +0.010 \end{array}$$

It can be minimally 0.188″, with its maximum size being 0.198″. The two tolerance ranges being in the same direction (both plus) indicates that such a dimension should never slip into the opposite, into minus.

The counterbore's depth begins with the same nominal dimension, toleranced on the minus side:

$$0.188 \begin{array}{l} -0.005 \\ -0.010 \end{array}$$

Here the opening's depth can be a maximum of 0.188″, with its minimal size 0.178″.

The total tolerance buildup between the two parts can be figured out by comparing their two most extreme dimensions:

$$0.198 - 0.178 = 0.020''$$

which means that the maximum difference between the height of the punch and the depth of the pocket can be 0.020″. The minimal difference will be zero, since both start with the same nominal size, 0.188″.

Another depth-tolerancing method uses slightly tighter tolerance ranges for both the punch head and its counterbore, with the offset between their basic dimensions. In such a case, the height of the punch head is

$$0.188 \begin{array}{l} +0.005 \\ -0.000 \end{array}$$

while the depth of the counterbore is automatically lowered some 0.005″ or 0.010″ (the exact amount depends on the manufacturing practice of the particular shop). The depth of counterbored pocket then becomes

$$0.178 \begin{array}{l} +0.000 \\ -0.005 \end{array}$$

The total tolerance buildup, obtained through comparison of the two most extreme sizes, will come out as

$$0.193 - 0.173 = 0.020''$$

The second punch head, shown in Fig. 4-3*B,* is dimensioned and toleranced similarly. Its tolerance range has already been tightened.

**Die button,** or **die bushing,** is shown in Fig. 4-4 in several variations. The first, *A* version, is used and dimensioned for piercing of a single opening, while the *B* style may produce the inner and outer diameter at the same time. The headless die button is often used with a lighter type of work or with parts such as wave washers. The absence of the heel may reduce the cost of such a bushing's manufacture and assembly, but the disadvantage of such an unsecured press fit may be considerable.

Again, the tightness of manufacturing tolerance of the cutting surface may be observed. Two-tenths of an inch (0.0002) variation from the nominal size can be viewed as precision work.

Every die bushing has an opening into which a punch slides when cutting the sheet-metal material. Such an opening is absolutely

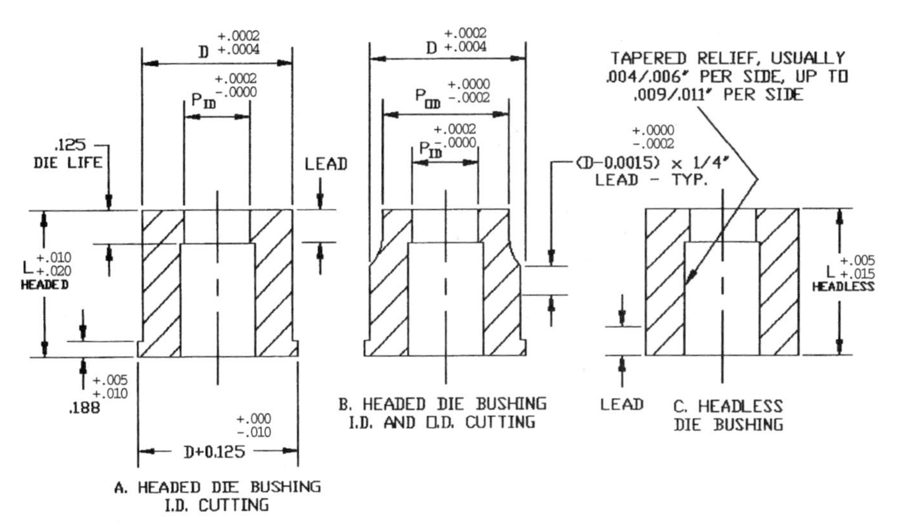


**Figure 4-4** Die button.

straight and precisely finished. It is called the "die life" (or "land"), and it is the amount of the die height which can be used up for subsequent sharpenings. The height of the die life depends on the number of pieces the die has to produce or on the number of expected sharpenings during the die button's existence in the die.

The height of this area is a debatable subject. In order to prolong the life of a die, a considerable die-life size may be chosen, which may be expected to provide for many sharpenings afterward. But, if such a tight portion is excessively long, the piercings, leaving the die through this channel, may get packed there, perhaps unable to move forward. Such a condition may endanger not only the particular punch and die but the whole die assembly as well.

Usually a ⅛″ length of die life is specified; rarely a greater size can be found.

At the other end of the die life the opening enlarges, turning into a clearance hole, through which the slugs leave the die. Usually this enlargement has a form of draft, most often in the vicinity of 1.5 to 2° taper.

**Variations of punch and die cutting diameter.** As already mentioned, the cutting portion of the punch is always the size of the opening to be pierced. The size of opening in the die amounts to a total of the punch size, plus metal-cutting clearance.

Metal-cutting clearance is the difference between the size of the punch and that of the die. This term is not to be confused with the manufacturing tolerance, which in this case is +0.0002/−0.0000. Metal-cutting clearance is discussed in Sec. 6-5.

Therefore, the inside-die diameter, or $P_{ID}$, will be

$$P_{ID} = (P_{PUNCH} + \text{CLEARANCE}) \begin{matrix} +0.0002 \\ -0.0000 \end{matrix}$$

However, should a die be used to cut both diameters, inner and outer, such as the one shown in Fig. 4-4*B*, its outer edge will take upon itself the function of a punch, with its corresponding die-bushing part being shifted into the punch-plate location. The outside-die dimension, or $P_{OD}$, will then become the exact size of the opening to be punched, with the metal-cutting clearance added to its corresponding die member, held in the punch plate. A reversed manufacturing tolerance of +0.0000/−0.0002 will then be assigned to such a punch. Its opposite die part would be dimensioned as

$$P_{DIE-pt} = (P_{OD} + \text{CLEARANCE}) \begin{matrix} +0.0002 \\ -0.0000 \end{matrix}$$

Mounting methods of guide bushings are similar to mounting methods for punches. Therefore, the earlier description of the procedure applies here too.

The height of the bushing in Fig. 4-5*A* should be equal to the thickness of the die block; the bushing shown in Fig. 4-5*B* must be higher than the die block to allow for dual cutting.

**Miscellaneous notes.** Other than round punches should be well aligned with their respective bushings and either part of the assembly should not rotate or otherwise deviate during the die function.

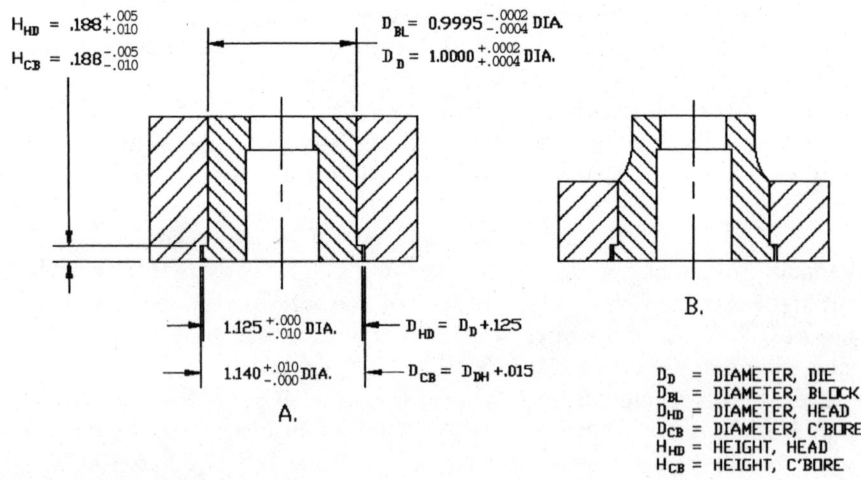


**Figure 4-5** Die mounting.

This, along with the correctness of the initial installation, is assured by keying the punch or die's head and placing a standard keyway against this surface.

Key flat portions are obtained by grinding the punch head all the way toward the shank diameter, as shown in Fig. 4-6. Some punches may have a single key flat; others may have two. With headless bushings, an undercut may be produced to serve the same purpose.

Another locating feature is a retaining notch in the shank of a punch (Fig. 4-6*D*). A screw inserted through a side opening in the punch plate will secure the punch against any rotational movement and against any vertical displacement as well.

The key flat portion is used in congruence with the cutting shape of a punch, where it indicates its orientation, as shown in Fig. 4-7*A*.

Die bushings utilized for cutting of openings of various shapes may sometimes be cut in half to allow for easier manufacture and easier alignment during installation. (See Fig. 4-7*C*.) Lately, however, with the emergence of EDM-wire machinery, punches and dies of all shapes can be produced with a single cut, where the thickness of the wire already provides for the metal-cutting tolerance.

A slug-ejecting device, shown in Fig. 4-7*B*, has a small, punchlike ejector pin already installed. The pin is continuously forced out by a spring pressure. During the cutting operation, the greater force of the press drives this pin to retreat back into the punch. As soon as its force is released, during the upward movement of the die, the pin

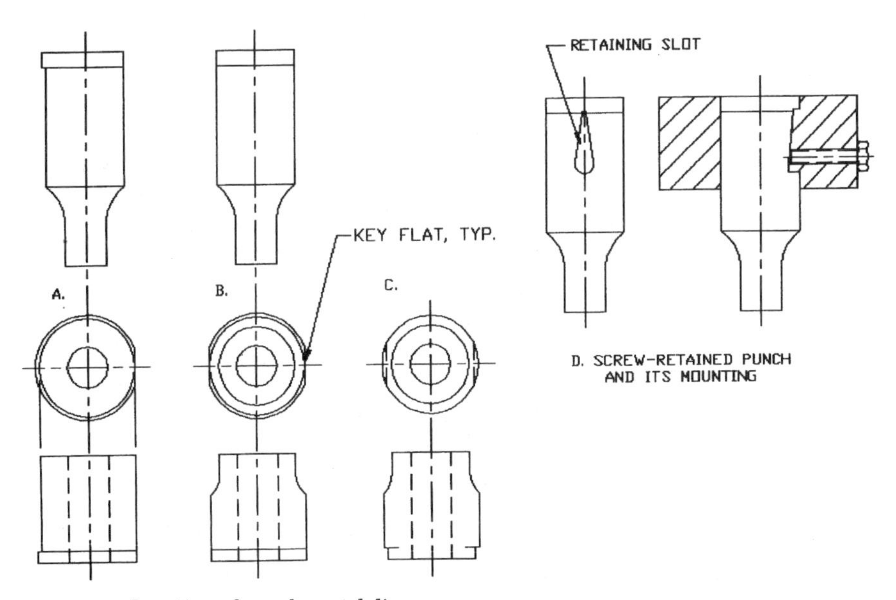


**Figure 4-6** Locating of punches and dies.

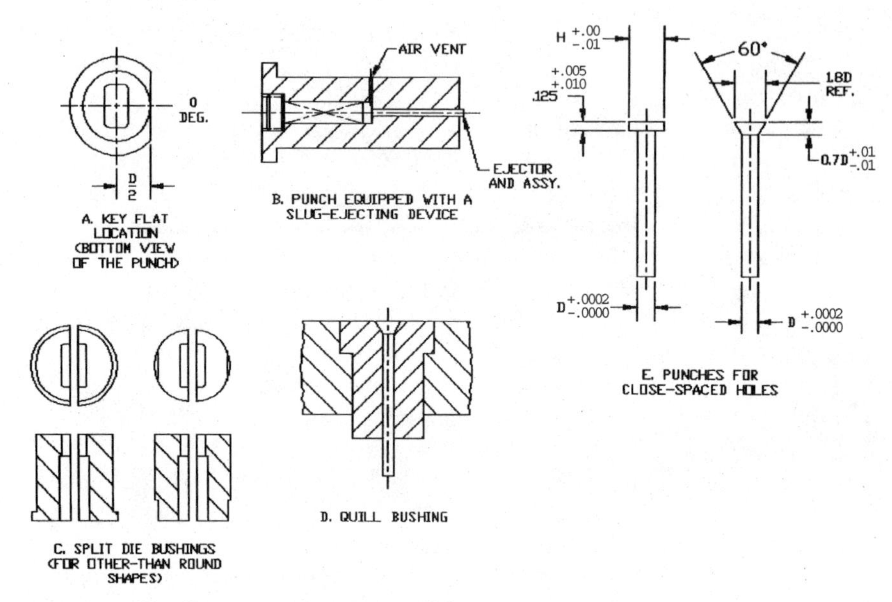


**Figure 4-7** Miscellaneous variations in tooling.

springs out, applying pressure against a slug, should it be sticking to the punch surface.

Punches for close-spaced openings (Fig. 4-7*E*) have quite small heads, so that they fit into congested areas. These punches are also used with quills (Fig. 4-7*D*), which actually are sleeves, or bushings, used to provide support to various fragile punches.

## 4-1-2. Pilots

In construction, pilots are similar to punches, with the only difference being in their smooth, radiused end. In the die, pilots provide for a guidance of the strip by sliding into at least two pierced openings, located at the extreme edges of the sheet-metal strip, and positioning, or fine-adjusting the surrounding material around their bodies (Fig. 4-8).

Pilots are always longer than any punches, to assure their contact with the strip prior to the occurrence of any cutting. Their diameter may be −0.003″ smaller than the diameter of the punch used for piercing pilot holes. Mounting of pilots utilizes the same procedure as that described for mounting of punches shown earlier in Fig. 4-3.

Spring pilots (Fig. 4-9) do not need precision mounting holes and are therefore cheaper to install. Aside from providing for the guidance of a strip, spring pilots may be used to support compression springs or to serve as lifting devices. The free end of the spring is usually contained in a simple pocket counterbored in the opposite plate.

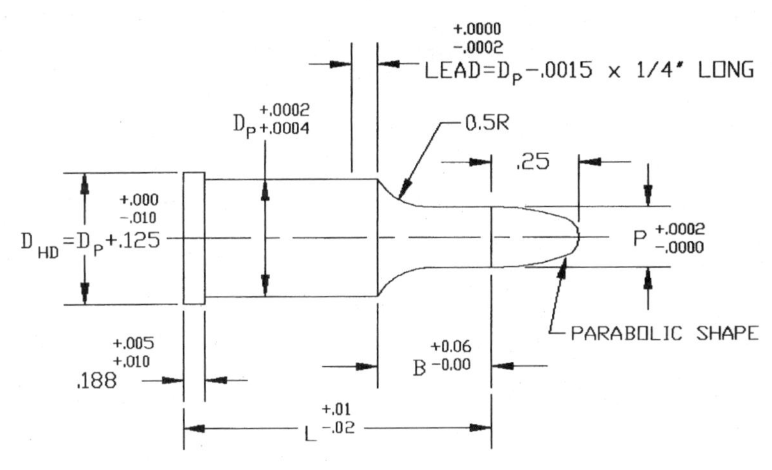


**Figure 4-8** Pilot detail.

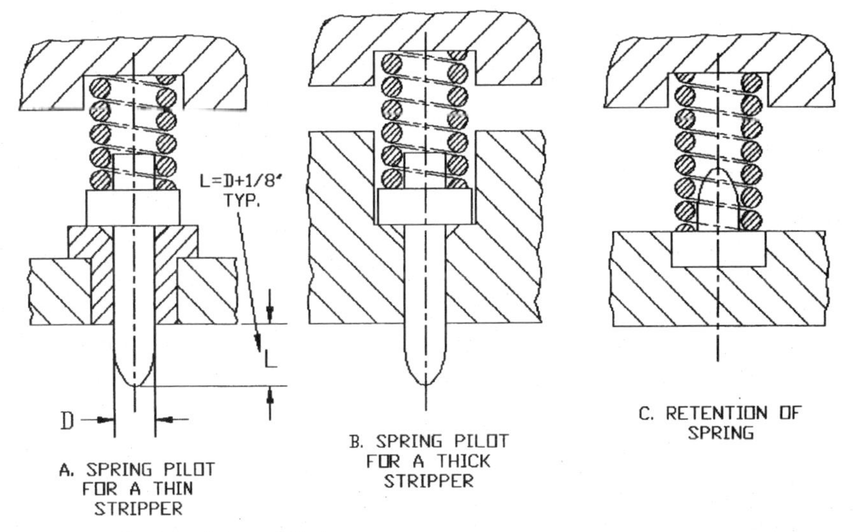


**Figure 4-9** Spring pilots and their use.

The length of a spring pilot tip should be equal to its diameter, plus ⅛″.

### 4-1-3. Guide bushings

Guide bushings (Fig. 4-10) are inserts in the stripper, guiding the punches and protecting the stripper plate from wear caused by the die operation. These bushings are hardened inserts, headed or headless, press-fitted into the plate, with slip-fit openings to contain the punches. Their countersink-shaped inner openings aid in the centering of the punch during the die operation.

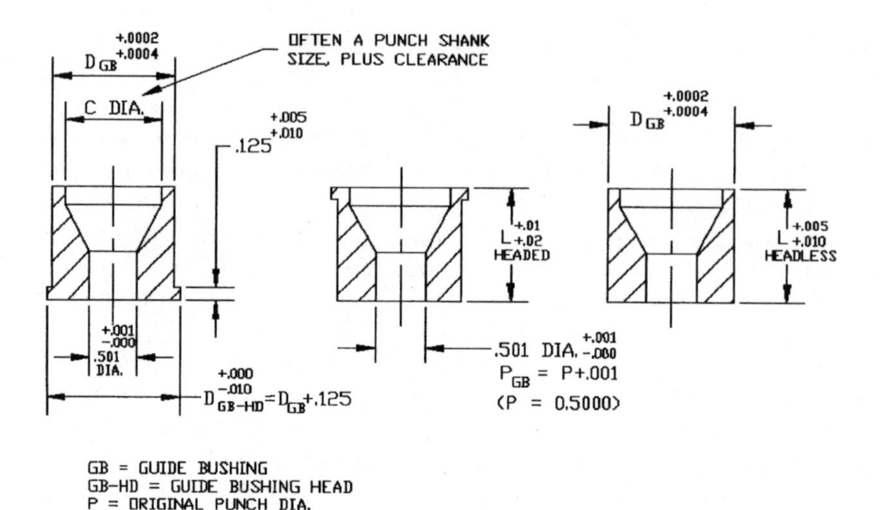


**Figure 4-10** Guide bushings.

The tolerance range of the guide bushing's inner openings is not as precise as that of cutting areas. The opening itself is usually made +0.001″ greater than the size of the punch, with tolerance of +0.001/−0.000.

Mounting of guide bushings in the stripper plate is the same as mounting of any other die member (Fig. 4-11). The head is always loose, sometimes even left outside the stripper's thickness. Bushings' heads are oriented around the stripper's upper surface for a majority of punching work. Only where greater stripping forces are expected,

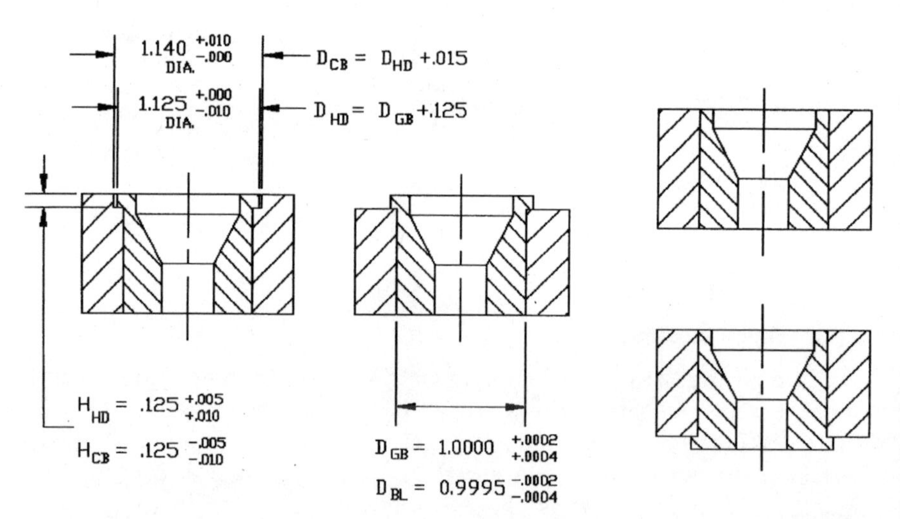


**Figure 4-11** Guide bushings' mounting.

the head may be turned upside down, toward the stripper block's bottom surface.

An alternative mounting method of punches, dies, and guide bushings (Fig. 4-12) uses only standard-size drills and reamers for this procedure. Where previously the same amount was added to the punch-body diameter for its head size (+0.125″), here the amount to be added corresponds with the availability of tooling, so that the counterbore will not be special in size. Such a counterbored pocket may be adapted to quite a narrow line of tooling, provided all punches and bushings are similarly altered to fit.

Here, the designer should be aware of the material cost and labor restrictions. If the step between the shank and head is too great, the blank size for that particular punch will also be greater, with subsequent increase in cost. The time needed for removal of a larger than usual chunk of metal will increase as well.

### 4-1-4. Knockouts, or knockout pins

Knockout pins, sometimes called kicker pins, push down the sheet-metal strip after the punching is done, when the upper die begins to move up and away from it (Fig. 4-13).

Similarly, as with the spring-loaded pin in the slug-ejecting device in Fig. 4-7*B*, the force of the spring-loaded knockout bar drives the knockout pins forward when the much greater force of the press is released.

The pins are small in diameter, and placed in close proximity to the punch cutting surface, they have proved very efficient.

The lifting type of knockout prevents the cut or formed part from falling below the upper surface of the die. This type of arrangement is

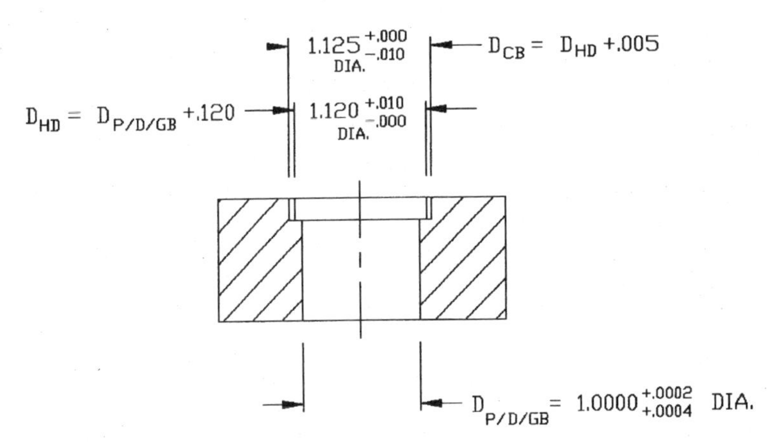


**Figure 4-12** Alternative mounting of punches and bushings.

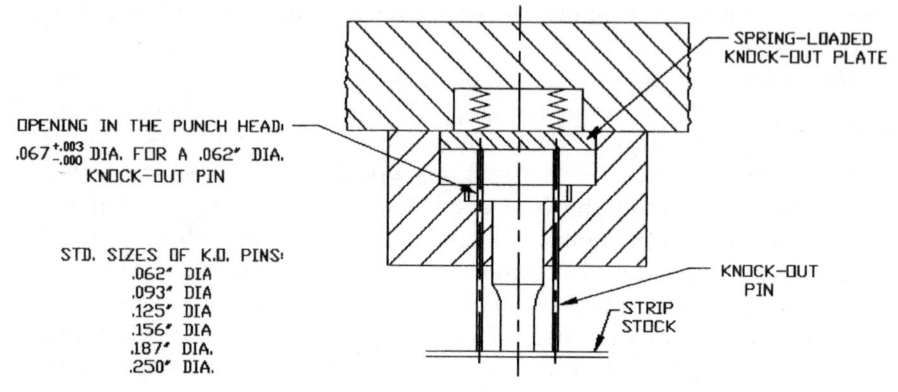


**Figure 4-13** Positive knockout system.

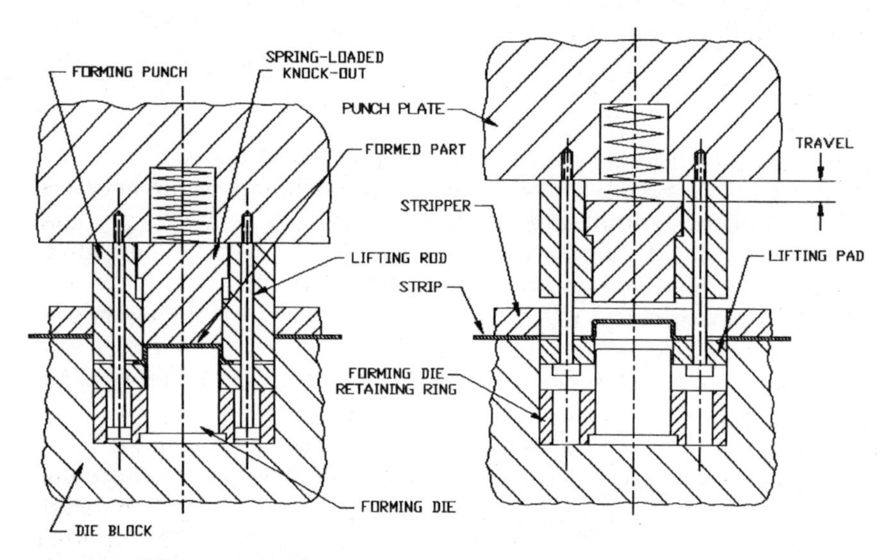


**Figure 4-14** Lifting-type knockout.

sometimes necessary where return blanking, flange forming, drawing, and similar operations are performed.

During the cup-forming operation, depicted in Fig. 4-14, the upper half of the die slides down, cutting the blank out of the sheet with the outer edge of its forming punch. The lifting pad moves along, attached to the punch plate with lifting rods. The whole array bottoms on the retaining ring, which secures the forming die in its location. Lifting rods' heads fit into the clearance openings of the retaining ring.

At the end of the forming sequence, when the upper half of the die moves up, the centrally located knockout forces the part out of the forming punch area. Lifting rods move along, pulling the lifting pad behind. When the lifting pad comes up flush with the die block, it provides a support to the formed cup before it gets removed from the die-working area.

### 4-1-5. Strippers

Stripping of parts off the face of the tooling is a complex problem, influenced by the thickness of material and its type, by the surface finish of the strip, and by the surface condition of the tooling as well.

The stripping of parts is further complicated by the prevailing manufacturing procedures, since all conventional metalworking machinery leaves circumferential grooves in the surface of a machined part. The sheet-metal material, forced by the pressure of tooling, may sometimes be coerced into fitting within these grooves in some sensitive sections and thus may generate a serious stripping problem.

For this reason, all problem-prone surfaces should obtain their final finish by some longitudinal grinding or polishing process, which may level these circumferential grooves and perhaps even replace them with slight lengthwise arrangements.

Another influence detrimental to parts' stripping is the emergence of vacuum between the cut metal and its tooling. This problem is discussed later in this chapter, with some possible remedial tooling alterations shown later in Fig. 4-35.

Strippers, as used in the die work, are either stationary (nonmoving) or spring-loaded (moving). Stationary strippers are low in cost when compared to spring strippers. Therefore, spring-backed stripping arrangements should be used with thin, fragile punches, where the immediate stripping action may prevent their breakage.

Spring strippers are of advantage also where additional flattening or material-retaining function is needed or considered beneficial. Such retaining action is usually utilized in drawing, flanging, or other forming operations.

Stationary strippers are provided with a milled channel made to accommodate the strip material. It retains the strip in a fixed location, preventing it from moving anywhere, up, down, or sideways, from bulging, or from swellings of any type. Naturally, this type of stripper is not adequate where the height of a part is increased during the die operation, such as the height of drawn, formed, embossed, or flanged parts.

The stationary stripper (Fig. 4-15) is attached to the die block, using the same screws and dowel pins as those necessary for attaching the die block to the die shoe. This way, a single set of dowel pins provides for the alignment between all plates, and a single set of screws is used for their attachment.

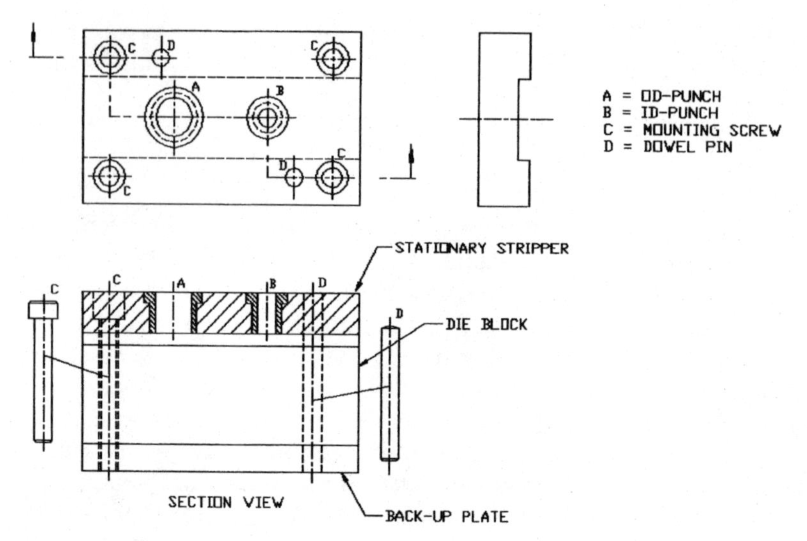


**Figure 4-15** Stationary stripper.

The holes containing dowel pins must be precision-reamed throughout all plates (shown later in Fig. 4-41). But the holes for screws cannot be tapped all the way through, as a misalignment, binding, and a host of other difficulties in assembly will be encountered. Openings for screws must be relief openings all the way through the blocks, no matter what their number or height should be, with only the final block being tapped, as shown later in Fig. 4-40.

Spring strippers (Fig. 4-16) are utilized where an increase in the height of a part is encountered. They also provide for much firmer stripping action, while acting partially as spring pads during the cutting, forming, or drawing activity of the die.

Spring strippers are attached to the punch plate, which makes them slide along with the movements of the ram.

**Strippers for drawn parts.** A stripper for thin-walled drawn parts is shown in Fig. 4-17*A*. This stripper is made of four circular segments, held together with a wire ring placed along their circumference.

The part, when forced by the action of a punch through such a stripping arrangement, coerces the segments to yield to its mass. This is accomplished with the assistance of the segments' central countersunk shape, shown in an enlarged detail. A minute movement of the whole assembly away from the center and against the holding force of the wire may be expected as well.

As the drawn shell passes through, its way back is blocked by the sharp edges of the opening in the segment ring.

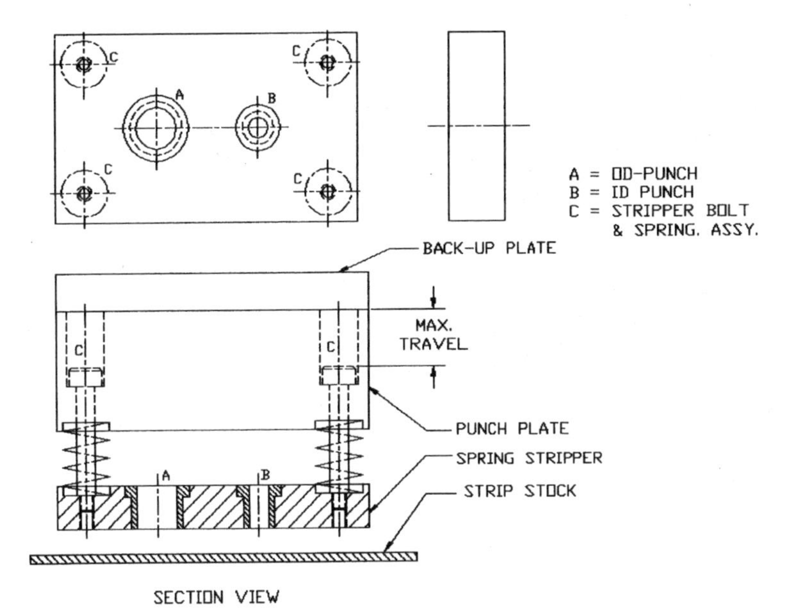


**Figure 4-16** Spring stripper.

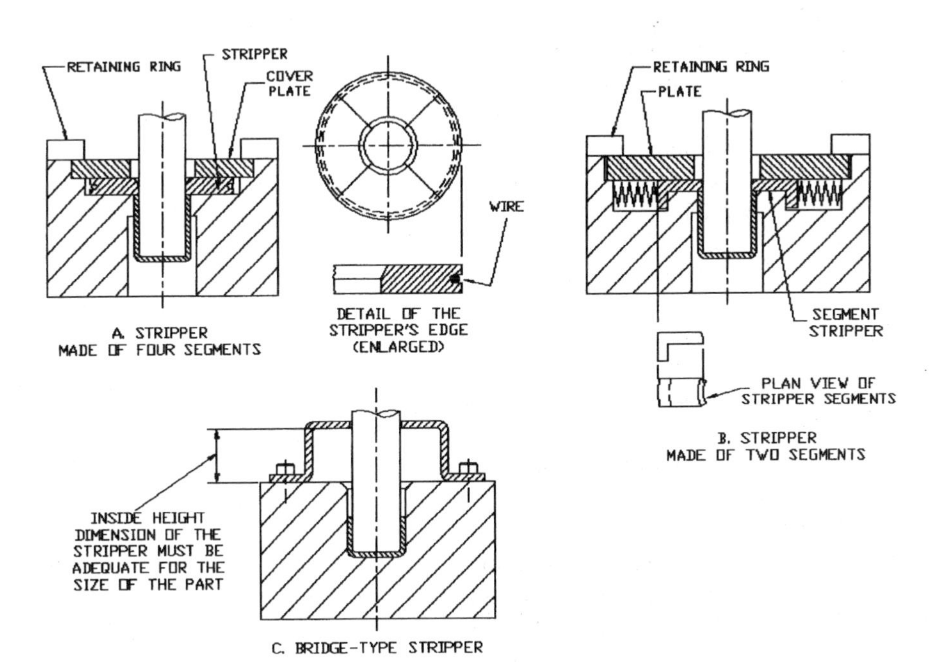


**Figure 4-17** Miscellaneous stripping arrangements.

The stripper in Fig. 4-17*B* is somewhat more positive in its action. Here the ring is replaced by two separate segments (shown in detail below the illustration), sliding toward the central punch under the force of springs. The material is pushed through, aided in its entry by the slant in the segments' surface. They immediately snap back, toward the body of the punch, allowing for no return of the shell.

This arrangement is found quite useful with thick-walled parts; thin shells may become torn by the stripping action applied to such a small area of their circumference. For thin-walled parts, a full ring of segments would be more appropriate, provided their construction allowed for a more controlled sliding-away and coming-back action.

Bridge-type stripper (Fig. 4-17*C*) strips securely and inexpensively all kinds of shells, regardless of their stock thickness or type of material.

**Minimal thickness of stripper plates.** The thickness of a stripper plate can be calculated by using the formula

$$h = \frac{1}{3}W + 2t \tag{4-2}$$

where $h$ = minimum stripper plate thickness
$W$ = stock width
$t$ = stripped material thickness

The result should be rounded up to the nearest fractional dimension in the eighths range, such as 0.375″, 0.50″, 0.625″, 0.750″, etc.

### 4-1-6. Stock guides

Stock guides can be of various forms and shapes. A set of pins, a set of blocks positioned alongside a strip, can be considered stock guides. Often, continuous, channel-type guides (shown in Fig. 4-18*A*) are used in the progressive die work. These are constructed so as to prevent the strip not only from moving aside but from following the upper die member in its movement during the die operation. This type of guiding arrangement, however, does not protect the sheet from bulging up when being pulled on by any of the punches. Such protection can be provided by substituting the side rails with a stationary stripper (Fig. 4-18*B*).

The width of all these stock-guiding channels should be equal to the width of the sheet strip, plus 0.010 to 0.020″. This increase in channel size is necessary to provide for the variation in the strip width and may be overcome by the locating activity of pilots.

The roller stock pusher (Fig. 4-19) enhances its stock-guiding capacity by combining it with greater accuracy. The hardened roller,

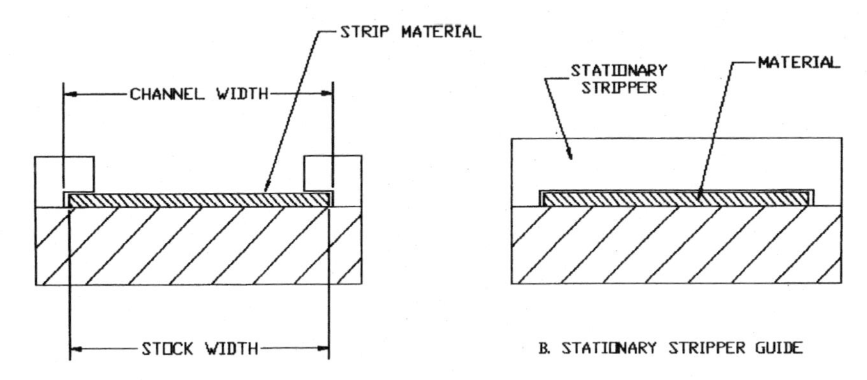


**Figure 4-18** Stock guides.

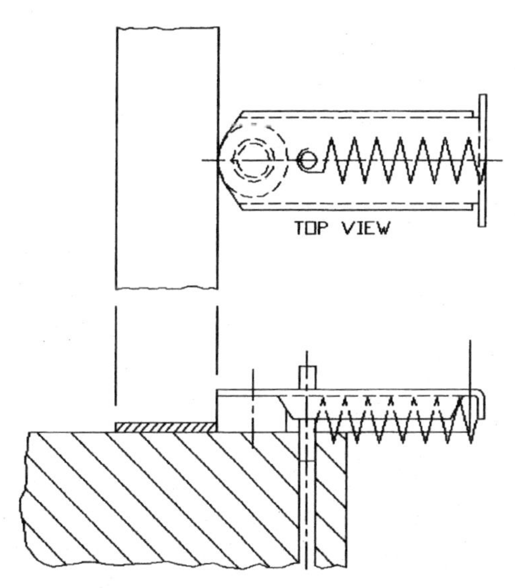


**Figure 4-19** Roller stock pusher. (*Printed with permission from Applied Mechanics Corp., Grand Rapids, MI.*)

contained at the tip of the unit, contacts the edge of a material and allows it to slide past by rotating along with its movement. The roller is held against the material's edge by the force of a spring, which allows for a width variation of the strip, rendering any additional adjustment needless.

The material-positioning device (Fig. 4-20) is mounted on the upper half of the die, and it moves up and down with the movement of the ram. During the downstroke, the long arm trips over the edge of a

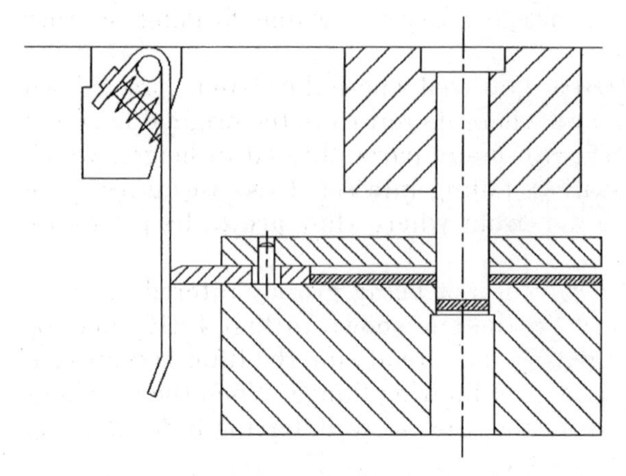

**Figure 4-20** Material positioner. (*Printed with permission from Applied Mechanics Corp., Grand Rapids, MI.*)

sliding block and pushes it toward the edge of the material. The sliding block, restricted in its movement by a pin, is sandwiched between the die block and the stripper. The tension adjustment of the slide-pushing arm's pressure is in the range of 2 to 20 lb.

### 4-1-7. Stock supports, stock lifters

Stock-supporting rails, as used in Fig. 4-21*A*, allow for the strip's travel above the die surface. This may be found quite helpful where

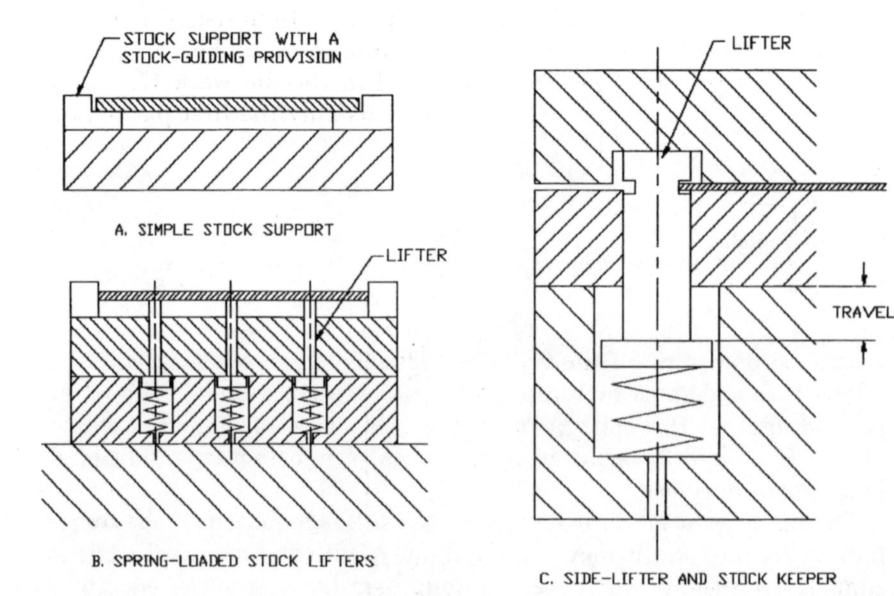


**Figure 4-21** Stock support, stock lifters.

the height of the strip increases because of drawing, forming, or other height-altering process.

Stock lifters (Fig. 4-21*B*) are utilized where the strip is forced down during the die operation and where its return to the original height is desired. Such need arises with many parts, altered in height, which travel from station to station, falling into relief recesses during the operational cycle of the die, from where they are to be pulled up again.

Where the height of a part is not being grossly altered, a lifter-retainer combination, such as the one shown in Fig. 4-21*C*, may be used. With the upper half of the die already up, the lifter is restricted from following along by its travel-limiting flange. When the die slides down, it exerts a pressure on the lifter as well, forcing it down, along with the strip it retains.

An adjustable version of this type of lifting device is shown in Fig. 4-22. By turning the slotted head of the unit, its position with regard to the sheet-metal strip as well as its height can be adjusted. This type of device also takes various sizes of heads, which makes its adjustability still more versatile.

### 4-1-8. Stops

Strip material, when first being guided into the die, must stop somewhere for the sequence of die operations to begin successfully. It is obvious that the strip should not go as far as the forming tool, which may need some preblanking work performed at the beginning. Advancing the strip too far may lead to greater than usual wear and tear of the tooling and its subsequent misalignment and breakage.

For that purpose, stops are introduced in the die work. The first stop, which the strip meets on its way, is usually the first pierce and

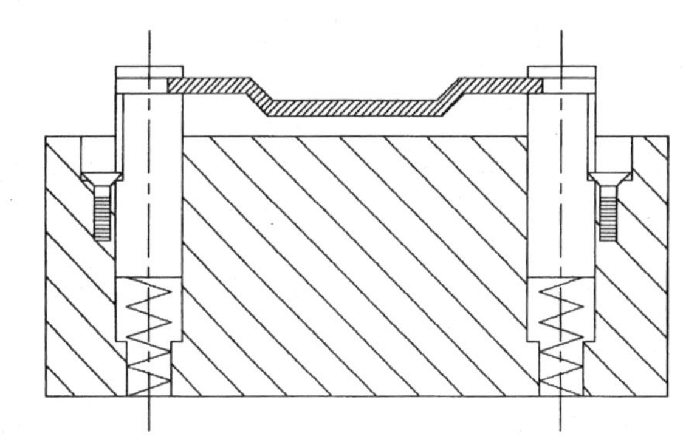

**Figure 4-22** Adjustable stock lifters. (*Printed with permission from Applied Mechanics Corp., Grand Rapids, MI.*)

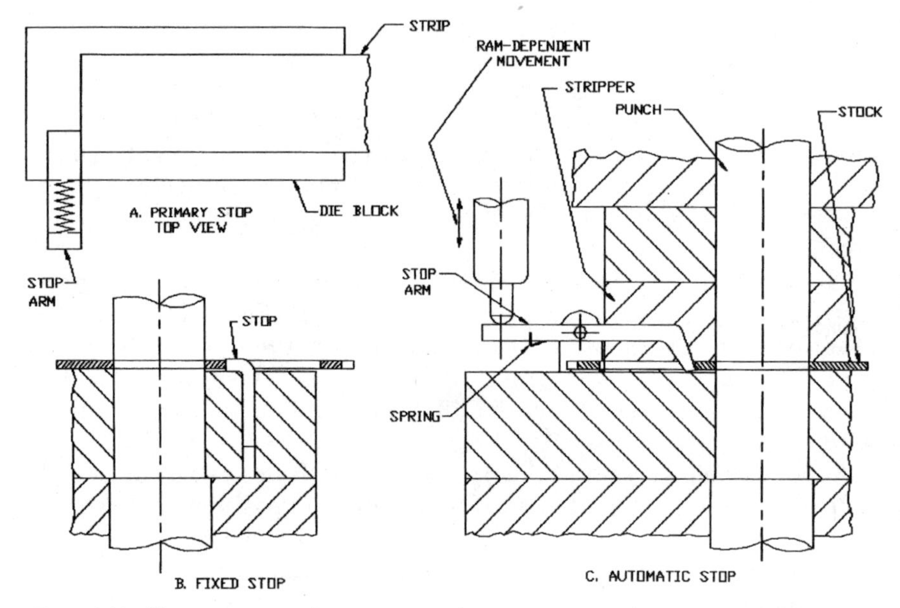


**Figure 4-23** Stop arrangements.

blank locator, which navigates the strip in such a way that all cutting is included prior to its arrival at forming and other stations.

The arrangement shown in Fig. 4-23*A* has a stop arm placed in the path of an advancing strip. On reaching the edge of the stop plate, the strip is automatically positioned under the vital punches, and the whole stop assembly may be pulled out of its way. This little device, when spring-loaded, snaps forward as the end of the strip leaves the die and is there ready to stop the next strip to be inserted.

A fixed stop is shown in Fig. 4-23*B*, where the material can bypass the stop pin's registration surface only by being lifted up above its level.

The automatic stop in Fig. 4-23*C* is a device which slides up and down along with the movement of the ram and either

- Forces the nose of the stop lever up, to release its engagement of the strip for the latter's progression (during the downward movement of the ram).
- Releases its pressure on the lever, thus allowing its nose to come down, pushed by a force of a spring. In such a position, the lever is ready for registration and retainment of the advancing strip (during the upward movement of the ram).

A similar device attacheable to the surface of the die block is shown in Fig. 4-24. It is activated by a spring, which forces the gauge's nose

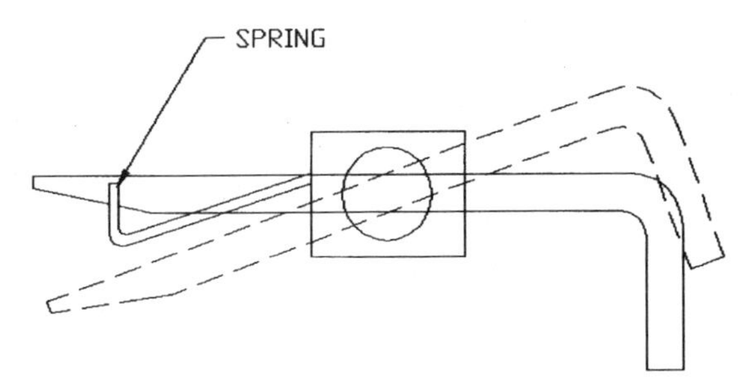


**Figure 4-24** Automatic stop and latch. (*Printed with permission from Applied Mechanics Corp., Grand Rapids, MI.*)

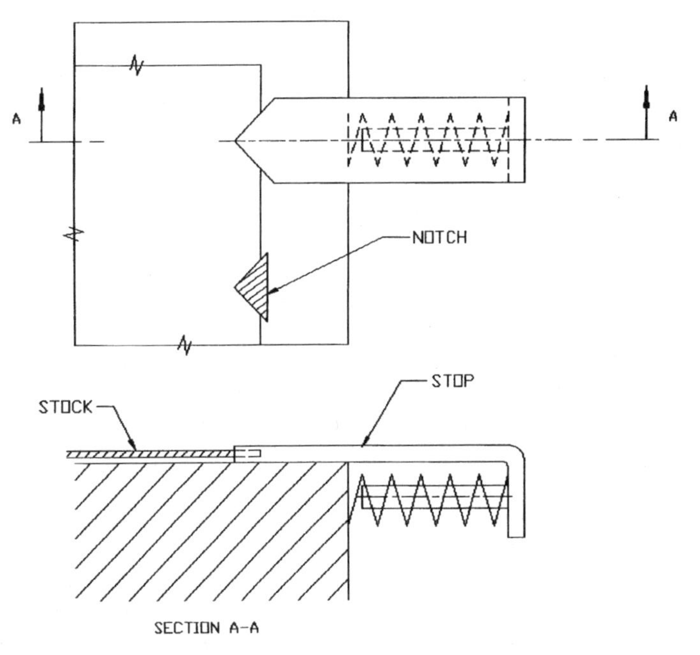


**Figure 4-25** V-notch stop.

toward the surface of the die block, holding it down to register the advancing strip.

A V-notch stop (Fig. 4-25) engages a V-notch cut in the side of the advancing material.

The nose of this device rides on the edge of the strip, snapping forward whenever a notch is encountered. When filling the notch, the spring force behind the latch pushes the whole strip material toward the opposite side of the channel, thus locating it under a punch.

This type of stop is in its origin a finger stop, shown in Fig. 4-26, which is similarly positioned against the strip's edge, where it pro-

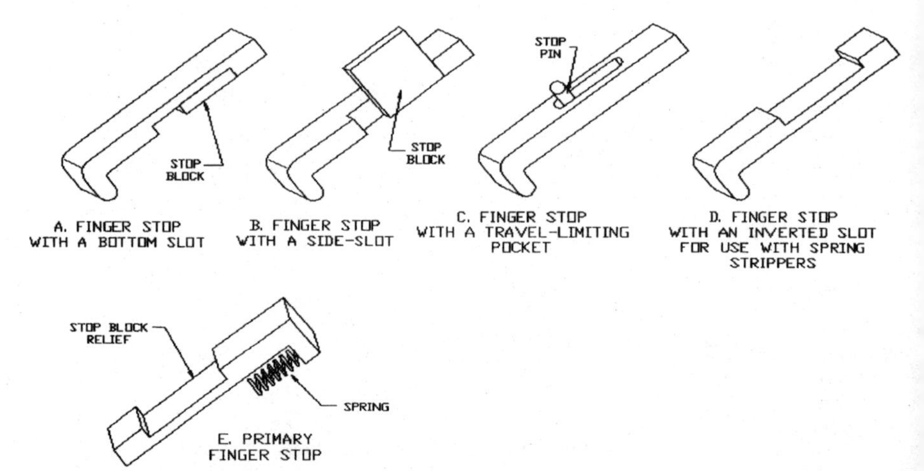


**Figure 4-26** Finger stops.

vides the force needed to push the material against the opposite side. The movement of the finger stop is controlled by a travel-limiting block or pin, positioned to fit within the relief slot in the stop body itself.

Various alterations of the travel-limiting slot location provide for a wide variety of applications. The side-located slot may be used with the travel-limiting function of a side block. The block with a pocket utilizes a pin to control the amount of its movement.

Some miscellaneous stopping ideas are demonstrated in Fig. 4-27. The material-deflecting pin in Fig. 4-27*B* makes the material slip over its rounded head, deflecting it down, where it leaves the die under an angle. The strip, already perforated, is easily averted from its straight path.

A material stop, allowing for eccentric positioning, is shown in Fig. 4-28. The whole unit can be rotated around a counterbore and secured in its final position by a steel ball emerging from its side. The steel

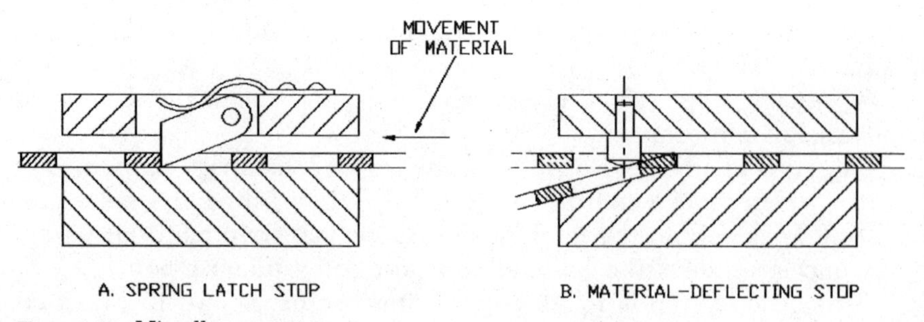


**Figure 4-27** Miscellaneous stops.

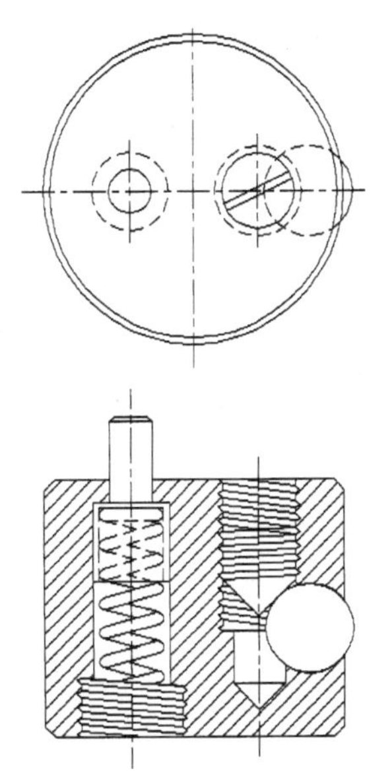

**Figure 4-28** Eccentric gauge stop. (*Printed with permission from Applied Mechanics Corp., Grand Rapids, MI.*)

ball is pushed out by the movement of a setscrew's cone point, which also retains it in the attained position.

The progressing strip stock is stopped in its movement on encountering the spring-loaded pin. When pushed down, the pin does not obstruct its advancement, provided there are no openings in its path.

A cam-operated slide (Fig. 4-29) delivers a fixed amount of adjustment to the sheet-metal strip. Attached to the upper half of the die, it moves down with the ram, driving the slide block toward the edge of

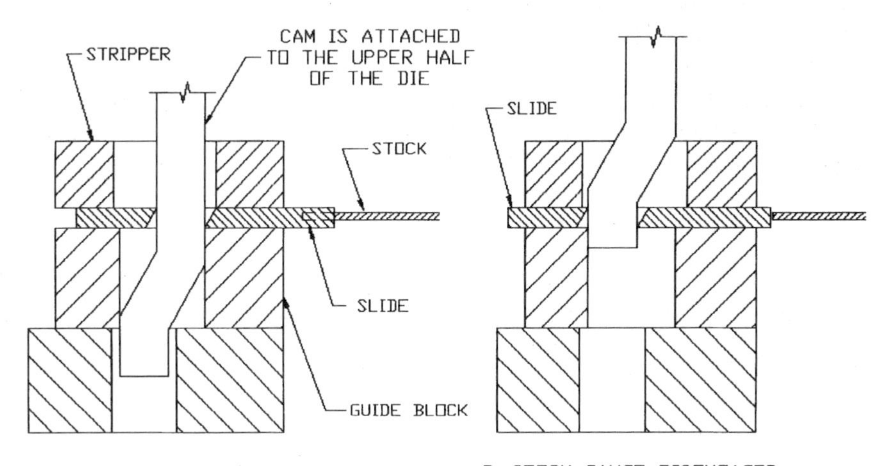


**Figure 4-29** Cam-operated stop.

the material (in its lowest position) and away from the strip (when moving up).

Electronically controlled material gauging units have a highly controllable area of function. Good gauging properties of this type of equipment can be made consistent in spite of variability of material thicknesses and sheet width.

### 4-1-9. Pressure pads

Pressure pads are actually small localized strippers, which operate on a slightly different basis. Instead of stripping parts off the tooling, they eject the pieces by pushing them out of the tooling.

By bringing the upper part of the die down on the work, as illustrated in Fig. 4-30, the spring pad is squeezed to its utmost position, up or down, whichever is appropriate. So restrained, the pad exerts a holding force on the sheet strip, not allowing the material to move or to be pulled along by the forming or bending action.

During the upstroke of the press, the pad, forced by the action of its spring, shoots forward, ousting the formed part from within the forming tool. Its opposite member, a forming block, provides the supportive action and may or may not contain some ejecting arrangement, either within its construction or beside it.

The necessity of additional spring ejecting force is determined by the shape of a part, by the depth of the formed section, material type and condition, aside from other influences.

The retention of the pad in a floating position is provided by shoulder screws. (Figure 4-31 shows pad retainers.) These not only secure the block in a relatively fixed position, they also control the length of

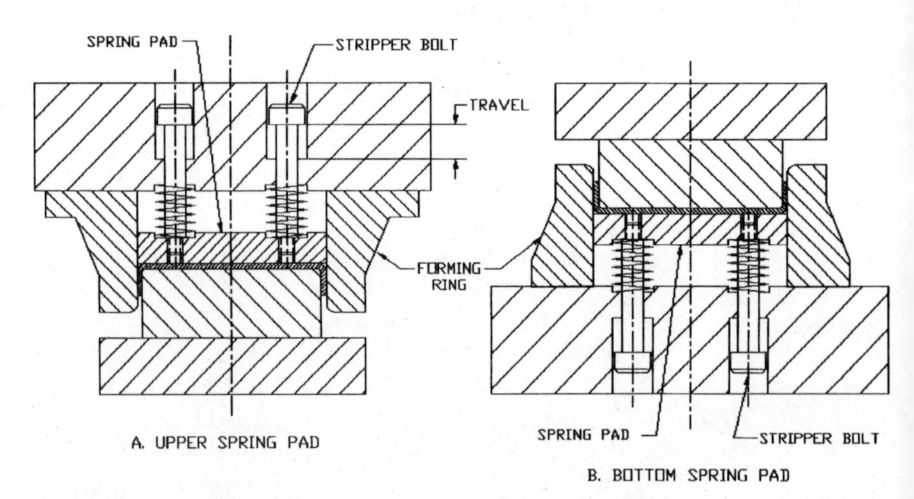


**Figure 4-30** Pressure pads.

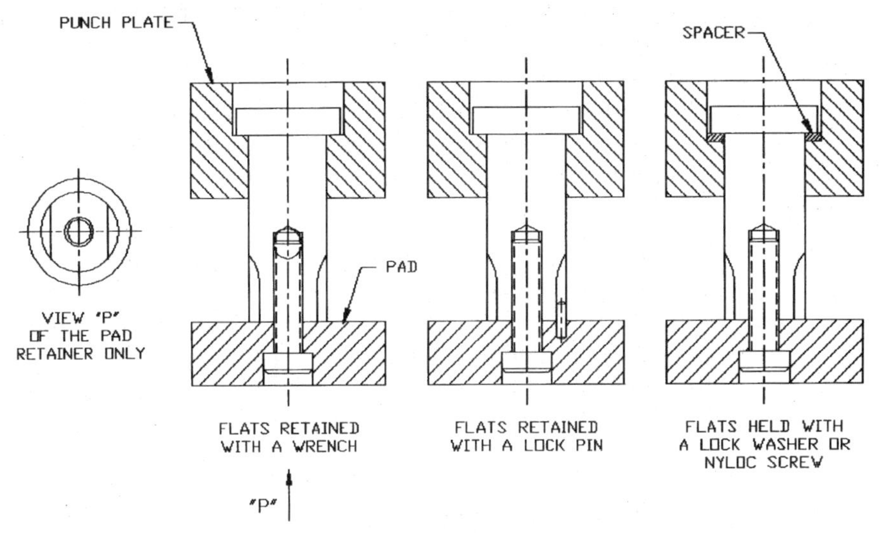


**Figure 4-31** Pad retainer. (*Printed with permission from Applied Mechanics Corp., Grand Rapids, MI.*)

its travel by the amount of space between the counterbore's surfaces and the bottoms of their heads. This space should be equal to the required travel distance, plus the life of the die.

If necessary, bolt heads may be shimmed for the proper distance, with some shims being removed during each sharpening.

In replacement of stripper bolts, which are used for the retention of spring-loaded blocks, pad retaining studs may be utilized. These are firmly attached to the pad by a socket head cap screw. Flat portions on their body diameter serve in assembly as an area for wrench application or as key flats protecting against rotation.

### 4-1-10. Nests

Some operations, especially shape altering or finishing operations such as shaving, restriking, precise forming, or drawing, require nests (Fig. 4-32).

Solid-block nests are utilized in empty stations of a progressive die, where they accommodate the formed part during the work cycle of a die, so that it will not be smashed and flattened by its downstroke action. The part is still attached to the strip on its way to the final blanking station (Fig. 4-32*A*).

During the upstroke, the strip material is forced up as well by the action of stock lifters and stock feeding devices. By such movement, it pulls the nested objects along on its way through the die.

Some nests, like those for shaving, can be adjustable to allow for easy removal and insertion of parts. Their moving portions should fit

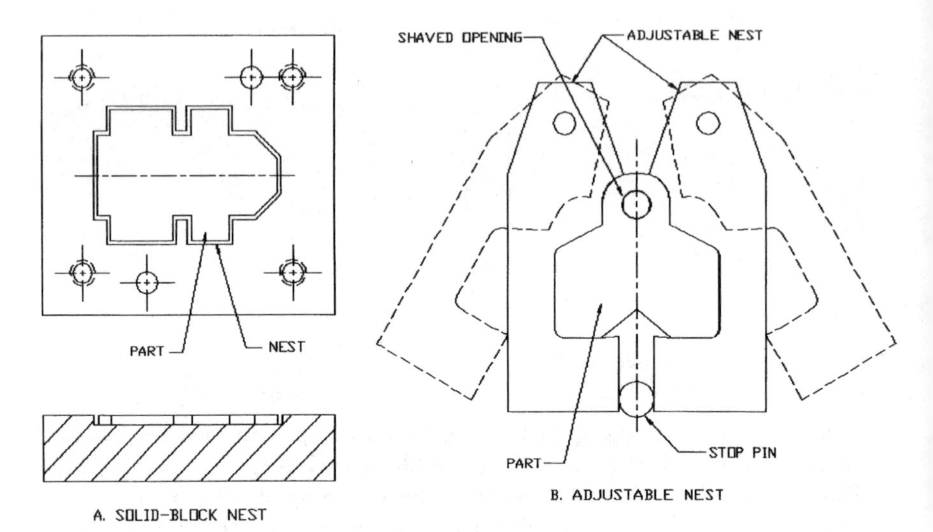


**Figure 4-32** Nests.

the outline of the part either all the way around or just in certain portions of it. This method of part locating will provide the necessary accuracy in positioning, where—for example—an opening has to be shaved and the shaving operation is to remove only a few thousandths of material per diameter.

### 4-1-11. Ejecting of parts

Spring-loaded stock lifters, pressure pins, or pressure pads may all be used as ejectors of finished parts, wherever these are not cut off in the last operation. The action of ejectors may be aided by a slant in the die surface, inclination of the press bed, or the application of compressed air.

**Air ejecting.** Air ejecting (Fig. 4-33) is used with formed cups and other formed shapes. It may also be used with flat washers and similar flat parts. Sometimes an inclination of the press bed is used in conjunction with an air-ejecting device to get the parts out of the working area quickly. First they are blown off the die block and, landing on the slant, are guided down into a drum, standing aside.

**Other methods of parts ejecting.** There are several other methods of getting the finished parts out of the die. Some parts are cut off during the last operation; others are blanked and sent down through the die opening. Drawn parts are usually forced down in the last drawing station unless their flanges are to be trimmed, in which case the (last)

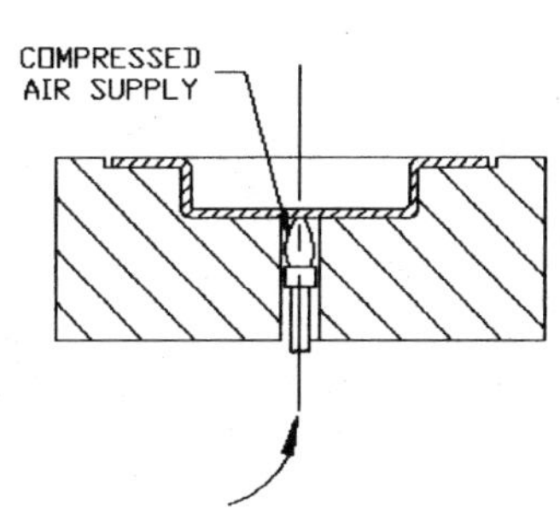


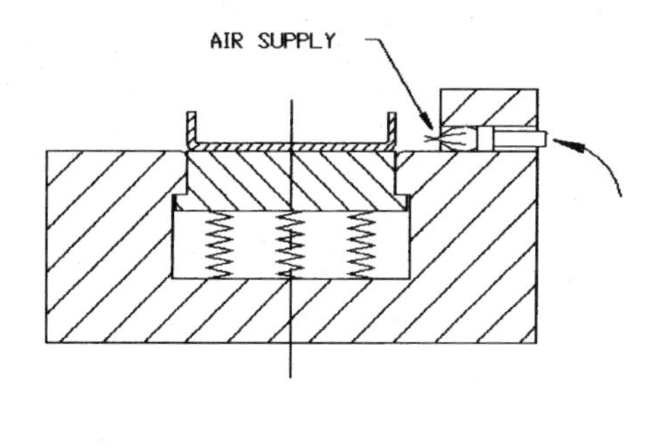


**Figure 4-33** Air ejecting.

trimming station disposes of them either through the die opening or by blowing them off the face of the die block with air.

Round blanks, such as washers or slugs, are most often disposed of by being forced down through the opening in the die bushing. For that reason, the inner diameter of the die opening increases in size, with the hole cross section tapered, usually under an angle of 1.5 to 2°, or 0.004/0.006″ per side, up to 0.009/0.011″ per side.

The continuation of this opening through all subsequent blocks is achieved by increasing its size +1⁄64″ with every new block it encounters.

If the opening for slug removal is to be straight, its diameter should be made 1⁄64″ greater than the diameter of the cutting portion. A plus-tolerance range in the vicinity of +0.002/−0.000 should be chosen for such a hole (see Fig. 4-34*B*).

A problem with small slugs may sometimes be encountered, as they tend to stick together and block up the passage. For that reason, slanted relief openings are placed in their path, drilled under an angle through the backup plate, as shown in Fig. 4-34*C*.

Drawn parts, as already mentioned, are often sent out of the die through the opening in the drawing die itself. The fact that the drawing portion of the die bushing consists of but a short section of its total length, with an increase in diametral size afterward, only aids the ejection of parts. (See earlier Fig. 4-17 for illustration.)

However, pieces of metal stick not only to each other but to their respective tooling as well, which is often due to the emergence of vacuum between the two. The tool, being pressed hard toward the material sheet, squeezes out all the air from the place of contact, attaching the cut shape to its surface. Such parts are very difficult to strip, and sometimes only the alteration of tooling can alleviate this problem.

With large-sized punches, the inner portion of the tool may be removed, as shown in Fig. 4-35*A*. Such a recess will limit the possible entrapment of air to the area of the ring of contact, which—with the

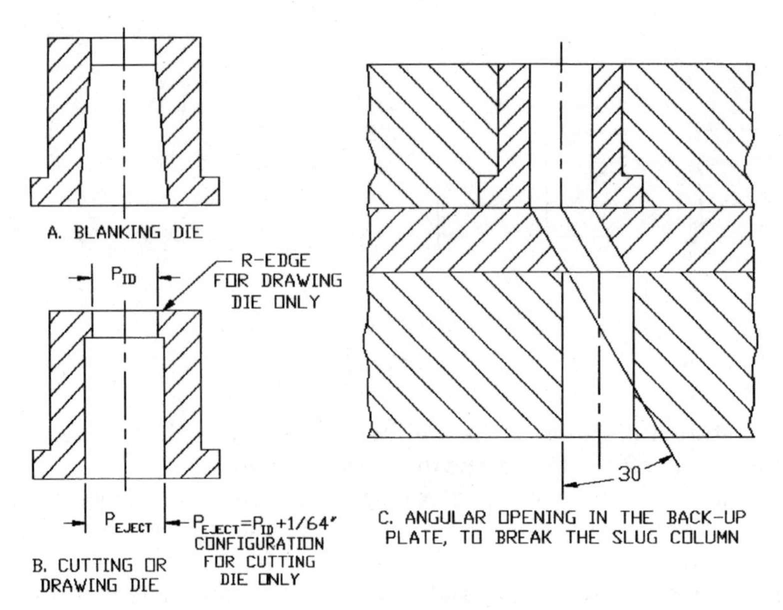


**Figure 4-34** Part ejecting through the die opening.

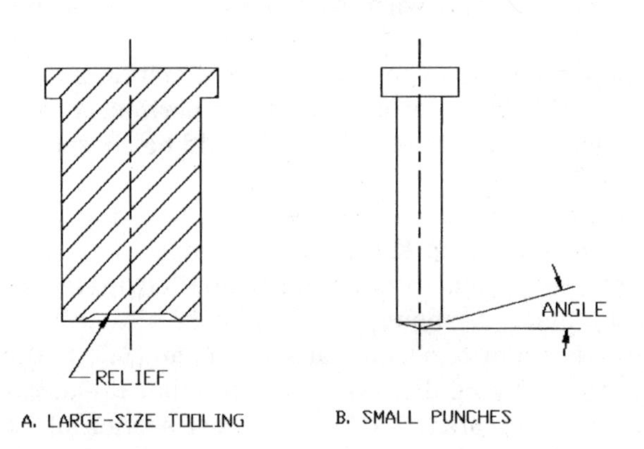


**Figure 4-35** Punch face alterations.

slightest bending of the cut metal—will lose its holding power altogether.

The cutting face of small punches may be shaped under an angle, as shown in Fig. 4-35*B*. Such an angle should be in a range of 1 to 2°, which should suffice to prevent vacuum-caused attachment.

However, such tools are more difficult to sharpen; because of this complication their use cannot be recommended without reservation.

### 4-1-12. Safety

Die safety is of the utmost importance to the security of the operating personnel, to the functionality of the expensive press equipment, and to the life of the die itself.

Metal stamping dies are very expensive devices, their design and manufacture being complex and demanding. Such tools must be protected from breakage or malfunction by all means, especially since their failure may bring about the deterioration of the press equipment in which they are running.

Safety of dies is controlled most often by limit switches, which are positioned to detect any misfeed situation, buckling of material, or other failures of the die's function. Limit switches may be used as safety stops, feed-control devices, or a means of additional control of the die function.

## 4-2. Mounting of Blocks

All the punches and dies described earlier are assembled within their respective blocks. The blocks themselves are firmly and with great precision attached to their supporting backup plates and to their die shoes. For mounting of blocks, socket head cap screws are most often utilized. The precision alignment is achieved by at least two dowel pins per assembly.

Usually, with smaller blocks, four screws and two dowel pins will suffice. With larger blocks, six or more screws and four dowels should be used.

### 4-2-1. Die block

The die block (Fig. 4-36) is made of high-quality steel, hardened and precision ground to exact sizes.

Die blocks, running in dies with a stationary stripper, would not contain tapped holes for screws; in such a situation the screws are inserted off the stripper's top surface, with counterbores in the stripper plate, relief holes through all the blocks, and a final tapped hole in the bottom die shoe.

With a spring stripper, which is not attached to the die block at all, the screws are driven in from the opposite direction, which is the bottom surface of the lower die shoe. In such a case, the die block contains tapped holes, with relief openings through all the adjoining plates and a counterbore for the screw heads in the die shoe.

When dimensioning the die block for manufacturing purposes, all the dimensions should go off one opening, which should be the only one to be dimensioned off the block's edge. Cumulative dimensioning

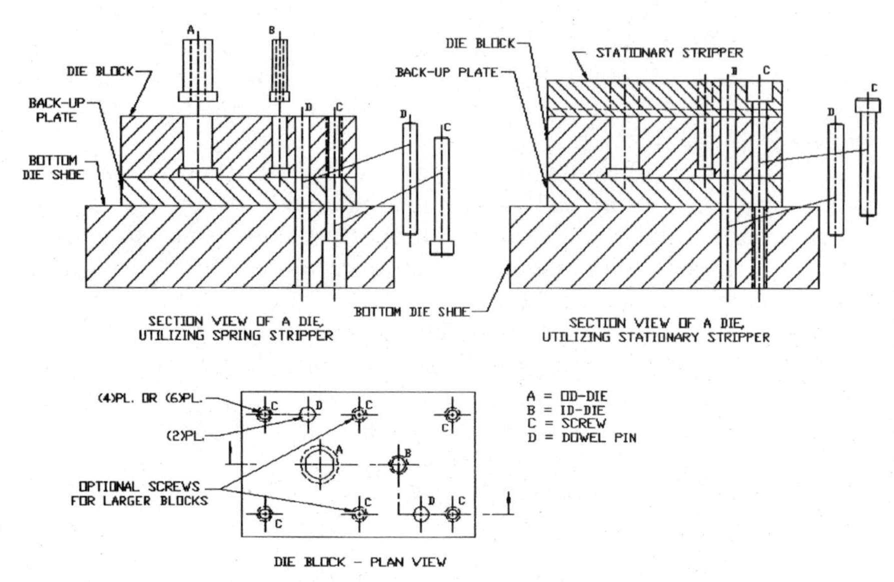


**Figure 4-36** Die block section.

must always be avoided, as such a method creates a cumulative tolerance buildup as well.

However, when generating a program for block-drilling NC machinery, all locations should be given off the first precision-finish hole, which is the dowel pin. The first dowel pin hole should be established as a zero location for NC equipment purposes.

A common range of general tolerances is added for illustration (see Fig. 4-37). However, the precision-finish openings containing die bushings and dowels, must be dimensioned as shown earlier in Figs. 4-5 and 4-12. Dowel pin mounting dimensions are given later in Fig. 4-41, screw mounting in Fig. 4-40.

The positioning of the first hole off the block's edge need not be overly precise. The surface roughness, along with the variation in straightness of the edge, may be a source of headaches for the inspection department, and yet the accuracy of this location is far from important for proper die manufacturing.

However, the distances between various openings, especially those of die (punch) and dowel pin holes, are of the utmost importance, and their precise locations cannot be overemphasized.

Occasionally an idea is expressed in the technical literature, stating that the designer should not make an effort to come up with dimensions of the hole-mounting locations, as the toolmaker should think up this part and make it work. It certainly is a debatable claim, as the toolmaker will certainly need much longer to arrive at the right

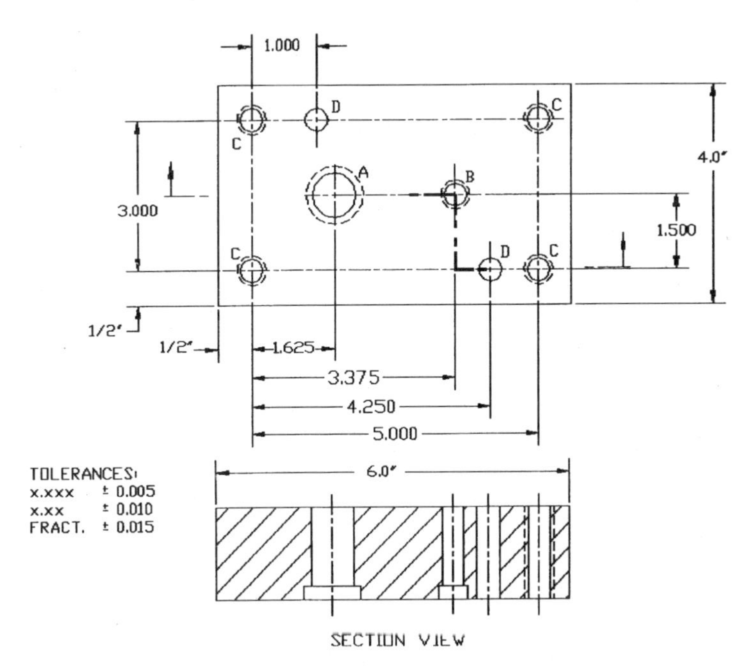


**Figure 4-37** Die block dimensioning.

solution than a worker with pencil and paper. Perhaps in some cases the toolmaker may even be forced to use the old trial-and-error procedure! Well—we all know that the designing process is much cheaper on paper than in steel.

### 4-2-2. Punch plate

The punch plate is designed, dimensioned, and manufactured similarly to the die block. There is one difference though, when considering the view location: The die block is always viewed off its top surface, whereas the punch plate is usually seen from below.

For the purpose of accuracy, when using NC block-cutting equipment provision, both of these plates should be viewed and manufactured from the same location, from the top view of the die.

The punch plate, since it provides the support for all punch shanks, must be adequately thick but not in excess of what is necessary, in order not to increase the weight of a die. The correct thickness of this plate is in the range of 1.5*D*, with *D* being the diameter or the largest dimension of the biggest punch the plate should accommodate.

The mounting procedure of the punch plate to the die shoe is as shown in Fig. 4-38. Screws are inserted through the die shoe, where their heads are embedded. The upper backup plate has only clearance

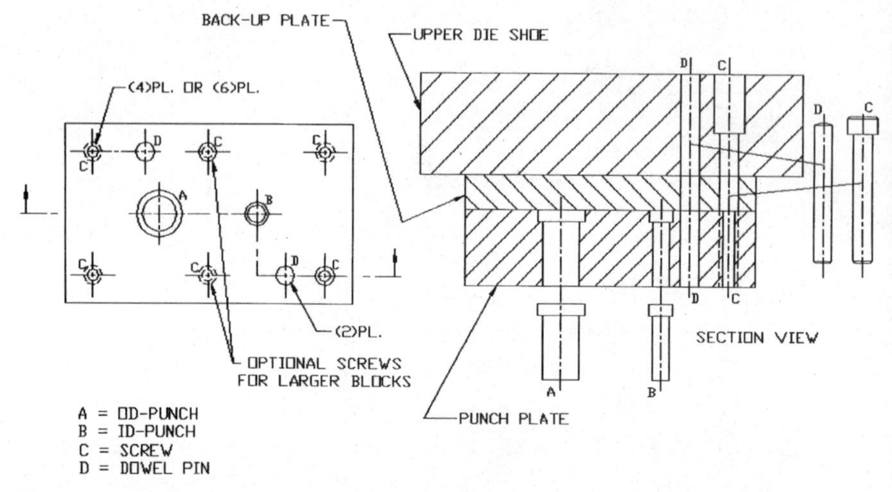


**Figure 4-38** Punch plate section.

openings for them, and the tapped hole going through the punch plate. This arrangement is the same whether a spring stripper is attached or not.

For dimensioning of punch plates, refer to the description for the die block, shown earlier in Fig. 4-37.

### 4-2-3. Backup plates

Both backup plates are made of hardened steel, 3⁄8″ thick for general work, 1⁄2″ thick for heavy-duty jobs.

Backup plates contain only clearance openings for screws and precision-finished openings for dowels. In slug-producing stations, however, they must have openings for the slug disposal; in blank-producing stations, such openings are used for parts' disposal.

## 4-3. Machining of Blocks

All the material to be removed in order to produce holes in the blocks has to be drilled away in stages. It is not possible to take a 5⁄8″ diameter drill and drill a 5⁄8″ diameter opening in one pass. This may be done in wood or in other soft material, but certainly not in metal and especially not in tool steel.

### 4-3-1. Drilling of holes

Every opening in a metal block must first be spot-drilled to establish its precise location for further processing. Following in a sequence, a rough drill is utilized, usually only slightly larger in size, let's say 5/32″ diameter. After this drill, another, larger drill (or subsequent drills) is used, to almost finish the diametral size of the opening, and then the jig bore reamer, a reamer, or a boring tool, or any other arrangement that provides a fine finish.

With counterbores, the process is similar: first the spot drill, followed by a roughing drill and all subsequent drills. Then comes the counterbore or end mill for the pocket, followed by a reamer or any other fine-finishing tool for the through hole.

The sequence of counterboring operation is true with one exception: when using a sub-land drill for making clearance screw holes, counter-boring operation should be performed first.

**Drilling Sequences for Various Openings**

| Hole dia, in | First drill | Second drill, in | Third drill, in | Fourth drill, in | Finishing tool, in |
|---|---|---|---|---|---|
| 1/8 | Spot | 3/32 | | | 1/8 reamer |
| 1/4 | Spot | 5/32 | 7/32 | | 1/4 reamer |
| 5/16 | Spot | 3/16 | 9/32 | | 5/16 reamer |
| 3/8 | Spot | 3/16 | 17/64 | 11/32 | 3/8 reamer |
| 7/16 | Spot | 3/16 | 9/32 | 13/32 | 7/16 reamer |
| 1/2 | Spot | 3/16 | 5/16 | 15/32 | 1/2 reamer |

With additional sizes, a fifth (later a sixth) drill may be added. The removal of material must be done in stages; otherwise the wear of tooling will result, along with poor-quality openings and their surface, overheated blocks, and subsequent tool breakage.

Drilling operation, once it exceeds the length of 2 to 3 times the hole diameter, is considered deep-hole drilling.

**An important detail:** In a sequence of operations, the last drill before the reamer should be −0.030″ per diameter smaller than the actual reamer's size. For counterbored holes, the difference between the last drill and the counterbore diameter should be 0.060″ per diameter.

### 4-3-2. Feeds and speeds

Feeds and speeds, useful for cutting of steel material, depend on the type of that material, its hardness, as well as the operation performed and the tool used.

*Surface feed,* or *tool feed,* is the rate at which the tool moves with every revolution of its spindle. In vertical drilling, surface feed is the movement of the tool down; in milling of pockets, it is the linear movement along a given path of work. Surface feed is usually specified in inches per minute or inches per revolution.

*Cutting speed* is actually the spindle speed, specified in either revolutions per minute (rpm) or surface feet per minute (sfpm).

**Some basic surface feeds:**

| | |
|---|---|
| D2–D6 material | 30 sfpm |
| O1 steel | 45 sfpm |
| Low-carbon steel | 65 sfpm |

Using the above values, different cutting processes should be limited to their percentages, as follows:

| | |
|---|---|
| Drill | 100% |
| Reamer | 70% |
| Counterbore | 110% |
| End mill | 110% |
| Bore | 200% |

Where a need to recalculate surface speed in rpms into inches per revolution arises, the formula is

$$\text{rpm} = \frac{3.8 \times \text{sfpm}}{\text{tool dia.}} \tag{4-3}$$

and, subsequently,

$$\text{sfpm} = \frac{\pi \times \text{rpm} \times \text{tool dia.}}{12} \tag{4-4}$$

All feeds for fluted tooling should be checked for the true number of revolutions per each of the flutes, which should never be less than a minimum of

| | |
|---|---|
| End mill | 0.001 in per rpm per 1 tooth |
| Reamer | 0.001 in per rpm per 1 tooth |
| Counterbore | 0.0015 in per rpm per 1 tooth |

Spot drills' speeds and feeds should be in the vicinity of 480 rpm and 3.5 in/min. Tap drill: 200 rpm, 12.5 in/min. For countersink values, drill speeds and feeds should be used.

### 4-3-3. Miscellaneous notes on machining

Openings which are to be precisely finished, such as punch and die retaining holes and dowel pin holes, should be rough-cut and almost finished prior to heat treatment, leaving only about 0.008″ per diameter to be removed afterward. Such a small amount of material should not present too great a problem, even when being removed off a hardened block. And yet many possible distortions, warping, deviations from straight line, and other defects caused by the heat treatment will be removed.

When drilling a block, operations should be grouped according to the tooling used. This way all spot drilling should be done to all openings at the same time. The same way, all the rough drilling should be performed together, followed by a sequence of finishing operations.

## 4-4. Mounting Hardware

All blocks are secured together with the aid of screws, the choice most often being socket head cap screws. With spring-and-bolt arrangements, shoulder screws are usually used to support the spring against misplacement or buckling with its shank, hold the spring-loaded block with its threaded portion, and control the length of the movement with the location of its head in the clearance pocket.

Block mounting hardware (Fig. 4-39) should never be positioned too close to the edge of the block, as bulging of its side may occur, along with other distortions. The minimal recommended distance off the edge should be as shown in Table 4-1.

**Attaching several blocks together.** When attaching several blocks together, screws must always be affixed to the last block of the whole

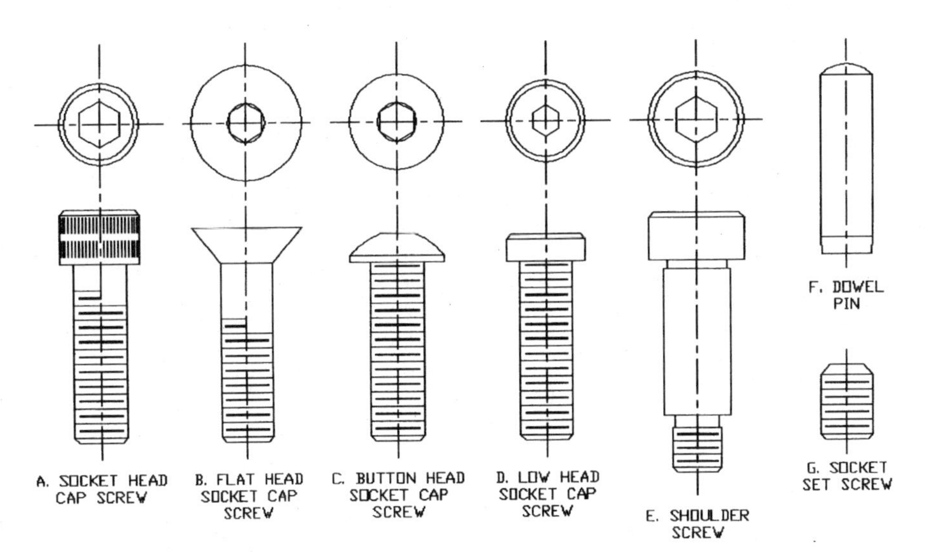


**Figure 4-39** Mounting hardware.

**TABLE 4-1 Minimum Distance between the Center of a Screw and the Edge of the Mounting Block**

| Type of installation | Material of the mounting block | |
|---|---|---|
| | Aluminum, CRS | Other alloys |
| Socket head cap screw | 1.7 diameters* | 2 diameters* |
| Flat head socket cap screw | 2 diameters* | 2.5 diameters* |

*Add diametral tolerance of the screw to the amount shown.

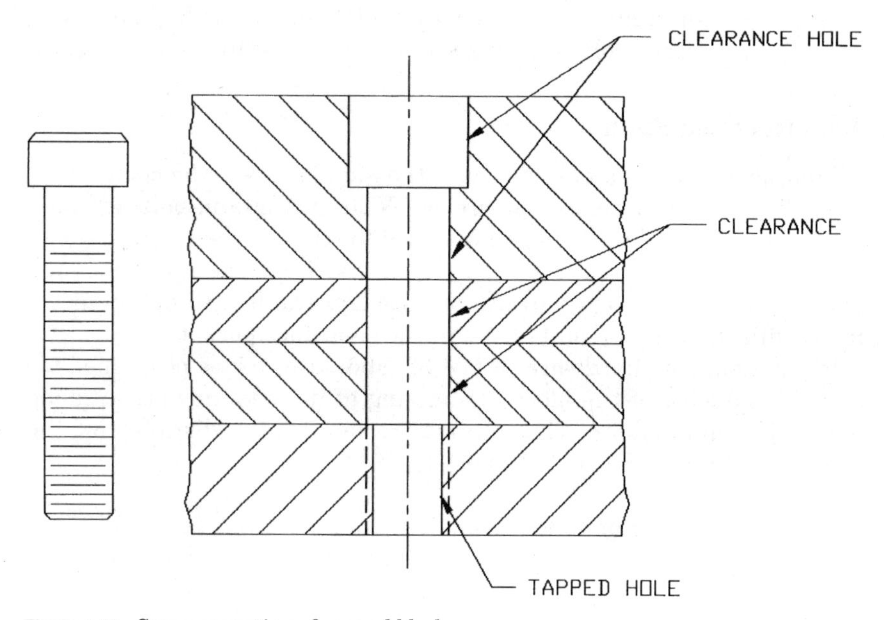


**Figure 4-40** Screw mounting of several blocks.

sandwich, which must be the only one to have tapped openings (Fig. 4-40). All openings in blocks in between, no matter how many of them there may be, must be clearance holes.

Should the holes in all blocks be tapped instead, it will be impossible to align their threads against the threaded portion of the screw in assembly. This will create gaps between the blocks, with binding and breakage of the tapped areas, provided, of course, the blocks are assembled at all.

The clearance openings for the head of the socket head screws must be counterbored to a minimal depth of

$$\text{Screw head thickness} + \text{die life} + \tfrac{1}{16}''$$

**Length of thread needed.** A screw needs to engage at least $2\frac{1}{2}$ of its threads to be considered holding the object. However, the designer need not be limited by such a claim, especially where enough space in blocks is provided for a longer threaded portion and where a few more turns of a wrench are not out of place.

The proven and acceptable lengths of a threaded portion should be the greater of these values:

UNC threads  Screw dia. + ½″ or ½(screw length)

UNF threads  1½(screw dia.) + ½″ or ⅜(screw length)

Tap drills, which always precede the tapping operation in metal machining (in plastic or wood, self-cutting screws are often used), should be used to drill the opening through the whole thickness of the block whenever possible. This procedure should be considered even in cases where the tapped portion of the hole is much shorter than the block itself.

The reason for this claim is the difficulty of producing a blind hole in metal, where chips often get locked inside and obstruct the performance of the drill and where air pockets may be encountered to further jeopardize the work.

**Dowel pins' assembly (Fig. 4-41).** Dowel pins are used for securing the blocks in a fixed, precise position relative to each other. Screws are but a means of attachment, whereas dowels provide for all the alignment needed.

For that reason, dowel pins must run through the whole length of the assembly of blocks, and in each of them the holes must be precision-finished. However, the dowels don't have to be surface-to-surface long, as a precaution against their protruding through one end or the other.

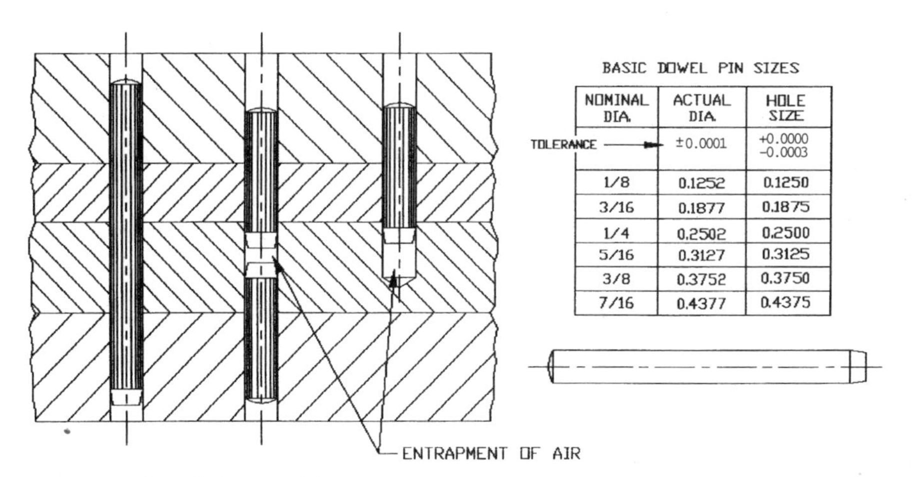

BASIC DOWEL PIN SIZES

| NOMINAL DIA. | ACTUAL DIA. | HOLE SIZE |
|---|---|---|
| TOLERANCE | ±0.0001 | +0.0000 / −0.0003 |
| 1/8 | 0.1252 | 0.1250 |
| 3/16 | 0.1877 | 0.1875 |
| 1/4 | 0.2502 | 0.2500 |
| 5/16 | 0.3127 | 0.3125 |
| 3/8 | 0.3752 | 0.3750 |
| 7/16 | 0.4377 | 0.4375 |



**Figure 4-41** Dowel pin assembly.

At least all dowel-pin openings in the two extreme blocks must be press-fit. The holes through the blocks in between may sometimes be only tight-fit. However, this consideration depends on the type of die, its function, and the common practice of that particular shop.

When looking at a block from the top view, the dowel pin's location should be at two extreme ends of the block to provide the broadest range of application. As already mentioned, dowels (and perhaps screws as well) should be spaced unequally, so that no block can be turned around and assembled the wrong way.

As a precaution, dowels should never be assembled into a blind hole, for a lack of ventilation at the bottom of it would create an air pocket, impairing the dowel's assembly and stability of performance.

## 4-5. Jigs and Fixtures

Jigs and fixtures can be considered a part of die building and die making, and almost every toolmaker displays a distinct ingenuity in devising and using them. Most often, jigs and fixtures are used to clamp down various parts, hold down the blocks, etc. The most widely used types of clamps are shown in Fig. 4-42.

A jig is actually a clamping device, holding a part in the position where it can be drilled, cut, formed, or otherwise altered. The jig may be assembled from all kinds of adjustable elements, but it must have a positive nesting arrangement, where the part may be placed and correctly oriented against the tool.

A drill jig should be made to stand on four legs. To leave the jig arrangement sitting flat on the table is not advisable, as the chips will have no place to go, while the table, acting like a plate, will make it difficult to orient the part against the tool.

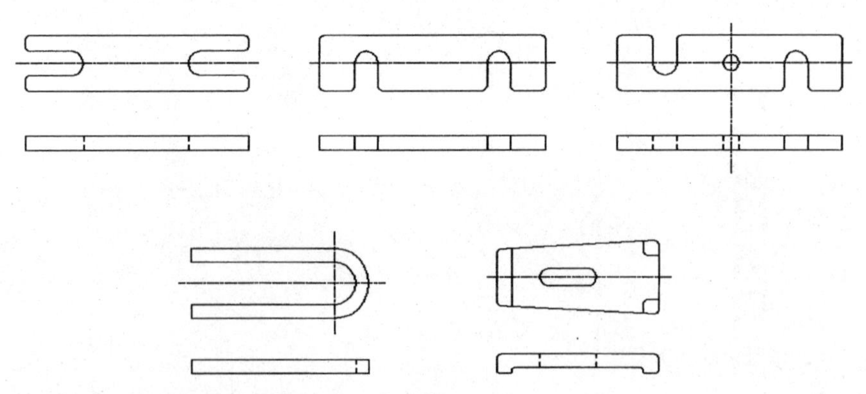

**Figure 4-42** Clamps used in fixturing of work.

Every jig drill should be received in a bushing, to guide its movement and protect it from breakage. Drill bushings should be made of hardened tool steel. They don't have to be contained in a block; clamping them in a proper location will serve the purpose as well.

There are many types of jigs: indexing jigs, vise jigs, plate jigs, universal jigs, and tumbling jigs.

Some other clamping and fixturing arrangements included here should be studied for their versatility and genuineness (Fig. 4-43). Most of these combinations were arrived at through many years of experience of many toolmakers, engineers, and designers.

The *A* method of fixturing shown in Fig. 4-43 displays a continuous spiral rise similar to that used on automatic lathes. Such a cam will lock securely at any point of its location and remain locked even during cutting. A 90° movement of the handle is quite adequate, but if a provision for greater material thickness is required, a 180° angle of movement is more appropriate.

In the cross section of a wedge clamp, shown in Fig. 4-44*A*, the push rod and a hand knob are connected to the horizontal mechanism of a wedge. A vertical stud, sliding in a steel bushing, leans against the surface of the wedge, and it is moved up and down by the horizontal movement of the whole part.

An alternate device, equipped with a spring plunger, is shown in Fig. 4-44*B*. Here a large hatlike top protects the plunger from being cut into. Air vents are provided at *A* and *B* locations.

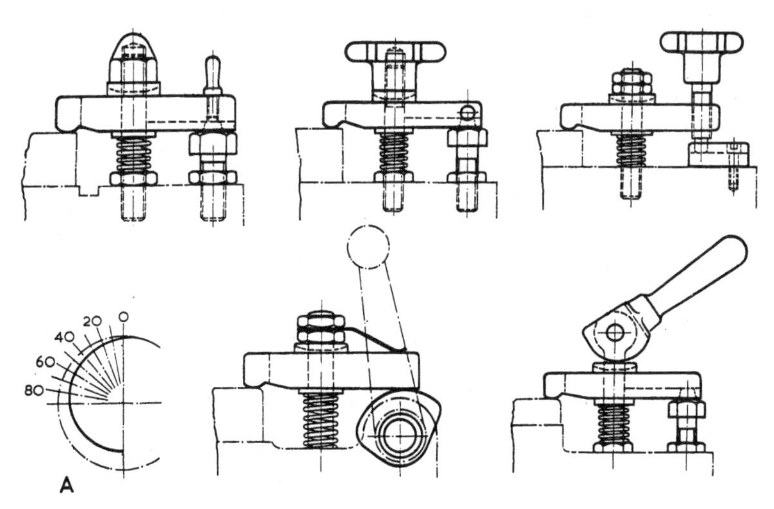


**Figure 4-43** Typical clamping devices used for jigs and fixtures. (*From H.C. Town,* "Technology of the Machine Shop," *1951. Published by Longmans, Green & Co.*)

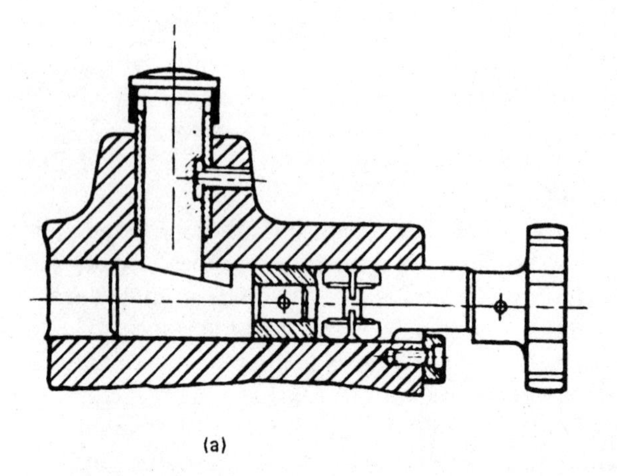


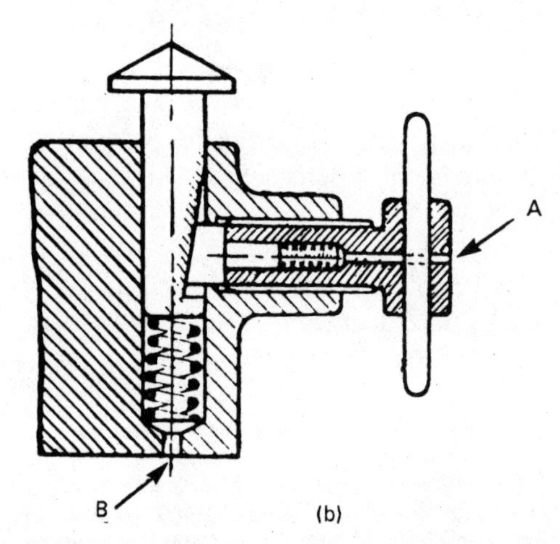


**Figure 4-44** (*a*) Wedge clamp. (*b*) Supporting plunger. (*From H.C. Town,* "Technology of the Machine Shop," *1951. Published by Longmans, Green & Co.*)

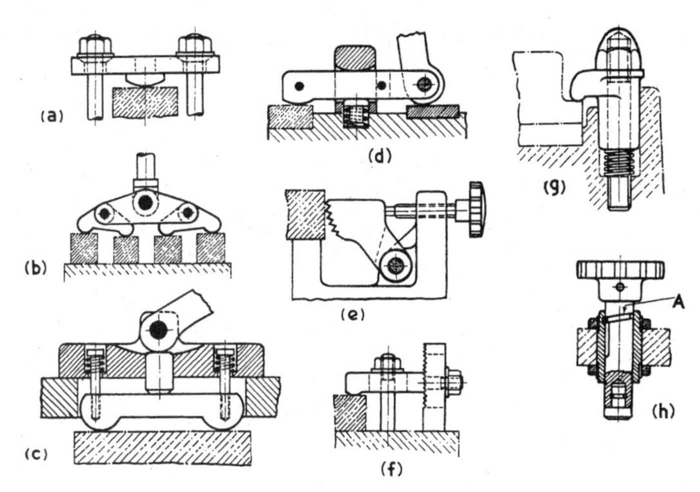


**Figure 4-45** Equalizing and other rapid clamping devices. (*From H.C. Town,* "Technology of the Machine Shop," *1951. Published by Longmans, Green & Co.*)

A strap, shown in Fig. 4-45*A*, is attached to the workpiece by two bolts. The whole assembly may experience quite a strain if the work surface is not flat. A pressure pad with rounded end is used to alleviate this problem. Also spherical-end washers and slots in the clamp may provide for the free movement, while not impairing its function. The disadvantage of such a clamping idea is that if a pressure is applied at a single point only, the working surface beneath may be marred.

A better arrangement is shown in Fig. 4-45*C*, where the pressure generated by the cam is transferred to the central stud, which subsequently presses against a strap with only two points of contact with the work surface.

The arrangement in Fig. 4-45*B* has a central rod mounted to a set of pivoted levers, which provide for the equal distribution of the vertical type of pressure.

A cam-operated sliding lever clamp is pictured in Fig. 4-45*D*. Here the centrally located support presses the arm against the upper surface of the guiding slot. Locking of the assembly is achieved through the movement of a cam-actuated lever arm.

Figure 4-45*E* has a serrated portion of the clamp leaning against the workpiece and tightened by a hand knob. The angle of serrations is of essence here, being established at 45°. With such an arrangement, the vibrations of the work process will not release the grip but rather push downward on the serrated jaw.

Figure 4-45*F* is a commercially available Hamber clamp. The heel-supporting piece is provided with accurately machined serrations, which allow for a rapid adjustment with variation in height.

A vertical clamp, shown in Fig. 4-45*G*, holds the workpiece by a pressure of its nose. The advantage of such clamping is the small amount of space it demands.

The benefit of two-way clamping (or multiple clamping) is the fact that even though it banks on two or more points, its clamping actuation is done from one position only. Some examples of this type of device are shown in Figs. 4-46 and 4-47. Additional clamping ideas are included in Fig. 4-48.

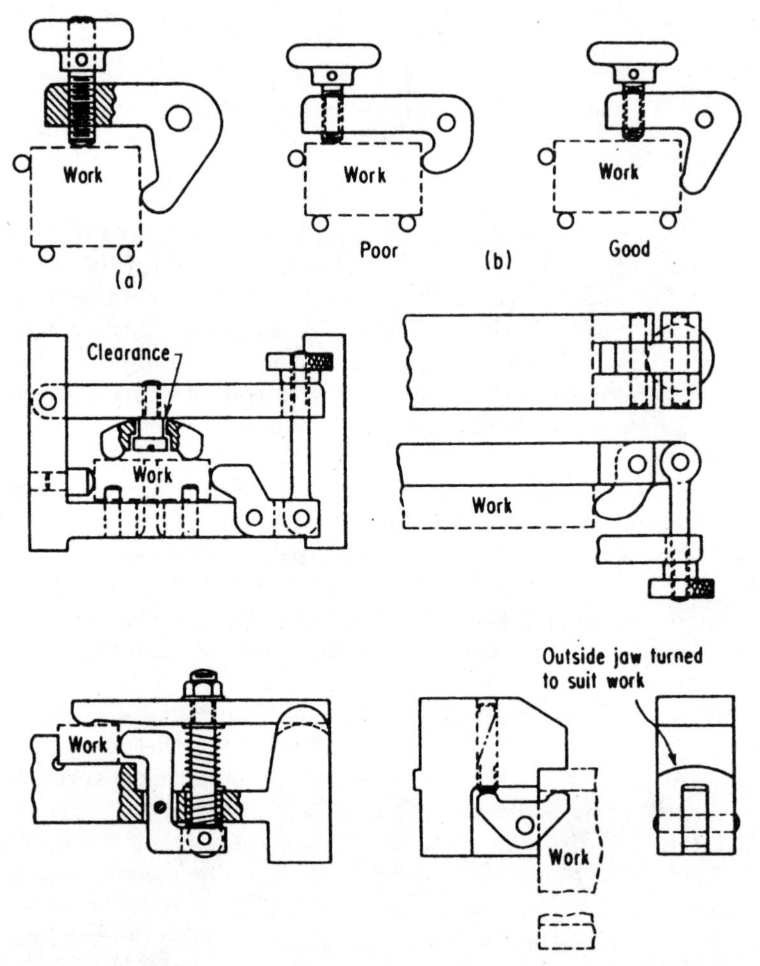


**Figure 4-46** Two-point clamping devices. (*From Frank W. Wilson,* "Handbook of Fixture Design," *1962. Reprinted with permission from The McGraw-Hill Companies.*)

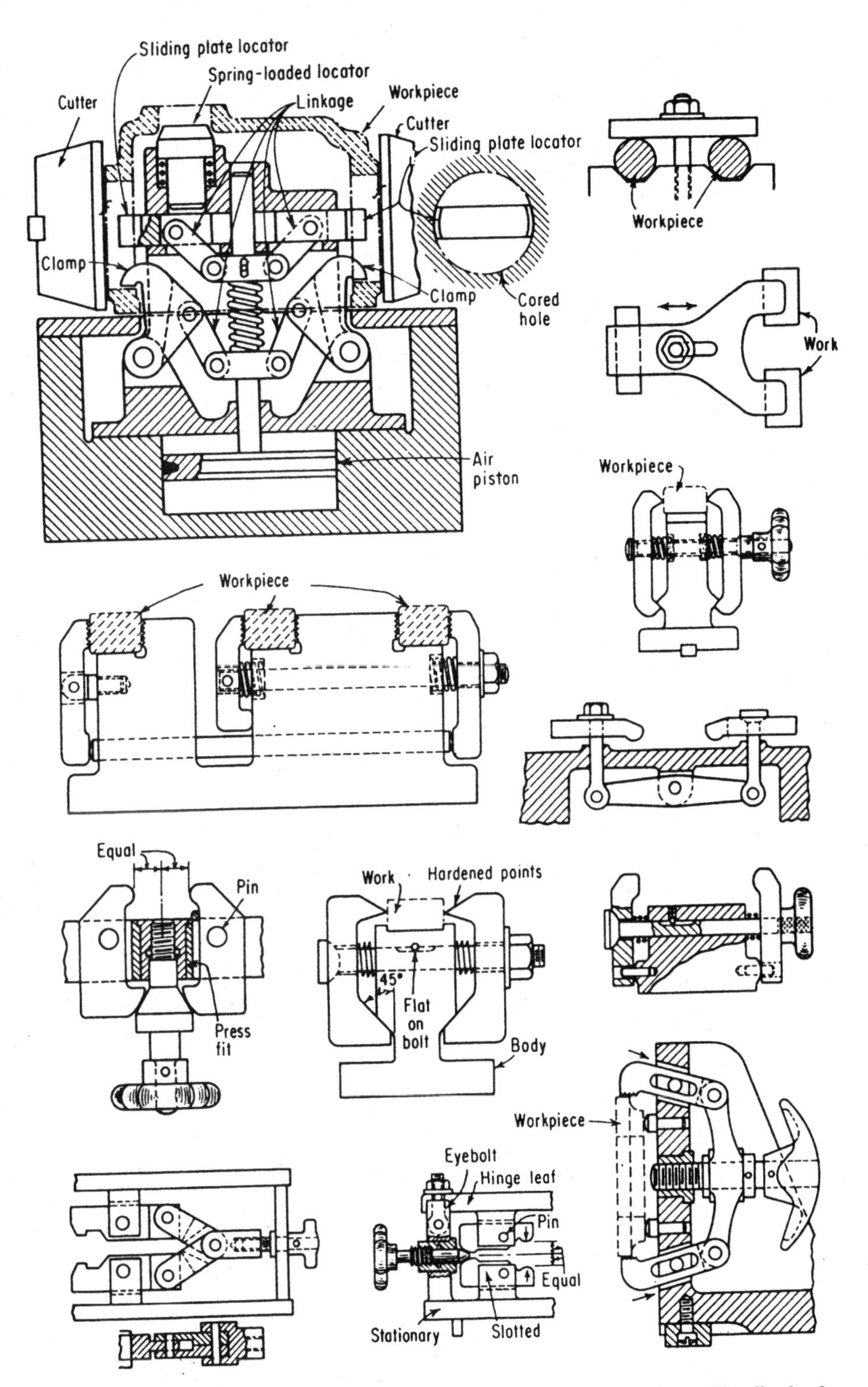


**Figure 4-47** Centrally actuated clamping devices. (*From Frank W. Wilson,* "Handbook of Fixture Design," *1962. Reprinted with permission from The McGraw-Hill Companies.*)

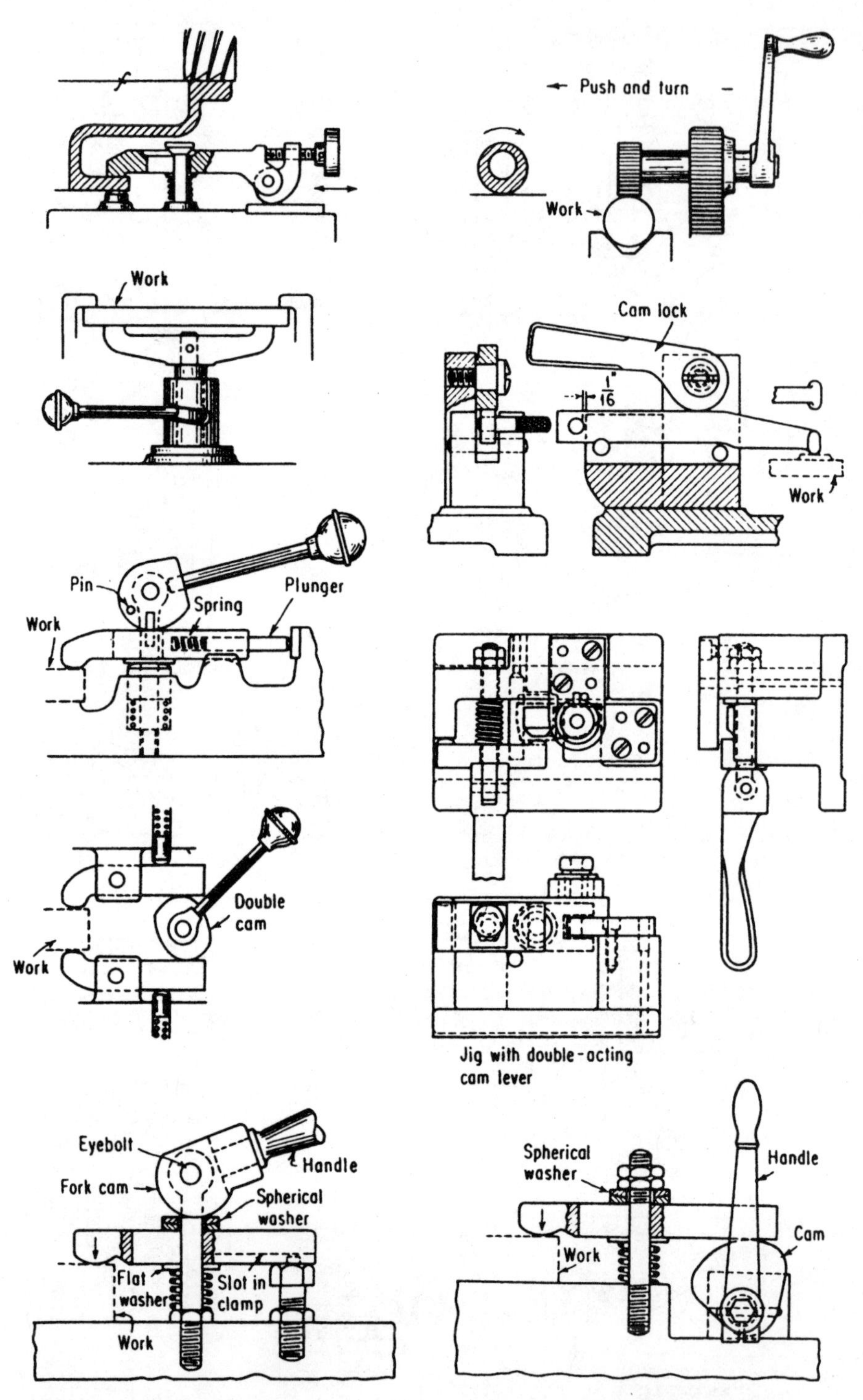


**Figure 4-48** Types of cam-actuated clamps. (*From Frank W. Wilson,* "Handbook of Fixture Design," *1962. Reprinted with permission from The McGraw-Hill Companies.*)

# Chapter 5

# Metalworking Machinery

## 5-1. Metalworking Machinery

The function of all metalworking machinery can be described as that of a force-producing mechanism which delivers its required output in a certain period of time with a predetermined amount of accuracy. The amount of the force and its method of application are not always consistent: Rolling machines apply a constant and uniform force toward a bar of steel, while drop hammers produce a single blow in a unit of time. The same amount of force may be divided into many smaller hits and distributed over a longer time period by the pneumatic hammer.

When classifying metalworking machinery according to the way it functions, there are two basic types of distinction:

- *Machines with linear movement of their tools,* such as presses, certain rolling machines, wire- and bar-drawing and stretching machines
- *Machines utilizing nonlinear movement of tooling,* such as bending and rolling machines, the rolling application being either longitudinal or cross rolling in type

In further evaluation of metalworking machinery, the scope of this problem is restricted to presses.

### 5-1-1. Presses, according to their function

Presses can be evaluated with regard to their function as

- *Energy-producing machines,* with the application of this energy being abrupt and instantaneous. All the machine's energy storage is depleted at the end of its work cycle. Such machines are hammers, which use a free fall principle as a basis of their function. However, sometimes their efficiency is increased by an addition of steam or pressurized air. The hammers' energy source is temporarily disconnected at the time the ram is released for an operating cycle, which consists of its falling down on the workpiece.

- *Force-producing machines,* which operate by generating a considerable amount of force, independent of the position of the ram. Hydraulic machinery is the main representative of this category.
- *Stroke-controlled machines,* or presses, the function of which depends on the movement and location of the ram. During the work cycle, the ram is always in contact with the source of its power. These presses can be further divided as
  - Presses driven by a crank or eccentric drive. *Simple crank drives* are the most commonly used types, with *extended crank drives* in knuckle-joint presses.
  - Presses driven by a cam. *Cam-driven systems* are used in presses with less tonnage.

Each of these groups has certain advantages and disadvantages closely connected with the type of process they represent. For example, hydraulic presses have fewer operating parts, which brings the cost of their repairs down. However, should these machines be in need of any repair, such a procedure will be very demanding. With mechanical presses their breakdowns are visually detectable, but a complete knowledge of the circuit is required to find a problem within a hydraulic system. Also the tolerance range of hydraulics is not as impressive as that of mechanical presses, with the latter group also being faster. Fortunately, each type's usefulness is given by the type of its applications, and that's where their function, along with its advantages and drawbacks, fits. (See Figs. 5-1 through 5-6.)

### 5-1-2. Presses, according to their energy supply

There are several categories of machines, their main difference being in the working media they use. We recognize:

- *Mechanical presses,* where the work force is supplied by some mechanical means, such as a cam or lever
- *Hydraulic presses,* utilizing the pressure of water or other fluid media
- *Steam presses,* using pressurized steam

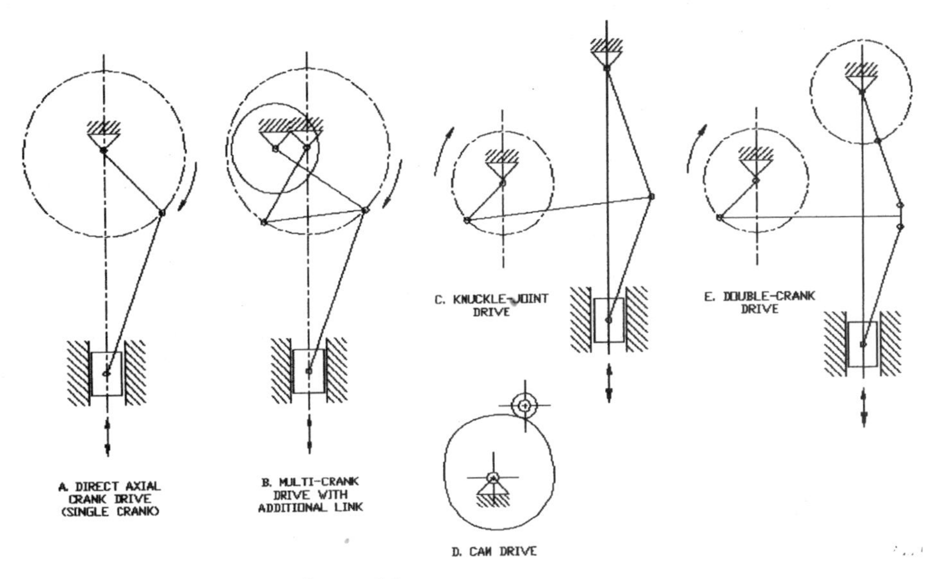


**Figure 5-1** Various types of press drives.

- *Pneumatic presses,* operating with the aid of pressurized air
- *Electromagnetic presses,* where the ram is moved by the application of electromagnetic force

### 5-1-3. Presses, according to their construction

Variations due to the construction of presses allow for the distinction, based on types of press frames, as

C-frame presses or gap-frame

Closed-frame presses or O-frame

C-frame construction is often used with smaller-capacity presses. Their main advantage lies in the easily accessible work area, which accounts for shorter setup and adjustment times. This advantage is perhaps outweighed by their faults, mostly attributable to the shape of their frame, whose construction is likely to suffer from deflection under load. However, in current machine building, heavy tie rods and other reinforcements are used to secure the machine's sturdiness and accuracy.

These two groups of frame-dependent classification can be further divided into presses with a single column, double column, and pillar-supported presses (Fig. 5-7).

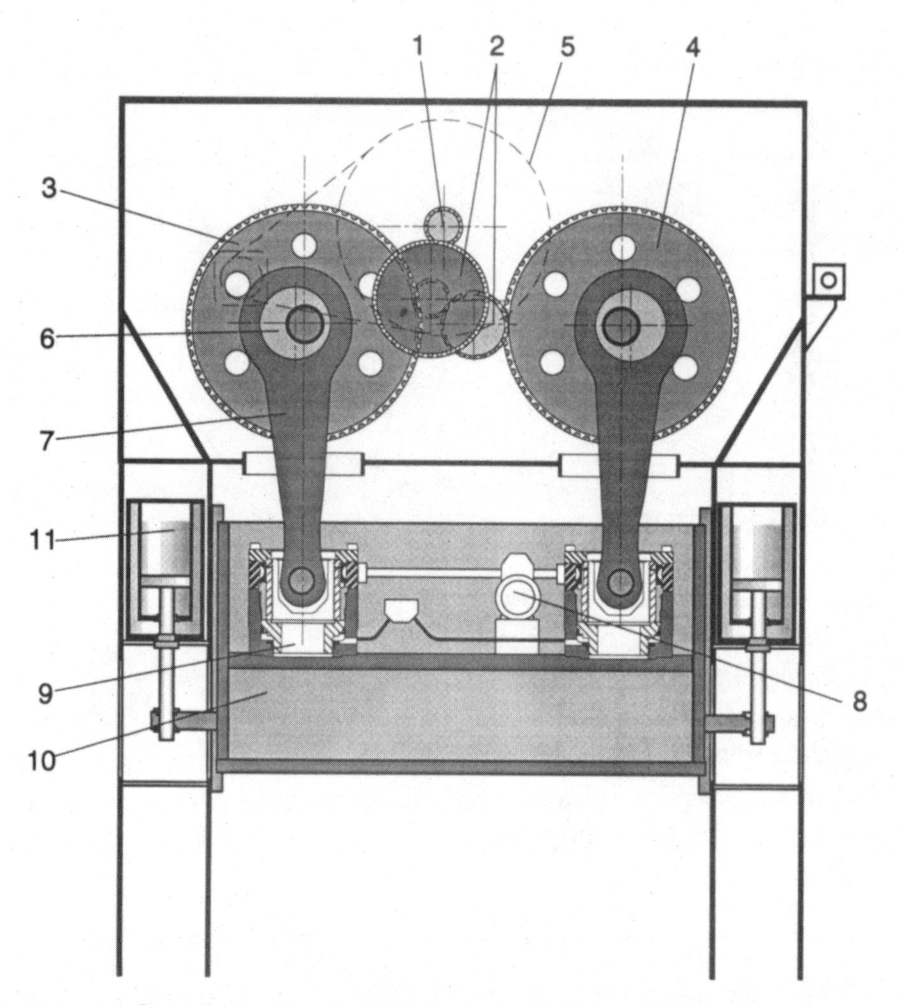


**Figure 5-2** Press drive principle: (1) Main shaft. (2) Pinion. (3) Gear. (4) Gear. (5) Driving belt. (6) Eccentric shaft. (7) Connection. (8) Slide adjustment. (9) Ram-retaining assembly. (10) Ram. (11) Counterweight. (*Reprinted with permission from Müller Weingarten AG, Germany.*)

Additionally, according to their width-to-height ratio and to the deviation from vertical, presses can be further categorized as vertical and horizontal presses, inclined or inclinable presses, and adjustable-bed presses (Fig. 5-8). Figures 5-9 through 5-16 show various presses.

The type of press drive used may introduce an additional variation. Figure 5-17 shows a *single-point system of drives,* in which the rod connecting the ram to the crankshaft is only one; a *two-point system,* in which the ram is tied to the crankshaft in two places; and a *four-point system* of drive variation, tied at four places. Most often the

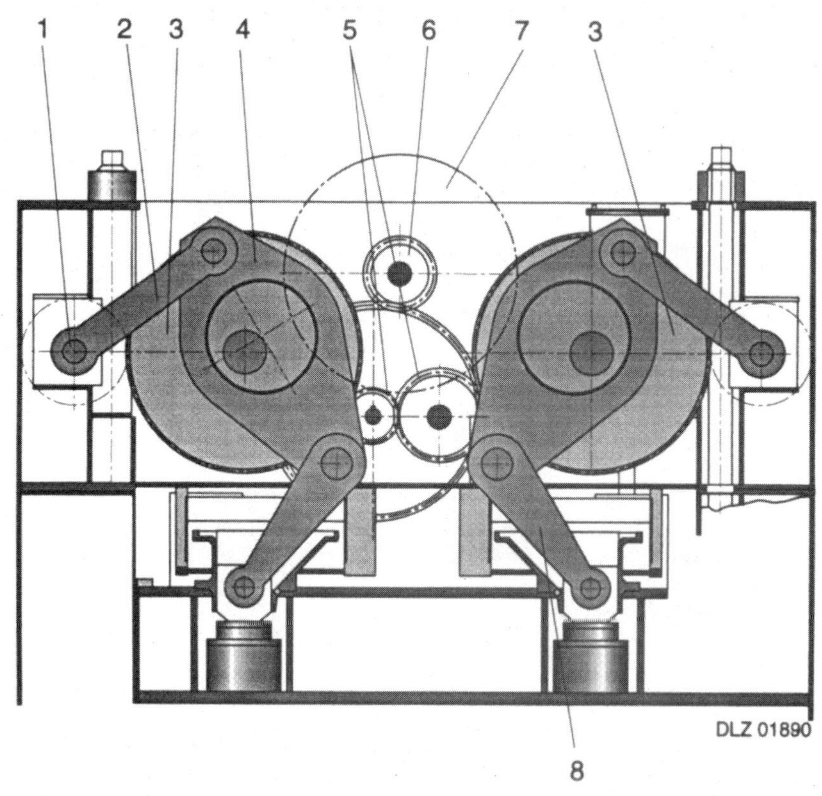


**Figure 5-3** Joint mechanism: (1) Fixed point. (2) Connecting link. (3) Eccenter. (4) Rocker. (5) Gear assembly. (6) Main shaft. (7) Flywheel. (8) Ram connection. (*Reprinted with permission from Müller Weingarten AG, Germany.*)

driving force is delivered from the upper area of the press, but sometimes even bottom-located driving systems may be used.

## 5-2. Parts of the Press

Presses consist of several essential components, assembled within the mass of their frames. Years ago, the presses were simple and straightforward, but also limited in their function. Today's presses contain electronic controls (some may even have certain programming capacities), and altogether, they can work miracles. These previously rigid and unyielding machines can now be set and adjusted to provide for any whim and fancy of a demanding toolmaker, shop supervisor, or engineer.

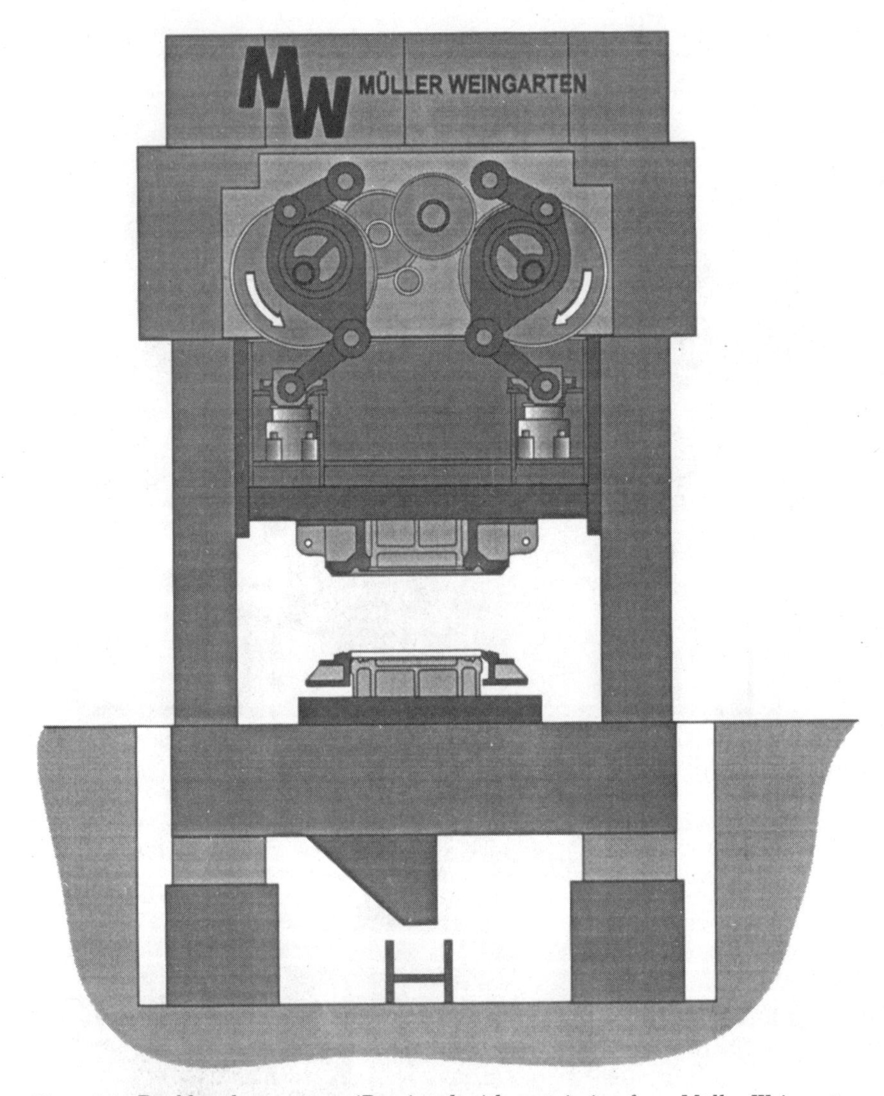


**Figure 5-4** Double-column press. (*Reprinted with permission from Müller Weingarten AG, Germany.*)

### 5-2-1. Press frame

The frame of the press must be sturdy and rigid in construction, which alone influences the functional accuracy of the machine. With weak frames, the deflection and later deformation of the mounting surfaces may occur, producing a subsequent misalignment of main parts, and possible breakage.

The sturdier and more rigid and stiff the frame is, the less the press mechanisms are affected by their own function.

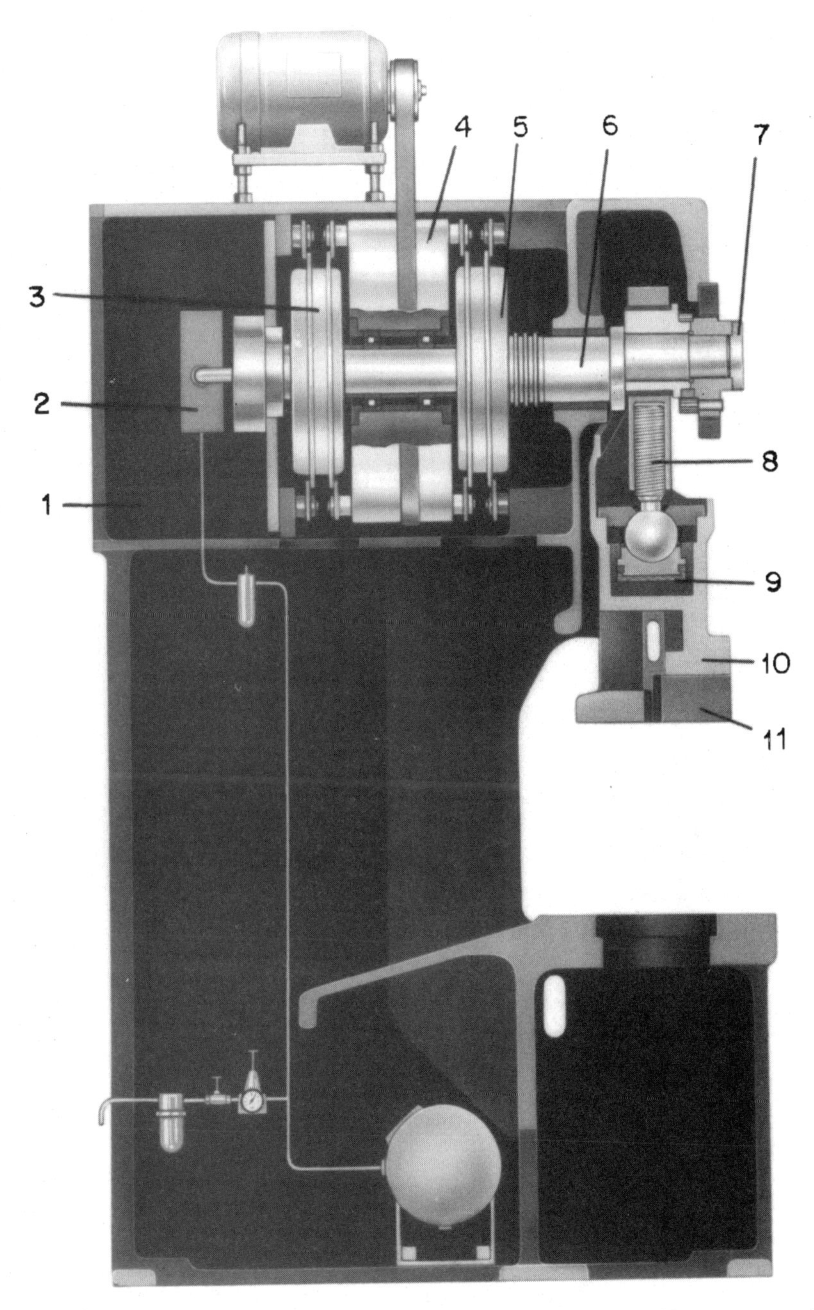


**Figure 5-5** Sectional view of the press mechanism without reduction gears: (1) Frame. (2) Control valve. (3) Clutch and brake. (4) Flywheel. (5) Clutch and brake. (6) Eccentric shaft. (7) Slide stroke adjustment. (8) Slide adjustment. (9) Overload safety device. (10) Slide. (11) Clamping block. (*Reprinted with permission from Müller Weingarten AG, Germany.*)

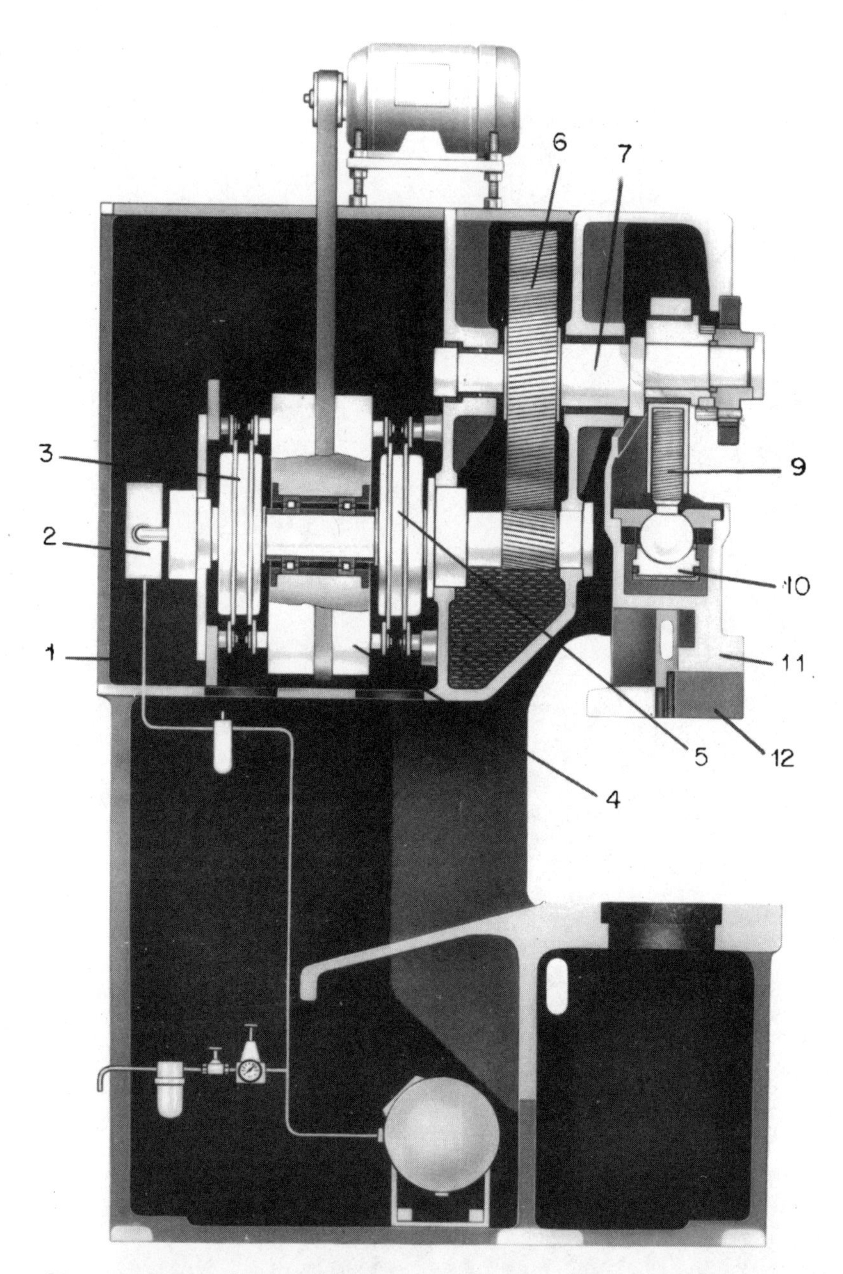


**Figure 5-6** Sectional view of the press mechanism with reduction gears: (1) Frame. (2) Control valve. (3) Clutch and brake. (4) Flywheel. (5) Clutch and brake. (6) Reduction gearing. (7) Eccentric shaft. (8) Slide stroke adjustment. (9) Slide adjustment. (10) Overload safety device. (11) Slide. (12) Clamping block. (*Reprinted with permission from Müller Weingarten AG, Germany.*)

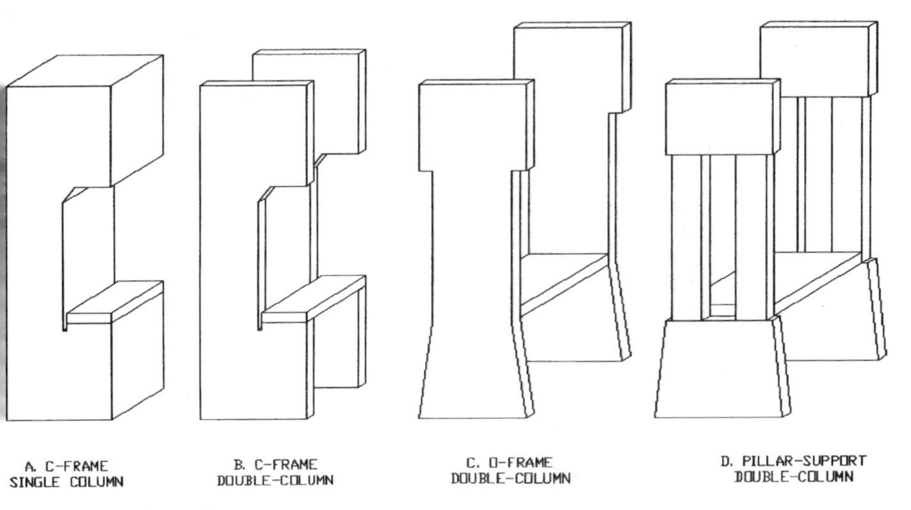


**Figure 5-7** Press frame types.

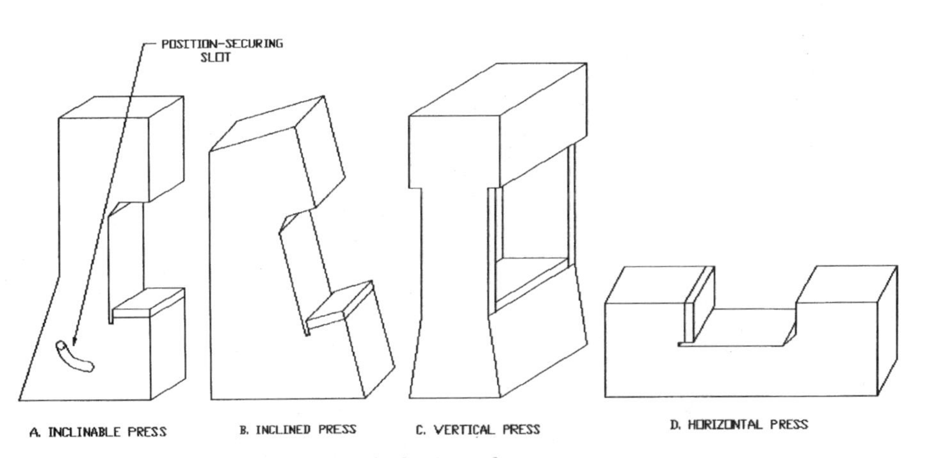


**Figure 5-8** Presses with respect to the horizontal.

The material of the frame is another aspect to be evaluated, for cast iron will succumb to deformation more readily than steel, since its modulus of elasticity is lower.

The best type of frame for press work is the fully enclosed O-type with the fewest openings in the sides and supports. All its flanges and openings should be provided with well-rounded corners, with no sharp or abrupt lines or connections of any kind.

Press frames are either cast or welded together from segments. Cast frames are used for smaller presses, their sides sometimes being

**Figure 5-9** Straight-side, high-speed automatic press. Frame: high-tensile cast iron, four-piece tie rod construction. Crankshaft: full eccentric, forged alloy steel. Eight-point gibbing. Flywheel or geared. 30 to 200 tons capacity, 60 to 500 strokes per minute. (*Reprinted with permission from Minster Machine Co., Ohio.*)

**Figure 5-10** Straight-side press. Frame: four-piece tie rod construction. Crankshaft: full eccentric, made of forged alloy steel. Square gibs guide the slide front to back and left to right. Single geared twin drive. 200 to 1000 tons capacity, up to 150 strokes per minute. (*Reprinted with permission from Minster Machine Co., Ohio.*)

**Figure 5-11** Open-front power press. Frame: gray cast iron. Optional steel plate welded construction. 630 to 4000 kN capacity, adjustable speed 20 to 60/min. (*Reprinted with permission from Müller Weingarten AG, Germany.*)

**Figure 5-12** Open-front power press. Notice the heavy tie rods. (*Reprinted with permission from Müller Weingarten AG, Germany.*)

**Figure 5-13** C-frame hydraulic press. Frame: welded. Capacity: 250 to 2500 kN. Working speed up to 8 to 38 mm/s. Maximum pressure 21 to 24.5 MPa. Electrohydraulic controls by Mannesmann Rexroth. (*Reprinted with permission from OSTROJ, Opava, Czech Rep.*)

**Figure 5-14** Three-axes, programmable transfer press with electronic transfer. (*Reprinted with permission from Müller Weingarten AG, Germany.*)

reinforced with additional plates or tie rods to increase their sturdiness in operation.

Welded frames may be produced either as a group of various assemblies welded together to form a single mass or may consist of various segments welded together but finally attached to each other by other means, the most common method of attachment being that of anchoring with ties, rods, and plates.

### 5-2-2. Bolster plate

Bolster plate, sometimes called press table, is positioned on top of the press bed. It is a heavy plate, ribbed with T slots (to receive T bolts in the assembly of a die), precision-aligned to the frame with dowel pins.

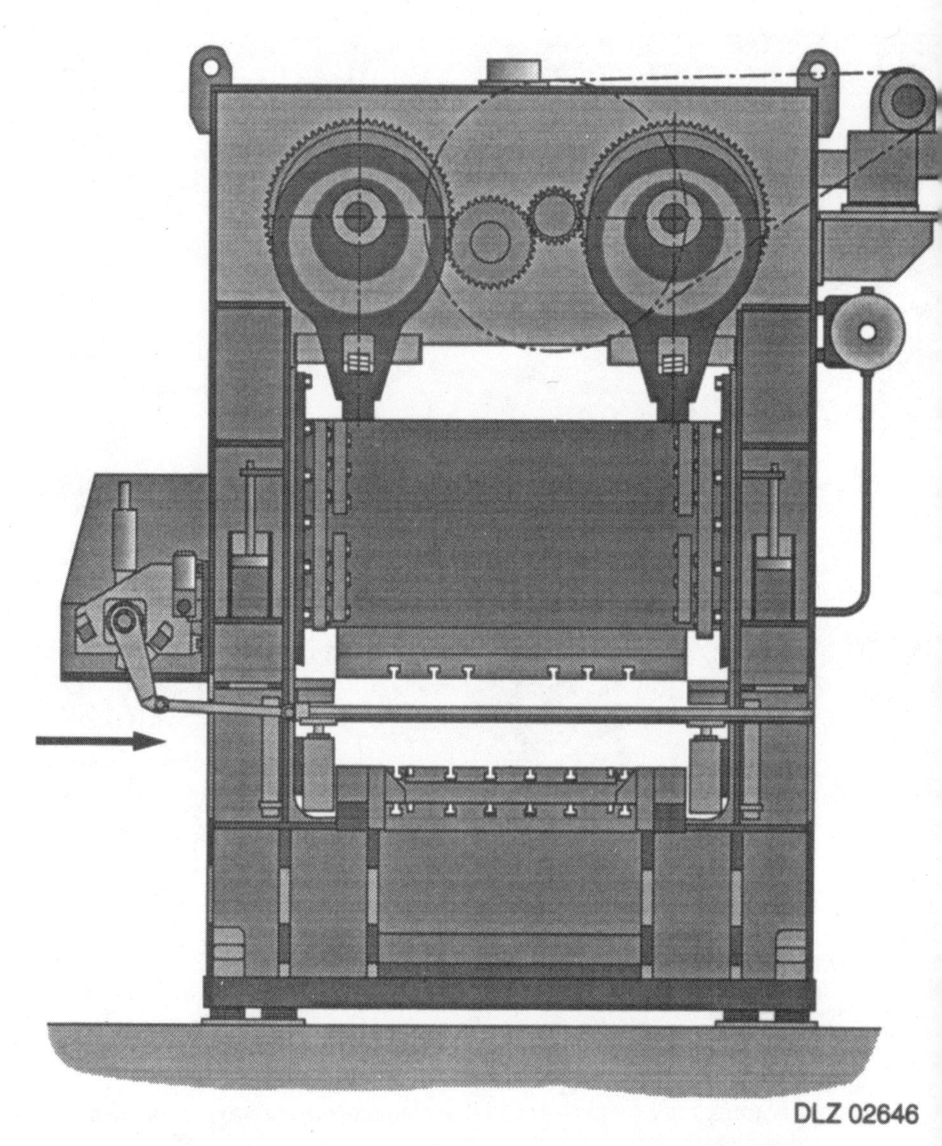


**Figure 5-15** Transfer press schematics. (*Reprinted with permission from Müller Weingarten AG, Germany.*)

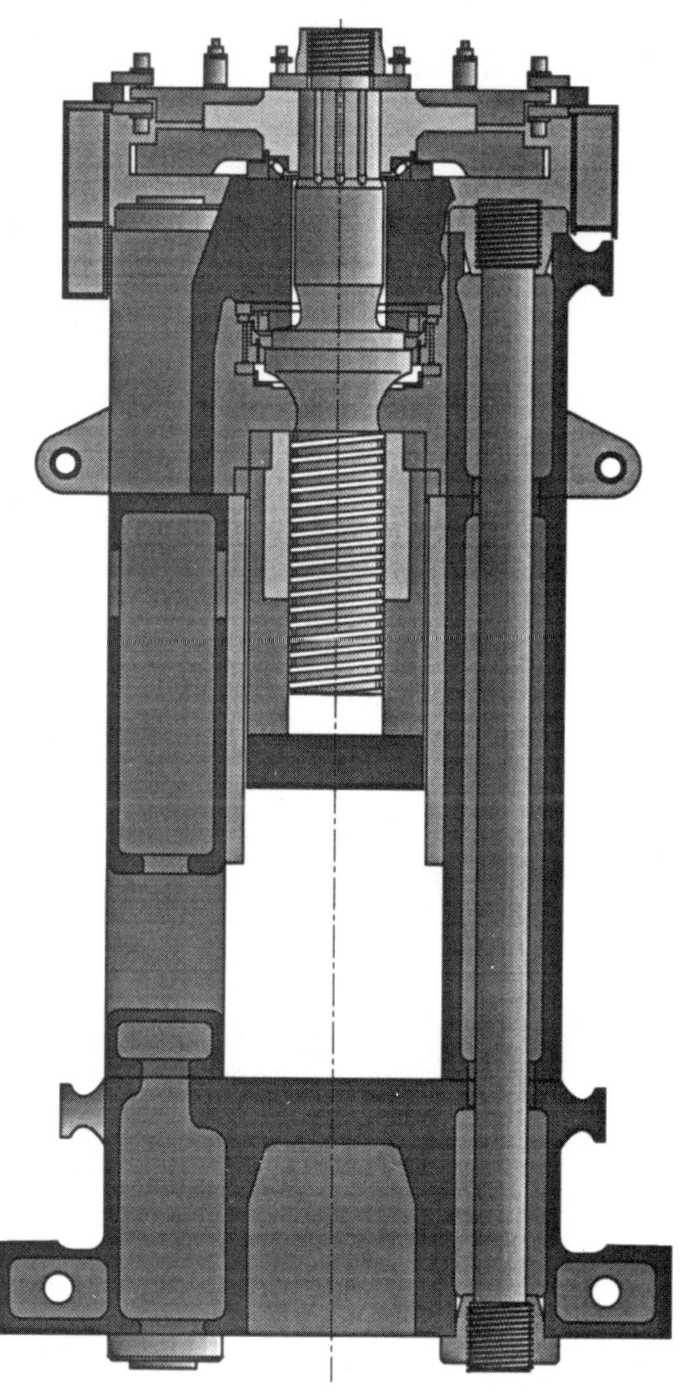

**Figure 5-16** Screw-press type, with motor-accelerated movement. 70,000 to 250,000 kN capacity. (*Reprinted with permission from Müller Weingarten AG, Germany.*)

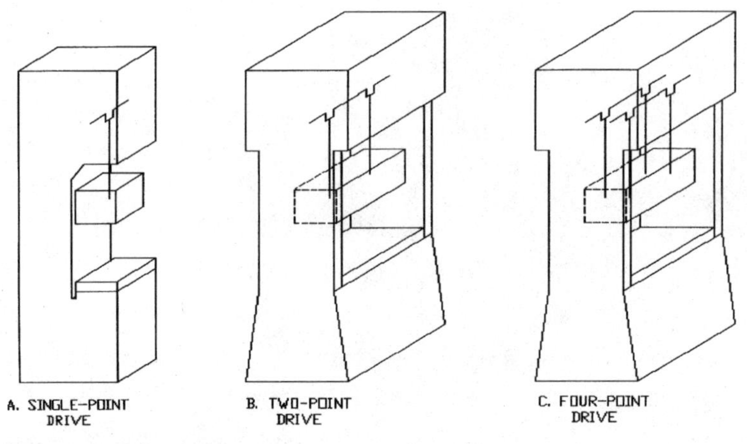


**Figure 5-17** Drive systems.

### 5-2-3. Ram

The ram, sometimes also called the slide, is the work-delivering part of the press, sliding up and down, as powered by the driving elements of the machinery.

### 5-2-4. Press drive

The power, which is used to drive mechanical presses, is transformed to the ram with the aid of either:

- Flywheel
- Flywheel and single-reduction gear
- Flywheel and multireduction gear arrangement

In all three types, the flywheel is the storage of energy, being incessantly run by a motor. During the press stroke, it temporarily slows down because of the transfer of a portion of its energy to the ram. Driven by the motor, the flywheel gains speed again in time for another stroke of the press.

Flywheel-only drive is most often used with small dies or where only light blanking is performed.

### 5-2-5. Crankshaft

To actuate the slide, crankshafts serve as a link between the press drive and the ram. In their basis, crankshafts can be of the following types:

- Crank type (crankpin)
- Eccentric

- Cam
- Knuckle joint
- Toggle mechanism
- Drag link

Contrary to all other types of crankshafts, knuckle joints can generate a tremendous force at the bottom of the stroke. For that reason they are used for compressing operations, which require short strokes at high pressures.

Toggles are actually levers, linked together. Their function may resemble that of a knuckle joint arrangement, but on close observation, toggles display a faster motion, with longer dwell intervals.

Because drag link mechanisms are slow in motion, they are utilized mostly for drawing operations.

### 5-2-6. Clutch and brake

The clutch, positioned between the flywheel and the press drive mechanism, controls the timing of the press by engaging and disengaging the drive shaft to the perpetually rotating flywheel. Clutches can be grouped into three main groups:

- Friction clutches
- Interlocking clutches (positive)
- Eddy-current clutches, operating through the influence of two magnetic fields

Brakes should stop the slide when the clutch is disengaged. Owing to the enormous mass of the whole machinery, the inertia of all its components can be of tremendous proportions, which calls for a good and dependable braking arrangement.

## 5-3. Press Operating Parameters

The energy of the press can be calculated by multiplying its average force by the distance through which it must be applied. The value of such energy is measured in inch-tons.

The power of the press is the amount of energy which must be exerted during a given time interval. Its unit of measure is the horsepower.

Some consider the secret of a good die-operating practice as depending on an adequate amount of pressure, be it the blankholding pressure or the stripping pressure. The selection and arrangement of springs, along with the selection of the press with adequate tonnage, are important here.

### 5-3-1. Tonnage

The value of press tonnage, as given by various press manufacturers, is based on a certain operating speed of the flywheel. Any difference from the flywheel's rpm's will alter the energy output of the press.

In a mechanical press, or other stroke-controlled machinery, the tonnage output will also vary with the position of the ram: When at its lowest, a drop in the press tonnage will be experienced. When the punching force is needed exactly at the moment the ram is at its lowest location, and the amount of stroke is comparatively short, the actual press force can be calculated as follows:

$$P_{act} = 3.5\sqrt{C_{dia}} \qquad \text{[tons]} \tag{5-1}$$

where $P_{act}$ = tonnage, actual, at the bottom of the stroke
$C_{dia}$ = diameter of crankshaft

This calculation and the information included in Table 5-1 is not applicable to end-wheel presses equipped with an overhanging crankpin. The preferred range of application is intended for a forged crankshaft, made of 0.45 percent carbon steel with higher elastic limits.

### 5-3-2. Shut height

Shut height of a press is the space reserved for the accommodation of the die. It is measured off the top of the bed to the bottom of the slide, with the stroke down and adjustment up.

### 5-3-3. Stroke

Stroke of the press is the dimensional variation of the slide's movement during the work cycle. The stroke must always be greater than the dimensional distance a die has to travel to operate properly.

## 5-4. Presses, According to Their Operation

After classifying presses according to various aspects of their construction and use, differentiating them according to the type of their operation is the final point of distinction of this type of equipment.

### 5-4-1. Single-action press

Single-action presses are used for general press work throughout the industry. The number of "actions" is given by the number of slides, or rams, operating on a common axis and mounted within the same frame.

**TABLE 5-1 Approximate Tonnage of Crankshaft Capacity at the Bottom of the Stroke**

| Crankshaft dia, in | Tons | |
|---|---|---|
| | Single-crank press | Double-crank press |
| 1.375 | 6 | |
| 1.5 | 7 | |
| 1.625 | 9 | |
| 1.75 | 10 | |
| 1.875 | 12 | |
| 2.0 | 14 | |
| 2.25 | 18 | |
| 2.5 | 22 | 22 |
| 2.75 | 26 | 26 |
| 3.0 | 31 | 31 |
| 3.5 | 43 | 43 |
| 4.0 | 56 | 56 |
| 4.5 | 71 | 71 |
| 5.0 | 88 | 88 |
| 5.5 | 106 | 106 |
| 6.0 | 126 | 126 |
| 6.5 | 150 | 150 |
| 7.0 | 180 | 180 |
| 7.5 | 215 | 215 |
| 8.0 | 255 | 255 |
| 9.0 | 345 | 345 |
| 10.0 | 440 | 450 |
| 11.0 | 545 | 650 |
| 12.0 | 665 | 900 |
| 13.0 | 790 | 1150 |
| 14.0 | 920 | 1400 |
| 15.0 | 1060 | 1700 |

### 5-4-2. Double-action press

A double-action press (Fig. 5-18), as its name implies, has two slides operating along the same axis yet independent in their movements. Where the first slide may be actuated by the usual means, the other slide has a different operating arrangement. Quite often, a cam-dependent movement tied to the first slide's function is utilized.

In double-action presses, the second slide, being an additional mechanism, is spaced around the first, original slide.

A double-action drawing combination has the drawing punch attached to the inner slide and the blankholder mounted to the second, outer slide. The whole operation can be adjusted to conform to a sequence, where the blankholder will come down first, retaining the part to be drawn, and only then the main slide with the drawing punch will follow. On coming up, the drawing punch leaves first,

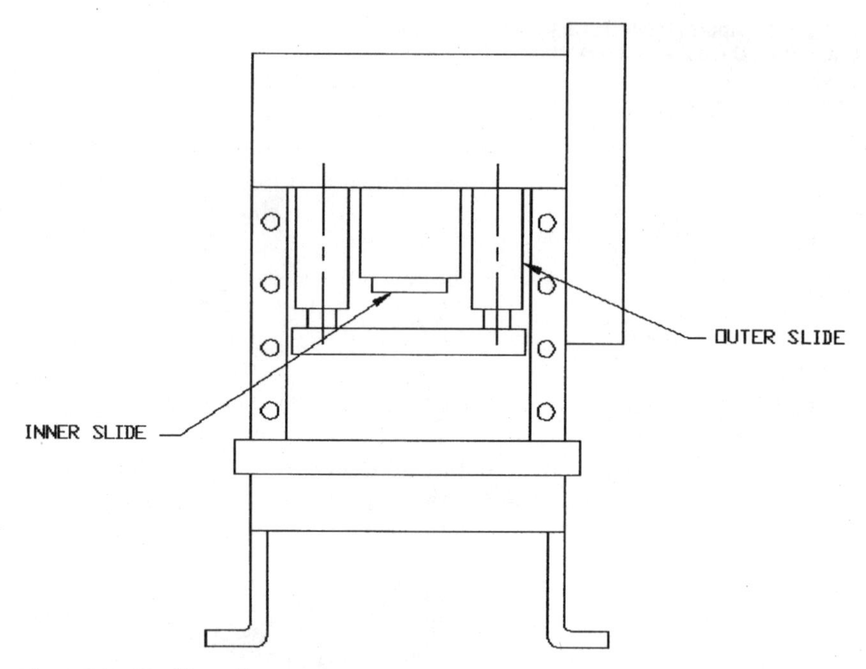


**Figure 5-18** Double-action press.

while the blankholder is still in its place, to retain the drawn part and prevent it from following up with the tool.

### 5-4-3. Triple-action press

A triple-action press has three slides, independent in their movement yet assembled within the same press frame. These presses are quite useful for complex drawing operations, where—for example—drawing has to be performed in more than one direction. In such a case, the first drawing punch and the blankholder may draw the part to the appropriate size and dwell to allow for the third slide's movement, which pulls the material in another direction.

### 5-4-4. Multislide press

Multislide presses may contain several slides assembled within a single press frame, with their range of operation along various axes. These slides may perform all kinds of work, either progressive or not. Such an arrangement is actually the same as if several presses were taken out of their shells and squeezed into a single machine frame.

### 5-4-5. Hydraulic press

Hydraulic presses are slower than mechanical presses, but their tonnage force is maintained constant throughout the stroke, no matter at which position the ram is located. Their total tonnage may be lowered for fragile die equipment. Double-action hydraulics can have the tonnage adjusted for both sections, for the punch holder and the blankholder as well.

Hydraulics cannot be overloaded, as they are protected by a combination of two separately adjustable relief valves. Die setup is easier, since they don't need to be adjusted for variation in material thickness. However, their motors must be larger than those of mechanical presses because they have no flywheel to store their energy.

Blanking operations can be detrimental to the hydraulic system, as the shock of puncturing the metal endangers its components. Their main range of application is for drawing, where hydraulics are an excellent choice, mainly because they maintain a constant pressure on the drawn part, progressing at more adaptable rates. Mechanicals enter the drawn part full-force and slow down toward the bottom of the stroke.

As mentioned previously, repairs of hydraulic presses, even though fewer, are much more complex and demanding, as the source of their breakdowns is hardly ever detectable visually.

## 5-5. Other Press-Room Machinery

There is a host of other press-room machinery needed to supplement the function of presses. Out of the whole array of equipment, several samples were selected for illustration (Figs. 5-19 through 5-22).

## 5-6. Electroerosive Machining

Electroerosive machining is a new machining method utilizing a bombardment of the metal material with the influence of electric current, accompanied by an appropriate cooling liquid. It is a process of metal removal, which can be used for production of cuts of various shapes, unattainable otherwise.

Electroerosive machining can be divided into two basic groups:

- Electrical-discharge machining (EDM), sometimes also called electrodestructive machining, which consists of cutting, pocket cutting, and grinding processes. The electrode is a metal wire or a thin metal strip, surrounded by the dielectric liquid. In the grinding application of this process, the electrode rotates around its own axis.

**Figure 5-19** Tongs feed unit, combined with a press. (*Reprinted with permission from Müller Weingarten AG, Germany.*)

- Electrochemical machining (ECM) is used for production of pockets in the metal material, or for grinding, polishing, and similar finishing operations. The process uses a rotating electrode, and the coolant is of an erosive nature, the action of which speeds up the removal of material while slowing down the wear of electrode. However, the accuracy of this method is less than that of EDM.

Because of the widespread utilization of EDM machining, only the first manufacturing method is treated in greater detail here.

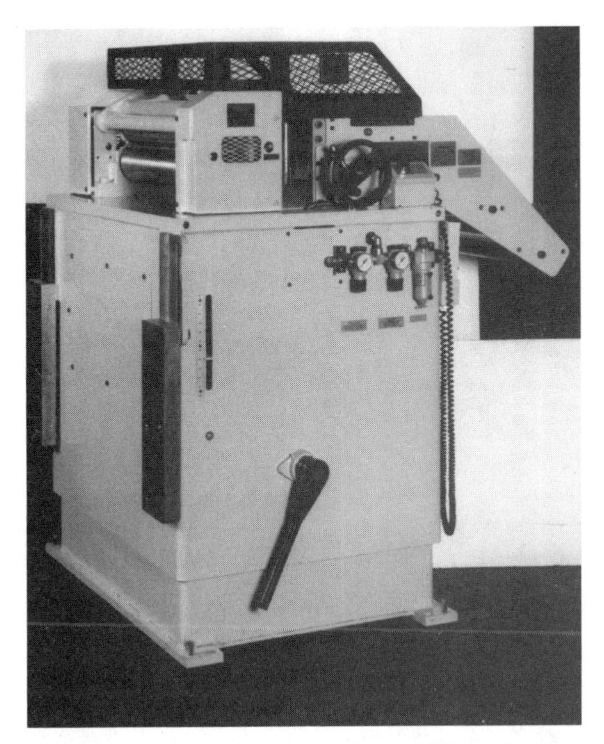

**Figure 5-20** Electric feed, servo-driven. Adjustable feed length during operation. Adjustable feed angle (dwell and position). No setup required for material thickness. *(Reprinted with permission from Minster Machine Co., Ohio.)*

EDM machining, or electrical-discharge machining, is used for production of a wide variety of cuts and shapes. In this process, a wire electrode is inserted through the metal workpiece, where by discharging electricity, it forms a localized arc between itself and the metal material. The electrode is made of either brass, copper, or tungsten. Where solid-block electrodes are used, these are usually made of copper or graphite. They wear quickly, for which reason they are often stepped in shape for cutting. A three-stage EDM electrode is shown in Fig. 5-23.

The EDM process is quite similar to a short circuit caused by touching the source of energy with a metal screwdriver. The electric spark so created erodes the metal surface in the point of contact. Because of the extremely high temperatures of the EDM arc, the eroded metal material literally melts away and evaporates, leaving a slight gap (Fig. 5-24). The dissolved metal is washed away by a coolant, which is usually a deionized water, petroleum, or oil. The scrap takes the

**Figure 5-21** Reels with expansion mandrels. Link-type expanded by hand crank, or wedge-type expanded by hydraulic cylinder. Pneumatic drag brake, hydraulic jog standard. (*Reprinted with permission from Minster Machine Co., Ohio.*)

shape of balls of compact metal dust. These sometimes get deposited back on the work surface, adhering to it as if welded.

The temperature between the electrode and the workpiece is enormous, with the arc being produced at high frequencies of many thousandths up to many ten-thousandths times per second. The dielectric coolant does not affect the spark, its influence being confined to the dissolved metal.

In this type of machining process, some overcut is always present. The depth into which the spark affects the metal may often be 0.004″ per side, which adds up to 0.008″ per diameter. This is the amount of additional and undesirable metal removal should the electrode's size be the same as the opening (see Fig. 5-24).

The surface condition of EDM cuts and the amount of overcut depend on the operating frequency and on the electric current going

**Figure 5-22** Straightener with eddy-current drive. (*Reprinted with permission from Minster Machine Co., Ohio.*)

through the EDM equipment. With the increase in frequency, the surface condition improves and the amount of overcut diminishes. With the increase in current, the overcut increases and the surface condition becomes worse.

Usually, the EDM cut is achieved in two phases: first a rough cut, followed by a finishing cut. An allowance for the second cut must be made in the size of the first opening. The finish obtained with the second cut is finer, even though the cutting process itself is faster. This is due to the much smaller amount of material being removed the second time.

EDM-produced cuts are not always straight because of the inclination of the electrode's surface, eroded by wear (Fig. 5-25). The arc,

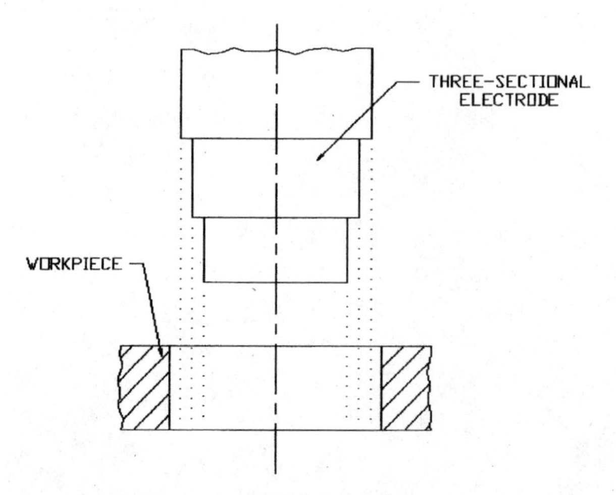


**Figure 5-23** Three-stage EDM electrode.

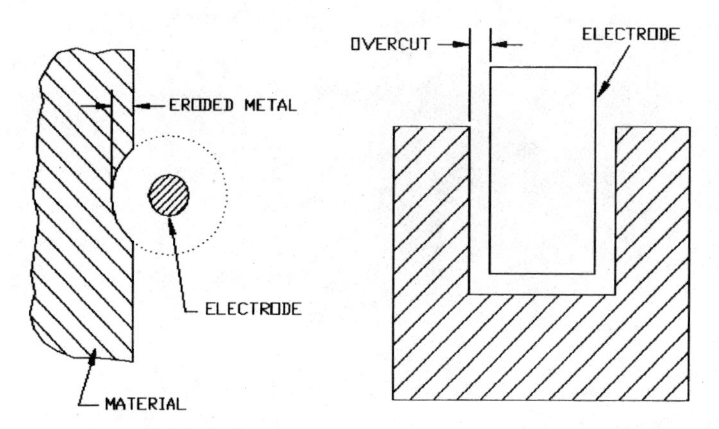


**Figure 5-24** EDM work process.

aside from removing the metal material from the workpiece as it should, also removes portions of the electrode in places of contact.

There are many advantages to the EDM process: Complicated shapes can be machined in a fraction of the time it used to take. Complex-shaped dies can be made in one piece instead of being sectioned to ease their machining. Metal blocks can be hardened prior to EDM cutting. Punch and die may often be made with the same cut, provided the amount of obtained tolerance (overcut) is acceptable.

A definite improvement of machining methods can be achieved with EDM undercutting capacities, as shown in Fig. 5-26. With a variable movement of a single electrode, multishaped undercuts can be attained, including slanted, angular surfaces, where the angle of

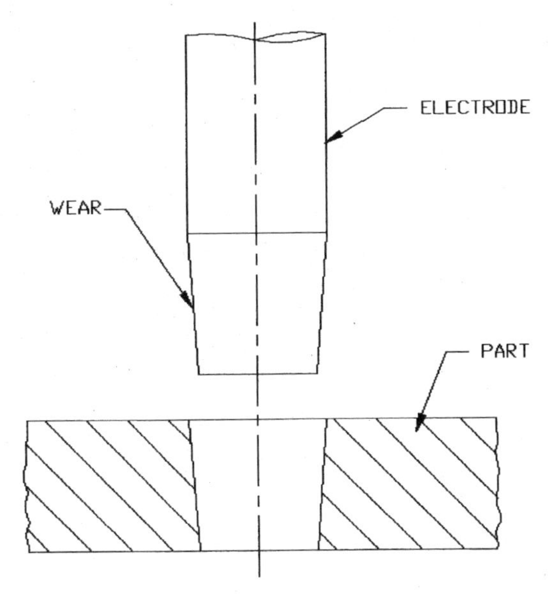


**Figure 5-25** Angular wear of the electrode.

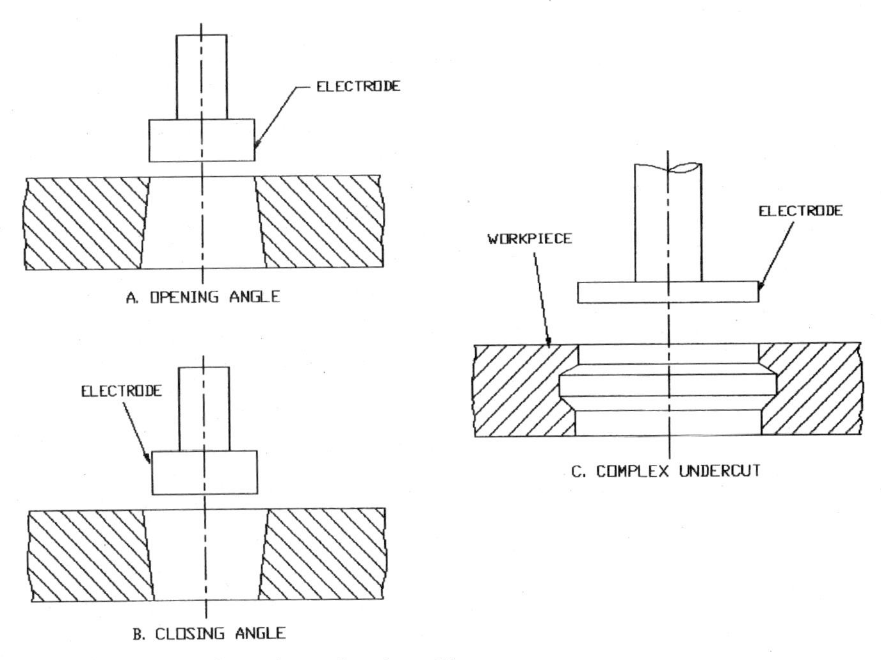


**Figure 5-26** Electrode cutting and undercutting.

inclination is either opening or closing toward the bottom of the block.

However, there are also disadvantages, the worst of them being the detrimental effect the EDM process has on the surface condition of the material. An underlying root of this problem is the continuous process of melting and solidification of the surficial layer of material. The effect of the extensive heat produced by the arc melts the material in its immediate vicinity, which is followed by its immediate cooling. This sequence is constantly repeated at great speed, and its effect on the surface of the material is enormous. First of all, a considerable brittleness of the upper layer develops, accompanied by a tendency to cracking. The depth of these cracks depends on the working temperature of the electrode: With higher temperatures, the depth of surficial fractures increases.

Underneath the first layer, the adjacent material is heated as well, but it does not melt, since its temperature does not reach such high levels. However, an alteration of its properties does occur, followed by changes in its structure and overall condition. The fatigue strength is decreased, cracks and microcracks appear, and a general degradation of the material follows.

In maraging steels, the effect of the EDM process on the surface condition is still more pronounced, in some cases almost detrimental to the quality and functionality of products made from these materials. Here, some tensile residual stresses of considerable proportions are introduced into the material, causing the formation of cracks and microcracks. With parts subjected to cyclic loading, such as springs, these imperfections are enhanced quite rapidly, progressing toward a fatigue-dependent premature failure of products.

Chapter

# 6

# Blanking and Piercing Operations

## 6-1. Metal-Cutting Process

Metal cutting is a process used for separating a piece of material of predetermined shape and size from the remaining portion of a strip or sheet of metal. It is one of the most extensively used processes throughout die and sheet-metal work. It consists of several different material-parting operations, such as piercing, perforating, shearing, notching, cutoff, and blanking.

In blanking, the piece is cut off from the sheet, and it becomes a finished part. In piercing, the cutout portion is scrap which gets disposed of while the product part travels on through the remainder of the die. The terminology is different here, though both processes are basically the same and therefore belong in the same category, which is the process of metal cutting (Fig. 6-1).

The actual task of cutting is subject to many concerns. The quality of surface of the cut, condition of the remaining part, straightness of the edge, amount of burr, dimensional stability—all these are quite complex areas of interest, well known to those involved in sheet-metal work.

Most of these concerns are based upon the condition of the tooling and its geometry, material thickness per metal-cutting clearance, material composition, amount of press force, accurate locating under proper tooling, and a host of additional minor criteria. These all may affect the production of thousands and thousands of metal-stamped parts.

With correct clearances between the punch and die, almost perfect edge surface may be obtained. This, however, will drastically change when the clearance amount increases, and a production run of rough-edged parts with excessive burrs will emerge from the die.

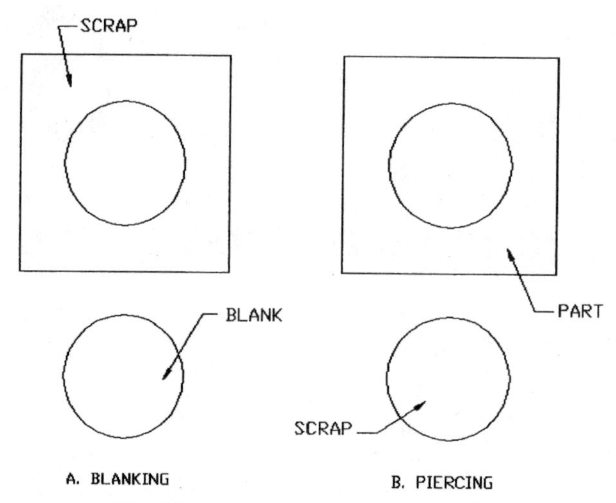


**Figure 6-1** Blanking and piercing differentiation.

Highly ductile materials, or those with greater strength and lower ductility, lesser thicknesses or greater thicknesses—these all were found similarly susceptible to the detrimental effect of greater than necessary clearances.

The literature recommends different tolerance amounts for cutting tools. Some claim $0.06t$ ($t$ = material thickness) to be sufficient for almost all applications. Others promote a $0.08t$ range, with $0.1t$ topping it off.

We already know from Chap. 2 that the shearing process consists of the punch moving toward the die opening, with sheet-metal material in between. The pressure applied to the strip causes the development of various compressive and tensile forces within the material. It begins to crack in the immediate vicinity of the edges of both cutting elements (Fig. 6-2). The progression of cracks finally results in separation of the cut area off the sheet, and the part is blanked (or pierced, perforated, etc.).

Naturally, a different type of separation must occur with a softer material than with its harder counterpart. The carbon content certainly has an influence on this process as well. Therefore, the tolerance range must have a provision to change not only with the stock thickness but with its composition as well.

As already mentioned, good condition of tooling is absolutely essential to the cutting process. We may have the most proper tolerance range between the punch and die, and yet the cut will suffer from imperfections if worn-out tools are used.

The wear of tooling (Fig. 6-3) considerably alters the cutting conditions, raising the demand for press force as well. Up to 50 percent

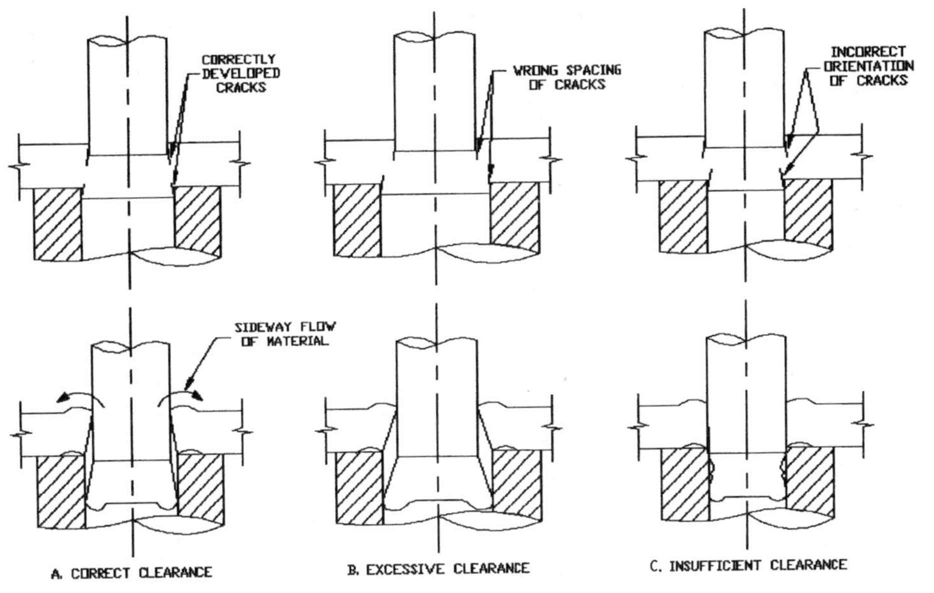


**Figure 6-2** Shearing of metal.

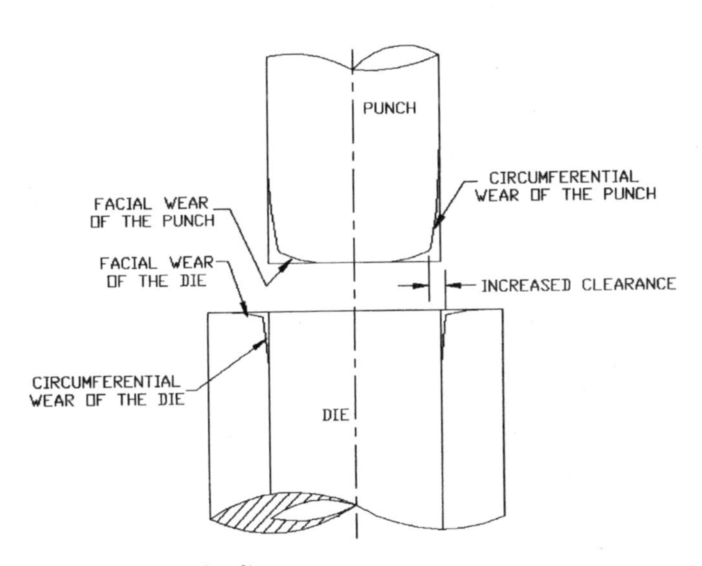


**Figure 6-3** Wear of tooling.

increase in tonnage has been observed with some parts. The punch-and-die clearance enlarges, with subsequent damage to the surface of the parts, which becomes rough and uneven. Excessive friction, followed by an increase in temperature during the die operation, only speeds up the destruction process.

## 6-2. Forces Involved in the Metal-Cutting Process

Aside from the press force acting upon the ram and applying strictly vertical pressure to the die and subsequently to the sheet-metal material, additional forces are involved in the metal-cutting process. As the punch enters the material, it pushes the bulk of it down through the opening in the die. However, a small portion of metal is forced sideways, as seen in Fig. 6-2. This flow, directed away from the cutting tool, is guided by the action of tensile and compressive forces within the cut metal, and is thus grain dependent: A different pattern of flow is seen along the grain than against it.

Such movement of material affects the structure of the sheet, especially in the immediate vicinity of the cutting station. Forced aside, the material becomes too crowded by such expansion in its content and it resorts to bulging through the only available outlet, through the surface of the sheet, which it deforms. In areas where piercing is more congested, the deformation progress is so widespread that the whole sheet becomes distorted, displaying either an excessive camber or waviness or any other variation from straightness.

The expansion of material pushes also against the body of a punch, applying a side-oriented force upon it. The punch is suddenly restricted in its movement by its squeeze, accompanied by an increase in friction. This force further attacks its stability, with slim and fragile tools often breaking under such a load.

The deformation of the cutoff portion of metal is often not so pronounced, which is probably due to its usually smaller size.

From Fig. 6-3 it is obvious that the flow of tensile and compressive forces resulting in the development of side-oriented and expansive shifts within the material is also a great contributor to the emergence of wear of the working surfaces. According to some, a side-oriented force generated by the cut material may amount to 2 to 20 percent of the total blanking force, with its marked dependency on the material thickness, its composition, and the amount of clearance between the cutting surfaces.

Additionally, forces within the cut material further influence the size of a sheared opening. On the complete retrieval of the punch, the bulging material slightly flattens out, its movement being oriented toward the empty space, which subsequently gets reduced in size.

Cutting clearances of up to $0.05t$ have been found to produce openings smaller than the size of the cutting punch.

As already mentioned, the punch on its way out of the cut material is restricted in movement by the emergence of frictional forces originating within the structurally altered material. For the punch to progress, a considerable force is needed to overcome this influence. This force, called a *stripping force,* may be calculated with regard to the material composition, its strength and thickness, the size of tooling, and its clearance.

Naturally, with increasing clearance between the punch and die arrangement, the amount of stripping force decreases. But the quality of the cut decreases along with it.

## 6-3. Alignment of Cutting Tools

Punches entering the material may or may not be absolutely concentric with the die opening below. Sometimes a shift from the mutual axis may be due to the assembly procedures; sometimes a minute movement in the frame of a press may cause a slight offset of the two centerlines, which ideally should match each other.

Even with perfect positioning, a long, unsupported, and unguided punch may be swayed aside by the movement of metal during the cutting operation or by its own off-center punching, or by an action of some other demanding operation within the die.

To alleviate this problem, punches must be guided in their movement unless their bulk is so great that they actually constitute the major portion of the die.

The guidance is provided through inserts in the stripper plate, which are appropriately called guide bushings. Slim punches should be further protected by quills, punch sleeves, or wraps and similar arrangements. Punches that have irregular shapes or those having their face area ground to an angle often utilize heels, which guide their progress during the cutting operation (Fig. 6-4).

Multipart retainers are an additional punch-guiding provision to a die. They resemble small, self-contained punch plates, and they come in various sizes and shapes and with different tool-retaining openings (Fig. 6-5). The whole unit, along with the punch or punches it holds, is secured to the holding plate with dowels and locked in this position by screws. The punch, equipped with a ball-retaining groove, is precision-located by a pressure of the spring-loaded ball.

Another method of tool guidance is that in which die shoes are aligned with precision guide pins. Four-pillar die sets were found to be the most accurately aligned instruments, surpassed only by subpress dies, which are actually considered small, self-contained, and self-aligned press units.

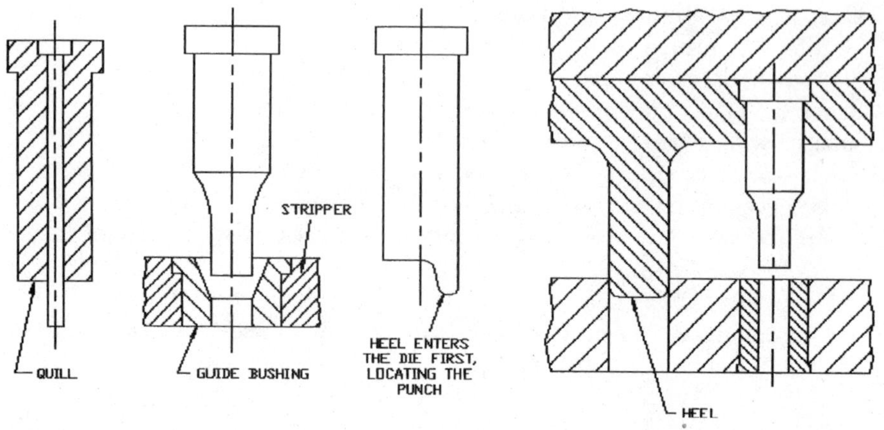


**Figure 6-4** Support and guidance of punches.

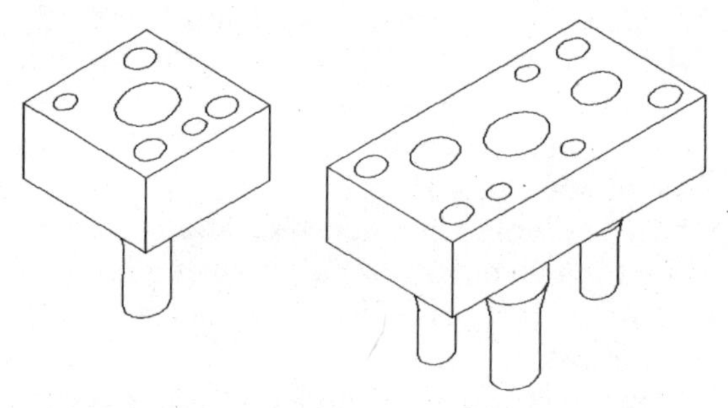

**Figure 6-5** Punch retainer.

Guide pin and pillar die sets are described in Secs. 3-1-2 and 3-1-3. Guide pins are precision-ground and fit into bushings of equal quality (Fig. 6-6). Their tolerance ranges are 0.0002 to 0.0004″, and their smooth function is aided by the lubricants retained in the grooves in the bushing.

Self-lubricating bushings (Fig. 6-7) are made of high-strength bronze material, where a lubricant is embedded throughout its structure. Such a lubricating arrangement usually lasts the entire life of the bushing.

However, the absolute of the die alignments is the ball bearing bushing (Figs. 6-8 and 6-9), which runs so tight that the effect of the side-oriented force on the tooling is almost eliminated.

## 6-4. Design of Cutting Tools

The shape, size, and operating sequence of cutting tools should be altered to achieve a trouble-free production of high-quality parts. Sometimes tooling shapes may be modified to become more sturdy

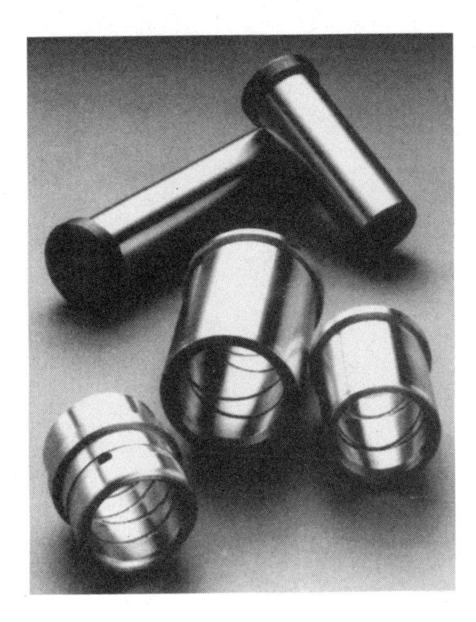

**Figure 6-6** Guide bushings and pins. Internal grooves are for distribution of lubricant. (*Reprinted with permission from LAMINA, Inc., Royal Oak, MI.*)

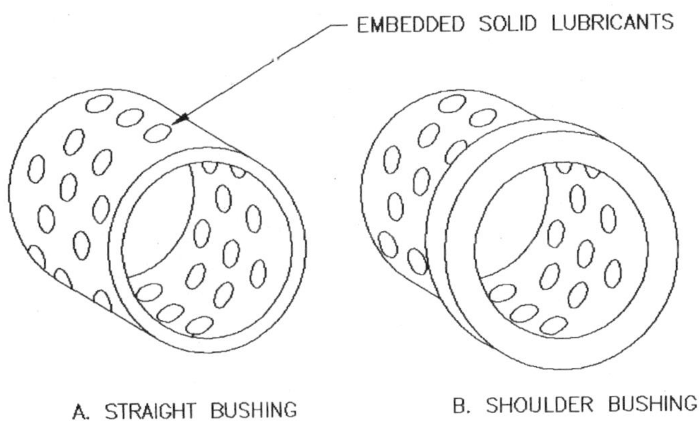


**Figure 6-7** Self-lubricating bushings.

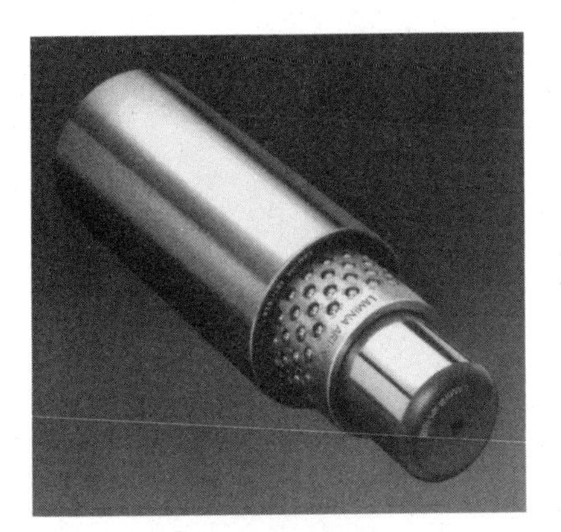

**Figure 6-8** Ball bearing bushing assembly. (*Reprinted with permission from LAMINA, Inc., Royal Oak, MI.*)

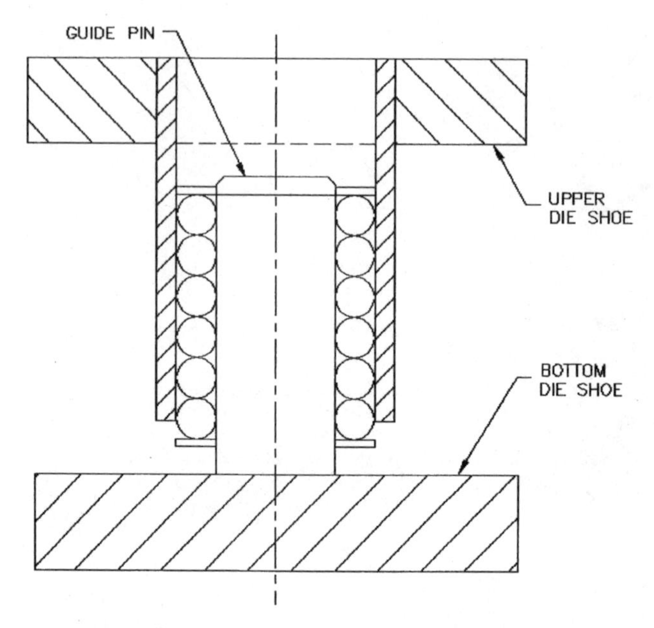


**Figure 6-9** Ball bearing bushing cross section.

and durable, with no resulting changes to the cut part whatsoever. During the die design stage, the operating sequences should be carefully scrutinized for possible defect-causing areas. By purposeful and cautious examination of all tooling and its combinations, perfect parts will certainly be produced.

### 6-4-1. Length of cutting tools

All cutting tools must be designed in such a way that the pressure applied toward their mass will not cause their buckling or outright distortion. For that reason, their lengths must be evaluated with regard to the necessity of their support.

**Cutting punches.** Basic length requirements for cutting punches made of heat-treated steels are

$$\frac{L}{\sqrt{I_{min}/A}} > 2.385 \qquad (6\text{-}1)$$

where $I_{min}$ = moment of inertia, minimal ($I = \frac{1}{2}mr^2$ for cylinders)
$L$ = length of the punch (see Fig. 6-10)
$A$ = area of the punch cross section

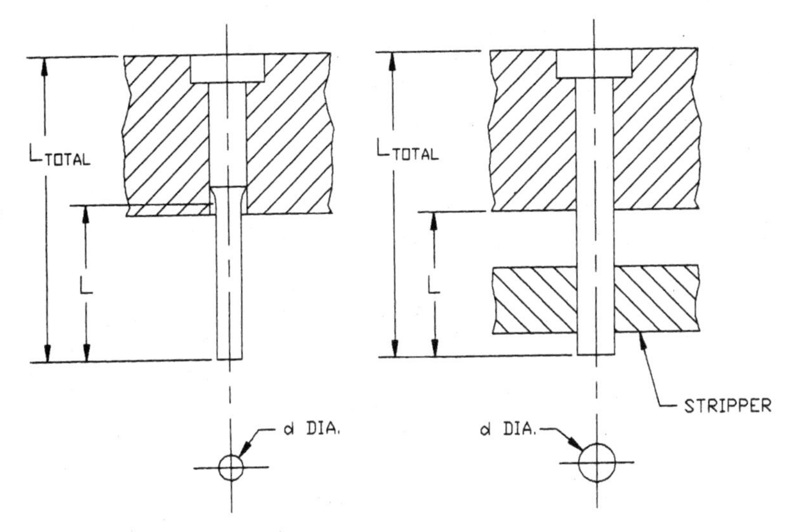


**Figure 6-10** Length of a punch.

For non-heat-treated steel material the result of formula (6-1) must be greater than 2.785.

A mean value of pressure force should be used in all calculations, even though the cutting stress varies with its distance off the cutting edge. Mean compression strength $S_{MC}$ is therefore expressed as a ratio of the cutting pressure and the area of the punch face:

$$S_{MC} = \frac{P}{A} = \frac{\pi d t S_{SH}}{\pi d^2/4} \tag{6-2a}$$

where $P$ = pressure, cutting, lb
$A$ = area of the punch face, in$^2$
$S_{SH}$ = shear strength of the punch material
$t$ = material thickness
$d$ = punch diameter

An adjusted expression of Eq. (6-2*a*) is

$$\frac{t}{d} \leq \frac{S_{S-\lim}}{4S_{SH}} \tag{6-2b}$$

where $S_{S-\lim}$ = allowed pressure limit against the punch support

Critical buckling pressure may be figured using Euler's formula:

$$P_{crit} = \frac{\pi^2 EI}{4L^2} \qquad (6\text{-}3a)$$

where $E$ = modulus of elasticity
$I$ = moment of inertia (minimal)
$L$ = length of the punch (see Fig. 6-10)

Further the critical pressure should be made equal to

$$P_{crit} = CP \qquad (6\text{-}3b)$$

where $C$ = safety factor; $C = 2$ to 3 for heat-treated steel; $C = 4$ to 5 for non-heat-treated steel

Subsequently the critical length of an unguided round punch can be calculated:

$$L_{crit} = \sqrt{\frac{\pi^2 EI}{4\,CP}} \qquad (6\text{-}4a)$$

The slenderness ratio $(L/d)_{min}$ may be calculated by using the formula

$$(L/d)_{min} = \frac{\pi}{8}\sqrt{\frac{E}{S_{S-lim}}} \qquad (6\text{-}5)$$

The critical buckling pressure for the guided punch is given by the relationship

$$P_{crit} = \frac{2\pi^2 EI}{2} \qquad (6\text{-}3c)$$

and

$$P_{crit} \le CP \qquad (6\text{-}3d)$$

The maximum length allowed for a round, guided punch will be

$$L_{crit} = \sqrt{\frac{2\pi^2 EI}{nP}} \qquad (6\text{-}4b)$$

where $n$ = exponent of strain hardening tendency of sheet-metal material, 0.2 to 0.5 in normal conditions, not involving superplasticity or ultrasound

The critical slenderness ratio will subsequently become

$$\left(\frac{L}{d}\right)_{\min} = \frac{\pi E}{S_{S-\lim}} \tag{6-6}$$

**Die bushings.** Die bushings may be compared to a thick cylinder, with an equally distributed pressure against the cutting opening. A rough height of the die button may then be calculated with the formula:

$$h \geq \sqrt[3]{P_{\max}} \tag{6-7}$$

Height of the round die with a round opening may be assessed with the aid of Eq. (6-8*a*). The bending tension $S_{\mathrm{SB}}$ may be calculated as

$$S_{SB} = \frac{1.5\,P}{h^2}\left(1 - \frac{d_1}{1.5\,d_2}\right) \tag{6-8a}$$

where $P$ = pressure, cutting, lb. Other values are per Fig. 6-11.

A groomed height of the die will become

$$h \geq \sqrt{\frac{1.5\,P}{S_{SB-\lim}}\left(1 - \frac{d_1}{1.5\,d_2}\right)} \tag{6-9a}$$

where $S_{SB\text{-lim}}$ = allowed bending pressure limit.

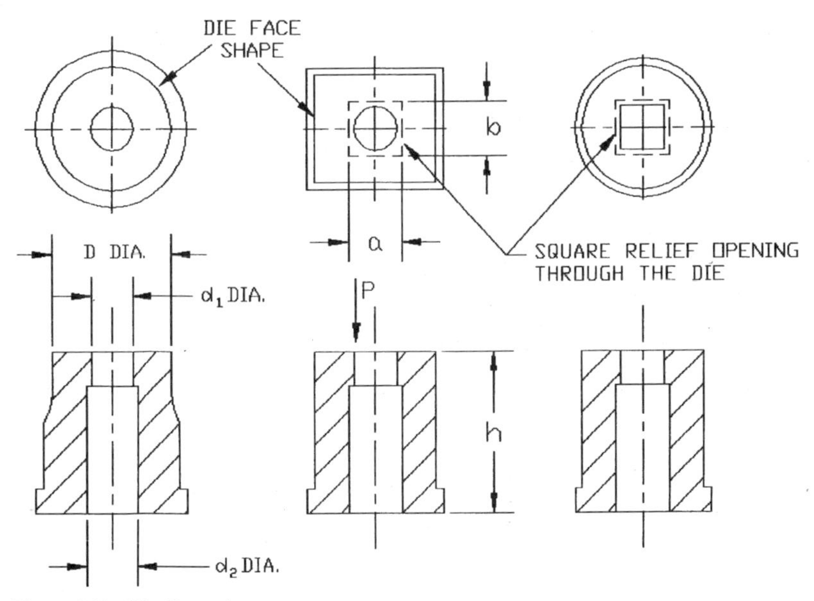


**Figure 6-11** Die dimensions.

A relationship must apply:

$$S_{SB} \leq S_{SB-\lim} \tag{6-8b}$$

For a square or rectangular die with a round cutting opening (see Fig. 6-11), the bending stress $S_{SB}$ will be

$$S_{SB} = \frac{1.5\,P}{h^2} \tag{6-8c}$$

and the height

$$h = \sqrt{\frac{1.5\,P}{S_{SB-\lim}}} \tag{6-9b}$$

With rounded punches with a square or rectangular cutting opening, the formula will apply:

$$S_{SB} = \frac{3\,P}{h^2}\,\frac{b/a}{1 + b^2/a^2} \tag{6-8d}$$

or

$$h \geq \sqrt{\frac{3\,P}{S_{SB-\lim}}\,\frac{b/a}{1 + b^2/a^2}} \tag{6-9c}$$

A side-oriented force against the die button may be evaluated as

$$P \doteq P_N + P_F + P_C \tag{6-10}$$

where $P_N$ = value of the side-oriented material pressure, equal to approximately 0.1 to 0.4 of the cutting force $P$
$P_F$ = force of friction between the die button and its support. A coefficient of friction $\mu = 0.15$ may be used with steel
$P_c$ = force depending on the pressure needed to push the slugs through the die. This force is

$$P_c = P_{c-\mathrm{act}}\,lt \tag{6-11}$$

where $P_{c\text{-act}}$ = pressure needed to push the slugs through the die
l = linear length of the cut opening
$t$ = material thickness

**Dowel pins.** Dowel pins are used to secure blocks against side shifting. If a side-acting force (or shear force) is applied toward their length, it may deform them, with subsequent damage to the block, cutting tools, and the whole die. The maximum allowed amount of such a force may be taken as

$$P_{max} = 0.4\, S_{C-\lim} D^2{}_{DP} \sqrt{\frac{S_{SB-\lim}}{S_{C-\lim}}} \tag{6-12}$$

where $D_{DP}$ = diameter of dowel pin
$S_{C\text{-lim}}$ = allowed pressure limit against the dowel pin support
$S_{SB\text{-lim}}$ = allowed bending pressure limit

The length of the dowel pin $L_{DP}$ will be as given by the relationship

$$L_{DP} = 0.77\, D_{DP} \sqrt{\frac{S_{SB-\lim}}{S_{C-\lim}}} \tag{6-13}$$

### 6-4-2. Shape of cutting tools

With round tools, their face surface may sometimes be tapered, which will reduce the cutting force needed for such an operation otherwise. The reasoning behind this statement is logical: As the punch cuts the metal in stages, the smaller amounts of press force are used, spread over an extended period of time.

**Inclined punch or die surface.** Applying a shear to the tool's face is one of the methods used to reduce the necessary cutting pressure. The punch is simply ground under an angle, which would allow for its longest portion to begin cutting first, with the rest of the tool to follow (Fig. 6-12).

Such an arrangement is unfortunately very crude, since the shearing process is not centralized around the tool axis, but rather side

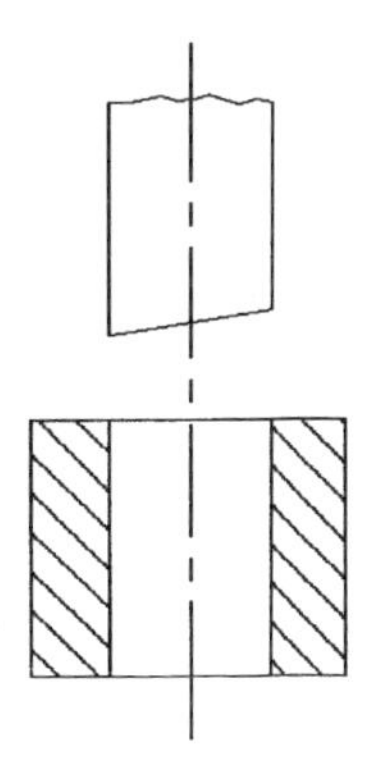

**Figure 6-12** Sheared punch.

shifting away from the center of the tool. A better approach is to shear the punch (or die) toward the center from all sides, as shown in Fig. 6-13. The inclination of surface, being center-oriented, will produce a centered cutting process, with cutting forces evenly spaced around the punch.

Either the punch or the die can have its cutting surfaces sloped, the amount of shear to be equal to the material thickness, with some thin materials using 1.5*t*.

Since the cut-out part or slug will always somewhat resemble the shape of the punch, it is advisable to apply the shear to that part of the tooling, which will be cutting scrap. This way the punch will be sheared if piercing, and the die will be sheared when used for blanking.

The *C method* of tool face alteration, shown in Fig. 6-13, where a slight inclination is produced on the die button, is considered preferable, since the punch here enters the material at both ends of the cut simultaneously, centering itself within the die opening.

**Inclined punches used to retain the material.** In some cases the slanted face surface of punches may be utilized as a stock holder as well. A sample of such an arrangement is shown in Fig. 6-14.

**Radiused tooling for exact cutting.** Rounding of the punch or die cutting edge is used where exact cuts are desired. This process utilizes the plastic deformation of material by pulling it over the rounded edge and severing it off afterward.

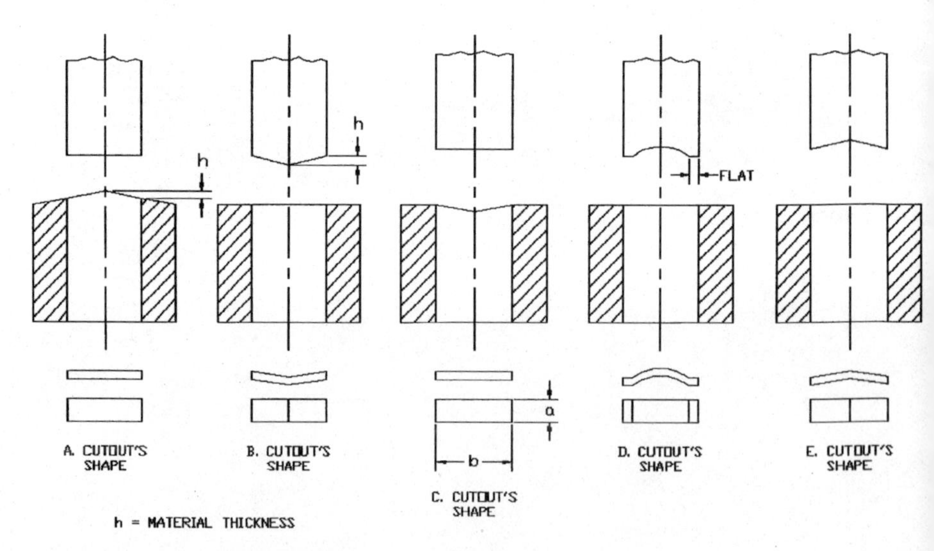


**Figure 6-13** Punch and die shear.

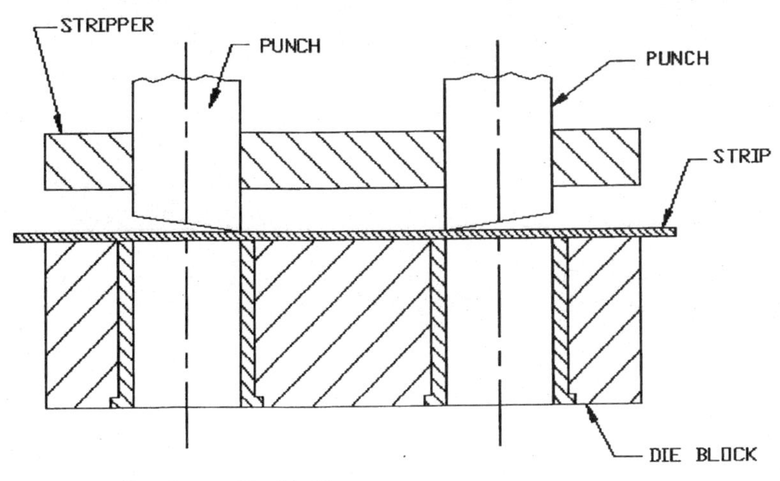


**Figure 6-14** Punches as blankholders.

Again, a rounded punch should be used for blanking and a rounded die for piercing operations. The tooling clearance can be calculated using Eq. (6-14).

$$CL_{\text{side}} = ct\sqrt{S_s} \tag{6-14}$$

where $S_s$ = shear strength of material. See Table 6-3
$t$ = material thickness
$C$ = constant, for material thickness up to 0.125″; $c$ = 0.000174; for material thickness over 0.125″; $c$ = 0.000133

In some cases, where a perfect finish of the cut edge is demanded, a die with a negative amount of cutting clearance can be used, as shown in Fig. 6-15. In such a configuration, the punch is actually bigger than the die opening through which the pierced part is squeezed.

**Shaped punch.** A method of grinding the cutting surface of a punch into a cone-shaped form may be used where some heavy piercing or piercing of hard material is to be done (Fig. 6-16). On coming to contact with the metal material, the point pushes down on it, holding it in a fixed position. Some stretching of the sheet is present when the metal is pulled down by the punch before it is sheared off on contact with the die. A certain amount of thinning of the cut part's edge surface may be expected.

**A half-twist shape.** A spiral shape added to the punch (from its two opposite sides, as shown in Fig. 6-17) may be used to offset the cutting pressure somewhat. As the punch slides down, its cutting is done

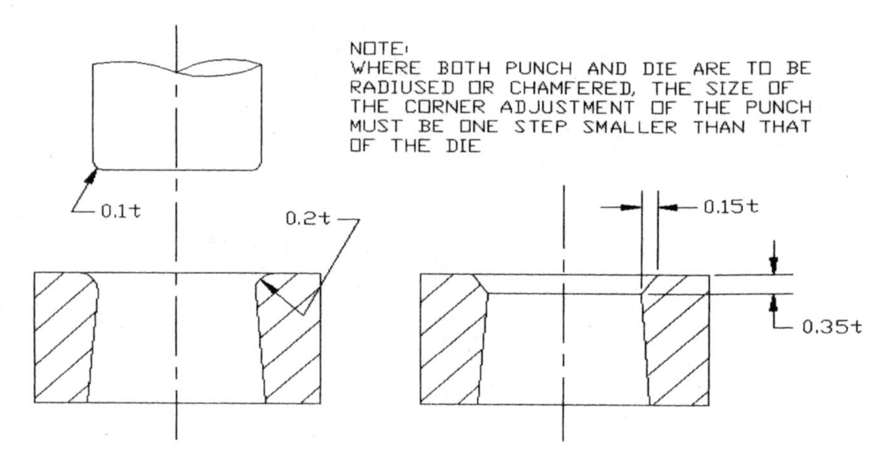


**Figure 6-15** Shapes of cutting die.

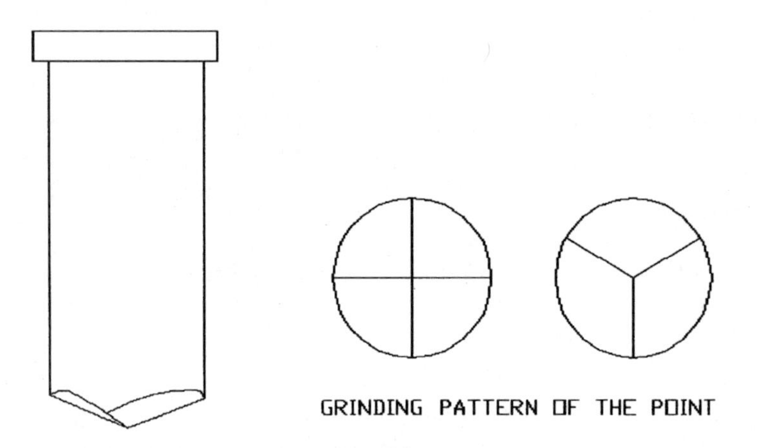


**Figure 6-16** Grinding of the cutting portion.

in stages, following the inclined line of the spiral shape as it comes into contact with the material and the die.

Of course, all these methods of cutting surface alteration have one disadvantage in common: Such tools are quite difficult to regrind, when becoming dull in service. For that reason, flat-faced punches and dies are used more frequently, with shaped tooling being reserved for special applications.

### 6-4-3. Combined tooling

Sometimes, where necessary or appropriate, two operations may be combined. Often this task is achieved by assembling two different tooling elements to make them perform two operations at the same shot.

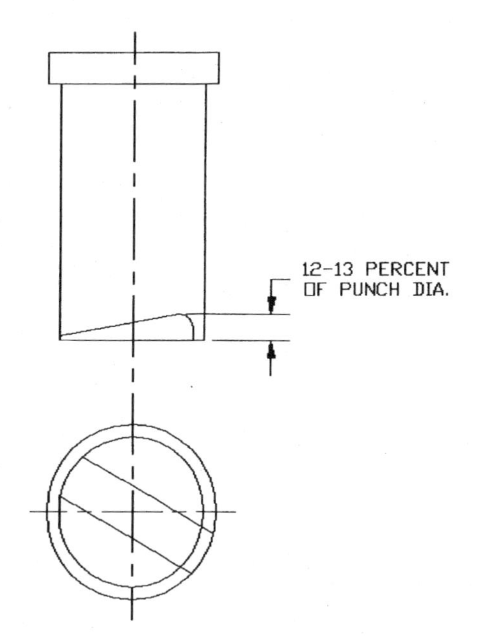


**Figure 6-17** Spiral end of a punch.

**Die insert to allow for shape cutting.** Where a tool of complex shape would be needed, a combination of two or more simple segments may produce the cut of the same shape.

To build a punch to cut the part from Fig. 6-18 in a single stroke of the press would not be difficult. But to make a corresponding die is a completely different task. Where the punch will have the material removed, the die needs the same amount of it added.

As depicted in Fig. 6-18, the needed metal is added in the form of an insert attached to the bulk of the die with a screw.

**Drawing (or forming) punch with a cutoff.** The part shown in Fig. 6-19*a* is drawn and cut off in a single operation. During the work cycle, the material is first drawn into the required depth, equal to the height of the drawing portion of the punch. When the trimming edge comes into contact with the die, the part is cut off.

**Forming and cutting tool.** The punch shown in Fig. 6-19*b* first pierces the opening in the center; then its wider body pulls the metal along on its way down. The spring-loaded lifting pad pushes the finished part up for removal once the punch is on its way up.

**Piercing and coining of the material.** Parts may be pierced first and pressed into a predetermined die shape afterward, as shown in Fig. 6-20*a*. Often such an elaborate arrangement may be superseded by sim-

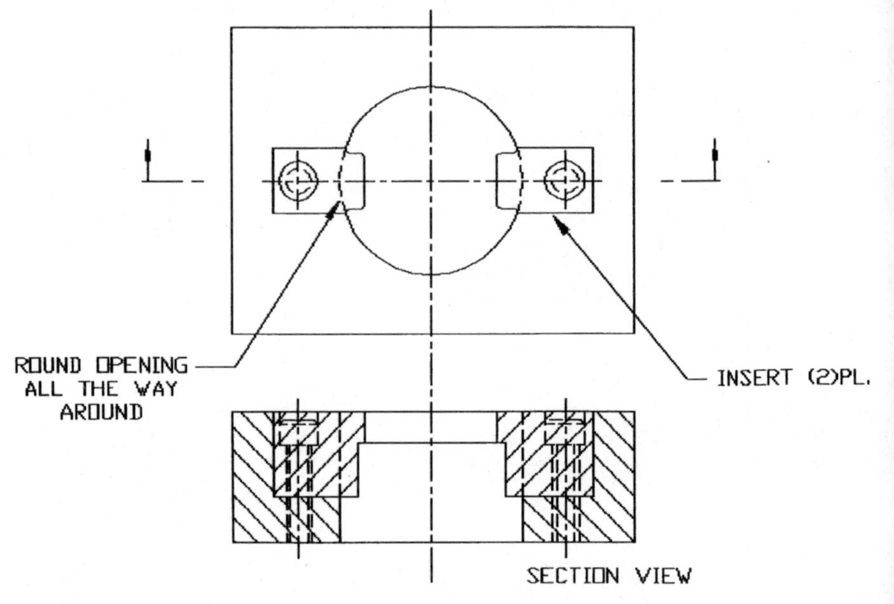


**Figure 6-18** Die with an insert.

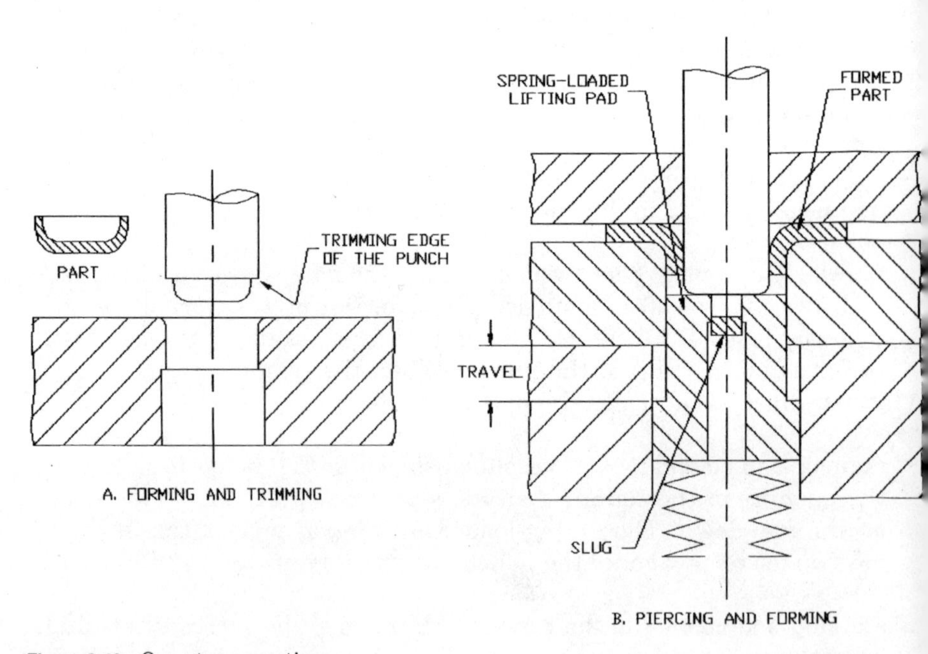


**Figure 6-19** One-step operations.

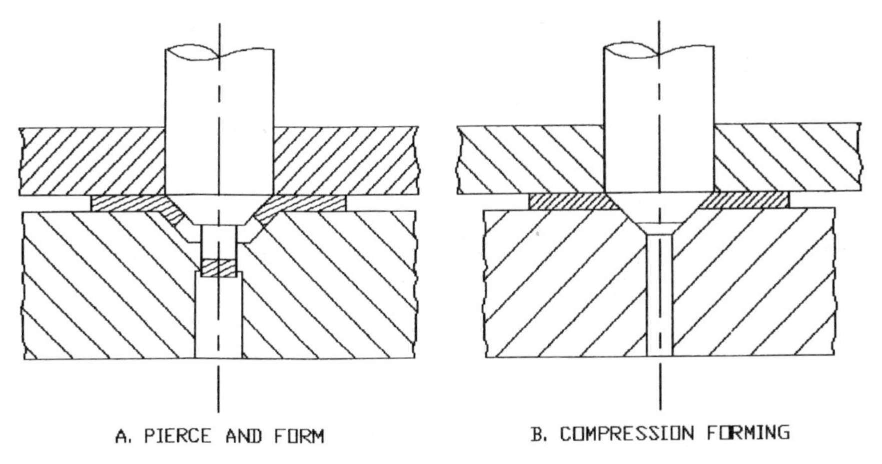


**Figure 6-20** Forming of countersunk shape.

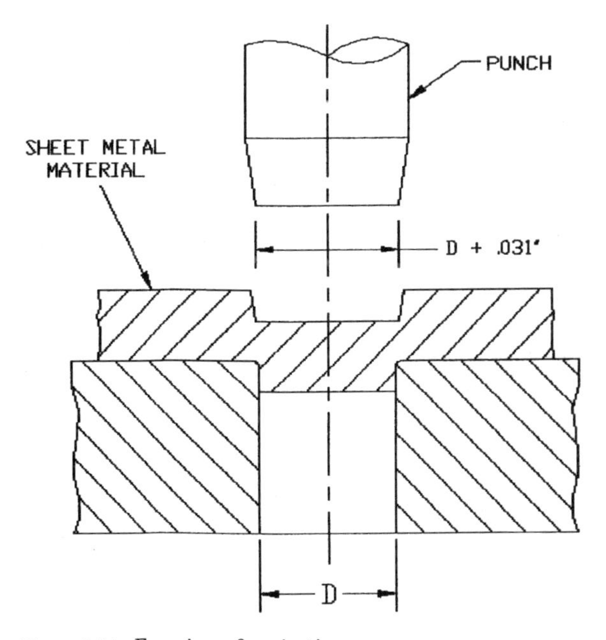


**Figure 6-21** Forming of projections.

ple compression forming, shown in Fig. 6-20*b*. Here a punch enters the precut material, compressing sides of the opening to form a countersink.

Another type of compression forming is shown in Fig. 6-21. In this case, the material is unsupported in the die opening, with the punch being used for indentation purposes only.

**Inclined pressure pad.** A pressure pad with the surface of contact under an angle is used for retention of material to be cut, as shown in Fig. 6-22.

The material is not only held down by the pressure of the pad. As the cutting progresses, it is also prevented from shifting around the cutting tool.

**Curling done in a single operation.** A one-step punch and curling tool may be used for cutting the blank and forming it to the shape shown in Fig. 6-23.

As the punch descends toward the stock, it first pierces through with its long sharp nose. While the cutting progresses, the portion of already parted material begins to imitate the shape of the punch and curls underneath it. When the flat section of the punch finally trims the part off, the curl is already formed.

### 6-4-4. Sequence of operations

In order to reduce the press force necessary for cutting, sequential cutting in stages may be used. This method is quite helpful where too many cuts are to be produced in a single die. Each punch (as shown in Fig. 6-24) is made slightly shorter than the previous one, so that they engage the material at different times. The difference in height is recommended to be equal to such thickness of material the punch must penetrate in order to separate the cutout from the strip. Such a distance is considered $0.33t$ by some.

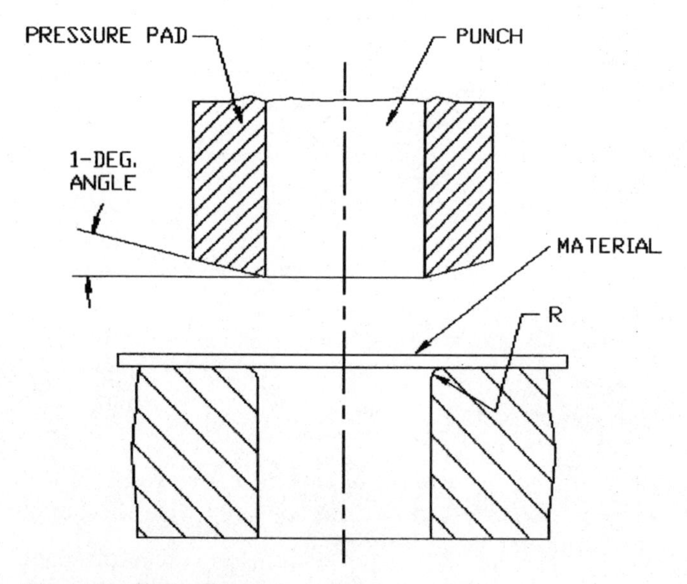


**Figure 6-22** Inclined pressure pad.

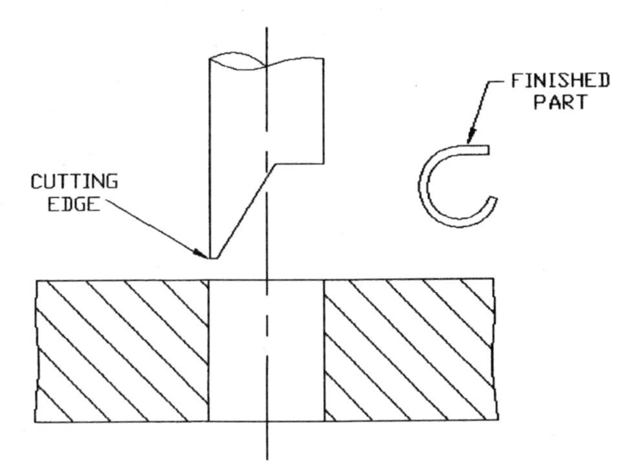


**Figure 6-23** Cutoff and curling.

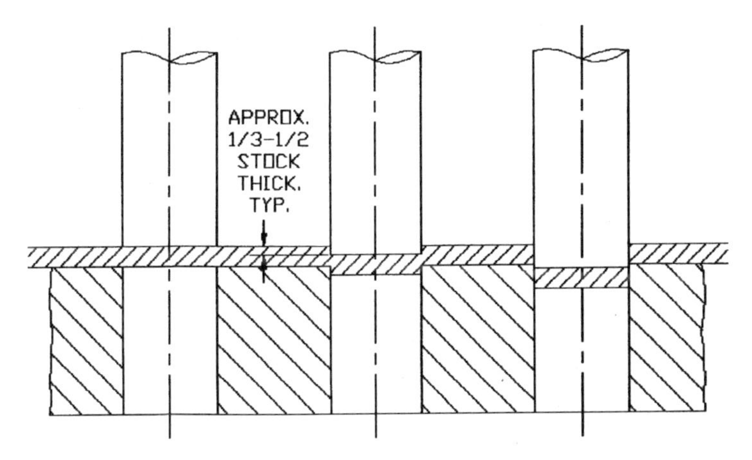


**Figure 6-24** Offset punching.

To calculate the exact cutting depth range for various materials, refer to Sec. 6-8, "Minimum Punch Diameter." Additionally, Table 6-1 depicts the percentage of material penetration needed to achieve severance off the strip.

The approach of sequence punching should be used where piercing of small holes occurs simultaneously with cutting of a large-sized opening. The flow of metal in such cases may be overwhelming if all punches are allowed to descend on the material at the same time. Because of the stress produced within the strip material by the activity of the large punch, all thin punches will certainly have a tendency to break.

As shown in Fig. 6-25, the large punch (or punches) should be long enough to enter the material first. When it penetrates the strip far

**TABLE 6-1 Penetration of Material**

The percentage of stock thickness that must be penetrated by the punch in order to consider the part cut off, or severed off the sheet.

| Material | % Under 0.156″ stock thickness | % Over 0.156″ stock thickness |
|---|---|---|
| Steel: | | |
| Soft | 45–60 | 35–45 |
| Medium-hard | 35–50 | 20–35 |
| Hard | 20–35 | 10–20 |
| Aluminum: | | |
| Soft | 45–65 | 45 |
| Hard | 30–50 | 30 |
| Brass: | | |
| Soft | 50–60 | 50 |
| Hard | 20–30 | 20 |

Note: Thinner materials should use the higher percentage amounts, while thicker materials should use the lower percentage values.

enough to consider the part cut off, thinner punches may be allowed to follow.

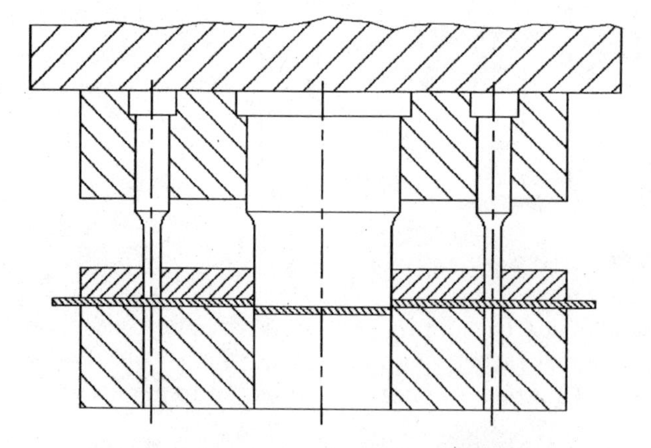

**Figure 6-25** Combination of small and large punches.

## 6-5. Cutting Clearances

The amount of cutting clearance between the punch and the die is of great importance in all sheet-metal work. It is usually given as a percentage of the thickness of cut material, as shown in Table 6-2.

The cutting clearance is *always* added to the die *bushing* of the par-

**TABLE 6-2 Cutting Clearances in Percentages of Material Thickness per Side**

|  |  |
|---|---|
| All materials, up to 0.001″ thickness | 0.0001″ per side |
| All materials, 0.001–0.025″ thick | 5% per side |
| Blue steel, up to 0.025″ thick | 5–8% per side |
| Copper, brass, soft steel, 0.025″ thick and up | 5% per side ID<br>8% per side OD |
| Medium hard steel, 0.025″ thick and up | 6% per side ID<br>9% per side OD |
| Stainless steel, carbon steel, Cr-Ni steel, 0.025″ thick and up | 8% per side ID<br>10% per side OD |

Clearances for aluminum to be used according to its comparable hardness.

ticular cutting station. The punch, as stated previously, has the exact size of the hole to be cut and a tolerance of

$$\begin{matrix} +0.0002 \\ -0.0000 \end{matrix}$$

The die opening, which is to contain the punch, will use its size, add the amount of cutting clearance to it, and attach a tolerance range, most often similar to that of the punch, or

$$\begin{matrix} +0.0002 \\ -0.0000 \end{matrix}$$

As an example, with the punch size of

$$1.0000 \begin{matrix} +0.0002 \\ -0.0000 \end{matrix} \text{ DIA}$$

the die opening at 8 percent cutting clearance will be

$$1.0800 \begin{matrix} +0.0002 \\ -0.0000 \end{matrix} \text{ DIA}$$

Even though the correct cutting clearance is recommended to be somewhere between $0.01t$ and $0.10t$, some manufacturers, especially those producing large-sized sheet-metal parts, use clearances up to $0.6t$. Such a gap is excessive, and it not only produces rough cuts with severe burrs, it also may cause a distortion of material in its immediate vicinity. This portion of material, once affected by such gross structural changes from within, becomes an obstacle to all successive operations done on that part.

In order to evaluate the possibility of the material deformation, Eq. (6-15) may be used for all sheets thinner than 0.093″.

$$f_L = ct\sqrt{S_s} \tag{6-15}$$

where $f_L$ = deformation, maximum limit
$t$ = thickness of pierced material
$S_s$ = shear strength of the material
$c$ = correction factor

The correction factor $c$ may be used from within the range of

0.00004 for the best surface quality of the cut

0.00030 for the least cutting force and least surface quality

Material testing has proved that all deformation limits above $0.3t$ not only cause an excessive distortion of the cut, they also exhaust the plasticity of the material in the surrounding area while altering some tensile properties as well.

**Calculated cutting clearances.** Cutting clearances $CL$ may also be calculated, using Eq. (6-16).

$$CL_{\text{side}} = ct\sqrt{S_s} \tag{6-16}$$

where $t$ = material thickness
$S_s$ = shear strength of material; see Table 6-3
$c$ = constant, for materials up to 0.125″ thick, $c$ = 0.00012 to 0.0008, for materials above 0.125″ thick, $c$ = 0.00017 to 0.0010

## 6-6. Punching and Blanking Pressure

The amount of force needed to punch out an opening in sheet metal has to be calculated in order to determine the size of a press to use.

Somewhere at the beginning stage of die design, the total amount of pressure, or tonnage, needed for carrying out all die operations has to be determined. On the basis of this tonnage, the appropriate press should be chosen, the data sheet of which supplies information concerning the maximum die size, shut height, and stroke length for further advancement of the die design.

The press tonnage should be figured out quite accurately, as the choice of a correct press size is relevant. Should a press of too low a ton-

**TABLE 6-3 Shear Strength of Material, Lb/In$^2$**

| Material | Shear strength |
|---|---|
| Tin, rolled sheet | 5,000 |
| Zinc: | |
| Rolled sheet | 18,000 |
| Hard rolled | 20,000 |
| Copper | 25,000 |
| Brass | 35,000 |
| Bronze | 35,000 |
| Aluminum: | |
| Soft | 10,000 |
| Half-hard | 20,000–30,000 |
| Full-hard | 40,000 |
| Steel: | |
| 0.1% carbon, hot rolled | 35,000 |
| 0.1% carbon, cold rolled | 45,000 |
| 0.2% carbon, hot rolled | 45,000 |
| 0.2% carbon, cold rolled | 55,000 |
| 0.3% carbon, hot rolled | 55,000 |
| 0.3% carbon, cold rolled | 65,000 |
| Stainless steel | 70,000–100,000 |
| High-carbon steel | 75,000–85,000 |

NOTE: 2,000 lb = 1 ton.

nage be chosen, excessive stresses may be created during the operating process, often resulting in some sort of breakage. With too powerful a press for the given job, its extra tonnage will be inefficiently wasted.

Naturally, an ideal choice would be a press having the exact or just very slightly higher tonnage than whatever's needed.

Tonnage can be calculated by using Eq. (6-17). It is assumed that no shear to either punch- or die-cutting surface is applied.

$$P_{BL\,(\text{lbs})} = LtS_s \tag{6-17a}$$

$$P_{BL\,(\text{tons})} = P_{\text{lbs}} \cdot 2000 \tag{6-17b}$$

where $L$ = total length of a cut (or cuts), in
$t$ = material thickness
$S_s$ = shear strength of material, per Table 6-3

The length of a cut should be a total length of all cut edges, be it straight lines or circumferential values.

**Inclined cutting surfaces.** To obtain a cutting force, needed for the application of sheared punches and dies, a different set of formulas have to be used. For cutting with a sheared punch, such as the one shown earlier in Fig. 6-12, where the shape of the punch is rectangular, the following equation is applicable:

$$P_{BL} = t\left(a + 2C_1 \frac{bt}{h}\right) S_s C_2 \tag{6-18a}$$

where $t$ = material thickness
$a$ = shorter side of the rectangular cut
$b$ = longer side of the rectangular cut
$h$ = height of the tool's inclination
$S_s$ = shear strength of the material
$C_1$ = amount of material penetration, from Table 6-1
$C_2$ = condition of cutting surfaces, ranging between 1.2 and 1.5

A rectangular cutout, using a punch with shear applied to its two sides, such as the one shown earlier in Fig. 6-13*E*, will utilize the formula

$$P_{BL} = 2t\left(a + C_1 \frac{bt}{h}\right) S_s C_2 \tag{6-18b}$$

A circular opening, with the shear applied to the die, as shown earlier in Fig. 6-13*A*, where the height of the inclined portion is equal to the material thickness, can be calculated as follows:

$$P_{BL} = \frac{2}{3}\pi dt\, S_s C_2 \tag{6-19}$$

where $d$ = punch diameter and $h = t$.

**Finding the center of pressure of a blanking station.** Where a blanking die is too large, it would better be positioned on the same axis as the ram of a press. This way the center of the press force and the center of its distribution throughout the blanking station will coincide. To position a powerful blanking station slightly off the center may result in greater than usual wear of die bushings, caused by the die's inclination in the direction of the lesser support.

In order to place a complicated shape dead on center, first the center must be located. The method of calculating the center of pressure of an irregular shape is demonstrated on the sample shown in Fig. 6-26.

First, the shape must be broken down into single lines without considering the geometrical entities they form. *X-Y* axes should be positioned in the corner of a shape to establish the zero position.

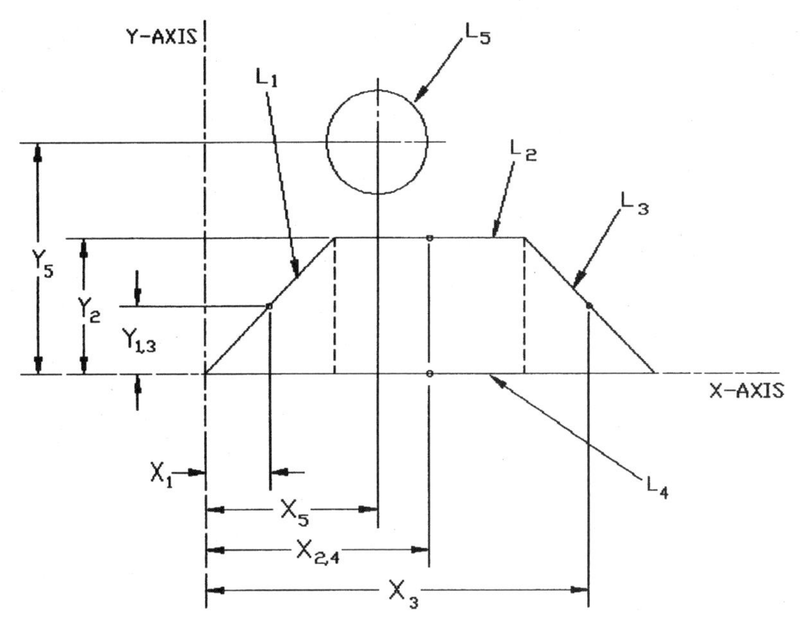


**Figure 6-26** Center of pressure.

The length of each line is to be calculated, along with the distance of its center off the zero point in both $x$ and $y$ dimension. The length of each line should be called $L$, and its subscript determines the sequential number of the line.

The center of shape should then be calculated separately along each axis. The formulas are as follows:

$$X_{\text{total}} = \frac{L_1X_1 + L_2X_2 + \cdots + L_nX_n}{L_1 + L_2 + \cdots + L_n} \tag{6-20}$$

and

$$Y_{\text{total}} = \frac{L_1Y_1 + L_2Y_2 + \cdots + L_nY_n}{L_1 + L_2 + \cdots + L_n} \tag{6-21}$$

## 6-7. Stripping Pressure

A stripping pressure calculation helps to determine the correct amount of the spring pressure a spring-loaded stripper must produce. It usually varies between 3 and 20 percent of the blanking pressure and can be figured out using Eq. (6-22):

$$P_s = 3.5\,Lt \tag{6-22}$$

where all values are the same as with the blanking pressure.

The amount of delivered stripping pressure depends mainly on the proper design and proper function of springs, which are supporting the stripper's mass. The second influential factor is the thickness of processed material, which governs the demand for stripping pressure approximately as shown.

| Stock thickness up to, in | Stripping pressure as percentage of cutting pressure |
|---|---|
| 0.042 | 3–8 |
| 0.093 | 8–10 |
| 0.156 | 10–13 |
| 0.250 | 13–20 |

The calculation above is but an approximation of the actual pressure needed to strip the part. The precise amount is very difficult to establish, since it is influenced by too many variables. The condition of the tooling, cutting clearance, type of material, and lubrication of tooling are just several out of many factors influencing the amount of stripping pressure needed.

Sheared punches may reduce the blanking pressure, but they have no effect on the stripping pressure requirements. However, staged punching, where the height of cutting tools is offset, will produce a decrease in demand of stripping pressure. Two levels of punches would halve the amount of stripping pressure otherwise needed. Three levels of punches will use up one-third of the pressure, etc.

## 6-8. Scrap and Hole Size Recommendations

Blanks, when positioned on a strip, certainly should not be spaced too far apart, for a large waste of material will result. However, placing them too close to each other may create a different kind of waste, that of a ruined strip, ruined tooling, and ruined die. The proper spacing is very important, and its amount depends on the material thickness above everything else (Figs. 6-27 and 6-28).

The proper spacing should be established by using Table 6-4, where the values are given with regard to the material thickness.

An additional distinction according to the cut material is provided in Table 6-5.

Openings, when punched too close to the edge, may produce bulging and distortion. However, where the holes must appear closer to the edge than recommended and a design change cannot be enforced,

**TABLE 6-5 Scrap Allowance for Blanking**

| Material | When length of skeleton segment between blanks or along edge is equal to or less than $2T$ | | | When length of skeleton segment between blanks or along edge is greater than $2T$ | | |
|---|---|---|---|---|---|---|
| | Thickness of stock $T$, in | Edge of stock to blank, in | Between blanks, same row, in | Thickness of stock $T$, in | Edge of stock to blank, in | Between blanks, same row, in |
| Cloth, paper (bond, manila, red rope, etc.) | All | $\frac{3}{32}$–$\frac{5}{32}$ | $\frac{3}{32}$–$\frac{5}{32}$ | All | $\frac{3}{32}$–$\frac{5}{32}$ | $\frac{3}{32}$–$\frac{5}{32}$ |
| Felt, leather, soft rubber | Under 0.062<br>Above 0.062 | $\frac{1}{16}$<br>$T$ | $\frac{1}{16}$<br>$T$ | Under 0.062<br>Above 0.062 | $\frac{1}{16}$<br>$T$ | $\frac{1}{16}$<br>$T$ |
| Hard rubber, celluloid, Protectoid | All | $0.4T$* or 0.040 min | $0.4T$* or 0.040 min | All | $0.4T$* or 0.040 min | $0.4T$* or 0.040 min |
| Metals, general: | | | | | | |
| Standard strip stock | Under 0.021<br>0.022–0.055<br>Above 0.055 | 0.050<br>0.040<br>$0.7T$ | 0.050<br>0.040<br>$0.7T$ | Under 0.044<br><br>Above 0.044 | 0.050<br><br>$0.9T$ | 0.050<br><br>$0.9T$ |
| Extra-wide stock and weak scrap skeleton | Under 0.042<br>Above 0.042 | 0.060<br>$1.4T$ | 0.050<br>$1.2T$ | Under 0.033<br>Above 0.033 | 0.060<br>$1.8T$ | 0.050<br>$1.6T$ |
| Stock run through twice | Under 0.042<br>0.043–0.055<br>Above 0.055 | 0.060<br>$1.4T$<br>$1.4T$ | 0.050†<br>0.040<br>$0.7T$ | Under 0.033<br>0.034–0.044<br>Above 0.044 | 0.060<br>$1.8T$<br>$1.8T$ | 0.050†<br>0.040<br>$0.9T$ |
| Stock run through twice and first and second rows of blanks interlock | Under 0.042<br>Above 0.042 | 0.060<br>$1.4T$ | 0.050*<br>$1.4T$† | Under 0.033<br>Above 0.033 | 0.060<br>$1.8T$ | 0.050*<br>$1.8T$† |

**TABLE 6-5 Scrap Allowance for Blanking (Continued)**

| Material | When length of skeleton segment between blanks or along edge is equal to or less than $2T$ | | | When length of skeleton segment between blanks or along edge is greater than $2T$ | | |
|---|---|---|---|---|---|---|
| | Thickness of stock $T$, in | Edge of stock to blank, in | Between blanks, same row, in | Thickness of stock $T$, in | Edge of stock to blank, in | Between blanks, same row, in |
| Mica, Micanite, phenol fabrics, and phenol fiber | All | $0.6T$* or 0.060 min | $0.6T$* or 0.060 min | All | $0.6T$* or 0.060 min | $0.6T$* or 0.060 min |
| Permalloy | All | 0.060 | 0.060 | All | $T$‡ | $T$‡ |
| Pressboard, asbestos board | Under 0.031<br>Above 0.031 | 1/16<br>$2T$ | 1/16<br>$2T$ | Under 0.031<br>Above 0.031 | 1/16<br>$2T$ | 1/16<br>$2T$ |
| Steel (silicon, spring, stainless) | Under 0.042<br>Above 0.042 | 0.060 min<br>$1.4T$ | 0.060 min<br>$1.4T$ | Under 0.033<br>Above 0.033 | 0.060 min<br>$1.8T$ | 0.060 min<br>$1.8T$ |
| Vulcanized fiber | All | $0.8T$* or 0.080 min | $0.8T$* or 0.080 min | All | $0.8T$* or 0.080 min | $0.8T$* or 0.080 min |

*Allow 0.060 between blanks at first and second rows.

†Allowance between blanks in same row and also between blanks of first and second rows.

‡When the blank edge parallels the edge of stock or between blanks more than four times the thickness, the allowance of 1.8 thickness of stock applies.

SOURCE: Frank W. Wilson, *Die Design Handbook,* New York, 1965. Reprinted with permission from The McGraw-Hill Companies.

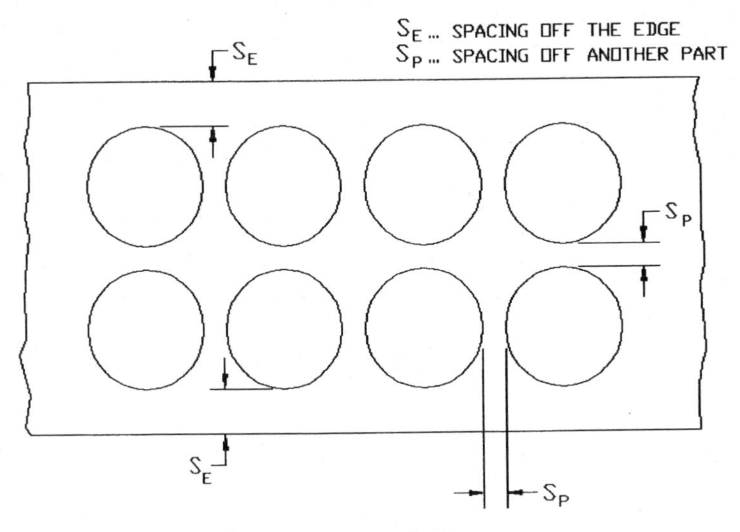


**Figure 6-27** Distances between parts on sheet.

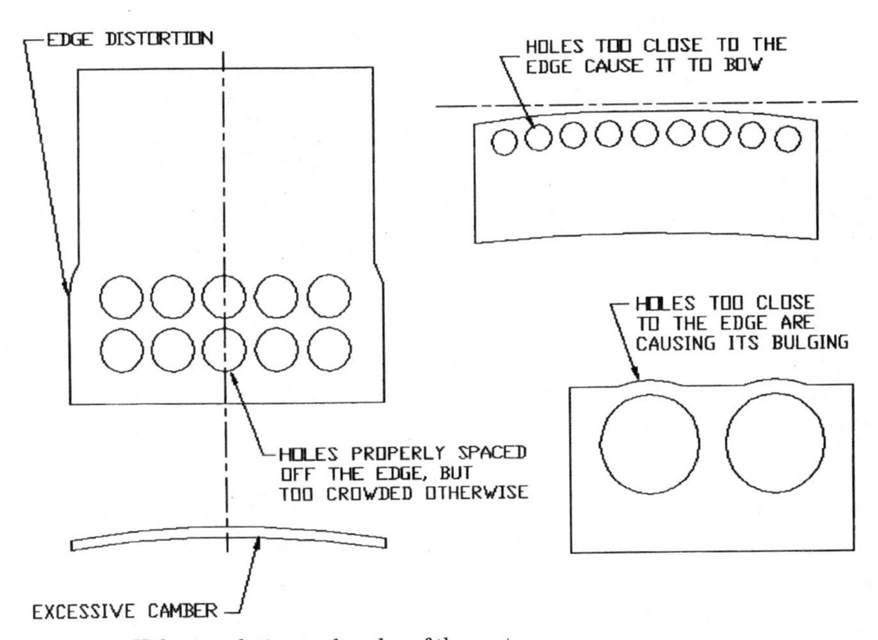


**Figure 6-28** Holes in relation to the edge of the part.

**TABLE 6-4 Scrap Allowance in Multiple Parts**

| Material thickness | Allowance ranges for metals | |
|---|---|---|
| Up to 0.031″ thick | $S_p$ = 0.031″ | $S_e$ = 0.062″ |
| 0.031–0.062″ thick | $S_p = t$ | $S_e$ = 0.062″ |
| 0.062″ and up | $S_p = t$ | $S_e = t$ |

where $t$ = material thickness
$S_p$ = space between parts
$S_e$ = space off the edge

The minimal $S_p$ value may change with different materials:

| | |
|---|---|
| Medium-hard steel | 0.031″ |
| Hard steel | 0.035″ |
| Bronze, hard brass | 0.045″ |
| Other brass | 0.048″ |
| Aluminum | 0.055″ |
| Plastics | 0.062–0.078″ |

Considering these variations in the minimal required distance between parts, the $S_p$ and $S_e$ values will have to be adjusted for the percentage difference attributable to the material change.

punching the openings first and subsequently producing the edge cut-off will certainly eliminate the part's tendency to distortion.

**Minimum punch diameter.** The old saying that the minimum hole diameter should be equal to the material thickness may not be true in all cases. The reason for a discrepancy is that this rule often ignored additional variables such as the material shear strength and the compressive strength of the punch. Actually, the minimum punch diameter must be within such a range that the compressive strength of the punch material will be greater than the force needed to punch the opening.

The maximum allowable compressive stress $S_c$ of the punch depends on the type of material used, its hardness, and other qualities. A tool grade steel, oil hardened and shock resistant, will take about 300,000 lb/in$^2$ before it will break. Therefore, a safe range of 200,000 lb/in$^2$ can be easily assumed, with regard to the extended tooling life.

With the shear strength of the material known, the punch diameter can be easily figured out by using the graph in Fig. 6-29. Otherwise for exact calculation, Eq. (6-23) must be used to evaluate the maximum compressive stress on the punch

$$S_c = \frac{tS_sL}{A} \tag{6-23}$$

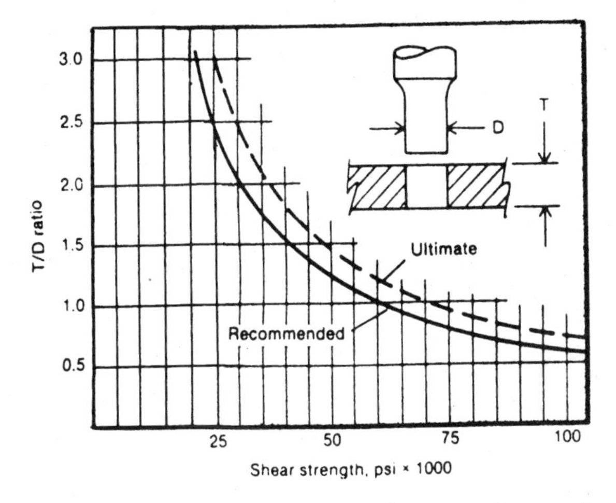


**Figure 6-29** Ratio of material thickness to the punch diameter as a function of shear strength. (*From O. D. Lascoe,* "Handbook of Fabrication Processes," *published by ASM International. Reprinted with permission from ASM International, Materials Park, OH.*)

where $S_c$ = compressive stress in the punch
$S_s$ = shear strength of the punched material
$t$ = material thickness
$L$ = length of the cut, in
$A$ = cross-sectional area of the punch, calculated per condition in Fig. 6-30

The result should be compared with the chart in Fig. 6-31. The minimal hole size should be read off the chart, with regard to the given material thickness.

To use the graph from Fig. 6-31, the thickness of the material has to be located on the vertical scale. Follow its line horizontally up to the lower edge of the area representing the material to be pierced. From the point of intersection, the corresponding punch diameter size can be read below.

The upper edge of each material's area represents the breaking range of the punch.

An overloaded punch will expand in size as its elastic limit is exceeded. An increased stripping force will have to be applied against such a tool, which will actually cause its breakage. Some punches may buckle under a load, breaking afterward.

Supported punches such as those shown in Figs. 6-32 and 6-4 will not fail as described. These tools would perform well up to a thickness-to-diameter ratio of 2:1 when piercing the mild steel.

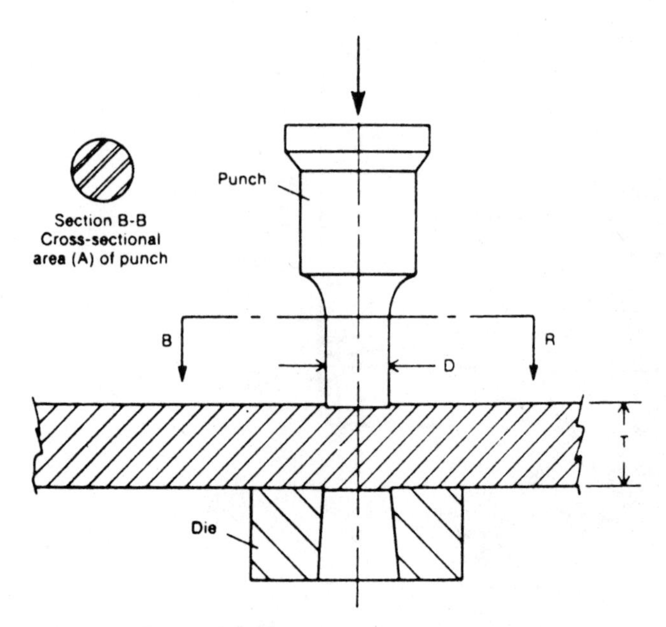


**Figure 6-30** Punch diameter to material thickness. (*From O. D. Lascoe,* "Handbook of Fabrication Processes," *published by ASM International. Reprinted with permission from ASM International, Materials Park, OH.*)

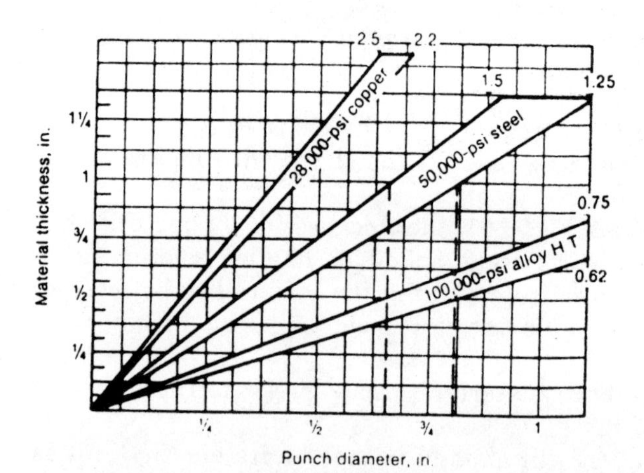


**Figure 6-31** Minimum punch diameter (hole size) as a function of material thickness. (*From O. D. Lascoe,* "Handbook of Fabrication Processes," *published by ASM International. Reprinted with permission from ASM International, Materials Park, OH.*)

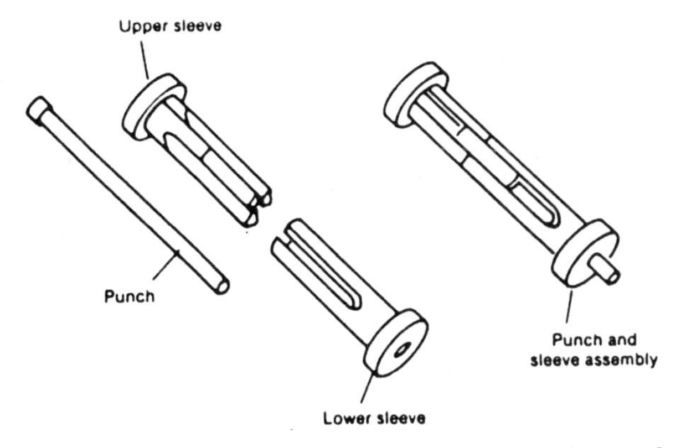


**Figure 6-32** Guided punch assembly (Durable Punch and Die Co.). (*From O. D. Lascoe,* "Handbook of Fabrication Processes," *published by ASM International. Reprinted with permission from ASM International, Materials Park, OH.*)

## 6-9. Practical Advice and Restrictions

All kinds of seemingly small details, trivial observations, and minor facts may often alleviate great problems out there in production. Such little notions are included here for the benefit of the reader.

**Punching of small notches with larger punches.** Where notches of various shapes have to be produced in a die, designers and toolmakers should never be tempted to take shortcuts. For a V notch, often a square punch is taken; for a half-round, a full-round punch is used (Figs. 6-33 and 6-34).

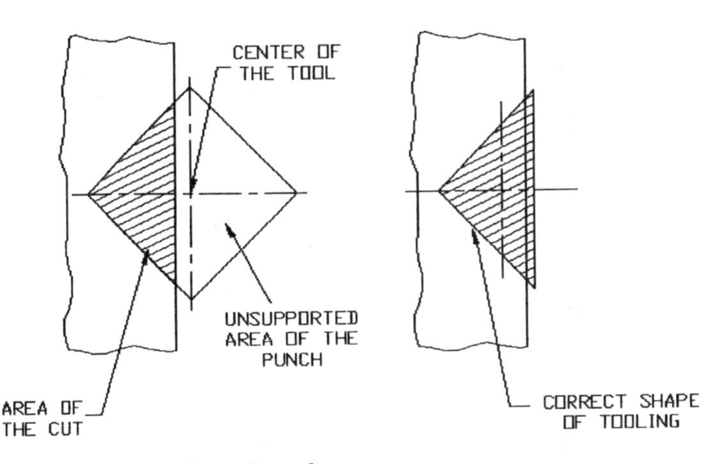


**Figure 6-33** 90° notching of an edge.

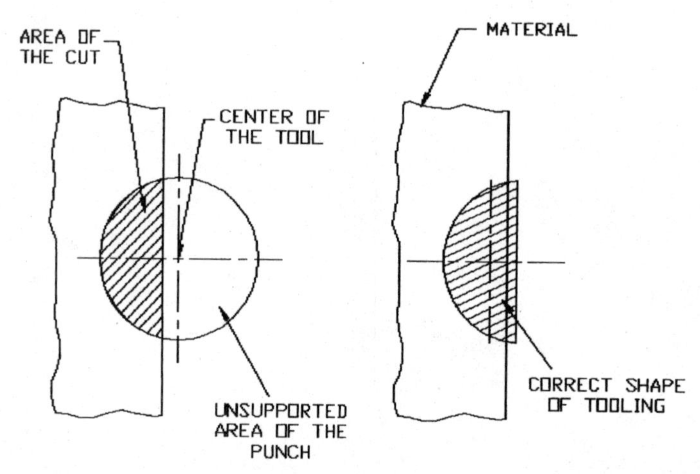


**Figure 6-34** Half-round notching of an edge.

Such solutions usually create more problems than they solve. To be adequately utilized, where no damage during a prolonged die operation is caused to it, a punch must cut through the material with at least 75 percent of its cutting surface area. Anything less than that will cause it to sway aside, followed by a tendency to buckling, bending, and subsequent breakage.

For that reason, all special shapes should either utilize special-shaped tooling or be performed in a sequence that provides for the utilization of at least three-quarters of their cutting area.

**Long and narrow tooling.** Using long and narrow tooling along the edge of a strip should be avoided wherever possible. Such punches, even though often cutting with a major percentage of their total punch area, will tend to be negatively affected by this operation, especially if the center of their shanks is off the sheet. An alteration of die design, with a larger distance between the edge of the cut and that of the sheet, should be chosen as shown in Fig. 6-35.

The long and narrow punch is always more secure if the whole surface of the tool is used for cutting.

**Cutting of shapes.** Where a complex shape is to be produced in the die, with the final cleanup of scrap to be provided at the end, the designer should beware the dangers not only of unsupported cutting but also of inadequate scrap removal.

In Fig. 6-36, the three rectangular cuts produced with tool (*A*) provide for the main ribbing of a part in the first station. Moving on, the central section is removed by using a rectangular punch (*B*). The

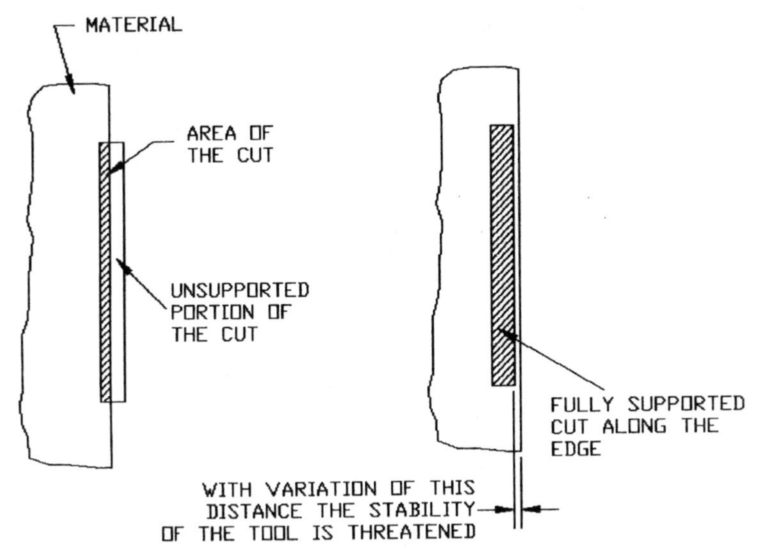


**Figure 6-35** Longitudinal cutting along an edge.

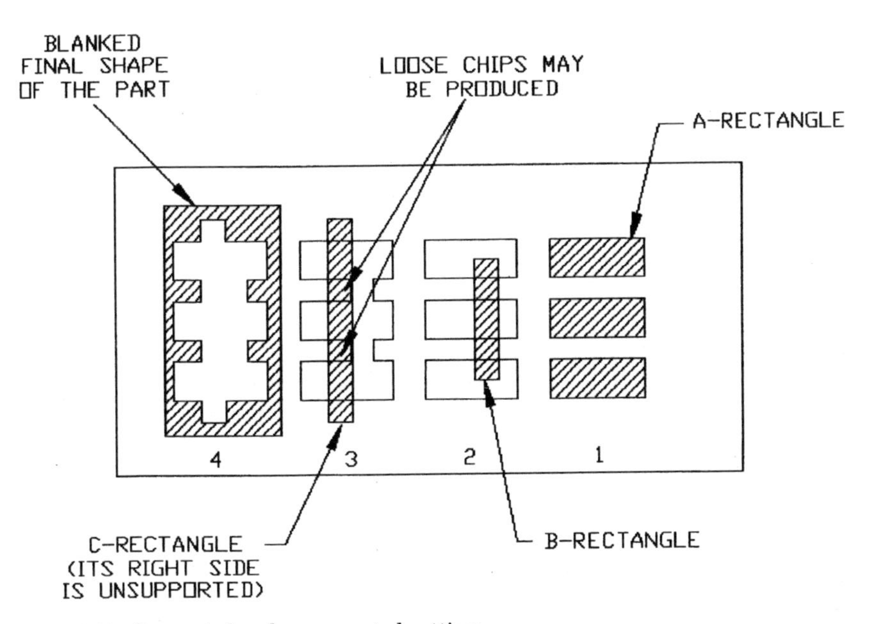


**Figure 6-36** Supported and unsupported cutting.

remaining portion is cut off in station 3 by a rectangular tool (*C*). The final blanking is performed in station 4.

The cut provided in die section 2 will certainly remove the material, with the load on the punch evenly distributed and therefore acceptable. On the sheet's arrival in the next station, however, with a portion of wall between horizontal ribs already removed, the remainder of that section, only partially attached to the sheet, may be sticking out either way with dependence on the type of material and stripping arrangement. Also, such small sections will certainly not provide the needed support to the long punch (*C*), which will be cutting with its left edge only.

The material removed in station 3 may further complicate the die operation, should some loose chips fail to leave through the opening in the die, as may sometimes happen. These small pieces may remain scattered over the die surface, and threaten its function, endangering the quality of parts. Another possibility of loose chips lying around the die surface, such as those created by piercing of 45° notches with improper tooling, has to be observed (Fig. 6-37). These chips of scrap may obstruct the work of a die by impairing its function or by attacking the surface of a part.

**Cutting of plastic.** Plastics, laminated materials, phenolics, and rubber can all be punched, blanked, or sheared in a die. Naturally, the rules for such production are slightly different from those for cutting of sheet-metal pieces.

The clearance between the punch and die should be an absolute minimum, especially when punching cold material: 1 to 2 percent per

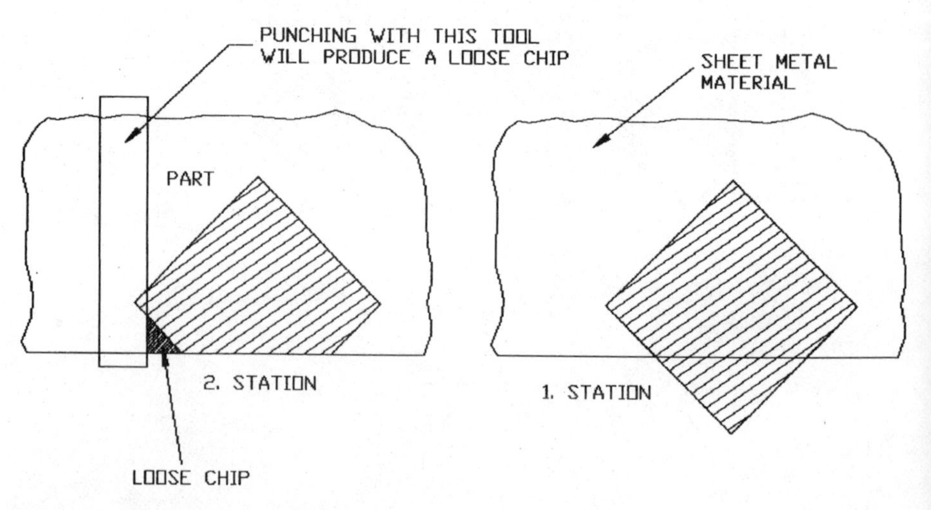


**Figure 6-37** Chips within the die.

side for each 0.031″ of material thickness will suffice. With some plastics, a slip fit between the punch and die may be needed.

The clearance between the punch and the stripper should be kept at an absolute minimum as well.

Some plastics or laminates may be punched cold; some need to be heated. It is advisable to check with the manufacturer for the particular material's preference.

The punch may often need to be made slightly smaller than the required size of the hole, as the opening often closes up on its retrieval. Approximately 0.001″ per diameter reduction in punch size is advisable.

Where rough edges of the punched opening are obtained, a shaving operation may be needed. Sometimes an alternative material-removing method should be chosen to alleviate such a problem.

Chapter

# 7

# Blank Calculation or Flat Layout

## 7-1. The Importance of Flat Layout or Blank Layout

The importance of an accurate flat layout has been stressed throughout the preceding text. In die work and any sheet-metal work in general, the importance of an accurate and dimensionally correct flat layout cannot be overemphasized. Many problems may be avoided if a full-sized or scaled layout is made first and the part's manufacture is evaluated on the basis of it, instead of referring to the bent-up drawing, which—after all—may not be manufacturable.

### 7-1-1. Flat layout development and calculation

When making a flat layout of a complex part, it always helps to start from a certain side, one that seems to be basic or the most complex, and unfold the remaining portions bend after bend.

A bracket, shown in Fig. 7-1, is simple enough to serve as an example of the unfolding technique. First, the *A* flange may be flattened out, to become flush with the vertical portion. This should be done visually, just by looking at the illustration and imagining the flange rotating around its pivot point, as if retained by hinges.

Next, the whole vertical segment should be folded down, to become a flat continuation of the horizontal section, as shown in Fig. 7-2.

Such a flattened portion may be sketched, adding other segments to it as they are unfolded. To provide the flat layout with dimensions, we start off a single corner and do all dimensioning with regard to that location. However, when checking the numbers later, we calculate

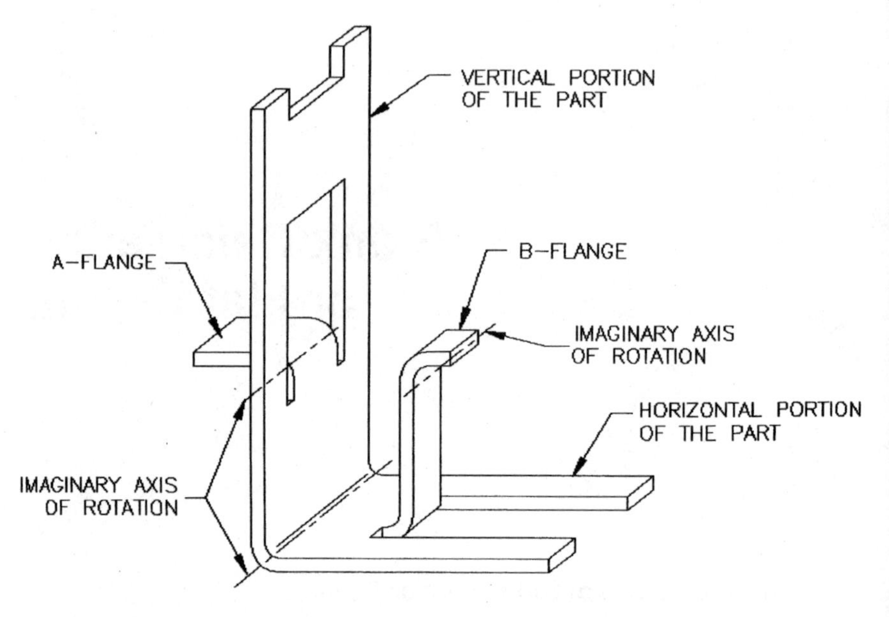


**Figure 7-1** Sample bracket.

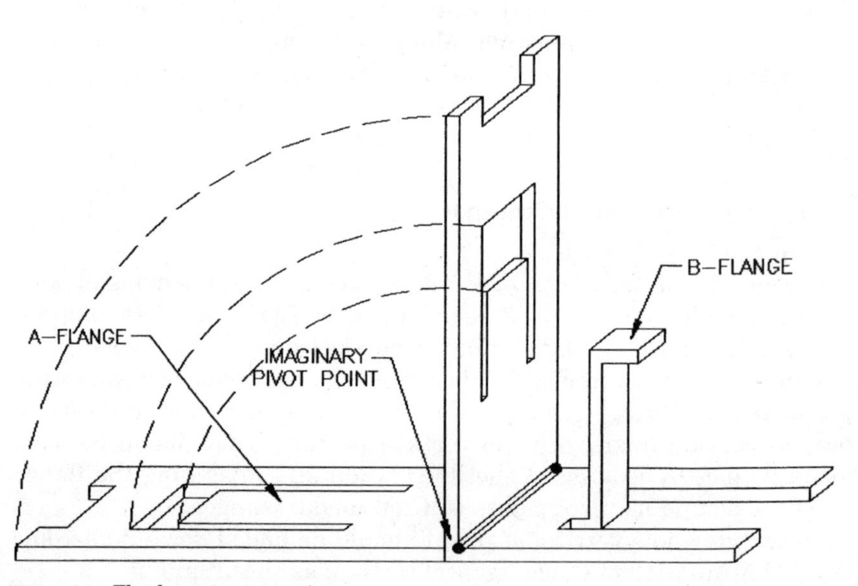


**Figure 7-2** Flat layout preparation.

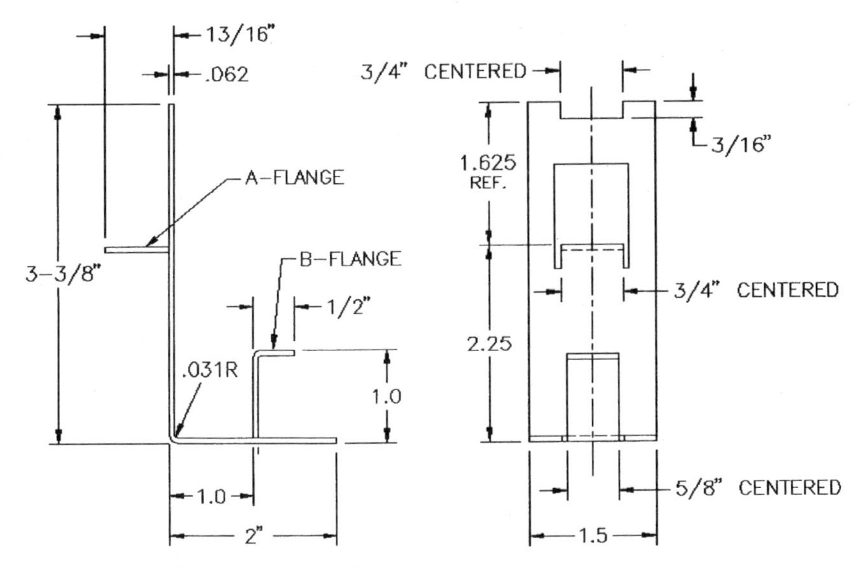


**Figure 7-3** Bracket, bent up with dimensions.

them off another location and see if the results will be the same. A sample of flat layout for the bracket shown previously is included in Fig. 7-4.

When calculating the dimensions in flat, flange *A* may be assessed off the top of the vertical section (see Fig. 7-3) by subtracting its height from the overall dimension, or

$$3.875 - 2.25 = 1.625$$

This dimension is included in Fig. 7-3 as a reference added for flat layout purpose only. On an actual drawing such a dimension will not be appropriate, as it is already expressed by the difference between 3.875 and 2.25.

Since the depth of relief slots is not indicated, it is probably not overly relevant, and in such a case we can make these cuts as deeply as necessary. Usually, a depth equivalent to the distance measured off the outside surface to the center of the bend radius plus one material thickness, with up to $2t$, will suffice, as shown in Fig. 7-5.

The width of the relief slot is not specified either, which similarly allows for a variation. Usually a size of at least one material thickness is appropriate.

The depth of the relief slot's bottom must therefore be calculated by using the dimensions

$$1.625 + \text{OuterRad} + 1.5t = 1.625 + 0.093 + 0.093 = 1.812$$

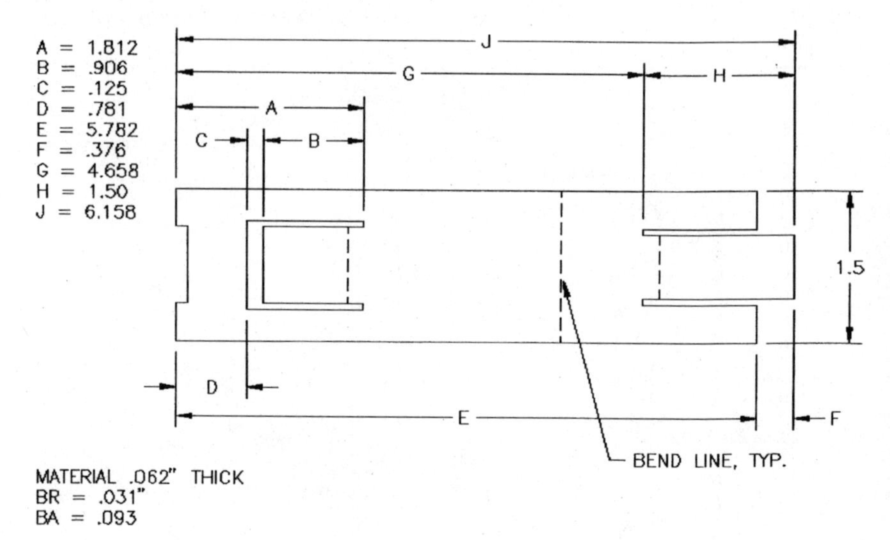


**Figure 7-4** Bracket, flat layout.

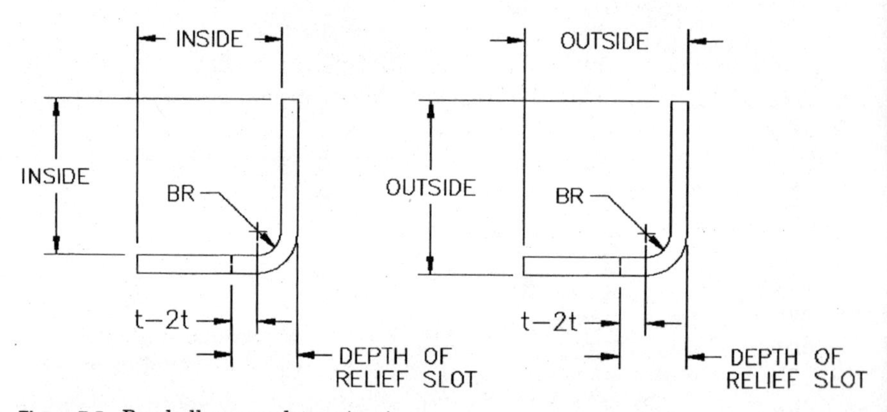


**Figure 7-5** Bend allowance determination.

The length of the flange can be calculated by adding the depth of the relief slot measured off the top of the flange ( = 0.187) plus the length of the flange ( = 0.812) and subtracting one bend allowance, or

$$0.187 + 0.812 - 0.093 = 0.906$$

Bend allowance may be taken off the charts in Chap. 8, "Bending and Forming Operations." Table 7-1 is an additional bend allowance chart, widely used in sheet-metal fabrication, where press-brake bending with V-type inserts is prevalent. The chart is added to permit a wider range of comparison of dimensions in flat, which sometimes are advisable to calculate using several different techniques.

**TABLE 7-1 Bend Allowance Chart for Sheet-Metal Material, Press-Brake Bending**
Use the outside dimensions of each bend and subtract the bend allowance shown

| Bend | Material thickness, in | | | | | | | | | | | | |
|---|---|---|---|---|---|---|---|---|---|---|---|---|---|
| radii | .016 | .021 | .032 | .040 | .054 | .062 | .068 | .078 | .091 | .125 | .156 | .187 | .250 |
| 1/64 | .027 | .033 | .048 | .058 | .076 | .087 | .094 | .107 | | | | | |
| 1/32 | .034 | .040 | .055 | .065 | .083 | .093 | .101 | .114 | .131 | | | | |
| 3/64 | .041 | .047 | .062 | .072 | .090 | .100 | .108 | .121 | .138 | .182 | | | |
| 1/16 | .047 | .054 | .068 | .078 | .096 | .107 | .114 | .127 | .144 | .188 | .228 | | |
| 5/64 | .054 | .061 | .075 | .085 | .103 | .114 | .121 | .134 | .151 | .195 | .235 | .275 | |
| 3/32 | .061 | .067 | .081 | .092 | .110 | .120 | .128 | .141 | .157 | .201 | .242 | .282 | .363 |
| 7/64 | .067 | .074 | .088 | .099 | .117 | .127 | .135 | .148 | .164 | .208 | .248 | .289 | .370 |
| 1/8 | .074 | .081 | .095 | .105 | .123 | .134 | .142 | .154 | .171 | .215 | .255 | .295 | .377 |
| 9/64 | .081 | .087 | .101 | .112 | .130 | .140 | .148 | .161 | .178 | .222 | .262 | .302 | .383 |
| 5/32 | .088 | .094 | .108 | .119 | .137 | .147 | .155 | .168 | .185 | .229 | .269 | .309 | .390 |
| 11/64 | .094 | .100 | .115 | .125 | .143 | .153 | .161 | .174 | .191 | .235 | .275 | .315 | .397 |
| 3/16 | .101 | .107 | .122 | .132 | .150 | .160 | .168 | .181 | .198 | .242 | .282 | .322 | .404 |
| 13/64 | .108 | .114 | .128 | .139 | .157 | .167 | .175 | .188 | .205 | .249 | .289 | .329 | .410 |
| 7/32 | .114 | .121 | .135 | .145 | .163 | .174 | .181 | .194 | .211 | .255 | .295 | .335 | .417 |
| 15/64 | .121 | .128 | .142 | .152 | .170 | .181 | .188 | .201 | .218 | .262 | .302 | .342 | .424 |
| 1/4 | .128 | .134 | .149 | .159 | .177 | .187 | .195 | .208 | .225 | .269 | .309 | .349 | .431 |
| 17/64 | .134 | .141 | .155 | .165 | .183 | .194 | .202 | .215 | .231 | .275 | .315 | .355 | .437 |
| 9/32 | .141 | .148 | .162 | .172 | .190 | .201 | .208 | .221 | .238 | .282 | .322 | .362 | .444 |
| 5/16 | .155 | .161 | .175 | .186 | .204 | .214 | .222 | .235 | .252 | .296 | .336 | .376 | .457 |
| 11/32 | .168 | .174 | .189 | .199 | .217 | .227 | .235 | .248 | .265 | .309 | .349 | .389 | .470 |
| 3/8 | .182 | .188 | .202 | .213 | .231 | .241 | .249 | .262 | .279 | .323 | .363 | .403 | .484 |
| 13/32 | .195 | .201 | .216 | .226 | .244 | .254 | .262 | .275 | .292 | .336 | .376 | .416 | .498 |
| 7/16 | .208 | .215 | .229 | .239 | .257 | .268 | .275 | .288 | .305 | .349 | .389 | .429 | .511 |
| 15/32 | .221 | .228 | .242 | .253 | .271 | .281 | .289 | .302 | .318 | .362 | .403 | .443 | .524 |
| 1/2 | .235 | .242 | .256 | .266 | .284 | .295 | .302 | .315 | .332 | .376 | .416 | .456 | .538 |

To use this chart, outside dimensions of each bend must be added up, with one bend allowance per bend subtracted from the total. The difference between the outside and inside measurements is shown in Fig. 7-5.

Where the material thickness or bend radius is not included in Table 7-1, such information may be calculated by using the formula

$$BA = 2t - \left[\frac{\pi}{2}(0.45t + BR) - 2BR\right]$$

For a so-called sharp bend, the formula will become

$$BA = 1.5t - \left[\frac{\pi}{2}(0.50t + BR) - 2BR\right]$$

where $BA$ = bend allowance
$BR$ = bend radius
$t$ = material thickness
$0.45t$ = 45 percent shrink allowance for a radiused bend
$0.50t$ = 50 percent shrink allowance for a sharp bend

Continuing with the evaluation of a flat layout shown in Fig. 7-4, dimensions *D, F, J, E* may be found to double each other by expressing the same information calculated off two different ends of the part. It sometimes pays to include these on the flat layout, especially where certain locations will have to be constantly recalculated off different ends. These dimensions may also serve as an efficient way of checking the flat layout, especially where the same person who drew and calculated it will have to check it.

Doubling of dimensions on a flat layout should not be considered inappropriate, since the layout is used for the purpose of constructing a die only. Sometimes, in order to calculate the exact location of the tooling, dimensions must be figured off various portions of the part, in which case it is helpful, as well as preventing error, if they are already included there and do not have to be recalculated separately each time the necessity arises.

The flat length of the flange *B* should be calculated the same way as described previously. Here, the difference between the outer edge of the horizontal portion and that of the flange should be figured starting off the bottom of the relief slot, which should be firmly established first. To verify the calculation, dimensions off the left edge of the blank should be used.

The bracket shown in Fig. 7-6 presents a slightly different problem. Here, the flange *A* contains an oval indentation for strengthening. Side flanges $B_1$ and $B_2$ show sharp-cornered cutouts of considerable size, which may cause problems in bending, because their bottom

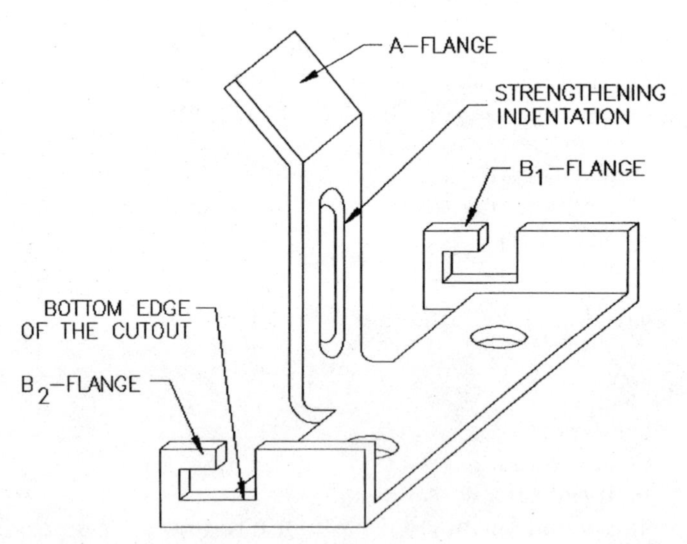


**Figure 7-6** Support bracket, bent up.

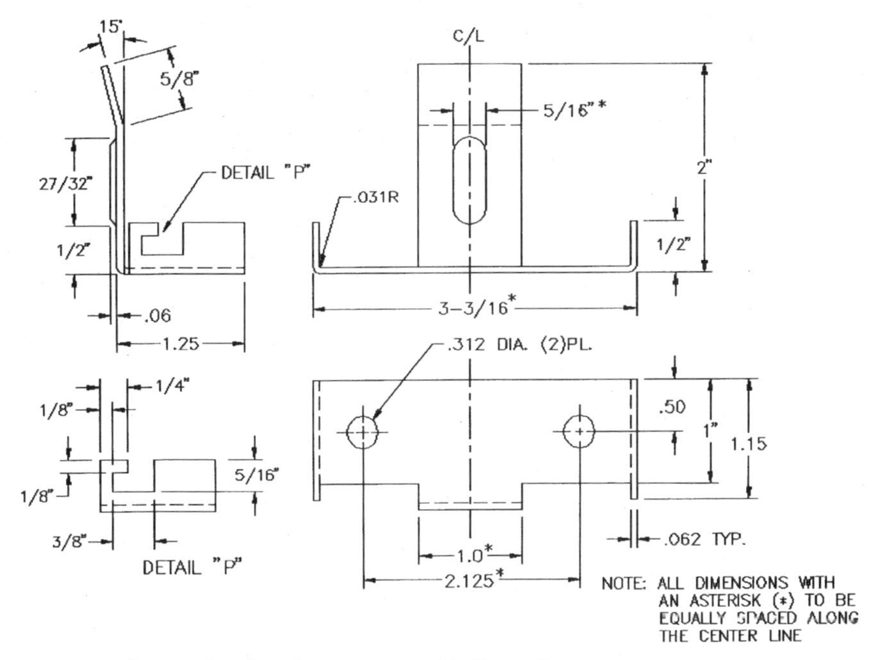


**Figure 7-7** Support bracket. Bent-up part with dimensions.

edge is located only 1.5$t$ off the center of bend radius. The dimensioned bent-up drawing of the part and its flat layout are included for study and comparison of results (see Figs. 7-7 and 7-8).

Notice that relief slots for the *A* flange are not necessary in this case, as the bend picks up much farther off the edge of the bottom portion of the part.

### 7-1-2. Phantom areas

A part in Fig. 7-9 containing phantom areas is shown for evaluation. These portions of unavailable material are not obvious from its bent-up drawing, but on observing the flat layout, shown in Fig. 7-11, it is certainly impossible to obtain the shaded corners of the two side flanges from the material given.

An alteration of the part or of the manufacturing process will have to be made. The extent of the change has to be examined with respect to the part's function, and with respect to the necessity of its outline.

When altering the part's design, the two relief slots will have to be made continuous, with the angular interruption removed, as shown dotted on the flat layout. If such a change is not possible, the middle flat portion will have to be used to provide the material for bottom

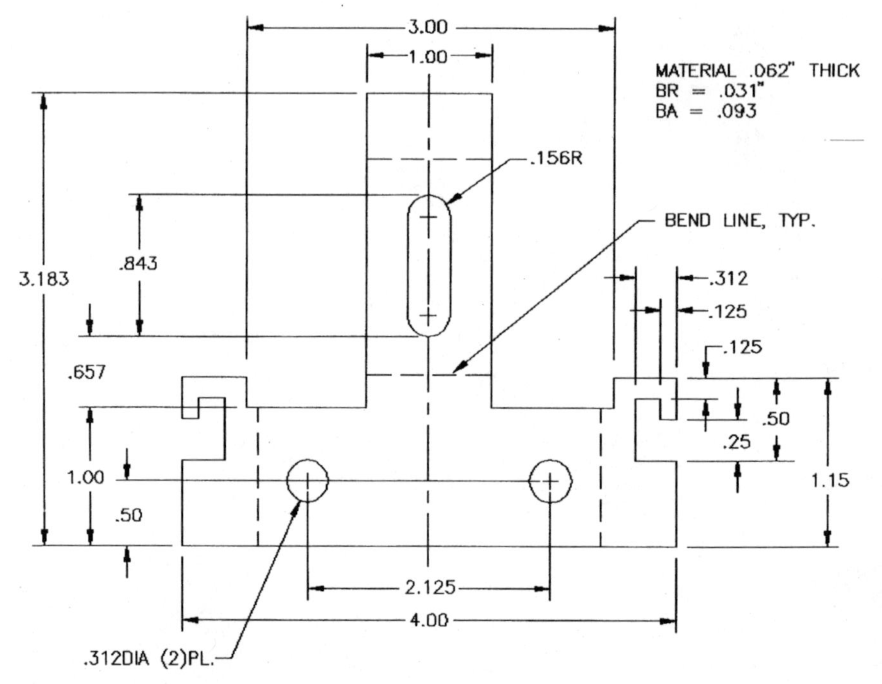


**Figure 7-8** Support bracket, flat layout.

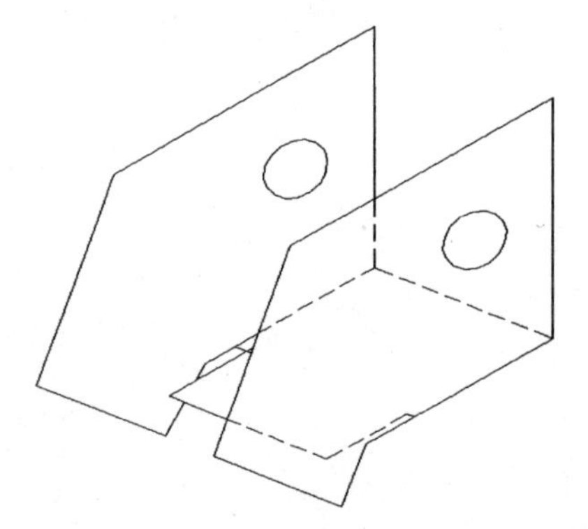

**Figure 7-9** Cover, sample part, bent up.

corners of side flanges, in which case the shaded area will be cut from the middle section. However, where an alteration of the manufacturing process should be resorted to, it will probably consist of welding the corners on or perhaps of welding the two sides to the basic flat area. It will be a highly inefficient and costly alteration.

A folded-up drawing of the part with dimensions and its flat layout are included for personal comparison and study. Note the width of the bottom flat portion being made 0.900 + 0.000/−0.015 on the flat layout in congruence with the folded-up drawing shown in Fig. 7-10,

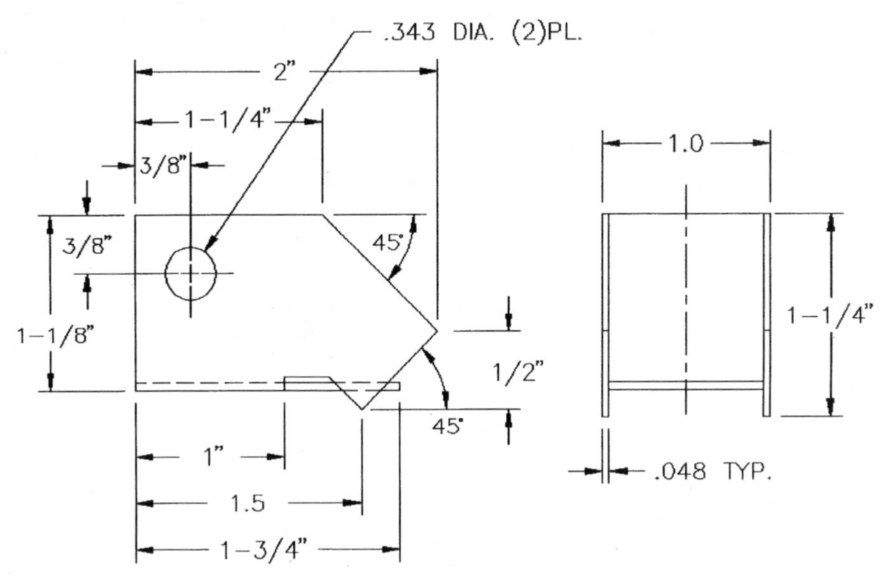


**Figure 7-10** Cover, bent-up part with dimensions.

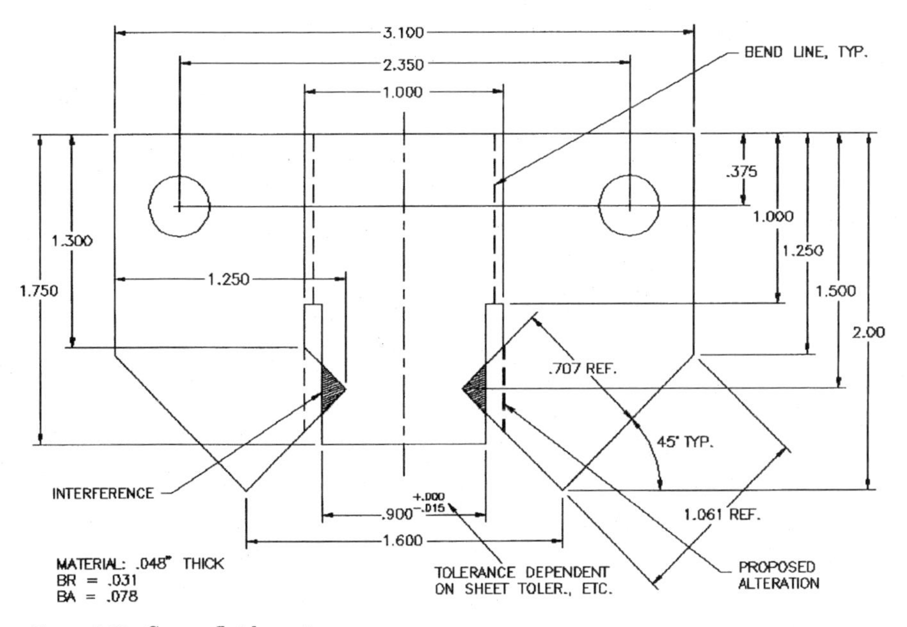


**Figure 7-11** Cover, flat layout.

where no gap is specified between the bottom and sides of the enclosure. The tolerance range in this case depends on several variables, such as

- Material thickness and its tolerance, especially its increase in thickness
- Tolerance requirements for the 1.0″ overall width of the part, as shown in Fig. 7-10
- The size of permissible gap between the edge of the flat bottom material and the inner surface of the two side flanges
- Bending operation tolerance range

Another phantom area containing a bracket is shown in Fig. 7-12. Here, the two side prongs are located on the same bend line, with the middle one offset backward. The offset prong has two side flanges, each pierced at two places. These two flanges, when observed on the flat layout shown in Fig. 7-13, cannot actually be made, as there is not enough material for their width.

If these flanges are located higher up, plenty of stock will be available once their height exceeds the top edge of the two side prongs. However, let's assume their location is firm and must remain as such.

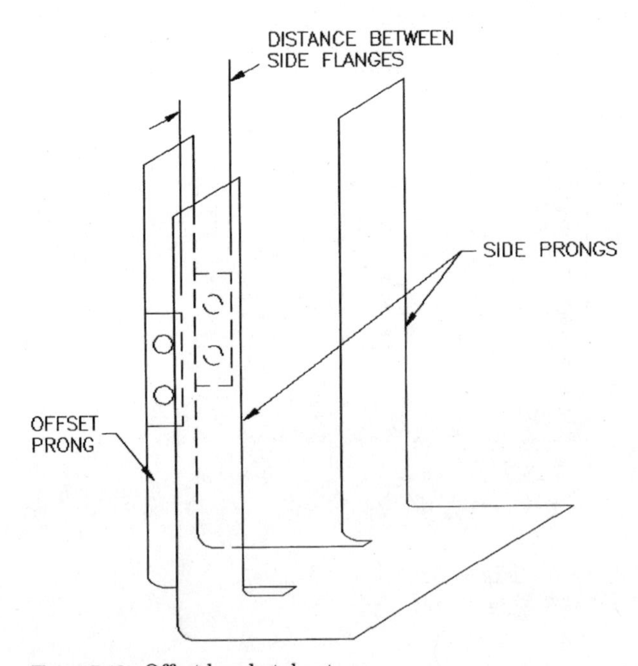


**Figure 7-12** Offset bracket, bent up.

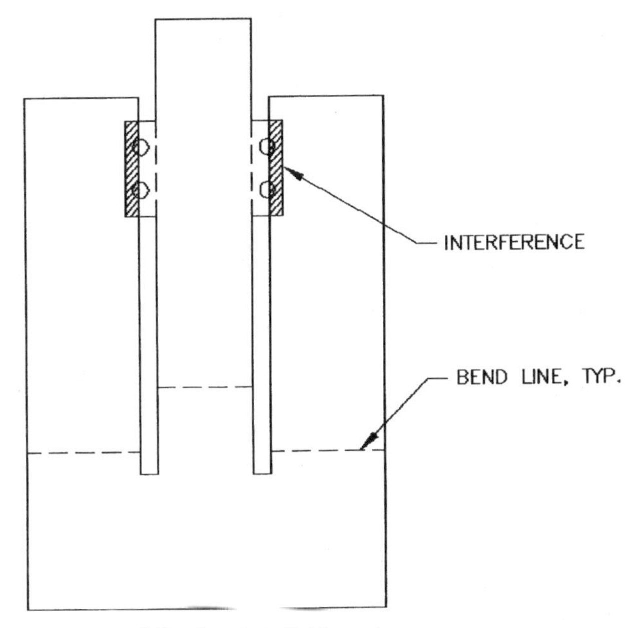


**Figure 7-13** Offset bracket, flat layout.

There seems to be no chance of producing the material for these flanges from anywhere unless it is taken from the material of the two prongs, as shown in Fig. 7-14*A*. Provided the sharp-edged cutout and the difference in width of the prongs' top will not impair the overall function of the part, this may become a solution to a given problem. Where the sharp edge may be objectionable, it may be either rounded or chamfered, as shown by dotted lines.

Tooling for such a design alteration will be quite costly, for which reason an overall widening of the relief slot, shown in Fig. 7-14*B*, may be more advantageous.

Naturally even this solution may not suffice, and another alternative has to be sought after. Shown in Fig. 7-15 is a different method of changing the given design. Here, the top of the middle prong is extended farther and spread out sideways, the extension being equal to the offset of two small side flanges, as shown in the folded view.

Such a design change may seem to do the job of placing the two side flanges in an appropriate location, but at a very steep price, for the part is much higher now, with its width increased as well. Aside from the obvious wastage of material, an additional bending operation is included. Also a possible need for two small tabs *P* may arise if the two narrow strips-turned-flanges are to be secured to the body of the middle prong. Such a design is not only impractical, it is outright clumsy.

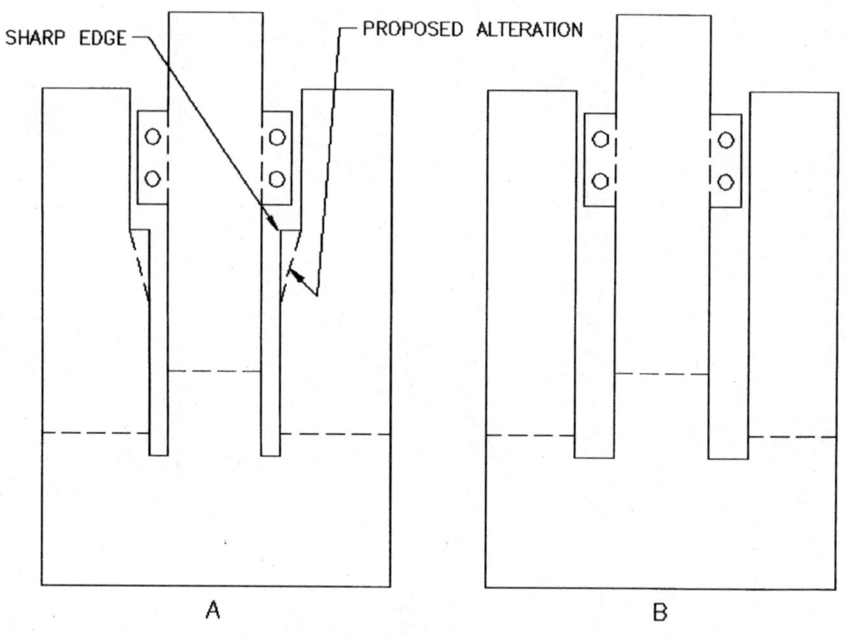


**Figure 7-14** Offset bracket, flat layout variation.

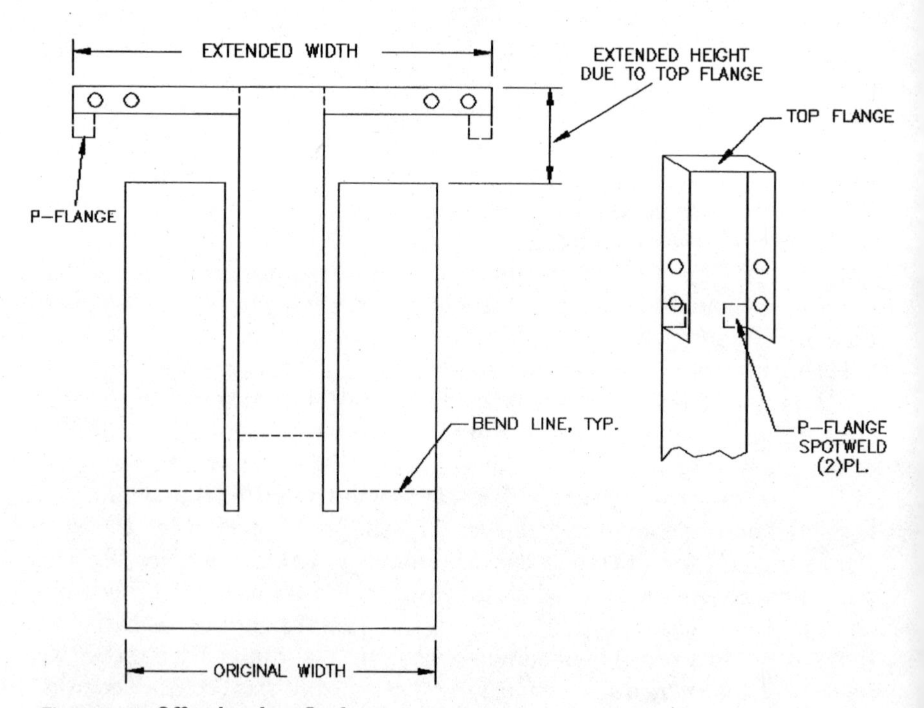


**Figure 7-15** Offset bracket, flat layout variation.

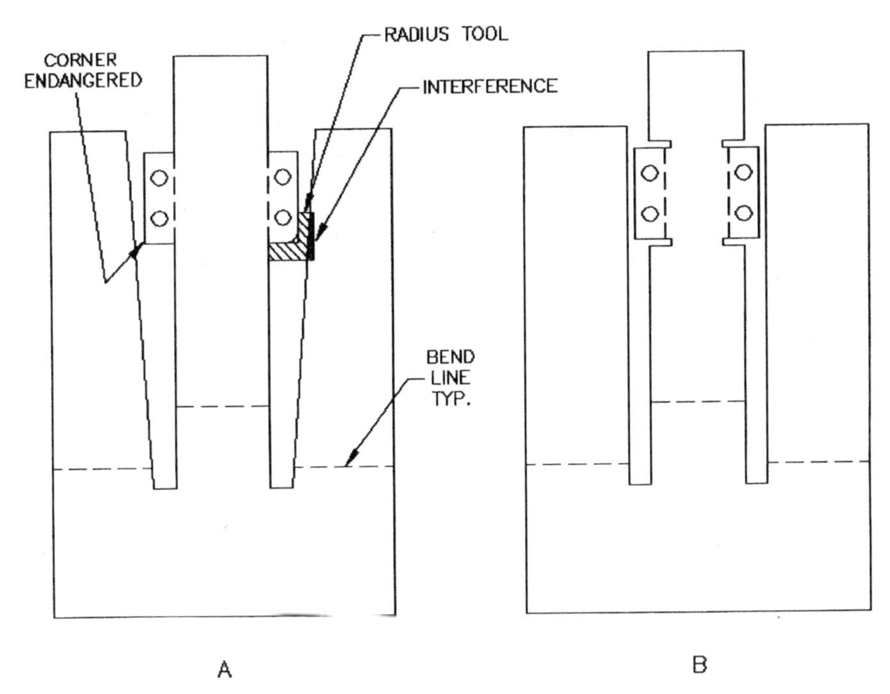


**Figure 7-16** Offset bracket, flat layout variation.

In Fig. 7-16 is shown another alteration of the given flat layout, where the relief cuts of the two side prongs have been slanted. It is a slightly less practical variation of the Fig. 7-14*A* and *B* method. The tool, cutting the angled portion, may not be narrow enough to miss the corner of each side flange, which may result in an involuntary and also uneven chamfer of its edge. True, the edge could have been previously rounded off, but a radius punch, if used, will certainly interfere with the material of side prong, unless it is made very thin, in which case it will frequently break.

Figure 7-16*B* depicts a halfway method, where the material for side flanges is taken off the area between the prongs by sinking the flanges into the material of the middle prong. This solution can be used where the horizontal distance between the flanges, as shown in Fig. 7-12, can be altered without impairing the function of the part.

## 7-2. Details of Flat Layout

When making a flat layout, there are certain areas where paying attention to the detail saves a great many headaches later, when the tooling is made in steel. These areas of interest are mainly corner relief slots, transitions between radii and straight surface lines, holes and their location off bends, etc. Many of these problem-prone areas may be successfully produced by a simple alteration of the manufacturing sequence of operations, but there are cases where such a solution would not be adequate.

Naturally, the importance of the positioning of parts with regard to the material grain is vital to the functionability of the product and to its manufacturability as well. Further, a correct assessment of the part's burr side and design of tooling in accordance with its location is of essence.

For example, to place a switch cover with the burr side up may tear the skin of the user, provided the switch is embedded within the plate. A similar effect would be achieved by reversing the burr side of the binder's opening-closing lever mechanism or that of any closure, hardware, and many other items of daily personal and industrial use.

The location of a burr with reference to the functional side of the part may cause the final product to either make it or break it, as they say. Therefore, a careful observation of the part and a correct assessment of its surfaces would be of essence where the location of the burr is not specified on the drawing.

### 7-2-1. Details of corners and relief slots

In order to eliminate the possibility of sharp corners some parts have their edges chamfered or rounded. This in itself seems like a logical solution when considering their functionability, but to actually produce such a corner is a different story. It is quite easy to chamfer an edge by using a square tool under 45° angle and hitting each corner prior to blanking its outline. We already know that we should never attempt to perform such a sequence of operations in reverse, and chamfer the corner after the outline is blanked out, since the punch, cutting with but a small portion of its total area, will have a tendency to buckling and breakage.

However, rounding of edges of parts is a completely different matter. With large-sized items, which demand a small, perhaps even dimensionally unspecified corner radius, this will not be such a problem. These parts may always be touched up individually later. But with small parts pouring from the die in quantity, a different solution will be needed where rounding of their corners is required.

As seen in Fig. 7-17*A*, a radius tool is used to produce a quarter-circle cut, with the rest of the material being chopped off by a rectangle. No matter which way the sequence of operations is arranged, a step between the cuts may be produced if the tooling is not sharp or tight enough, or where the detrimental qualities of the material, such as high ductility or brittleness, are present.

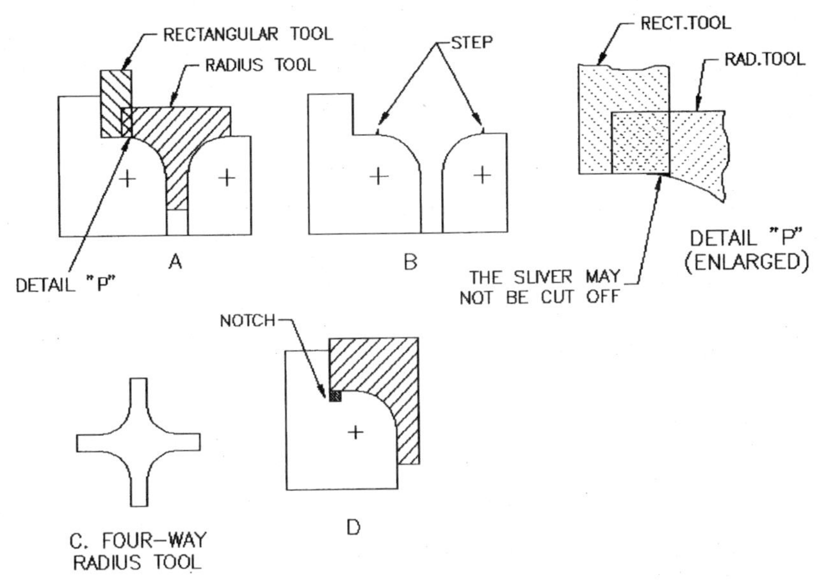


**Figure 7-17** Corner radii.

Where the tools overlap, a step may be present at either or both ends of the length of the overlap. This may be even more complicated if punching first with the rectangle and following up with the radius tool. A small sliver of material, shown in the enlarged detail *P*, may not be cut off properly. Because of its insignificant size, this little corner may fit within the tooling tolerance range, and instead of being cut, it will be squeezed down, where it will aid in the formation of the burr.

A four-way radius tool, often used in sheet-metal practice for providing radii at various locations, is as shown in Fig. 7-17*C*.

An alternative to detail *A* cutting is presented in detail *D* of the same illustration. Here the whole area is removed by a single punch, shaped to suit the required size and contour. However, even though the resulting cut will be perfect, a tool like this is highly specialized, usable for one particular operation only. Further, if a relief notch is to be produced in the corner, an additional hit, removing that small section of material, will be needed.

*Corners of sheet-metal parts* may often pose a problem, especially where joining two flanges together, as seen in Fig. 7-18*A*. Here the two edges of a box-type shape must be very close, almost touching. Some may attempt to pattern the relief slots as shown in detail *B*. This solution may often be used with different sizes of bend radii, provided the sharp corner may be filed off where it should exceed the shape of the part. A method of replacing such a complicated relief slot pattern with a single round opening, shown in detail *C*, may often be

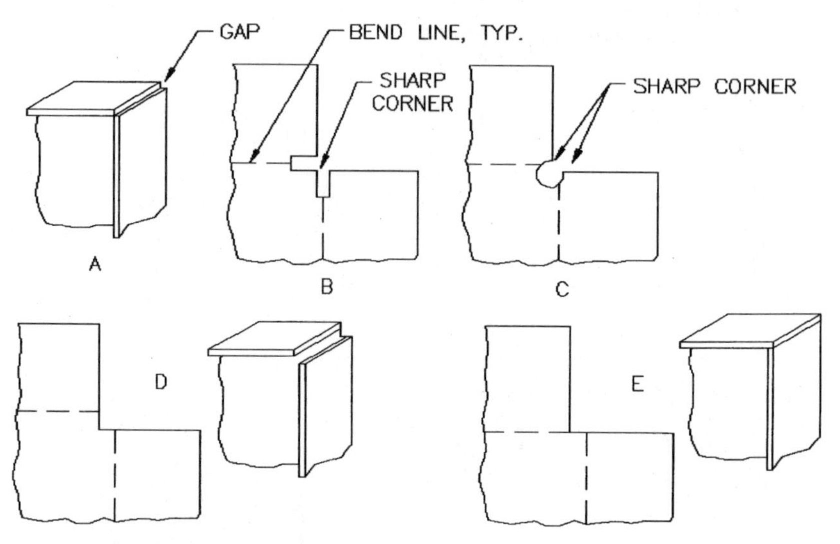


**Figure 7-18** Corner design variations.

a better solution. The sharp corner will be shifted into another location, with its size and sharpness dependent on the distance of the round tool off the part's edge.

Another manufacturing method of producing a corner relief is to offset both cuts slightly below the bend line, as shown in Fig. 7-18*D*. Such an arrangement will certainly work well with parts where a sharp corner is permissible. An alternative of the same joining method is shown in *E*, where one side of the box is bent to fit under the other flange.

Face flanges present a slightly different challenge to the avid die or sheet-metal designer. Here the corner may be finished either by joining two perpendicular flanges or by providing a 45° cut in the middle, as shown in Fig. 7-19*A* and *D*, respectively.

The method of corner finishing depicted in detail *A* will always display a gap of some kind, especially where larger bend radii are involved. With large-sized parts, these gaps may be filled in and ground smooth, but with small, mass-produced items, such a finishing procedure is highly unreasonable. A 45° cut is more appropriate in a case like this.

The amount of space between the adjoining flanges shown in *D* may be diminished to a negligible size by a correct assessment of the depth of relief cut. The variation of such a space will be achieved by moving the peak of the cut with reference to the face bend line, as shown in *E* and *F*.

An easy method of evaluating the gap size is to calculate the exact position of the edge of each flange and determine the size of the space

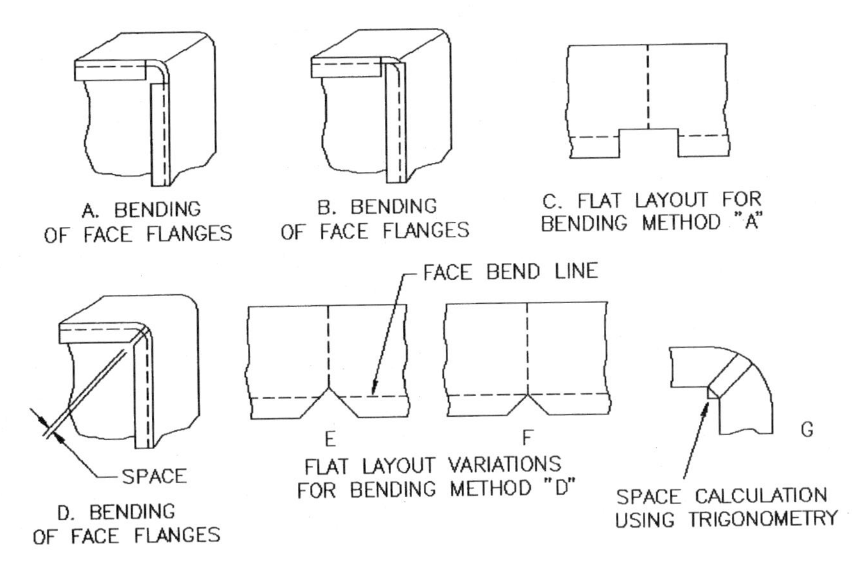


**Figure 7-19** Corner design variations.

between them by using basic trigonometry. Naturally, the bend allowance for such a bend must be divided equally between both sides of a part. Where the peak of a relief cut will end at the bend line, a slight bulging of the material around that area may be expected after bending.

Sometimes, if the closest possible contact between the two flanges is required, some bend reliefs are purposely made smaller than necessary, allowing for tearing of the material in bending. This is done especially with large parts, where these areas may be sanded smooth later on.

### 7-2-2. Holes versus bends

The location of openings with reference to the bend line is of considerable importance in sheet-metal practice. The metal material, when subjected to stretching and compressing during the bending process, has a tendency to alter the shape of a hole, changing its contour from round to elliptical. This may be prevented where it is possible to bend the part first and pierce the holes afterward. However, such an alteration usually involves a side action of the die, utilizing cams, which is not always the most desirable method of solving this problem.

In some cases, the holes in question may be pierced oval in shape, so that when being tugged upon by the strain of the bending process, their elongation will actually make them round. See Fig. 7-20.

The minimal distance of the edge of an opening off the bend should equal $2t$, or two material thicknesses, measured off the center of the

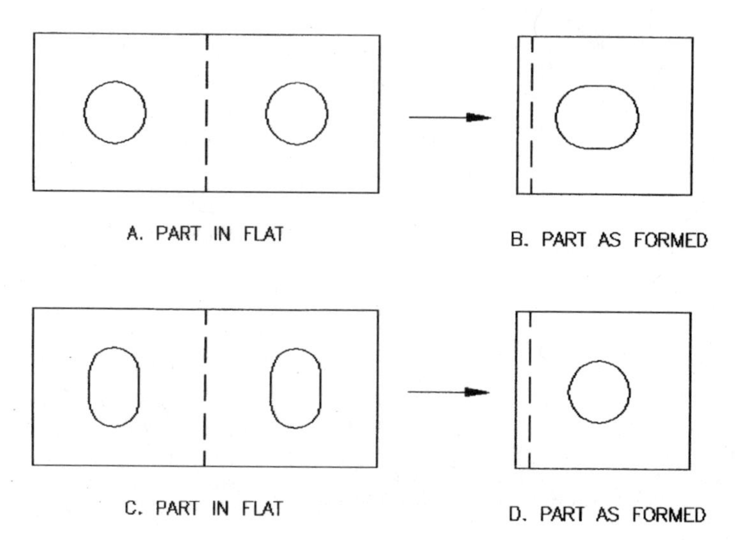


**Figure 7-20** Shape of openings altered by bending.

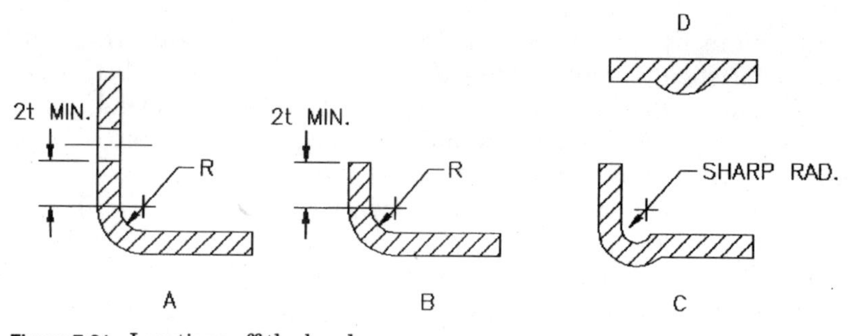


**Figure 7-21** Locations off the bend.

corner radius. In the same way the edge of the part should be at least two material thicknesses off the edge of the radius, as shown in Fig. 7-21.

Where a sharp corner bend is absolutely necessary, perhaps to serve as a relief for the corner of an adjoining part, it may be produced by restriking the bent-up part with the intention of sharpening the inner radius. Since during such a process the material of the edge will be diminished in thickness, slight bulging of the bend line may be needed prior to the bending operation, as shown in Fig. 7-21*D*.

**Relief openings.** To punch relief openings in metal, such as those providing an access for the assembly tool (see Fig. 7-22*A*), or clearance holes for screws or other inserts (see Fig. 7-22*B*), it pays to make these holes as large as possible. Larger openings not only provide for

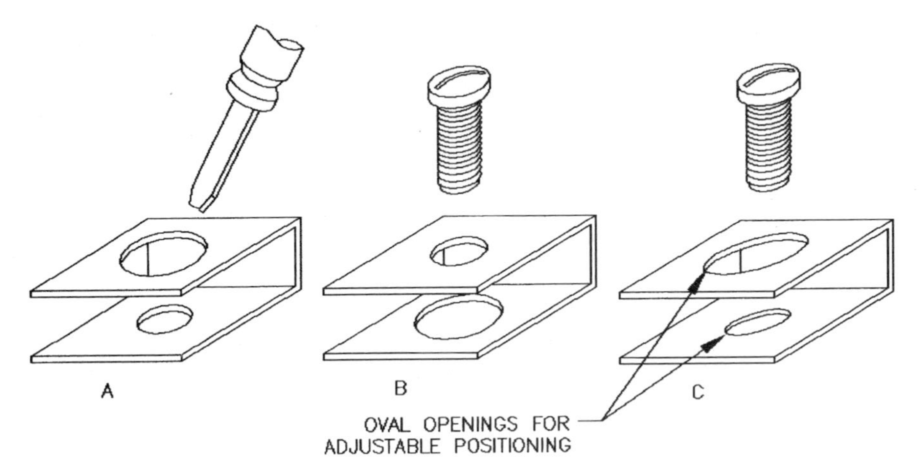


**Figure 7-22** Access openings.

a positive relief action, they are not easily affected by any dimensional discrepancy, which in bent-up parts may sometimes be seen.

Openings where the location of their center with respect to the corresponding part is outright questionable, as with segments of large-sized assemblies or constructions, should contain oval holes instead of rounds, with the longer side of the shape positioned along the path of the expected dimensional discrepancy (see Fig. 7-22*C*).

### 7-2-3. Extruded openings and bosses

The size and depth of extruded shapes should be calculated the same way a bend would be assessed. After all, that's what these contours are, when observing their cross-sectional outlines: They are bent-up, sometimes slightly drawn shapes.

Most extruded openings must first be pierced in order to provide enough space for the extruding punch, which is to form the edges afterward. To calculate the size of the basic opening, we must assess the length of the bent-up portion, apply the necessary bend allowance, and subtract the result from the size of the extruded shape, or

$$d = 2[B - (A + h - BA)]$$

where $BA$ = bend allowance

All other values are given in Fig. 7-23.

This calculation does not take into consideration any drawing tendency of the material, considering the obtained shape to be attributable strictly to bending.

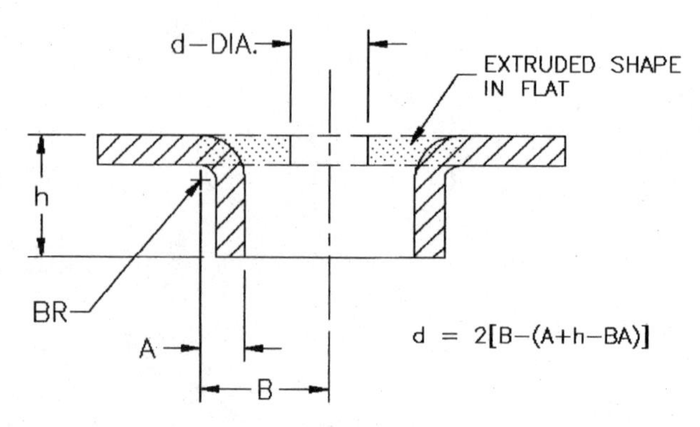


**Figure 7-23** Extruded openings.

Often such a calculation may be used when determining the largest possible height *h,* obtained from the smallest possible pierced opening. In such a case, the diametral size of the original opening with respect to the diameter of the final shape (2*B* diameter) are the factors limiting the height of the extruded shape.

Bosses and dimples may be evaluated the same way; however, these indentations often contain some stretching of material, with subsequent thinning of some sections.

### 7-2-4. Dutch bends and joggles

Dutch bends, or 180° bends, are used as hems in sheet-metal practice (Fig. 7-24). Such bends are actually doubled 90° bends, with the outer surface of the material forming a whole half circle around the center of the bend. There is almost no bend radius with this type of bending.

To calculate the size of a hemmed part in flat, the chart shown in Table 7-2 should be used. Table 7-3 gives the minimal length of hem *A.*

*Joggles* are offsets in height, connected by a short piece of angular strip of the same material, as shown in Fig. 7-25. Joggles are used where an offset for another part of the assembly is to be provided or where a greater flatness of the actual contact surface is desired. Joggles may also be used to stiffen the part or to provide for collection

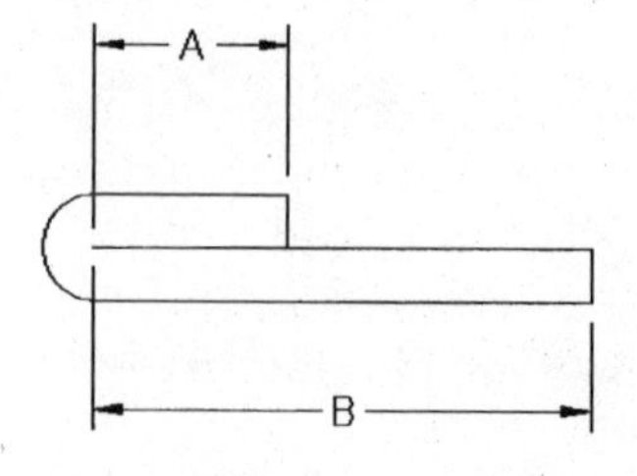


**Figure 7-24** Dutch bend (hem).

**TABLE 7-2 Hemming Dimensions in Flat**

| Material thickness, in | Total length in flat, in |
|---|---|
| .030 | $A + B + .050''$ |
| .036 | $A + B + .060''$ |
| .047 | $A + B + .080''$ |
| .062 | $A + B + .090''$ |
| .093 | $A + B + .140''$ |
| .109 | $A + B + .160''$ |
| .125 | $A + B + .190''$ |

**TABLE 7-3 Minimum Length of the Hemming Flange**

| Material thickness, in | Minimum $A$ dimension, in |
|---|---|
| .030 | .250 |
| .036 | .250 |
| .047 | .250 |
| .062 | .250 |
| .093 | .375 |
| .109 | .375 |
| .125 | .375 |
| .187 | .625 |
| .250 | .750 |

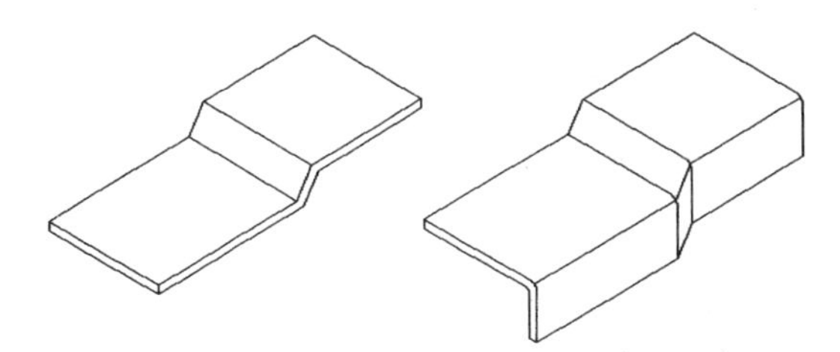

**Figure 7-25** Joggles.

of liquids, in which case they take the shape of a shallow cuplike indentation within the flat surface.

Joggles, indentations, bosses, and other height-altering additions should use the same recommended bend radii as regular bent-up parts. Bend allowances for joggles should be taken in the vicinity of 33 percent of the material thickness. As an example, a 0.060″ high joggle in 0.048″ thick cold-rolled steel will use a +0.014″ bend allowance.

## 7-2-5. Stretching of densely perforated sheets

The problem of shape alteration of densely perforated surfaces has been addressed in preceding chapters. We all know by now that the piercing process cuts across the strains of material, causing them to bulge around the cut, with subsequent extension in length and dimensional distortion of the part. Such a variation in size of heavily perforated surfaces may often be considerable, and a sequence of operations should be altered to accommodate such differences.

Where *A* holes, shown in Fig. 7-26, are to be reasonably accurately spaced off the bottom of their flanges, the bottom rectangular surface must be perforated first. Naturally, the strain produced by such intense piercing will cause the material to stretch and perhaps even to bulge. If the holes and the shape of the narrow flange are provided prior to the perforation, their location will definitely be thrown off. Therefore, these cuts must follow after the perforation is finished.

Removing the material around the flanges prior to perforation and hoping for the gap to provide the needed relief for stresses introduced by the process of perforating will not help the situation either, even though a partial relief may certainly be counted upon (see Fig. 7-26*A*). The middle band, within which the two *A* holes are located, will still remain affected by the movement of material during piercing.

But even perforating first and removing the material around the narrow flanges afterward may not be as easy as it seems. As shown in

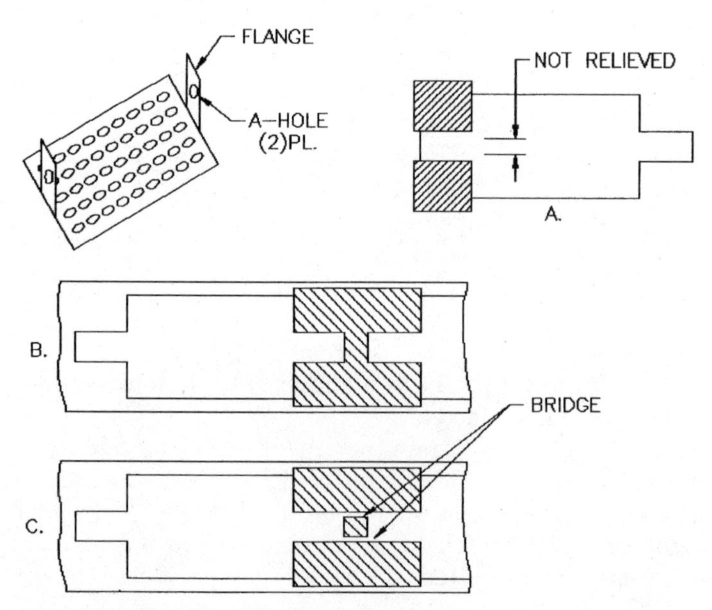


**Figure 7-26** Stretching of perforated surface.

Fig. 7-26*B,* the lengthwise positioned perforated part will stretch and perhaps bulge. When relieved by the removal of the hatched area, the cut itself may not be positioned properly, as some shifting due to the material expansion in size may be encountered. With dependence on the type of material and length of flanges, the narrow flanges may further obstruct the movement of the sheet through the die.

Should the flanges be kept attached by small bridges of material, as shown in Fig. 7-26*C,* the still-connected parts, expanded lengthwise, will tend to bulge in that area, complicating the dimensional stability of *A* holes.

Perhaps the greatest dimensional stability may be obtained by placing the parts on the strip perpendicularly to the previous design, as shown in Fig. 7-27. By perforating first, an extension in size will be encountered. This dimensional variation will affect the flatness of a surface only marginally, as the parts will have plenty of space to expand outward, bulging the sides of the strip. With the metal removal around the flanges, this tendency will be only encouraged, and when the two *A* holes are pierced, their location should be reasonably accurate.

## 7-3. Flat Layout and Its Additional Uses

A correct and accurate flat layout has additional advantages in sheet-metal practice. Reproduced several times and cut out, it may serve to show a rough positioning of parts on the sheet. Their location may be

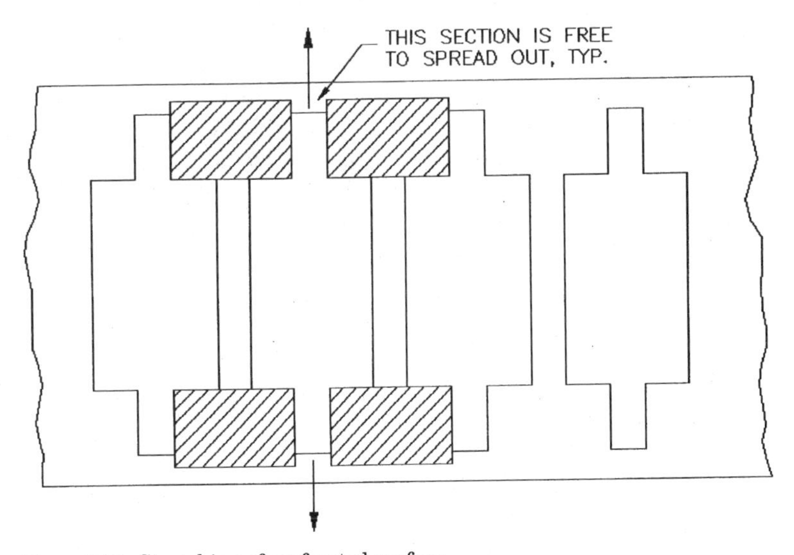


**Figure 7-27** Stretching of perforated surface.

altered without wasting time on sketching of every design alternative.

Placed over the finished piece, flat layout may act as a method of quick evaluation of the product's dimensional correctness. Pasted over the part, it may be used as a template in setting up some preliminary operation. In some instances, where a limited number of pieces is to be experimentally altered, a good flat layout could be used to show the exact location of areas of interest.

In the product designing field, flat layout may be used as a visual aid when evaluating the part's shape and its interference with other elements of an assembly.

Flat layout drawn on mylar may be used for a comparison with other similar parts and their dimensional evaluation. This method of assessment is of great use especially where many similar parts are being made. Instead of checking various drawings step by step, such a quick method of evaluation can be used to preselect certain parts visually.

Where a computerized drafting system is employed, flat shapes of all parts should be placed within the same file, allocating each layout to a certain layer and positioning them all at the same distance off the origin. A quick method of comparison, unsurpassed by anything else, may be achieved this way. Computerized drafting, as opposed to a manually drawn sketch, may allow for a quick comparison of the main dimension without pulling large sheets of papers and trying to find the way through the maze of already bent-up shapes. Here, having all the parts already on the screen in flat, a visual comparison of the basic shapes, followed by a simple measurement of any questionable dimension, will ascertain quickly and efficiently any similarities or variations, if present.

The last method is useful especially in shops where a multitude of similar parts are being produced. Often a single die, perhaps with a slight alteration, may serve a purpose of producing a completely different part, without the expense of building another tool.

Chapter

# 8

# Bending and Forming Operations

## 8-1. Stress, Strain, Elongation, Compression

Bending and forming are quite similar operations, the only difference being the presence of the drawing action in forming. In bending, a portion of the part is flexed along a straight line until a bend is obtained. In forming, the bend line may be curved, circular, or otherwise shaped. The variability of the bend line contour causes the material to either expand on one side or be compressed elsewhere.

There are basically two types of forming operations, one producing shrink flanges and the other used for making stretch flanges. In shrink flanges, as the name implies, the flange material must be squeezed or compressed during forming, whereas stretch flanges must be stretched and subsequently thinned (Fig. 8-1).

The evaluation of the flange type should be pursued by observing the flat layout of a part, which clearly shows where the material for each flange may be taken from.

Bending, even though generally considered a draw-free method of obtaining flanges, does contain some minute amounts of drawing action as well. Drawing-type movement of material can be found along the circumference of the bend, where a small amount of stretching and shrinking can be seen. Some may consider this to be a lengthwise shift of various material layers with respect to each other, but the expansion or contraction of material attributable to its increase in linear length, with subsequent change in thickness (see Fig. 8-2) is certainly a minute amount of drawing action. True it is an occurrence so greatly limited in scope that we may consider it negligible.

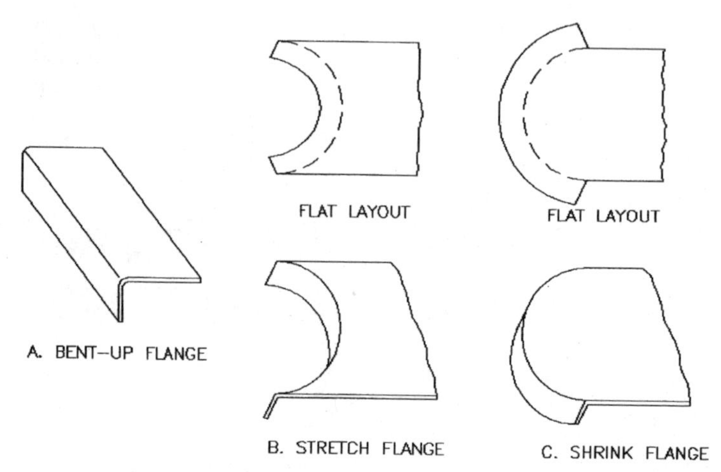


**Figure 8-1** Types of flanges.

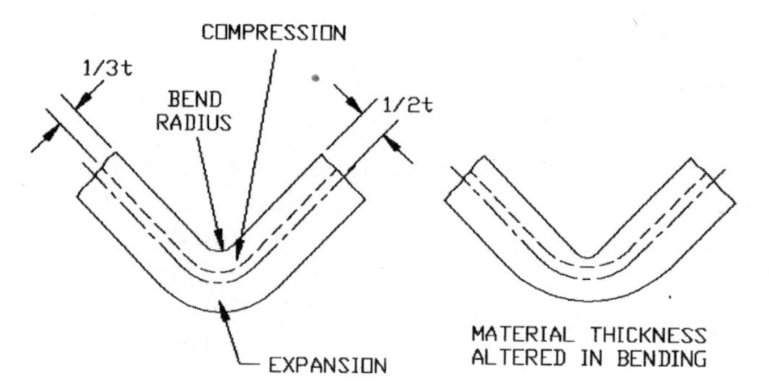


**Figure 8-2** Neutral axis location in pure bending operation.

The inside portion of the bend usually does not display any considerable changes because its outline is restricted by the contact with tooling.

The cross section of the bend and flange, in order to be considered a result of the bending operation, must maintain the same thickness as the rest of the sheet out of which it was made. An ideal situation can be found where spring-hard sheet-metal material, which—bent freely, preferably by hand, without the use of any bend-enforcing tools—forms an extremely shallow radius, returning to its original shape immediately on cessation of the applied force.

With softer metals, the tendency to succumb to the bending force is greater, and these materials do not always return to their original form and shape. The amount of spring-back, or back-returning force within the metal, is less in such ductile materials, making them easily altered by the application of force.

For the above reasons, bending of stiff and hard materials should utilize more acute angles than required, as the bent-up flange will always have a tendency to return to its original shape or to spring-back (Fig. 8-3).

Another method used for minimizing of the spring-back effect is the application of a die force against the formed part, or bottoming. This action, being in kind a coining operation, forcibly returns the bent-up portion to the desired location, while at the same time interlocking the free layers of material. Such a bend is firm and rigid, secured against any spring-back tendency of the material. The strain hardening, which occurs in the material owing to cold working of both the bending and coining operations, increases the strength of that section.

## 8-2. Bend Radius

All bending and forming of sheet metal is considerably affected by two important factors:

1. Amount of bend radius
2. Size of bend angle

With a generous bend angle, such as the one shown in Fig. 8-4*A*, any material can be formed with probably no great problems encountered, as there would be less difference between the material in flat and that already formed.

With an obtuse angle, the bend radius may be specified as "sharp," and it may actually be obtained as such, for a radius in such a loose angle is extremely hard to measure, and a sharp line left by the tooling at the inner section of the bend may be considered a sharp radius.

However, as the angle becomes less obtuse, many difficulties begin to surface. The amount of bend radius becomes more critical, the material hardness more detrimental to the success of the bending process, the material thickness more commanding. With angles under 90°, or acute angles, all these influences are still more intensified.

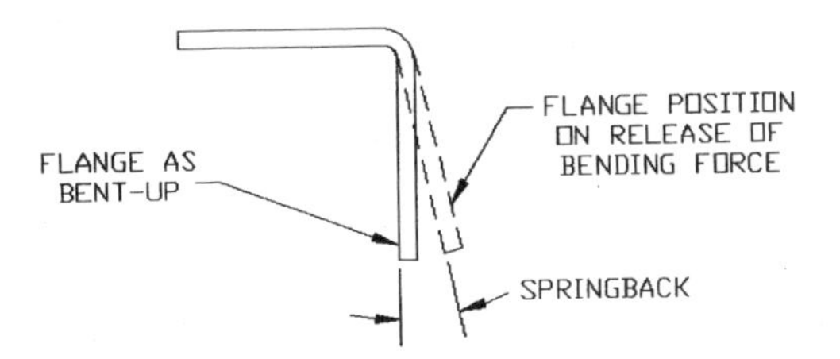


**Figure 8-3** Spring-back of the bent-up flange.

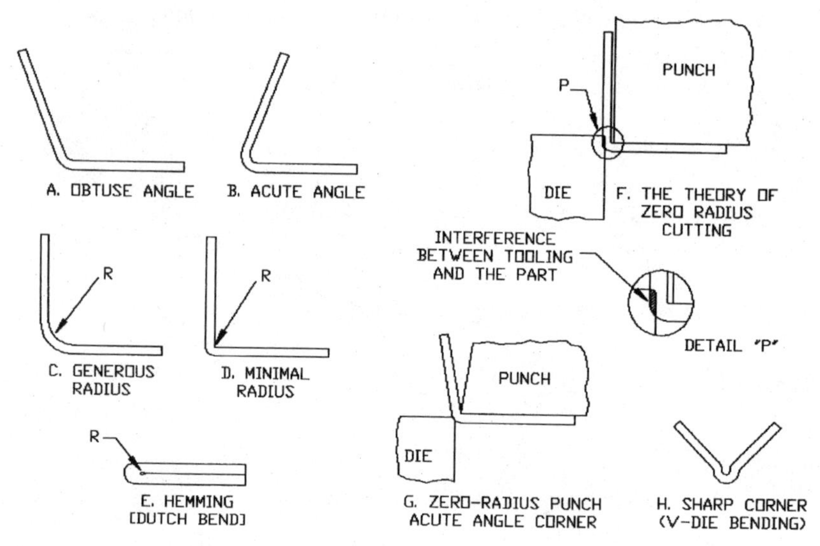


**Figure 8-4** Various types of bends.

Bend radius affects the success of the bending operation the most. It is dependent on the material thickness and material properties, such as its formability and hardness, aside from other small but not negligible influences. The smallest attainable bend radii for various materials are listed in Tables 8-1, 8-2, and 8-3.

The bend radius in bending or forming operations always pertains to the *inside radius* of the bend.

The size of bend radius varies with cold bending, as opposed to bending with addition of heat. It is obvious that with a heated metal, its properties change, and sharper bends may be attained in previously unyielding materials. Also the required bending force will be lessened and spring-back values altered, their occurrence being dependent on the amount of heat supplied.

The minimum bend radius may also be calculated using Eq. (8-1).

$$BR_{min} = kt \tag{8-1}$$

where $t$ = thickness of material

$k$ = coefficient from Table 8-4

The "sharp bend" so often specified on the drawings actually cannot be produced in sheet-metal bending (see Fig. 8-4*F*). With any material other than modeling clay, should it be exposed to a strain of such excessive bending demand, it will end up fractured or outright torn off.

Sheet-metal material, even where bent at the sharpest corner possible, will still maintain a trace of a radius. This will offset the flange

**TABLE 8-1 Minimum Bend Radii for 90° Cold Bends in Steel and Aluminum Alloys**

| Material type | 0.015 | 0.031 | 0.062 | 0.125 | 0.187 |
|---|---|---|---|---|---|
| | Sheet thickness, in | | | | |
| | Fractions of material thickness | | | | |
| Steel: | | | | | |
| 1020–1025 | 2 | 1½ | 1 | 1 | 1 |
| 4130, 8630 | 2 | 2 | 1½ | 1½ | 1½ |
| 1070, 1095 | 3–3½ | 3 | 2½ | 2½ | 2½ |
| Aluminum: | | | | | |
| 1100-0, H12, H14, H18; 2024-0; ALCLAD 2014-0; 3003-H12, H14, H32; 3000-0; 5005-0, H12, H14, H32, H34; 5050-0, H32, H34; 5052-0, H36; 5086-0; 5357-0, H32, H34; 5454-0; 5457-0; 5557-0; 6061-0 | 0 | 0 | 0 | 1 | 1 |
| 1100-H16; 5052-H34; 5456-0; 7075-0; 7178-0 | 0 | 0 | 0–1 | 1–1½ | 1–2 |
| 3004-H34; 5050-H34; 5052-H34; 5086-H32, H34 | 0 | ½–1 | 1–1½ | 1–1½ | 1–2 |
| 3003-H16; 5005-H16, H36; 5050-H36; 5154-H34, H36; 5454-H32, H34; 5456-H36; 5357-H36; 6061-T4, T42 | 0–1 | 0–1 | ½–1½ | 1–2 | 1½–3 |
| 1100-H18; 5052-H36; 5083-H323, H324; 5456-H323, H343; 6061-T6 | 0–1 | ½–1½ | 1–2 | 1½–3 | 2–4 |
| 3003-H18, H38; 5005-H18, H38; 5052-H38; 5357-H38; 5457-H38; 5557-H28 | ½–1½ | 1–2 | 1½–3 | 2–4 | 3–5 |
| 2024-T3, T4 | 1½–3 | 2–4 | 3–5 | 4–6 | 4–6 |
| ALCLAD 2014-T3, T4; 5154-H38 | 1–2 | 1½–3 | 2–4 | 3–5 | 4–6 |
| ALCLAD 2014-T6; 2024-T36; 075-T6; 7178-T6 | 2–4 | 3–5 | 3–5 | 4–6 | 5–7 |

**TABLE 8-2 Minimum Bend Radii for 90° Cold Bends for Stainless Steel**

| Material type | Temper | Up to 0.050 | 0.051–0.187 |
|---|---|---|---|
| | | Sheet thickness, in. Values are fractions of stock thickness | |
| 301, 302, 304, 316 | Annealed | 90° and 180° = ½ | 90° and 180° = 1 |
| 301, 302 | ¼ hard | 90° and 180° = ½ | 90° and 180° = 1 |
| 301, 302 | ½ hard | 90° and 180° = 1 | 90° and 180° = 1 |
| 301, 302 | Full hard | 90° and 180° = 2 | |
| 316 | ¼ hard | 90° and 180° = 1 | 90° and 180° = 1 |

**TABLE 8-3 Minimum Bend Radii for Copper Alloys**

| Material type | Material thickness, in | Material condition | Bend across the grain | Bend at 45° to the grain | Bend parallel with the grain |
|---|---|---|---|---|---|
| Copper | 0.021 | Half hard | 0.031 | 0.031 | 0.046 |
| Brass, yellow | 0.021–0.062 | Half hard | Sharp | Sharp | 0.015 |
| Brass, yellow | 0.042 | Full hard | Sharp | 0.015 | 0.031 |
| Brass, yellow | 0.042 | Spring temper | 0.046 | 0.218 | 0.218–0.250 |
| Brass, red | 0.021–0.046 | Half hard | Sharp | Sharp | 0.015 |
| Brass, red | 0.042 | Full hard | 0.015 | 0.031 | 0.093 |
| Brass, red | 0.042 | Spring temper | 0.062 | 0.187 | 0.437–0.500 |
| Phosphor bronze, 5% | 0.021–0.062 | Half hard | Sharp | Sharp | 0.015 |
| Phosphor bronze, 5% | 0.042 | Full hard | 0.062 | 0.062 | 0.125 |
| Phosphor bronze, 5% | 0.042 | Spring temper | 0.093 | | |
| Phosphor bronze, 8% | 0.021–0.062 | Half hard | Sharp | Sharp | 0.015 |
| Phosphor bronze, 8% | 0.042 | Full hard | 0.031 | 0.125 | |
| Phosphor bronze, 8% | 0.042 | Spring temper | 0.093 | 0.250 | 0.437–0.500 |
| Cartridge brass, 70% | 0.021–0.046 | Half hard | Sharp | Sharp | 0.015 |
| Cartridge brass, 70% | 0.042 | Full hard | 0.015 | 0.031 | 0.046 |
| Cartridge brass, 70% | 0.042 | Spring temper | 0.062–0.125 | 0.218 | 0.218–0.250 |

**TABLE 8-4 Coefficient *k* of the Minimum Bend Radius**

| | Material condition | | | |
|---|---|---|---|---|
| | Annealed | | Heat-treated | |
| | Relationship bend/grain | | Relationship bend/grain | |
| Material | $\perp$ | $\parallel$ | $\perp$ | $\parallel$ |
| Deep drawing steel | 0.0 | 0.2 | 0.2 | 0.5 |
| Steel, AISI 1010, 1040 | 0.1 | 0.5 | 0.5 | 1.0 |
| Steel, AISI 1015, 1020 | 0.2 | 0.6 | 0.6 | 1.2 |
| Steel, AISI 1049 | 0.3 | 0.8 | 0.8 | 1.5 |
| Steel, AISI 1064 | 0.7 | 1.3 | 1.3 | 2.0 |
| Copper | 0.1 | 0.2 | 1.0 | 2.0 |
| Brass | 0.0 | 0.2 | 0.4 | 0.8 |
| Zinc | 0.5 | 1.0 | | |
| Aluminum | 0.0 | 0.2 | 0.3 | 0.8 |
| Aluminum, hard | 1.0 | 1.5 | 3.0 | 4.0 |
| Electron, Mg-metal | 2.0 | 3.0 | 7.0 | 9.0 |
| Titanium | 0.5 | 1.0 | 3.0 | 5.0 |

SOURCE: Svatopluk Černoch, *Strojně technická příručka,* 1977. Reprinted with permission from SNTL Publishers, Prague, CZ.

away from the body of the forming punch, as shown in detail *P.* Such a gap will shift the flange into the way of the forming die, which—on interference—will either shear that portion off or will make an attempt at drawing or ironing that wall. Even a 180° bend, such as a dutch bend or hem, shown in Fig. 8-4*E,* will contain a slight bend radius unless completely flattened out by a coining operation.

Where a punch with no corner radius and with a corner angle under 90° should be used to produce a sharp corner in sheet metal, such a tool will certainly cut more easily through the metal than form it (see Fig. 8-4*G*).

A method of producing a sharp corner in a V die is shown in Fig. 8-4*H.* Here the corner of a part is extended beyond the outline of both flanges, leaving a gap where the radius-forming material would normally be.

Another obstacle in sharp-edge bending is the condition of the edges of a material. Where the sides of the sheet were cut (or sheared) with dull instruments, these edges will display a greater tendency to cracking. The ductility of the material naturally affects this condition, but generally, unless the edges of the part are smooth, preferably deburred, or otherwise finished, cracking may be experienced with very sharp corners.

The *bending punch and die radius* determines the size of the plasticized area created by the bending process. It influences the dimensional accuracy of the part, as well as the amount of spring-back. Values of forming die radii should be approximately

$$R_{\text{die}} = 0.5t \text{ for materials of greater ductility}$$

$$R_{\text{die}} = 0.8t \text{ for materials of lesser ductility}$$

These values are for *across the grain* forming only. Where *along the grain* forming cannot be avoided, the radii should be 20 to 25 percent greater.

The radius of a forming punch should be made in congruence with the following:

$$R_{\text{punch}} = 2t \text{ for thicknesses under } 0.125''$$

$$R_{\text{punch}} = \text{chamfered for thicknesses above } 0.125'' \text{ thick}$$

In the U type of bending, the space between punch and die is significant for the success of the bending procedure. Sometimes, in an attempt to eliminate the spring-back, this space may be diminished toward the bottom of the forming area (see Fig. 8-5*C*). Such an alteration results in increased plastic flow within the material, which in turn limits the occurrence of spring-back.

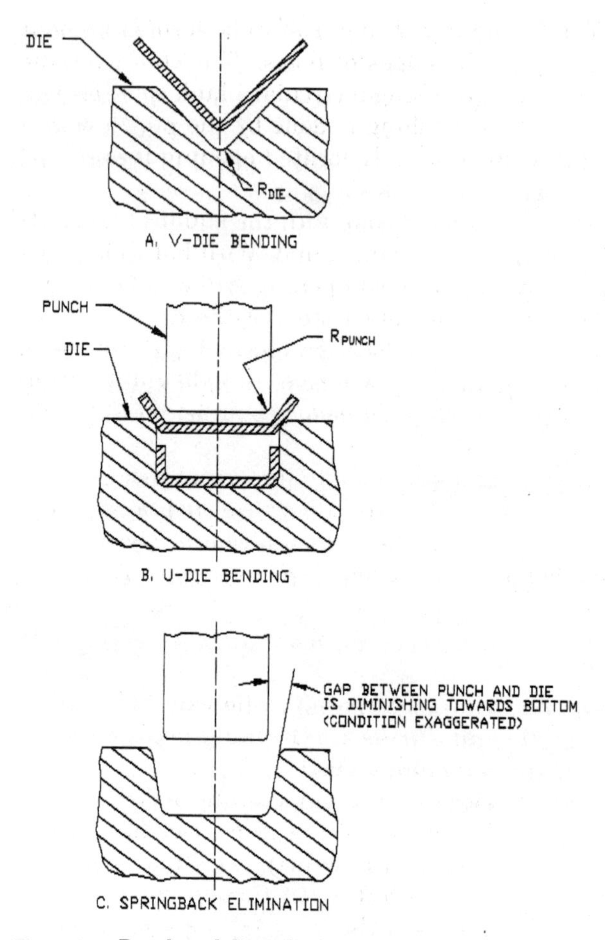


**Figure 8-5** Punch and die radii determination.

The space between the forming punch and die should follow the approximate guidelines:

$$v = 1.05 \text{ to } 1.15t \qquad \text{for steel}$$

$$v = 1.00 \text{ to } 1.10t \qquad \text{for nonferrous metals}$$

The radius of punch and die (see Fig. 8-37, later) should be

$$R_E = (2 \text{ to } 6)t \text{ for nonmoving edges}$$

$$R_P = \text{according to the requirements of the part}$$

$$R_D = R_P + (1.2 \text{ to } 1.25)t$$

The *edge formability* is the ability of material to be formed without fracturing or thinning around the edges of holes. This characteristic is experimentally assessed by stretching a circular blank containing a round hole in the middle. The stretching is done by the punch with a flat bottom, and the edge of the centrally located opening is observed for the appearance of cracks.

Interestingly enough, it was noticed that with the equally dispersed punch pressure, the hole does not remain round. With hot-rolled low-carbon steel used for drawing, the largest opening size was found at a 45° angle off the grain line. In cold-rolled steel of the drawing type, the largest diameter was found at the location parallel with the grain direction or at 0° off the grain line, whereas in cold-rolled high-strength low-alloy steels the *smallest diameter* could be found at 90° off the grain line.

Edge formability is going to be increasingly important owing to the escalated use of high-strength steel. This is because such a mechanical property of the material is crucial in evaluation of sheet-metal behavior in forming and bending operations.

## 8-3. Neutral Axis in Bending

The neutral axis of the material is supposed to do exactly what its name implies: to remain neutral during the bending process and to conform to neither side of the altered material.

However, the neutral axis does not remain totally neutral, as it shifts slightly in the forming operation (Fig. 8-6). Because of the deformation of the bent-up material, the neutral axis moves toward the center of the bend radius. This shift, even though small in size, may be of importance in some areas of the industry, for which reason it is included in the following. Its amount is based on the ratio of the bend radius and material thickness:

| $R/t$ | 0.1 | 0.25 | 0.5 | 1.0 | 2.0 | 3.0 | 4.0 | 6.0 | 10.0 |
|---|---|---|---|---|---|---|---|---|---|
| $v$ | 0.0125 | 0.0135 | 0.015 | 0.0165 | 0.0175 | 0.018 | 0.0185 | 0.019 | 0.019 |

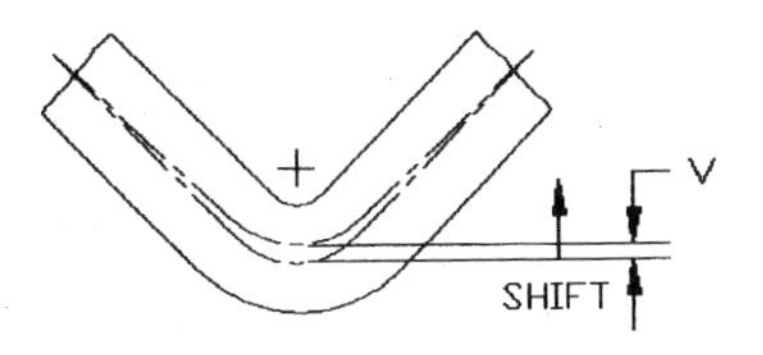


**Figure 8-6** Shift of the neutral axis.

where $v$ is the change in the neutral axis location, directed toward the origin of the bend radius (see Fig. 8-6).

For purposes of calculation of the flat size of a part, the location of neutral axis, presented here as an ingredient of various formulas and tables, has been altered with each type of bending operation performed. These percentile values are based on actual tests and years of experience of toolmakers, designers, and engineers.

## 8-4. Types of Bending Operations

Bending of sheet metal can be accomplished through utilization of several manufacturing processes. First of all, the distinction can be made as far as the bending part's support is concerned: There is supported bending and unsupported bending.

*Unsupported bending* is similar to the process of stretching, where a flat piece of metal, retained in a die, stretches along with the application of tool pressure.

U-die and V-die bending are both considered unsupported bending processes at their beginning stages, as shown in Fig. 8-7. As the bending process continues, and the material is pulled down into the recess, all the way down, the bending becomes supported, as shown later in Fig. 8-9.

Other unsupported bending is wipe bending and rotary bending, shown in Fig. 8-8.

*Supported bending* may be considered any bending where a pad, usually spring-loaded, is included for support of the formed part (see Fig. 8-9*C*). Similarly, the finishing stages of U- and V-die forming are considered supported bending.

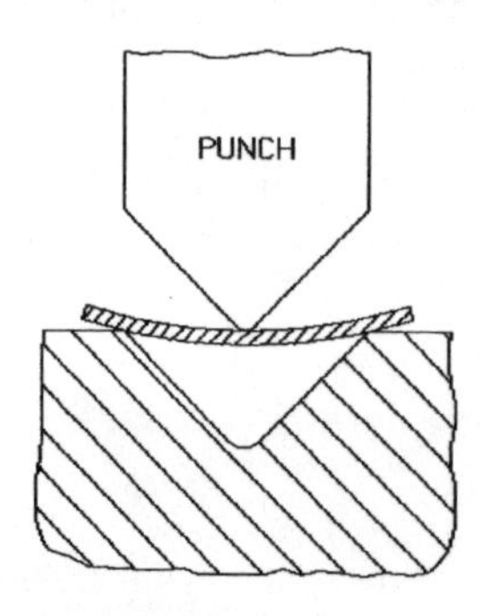


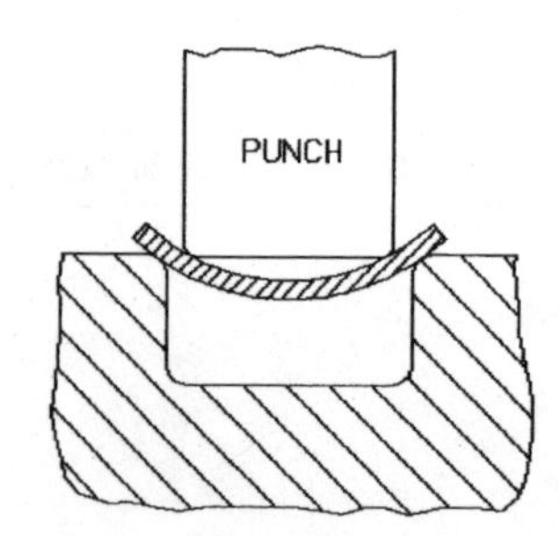


**Figure 8-7** Unsupported bending.

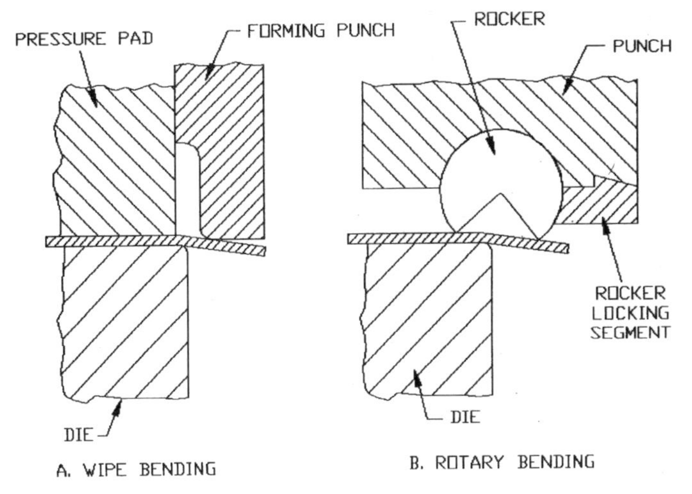


**Figure 8-8** Unsupported bending.

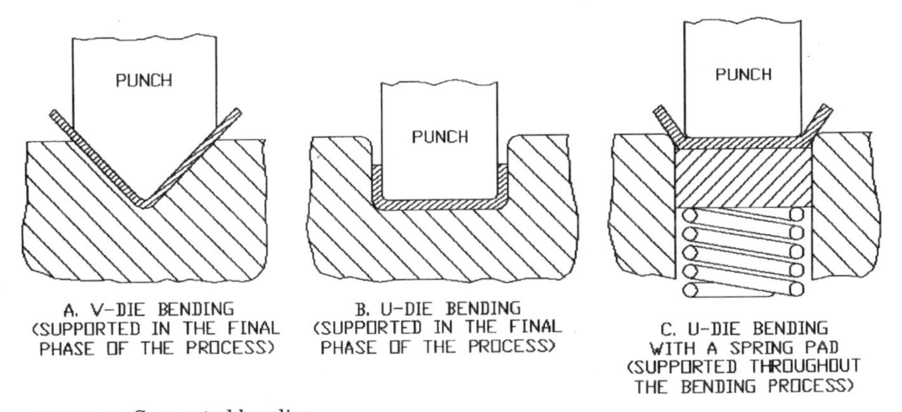


**Figure 8-9** Supported bending.

Supported bending has one advantage over the other methods, the added benefit of coining. As already mentioned, after a part is formed, the material's tendency to return to its original shape may be prevented by overbending its flanges and coining, or bottoming. Overbending consists of bringing the flange to a more acute angular distance than necessary. Coining afterward applies a squeezing pressure to the strains of material, securing them in their present location.

Another category of bending, representing its own group within metal-bending processes, is rotary bending, and its predecessor, a roller bending process. Rotary bending is a patented process, developed by R. J. Gargrave, president of Ready Tools, Inc., Dayton, Ohio. As may be seen from Fig. 8-8*B*, the rocker engages the material strip, and by being forced downward, it forces the strip to conform to the die underneath it.

### 8-4-1. V-Die bending

Even though V-die bending is probably the most inaccurate of all bending processes, it is widely used throughout the industry. The reasons are obvious: The tooling is simple and may be used for more than one flange and for more than one part.

During V-die bending, the punch slides down, coming first to a contact with the unsupported sheet metal. By progressing farther down, it forces the material to follow along, until finally bottoming on the V shape of the die. As may be observed, at the beginning of this process, the sheet is unsupported, but as the operational cycle nears its end, the bent-up part becomes totally supported while retained within the space between the punch and die (Fig. 8-10).

According to the difference of punch and die radii, there are two conditions of their alignment:

- With fully closed punch and die operation, the bottom radius of the die is equal to the amount of punch radius plus the thickness of the material.

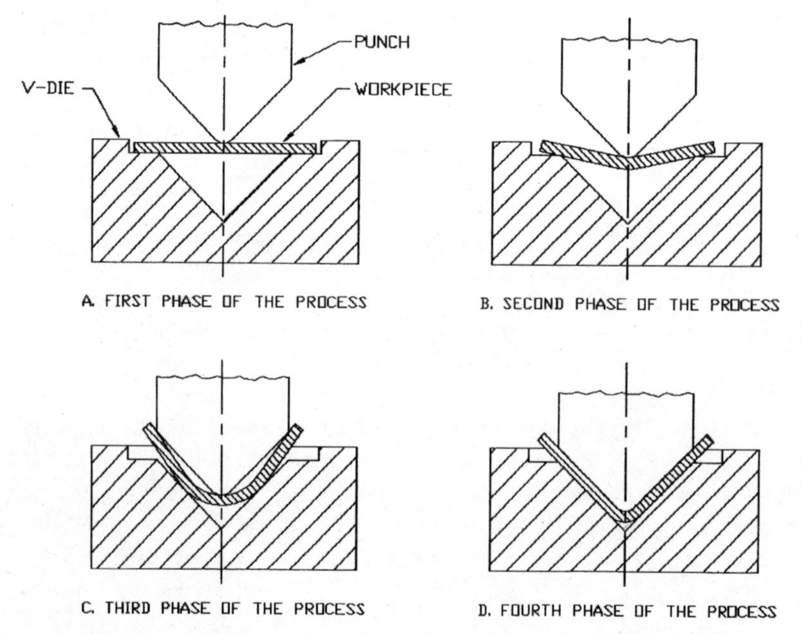


**Figure 8-10** V-die bending.

- With semiclosed punch and die operation, the radius of the die is smaller than the amount of combined punch radius and metal thickness. Usually such a die radius is actually a sharp-corner connection between the two slanted surfaces of the die, or a sharp V (Fig. 8-10).

To calculate the length of a piece in flat, when only bent-up dimensions are given (which is the usual way of dimensioning sheet-metal parts), various formulas that follow may be utilized. These calculations are quite similar, their only difference being the anticipated variation of the location of the neutral axis with respect to the bending or forming process used.

Calculation of the bend allowance for a soft steel material, using a V-die bending or forming process, shifts the neutral axis into one-third of the material thickness. The formula is

$$BA_{\text{V-die}} = C\left(R_{\text{in}} + \frac{t}{3}\right) \tag{8-2}$$

where $BA$ = bend allowance (see also Table 8-5)
$R_{\text{in}}$ = inside radius of the bend
$t$ = material thickness
$C$ = constant, depending on the angle of the bend. For 90° bends, this value is 1.5708. For bends of different angularity, see Table 8-6.

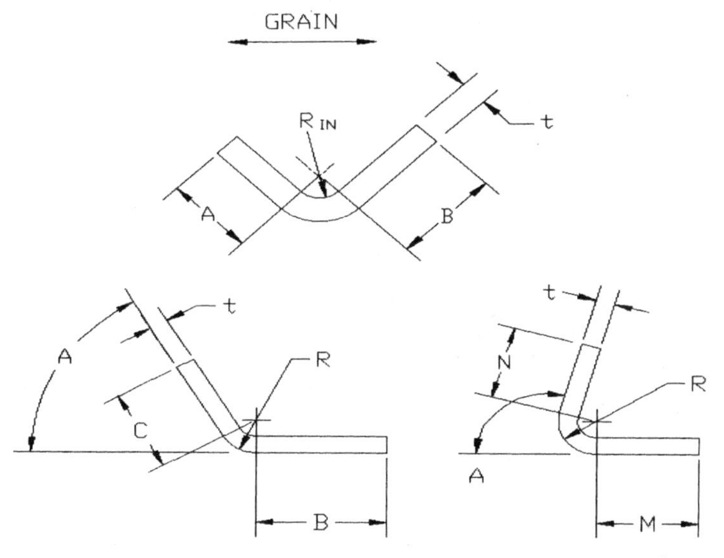


**Figure 8-11** 90° V-die bend calculation for bends across the grain only.

**TABLE 8-5 Bend Allowance Chart for 90° Bends in Annealed Steel V-Die Bending**

| Stock thickness, in | Bend radius, in | | | | | | | | | | | | | | | | |
|---|---|---|---|---|---|---|---|---|---|---|---|---|---|---|---|---|---|
| | 0.015 | 0.031 | 0.062 | 0.093 | 0.125 | 0.156 | 0.187 | 0.218 | 0.250 | 0.281 | 0.312 | 0.343 | 0.375 | 0.406 | 0.437 | 0.468 | 0.500 |
| 0.015 | 0.031 | 0.057 | 0.105 | 0.154 | 0.204 | 0.253 | 0.302 | 0.350 | 0.401 | 0.449 | 0.498 | 0.547 | 0.597 | 0.646 | 0.694 | 0.743 | 0.793 |
| 0.018 | 0.033 | 0.058 | 0.107 | 0.156 | 0.206 | 0.254 | 0.303 | 0.352 | 0.402 | 0.451 | 0.500 | 0.548 | 0.598 | 0.647 | 0.696 | 0.745 | 0.795 |
| 0.024 | 0.036 | 0.061 | 0.110 | 0.159 | 0.209 | 0.258 | 0.306 | 0.355 | 0.405 | 0.454 | 0.503 | 0.551 | 0.602 | 0.650 | 0.699 | 0.748 | 0.798 |
| 0.031 | 0.040 | 0.065 | 0.114 | 0.162 | 0.213 | 0.261 | 0.310 | 0.359 | 0.409 | 0.458 | 0.506 | 0.555 | 0.605 | 0.654 | 0.703 | 0.751 | 0.802 |
| 0.036 | 0.042 | 0.068 | 0.116 | 0.165 | 0.215 | 0.264 | 0.313 | 0.361 | 0.412 | 0.460 | 0.509 | 0.558 | 0.608 | 0.657 | 0.705 | 0.754 | 0.804 |
| 0.048 | 0.049 | 0.074 | 0.123 | 0.171 | 0.221 | 0.270 | 0.319 | 0.368 | 0.418 | 0.467 | 0.515 | 0.564 | 0.614 | 0.663 | 0.712 | 0.760 | 0.811 |
| 0.054 | 0.052 | 0.077 | 0.126 | 0.174 | 0.225 | 0.273 | 0.322 | 0.371 | 0.421 | 0.470 | 0.518 | 0.567 | 0.617 | 0.666 | 0.715 | 0.763 | 0.814 |
| 0.062 | 0.056 | 0.081 | 0.130 | 0.179 | 0.229 | 0.278 | 0.326 | 0.375 | 0.425 | 0.474 | 0.523 | 0.571 | 0.622 | 0.670 | 0.719 | 0.768 | 0.818 |
| 0.075 | 0.063 | 0.088 | 0.137 | 0.185 | 0.236 | 0.284 | 0.333 | 0.382 | 0.432 | 0.481 | 0.529 | 0.578 | 0.628 | 0.677 | 0.726 | 0.774 | 0.825 |
| 0.093 | 0.072 | 0.097 | 0.146 | 0.195 | 0.245 | 0.294 | 0.342 | 0.391 | 0.441 | 0.490 | 0.539 | 0.587 | 0.638 | 0.686 | 0.735 | 0.784 | 0.834 |
| 0.109 | 0.081 | 0.106 | 0.154 | 0.203 | 0.253 | 0.302 | 0.351 | 0.400 | 0.450 | 0.498 | 0.547 | 0.596 | 0.646 | 0.695 | 0.744 | 0.792 | 0.842 |
| 0.125 | 0.089 | 0.114 | 0.163 | 0.212 | 0.262 | 0.310 | 0.359 | 0.408 | 0.458 | 0.507 | 0.556 | 0.604 | 0.655 | 0.703 | 0.752 | 0.801 | 0.851 |
| 0.140 | 0.097 | 0.122 | 0.171 | 0.219 | 0.270 | 0.318 | 0.367 | 0.416 | 0.466 | 0.515 | 0.563 | 0.612 | 0.662 | 0.711 | 0.760 | 0.808 | 0.859 |
| 0.156 | 0.105 | 0.130 | 0.179 | 0.228 | 0.278 | 0.327 | 0.375 | 0.424 | 0.474 | 0.523 | 0.572 | 0.620 | 0.671 | 0.719 | 0.768 | 0.817 | 0.867 |
| 0.187 | 0.121 | 0.147 | 0.195 | 0.244 | 0.294 | 0.343 | 0.392 | 0.440 | 0.491 | 0.539 | 0.588 | 0.637 | 0.687 | 0.736 | 0.784 | 0.833 | 0.883 |
| 0.203 | 0.130 | 0.155 | 0.204 | 0.252 | 0.303 | 0.351 | 0.400 | 0.449 | 0.499 | 0.548 | 0.596 | 0.645 | 0.695 | 0.744 | 0.793 | 0.841 | 0.892 |
| 0.218 | 0.138 | 0.163 | 0.212 | 0.260 | 0.310 | 0.359 | 0.408 | 0.457 | 0.507 | 0.556 | 0.604 | 0.653 | 0.703 | 0.752 | 0.801 | 0.849 | 0.900 |
| 0.234 | 0.146 | 0.171 | 0.220 | 0.269 | 0.319 | 0.368 | 0.416 | 0.465 | 0.515 | 0.564 | 0.613 | 0.661 | 0.712 | 0.760 | 0.809 | 0.858 | 0.908 |
| 0.250 | 0.154 | 0.180 | 0.228 | 0.277 | 0.327 | 0.376 | 0.425 | 0.473 | 0.524 | 0.572 | 0.621 | 0.670 | 0.720 | 0.769 | 0.817 | 0.866 | 0.916 |

**TABLE 8-6 Bend Allowance Constants for Bends Across the Grain Only**

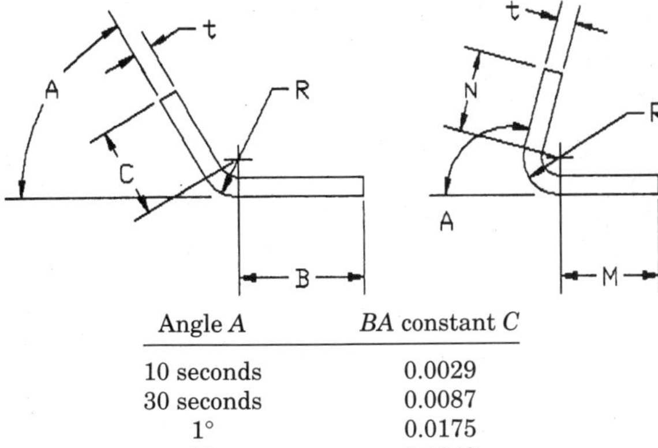


| Angle $A$ | $BA$ constant $C$ |
|---|---|
| 10 seconds | 0.0029 |
| 30 seconds | 0.0087 |
| 1° | 0.0175 |
| 2° | 0.0349 |
| 3° | 0.0524 |
| 4° | 0.0698 |
| 5° | 0.0873 |
| 10° | 0.1745 |
| 15° | 0.2618 |
| 20° | 0.3491 |
| 25° | 0.4363 |
| 30° | 0.5236 |
| 35° | 0.6109 |
| 40° | 0.6981 |
| 45° | 0.7854 |
| 50° | 0.8728 |
| 55° | 0.9599 |
| 60° | 1.0472 |
| 65° | 1.1345 |
| 70° | 1.2217 |
| 75° | 1.3090 |
| 80° | 1.3963 |
| 85° | 1.4835 |
| 90° | 1.5708 |

Values may vary with different types of material. For bends smaller than 90° use the constants given. For bends greater than 90° combine the constant for 90° with the constant for the angular excess.

The bend allowance per Eq. (8-2) would be further used to calculate the total length of the part, $L_{total}$ (see Fig. 8-11), as follows:

$$L_{total} = BA + A + B \tag{8-3a}$$

Numerous adaptations of V-die bending exist throughout the manufacturing field. Some examples are shown in Fig. 8-12.

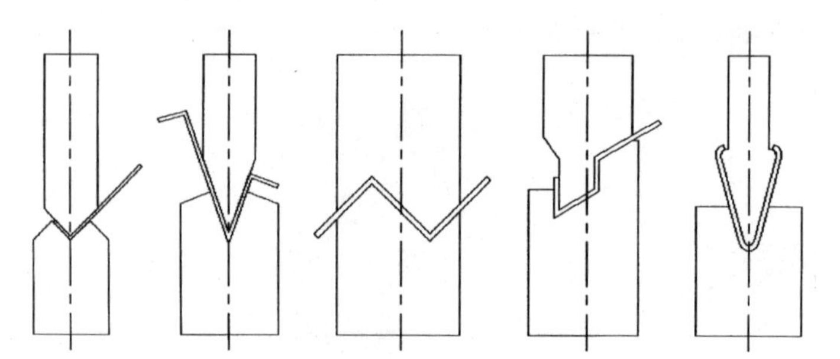

Figure 8-12 V-die bending samples.

### 8-4-2. U-Die bending

In this type of bending, the process begins with a sheet of metal positioned over a U-shaped opening or an insert of such a shape. As the punch comes down, it contacts the sheet material first and pulls it along on further descent, forcing it into the U-shaped opening.

Calculation of the bend allowance for a bent or formed flange, made of hard steel, considers the neutral axis to be located in the middle of the material thickness. Such a formula is as follows:

$$BA = C\left(R_{in} + \frac{t}{2}\right) \tag{8-4}$$

where $R_{in}$ = inside radius of the bend
$t$ = material thickness
$C$ = constant, depending on the angle of the bend. For 90° bends, this value is 1.5708. For bends of different angularity, see Table 8-6.

The bend allowance is further used the same way as shown in Eq. (8-3*a*), where all abbreviations are based on Fig. 8-13. A chart of calculated bend allowance values is shown in Table 8-7.

The bend allowance formula for soft steel, formed or bent in a U die, or for a condition where the metal is drawn over the edge of either punch or die, is Eq. (8-5).

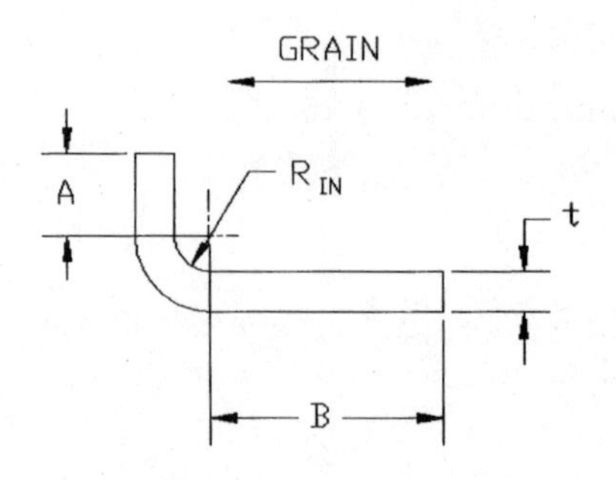


Figure 8-13 90° bend calculation.

**TABLE 8-7 Bend Allowance Chart for 90° Bends in Hardened Steel for Bending, Forming, Drawing Calculations, Including Forming of Round Stock**

| Stock thickness, in | Bend radius, in | | | | | | | | | | | | | | | | |
|---|---|---|---|---|---|---|---|---|---|---|---|---|---|---|---|---|---|
| | 0.015 | 0.031 | 0.062 | 0.093 | 0.125 | 0.156 | 0.187 | 0.218 | 0.250 | 0.281 | 0.312 | 0.343 | 0.375 | 0.406 | 0.437 | 0.468 | 0.500 |
| 0.015 | 0.035 | 0.060 | 0.109 | 0.158 | 0.208 | 0.257 | 0.306 | 0.354 | 0.404 | 0.453 | 0.502 | 0.551 | 0.601 | 0.650 | 0.698 | 0.747 | 0.797 |
| 0.018 | 0.038 | 0.063 | 0.112 | 0.160 | 0.210 | 0.259 | 0.308 | 0.357 | 0.407 | 0.456 | 0.504 | 0.553 | 0.603 | 0.652 | 0.701 | 0.749 | 0.800 |
| 0.024 | 0.042 | 0.068 | 0.116 | 0.165 | 0.215 | 0.264 | 0.313 | 0.361 | 0.412 | 0.460 | 0.509 | 0.558 | 0.608 | 0.657 | 0.705 | 0.754 | 0.804 |
| 0.031 | 0.048 | 0.073 | 0.122 | 0.170 | 0.221 | 0.269 | 0.318 | 0.367 | 0.417 | 0.466 | 0.514 | 0.563 | 0.613 | 0.662 | 0.711 | 0.759 | 0.810 |
| 0.036 | 0.052 | 0.077 | 0.126 | 0.174 | 0.225 | 0.273 | 0.322 | 0.371 | 0.421 | 0.470 | 0.518 | 0.567 | 0.617 | 0.666 | 0.715 | 0.763 | 0.814 |
| 0.048 | 0.061 | 0.086 | 0.135 | 0.184 | 0.234 | 0.283 | 0.331 | 0.380 | 0.430 | 0.479 | 0.528 | 0.576 | 0.627 | 0.675 | 0.724 | 0.773 | 0.823 |
| 0.054 | 0.066 | 0.091 | 0.140 | 0.188 | 0.239 | 0.287 | 0.336 | 0.385 | 0.435 | 0.484 | 0.533 | 0.581 | 0.631 | 0.680 | 0.729 | 0.778 | 0.828 |
| 0.062 | 0.072 | 0.097 | 0.146 | 0.195 | 0.245 | 0.294 | 0.342 | 0.391 | 0.441 | 0.490 | 0.539 | 0.587 | 0.638 | 0.686 | 0.735 | 0.784 | 0.834 |
| 0.075 | 0.082 | 0.108 | 0.156 | 0.205 | 0.255 | 0.304 | 0.353 | 0.401 | 0.452 | 0.500 | 0.549 | 0.598 | 0.648 | 0.697 | 0.745 | 0.794 | 0.844 |
| 0.093 | 0.097 | 0.122 | 0.170 | 0.219 | 0.269 | 0.318 | 0.367 | 0.415 | 0.466 | 0.514 | 0.563 | 0.612 | 0.662 | 0.711 | 0.759 | 0.808 | 0.858 |
| 0.109 | 0.109 | 0.134 | 0.183 | 0.232 | 0.282 | 0.331 | 0.379 | 0.428 | 0.478 | 0.527 | 0.576 | 0.624 | 0.675 | 0.723 | 0.772 | 0.821 | 0.871 |
| 0.125 | 0.122 | 0.147 | 0.196 | 0.244 | 0.295 | 0.343 | 0.392 | 0.441 | 0.491 | 0.540 | 0.588 | 0.637 | 0.687 | 0.736 | 0.785 | 0.833 | 0.884 |
| 0.140 | 0.134 | 0.159 | 0.207 | 0.256 | 0.306 | 0.355 | 0.404 | 0.452 | 0.503 | 0.551 | 0.600 | 0.649 | 0.699 | 0.748 | 0.796 | 0.845 | 0.895 |
| 0.156 | 0.146 | 0.171 | 0.220 | 0.269 | 0.319 | 0.368 | 0.416 | 0.465 | 0.515 | 0.564 | 0.613 | 0.661 | 0.712 | 0.760 | 0.809 | 0.858 | 0.908 |
| 0.187 | 0.170 | 0.196 | 0.244 | 0.293 | 0.343 | 0.392 | 0.441 | 0.489 | 0.540 | 0.588 | 0.637 | 0.686 | 0.736 | 0.785 | 0.833 | 0.882 | 0.932 |
| 0.203 | 0.183 | 0.208 | 0.257 | 0.306 | 0.356 | 0.404 | 0.453 | 0.502 | 0.552 | 0.601 | 0.650 | 0.698 | 0.748 | 0.797 | 0.846 | 0.895 | 0.945 |
| 0.218 | 0.195 | 0.220 | 0.269 | 0.317 | 0.368 | 0.416 | 0.465 | 0.514 | 0.564 | 0.613 | 0.661 | 0.710 | 0.760 | 0.809 | 0.858 | 0.906 | 0.957 |
| 0.234 | 0.207 | 0.232 | 0.281 | 0.330 | 0.380 | 0.429 | 0.478 | 0.526 | 0.576 | 0.625 | 0.674 | 0.723 | 0.773 | 0.822 | 0.870 | 0.919 | 0.969 |
| 0.250 | 0.220 | 0.245 | 0.284 | 0.342 | 0.393 | 0.441 | 0.490 | 0.539 | 0.589 | 0.638 | 0.686 | 0.735 | 0.785 | 0.834 | 0.883 | 0.931 | 0.982 |

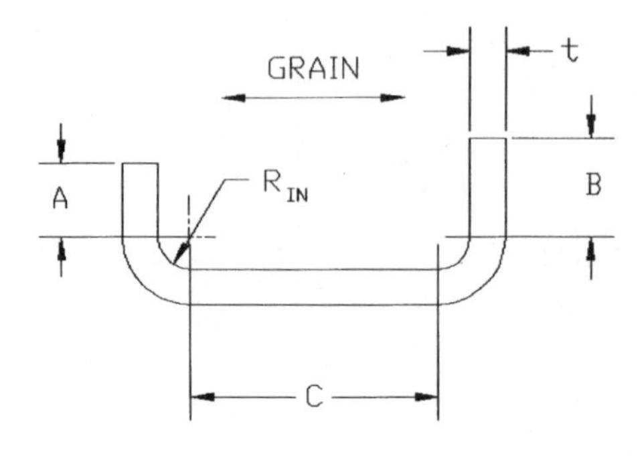


**Figure 8-14** U-type bend calculation.

$$BA_{U\text{-die}} = C\left(R_{in} + \frac{t}{4}\right) \tag{8-5}$$

The bend allowance is further used to calculate the total length of the part, using the altered formula (8-3*a*):

$$L_{\text{total}} = BA + A + B + C \tag{8-3b}$$

The minimum length of a U-forming punch should be equal to the inner length of the formed part, plus $2t$.

Three segments of a bending process, shown in Fig. 8-15, depict an unsupported forming in the first stage, followed by a stretching of the material due to the punch pressure in the second stage. The last operation, bottoming, shows the tendency of the bulging bottom to allocate the excess material within the corners of the part. A subsequent bending of flanges toward the body of a punch may be observed.

In U-shaped die forming, the difference in the outcome is introduced with an addition of a spring-loaded pressure pad underneath the part (see Fig. 8-16). The blank, when pulled by the punch into the die opening, is firmly supported by the pressure pad already at the beginning of the forming operation. This way the punch cannot stretch the material, leaving the bottom of the part flat.

When the punch-metal-pad sandwich finally bottoms, the formed part remains the same, with no bulging or distortion of any kind.

### 8-4-3. Wipe bending

In the wipe bending method of producing flanges, the blank is retained in a fixed position by the spring-loaded pressure pad (Fig. 8-17). The forming punch comes down toward the exposed flange and bends it during its further descent.

The formula to use for this type of bend allowance is as follows:

$$BA_{\text{wipe}} = C\left(R_{\text{in}} + \frac{t}{5}\right) \tag{8-6}$$

**TABLE 8-8 Bend Allowance Chart for 90° Bends in Annealed Steel Wipe Bending or U-Die Bending, Across the Grain**

| Stock thickness, in | Bend radius, in | | | | | | | | | | | | | | | | |
|---|---|---|---|---|---|---|---|---|---|---|---|---|---|---|---|---|---|
| | 0.015 | 0.031 | 0.062 | 0.093 | 0.125 | 0.156 | 0.187 | 0.218 | 0.250 | 0.281 | 0.312 | 0.343 | 0.375 | 0.406 | 0.437 | 0.468 | 0.500 |
| 0.015 | 0.029 | 0.055 | 0.103 | 0.152 | 0.202 | 0.251 | 0.300 | 0.348 | 0.399 | 0.447 | 0.496 | 0.545 | 0.595 | 0.644 | 0.692 | 0.741 | 0.791 |
| 0.018 | 0.031 | 0.056 | 0.104 | 0.153 | 0.203 | 0.252 | 0.301 | 0.350 | 0.400 | 0.448 | 0.497 | 0.546 | 0.596 | 0.645 | 0.694 | 0.742 | 0.792 |
| 0.024 | 0.033 | 0.058 | 0.107 | 0.156 | 0.206 | 0.254 | 0.303 | 0.352 | 0.402 | 0.451 | 0.500 | 0.548 | 0.598 | 0.647 | 0.696 | 0.745 | 0.795 |
| 0.031 | 0.036 | 0.061 | 0.110 | 0.158 | 0.209 | 0.257 | 0.306 | 0.355 | 0.405 | 0.454 | 0.502 | 0.551 | 0.601 | 0.650 | 0.699 | 0.747 | 0.798 |
| 0.036 | 0.038 | 0.063 | 0.112 | 0.160 | 0.210 | 0.259 | 0.308 | 0.357 | 0.407 | 0.456 | 0.504 | 0.553 | 0.603 | 0.652 | 0.701 | 0.749 | 0.800 |
| 0.048 | 0.042 | 0.068 | 0.116 | 0.165 | 0.215 | 0.264 | 0.313 | 0.361 | 0.412 | 0.460 | 0.509 | 0.558 | 0.608 | 0.657 | 0.705 | 0.754 | 0.804 |
| 0.054 | 0.045 | 0.070 | 0.119 | 0.167 | 0.218 | 0.266 | 0.315 | 0.364 | 0.414 | 0.463 | 0.511 | 0.560 | 0.610 | 0.659 | 0.708 | 0.756 | 0.807 |
| 0.062 | 0.048 | 0.073 | 0.122 | 0.170 | 0.221 | 0.269 | 0.318 | 0.367 | 0.417 | 0.466 | 0.514 | 0.563 | 0.613 | 0.662 | 0.711 | 0.759 | 0.810 |
| 0.075 | 0.053 | 0.078 | 0.127 | 0.176 | 0.226 | 0.274 | 0.323 | 0.372 | 0.422 | 0.471 | 0.520 | 0.568 | 0.619 | 0.667 | 0.716 | 0.765 | 0.815 |
| 0.093 | 0.060 | 0.085 | 0.134 | 0.183 | 0.233 | 0.282 | 0.330 | 0.379 | 0.429 | 0.478 | 0.527 | 0.575 | 0.626 | 0.674 | 0.723 | 0.772 | 0.822 |
| 0.109 | 0.066 | 0.091 | 0.140 | 0.189 | 0.239 | 0.288 | 0.337 | 0.385 | 0.436 | 0.484 | 0.533 | 0.582 | 0.632 | 0.681 | 0.729 | 0.778 | 0.828 |
| 0.125 | 0.073 | 0.098 | 0.146 | 0.195 | 0.245 | 0.294 | 0.343 | 0.392 | 0.442 | 0.490 | 0.539 | 0.588 | 0.638 | 0.687 | 0.736 | 0.784 | 0.834 |
| 0.140 | 0.079 | 0.104 | 0.152 | 0.201 | 0.251 | 0.300 | 0.349 | 0.397 | 0.448 | 0.496 | 0.545 | 0.594 | 0.644 | 0.693 | 0.741 | 0.790 | 0.840 |
| 0.156 | 0.085 | 0.110 | 0.159 | 0.207 | 0.258 | 0.306 | 0.355 | 0.404 | 0.454 | 0.503 | 0.551 | 0.600 | 0.650 | 0.699 | 0.748 | 0.796 | 0.847 |
| 0.187 | 0.097 | 0.122 | 0.171 | 0.220 | 0.270 | 0.318 | 0.367 | 0.416 | 0.466 | 0.515 | 0.564 | 0.612 | 0.662 | 0.711 | 0.760 | 0.809 | 0.859 |
| 0.203 | 0.103 | 0.128 | 0.177 | 0.226 | 0.276 | 0.325 | 0.373 | 0.422 | 0.472 | 0.521 | 0.570 | 0.619 | 0.669 | 0.717 | 0.766 | 0.815 | 0.865 |
| 0.218 | 0.109 | 0.134 | 0.183 | 0.232 | 0.282 | 0.331 | 0.379 | 0.428 | 0.478 | 0.527 | 0.576 | 0.624 | 0.675 | 0.723 | 0.772 | 0.821 | 0.871 |
| 0.234 | 0.115 | 0.141 | 0.189 | 0.238 | 0.288 | 0.337 | 0.386 | 0.434 | 0.485 | 0.533 | 0.582 | 0.631 | 0.681 | 0.730 | 0.778 | 0.827 | 0.877 |
| 0.250 | 0.122 | 0.147 | 0.196 | 0.244 | 0.295 | 0.343 | 0.392 | 0.441 | 0.491 | 0.540 | 0.588 | 0.637 | 0.687 | 0.736 | 0.785 | 0.833 | 0.884 |

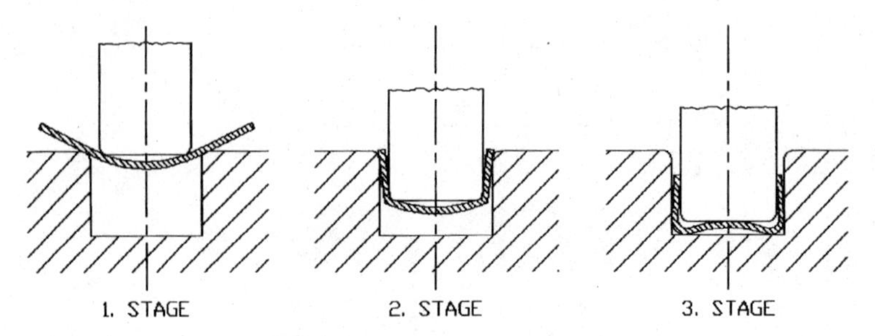


**Figure 8-15** U-die bending process.

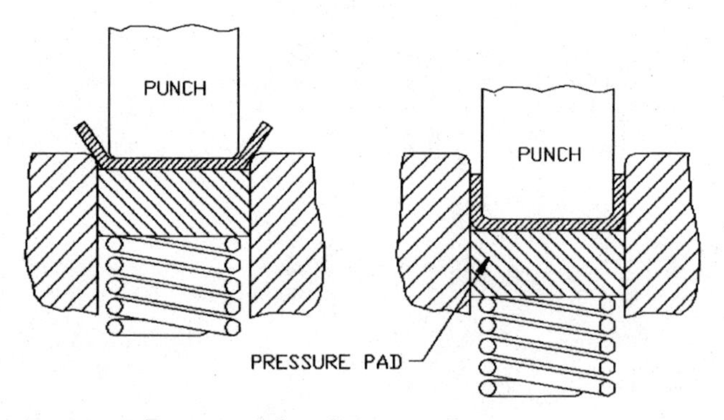


**Figure 8-16** Pressure pad application in forming.

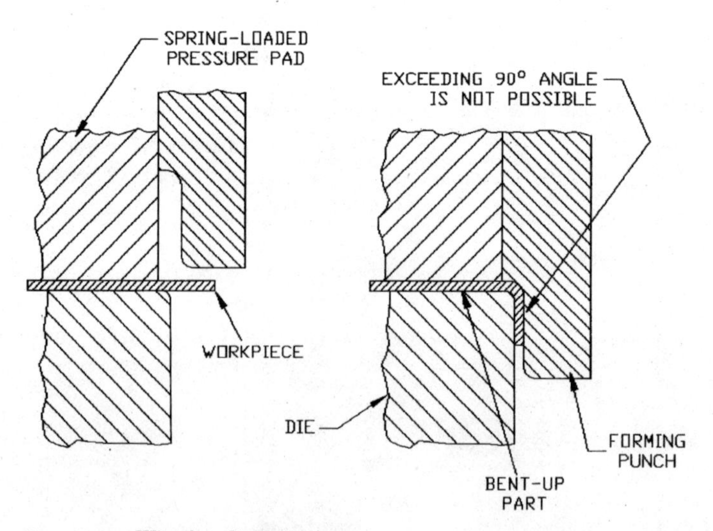


**Figure 8-17** Wipe bending process.

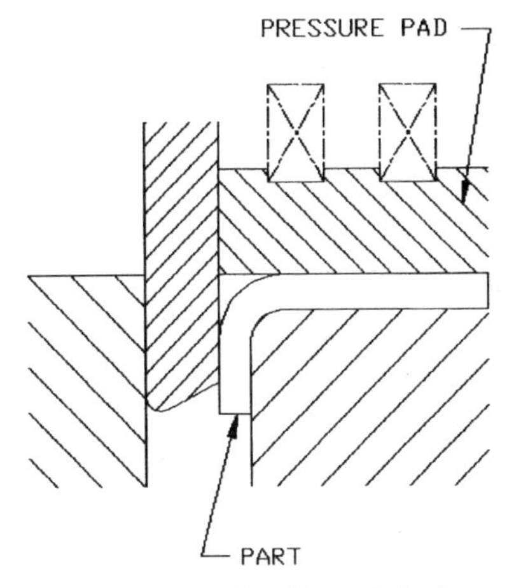


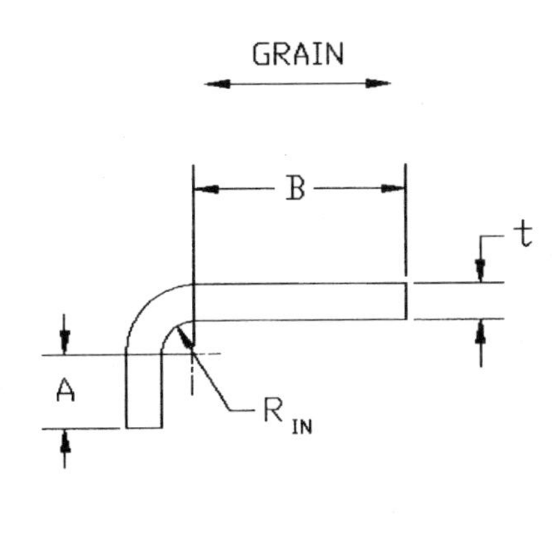


**Figure 8-18** Wipe bending calculation.

where $R_{in}$ = inside radius of the bend
$t$ = material thickness
$C$ = constant, depending on the angle of the bend. For 90° bends, this value is 1.5708. For bends of different angularity, see Table 8-6.

The bend allowance is further used to calculate the total length of the part, as shown in Eqs. (8-3*a*) or (8-3*b*). A chart of calculated bend allowance values is shown in Tables 8-9 and 8-8.

### 8-4-4. Rotary bending

Rotary bending has several advantages over traditional types of bending. Not only does it utilize less bending force, it also does not need a pressure pad for retention of the material, as the rocker provides for it automatically (Fig. 8-19).

As the tool comes down, the rocker lands on the material, positioning itself with one edge over the die and the other over the gap. Coming farther down, its pressure bends down the flange, but it does not stop at 90°; it continues farther to attain a 2° overbend as a protection against spring-back.

The rocker is shaped like a cylinder with an angular portion of its entire length cut out. Both the rocker and the punch plate are usually made of air-hardening tool steel, the rocker being around 50 Rockwell hardness Scale C. The punch plate should be harder, up to 60 Rockwell hardness Scale C. The die surface must be slanted, the

**TABLE 8-9 Bend Allowance Chart for 90° Bends in Annealed Steel Wipe Bending with a Spring Pad, Across the Grain**

| Stock thickness, in | Bend radius, in | | | | | | | | | | | | | | | | |
|---|---|---|---|---|---|---|---|---|---|---|---|---|---|---|---|---|---|
| | 0.015 | 0.031 | 0.062 | 0.093 | 0.125 | 0.156 | 0.187 | 0.218 | 0.250 | 0.281 | 0.312 | 0.343 | 0.375 | 0.406 | 0.437 | 0.468 | 0.500 |
| 0.015 | 0.028 | 0.053 | 0.102 | 0.151 | 0.201 | 0.250 | 0.298 | 0.347 | 0.397 | 0.446 | 0.495 | 0.543 | 0.594 | 0.642 | 0.691 | 0.740 | 0.790 |
| 0.018 | 0.029 | 0.054 | 0.103 | 0.152 | 0.202 | 0.251 | 0.299 | 0.348 | 0.398 | 0.447 | 0.496 | 0.544 | 0.595 | 0.643 | 0.692 | 0.741 | 0.791 |
| 0.024 | 0.031 | 0.056 | 0.105 | 0.154 | 0.204 | 0.253 | 0.301 | 0.350 | 0.400 | 0.449 | 0.498 | 0.546 | 0.597 | 0.645 | 0.694 | 0.743 | 0.793 |
| 0.031 | 0.033 | 0.058 | 0.107 | 0.156 | 0.206 | 0.255 | 0.303 | 0.352 | 0.402 | 0.451 | 0.500 | 0.549 | 0.599 | 0.647 | 0.696 | 0.745 | 0.795 |
| 0.036 | 0.035 | 0.060 | 0.109 | 0.157 | 0.208 | 0.256 | 0.305 | 0.354 | 0.404 | 0.453 | 0.501 | 0.550 | 0.600 | 0.649 | 0.698 | 0.746 | 0.797 |
| 0.048 | 0.039 | 0.064 | 0.112 | 0.161 | 0.211 | 0.260 | 0.309 | 0.358 | 0.408 | 0.456 | 0.505 | 0.554 | 0.604 | 0.653 | 0.702 | 0.750 | 0.800 |
| 0.054 | 0.041 | 0.066 | 0.114 | 0.163 | 0.213 | 0.262 | 0.311 | 0.359 | 0.410 | 0.458 | 0.507 | 0.556 | 0.606 | 0.655 | 0.703 | 0.752 | 0.802 |
| 0.062 | 0.043 | 0.068 | 0.117 | 0.166 | 0.216 | 0.265 | 0.313 | 0.362 | 0.412 | 0.461 | 0.510 | 0.558 | 0.609 | 0.657 | 0.706 | 0.755 | 0.805 |
| 0.075 | 0.047 | 0.072 | 0.121 | 0.170 | 0.220 | 0.269 | 0.317 | 0.366 | 0.416 | 0.465 | 0.514 | 0.562 | 0.613 | 0.661 | 0.710 | 0.759 | 0.809 |
| 0.093 | 0.053 | 0.078 | 0.127 | 0.175 | 0.226 | 0.274 | 0.323 | 0.372 | 0.422 | 0.471 | 0.519 | 0.568 | 0.618 | 0.667 | 0.716 | 0.764 | 0.815 |
| 0.109 | 0.058 | 0.083 | 0.132 | 0.180 | 0.231 | 0.279 | 0.328 | 0.377 | 0.427 | 0.476 | 0.524 | 0.573 | 0.623 | 0.672 | 0.721 | 0.769 | 0.820 |
| 0.125 | 0.063 | 0.088 | 0.137 | 0.185 | 0.236 | 0.284 | 0.333 | 0.382 | 0.432 | 0.481 | 0.529 | 0.578 | 0.628 | 0.677 | 0.726 | 0.774 | 0.825 |
| 0.140 | 0.068 | 0.093 | 0.141 | 0.190 | 0.240 | 0.289 | 0.338 | 0.386 | 0.437 | 0.485 | 0.534 | 0.583 | 0.633 | 0.682 | 0.730 | 0.779 | 0.829 |
| 0.156 | 0.073 | 0.098 | 0.146 | 0.195 | 0.245 | 0.294 | 0.343 | 0.391 | 0.442 | 0.490 | 0.539 | 0.588 | 0.638 | 0.687 | 0.735 | 0.784 | 0.834 |
| 0.187 | 0.082 | 0.107 | 0.156 | 0.205 | 0.255 | 0.304 | 0.352 | 0.401 | 0.451 | 0.500 | 0.549 | 0.598 | 0.648 | 0.696 | 0.745 | 0.794 | 0.844 |
| 0.203 | 0.087 | 0.112 | 0.161 | 0.210 | 0.260 | 0.309 | 0.358 | 0.406 | 0.456 | 0.505 | 0.554 | 0.603 | 0.653 | 0.702 | 0.750 | 0.799 | 0.849 |
| 0.218 | 0.092 | 0.117 | 0.166 | 0.215 | 0.265 | 0.314 | 0.362 | 0.411 | 0.461 | 0.510 | 0.559 | 0.607 | 0.658 | 0.706 | 0.755 | 0.804 | 0.854 |
| 0.234 | 0.097 | 0.122 | 0.171 | 0.220 | 0.270 | 0.319 | 0.367 | 0.416 | 0.466 | 0.515 | 0.564 | 0.612 | 0.663 | 0.711 | 0.760 | 0.809 | 0.859 |
| 0.250 | 0.102 | 0.127 | 0.176 | 0.225 | 0.275 | 0.324 | 0.372 | 0.421 | 0.471 | 0.520 | 0.569 | 0.617 | 0.668 | 0.716 | 0.765 | 0.814 | 0.864 |

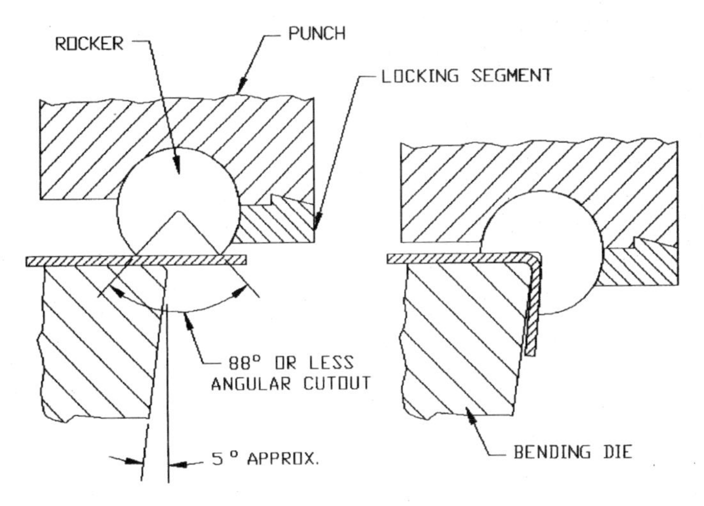


**Figure 8-19** Rotary bending process.

angle of incline being greater than the amount of overbend to provide clearance for the flange (see Fig. 8-19).

One drawback of this type of forming is the assembly's sensitivity to the variation of stock thickness.

**Bending with rotary inserts.** This type of bending is actually an alternative of U-die application, with rotary inserts placed at the corners of a die (Fig. 8-20). The inserts are spring-loaded, allowing the material to land upon them, and retreating under the press force. On release, these rotary inserts force the part up.

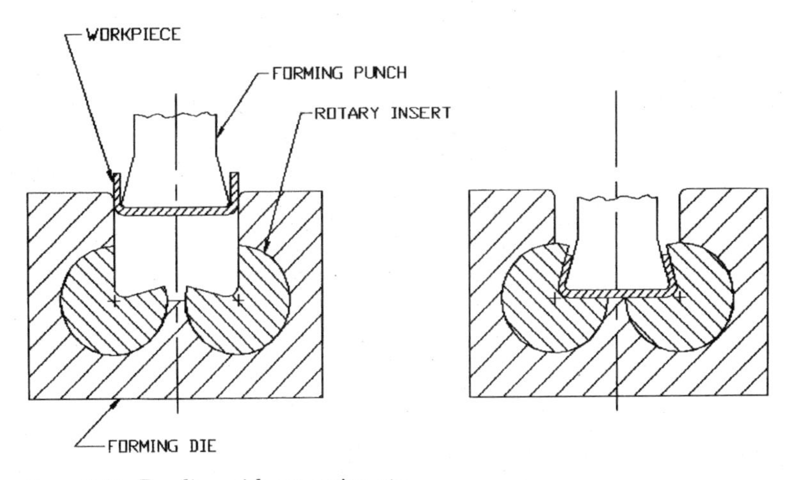


**Figure 8-20** Bending with rotary inserts.

One advantage of this process is the possibility of overbending the flanges as a protection against spring-back occurrence.

**Bending with a pivoted roller.** The roller is attached to the punch plate by a pin. As the upper section of the die slides down, the roller engages the material, forcing it down and under the nose in the die block (see Fig. 8-21).

### 8-4-5. Bending with flexible tooling

Bending with flexible tooling utilizes rubber or urethane forming pads instead of hard-metal tooling (Figs. 8-22 to 8-24). Its advantage lies in the possibility of forcing the flexible tooling material to fill gaps and undercuts, taking the in-between sheet-metal strip along.

Forces necessary for bending with flexible tooling are higher than those needed for conventional bending methods. These differences have not yet been assessed because of the great amount of variables involved. The strain-hardening tendency of the material will have to be evaluated in comparison with the elastic properties of the flexible forming material, aside from other influencing aspects.

With V-die bending utilizing elastic material, the bending process is rather different from the one using hard tooling. First, the sheet-metal material is compressed by the nose of a tool, elongating tangentially under its pressure (see Fig. 8-25). The pressure of the elastic pad is restricted to quite a small area immediately surrounding the tool impression.

The radius of the bent-up part usually ends up being larger than that of the tooling. Enlarging the radius of the V die does not readily

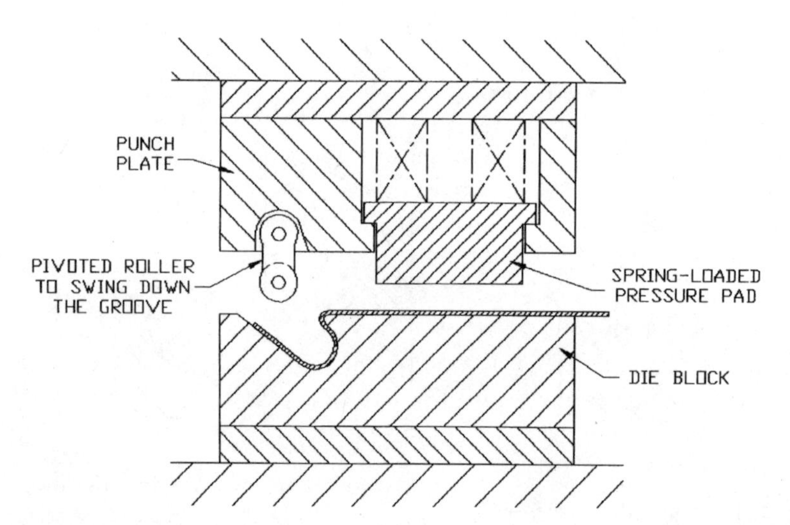


**Figure 8-21** Roller-type bending die.

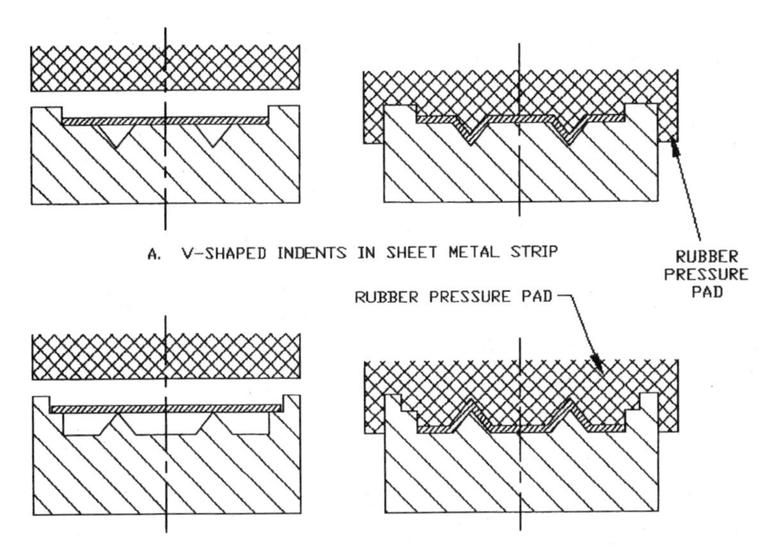


**Figure 8-22** Bending with flexible tooling.

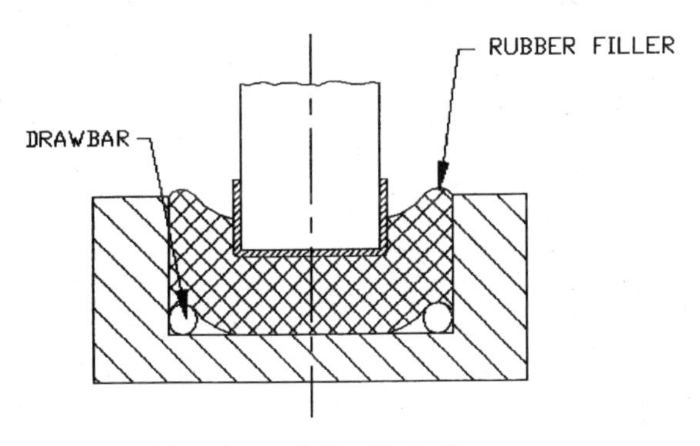


**Figure 8-23** Forming with flexible tooling.

solve the problem, as a considerable enlargement is needed for the part to follow the shape of the punch. However, there are no directions to follow in such an undertaking, as all the work on a subject has been arrived at experimentally, with not enough data collected yet.

### 8-4-6. Miscellaneous bending techniques

Various bending calculations that do not fit into any described category are presented here for possible evaluation and use. They have proved quite accurate for certain materials and applications.

1. *Bending of soft copper and soft brass.* The formula to calculate bend allowance is Eq. (8-7). Its application is the same as previously described.

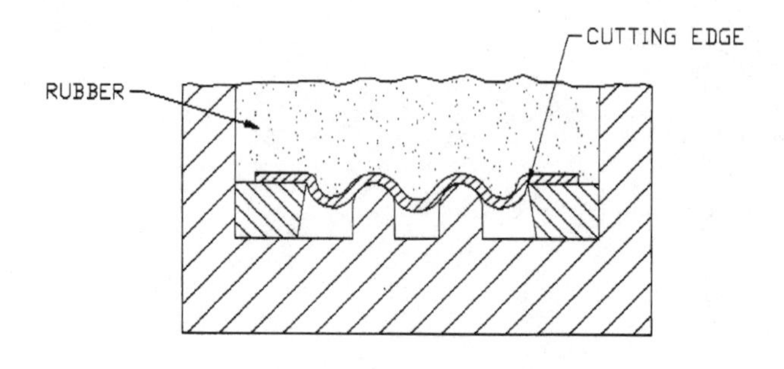


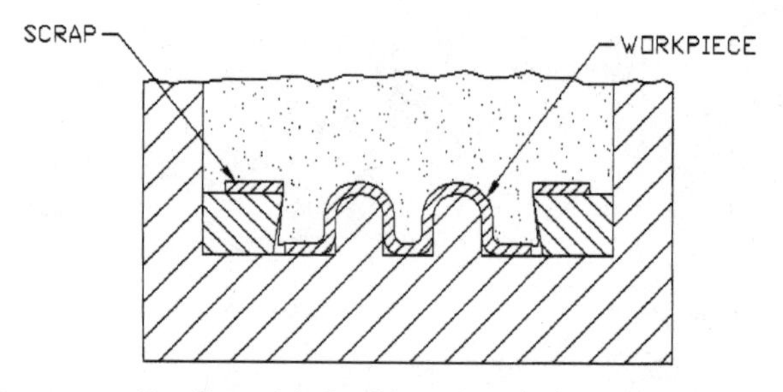


**Figure 8-24** Bending with flexible tooling.

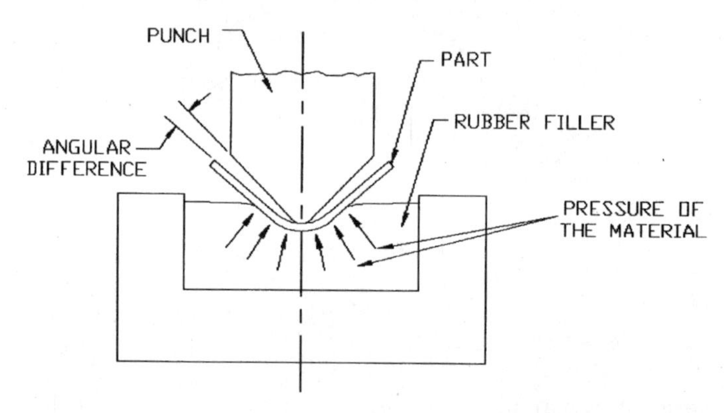


**Figure 8-25** V bending with elastic tooling.

$$BA_{Cu} = CR_{in} + (0.55t) \tag{8-7}$$

where $R_{in}$ = inside radius of the bend
$t$ = material thickness
$C$ = constant, depending on the angle of the bend. For 90° bends, this value is 1.5708. For bends of different angularity, see Table 8-6.

Further calculation of the total length of the part $L_{total}$ is performed the same way, as shown in Eqs. (8-3*a*) or (8-3*b*). A chart of calculated bend allowance values is shown in Table 8-10.

**TABLE 8-10 Bend Allowance Chart for 90° Bends in Annealed Copper and Brass, U-Die Bending**

| Stock thickness, in | Bend radius, in | | | | | | | | | | | | | | | | |
|---|---|---|---|---|---|---|---|---|---|---|---|---|---|---|---|---|---|
| | 0.015 | 0.031 | 0.062 | 0.093 | 0.125 | 0.156 | 0.187 | 0.218 | 0.250 | 0.281 | 0.312 | 0.343 | 0.375 | 0.406 | 0.437 | 0.468 | 0.500 |
| 0.015 | 0.032 | 0.057 | 0.106 | 0.154 | 0.205 | 0.253 | 0.302 | 0.351 | 0.401 | 0.450 | 0.498 | 0.547 | 0.597 | 0.646 | 0.695 | 0.743 | 0.794 |
| 0.018 | 0.033 | 0.059 | 0.107 | 0.156 | 0.206 | 0.255 | 0.304 | 0.352 | 0.403 | 0.451 | 0.500 | 0.549 | 0.599 | 0.648 | 0.696 | 0.745 | 0.795 |
| 0.024 | 0.037 | 0.062 | 0.111 | 0.159 | 0.210 | 0.258 | 0.307 | 0.356 | 0.406 | 0.455 | 0.503 | 0.552 | 0.602 | 0.651 | 0.700 | 0.748 | 0.799 |
| 0.031 | 0.041 | 0.066 | 0.114 | 0.163 | 0.213 | 0.262 | 0.311 | 0.359 | 0.410 | 0.458 | 0.507 | 0.556 | 0.606 | 0.655 | 0.703 | 0.752 | 0.802 |
| 0.036 | 0.043 | 0.068 | 0.117 | 0.166 | 0.216 | 0.265 | 0.314 | 0.362 | 0.413 | 0.461 | 0.510 | 0.559 | 0.609 | 0.658 | 0.706 | 0.755 | 0.805 |
| 0.048 | 0.050 | 0.075 | 0.124 | 0.172 | 0.223 | 0.271 | 0.320 | 0.369 | 0.419 | 0.468 | 0.516 | 0.565 | 0.615 | 0.664 | 0.713 | 0.762 | 0.812 |
| 0.054 | 0.053 | 0.078 | 0.127 | 0.176 | 0.226 | 0.275 | 0.323 | 0.372 | 0.422 | 0.471 | 0.520 | 0.568 | 0.619 | 0.667 | 0.716 | 0.765 | 0.815 |
| 0.062 | 0.058 | 0.083 | 0.131 | 0.180 | 0.230 | 0.279 | 0.328 | 0.377 | 0.427 | 0.475 | 0.524 | 0.573 | 0.623 | 0.672 | 0.721 | 0.769 | 0.820 |
| 0.075 | 0.065 | 0.090 | 0.139 | 0.187 | 0.238 | 0.286 | 0.335 | 0.384 | 0.434 | 0.483 | 0.531 | 0.580 | 0.630 | 0.679 | 0.728 | 0.776 | 0.827 |
| 0.093 | 0.075 | 0.100 | 0.149 | 0.197 | 0.248 | 0.296 | 0.345 | 0.394 | 0.444 | 0.493 | 0.541 | 0.590 | 0.640 | 0.689 | 0.738 | 0.786 | 0.837 |
| 0.109 | 0.084 | 0.109 | 0.157 | 0.206 | 0.256 | 0.305 | 0.354 | 0.402 | 0.453 | 0.501 | 0.550 | 0.599 | 0.649 | 0.698 | 0.746 | 0.795 | 0.845 |
| 0.125 | 0.092 | 0.117 | 0.166 | 0.215 | 0.265 | 0.314 | 0.362 | 0.411 | 0.461 | 0.510 | 0.559 | 0.608 | 0.658 | 0.706 | 0.755 | 0.804 | 0.854 |
| 0.140 | 0.101 | 0.126 | 0.174 | 0.223 | 0.273 | 0.322 | 0.371 | 0.419 | 0.470 | 0.518 | 0.567 | 0.616 | 0.666 | 0.715 | 0.763 | 0.812 | 0.862 |
| 0.156 | 0.109 | 0.134 | 0.183 | 0.232 | 0.282 | 0.331 | 0.380 | 0.428 | 0.479 | 0.527 | 0.576 | 0.625 | 0.675 | 0.724 | 0.772 | 0.821 | 0.871 |
| 0.187 | 0.126 | 0.152 | 0.200 | 0.249 | 0.299 | 0.348 | 0.397 | 0.445 | 0.496 | 0.544 | 0.593 | 0.642 | 0.692 | 0.741 | 0.789 | 0.838 | 0.888 |
| 0.203 | 0.135 | 0.160 | 0.209 | 0.258 | 0.308 | 0.357 | 0.405 | 0.454 | 0.504 | 0.553 | 0.602 | 0.650 | 0.701 | 0.749 | 0.798 | 0.847 | 0.897 |
| 0.218 | 0.143 | 0.169 | 0.217 | 0.266 | 0.316 | 0.365 | 0.414 | 0.462 | 0.513 | 0.561 | 0.610 | 0.659 | 0.709 | 0.758 | 0.806 | 0.855 | 0.905 |
| 0.234 | 0.152 | 0.177 | 0.226 | 0.275 | 0.325 | 0.374 | 0.422 | 0.471 | 0.521 | 0.570 | 0.619 | 0.667 | 0.718 | 0.766 | 0.815 | 0.864 | 0.914 |
| 0.250 | 0.161 | 0.186 | 0.235 | 0.284 | 0.334 | 0.383 | 0.431 | 0.480 | 0.530 | 0.579 | 0.628 | 0.676 | 0.727 | 0.775 | 0.824 | 0.873 | 0.923 |

2. *Bending of half-hard copper, half-hard brass, and half-hard steel.* The formula to obtain the bend allowance is

$$BA_{1/2\text{Cu}} = CR_{\text{in}} + (0.64t) \tag{8-8}$$

To calculate the total length, use the bend allowance obtained in Eqs. (8-3*a*) or (8-3*b*). A chart of calculated bend allowance values is shown in Table 8-11.

3. *Bending of hard copper, bronze, cold-rolled steel, and spring steel.* The bend allowance formula is

$$BA_{\text{spring}} = CR_{\text{in}} + (0.71t) \tag{8-9}$$

Use Eqs. (8-3*a*) or (8-3*b*) to calculate the total length of the part. A chart of calculated bend allowance values is shown in Table 8-12.

4. *Curling-based bending* demands that the workpiece be accurately located within the die. When the bending arm slides against the part, the exposed flange is forced to follow its shape, as if in curling (see Fig. 8-26).

The burr location is of importance with this type of bending, as it should never be located leaning against the surface of the forming section.

5. *Cam-die forming.* The upper punch member shown in Fig. 8-27 rough-forms the part, while the additional forming section, forced by a movement of a cam, slides against the part, finish-forming its side. The movement of the cam is guided over adjustable inserts, placed in strategic locations, and attached to the surrounding blocks. This way the wear of main blocks is prevented, with only these sections being interchanged whenever appropriate.

## 8-5. Spring-back

The amount of spring-back depends on mechanical properties of the metal material. It increases with greater yield strength or with the material's strain-hardening tendency. Cold working and heat treatment both increase the amount of spring-back. It may be decreased with the reduction of elastic modulus.

It is obvious that spring-back is the amount of elastic distortion a material has to go through before it becomes permanently deformed, or formed. It is the amount of elastic tolerance, which is to some extent present in every material, be it a ductile, annealed metal or a hard-strength maraging steel. Naturally, in ductile materials, the spring-back is much lower than in hard metals.

Spring-back occurs in all formed or bent-up parts on release of forming pressure and withdrawal of the punch. The material, previously

**TABLE 8-11 Bend Allowances for 90° Bending in Half-Hard Steel, Brass, or Copper Wipe Bending**

| Stock thickness, in | Bend radius, in | | | | | | | | | | | | | | | | |
|---|---|---|---|---|---|---|---|---|---|---|---|---|---|---|---|---|---|
| | 0.015 | 0.031 | 0.062 | 0.093 | 0.125 | 0.156 | 0.187 | 0.218 | 0.250 | 0.281 | 0.312 | 0.343 | 0.375 | 0.406 | 0.437 | 0.468 | 0.500 |
| 0.015 | 0.033 | 0.058 | 0.107 | 0.156 | 0.206 | 0.255 | 0.303 | 0.352 | 0.402 | 0.451 | 0.500 | 0.548 | 0.599 | 0.647 | 0.696 | 0.745 | 0.795 |
| 0.018 | 0.035 | 0.060 | 0.109 | 0.158 | 0.208 | 0.257 | 0.305 | 0.354 | 0.404 | 0.453 | 0.502 | 0.550 | 0.601 | 0.649 | 0.698 | 0.747 | 0.797 |
| 0.024 | 0.039 | 0.064 | 0.113 | 0.161 | 0.212 | 0.260 | 0.309 | 0.358 | 0.408 | 0.457 | 0.505 | 0.554 | 0.604 | 0.653 | 0.702 | 0.750 | 0.801 |
| 0.031 | 0.043 | 0.069 | 0.117 | 0.166 | 0.216 | 0.265 | 0.314 | 0.362 | 0.413 | 0.461 | 0.510 | 0.559 | 0.609 | 0.658 | 0.706 | 0.755 | 0.805 |
| 0.036 | 0.047 | 0.072 | 0.120 | 0.169 | 0.219 | 0.268 | 0.317 | 0.365 | 0.416 | 0.464 | 0.513 | 0.562 | 0.612 | 0.661 | 0.709 | 0.758 | 0.808 |
| 0.048 | 0.054 | 0.079 | 0.128 | 0.177 | 0.227 | 0.276 | 0.324 | 0.373 | 0.423 | 0.472 | 0.521 | 0.570 | 0.620 | 0.668 | 0.717 | 0.766 | 0.816 |
| 0.054 | 0.058 | 0.083 | 0.132 | 0.181 | 0.231 | 0.280 | 0.328 | 0.377 | 0.427 | 0.476 | 0.525 | 0.573 | 0.624 | 0.672 | 0.721 | 0.770 | 0.820 |
| 0.062 | 0.063 | 0.088 | 0.137 | 0.186 | 0.236 | 0.285 | 0.333 | 0.382 | 0.432 | 0.481 | 0.530 | 0.578 | 0.629 | 0.677 | 0.726 | 0.775 | 0.825 |
| 0.075 | 0.072 | 0.097 | 0.145 | 0.194 | 0.244 | 0.293 | 0.342 | 0.390 | 0.441 | 0.489 | 0.538 | 0.587 | 0.637 | 0.686 | 0.734 | 0.783 | 0.833 |
| 0.093 | 0.083 | 0.108 | 0.157 | 0.206 | 0.256 | 0.305 | 0.353 | 0.402 | 0.452 | 0.501 | 0.550 | 0.598 | 0.649 | 0.697 | 0.746 | 0.795 | 0.845 |
| 0.109 | 0.093 | 0.118 | 0.167 | 0.216 | 0.266 | 0.315 | 0.363 | 0.412 | 0.462 | 0.511 | 0.560 | 0.609 | 0.659 | 0.708 | 0.756 | 0.805 | 0.855 |
| 0.125 | 0.104 | 0.129 | 0.177 | 0.226 | 0.276 | 0.325 | 0.374 | 0.422 | 0.473 | 0.521 | 0.570 | 0.619 | 0.669 | 0.718 | 0.766 | 0.815 | 0.865 |
| 0.140 | 0.113 | 0.138 | 0.187 | 0.236 | 0.286 | 0.335 | 0.383 | 0.432 | 0.482 | 0.531 | 0.580 | 0.628 | 0.679 | 0.727 | 0.776 | 0.825 | 0.875 |
| 0.156 | 0.123 | 0.149 | 0.197 | 0.246 | 0.296 | 0.345 | 0.394 | 0.442 | 0.493 | 0.541 | 0.590 | 0.639 | 0.689 | 0.738 | 0.786 | 0.835 | 0.885 |
| 0.187 | 0.143 | 0.168 | 0.217 | 0.266 | 0.316 | 0.365 | 0.413 | 0.462 | 0.512 | 0.561 | 0.610 | 0.658 | 0.709 | 0.757 | 0.806 | 0.855 | 0.905 |
| 0.203 | 0.153 | 0.179 | 0.227 | 0.276 | 0.326 | 0.375 | 0.424 | 0.472 | 0.523 | 0.571 | 0.620 | 0.669 | 0.719 | 0.768 | 0.816 | 0.865 | 0.915 |
| 0.218 | 0.163 | 0.188 | 0.237 | 0.286 | 0.336 | 0.385 | 0.433 | 0.482 | 0.532 | 0.581 | 0.630 | 0.678 | 0.729 | 0.777 | 0.826 | 0.875 | 0.925 |
| 0.234 | 0.173 | 0.198 | 0.247 | 0.296 | 0.346 | 0.395 | 0.443 | 0.492 | 0.542 | 0.591 | 0.640 | 0.689 | 0.739 | 0.788 | 0.836 | 0.885 | 0.935 |
| 0.250 | 0.184 | 0.209 | 0.257 | 0.306 | 0.356 | 0.405 | 0.454 | 0.502 | 0.553 | 0.601 | 0.650 | 0.699 | 0.749 | 0.798 | 0.846 | 0.895 | 0.945 |

**TABLE 8-12 Bend Allowance Chart for 90° Bend in Full-Hard and Spring Temper Materials U-Die Bending**

| Stock thickness, in | Bend radius, in | | | | | | | | | | | | | | | | |
|---|---|---|---|---|---|---|---|---|---|---|---|---|---|---|---|---|---|
| | 0.015 | 0.031 | 0.062 | 0.093 | 0.125 | 0.156 | 0.187 | 0.218 | 0.250 | 0.281 | 0.312 | 0.343 | 0.375 | 0.406 | 0.437 | 0.468 | 0.500 |
| 0.015 | 0.034 | 0.059 | 0.108 | 0.157 | 0.207 | 0.256 | 0.304 | 0.353 | 0.403 | 0.452 | 0.501 | 0.549 | 0.600 | 0.648 | 0.697 | 0.746 | 0.796 |
| 0.018 | 0.036 | 0.061 | 0.110 | 0.159 | 0.209 | 0.258 | 0.307 | 0.355 | 0.405 | 0.454 | 0.503 | 0.552 | 0.602 | 0.651 | 0.699 | 0.748 | 0.798 |
| 0.024 | 0.041 | 0.066 | 0.114 | 0.163 | 0.213 | 0.262 | 0.311 | 0.359 | 0.410 | 0.458 | 0.507 | 0.556 | 0.606 | 0.655 | 0.703 | 0.752 | 0.802 |
| 0.031 | 0.046 | 0.071 | 0.119 | 0.168 | 0.218 | 0.267 | 0.316 | 0.364 | 0.415 | 0.463 | 0.512 | 0.561 | 0.611 | 0.660 | 0.708 | 0.757 | 0.807 |
| 0.036 | 0.049 | 0.074 | 0.123 | 0.172 | 0.222 | 0.271 | 0.319 | 0.368 | 0.418 | 0.467 | 0.516 | 0.564 | 0.615 | 0.663 | 0.712 | 0.761 | 0.811 |
| 0.048 | 0.058 | 0.083 | 0.131 | 0.180 | 0.230 | 0.279 | 0.328 | 0.377 | 0.427 | 0.475 | 0.524 | 0.573 | 0.623 | 0.672 | 0.721 | 0.769 | 0.819 |
| 0.054 | 0.062 | 0.087 | 0.136 | 0.184 | 0.235 | 0.283 | 0.332 | 0.381 | 0.431 | 0.480 | 0.528 | 0.577 | 0.627 | 0.676 | 0.725 | 0.773 | 0.824 |
| 0.062 | 0.068 | 0.093 | 0.141 | 0.190 | 0.240 | 0.289 | 0.338 | 0.386 | 0.437 | 0.485 | 0.534 | 0.583 | 0.633 | 0.682 | 0.730 | 0.779 | 0.829 |
| 0.075 | 0.077 | 0.102 | 0.151 | 0.199 | 0.250 | 0.298 | 0.347 | 0.396 | 0.446 | 0.495 | 0.543 | 0.592 | 0.642 | 0.691 | 0.740 | 0.788 | 0.839 |
| 0.093 | 0.090 | 0.115 | 0.163 | 0.212 | 0.262 | 0.311 | 0.360 | 0.408 | 0.459 | 0.507 | 0.556 | 0.605 | 0.655 | 0.704 | 0.752 | 0.801 | 0.851 |
| 0.109 | 0.101 | 0.126 | 0.175 | 0.223 | 0.274 | 0.322 | 0.371 | 0.420 | 0.470 | 0.519 | 0.567 | 0.616 | 0.666 | 0.715 | 0.764 | 0.813 | 0.863 |
| 0.125 | 0.112 | 0.137 | 0.186 | 0.235 | 0.285 | 0.334 | 0.382 | 0.431 | 0.481 | 0.530 | 0.579 | 0.628 | 0.678 | 0.726 | 0.775 | 0.824 | 0.874 |
| 0.140 | 0.123 | 0.148 | 0.197 | 0.245 | 0.296 | 0.344 | 0.393 | 0.442 | 0.492 | 0.541 | 0.589 | 0.638 | 0.688 | 0.737 | 0.786 | 0.835 | 0.885 |
| 0.156 | 0.134 | 0.159 | 0.208 | 0.257 | 0.307 | 0.356 | 0.404 | 0.453 | 0.503 | 0.552 | 0.601 | 0.650 | 0.700 | 0.749 | 0.797 | 0.846 | 0.896 |
| 0.187 | 0.156 | 0.181 | 0.230 | 0.279 | 0.329 | 0.378 | 0.427 | 0.475 | 0.525 | 0.574 | 0.623 | 0.672 | 0.722 | 0.771 | 0.819 | 0.868 | 0.918 |
| 0.203 | 0.168 | 0.193 | 0.242 | 0.290 | 0.340 | 0.389 | 0.438 | 0.487 | 0.537 | 0.586 | 0.634 | 0.683 | 0.733 | 0.782 | 0.831 | 0.879 | 0.930 |
| 0.218 | 0.178 | 0.203 | 0.252 | 0.301 | 0.351 | 0.400 | 0.449 | 0.497 | 0.547 | 0.596 | 0.645 | 0.694 | 0.744 | 0.793 | 0.841 | 0.890 | 0.940 |
| 0.234 | 0.190 | 0.215 | 0.264 | 0.312 | 0.362 | 0.411 | 0.460 | 0.509 | 0.559 | 0.608 | 0.656 | 0.705 | 0.755 | 0.804 | 0.853 | 0.901 | 0.952 |
| 0.250 | 0.201 | 0.226 | 0.275 | 0.324 | 0.374 | 0.423 | 0.471 | 0.520 | 0.570 | 0.619 | 0.668 | 0.716 | 0.767 | 0.815 | 0.864 | 0.913 | 0.963 |

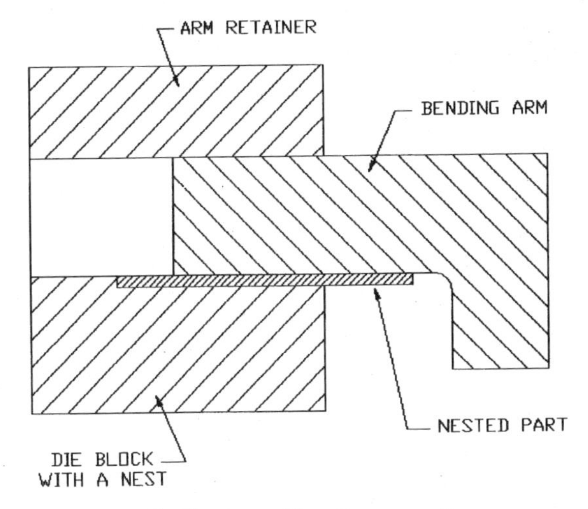


**Figure 8-26** Curling type of bending operation.

held in a predetermined arrangement by the influence of these two elements, is suddenly free from outside restriction and immediately makes an attempt to return to its original shape and form (Fig. 8-28).

The angle of the bent-up flange $\alpha_B$ is greater than that altered by spring-back $\alpha_S$. The same way, the radius $R_B$ increases on becoming affected by the spring-back $R_S$. However, the lengthwise portion $W$, which is the length of the arc, remains the same. Its relationship to other areas of significance is given as

$$W = \alpha_B\left(R_B + \frac{t}{2}\right) = \alpha_S\left(R_S + \frac{t}{2}\right) \tag{8-10}$$

From this relationship, a spring-back factor $K$ can be obtained:

$$K = \frac{R_B + \dfrac{t}{2}}{R_S + \dfrac{t}{2}} \tag{8-11}$$

Usually spring-back can be found between 0.9 and 1.0 for bends, using small bend radii. When the bend radius increases, the spring-back diminishes.

Equation (8-11) was proved true for bends with large bend radii or for those with small bend angles. However, with small bend radii, it may be considered valid only if the bend angle has a greater than 45° bending angle. For small bending angles and sharp bend radii, the spring-back is usually quite large.

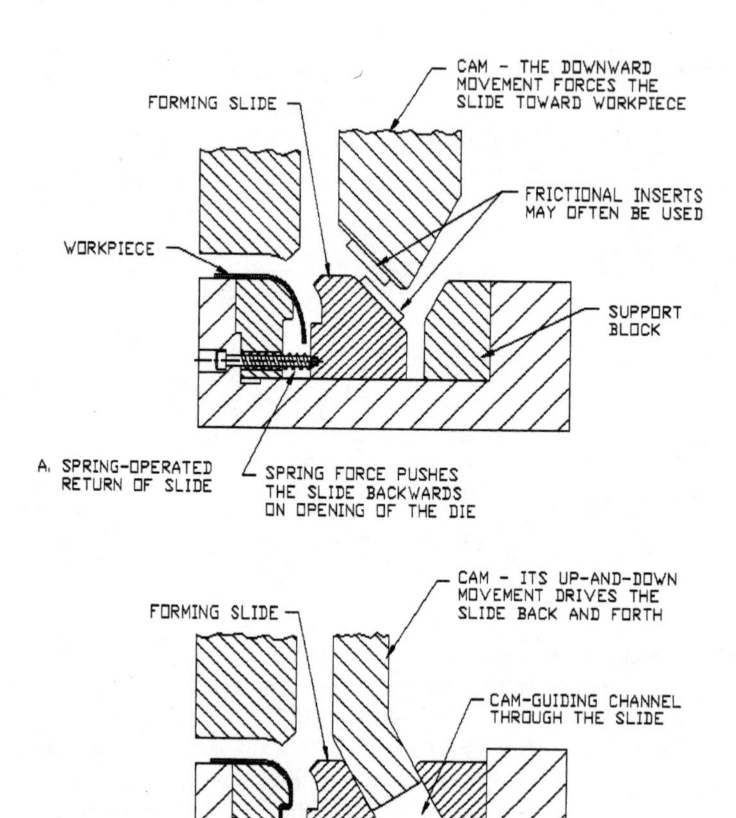


**Figure 8-27** Two types of cam movement.

Values of spring-back for steel are shown in Table 8-13.

To overcome the undesirable effect of spring-back, flanges are often bent more than desired, so that when the bent-up flange slightly returns to its previous position, or springs back, it finds itself within the limits of drawing or other requirements.

To diminish the spring-back effect, sometimes the enlargement of the bend radius may be of service. An additional benefit may be found in bottoming with the forming punch, or coining of the material.

A greater tendency toward spring-back even in annealed materials may be observed where the U die has too sharp corners at the bottom. The effect of such an unnatural junction distorts the material structure, resulting in bowing of the bottom surface, which in turn increases the spring-back tendency of the flanges.

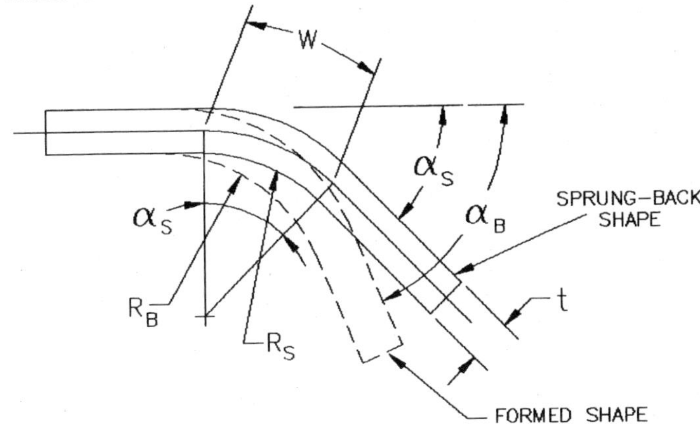


**Figure 8-28** Spring-back.

**TABLE 8-13 Spring-back Values for Steel Material**

| Angle of the bend | Condition of material | Ratio $R/t$ | | | | |
|---|---|---|---|---|---|---|
| | | 1 | 2 | 4 | 6 | 10 |
| 30° | Annealed | 1 | 1 | 3 | 4 | 7 |
| 30° | Hard | 0 | 2 | 5 | 8 | 14 |
| 60° | Annealed | 0 | 1 | 2 | 3 | 5 |
| 60° | Hard | 0 | 1 | 3 | 5 | 8 |
| 90° | Annealed | 0 | 0 | 1 | 2 | 4 |
| 90° | Hard | −1 | 0 | 2 | 3 | 7 |
| 120° | Annealed | −1 | 0 | 0 | 1 | 3 |
| 120° | Hard | −1 | 0 | 1 | 2 | 4 |

SOURCE: Svatopluk Černoch, *Strojně technická příručka,* 1977. Reprinted with permission from SNTL Publishers, Prague, CZ.

Spring-back may also be eliminated by the inclusion of a curved pressure pad, as shown in Fig. 8-29. On ejection of the part from the die, the slightly bowed bottom has a tendency to spring back toward its original flatness. This pulls the sides of the part toward the center.

## 8-6. Surface Flatness after Bending

In bending, as in drawing, it is sometimes quite difficult to produce a flat surface on a part. Any portion that has to be flat, unless it is very small in area, tends to become warped, bowed, inclined, or otherwise

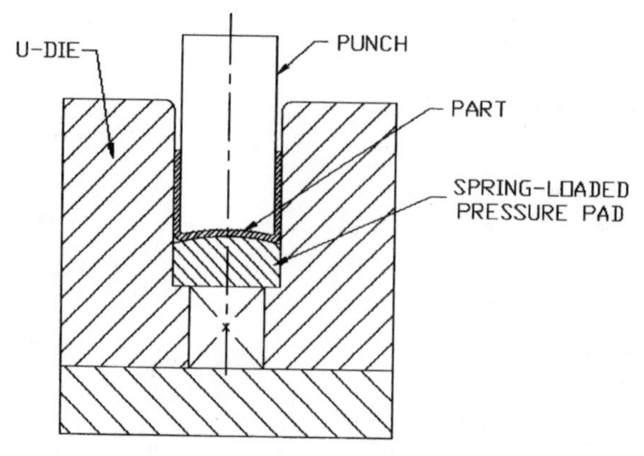


Figure 8-29 Spring-back elimination method.

distorted, but not flat. Such a surface, when pressed upon by hand, snaps back and forth, like an oilcan, from which this occurrence took its name: an oilcan effect.

In order to prevent this from happening, ribs, joggles, and other strengthening indentations were introduced, to provide for the dimensional as well as functional stability of parts. Objects that have to stand on a flat surface, like containers or cans, have their bottoms either bent or drawn in, keeping but a narrow rim of flat surface to stand on. Some products are offset, with a small ridge to provide for all the necessary flatness.

Where the supposedly flat surface of an object is distorted or warped, no amount of hammering or presswork will make it straight again. This is due to the mechanical properties of the material, which does not allow for any permanent alteration unless the elastic limit of the material is exceeded.

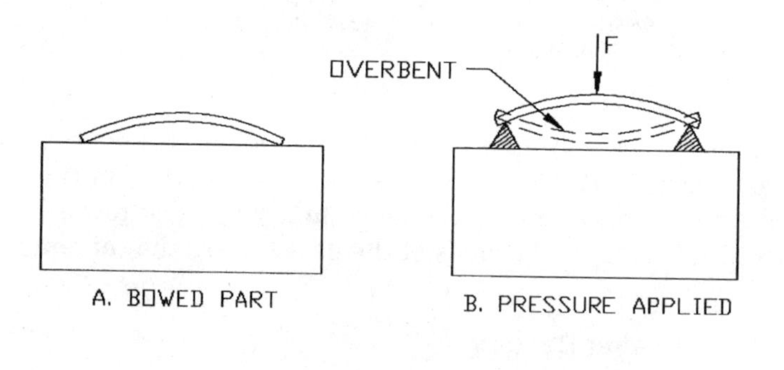


Figure 8-30 Straightening of sheet metal.

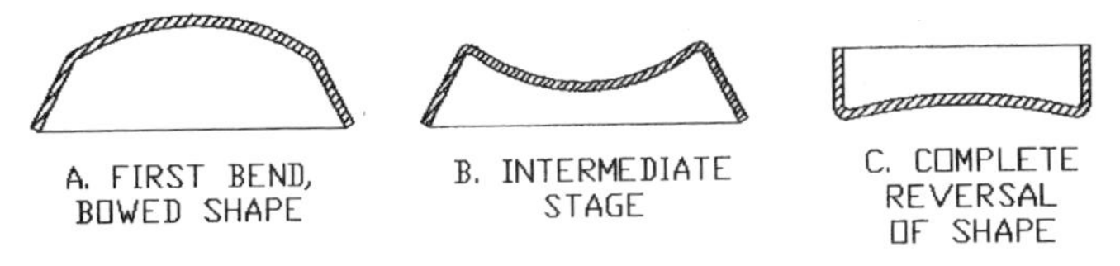


**Figure 8-31** Method of bending and rebending for accuracy.

Part *A* in Fig. 8-30 cannot be straightened by any amount of pressure applied from above. In order to flatten this surface, it must be reversed and supported on two extreme ends, as shown in *B*. In such a position, even a minute force will produce the flattening effect.

Often, where parts cannot be formed to hold their shape, a double bending method is used (Fig. 8-31). In such a case, the bend is first produced in the opposite direction to that desired. The bend is then reversed until a correctly shaped product is obtained.

## 8-7. Forming

Metal forming is a process totally dependent on the influence of outside tensile forces against the structure of the material. The resulting permanent deformation is called forming. The force-exerting instrument is the punch, which by pulling the sheet-metal material along, makes it enter the die, where it is compelled to take upon itself the impression of the assembly.

The decision if the part is to be formed or drawn is usually based on the evaluation of its shape and dimensional requirements. Drawing is utilized for those parts made of thicker materials or for those with vertical (or slightly inclined) walls and sharp corners at the bottom.

Since a forming die may often be instrumental in the formation of wrinkles or cause development of excessive tensile strains in the material, tearing the part in the process, drawing is often resorted to in such cases.

### 8-7-1. Forming of singular recesses

Singular recesses in the flat metal sheet are usually formed by stretching or drawing. Forming is reserved for parts with smooth connection of contours, without excessively sharp edges. Several samples of stretched parts are shown in Fig. 8-32.

The maximum amount of stretch for a given material depends on its distribution over the area of stretch. Naturally, the larger such an area is, the greater the maximum amount of stretch that can be obtained.

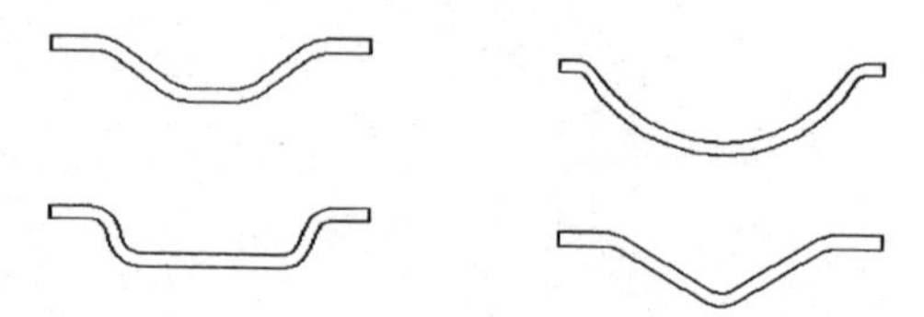


**Figure 8-32** Shapes of stretched parts.

To evaluate the strain with respect to the amount of stretch, Eq. (8-12) may be used:

$$s = \frac{L_s - L_o}{L_o} \tag{8-12}$$

where $s$ = stretch
All other values are in Fig. 8-33.

The linear distance of the drawn-in portion of the radius $R_{in}$ can be calculated:

$$R_{in} = \frac{1}{2}(D_o - D_s) \tag{8-13}$$

All calculations or measurements should be taken between mold lines, ignoring the radii of the edges. The mold line, as shown in Fig. 8-33, is an extension of the curved or linear portion, connecting with another line sharply, with no radius applied.

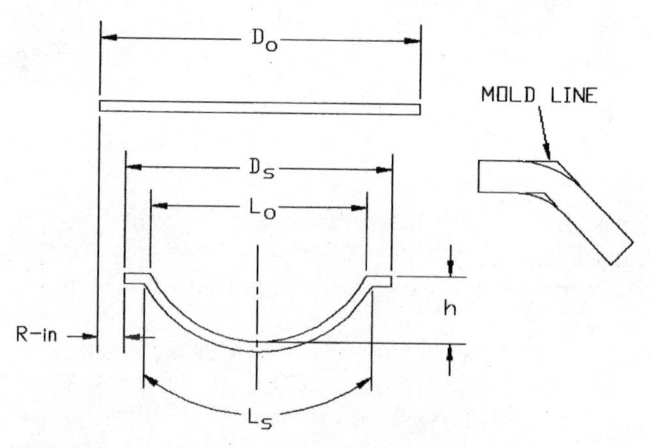


**Figure 8-33** Dimensional values of a stretched recess.

**TABLE 8-14 Stretch Values of Circular Recesses**

| Ratio $L_s/L_o$ | Ratio $h/L_o$ | Stretch % |
|---|---|---|
| 1.05 | 0.157 | 5 |
| 1.10 | 0.218 | 10 |
| 1.15 | 0.264 | 15 |
| 1.20 | 0.302 | 20 |
| 1.25 | 0.334 | 25 |
| 1.30 | 0.361 | 30 |
| 1.35 | 0.386 | 35 |
| 1.40 | 0.408 | 40 |
| 1.45 | 0.429 | 45 |
| 1.50 | 0.447 | 50 |

Round recesses can be assessed from Table 8-14. Their stretch values are based on the ratio of the depth $h$ to the length of the recess in flat $L_s$.

### 8-7-2. Stretch flange forming

Stretch flanges, when viewed from the top, form a concave curvature. These are flanges which on forming must be lengthened or stretched (Fig. 8-34). They may be compared to flanges surrounding a hole in sheet metal. Processes such as extruding, dimpling, and countersinking are all basically stretch flange forming.

The evaluation of the flange type should be pursued by observing the flat layout of a part, which clearly shows where the material for each flange may be taken from.

Owing to their stretching, the material thickness of stretch flanges decreases. This places a lateral strain on the material, with a resulting circumferential tension. The strain $l$ can be calculated from the

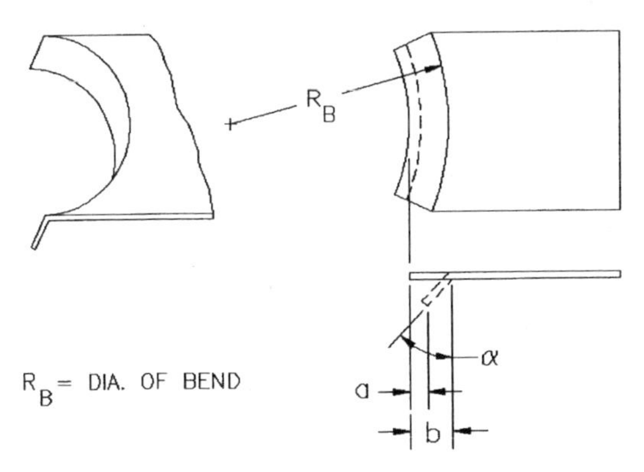


**Figure 8-34** Stretch flange.

ratio of the wall thickness $\Delta t$ and the amount of wall thickness change $t$ as follows:

$$\frac{\Delta t}{t} = -\frac{e}{2} \quad (8\text{-}14)$$

A minus sign attached to Eq. (8-14) depicts the variation in wall thickness, which in stretch flanges diminishes (−) and in shrink flanges increases (+).

From this relationship, other values may be assessed with the use of the formula

$$e = \frac{a}{R_B - b} \quad (8\text{-}15)$$

where all values are as shown in Fig. 8-35.

However, with the bend radius being too small, the value $a$ of flange movement may be approximated. In such a case, the material thickness is to be considered zero and the flange width constant. The formula to be used is then

$$a = b(1 - \cos\alpha) \quad (8\text{-}16)$$

Stretch flanges are limited by the amount of material from which they can draw for their development. Beyond such a limit, the flanges will crack around the edges. Too small a radius of curvature also adds to the problems, and it should be made as generous as possible, with a definite preference for straight lines.

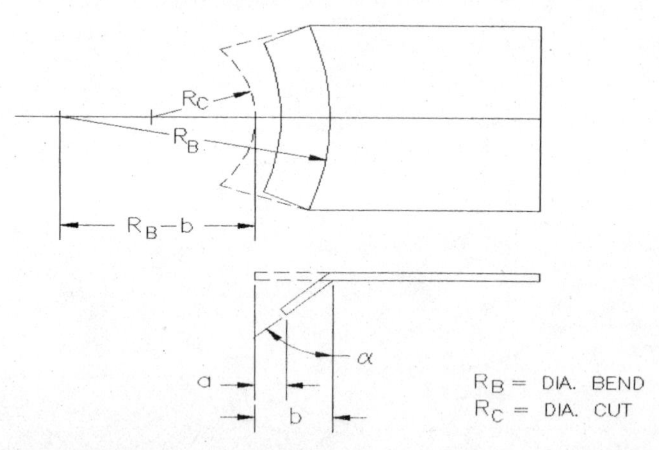


**Figure 8-35** Flange geometry.

The limits on the 90° flange are as given by the equation

$$e_{\text{limit}} = 1 - \frac{\cos \alpha}{R_B/b - 1} \tag{8-17}$$

### 8-7-3. Shrink flange forming

Shrink flanges are those which are reduced in length on forming, or shrunk. The shrunk flange, when viewed from the top, usually forms a convex line. The wall thickness of these flanges increases, which is caused by a circumferential compression during the forming process.

The lateral strain, acting within the flange material, can be calculated by using Eq. (8-14). Similarly as with the stretch flanges, other pertinent values may be assessed:

$$e = -\frac{a}{R_B + b} \tag{8-18}$$

where the values are as shown in Figs. 8-35 and 8-36.

And again, the value $a$ of flange movement may be approximated, using Eq. (8-16).

Shrink flanges are actually quite difficult to form from a flat blank. The material is often reluctant to succumb to such compressive stresses and has a tendency to wrinkle or buckle. With wider flanges, the tendency to wrinkling is increased.

Buckling is found controllable where the ratio of the flange width to the material thickness remains within a range of 3 to 4.

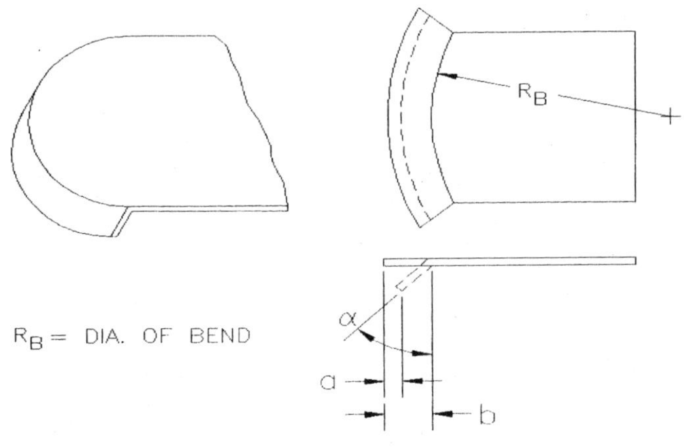


**Figure 8-36** Shrink flange.

## 8-8. Bending and Forming Pressure Calculations

Several formulas are utilized for calculation of bending and forming pressures. They may vary with the type of bending utilized.

1. *Bending in a V die, with rectangular cross section:*

$$P_V = \frac{k_V S W t^2}{L} \tag{8-19}$$

where $k_V$ = die opening factor, 0.75 to 2.5 (larger values are for smaller $R/t$ ratios and vice versa). A 1.33 value is used for a die opening of 8 times metal thickness.
$W$ = width of the bent-up portion
$L$ = distance between material supports (see Fig. 8-37)
$S$ = ultimate tensile strength, tons/in$^2$ (Table 8-15)

2. *Bending in a U die, equipped with a spring-loaded pressure pad:*

$$P_U = \frac{k_U S W t^2}{R_E + R_D + t} + P_{pad} \tag{8-20}$$

where $k_U$ = die opening factor, 0.4 to 1.0
$R_E$ = radius, die edge (see Fig. 8-38)
$R_D$ = radius, bottom of U channel
$P_{pad}$ = pressure of spring-loaded support

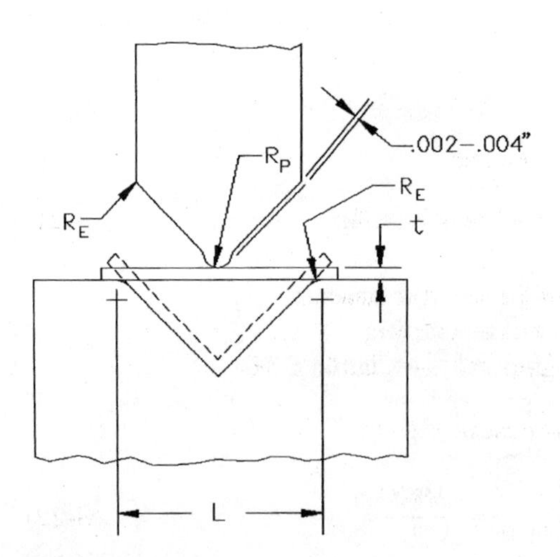


**Figure 8-37** V-die bending geometry.

**TABLE 8-15 Ultimate Tensile Strength of Materials**

| | Tons/in$^2$ |
|---|---|
| Steel, low carbon, 1025 | 30–51.5 |
| Steel, medium carbon, 1045 | 40–91 |
| Steel, high carbon, 1095 | 45–106.5 |
| Steel, stainless, 303 | 42.5–62.5 |
| Aluminum alloy, cold-worked | 6.0–31.5 |
| Aluminum alloy, heat-treated | 11.0–41.5 |
| Copper | 16–28.5 |
| Phosphor bronze | 20–64 |
| Zinc | 9.75–15.5 |

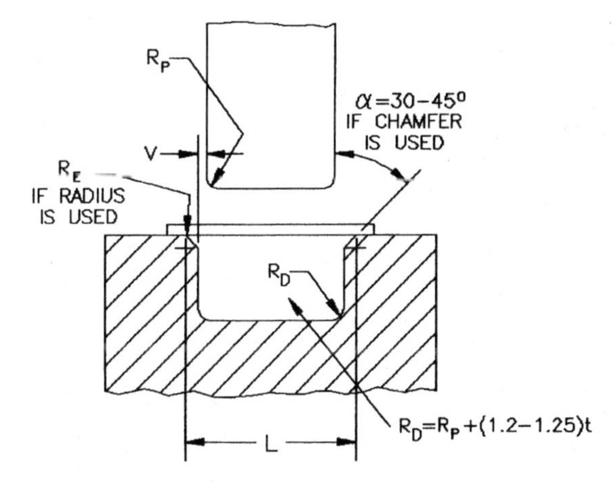


**Figure 8-38** U-die bending geometry.

3. *Bending with bottoming (coining):*

$$P_{\text{bottom}} = (2 \text{ to } 4)P = Ap \qquad (8\text{-}21)$$

where $P$ = bending pressure of particular process
$A$ = area of part, subjected to coining
$p$ = bending pressure, general (see Table 8-16)

4. *Wipe bending dies' pressure calculation:*

$$P_{\text{total-wipe}} = \frac{SWt^2}{L} \qquad (8\text{-}22)$$

**TABLE 8-16 Selected Bending Pressures in Tons/in²**

| | Material thickness, in | |
|---|---|---|
| | Under 0.125 | Over 0.125 |
| Steel, annealed | 38–62 | 62–87 |
| Steel, hard | 52–87 | 87–110 |
| Aluminum | 10–30 | 30–43 |
| Brass | 25–50 | 50–72 |

SOURCE: Svatopluk Černoch, *Strojně technická příručka,* 1977. Reprinted with permission from SNTL Publishers, Prague, CZ.

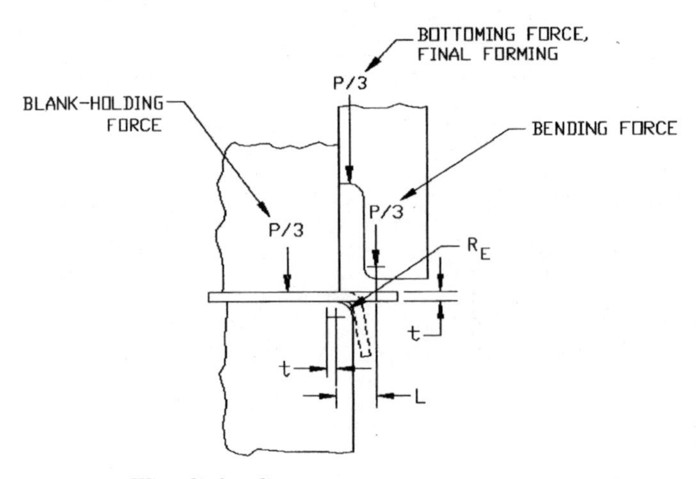


**Figure 8-39** Wipe die bending geometry.

where $L$ = distance between supports of the material (see Fig. 8-39)
$W$ = width of the bent-up portion
$S$ = ultimate tensile strength (Table 8-15)

Subsequently, each of the three forces acting upon the appropriate point in the assembly is one-third of the total force. These forces are: (1) force of blank holder; (2) bending force of the punch; (3) final bottoming force of the punch (see Fig. 8-39).

$$P_{1 \text{ or } 2 \text{ or } 3} = 0.333 \frac{SWt^2}{L} \tag{8-23}$$

5. *Calculation of the pressure involved in rotary bending is as follows:*

$$P_{\text{total}} = 2.25L \frac{SWt^2}{L} = L_P + t + R_E \tag{8-24}$$

where all values are as shown in Fig. 8-40.

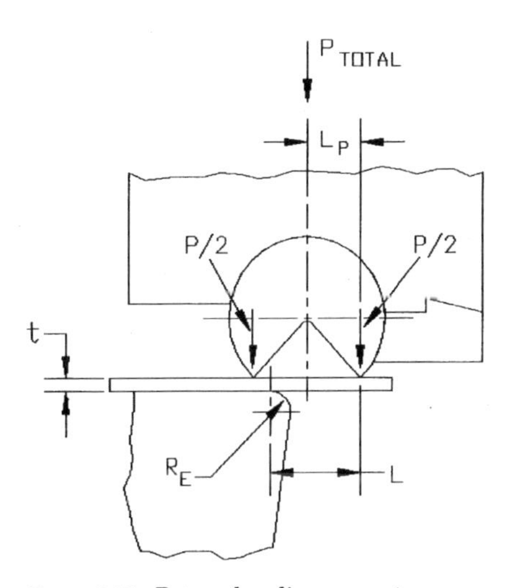


**Figure 8-40** Rotary bending geometry.

Chapter

# 9

# Drawn Parts

## 9-1. Drawing of Sheet Metal

Drawing means altering the structure of sheet-metal blank by extending its shape along the third axis *Z*. This is achieved by forcing the blank material to follow the movement of a punch, during which process the metal content of the flange changes its location by flowing into the body of the drawn part. The material, forced to overcome its elastic limit, succumbs to a plastic deformation and becomes permanently drawn, or extended. Figure 9-1 shows the forces involved in the cup-drawing process.

The blank is usually restricted from following the punch by having its edges confined between the surface of the die and the pressure pad.

The metal taken from the flange is used up against the increase in height of the part. A rather crude demonstration of this shift is depicted in Fig. 9-2. Here the whole segments of material are being displaced, flowing away from the flange toward the body of the shell.

A drawn shell may emerge from the drawing operation altered in either of the following ways:

- With the diameter of the flange diminished but the wall thickness unaltered (shift in Fig. 9-2).
- With the diameter of the flange unaltered but the wall thickness diminished. This occurs where the material is submitted to a shift of vertically situated layers, as opposed to horizontally oriented segments, shown in Fig. 9-2.
- Combination of the two possibilities.

The blank is a flat piece of sheet metal of uniform thickness, usually round. During the process of drawing its shape may change, with

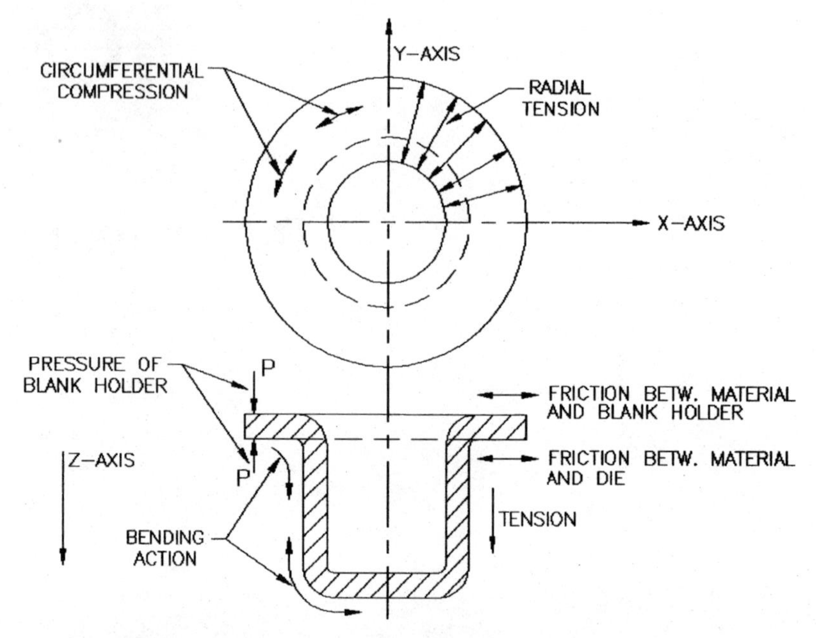


**Figure 9-1** Forces involved in cup-drawing process.

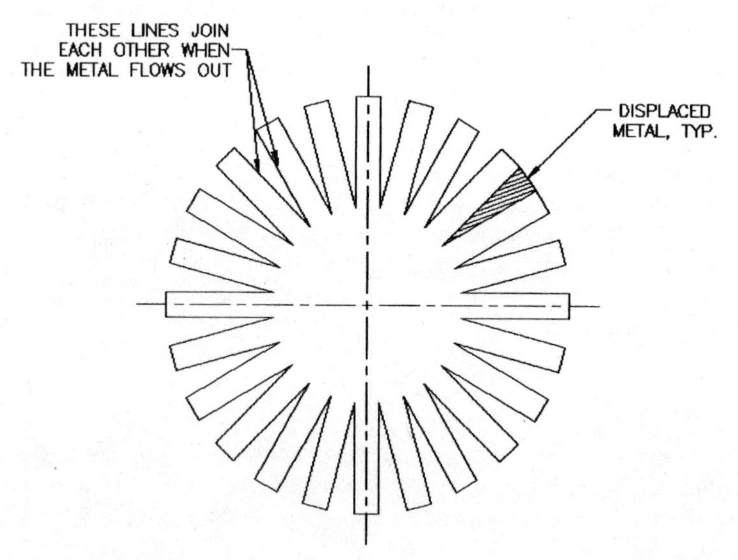


**Figure 9-2** Displacement of metal in drawing.

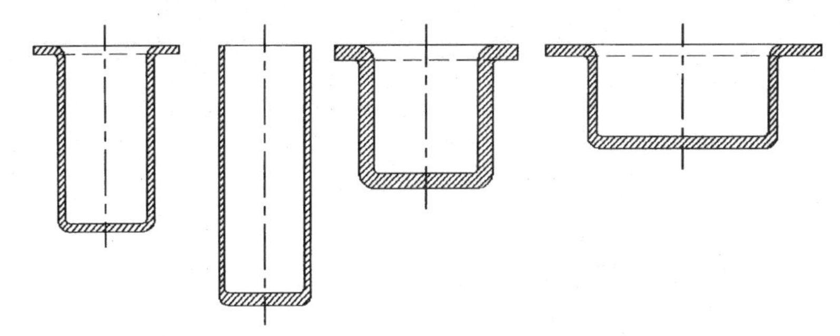

**Figure 9-3** Volumnar equality of shells made from the same blank.

its thickness sometimes changing as well, but its volumnar value will remain the same, should it be drawn into a short, thick-walled cylinder, or a tall, thin-walled shell (Fig. 9-3).

Consequently a correct size of the blank must be well assessed to provide for all the needed volume of material and yet not be excessive in size or thickness. Drawing from blanks the diameter of which is too small to their depth always poses a problem, as the thinning of their walls will be unreasonable and parts may emerge from the die distorted or outright fractured.

Numerous other influences controlling the outcome of the drawing process can be found within mechanical properties of the material, such as strength, ductility, elasticity, and even thickness of the drawn stock. Should these values be either inadequate or excessive, they certainly will have an effect on the drawn part, and either favorably or negatively alter the whole process and its outcome.

The total expansion in depth often cannot be attained in a single operation. No material, with the exception of a rubber band, has such elastic properties as to allow for its stretching into depths greater than certain limiting percentages, which are listed later in this chapter. If a part is drawn more than it can tolerate, its stretching will place such a strain on its structure that a permanent deformation followed by fractures and tearing will result.

Therefore, drawing into greater depths must be done in stages, with each operation to be performed within the limits set for that particular material. And each drawing pass should stretch the shell slightly more until a final drawn cup is produced (Fig. 9-4).

To produce a well-shaped and high-quality drawn part, the edges of both punch and die must be radiused, or chamfered; otherwise tearing of the product will occur. The radii should be quite liberal, ranging at least four times the material thickness, even though the part's blueprint does not call for them. Where a drawn product must have

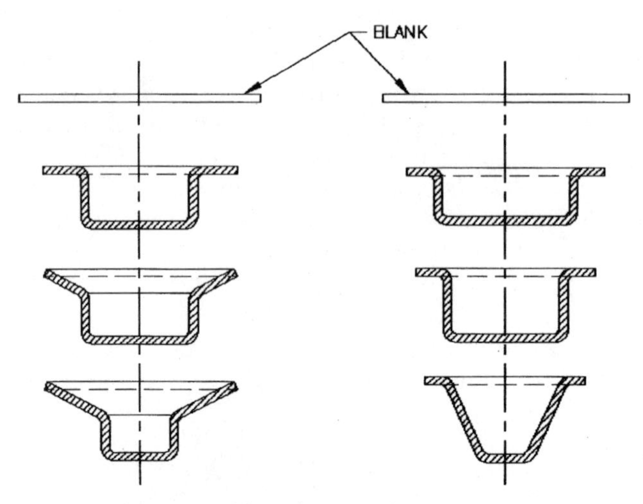


Figure 9-4 Sequence of drawing operations.

smaller than possible radii, these should be produced later, in the restriking operation.

The restriking process does not draw the shape any further; rather, it forces the already drawn product to conform dimensionally to the requirements, unacquirable otherwise. During restrike, radii may be produced smaller than those enforced by the requirements of the drawing process. Bottoms may be flattened (somewhat), or bulged, or sides of a shell may be straightened (Fig. 9-5).

Restriking differs from redrawing in that the punch does not attempt to extend the drawn shell, whereas redrawing is used strictly for deepening of the drawn portion.

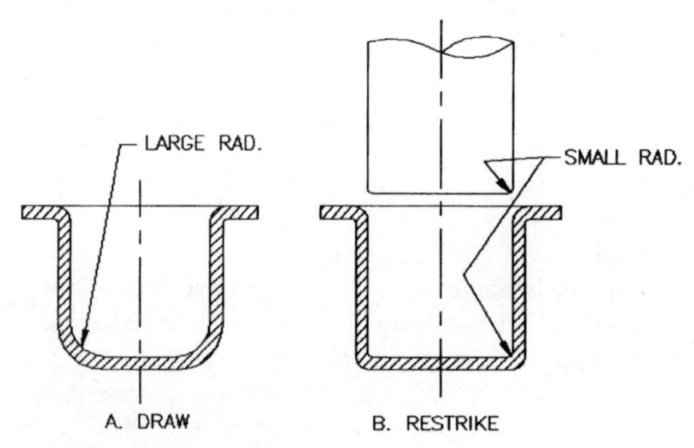


Figure 9-5 Restriking operation.

Repeated redrawing will produce strain hardening within the drawn material. After two or three drawing passes, some materials are hardened so greatly that the press force to overcome such an obstacle will have to be tremendous, and yet it may not achieve another extension in shape, as the hardened part may tear or rupture.

Therefore, annealing of the drawn material must be performed between various redrawing processes. Annealing brings the mechanical properties of metal back to its predrawing stage, or at least quite close to it. Additional drawing passes may then be performed without causing unnecessary disturbances of the part's structure.

Some materials, such as brass, copper, and some steel, has to be annealed between every drawing stage. The effect of their strain hardening is too massive, handicapping further drawing operations. This may be observed with a piece of soft copper wire, which—when bent up and down several times in succession—suffers from such strain hardening that it becomes quite rigid.

Aluminum wire, on the other hand, softens and tears readily, with the breakage occurring within the area of the bend. When drawn, aluminum can attain deeper shapes in fewer drawing passes and with less annealing in between. Naturally, not all aluminum grades perform equally, which makes the above statement applicable mainly to alloys designated for drawing purposes.

Drawing differs from other metalworking processes in that it totally exploits the elastic and plastic properties of materials. The flat blank, forced to alter its shape to comply with the tooling, wraps around the punch, tightly adhering to its surface. Drawn parts always conform to the shape of the punch, while the opening in the die is immaterial, as long as its size is adequate for the given stock thickness (Fig. 9-6).

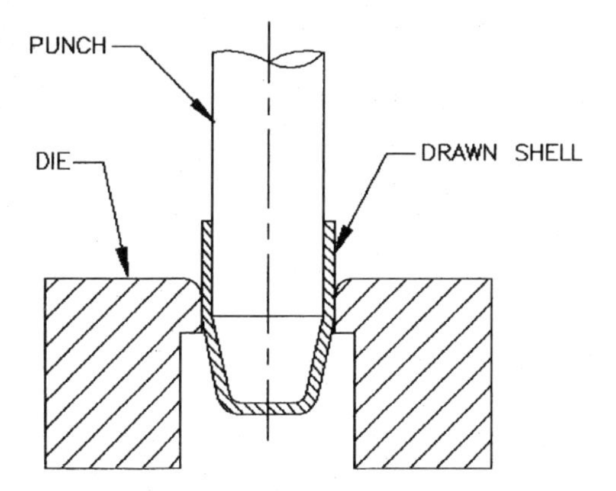


**Figure 9-6** Drawn material conforms to the shape of the punch.

## 9-2. Metal Movement in Drawing Operation

During the process of sheet-metal drawing, the metal of the blank is exposed to various influences, which lead to its alteration in shape and sometimes in thickness (see Fig. 9-1). A plastic deformation of the material can be seen in the area, exposed to the pressure of the blank holder, while the plastic deformation caused by the drawing punch face is minimal. The plastic deformation occurring within the flange is positive where the radial tension is concerned, and negative owing to compression in the tangential direction. The direction of deformation normal to the flange is at first negative, with the resulting thinning of walls. But at the diametral distance greater than

$$\text{Punch dia.} + 1.214R$$

the deformation becomes positive, with subsequent increase in thickness (Fig. 9-7).

Plastic deformation of the material may be enhanced or decreased by altering the amount of friction between the drawn part and its tooling. However, the influence of friction varies with its location within the drawing process. Friction between the material and the drawing die or blank holder causes the radial tension and the ultimate coefficient of cupping to increase, with subsequent restriction of the maximum possible depth of draw. Friction between the material and drawing punch exerts an opposite influence on the outcome by increasing the maximum possible depth of the draw as a consequence of increase in friction.

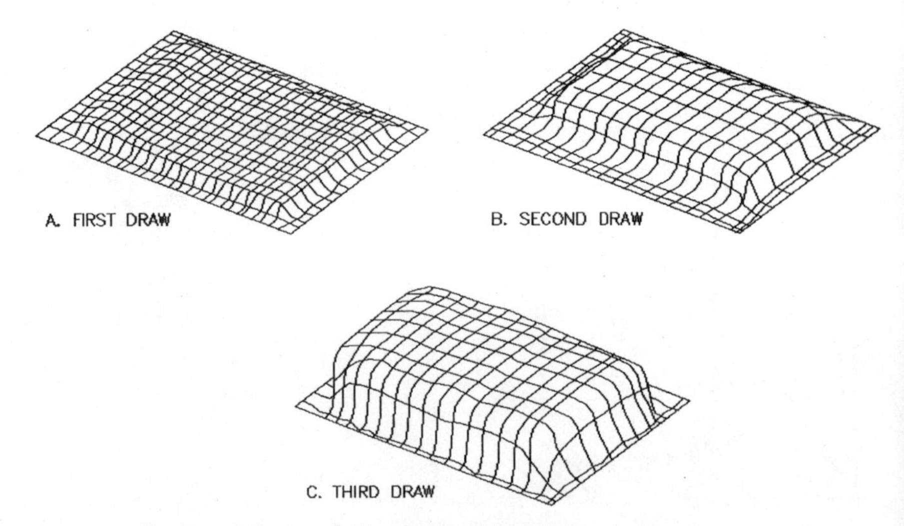


**Figure 9-7** Structural changes during drawing operation.

This scenario may sometimes be enhanced by frictional inserts in the punch or by roughing of its cylindrical surface. The alteration is efficient even as a prevention of the excessive plastic deformation, or occurrence of wrinkles.

Wrinkling of material can also be prevented by the inclusion of a blank holder within the drawing die arrangement. The blank holder not only prevents wrinkling of the flange, it also retains the blank, so that it may not be pulled into the die without being drawn.

Not all materials need the blank holder, though. Some thicker stock may be successfully drawn without being retained under pressure. However, deforming influences within the flange may develop, caused by the tension $\sigma$, the value of which depends on the cupping strain factor $E_c$. Where such tension is greater than the critical tension $\sigma_{crit}$, which is dependent on the thickness of stock, a blank holder is necessary.

## 9-3. Technological Aspects of Drawing Process

A drawn part is exposed to various technological influences, which affect every manufacturing process and its outcome. These are factors, including but not limited to the type of tooling, the type of manufacturing process, the amount of friction, speed of the process, temperature of the product and its tooling, and numerous other influences, exerting their control over the final product.

All these factors may affect the part and its manufacturing process either singularly or in a combination of two or more circumstances. For example, the drawing process itself will be affected by the amount of friction, which may give rise to the temperature of working surfaces, with subsequent wear of the tooling.

There are numerous small and large influences, all insidiously waiting to be omitted from the total assessment of the situation, so that they may exert their influence over the process unexpectedly and at the most inopportune moment. All these aspects have to be properly evaluated so that their span of influence is limited in scope and in magnitude as well.

### 9-3-1. Suitability of materials for drawing

A valuable contribution to the successful drawing process is a properly selected drawing material. The choice is governed mainly by its drawability, or rather by the portion of it regarding the deformation and its distribution. The value of deformation should be within 25 to 75 percent of the value of drawability.

Drawability of metals can be defined as their capacity to assume the predetermined shape without suffering any loss of stability, without fracturing or being otherwise distorted by the drawing process.

There are several theories on materials' drawability, all supported by a thorough testing. Erichsen's test was carried in accordance with PN/68/11-04400 (Polish Standard), where a punch ending with a ø20-mm ball was used. The resulting fracture occurrence was determined with 0.01-mm accuracy. Jovignot's test utilized a ø50-mm die with ø5-mm profile radius. The accuracy of these findings was also 0.01 mm. In the Swift test, a ø32-mm punch with a flat face was used. The Engelhardt-Gross test employed a ø20-mm punch against the ø52-mm blank. The Fukui test, using a conical cup, was performed with ø8- to 27-mm punches. Siebel-Pomp tested 80 × 80 mm samples with a central opening of ø12 mm.

Various testing methods established the drawability factor as a function of the mechanical makeup of the material, depending mainly on its strength, elastic/plastic properties, and chemical composition.

A certain lack of relationship between Erichsen's drawability index and the chemical composition of the material renders its influence meaningless. Tested materials ranged in the following values: 0.045 to 0.16 percent carbon, 0.24 to 0.48 percent manganese, 0.011 to 0.039 percent phosphorus, 0.005 to 0.03 percent sulfur.

Other findings proved the influence of phosphorus and manganese on drawability controversial. In these tests, the phosphorus content ranged between 0.01 and 0.25 percent, while manganese was included at 0.25 to 0.39 percent.

Still other tests found that the 0.015 to 0.025 percent of phosphorus was actually an improvement to drawability.

The most pronounced effect on the materials' drawability was considered normal anisotropic plasticity, where the actual tests fully confirmed previously obtained theoretical analyses. However, the lack of conformity between the theory and experiments in the case of strain-hardening influence on the drawability was too obvious.

Experimental findings also substantiated the difference in the drawability of materials of the same thickness, as based on the type of manufacturing process of the basic steel type. Materials stabilized by aluminum were found to have their elastic limits approximately 5 percent greater than those stabilized by titanium, whereas the drawability of titanium-stabilized steel was found 7 percent greater than that of the aluminum-stabilized steel.

**Normal anisotropy.** Suitability of drawing materials should also be evaluated on the basis of its coefficient of normal anisotropy $r$. This coefficient is a ratio of the actual deformation (or variation) within the metal to the variation in its thickness. The relationship can be defined as

$$r = \frac{\ln(b_0/b)}{\ln(t_0/t)} \tag{9-1}$$

where all values are as shown in Fig. 9-8.

Coefficient of normal anisotropy is a speculative value, comparing the behavior of the flange material with that in the drawn section and that located under the face of the drawing punch. A proper development of tangential and radial deformations within the flange and an attainment of low amounts of deformation within the drawn section of the shell are vital to the success of the drawing process. With higher values of coefficient of normal anisotropy, the material's formability will be greater, producing deeper draws and lowering the cupping strain factor.

Since the coefficient of normal anisotropy $r$ is grain orientation–dependent, its value is specified with respect to its variation from the grain line, which is considered at 0°. Subsequently, $r_0$ goes along the grain line, $r_{45}$ at 45°, and $r_{90}$ at 90° (see Fig. 9-9). The mean value of the normal coefficient of anisotropy $\bar{r}$ can be obtained from the formula

$$\bar{r} = \frac{r_0 + 2r_{45} + r_{90}}{4} \tag{9-2}$$

while the surficial anisotropy can be obtained with the formula

$$\Delta r = \frac{r_0 - 2r_{45} + r_{90}}{2} \tag{9-3}$$

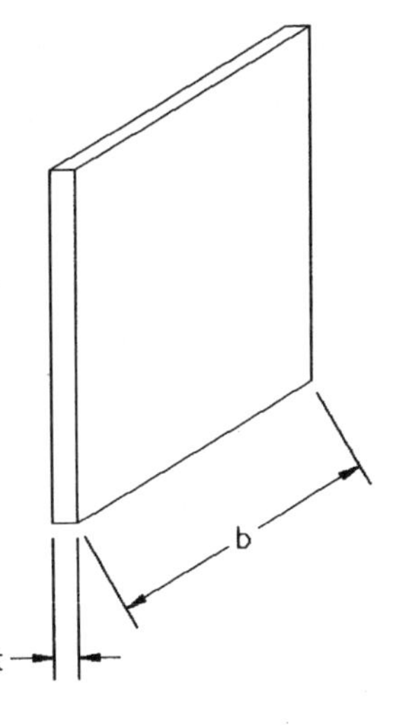


**Figure 9-8** Normal anisotropy values.

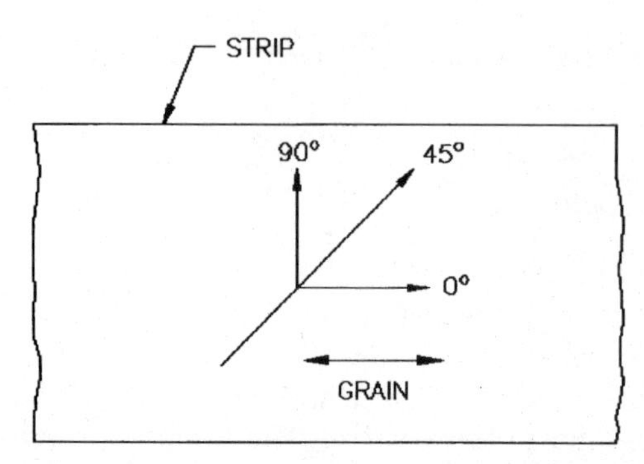


**Figure 9-9** Anisotropy in a plane (surficial).

The value of $\Delta r$ influences the variation in drawing results with respect to the grain line of the material, which present themselves as a variation in straightness of the upper edge of the drawn shell, as shown in Fig. 9-10.

Where the anisotropy $r$ would be greater in the direction of 0° and 90°, in that direction the material will be drawn to greater depths than along the 45° line, bringing the value of $\Delta r$ for 0° and 90° above zero.

Where the anisotropy at 45° exceeds the other directionally oriented values, the inequality in the surface will be the most pronounced along that line, and the value of $\Delta r$ will be driven under zero.

### 9-3-2. Severity of draw and number of drawing passes

The severity of the drawing operation may be expressed by the relationship of the blank diameter to the cup diameter. This ratio, often called a cupping ratio, allows for an assessment of the amount of drawing passes needed to produce a particular shell.

Where this ratio is exceeded, a fracture of the shell results, attributable to the exhaustion of drawing properties of the particular material. This means that from a blank of a certain size, only a certain cup diameter may be produced during a single drawing pass.

The severity of draw is calculated using Eq. (9-4):

$$K = \frac{D_0}{d} \qquad \text{or} \qquad M = \frac{d}{D_0} \tag{9-4}$$

where $K$ = severity of draw factor
$M$ = reverse value of severity of draw factor

Recommended values of $M$ to be used for the first drawing pass are $M = 0.48$ to $0.60$, with dependence on the drawn material properties.

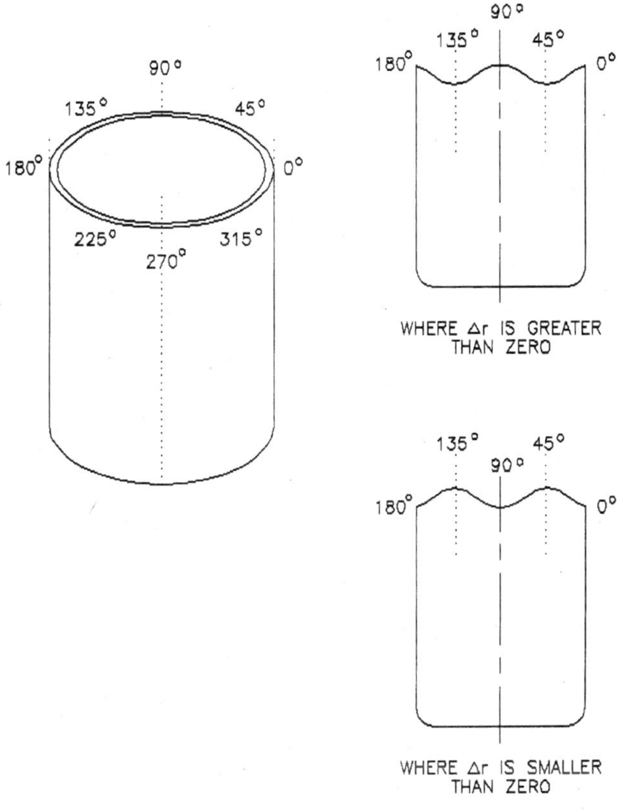


**Figure 9-10** Uneven edges of drawn shells due to surficial anisotropy $\Delta r$.

The CSN 22 7301 (Czech National Standard, similar to DIN) recommends a range from

$$M_{\min} = \frac{50 + 0.01D_0}{100} \tag{9-5a}$$

up to

$$M_{\max} = \frac{60 + 0.01D_0}{100} \tag{9-5b}$$

The advantage of using the severity of draw calculations at the early stages of the drawn shell evaluation is the immediate assessment of the number of drawing passes needed.

Subsequent drawing passes may use the CSN 22 7301 recommendations of the $M$-value range, starting at

$$M_{n-\min} = \frac{70 + 0.01d_{n-1}}{100} \tag{9-6a}$$

up to

$$M_{n-\max} = \frac{80 + 0.01d_{n-1}}{100} \tag{9-6b}$$

The total of all *M* coefficients for the particular drawing sequence is dictated by the geometry of all drawn and redrawn shapes and by the subsequent geometry of the finished shell. It may be calculated by using Eq. (9-7*a*):

$$M_{\text{total}} = M_1 \times M_2 \times M_3 \times \cdots \times M_n \tag{9-7a}$$

and subsequently

$$M_{\text{total}} = \frac{d_1}{D_0} \times \frac{d_2}{d_1} \times \frac{d_3}{d_2} \times \cdots \times \frac{d_n}{d_{n-1}} = \frac{d_n}{D} \tag{9-7b}$$

With a greater radius of the drawing die ranging between 8*t* and 15*t*, smaller values of the severity of the draw coefficient may be used. Subsequently, with smaller drawing die radii such as those ranging between 4*t* and 8*t*, larger coefficients are recommended.

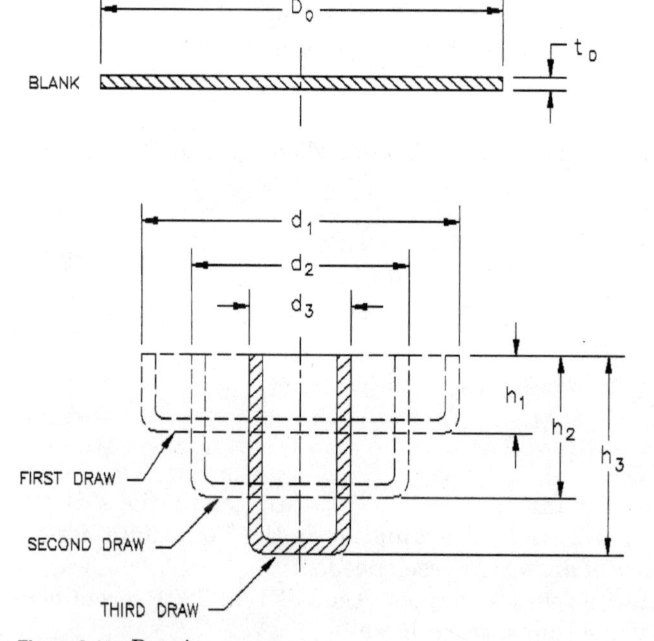


**Figure 9-11** Drawing sequence.

For metals low in ductility, such as brass and some harder grades of aluminum, the coefficient should be made purposely larger and lowered for more ductile materials.

### 9-3-3. Cupping strain factor $E_c$

The strain factor of the cupping operation shows the actual strain in the metal created by its elongation during the deep-drawing process. For evaluation of this type of stress, Eq. (9-8$a$) should be utilized, perhaps in replacement of the severity of draw calculations included in the preceding section.

$$E_c = \frac{h}{H} = \frac{2(D^2 - d^2)}{4d(D - d)} = \frac{\frac{D}{d} + 1}{2} \tag{9-8a}$$

Subsequently, to calculate the mean diameter of the cup $d$ with respect to the allowable amount of the cupping strain factor, an alteration of this formula will serve the purpose:

$$d = \frac{D}{2E_c - 1} \tag{9-8b}$$

The cupping strain factor, depicting the strain effect of cupping after the $n$-redrawing operation, is

$$E_{c,n} = \frac{h_n}{H_n} = \frac{t(D + d_n)}{2d_n t_n} \tag{9-9}$$

and the ultimate strain factor will be

$$E_{max} = 1 + \frac{e}{100} \tag{9-10}$$

where $e$ = maximum elongation at fracture, percent

Several recommended cupping ratios and their respective strain factors are shown in Table 9-1.

As each redrawing stage becomes progressively more impaired by the strain-hardening influence, the strain factor for each successive redrawing operation should always be smaller than the preceding one. Usually the first redrawing strain factor may be derived from the original $E_c$ value by calculating

$$E_{redraw} = (E_c)^x$$

with the exponent $x$ to be between 0.4 and 0.6. Usually a strain factor of 1.12 to 1.18 may be utilized in redrawing most materials.

**TABLE 9-1 Recommended Maximum Reductions for Cupping**

| Metal | Reduction in diam., % (max)* | Cupping ratio, $D/d$ | Strain factor, $E_{c,n}$ | Reduction in area, % (max)† |
|---|---|---|---|---|
| Aluminum alloys | 45 | 1.80 | 1.40 | 28 |
| Aluminum, heat-treatable | 40 | 1.60 | 1.30 | 23 |
| Copper, tombac | 45 | 1.80 | 1.40 | 28 |
| Brasses, high, 70/30, 63/37 | 50 | 2.00 | 1.50 | 33 |
| Bronze, tin | 50 | 2.00 | 1.50 | 33 |
| Steel, low-carbon | 45 | 1.80 | 1.40 | 28 |
| Steel, austenitic stainless | 50 | 2.00 | 1.50 | 33 |
| Zinc | 40 | 1.60 | 1.30 | 23 |

* = $100(1 - d/D)$.
† = $100(1 - a/A)$.
SOURCE: Frank W. Wilson, *Die Design Handbook,* New York, 1965. Reprinted with permission from The McGraw-Hill Companies.

In multioperational redrawing sequences, the total strain factor should be considered to be a multiple of the respective stress factors of all drawing operations.

This relationship may be expressed as

$$E_{c-\text{total}} = E_1 \times E_2 \times E_3 \times \cdots \times E_n \tag{9-11}$$

This means that should a total stress factor be $E_c = 1.4$, stress factors of various redrawing operations within the operational sequence should be chosen to equal their total multiple to 1.4.

The amount of stress factor value is mainly influenced by the ductility and strain hardening of the particular material. Where the total stress factor amount is reached sooner than the finished product is produced, annealing of the shell must be performed.

Singular strain factors—as may be seen from the formulas above—depend on ratios of the blank diameter to the shell diameter, or on the height of the drawn cup. The thickness of metal and the amount of friction within the particular drawing pass are also of importance in this process.

### 9-3-4. Reduction ratios

A shell may be drawn into a certain depth only without causing damage to its shape or structure. Where greater depths or reductions are required, subsequent drawing passes must be added. To determine the amount of reduction per given shell size, the following formulas should be used.

For the first operation die:

$$d_1 = \frac{B_1 \times D}{100 - 0.635D} \tag{9-12$a$}$$

For all redrawing dies:

$$d_2 = \frac{B_2 \times d_1}{100 - 0.635d_1} \qquad (9\text{-}12b)$$

$$d_3 = \frac{B_2 \times d_2}{100 - 0.635d_2} \qquad (9\text{-}12c)$$

etc.

where $D$ = blank diameter
$d_1$ = mean diameter of first shell
$d_2$ = mean diameter of second shell
$d_3$ = mean diameter of third shell
$B_1$ and $B_2$ = factors depending upon thickness of metal to be drawn, from Table 9-2

The reduction ratio $R_c$ may be calculated by using the following formula:

$$R_c = \frac{100(D - d)}{D} \qquad (9\text{-}13)$$

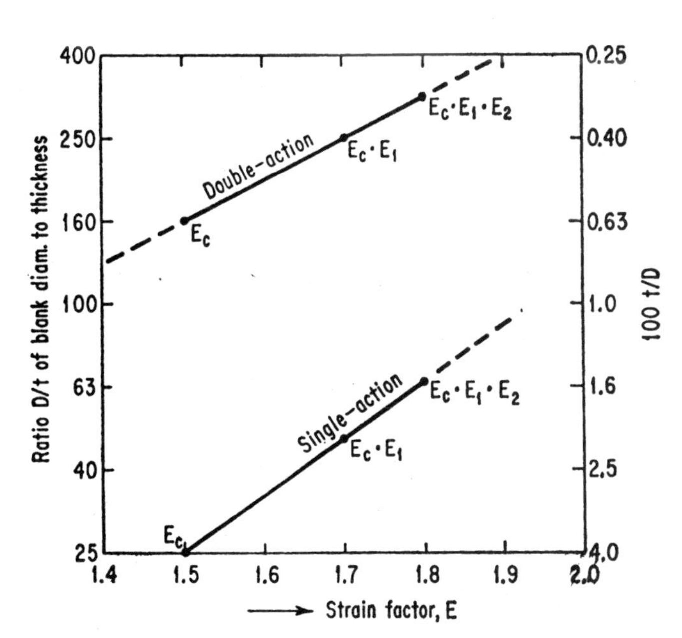


**Figure 9-12** Relationship between strain factors and blank dimensions for deep drawing with single- and double-action dies. (*From Frank W. Wilson, "Die Design Handbook," New York, 1965. Reprinted with permission from The McGraw-Hill Companies.*)

**TABLE 9-2 Factors $B_1$ and $B_2$**

| Stock thickness, in | First operation die $B_1$ value | Any redrawing die $B_2$ value |
|---|---|---|
| 0.015–0.018 | 61 | 74 |
| 0.021 | 58 | 73 |
| 0.022–0.024 | 56 | 72 |
| 0.027 | 54 | 71 |
| 0.031 | 50 | 70–71 |
| 0.062–0.109 | 47 | 70 |
| 0.125–0.250 | 51 | 65 |

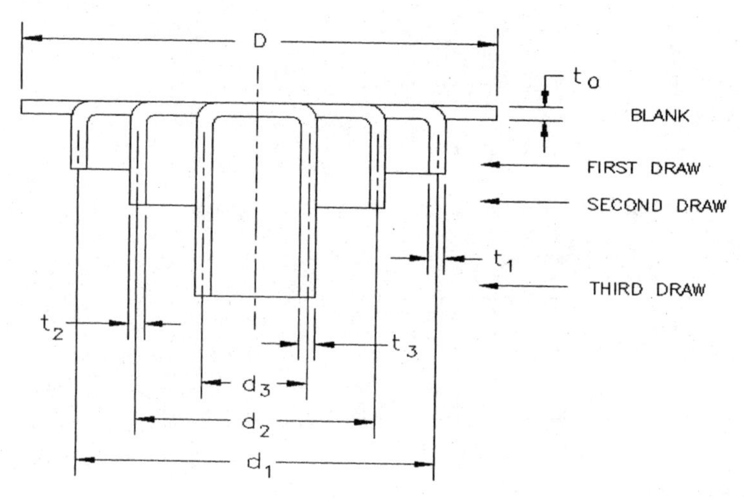


**Figure 9-13** Drawing sequence.

A maximum percentage of reduction for deep-drawing materials of various thicknesses is slightly higher than the values included previously. Intermediate annealing is to be utilized only when the shells become strain hardened or when cracks begin to form.

Table 9-3, giving the values of the maximum possible reduction, should be used for dies operating in hydraulic presses, where the pressure of the blank holder is constant. The percentages given here are recommended for drawing operations only where no ironing is involved. Should ironing of the shell be needed, the values shown in Table 9-3 must be reduced.

**Drawing of stainless-steel shells.** Drawing of stainless-steel shells may basically follow the same procedures as drawing of other materials. But a slight change in reduction formulas is necessary, for in the drawing process, stainless steel behaves differently from other materials.

**TABLE 9-3 Maximum Percentage of Reduction for Deep-Drawing Materials**

| Stock thickness, in | First drawing operation | Second drawing operation | Any additional drawing operation |
|---|---|---|---|
| 0.010 | 27 | 18 | 17 |
| 0.015 | 32 | 20 | 19 |
| 0.021 | 35 | 21 | 20 |
| 0.024 | 39 | 22 | 21 |
| 0.031 | 42 | 23 | 22 |
| 0.036 | 44 | 26 | 24 |
| 0.042 | 46 | 28 | 25 |
| 0.046 | 47 | 28 | 25 |
| 0.054 | 47 | 29 | 26 |
| 0.062–0.124 | 48 | 30 | 27 |
| 0.125–0.250 | 47 | 28 | 26 |

For example, a large reduction from the basic flat blank is possible for stainless steel of 18-8 type, but the subsequent drawing operations must be very moderate.

A chromium type 17-20 steel cannot be drawn into great depths from a blank, yet larger reductions may be obtained through succeeding redrawing operations.

Generally, chromium-nickel stainless steel strain hardens quite readily, for which reason more frequent anneals, combined with lower drawing speeds and better lubrication, are required.

The amount of reduction of the particular stainless-steel material may be calculated with the help of constants $B_{SS\text{-}1}$ and $B_{SS\text{-}2}$, listed in Table 9-4.

**TABLE 9-4 Factors $B_{SS\text{-}1}$ and $B_{SS\text{-}2}$**

| Material thickness, in | First operation die $B_{SS\text{-}1}$ value | | Any redrawing die $B_{SS\text{-}2}$ value | |
|---|---|---|---|---|
| | 18-8 SS | 17Cr SS | 18-8 SS | 17Cr SS |
| 0.015–0.018 | 61 | 68 | 81 | 77.5 |
| 0.021 | 58 | 65 | 80 | 76.5 |
| 0.022–0.024 | 56 | 63 | 80 | 75.5 |
| 0.027 | 54 | 60 | 79 | 74.5 |
| 0.031 | 50 | 56 | 77 | 74 |
| 0.062–0.109 | 47 | 53 | 75 | 73.5 |
| 0.125 | 51 | 51 | 65 | 65 |

Annealing between drawing stages is required for stainless. The preferred press equipment to be used is a double-action press or a single-action die with a drawing collar and an air cushion.

The formula for obtaining the maximum reduction in stainless-steel material to be used for the first operation die is

$$d_1 = \frac{B_{SS\text{-}1} D}{100 - 0.625D} \qquad (9\text{-}14a)$$

And for all succeeding redrawing dies:

$$d_2 = \frac{B_{SS\text{-}2} d_1}{100 - 0.625d_1} \qquad (9\text{-}14b)$$

and

$$d_3 = \frac{B_{SS\text{-}2} d_2}{100 - 0.625d_2} \quad \text{etc.} \qquad (9\text{-}14c)$$

where $D$ = blank diameter
$d_1$ = mean diameter of first shell
$d_2$ = mean diameter of second shell
$d_3$ = mean diameter of third shell, etc.
$B_{SS\text{-}1}$ and $B_{SS\text{-}2}$ = constants, depending upon the thickness of metal to be drawn, from Table 9-4

### 9-3-5. Strain hardening of material

The research has asserted that the strain-hardening coefficient has a dual effect on the formation of wrinkles in drawn material. First, it enhances the material's resistance to wrinkling through its supportive action toward the buckling modulus. Further, it affects the distribution of strain, caused by the action of drawing, while supporting the development of higher compressive stresses within the walls of shells.

To eliminate the effect of strain hardening, radial tension in the walls of the drawn part must be enhanced, and the die radius should

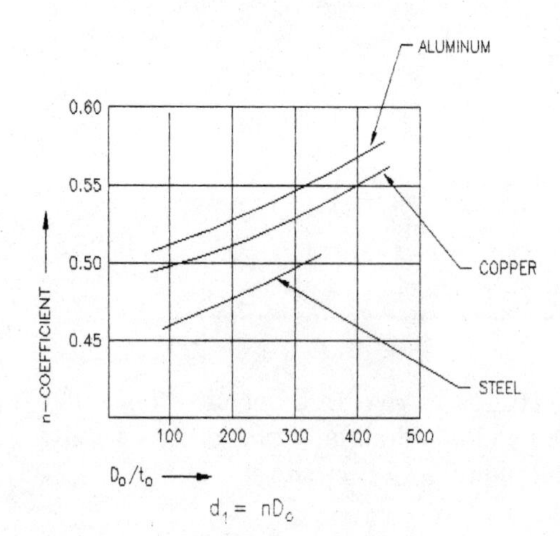


**Figure 9-14** Alternative calculation of blank reduction in drawing.

be decreased. Also the blank holder's pressure has to be increased, while lower-grade lubricants should be utilized.

Wrinkling of drawn parts usually begins quite close to the die radius; that's where the compressive stresses within the material are largest. Where the pressure of the blank holder is inadequate, these wrinkles will increase; but with higher blank-holding pressure, fracturing of metal covering the nose of a punch will occur.

In cold-rolled steels and high-strength low-alloyed materials the limitation in drawing depth due to fractures or wrinkling was established. In high-strength low-alloyed materials, the tendency to wrinkling is known, which—in order to be prevented—will demand higher blank-holding pressures applied against the blank.

In martensitic stainless steel, the strain-hardening tendency of the material decreases as the temperature rises. Yield stress can be lowered by the use of coarse-grained material, provided its surface layer is not as coarse as so-called orange-peel texture. Coarsely grained material also displays an increase in its ductility.

Uniformity of the strain distribution within the material, as well as a decrease of its strain-hardening tendencies, will be affected by choosing a material with quite a high carbon content, 0.1 percent and up. An equal distribution of the peak uniform strain depends also on the carbon content, most probably because of carbon's strengthening effect in strain-induced martensite materials.

With forming of deep stainless-steel shapes, using a 302 type of material, 50 to 60°C was attained in critical areas of the product. At such a temperature range, a strain of highest uniformity may be obtained in steels with 1.0 austenite stability factor.

Generally, slight variations in the strain rate values do not affect the strain-hardening tendency of the material. That tendency, along with the level of peak uniform strain and yield stress, can be increased by elevating the material's carbon and nitrogen content.

#### Strain hardening–related calculations

The strain factor of the material $E$, when equal to 1.0, is applicable to the annealed metal of the yield strength $S_o$. With strain-hardened metal, the $E_{max}$ value, which is the ultimate strain factor [see Eq. (9-10)] will enter the picture, accompanied by the ultimate tensile strength $S_u$. For purposes of calculation, these material data can be obtained for the mill.

A comparison of strain hardening, strain factor, tensile strength, and diametral reduction can be found in Fig. 9-15. The drawn shell's tensile strength $S_c$ may be obtained as

$$S_c = S_o \log_e E_c^n \tag{9-15}$$

where $n$ = strain-hardening exponent

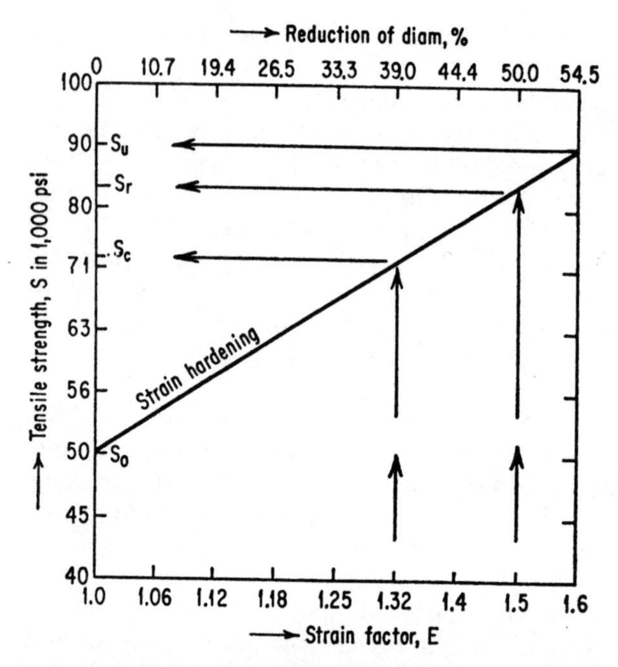


**Figure 9-15** Relationship between strain hardening, strain factor, and tensile strength. (*From Frank W. Wilson, "Die Design Handbook," New York, 1965. Reprinted with permission from The McGraw-Hill Companies.*)

The strain-hardening exponent $n$ determines the plastic limit or ultimate strength of the material usually not used in general drawing practice. Ultimate strength, with elastic limit, hardness, and yield point, increase when the strain factor of cupping $E_c$ increases.

With the cupping strain factor $E_c = 1.32$, a tensile strength of $S_c = 71{,}000$ lb/in² is encountered in the upper portion of the shell, which is the most strained area. Choosing a safe value for the total strain factor as $E_{c-\text{total}} = 1.5$ results in tensile strength in the cup's upper part of $S_c = 82{,}000$ lb/in². From these assumptions, an appropriate redrawing strain factor may be computed by using Eq. (9-11) as

$$E_1 = \frac{E_{c-\text{total}}}{E_c} = \frac{1.5}{1.32} = 1.14$$

The logarithmic relationship of the cupping ratio $D/d$ comes out as a straight diagonal line, shown in Fig. 9-16.

### 9-3-6. Wall thickness decrease or ironing

Some products can be designed with an intentional decrease in their wall thickness, to be attained during deep drawing. This decrease

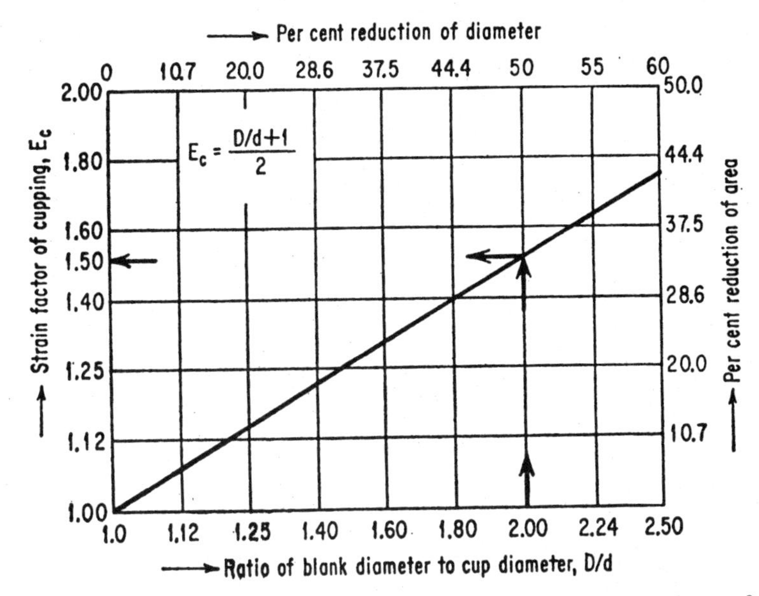


**Figure 9-16** Relationship between cupping ratio $D/d$ and the strain factor of cupping $E_c$, (logarithmic base). (*From Frank W. Wilson, "Die Design Handbook," New York, 1965. Reprinted with permission from The McGraw-Hill Companies.*)

may often be considerable, and often several drawing passes are needed if these shells are to be produced using either conventional processes or alternative manufacturing methods such as extruding or reverse drawing.

Shells of smaller diameters may be deep-drawn with an addition of thinning, or ironing of their walls. The number of necessary drawing passes depends on the ratio of the wall thicknesses before and after drawing, and it is influenced by the maximum possible deformation of the material. Equation (9-16) should be used to calculate deformation

$$F = 100 \frac{t_{n-1} - t_n}{t_{n-1}} (1 - k_t) \quad \% \tag{9-16}$$

where $F$ = allowed deformation per Table 9-5.

$t_n$ and $t_{n-1}$ = wall thickness per Fig. 9-17. The $n$ denotes the sequential number of the draw, 1, 2, 3,…,$n$

$k_t$ = thinning coefficient, obtained from the formula

$$k_t = 100 \frac{t_n}{t_{n-1}} \quad \% \tag{9-17}$$

**TABLE 9-5 Ironing-Related Deformation (*F*) (Percentages)**

| Material | First draw | | Second draw | |
|---|---|---|---|---|
| | *F* | *kt* | *F* | *kt* |
| Steel, annealed | 55–60 | 40–45 | 35–40 | 50–55 |
| Steel, medium hard | 35–40 | 55–60 | 25–30 | 65–70 |
| Aluminum | 55–60 | 35–40 | 40–50 | 50–55 |
| Brass | 60–65 | 30–40 | 50–60 | 40–50 |

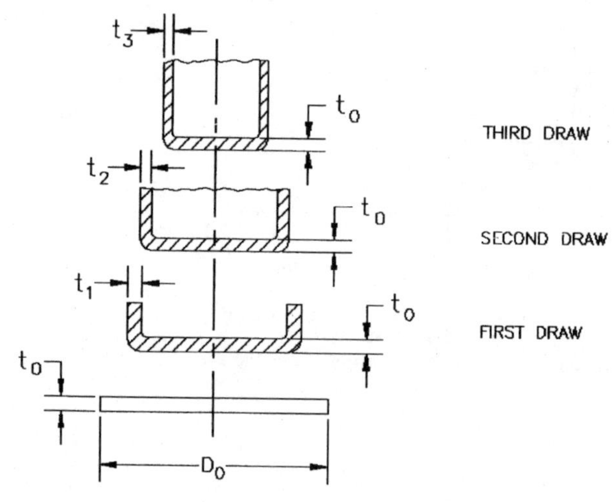


**Figure 9-17** Drawing sequence with thinning of the wall.

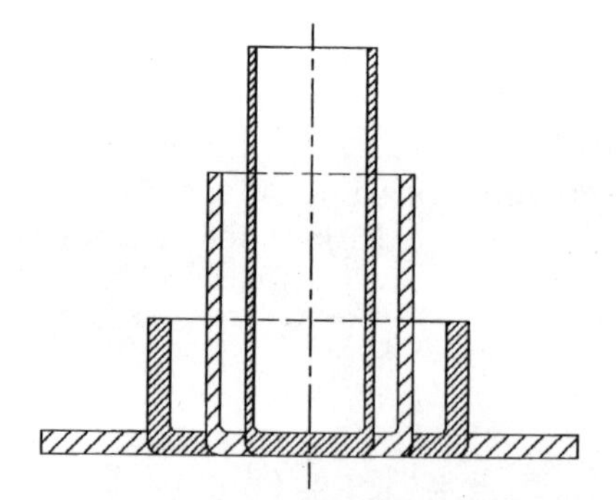

**Figure 9-18** Drawing operations with thinning of the wall.

Today's manufacturing methods allow for a combined drawing in a single die, using various inserts in succession, to achieve the desired decrease in the product wall (Fig. 9-19). Since the friction between

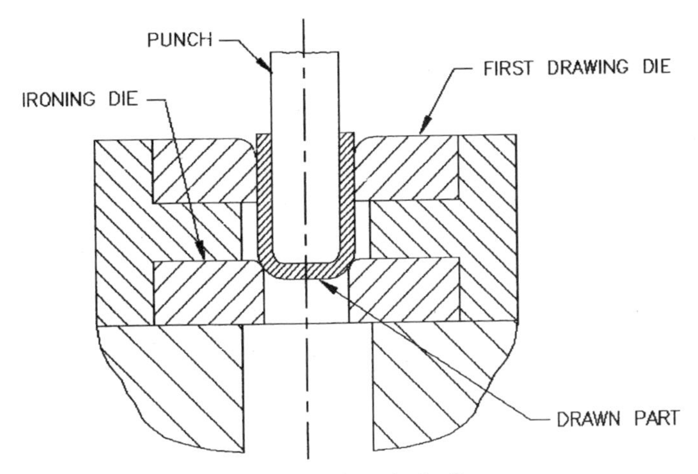


**Figure 9-19** Two drawing inserts in a single die.

material and tooling may be considerable in such a case, parts are coated prior to drawing with copper or phosphate coatings.

## 9-3-7. Forming of flanges

Various blanks may be drawn into impressive depths, provided a correct material type is chosen, the tooling design is sound, and the manufacturing procedures and the choice of press are acceptable. All drawn products can be placed in two groups:

- Drawn shells with flanges
- Drawn shells without flanges

Where flanges are to be produced on a drawn shell, the size of the blank must be adequate to suffice for their width, often leaving some additional material for trimming. Trimming of the drawn cup with a flange is usually inevitable, for the outer edge of the blank may become distorted by the drawing process (Fig. 9-20).

Trimming of the flange is performed in the last operation of the sequence, where the finished part is also ejected out of the die. Sometimes a pinch-trim of the flange is preferred, because of its speed and simplicity of operation (Fig. 9-21).

Where parts without flanges are to be drawn, the blank size for their production should be exact, with no material to be removed afterward. This way the material volume gets all used up in the drawing process. The finished shell may be ejected from the die in a last drawing stage, where it is dropped down during the return stroke of the punch (Fig. 9-22).

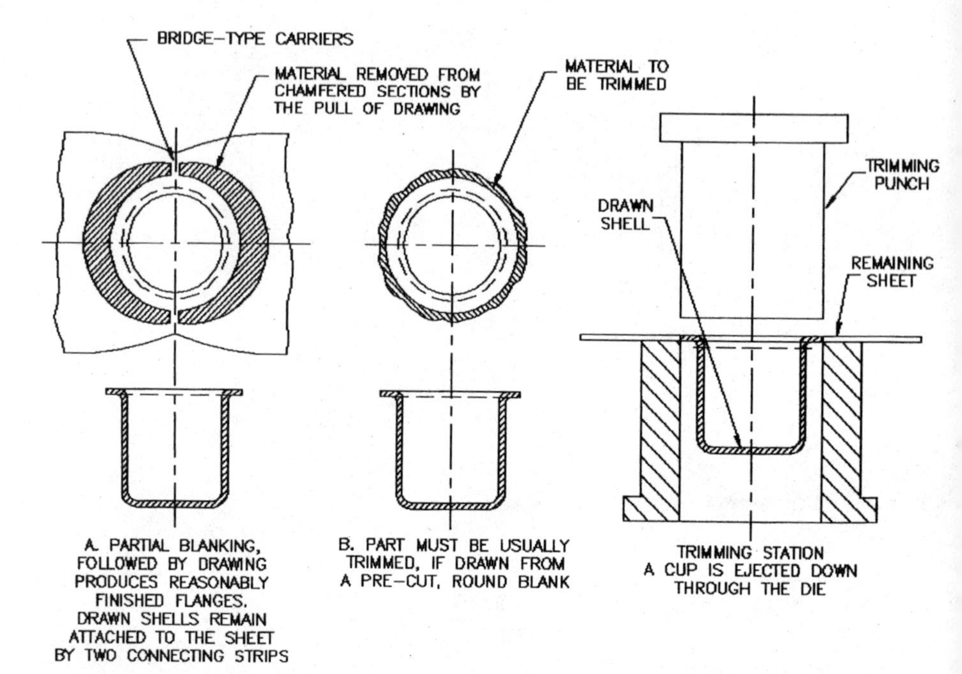


**Figure 9-20** Flange trimming.

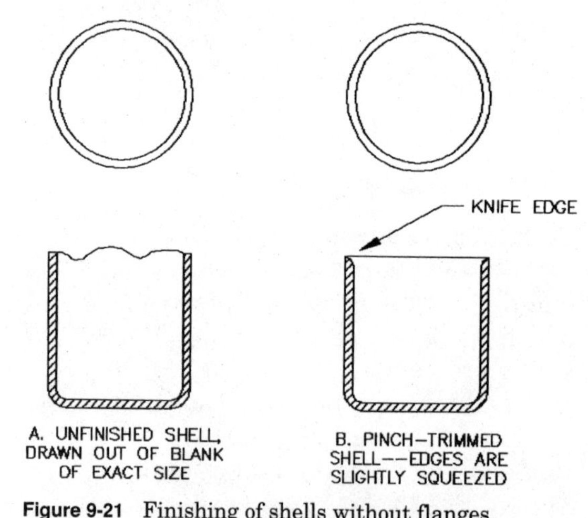


**Figure 9-21** Finishing of shells without flanges.

This type of ejecting method utilizes various types of strippers, restricting the formed cup from following the movement of the punch. Some samples of such stripping arrangements are shown in Sec. 4-1-5, "Strippers."

The method of forming flanges further forks into two procedures in their production. First, where the diametral reduction of the flange is

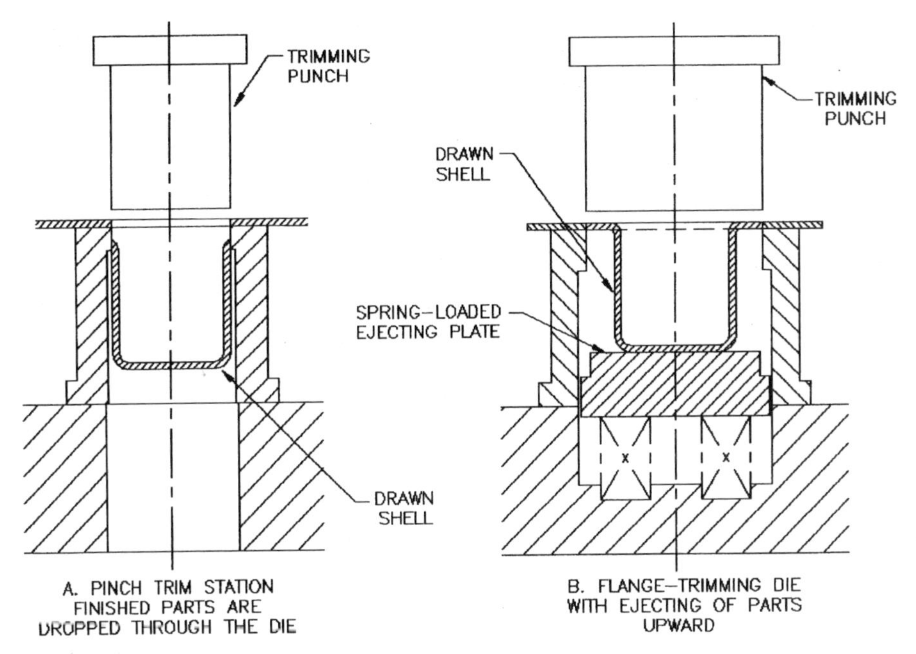


**Figure 9-22** Ejecting of drawn shells.

not of concern, the blank may be drawn into the desired shape and depth by drawing and redrawing until a finished part is produced. Each drawing stage slightly reduces the blank diameter, utilizing the material for the depth of the shell. The width of a flange is not affected, but its overall diameter diminishes.

Such a procedure subjects the material to a greater strain, created by excessive stretching of both the flange and the cup. The strain-hardening effect is increased, with more annealing required between redraws. Nevertheless, this is the most commonly used drawing practice (Fig. 9-23).

The second drawing method keeps the blank diameter intact while increasing the width of the flange with each redrawing pass (Fig. 9-24). This method should be used where the diameter of the flange is considerably larger than the diameter of the shell.

During the drawing process, the material is actually forced back into the flange with each decrease of the shell diameter. This is achieved by making a radius of the first drawing die as large as possible, with all subsequent radii reduced in size. Such an alteration promotes the upward flow of metal, keeping the wall thickness more uniform and creating less strain within the material structure, which subsequently decreases the number of annealings needed.

Care should be taken when adjusting the stroke of a press so that it does not exceed the required depth, for if the punch begins to draw

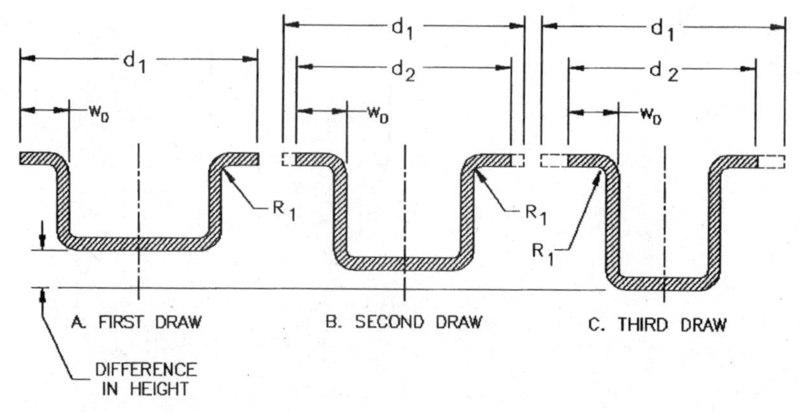


**Figure 9-23** First method of shell drawing.

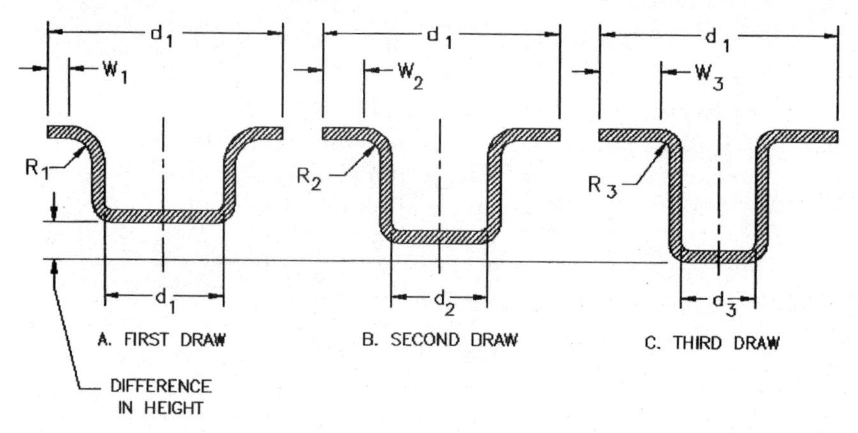


**Figure 9-24** Second method of shell drawing.

the material in each redrawing operation, it certainly will pull the material from the flange.

Shells with wide flanges often pose a problem in the drawing process, as the tendency to form wrinkles is sometimes difficult to overcome. Therefore, an adjustment in the drawing procedure should be made to allow for the allocation of metal to be drawn partially from the flange as well as from the body of the shell (Fig. 9-25).

Here the depth is produced in the first drawing operation, leaving the shell with tapered walls. The second draw straightens the sides of the cup, pushing the excess material into the flange, which turns wider. In the third drawing station the bottom diameter is drawn smaller in size, the excess material being allocated for a production of another inclination of the wall. The last drawing pass straightens the wall, pushing the excess material toward the flange.

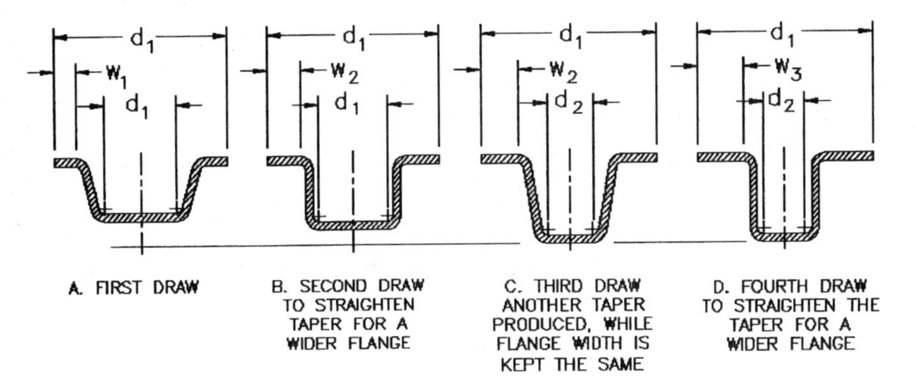


**Figure 9-25** Drawing of a shell with wide flanges.

### 9-3-8. Height of a shell

The height of the shell consists of a displaced metal taken mainly from the flange and more or less from the other areas of the blank. Where no thinning of walls is encountered, the bottom of the shell remains unaltered, with no metal being removed or added there.

The maximum height attainable from a given blank size can therefore be calculated. For the purpose of simplicity this evaluation is approximate, where the corner radii are neglected and the shell thickness is considered equal to that of the blank in all its cross-sectional areas. In such a case, the height will be

$$h = \frac{D^2 - d^2}{4d} \tag{9-18a}$$

where $h$ = height of shell
$D$ = blank diameter
$d$ = mean diameter of shell

The height of a shell subjected to $n$-redrawing operations may be calculated:

$$h_n = \frac{t(D^2 - d_n^2)}{4d_n t_n} \tag{9-18b}$$

Obviously, $d_n$ is the diameter of the shell after $n$-redraw and $t_n$ is the thickness of the wall then, while $t$ is the original thickness of the blank.

### 9-3-9. Drawing speed

While other die processes are not overly affected by the actual speed, the drawing operation is speed-dependent with respect to the material drawn. Where zinc is included in the material buildup, a slower drawing rate should be chosen. Such speed is also beneficial for drawing of austenitic stainless steel. With aluminum- and copper-based materials, greater speeds are possible.

Generally used drawing speeds for single-action and double-action dies fall within a range of the following values.

| Material | Single-action drawing, ft/min | Double-action drawing, ft/min |
|---|---|---|
| Steel | 60 | 35–55 |
| Stainless steel | — | 20–30 |
| Aluminum | 180 | 100 |
| Aluminum alloys | — | 30–40 |
| Copper | 150 | 85 |
| Brass | 200 | 100 |

## 9-4. Drawing of Thick-Walled Cylinders

Drawing of thicker materials is slightly different from drawing of thin-walled shells. First of all, these parts do not need a blank holder, as the thickness of material prevents the blank from collapsing under the tangential pressure. Single-action tooling may be utilized for production, which brings the cost of the die down. The disadvantage is the more demanding construction of the die and a larger value of the stroke.

The radius of the drawing die is of utmost importance, for it needs to produce a low coefficient of friction. One way to achieve such a scenario is to make the die radius of such a size and shape that it continuously pushes the drawn material against the punch.

In Fig. 9-26, the difference between small and large radius is shown pictorially. With a greater than necessary distance between the point of support $A$ and the edge of the blank, an additional bend in the flange is created. Such a bend not only increases the need of drawing pressure, it also exposes the material to wrinkling within the area of the flange.

**Shaped die edge.** Another technique for solving the drawing punch and die relationship is Pelczinski's method, which forms the die radius in such a fashion that the point of contact between the die and the drawn part is always restricted to the edge of the radius. Such a die edge modification is arrived at graphically, with the drawing operation divided into segments and with each of the segments evaluated separately for the location of the edge as seen in Fig. 9-27.

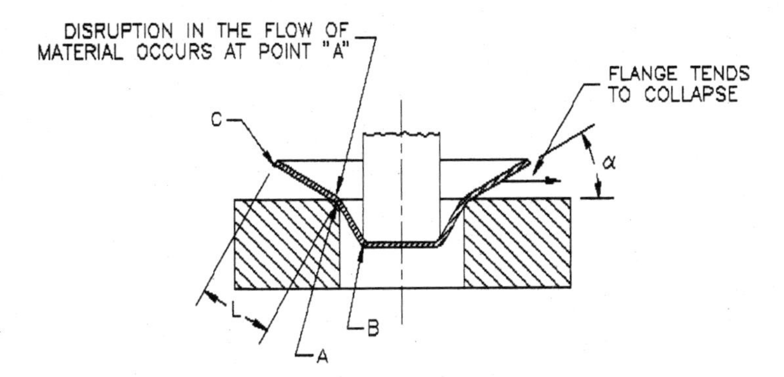


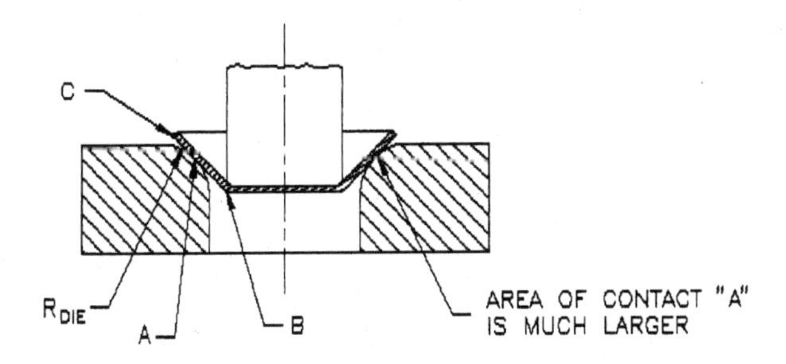


**Figure 9-26** Drawing of thick-walled cylinders.

This theory considers the wall thickness of the drawn part constant, in which case Eq. (9-19) applies:

$$\frac{D^2 - d^2}{4} = \alpha R d \pi + 2\pi R^2 (L - \cos \alpha) + L\pi (d + 2R \sin \alpha + L \cos \alpha) \tag{9-19}$$

The result helps to determine additional dimensions involved in this evaluation as

$$\cos \alpha = \frac{R^2 - (d/2)^2}{(D_0/2)^2 - (d/2)^2} \tag{9-20}$$

where all values are from Fig. 9-27.

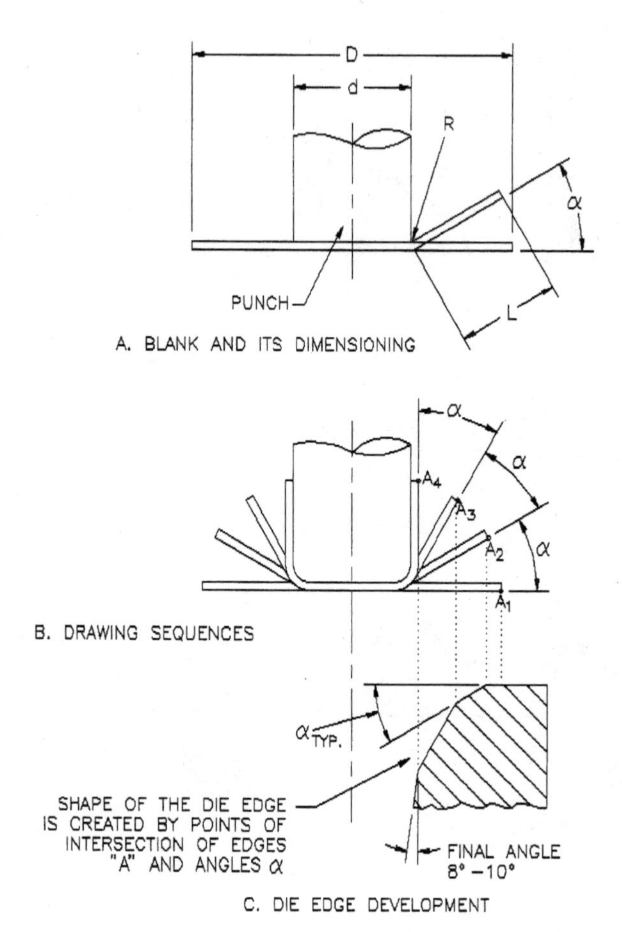


**Figure 9-27** Development of the drawing die radius.

The configuration of the die radius is shown in Fig. 9-27*b*, where all points corresponding with various locations of the bent-up flange are shown. The final angle is the continuation of the last inclined surface, usually in the vicinity of 8 to 10° with the vertical.

The drawing force diagram, if drawn, will not show any great deviation from the constant until the shell arrives at the area of 8 to 10° draft, at which point there will be an increase in the drawing pressure. Such a curve, if compared to that created by a large-radiused die, will clearly show a decreased demand for drawing pressure.

**Huygens' tractrix curve.** A different method of drawing edge design is presented here by including the tractrix curve technique. Tractrix is a line, shown as beginning at point $A$ (Fig. 9-28*a*) and created by an evolving catenary. This type of drawing edge construction method

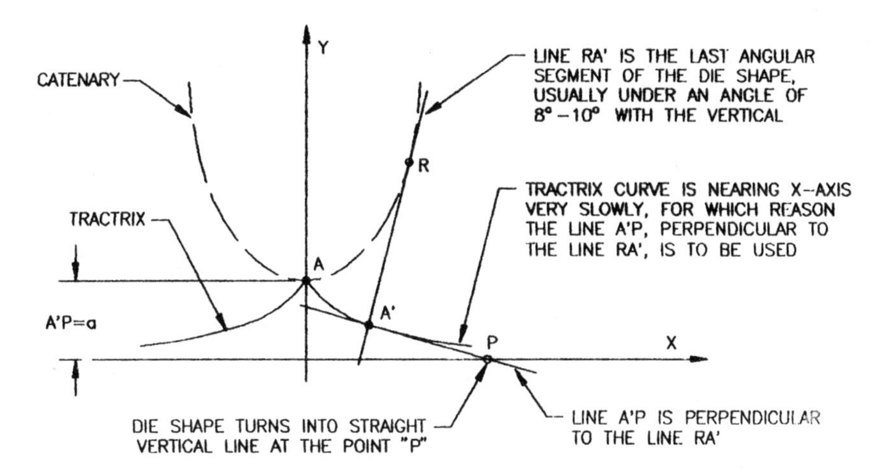


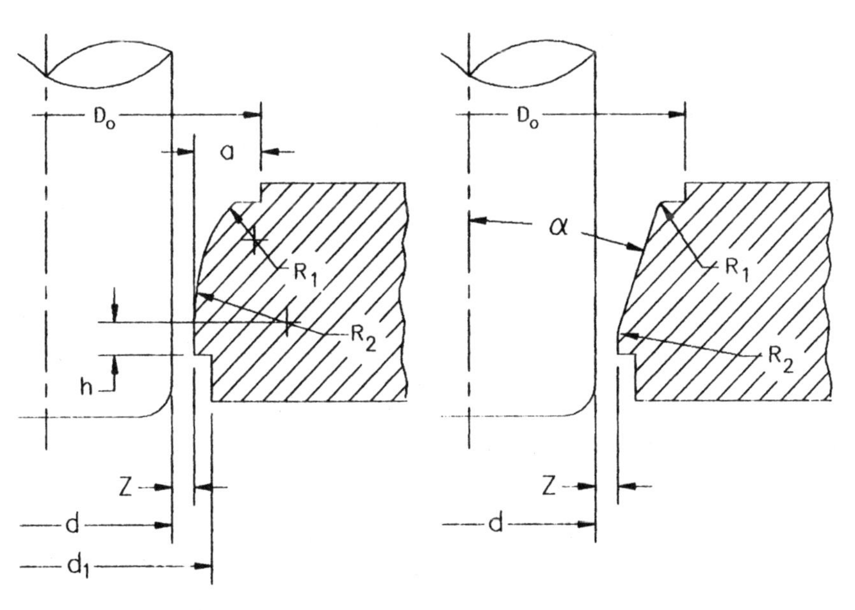


**Figure 9-28** Drawing edge development.

allows for the use of the smallest drawing pressures, while attaining the best severity of draw coefficient of 0.35. The tractrix curve is produced by a continuous tangential contact of the drawn part's edge with the die.

The equation form of the tractrix curve is

$$x = a \ln \frac{a + \sqrt{a^2 - y^2}}{y} - \sqrt{a^2 - y^2} \tag{9-21}$$

The tractrix's asymptote is the axis $x$, with the axis of rotation being $y$ and the radius of the curve being

$$R = a \operatorname{cotg} \frac{x}{y} \tag{9-22}$$

Geometry of the drawing die is shown in Fig. 9-28*b*. The radius here consists of two combined radii $R_1$ and $R_2$, out of which the first $R_1$ is the line of the tractrix. However, since the tractrix curve is nearing the $x$ axis—which actually is a centerline of the drawn part—only very slowly, a great height of such a die would be necessary, with a subsequent extended stroke of the press. For that reason, a tangent under an angle of $\alpha = 8$ to $10°$ with the vertical is attached to the end of the tractrix curve.

Practical parameters for this type of evaluation are as shown below:

$$a = \frac{D_0 - d_1}{2}$$

$$R_1 = 0.5a$$

$$R_2 = 2.3a$$

$$d_1 = d + 2z$$

$$z = (1.2 - 1.4)t_0$$

$h$ = between 0.040 and 0.400 in, depending on the thickness of the blank

Since the actual manufacture of such a complex curve may sometimes pose a problem, a simplified shape of the drawing die, shown in Fig. 9-28*c*, has been derived. Here the tractrix line is replaced by a straight angular segment, usually at angle $\alpha = 30°$ with the vertical. Other values may be used as

$$R_1 = 0.05\,D_0$$

$$R_2 = 5t_0$$

$$d = M(D_0)$$

where $M$ = coefficient of severity of draw

$$z = (1.2 - 1.4)t_0$$

The radius of the drawing punch should be made equal to the radius of the drawing die, or $R_{punch} = R_{die}$. For progressive redrawing operations, the edge of the punch is often chamfered at 35 to 45°.

The radius of the last redrawing punch of a sequence should be a minimum of (3 to 7) $t_0$, with the following variations for a given thickness of blank:

3 to $4t_0$ for blanks of 0.40 to 4.00 in diameter

4 to $5t_0$ for blanks of 4.00 to 8.00 in diameter

5 to $7t_0$ for blanks of 8.00 in diameter and up

The surface of the drawing punch must be smooth in order not to impair the drawing process in any way. Air vents are vital to the outcome as well.

## 9-5. Drawing of Square or Rectangular Shapes

When drawing a square or rectangular shape, the first that must be evaluated is the depth of the draw. If the depth of a part is not excessively deep, perhaps the same product may be obtained by a much easier manufacturing method, such as stretching the material. However, where greater depths are involved, the drawing process must be utilized.

Drawing of box-type objects is susceptible to various stresses within the material, which are far from being evenly distributed around the edge of a die. Plastic deformation at the level of the flange is present within the area of partial-cylindrical corners, as with round cylindrical shells. The sides of a box, which theoretically should contain only a radial tension, such as that imposed by simple bending, are actually quite stressed. This is due to the stress step between the two stress differences, the influence of which is considerable.

Shown in Fig. 9-29 is a sketch of a drawn cylindrical shape (Fig. 9-29*a*) and a box-type part (Fig. 9-29*b*). Main dimensions $R_b$, $R_d$, $R_c$, $h$, and $t_0$ are the same for both products. In both cases, the two-dimensional stress within the flange ($\sigma_1$ tension, $\sigma_3$ compression) is applicable. With the cylindrical shape, the tension is equally dispersed

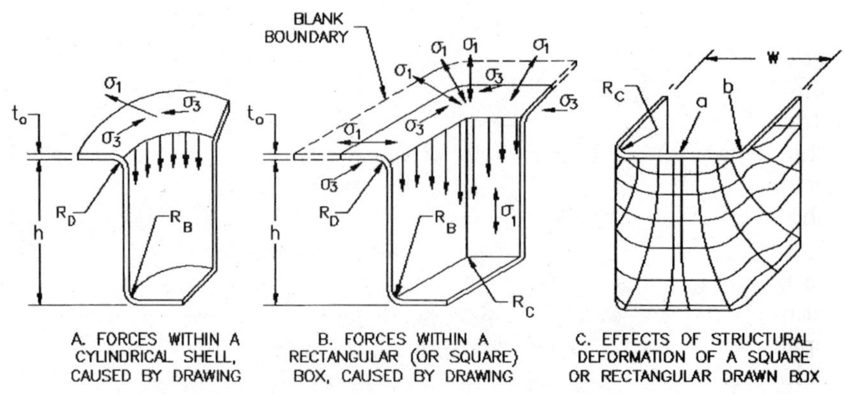


**Figure 9-29** Deformation of drawn shapes.

throughout the circumference, and its direction and value are consistent all around. The box-type drawn part suffers from inequalities of these influences, the directions and values of which change as they approach the corner: Their quantity diminishes, and their direction is diverted from their radial course toward the adjacent sides (see Fig. 9-29*c*).

The tension contained in the drawn body of each shell is directed along a single axis. However, in a square shell, its value reaches the upper limit in the corner, with decreasing tendencies along the straight sides.

Figure 9-29*c* depicts the lines of deformation within the part. It is obvious that the straight sides of the box suffer from certain bending influences, accompanied by complex deformations, which produce a compression tendency of the circumference of the part and an extension of the area of the bottom edge. Therefore, the widely accepted claim that the drawing of a square or rectangular shape is restricted to the corner areas is proved inaccurate.

Approximate values of the deformation, assessed in the middle of the upper edge (point $a$) and around the corner radius (point $b$) are recorded below. Values of the cylindrical shape are included for comparison.

| | Value of deformation | | |
|---|---|---|---|
| | Rectangular shape | | Cylindrical shape |
| Height of drawn shape | Point $a$ | Point $b$ | |
| $h = W$ | 25–30 | 45–50 | 55 |
| $h = 0.5W$ | 15–20 | 30–46 | 40 |
| $h = 0.3W$ | 5–8 | 15–30 | 31 |

With regard to the depth of draw, the tangential flow may reach quite high levels, exceeding columnar (buckling) limits of the drawn material. This may cause an appearance of wrinkles within the drawn body of the part. Columnar limits depend on the thickness of the wall and geometry of the die radius, whereas tangential stress is related to the radius of the corner of a part, plus wall thickness and the depth of draw, or $t/d$.

Alignment of the two different stress situations, such as those within the corner and straight side of a part, can be solved by the alteration of the part's geometry and by a partial decrease of the drawing space between the punch and die in straight areas. Another helpful solution is the installation of draw beads, which slow down the drawing process of these sections. An application of draw beads is shown in Fig. 9-30.

Draw beads may be positioned either on the punch side or on the die block. Their distance off the drawing edge should be approximately 1.187 in, and they may be added as a single row, or doubled, with a maximum of three rows (see Fig. 9-30). The exact number of draw beads per side of the rectangular or square box is determined on the basis of testing.

Naturally, the width of a flange is of importance in such drawing, and an exact blank layout of the part is essential for the success of the drawing operation.

To calculate widths and lengths of the straight, bent-up portion of a box-type shell, the formula for 90° bends may be used (see Sec. 8-4-2, "U-Die Bending," Table 8-7).

**Tolerance ranges.** For drawing of rectangular or square shells in one operation, the clearance between the punch and die should be equal to the thickness of the drawn material. This situation is sometimes referred to as "metal-to-metal."

A slightly larger than metal thickness corner clearance will greatly reduce the drawing pressure while reducing the strain on the shell as well.

Shells that require more than one drawing operation should use a clearance chosen with regard for the material being drawn. Basic guidelines for the proper choice of such a clearance are included in Table 9-6.

### 9-5-1. Drawing radii

Drawing of box-type shapes usually cannot be performed in one drawing pass, because the depth of the draw may be quite restrictive. However, using two successive dies may create another problem,

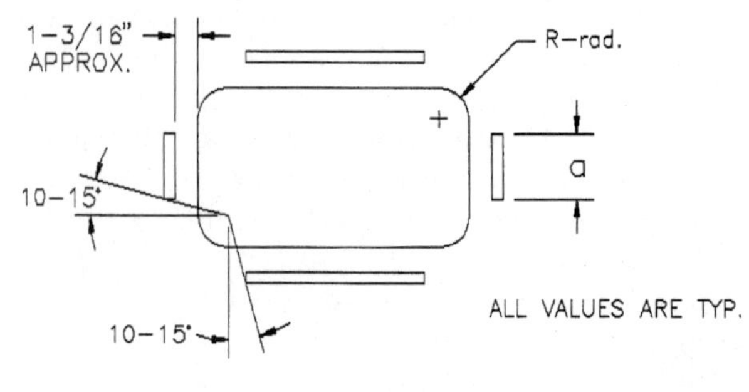


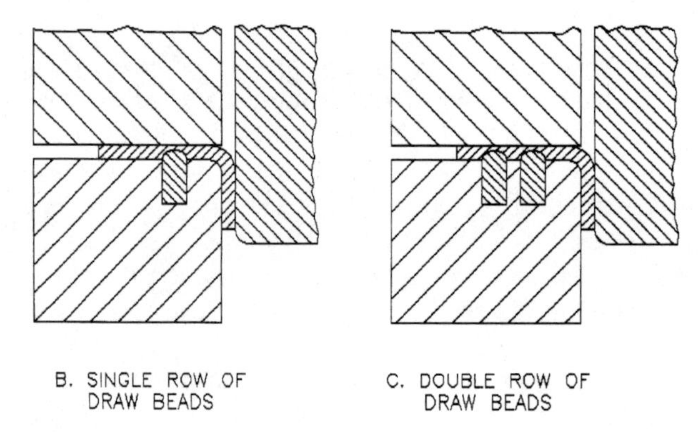


**Figure 9-30** Draw beads in rectangular or square parts drawing.

**TABLE 9-6 Clearance between the First Drawing Punch and Die for Square or Rectangular Shells**

| Material | Per-side clearance between punch and die |
|---|---|
| Aluminum | 1.25$t$ |
| Steel, deep drawing | (1.1 to 1.2)$t$ |
| Stainless steel | (1.75 to 2.5)$t$ |

$t$ = stock thickness.

especially where drawing steel material. The drawing pressure, being applied twice in succession to the corner areas, produces considerable stresses within these sections of the part. The sides of a box, often considered to be nothing but flat, bent-up sections, are actually vastly affected by the variation in the volume and orientation of stresses caused by the sudden breakage in the continuity of the shape.

Deformation, which affects the sides of the box, is directed toward the corner areas, making these susceptible to greater than usual wear, with the material of tooling being literally eaten away. This condition produces an increase of the gap between the tooling, with ensuing thickening of the drawn material in those sections. Often the corners of such boxes are so congested and stain-hardened that their condition impairs the second drawing process and annealing must be administered in between.

The relationship of two successive dies drawing a square or rectangular shell may be observed from Fig. 9-31.

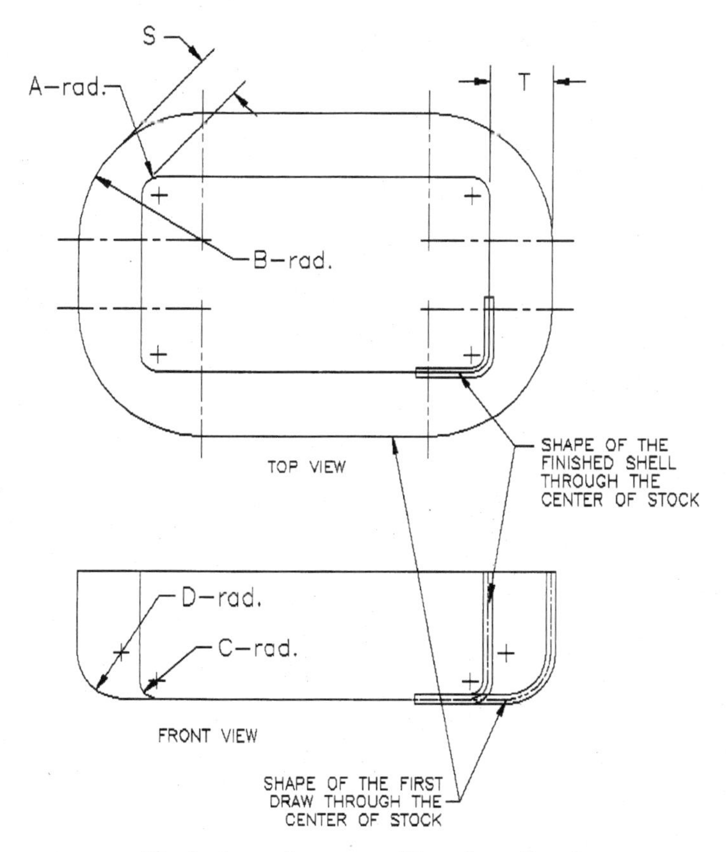


**Figure 9-31** Blank size and corner radii configuration for square or rectangular shells, produced in two drawing sequences.

$A$ = corner radii of the finished shell

$B = 5A$

$S$ = approximately 0.125 in to serve as a drawing edge

$T = 3A$ (where $A$ is less than 0.5 in), or 0.75 in (where $A$ is larger than 0.5 in)

$C$ = bottom radii of finished shell

$D = T + 0.25C$

The size of the drawn box corner radii may be figured with regard to the value of severity of draw $M$, or

$$M = \frac{2R_{\text{corner}}}{D} \tag{9-23}$$

Values of the coefficient $M$ for the first and second draw are given in Tables 9-7 and 9-8. It may be observed that the longer the straight side portion of the box-type drawn part is the lower a coefficient of severity of draw may be used.

### 9-5-2. Maximum depth of a drawn shell

The maximum depth of a box-type shell, attained in a single drawing pass, may be determined on the basis of the following variables:

- Type of material and its thickness
- Size of corner radius through the center of stock
- Size of the bottom radius
- Ratio between corner radii and length of straight portions of shortest side
- Width of the flange
- Blank shape and overall size of the shell
- Type of die and press

The size of corner radius is the main factor in determining the depth of a square or rectangular shell. The distance between the corner radii centers should be at least six times the value of a corner radius (see Fig. 9-32). If such a distance should be smaller than $6R$, the depth of the drawn shell has to be decreased proportionately.

The bottom (bend) radius along the straight sides and corners as well should be the same size as the corner radius or larger. The size of corner radius in relation to the depth of draw is shown in Table 9-9.

**TABLE 9-7 Severity of Draw Coefficient *M, Square or Rectangular Shells, First Drawing Pass Only***

| Length $a$* | Percentage ratio of thickness to blank diameter 100($t/D$) | | | | | |
|---|---|---|---|---|---|---|
| | 0.1–0.3 | 0.3–0.6 | 0.6–1.0 | 1.0–1.5 | 1.5–2.0 | Above 2.0 |
| 2$R$ | 0.550 | 0.540 | 0.520 | 0.500 | 0.475 | 0.450 |
| 5$R$ | 0.480 | 0.465 | 0.450 | 0.425 | 0.405 | 0.380 |
| 10$R$ | 0.360 | 0.345 | 0.330 | 0.305 | 0.285 | 0.260 |

*$a$ per Fig. 9-30.

**TABLE 9-8 Severity of Draw Coefficient *M, Square or Rectangular Shells, All Redrawing Operations***

| Length $a$* | Percentage ratio of thickness to blank diameter 100($t/D$) | | | | | |
|---|---|---|---|---|---|---|
| | 0.1–0.3 | 0.3–0.6 | 0.6–1.0 | 1.0–1.5 | 1.5–2.0 | Above 2.0 |
| 2$R$ | 0.760 | 0.750 | 0.740 | 0.730 | 0.715 | 0.700 |
| 5$R$ | 0.660 | 0.650 | 0.640 | 0.625 | 0.610 | 0.590 |
| 10$R$ | 0.500 | 0.490 | 0.475 | 0.455 | 0.435 | 0.415 |
| Last draw | 0.880 | 0.875 | 0.860 | 0.855 | 0.845 | 0.830 |

*$a$ per Fig. 9-30.

Another way of determining if a given part needs more than one drawing operation is to build a die for the last drawing station and try it.

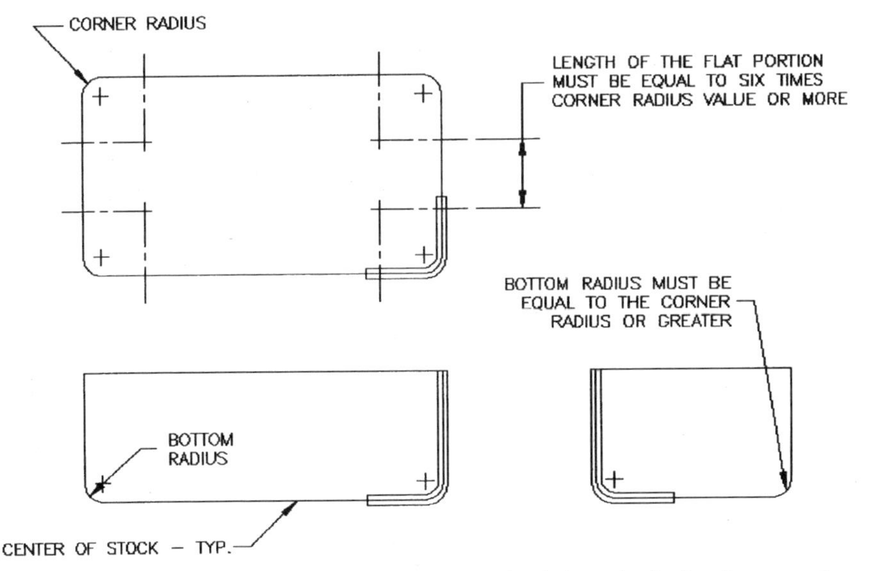


**Figure 9-32** Parameters for obtaining a maximum depth in a single drawing operation (square or rectangular shells only).

**TABLE 9-9 Maximum Depth of a Square or Rectangular Shell, Obtainable in a Single Drawing Pass, Using Deep-Drawing Stock**

| Corner radius through center of stock | Maximum drawing depth factor, (multiply corner radius by the number below) |
|---|---|
| Up to 0.187″ radius | 8 |
| 0.187″ to 0.375″ radius | 7 |
| 0.375″ to 0.50″ radius | 6 |
| 0.50″ to 0.750″ radius | 5 |
| Over 0.75″ radius | 4 |

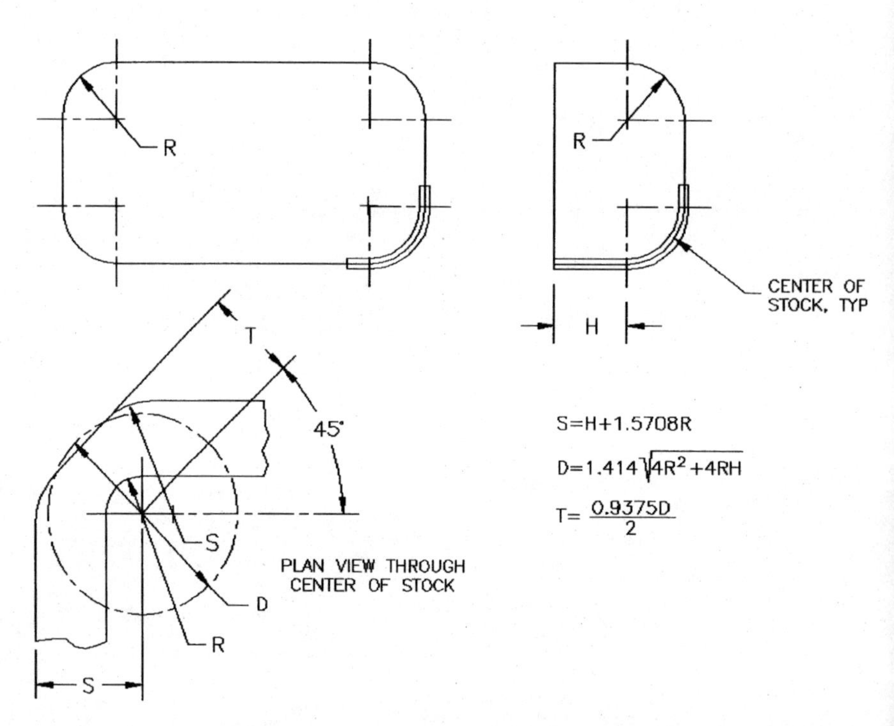


**Figure 9-33** Approximate size and shape of corners (in flat) for square or rectangular shells.

### 9-5-3. Approximate blank corner shape for square or rectangular shells whose width exceeds their depth

The method of obtaining an approximate shape of the corners of blank for square or rectangular shells is shown in Fig. 9-33. It may be used only for such drawn products whose width exceeds the depth and whose corner radius is equal to the bottom radius.

To find the length and width (in flat) of the sides of the box, bending formulas from Sec. 8-4-2 may be used.

This method of corner development should be used to obtain the shape and size of a trial blank. Formulas for the development of accurate shape of corners cannot be derived, owing to the involvement of too many factors. After drawing of the trial blank, small alterations to the blank outline may be needed.

### 9-5-4. Graphical method of blank development for square or rectangular shells whose depth is equal to or greater than their width

To distribute the metal equally during the drawing process, the calculated length of the blank must be diminished, while the width must be extended in the area of radii $V$ and $S$. The amount of decrease in length and increase in width cannot be calculated, as such formulas do not exist. Rather, a sketch shown in Fig. 9-34 should be utilized to graphically construct the shape of the blank. Values for different radii

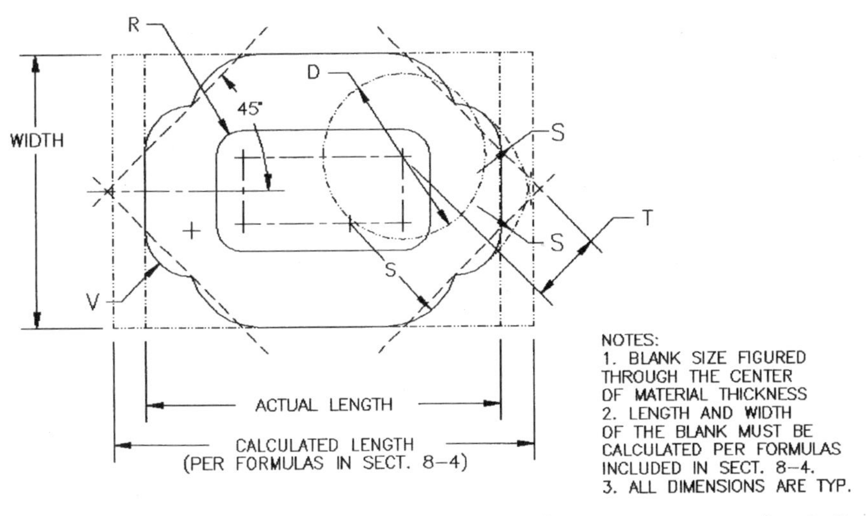


**Figure 9-34** Graphical method of blank development for square or rectangular shells whose depth is equal to or greater than their width.

shown should be calculated with the aid of formulas in Figs. 9-33 and 9-35 through 9-38.

### 9-5-5. Approximate blank corner shape for square or rectangular shells of various cross sections whose width exceeds their depth

The approximate shape of corners of blanks for square or rectangular drawn shells whose width exceeds their depth may be figured with the help of formulas and illustrations shown in Figs. 9-35 through 9-38. Blanks developed with these formulas consider the bottom radius of the shell to be zero. However, since it is nearly impossible to draw shells without bottom radii, these methods should serve only as guidelines, to quickly and efficiently determine the approximate shape of the trial blank corners.

## 9-6. Cylindrical Shell Blank Sizes

The displacement of metal in drawing operations varies along the shape of a shell. The flange is subject to the greatest alterations, while the bottom remains almost unchanged.

The metal flow during the drawing process promotes the increase in height of a part toward which it is applied. Whole segments shift away from the flange area into the body of the shell. The surface most

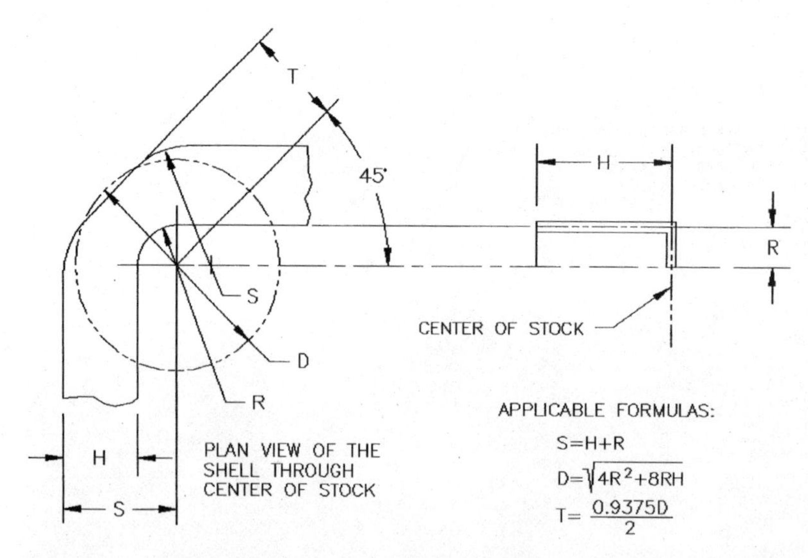


**Figure 9-35** Approximate blank corner shape for square or rectangular shells of various cross sections whose width exceeds their depth.

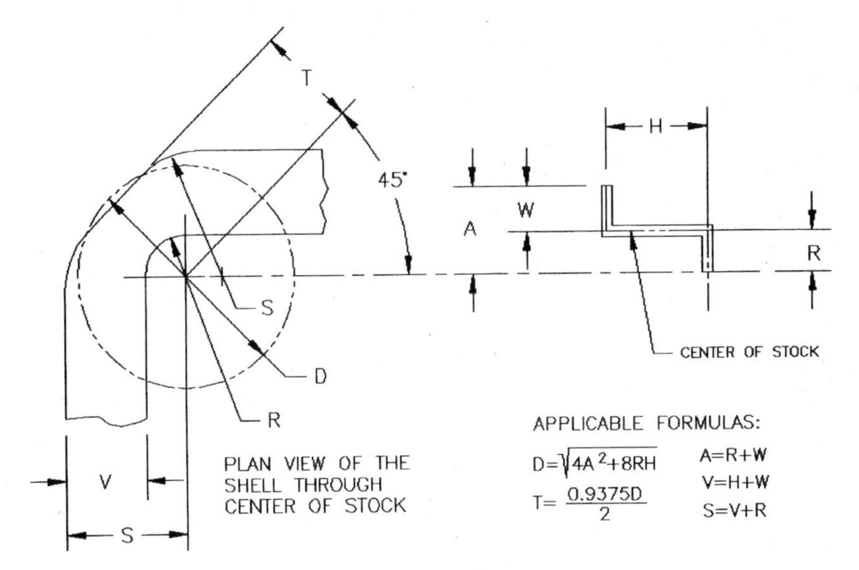


**Figure 9-36** Approximate blank corner shape for square or rectangular shells of various cross sections whose width exceeds their depth.

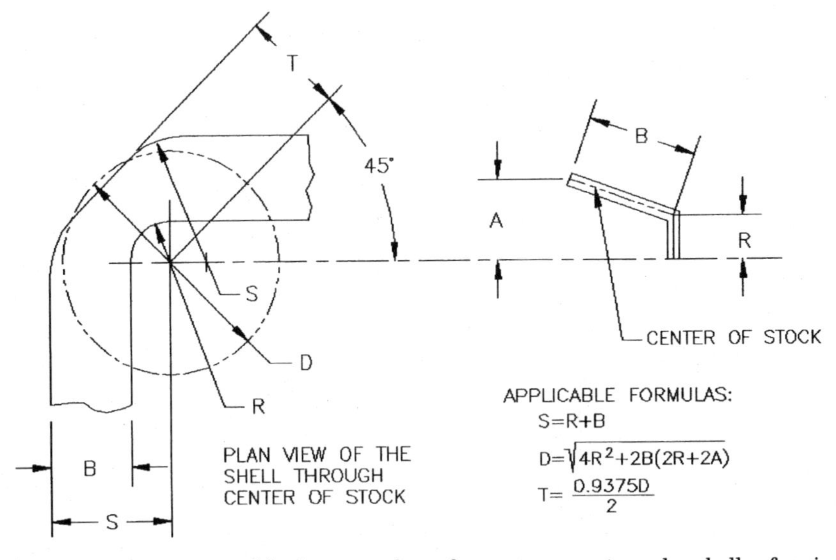


**Figure 9-37** Approximate blank corner shape for square or rectangular shells of various cross sections whose width exceeds their depth.

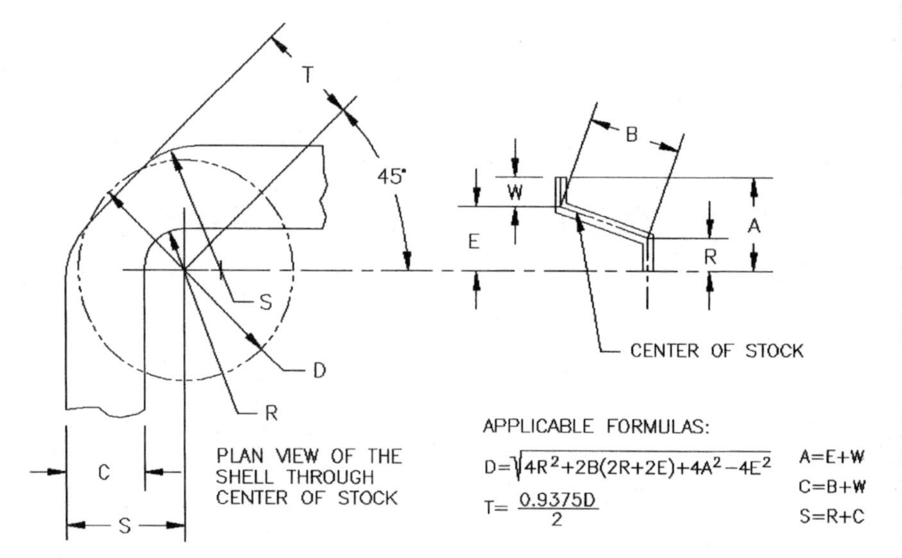


**Figure 9-38** Approximate blank corner shape for square or rectangular shells of various cross sections whose width exceeds their depth.

affected by such changes is that located farthest away from the shell body, which is the outer surface of the flange.

To calculate the basic size of a blank from which—through such transformation—a drawn cup may be obtained, the area of the part has to be assessed, which then will be projected into a diametral size of the blank.

Two methods of blank calculation, both applicable only to symmetrical shells, are described further. The first method is based on a theory that the area of any shape is given by the length of its profile, multiplied by the length of travel of its center of gravity.

**First method of blank calculation.** Lengths of line segments $L_1$, $L_2$, and $L_3$, as shown in Fig. 9-39, should be assessed along their neutral axis. Distances of their centers of gravity along $X$ axis, $X_1$, $X_2$, $X_3$ should be established. The formula to calculate the linear distance of the center of gravity of the shape $CG$ is

$$X = \frac{L_1X_1 + L_2X_2 + L_3X_3}{L_1 + L_2 + L_3} \tag{9-24}$$

As there is no need to calculate the distance of the center of gravity $CG$ along the $Y$ axis, it will not be attempted here.

The total length of the shape can be obtained by adding all segment lengths together. Multiplying this value by the length of the circular path of the center of gravity $CG$ can be done by using the formula

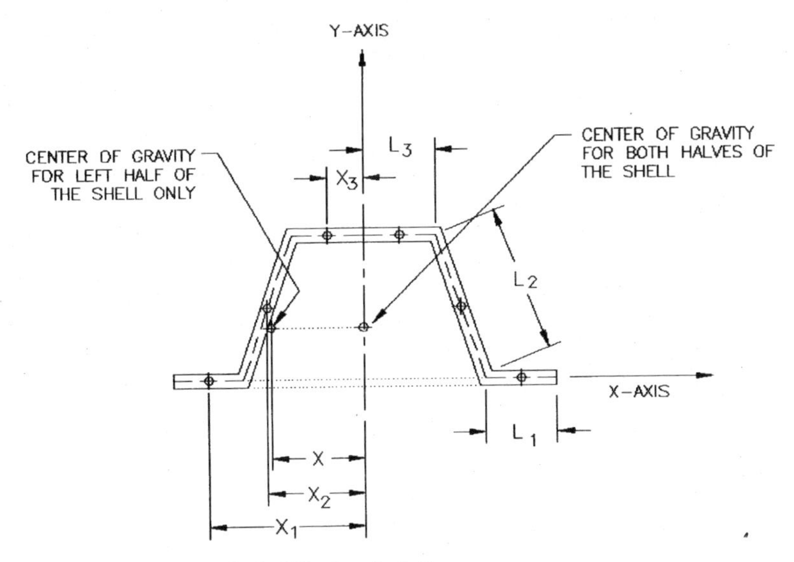


**Figure 9-39** First method of blank calculation.

$$A = 2\pi X(L_1 + L_2 + L_3) \tag{9-25}$$

From the result, a blank diameter may be acquired:

$$D = 2\sqrt{\frac{A}{\pi}} = \sqrt{8XL_{\text{total}}} \tag{9-26}$$

**Second method of blank calculation.** The second method of blank computation calculates each section of the drawn shell separately, adding their lengths up (Fig. 9-40).

Both of these methods give only an approximate size of the blank diameter, since to calculate its exact proportions is impossible. Too many variables influence the drawing process, making it more complex than any other manufacturing method. The movement of metal, which may produce thickening or thinning of various sections, a possibility of ironing, the variation in height, are a few factors out of many that expose the drawn part to so many influences that the total outcome is unpredictable.

Therefore, the blank size is usually chosen either slightly larger than necessary and trimmed afterward or an exact blank size may be considered, which is further adjusted in size after completion of a trial run.

The following set of calculations considers the size of blank diameter for simple cylindrical shells to be dependent on the ratio of the shell diameter to the corner radius $d/R$. It regards the blank to be of

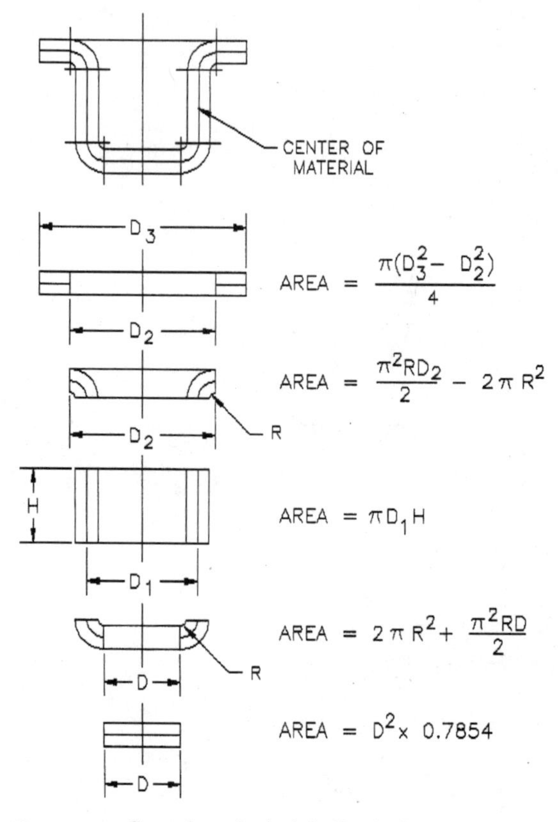


**Figure 9-40** Second method of shell calculation.

the same surface area as the finished shell.

Where $d/R = 20$ or more, the calculation to use is

$$D = \sqrt{d^2 + 4dh} \tag{9-27a}$$

With $d/R = 15$ to 20, the following formula should be used:

$$D = \sqrt{d^2 + 4dh} - 0.5R \tag{9-27b}$$

Where $d/R = 10$ to 15, the formula is

$$D = \sqrt{d^2 + 4dh} - R \tag{9-27c}$$

And where $d/R =$ below 10, the calculation becomes

$$D = \sqrt{(d - 2R)^2 + 4d(h - R) + 2\pi R(d - 0.7R)} \tag{9-27d}$$

Some recommend a formula for calculating all types of drawn shell blanks as

$$D = \sqrt{d^2 + 4dh - 3.44dR} \tag{9-27e}$$

where $D$ = blank diameter
$d$ = shell diameter
$h$ = height of shell
$R$ = radius of corner

For shells where some ironing is expected, resulting in thinner walls than the bottom surface, the following blank size calculation may be used:

$$D = \sqrt{d^2 + 4dh \frac{t_W}{t_B}} \tag{9-28}$$

where $t_B$ = thickness of bottom area or that of a blank
$t_W$ = thickness of wall

For an evaluation of more complex shell shapes, their cross-sectional outline should be dismembered to obtain simple sections whose areas can be calculated with the aid of formulas in Figs. 9-41 and 9-42. The total blank size is obtained by adding up all results.

## 9-7. Clearance between Punch and Die

If a gap between a punch and die is too generous, the material will not be in contact with both parts of tooling simultaneously. The drawing operation will be altered because of such a discrepancy, and the process will rather resemble stretching.

An insufficient clearance between the tooling will produce thinner walls, which is an effect called ironing. Ironing is sometimes used on purpose, where an improvement in the shell surface is needed. Too small a clearance also prevents the material from relocating freely within the thickness of the part, which promotes its compacting and compressing.

The proper clearance between the drawing punch and die should be in the vicinity of

$$z = \text{clearance} = 1.4t$$

or 40 percent greater space than the material thickness. Clearance may be smaller for flanged shells than for those having no flange and being pushed out through the die.

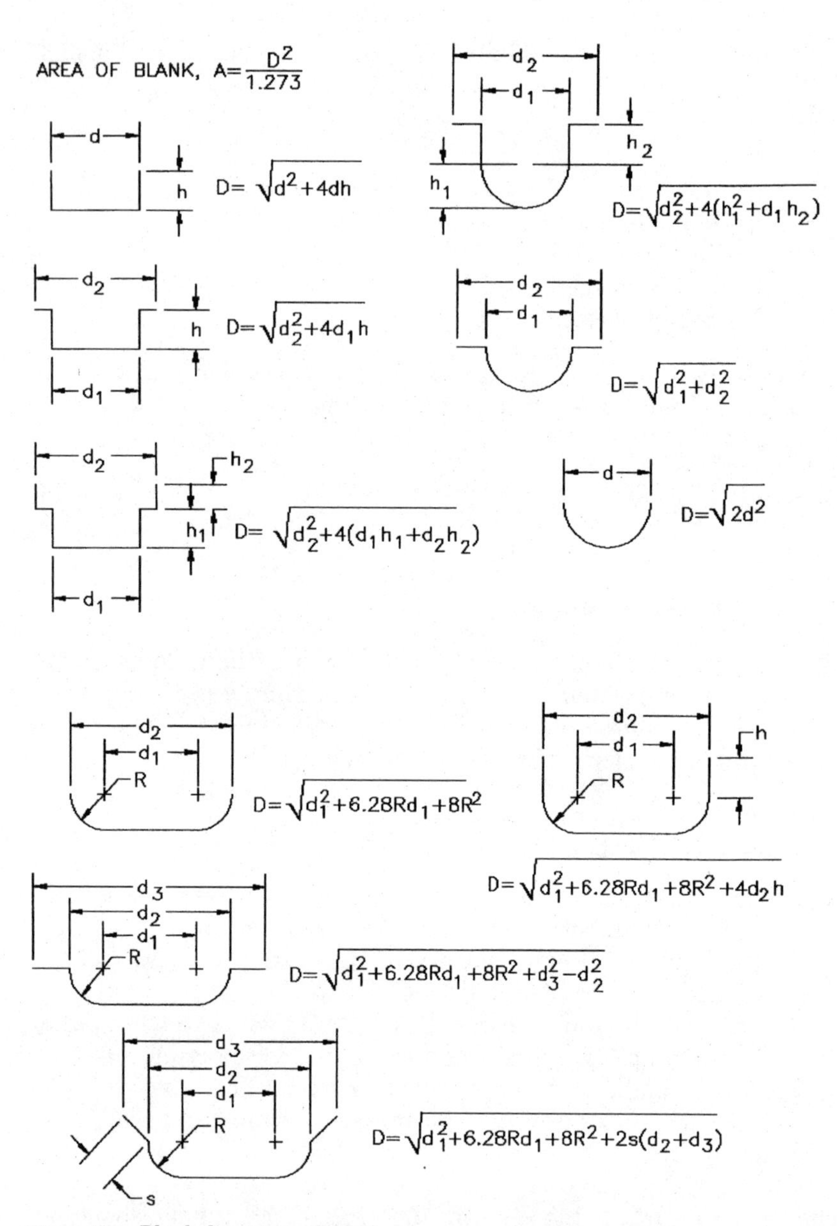


**Figure 9-41** Blank diameter calculation.

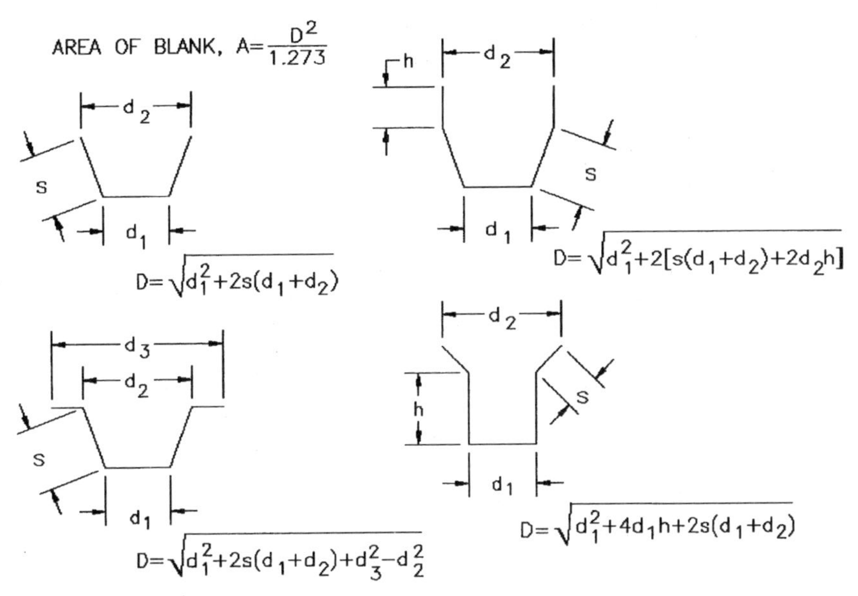


**Figure 9-41** (*Continued*)

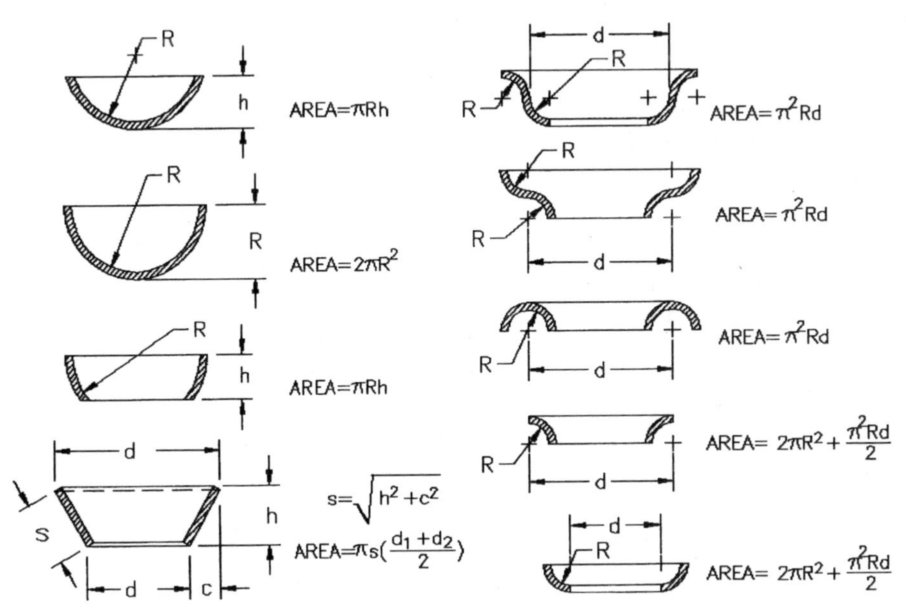


**Figure 9-42** Areas of drawn shells (through center of stock).

The CSN 22 7301 (Czech National Standard, similar to DIN) suggests

$$z = 1.2t_{max}$$

Oehler's recommendation is given by a formula

$$z = t_{max} + k\sqrt{10t_0} \tag{9-29}$$

where $k$ = 0.07 for steel
= 0.04 for nonferrous metals
= 0.02 for aluminum

The clearance, however, depends mainly on the material to be drawn. In Table 9-10, some basic guidelines are given, differentiating the material influence in drawing.

Tolerances, allowing for a slight ironing action, such as those using the gap size between $1.1t$ and $1.2t$, are often used within the industry. In such a case, the wall will come out slightly thinner, but its straightness and uniformity of thickness are usually beneficial to the outcome, even where a redrawing station has to be utilized.

Another evaluation of the ironing operation recommends the diameter of the ironing die to be

$$DIA_{die} = DIA_{punch} + 2t + 0.004 \text{ to } 0.008 \text{ in}$$

Such a clearance should provide a smooth finish or accurate outer diameter of the shell. Where a finer finish is needed, the diameter of the ironing die should be

$$DIA_{die} = DIA_{punch} + 2t + 0.002 \text{ in}$$

Ironing of the shell with the intention of reducing the material thickness necessitates drawing of the shell to the required inside dimensions, and ironing in a separate die. The ironing process considerably increases the pressure of the punch against the bottom of the

**TABLE 9-10 Clearance between the First Drawing Punch and Die for Cylindrical Shells**

| Material | Per-side clearance between punch and die |
|---|---|
| Aluminum | $1.25t$ |
| Steel, deep drawing | $(1.1 \text{ to } 1.25)t$ |
| Stainless steel | $(1.75 \text{ to } 2.25)t$ |

$t$ = stock thickness.

shell. For that reason, the diametral reduction of each redrawing station should be smaller than that used for redrawing with no ironing. One-half of the amount needed for drawing without ironing is recommended for ironing-oriented reduction.

Tolerances smaller than the material thickness are generally not recommended, since the ironing force will be excessive and tearing of the product may occur even though a process of long tubes' drawing tubes often utilizes clearances of up to 0.9*t*, or up to 10 percent smaller gaps than the material thickness.

## 9-8. Radius of Drawing Punch and Die

As already mentioned, the radius of the die section should be quite liberal in size. However, too large radii are not desirable either, as they enhance the material's tendency to wrinkle and fold.

The basic recommended sizes of tooling radii for drawing punches and dies are:

First drawing die radius: $(6 \text{ to } 10)t$

Redrawing die radius: $(6 \text{ to } 8)t$

Drawing punch values differ with the diameter of shell $d$ as follows:

For $d = 0.25$ to 4.0 in, $R_p = (3 \text{ to } 4)t$

For $d = 4.0$ to 8.0 in, $R_p = (4 \text{ to } 5)t$

For $d =$ above 8.0 in, $R_p = (5 \text{ to } 7)t$

The size of the punch radius is further influenced by the depth of the draw, percentage of reduction, and type of metal. For comparison, a chart of the recommended radii for a given thickness of the material is included in Table 9-11.

The continuity of the radius is not required to cover a whole 90° area of the corner. Variations of this portion are possible (see Fig. 9-43).

**TABLE 9-11 Radii of Drawing Punches and Dies**

| | Drawing edge radius, punch or die | |
|---|---|---|
| Stock thickness, in | Minimum, in | Maximum, in |
| 0.015–0.018 | 0.156 | 0.250 |
| 0.021–0.027 | 0.187 | 0.281 |
| 0.031–0.046 | 0.187 | 0.312 |
| 0.048–0.062 | 0.250 | 0.375 |
| 0.078–0.093 | 0.312 | 0.437 |
| 0.109–0.125 | 0.343 | 0.468 |

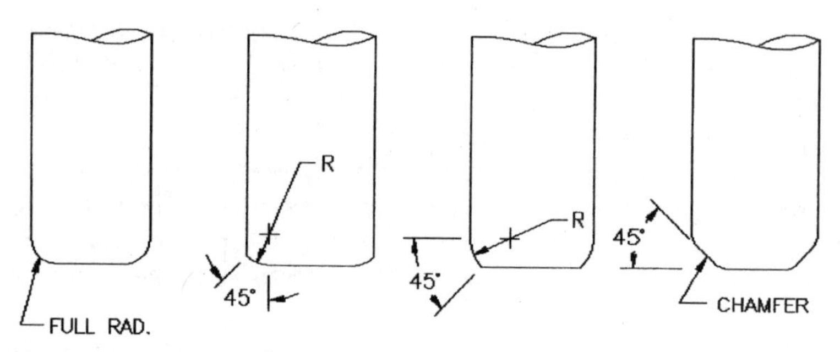


**Figure 9-43** Edge shapes of drawing punches.

The radius of the die has a direct influence on the drawability of the material, which it may influence in a positive or negative manner. Since the size of the die radius depends on its relationship to the thickness of the drawn material, it subsequently suggests that with a thicker blank, the influence of the die radius diminishes (Fig. 9-44).

Friction, as encountered in the drawing process, may often be considerable, because the surface of a part, as affected by the tangential compression, turns rough. To prevent this negative influence on the drawing operation and its outcome, the die radius, which is the fundamental factor delineating the success of a draw, must be as smooth as possible. Size of the die radius, if enlarged, decreases some frictional forces and enhances the drawing process, so that greater depths may be attained in one operation. However, with the enlargement of the radius, the area of the flange under the blank holder diminishes, affecting the stability of the drawing operation, with possible emergence of defects. Too small a die radius further causes tearing of the stock, while too large a radius may be the reason behind the appearance of wrinkles.

If an angle is selected in replacement of the die radius, it should be made equal to 60° off the horizontal for a simple push-through die. Such a type of die should be used only for a stock thickness greater than 0.062 in, or for shells smaller than 3.25 in diameter (see Figs. 9-45 and 9-46).

Angular die edge, where used for redrawing of cylindrical shells in a simple push-through die (Fig. 9-47) should be restricted to a stock thickness greater than 0.062 in or to shells smaller than 3.25 in diameter.

A beveled drawing die for redrawing, utilizing an inside blank holder (shown in Fig. 9-48), may be used for all material thicknesses. The angle, however, may vary (see Table 9-12 ).

A method of determining the size of corner angle and bottom radius for redrawing dies (shown in Fig. 9-49) bases the size of radii on the

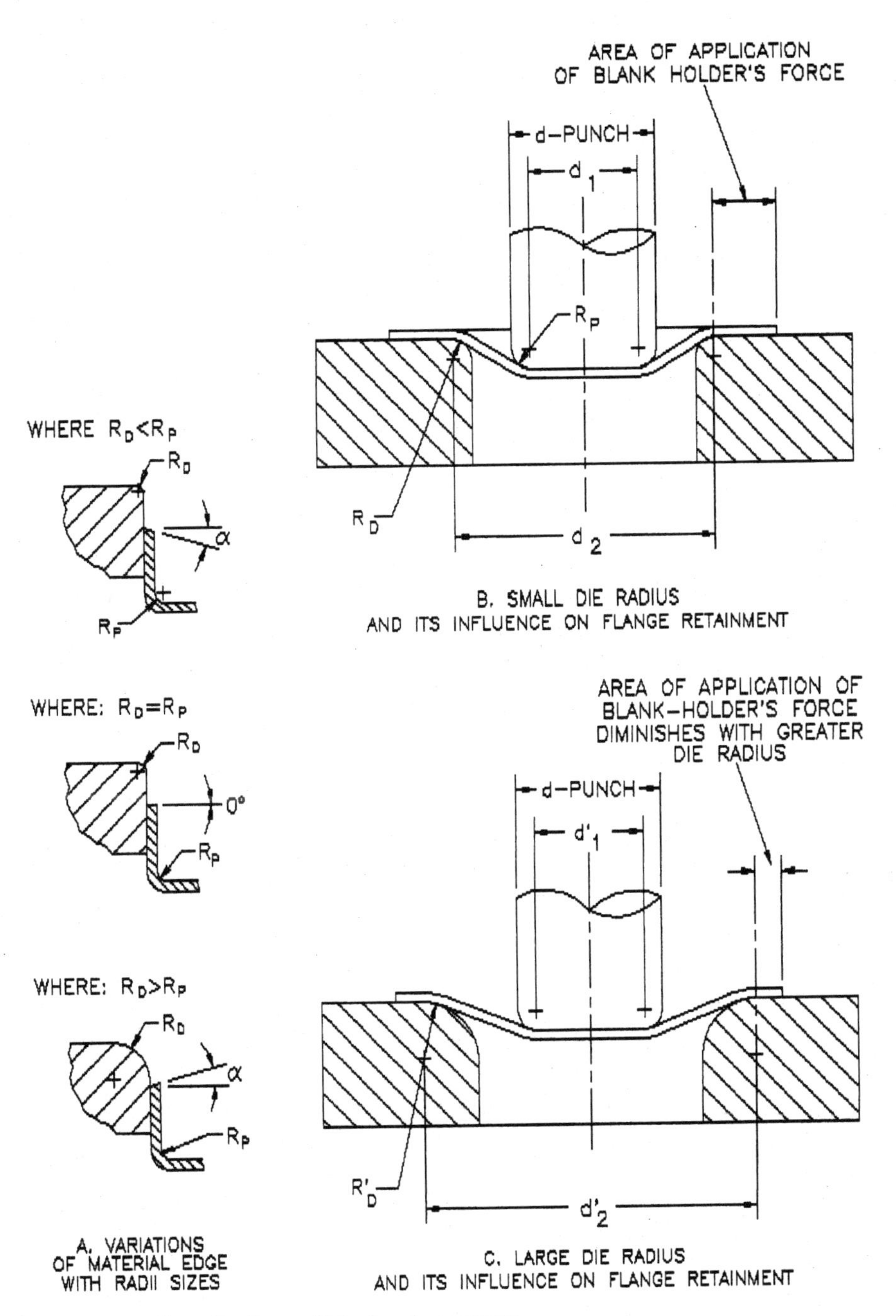


**Figure 9-44** Size of tooling radii in drawing process.

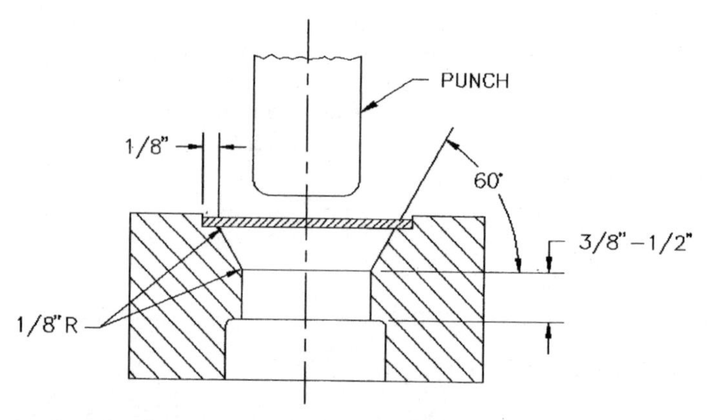


**Figure 9-45** Cross section of beveled drawing surface of a simple, push-through drawing die (dimensions are approximate).

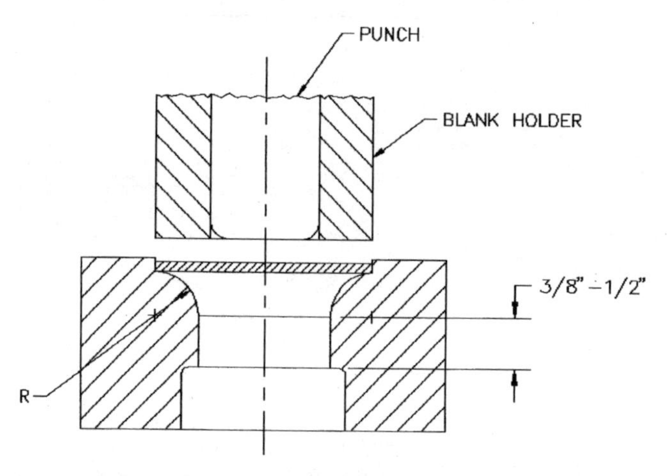


**Figure 9-46** Cross section of radiused drawing surface of a simple, push-through drawing die (dimensions are approximate).

material thickness as well. Here a shell requiring two preliminary drawing operations and a final finishing draw is shown. Where such a shell will need more than this many preliminary drawing passes, the same procedure for corner angle layout should be followed.

A mathematical method of evaluation of the proper size of the first drawing die radius follows.

$$R_{\text{die-first}} = \sqrt{0.8(D_0 - d)t_0} \tag{9-30a}$$

where $D_0$ = blank diameter
$d$ = inner diameter of shell
$t_0$ = thickness of blank

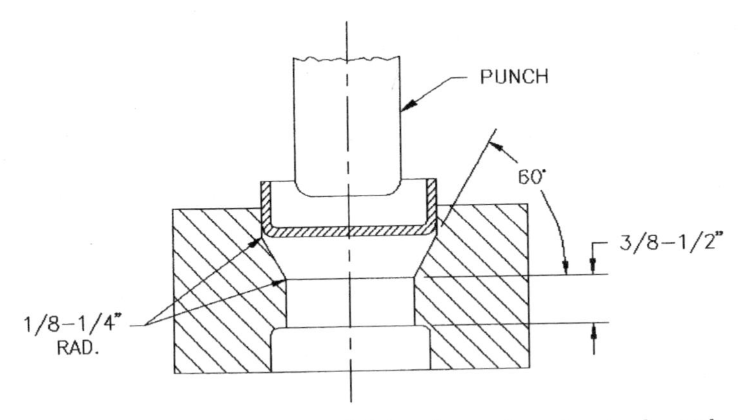


**Figure 9-47** Cross section of beveled drawing surface of a simple, push-through redrawing die (dimensions are approximate).

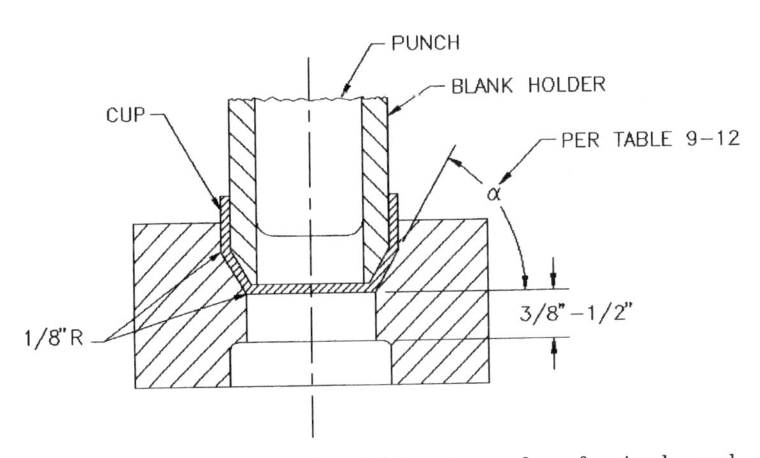


**Figure 9-48** Cross section of beveled drawing surface of a simple, push-through redrawing die with inner blank holder (dimensions are approximate).

**TABLE 9-12 Angle of Drawing Surface and Its Variation with Stock Thickness. Redrawing of Cylindrical Shells, Utilizing Inner Blank Holder**

| Stock thickness, in | Angle $\alpha$ value, degrees |
|---|---|
| 0.012–0.031 | 30 |
| 0.031–0.062 | 40 |
| Over 0.062 | 45 |

**TABLE 9-13 Angles of Redrawn Cylindrical Shells and Their Variation with Stock Thickness**

| Stock thickness, in | Angle $\alpha$ value, degrees |
|---|---|
| 0.012–0.031 | 30 |
| 0.031–0.062 | 40 |
| Over 0.062 | 45 |
| Stainless steel, all thicknesses | 45 |

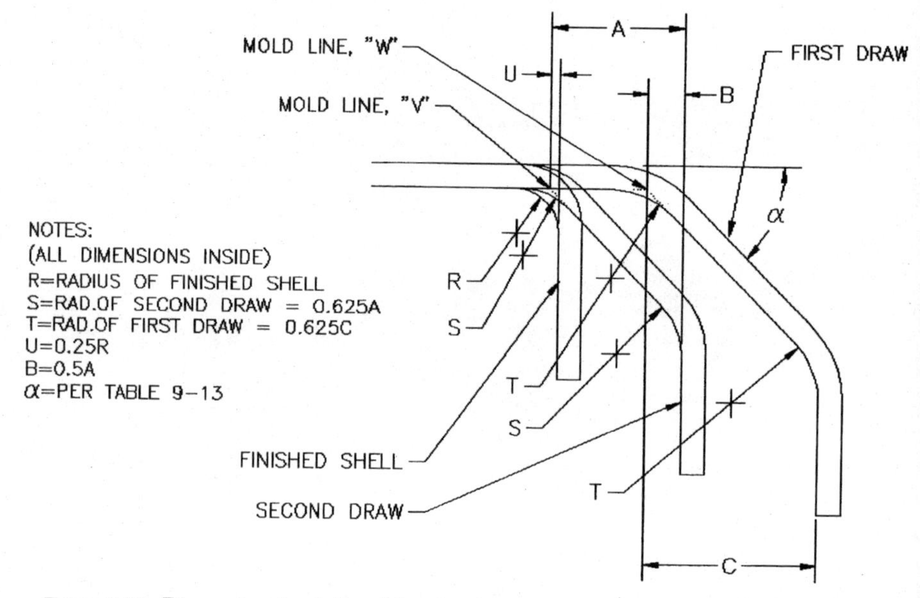


**Figure 9-49** Dimensional relationship of redrawn cylindrical shells (dimensions are approximate).

CSN 22 7301 recommends using the radius size of (6 to 10)$t$ for a single-pass drawing operation.

For all redrawing dies, the above formula becomes

$$R_{D-\text{redraw}} = \frac{d_1 - d_2}{2} - t_0 \tag{9-30b}$$

where $d_1$ = shell diameter before redrawing
$d_2$ = anticipated shell diameter after redrawing

The relationship of the blank thickness and diameter may also be expressed by the equation

$$\Delta t = \frac{100t_0}{D_0} \tag{9-31}$$

where $t_0$ = blank thickness
$D_0$ = blank diameter

If the result is $\Delta t > 2$, a blank holder is not necessary.

If $\Delta t < 1.5$, a blank holder must be used.

If $\Delta t$ is between 1.5 and 2, the decision is not substantiated and has to be evaluated on the basis of actual test.

Some recommended values of drawing die radii are shown in Fig. 9-50. Different hatched areas of this graph are to be used for the following applications:

Area $a$ = for flanged shells

Area $b$ = shells without flange

Area $c$ = shells having a frictional insert (such as draw beads)

Area $d$ = a continuation of area $b$

Area $e$ = for progressive tooling

This suggests that thin metals may be drawn without a blank holder only where the drawn shell is shallow and the coefficient $M$ of severity of draw is large.

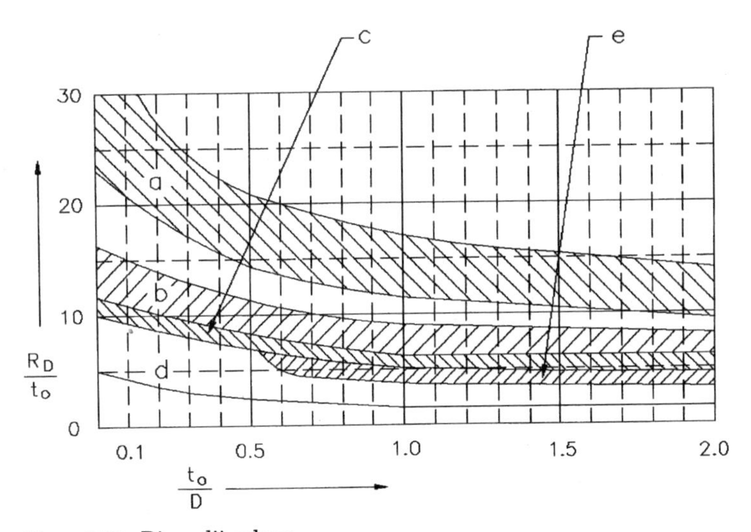


**Figure 9-50** Die radii values.

Additional methods of evaluation of the size and shape of the drawing die edge may be found in Sec. 9-4, "Drawing of Thick-Walled Cylinders." The *shaped die edge* and *tractrix curve edge* described there may be applied to drawing of regular cylindrical shells as well as to thick-walled cylinders.

## 9-9. Redrawing Operations

All redrawing may be divided into two groups:

- Redrawing, where the actions of both punch and die are oriented toward the whole length of the part (Fig. 9-51).
- Redrawing, where the punch exerts its pressure only at the bottom of the part, in which case the process is called reducing (Fig. 9-52*a*)

When redrawing, the part may be drawn into smaller diameter, greater depth, thinner wall, or a different shape. Some parts are reverse-drawn, which means that the drawing operation progresses in the direction opposite to that of the previous drawing operation, as shown later in Fig. 9-54.

Redrawing with no reduction in wall thickness is also called sinking. Redrawing with subsequent reduction of the wall thickness is ironing (Fig. 9-52).

Reducing affects mainly the contour of the part. It may be called necking, tapering, or closing. Some samples of reducing operations are included in Fig. 9-53.

All these operations affect the formation of the part cross section by reducing either a small portion of it or the whole length.

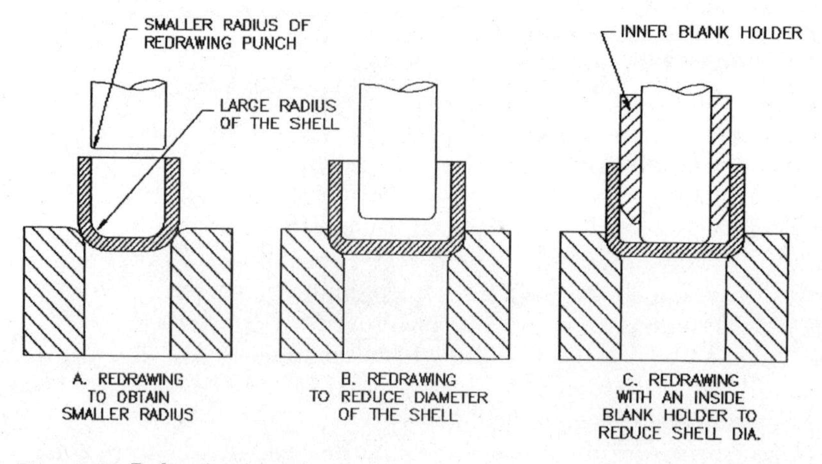


**Figure 9-51** Redrawing operation.

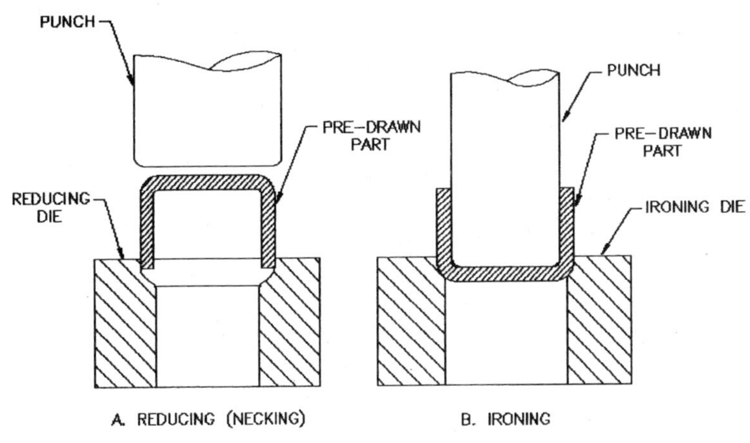


**Figure 9-52** Reducing and ironing.

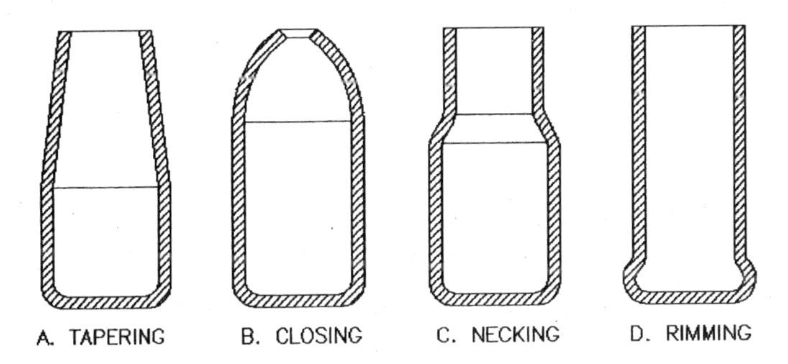


**Figure 9-53** Reducing operations.

### 9-9-1. Drawing inside out

Reversal of the drawing process is utilized for redrawing of existing shells, where greater elongation of the part is desired, while eliminating subsequent drawing operations, needed otherwise. This is achieved by doubly overcoming the material's elastic limit during the two drawing passes, each opposite the other.

Another advantage of inside-out redrawing is the avoidance of wrinkles, with an enhanced dimensional stability of the shell.

However, recommended die radii with respect to the stock thickness cannot be honored, as their value will be dictated by the difference between drawing diameters $d_1$ and $d_2$ (see Fig. 9-54). Figure 9-55 shows drawing of a half-round shape.

Reverse drawing is advantageous for thinner materials, where $2t/d_1 = 0.2$ and for larger ratios of the diameters' difference, $d_1/d_2 = 1.4$.

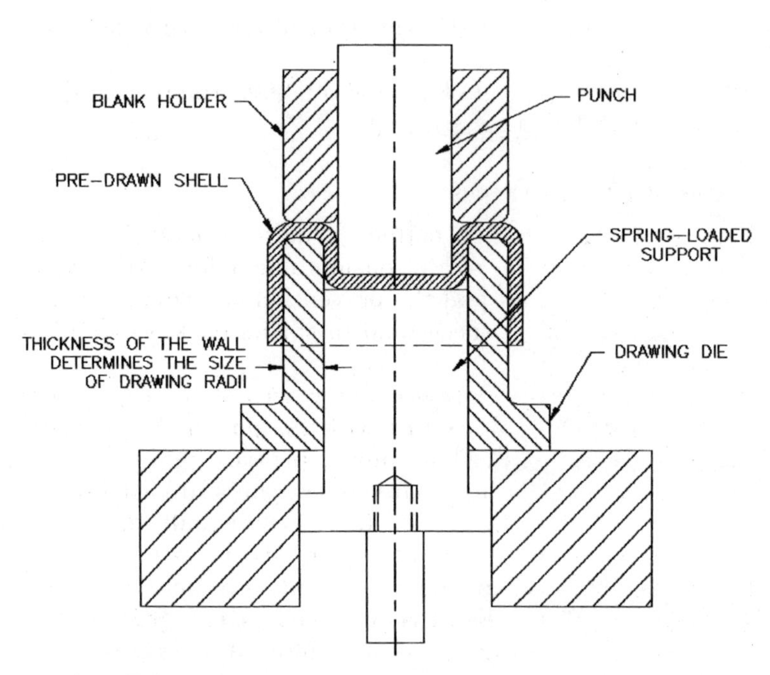


**Figure 9-54** Reverse drawing.

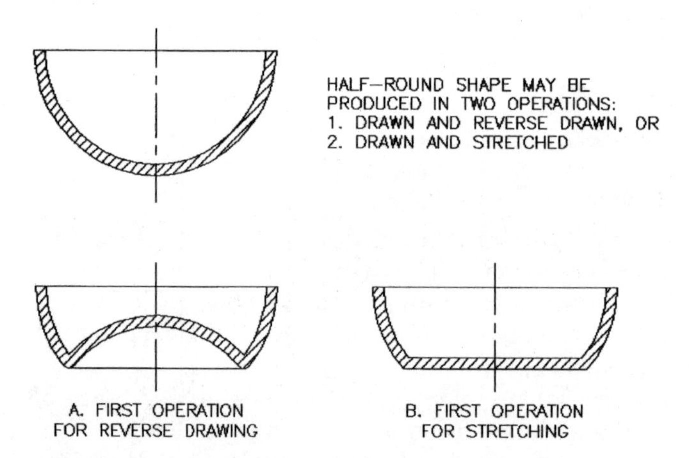


**Figure 9-55** Drawing of a half-round shape.

With thicker materials, $2t/d_1 = 0.1$, this type of redrawing is not recommended.

The drawing method shown in Fig. 9-56 combines the regular drawing process with that of reverse drawing.

### 9-9-2. Drawing of spherical shapes

Drawing of objects with spherical bottoms is more difficult than drawing of cylindrical shells. First of all, considerable deformation is generated in the curved portion, and the prevention of wrinkling is also impaired. Second, parts like these tend to spring back, especially if drawn in solid tooling.

There are quite intensive deformation processes within that portion of drawn material which is in contact with the punch. At the beginning of the drawing sequence, the punch presses against a large unsupported area of blank, which is free to become wrinkled anytime. At the point of contact *A,* in Fig. 9-57*a,* there is an equally distributed radial and tangential tension. But at the point of contact *B,* which can be found at the beginning of the die radius, only a radial tension may be found, along with a tangential compressive stress. Stretching of the material causes thinning of the wall and often may result in breakage.

In cases where the spherical radius of the bottom $R_{\text{sphere}} > d/2$ and the height of the cylinder is quite small, the stability of the part in

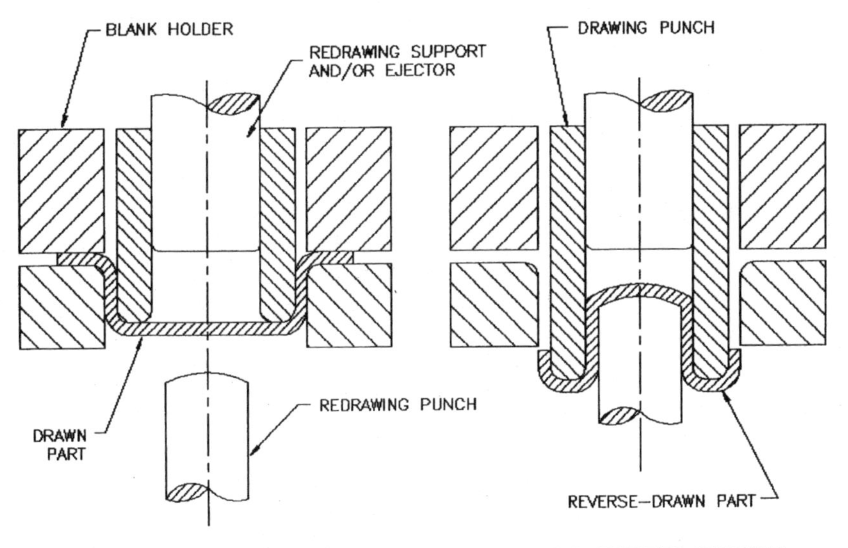


**Figure 9-56** Reverse drawing combined with regular drawing.

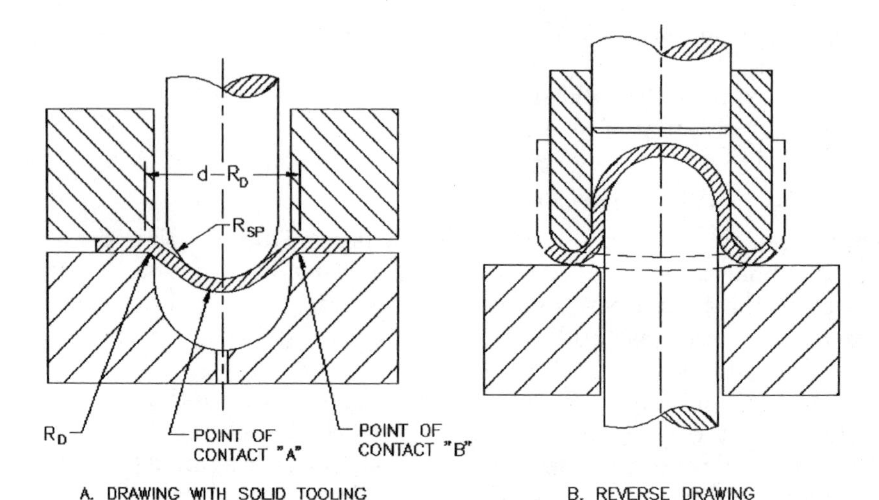


**Figure 9-57** Drawing of spherical shapes.

areas not retained by the tooling is endangered. Wrinkling may occur, and it may be eliminated only where the following relationship exists:

$$\frac{100R_{\text{sphere}}}{t_0} \geq 1.3$$

Some control over the drawing process may be achieved with reverse drawing or through the inclusion of draw beads.

The number of needed drawing passes to produce a spherical-bottom cylindrical shell may be assessed the same way as with regular cylindrical shells, provided the condition $R_{\text{sphere}} > d/2$ applies, where $M = d/D_0 = 0.71$ max, or $K = D_0/d = 1.4$ max.

The coefficient of severity of draw $M$ is constant and for that reason may not be reliable as a guideline in evaluation of the amount of needed drawing passes. Rather a percentage of blank thickness to blank diameter should be used. Where a single drawing pass is contemplated, as followed by a finishing draw, and a condition on left applies, the recommended drawing practice (on right) should be observed.

| | |
|---|---|
| $\frac{100t_0}{D_0} > 0.5$ | Either reverse drawing or regular drawing with a blank holder should be used |
| $\frac{100t_0}{D_0} < 0.5$ | Either reverse drawing or regular drawing with draw beads should be used |
| $\frac{100t_0}{D_0} > 3.0$ | Single-operation drawing with bottoming should be used. No blank holder is necessary. |

## 9-10. Types of Drawing Dies and Their Construction

Most often, drawn parts are cylindrical, with some marginal squares or rectangular shapes to be found. The drawing process, with dependence on its speed, the necessity of a blank holder, and the type of shell disposal, may be performed in either a single-action or a double-action die. Single-action dies are used for parts of greater material thickness. In double-action drawing, the blank holder is employed for the retention of the blank (Fig. 9-58).

### 9-10-1. Double-action dies

Double-action dies are two in kind: Those that push the finished shell down through the opening in the die and those that bottom at the end of the drawing operation (Fig. 9-59).

Thinner stock should be drawn in double-action dies, for it should be restricted from movement by the blank holder's pressure; otherwise it will wrinkle. Wrinkled parts, no matter how many times redrawn, remain wrinkled, and therefore defective. This is attributable to the material's elastic limit, which must be exceeded for the part to accept a new shape. With redrawing for removal of wrinkles, all the force is applied in the direction in which they were formed, which does not provide for surpassing of the material's elastic limit. This can be accomplished only by applying the pressure from the opposite direction.

The blank holder's pressure in double-action drawing is adjustable, so that it clamps the blank firmly, regardless of possible thickness

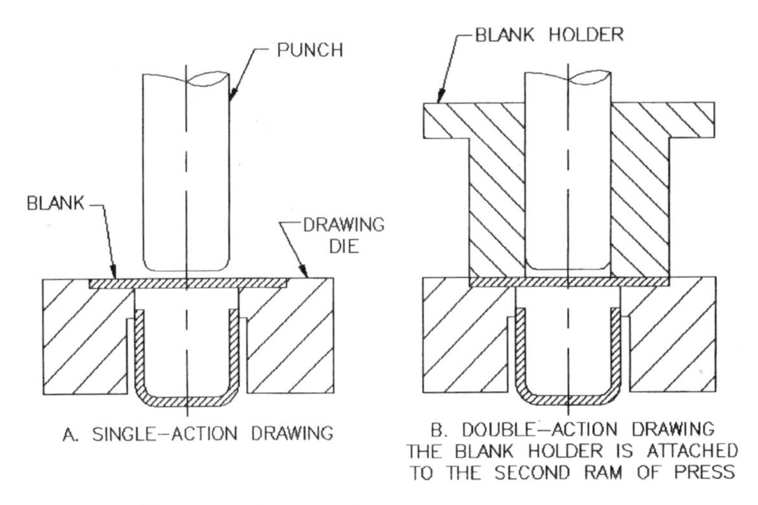


**Figure 9-58** Two types of cup forming.

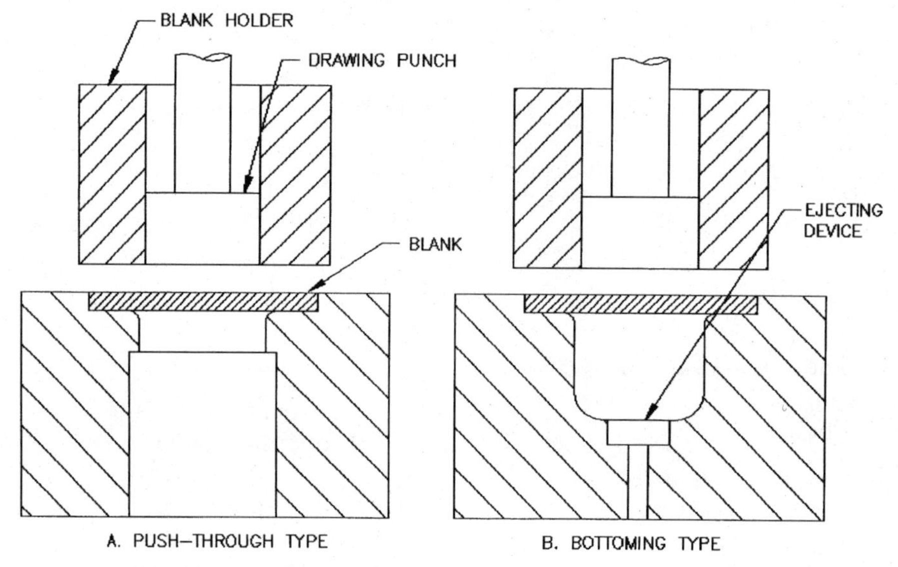


**Figure 9-59** Double-action drawing dies.

deviation. It secures the metal from any movement, prohibiting it from following the drawing punch and getting congested in the die.

### 9-10-2. Triple-action dies

Triple-action dies are used where additional die work is to be performed simultaneously with the basic drawing. Such dies may be used for blanking-embossing-drawing, or blanking-drawing-compressing, or blanking-drawing-drawing and similar arrangements (Fig. 9-60).

Triple-action dies are often used in replacement of bottoming double-action dies (see Fig. 9-59*B*), where the finished part has to be pushed up. In triple-action tooling the finished shell may still be disposed of through the bottom opening in the die, which often may be preferable.

### 9-10-3. Drawing with flexible tooling

Drawing with flexible tooling (such as rubber and fluids) often allows for deeper draws and greater reduction ratios than hard, inflexible punches and dies. The radius of the flexible die $R_D$ as seen in Fig. 9-61 is capable of altering its shape under the pressure of fluid $P_{FL}$. Such a shape may take the form of either higher or lower line, depending on the amount of pressure $P_{FL}$ which may be adjusted during the drawing process.

The required fluid pressure $P_{FL}$ may be calculated by using a formula

$$P_{FL} = \frac{2tS(R_P \cos \alpha + R_D \cos \beta)}{R_P^2 - R_D^2} \qquad (9\text{-}32)$$

where $t$ = material thickness
$S$ = tensile strength of drawn material
$R_P, R_D$ = instantaneous radii of the punch and die
$\alpha$ = angle of tangent at the point of transition of the wave into the flange
$\beta$ = angle of tangent at the point of transition of the wave into the vertical height of the shell

The result comes out in lb/in$^2$ or in tons/in$^2$ according to the tensile strength's denomination.

The radius of the curve of the die $R_D$ may be obtained:

$$R_D = \frac{R_P}{\cos \alpha + \cos \beta} - t \qquad (9\text{-}33)$$

The radius of the die $R_D$ has a considerable influence on the stability of the drawing edge, which in this case is adversely affected by the fact that the surface of the die is flexible.

Because of possible regulation of the pressure of fluid even during the drawing operation, flexible drawing tooling is unsurpassed by any other manufacturing method. It may be utilized for drawing of simple shapes, as well as for complex ones, such as spherical, conical, or

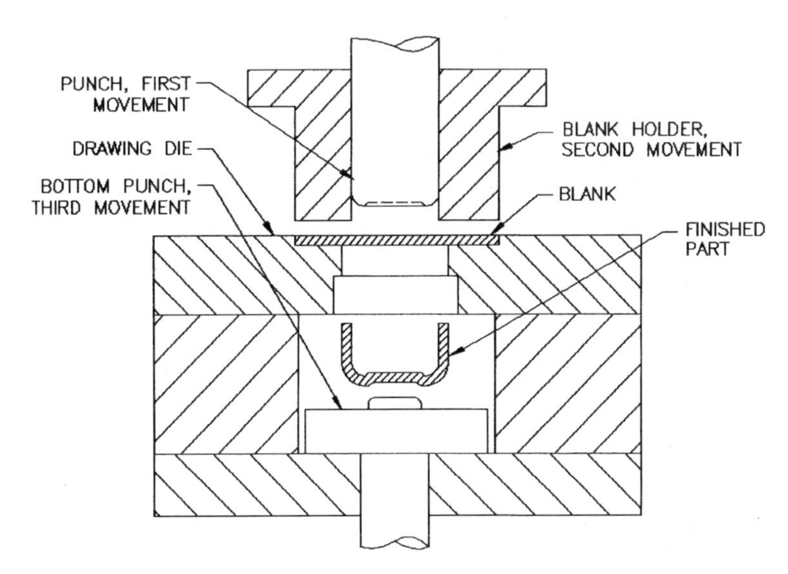


**Figure 9-60** Triple-action die.

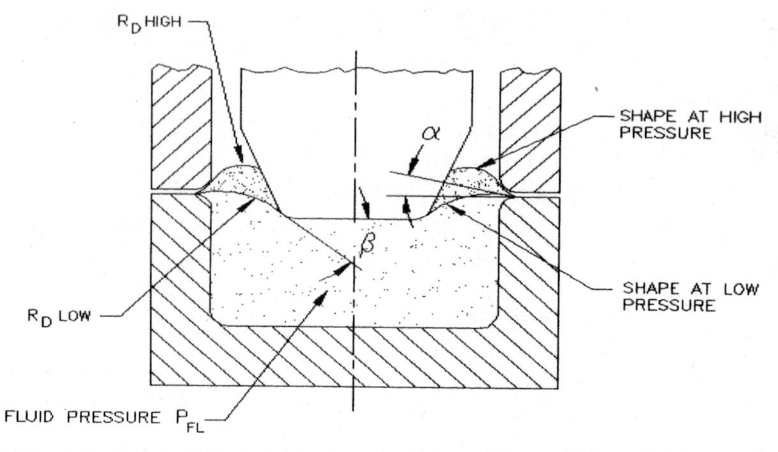


**Figure 9-61** Shape of the flexible drawing die and its variation with the pressure.

hyperbolic cross sections, which will otherwise be prone to the formation of wrinkles.

Drawing with flexible tools may be divided into two sections, where the first utilizes the Marform or Hydroform processes of drawing. In the Marform process (Fig. 9-62), the rubber insert serves as a blank holder, retaining the blank on the surface of the stationary punch. The hydroform process (Fig. 9-63) draws the part with the pressur-

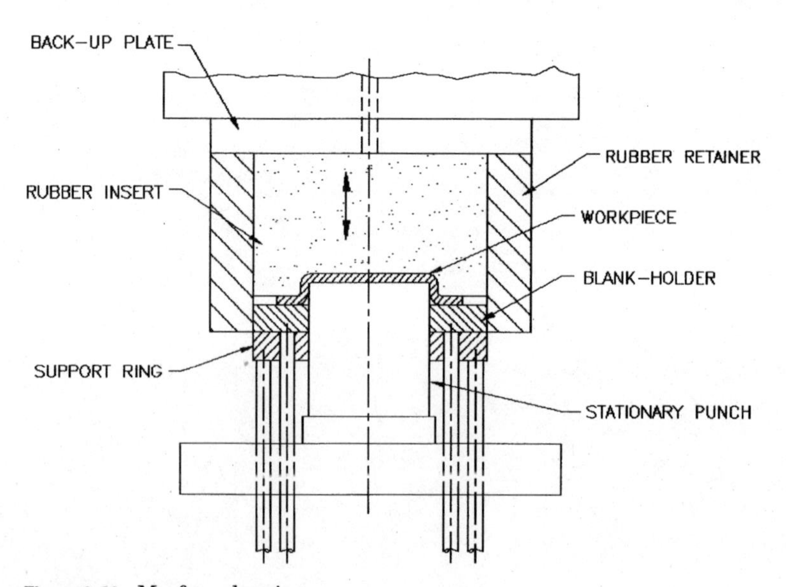


**Figure 9-62** Marform drawing process.

ized fluid. The hydrodynamic forming process and the hydromechanical drawing process are shown in Figs. 9-64 and 9-65.

The second group of drawing techniques using flexible tooling utilizes the Verson-Wheelon or Asea Quintus process (Fig. 9-66), where a flexible rubber sack exerts its pressure on the part. The drawn object

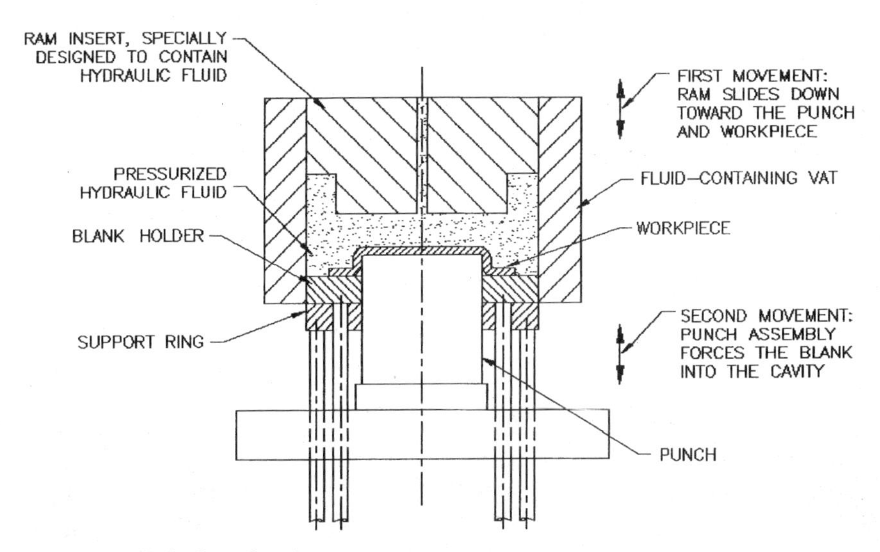


**Figure 9-63** Hydroform drawing process.

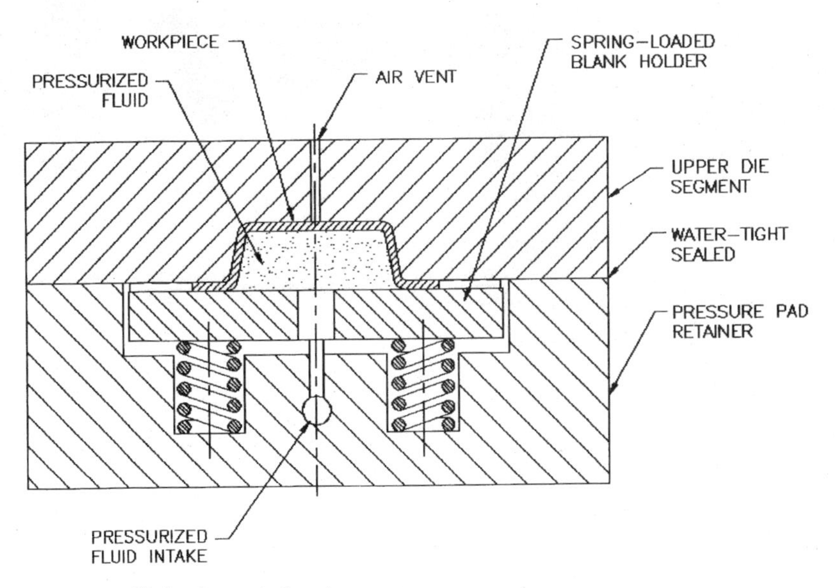


**Figure 9-64** Hydrodynamic forming process.

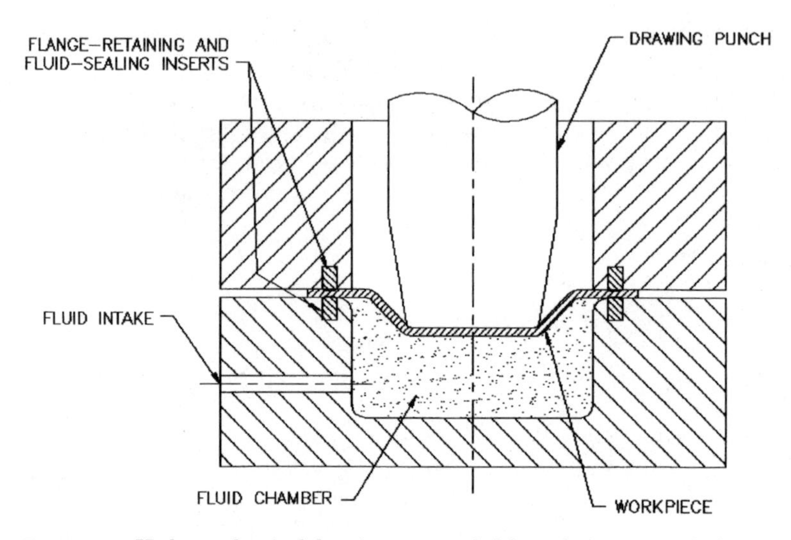


**Figure 9-65** Hydromechanical drawing process (with workpiece retention).

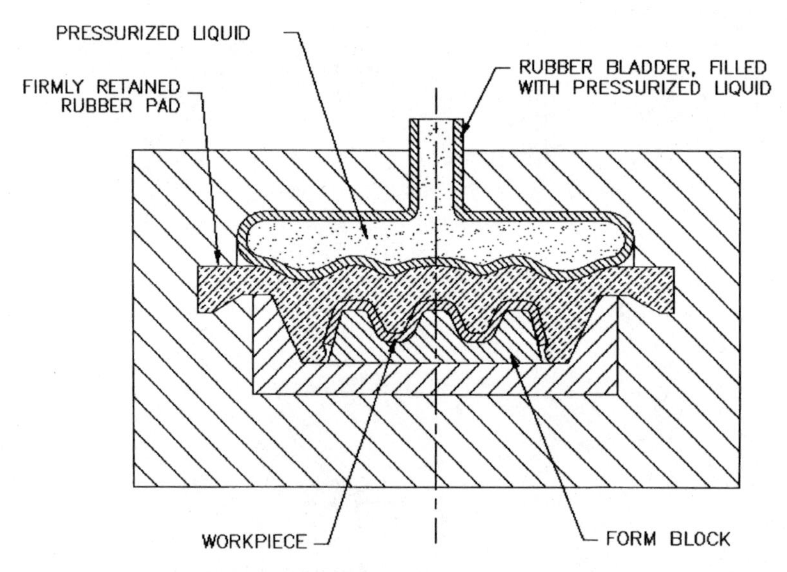


**Figure 9-66** Verson-Wheelon drawing process.

is wrapped around the punch, often without any assistance of a blank holder, which sometimes causes waves and wrinkles in the material.

The first group of drawing techniques regulates the pressure of the flexible element $P_{FL}$ by the counterpressure of the blank holder. Such a counterpressure is governed either by a regulating valve or by a microprocessor. It allows for a decreased drawing pressure at the beginning of the drawing operation, with subsequent adjustment to higher levels when the drawing process demands it.

The second group suffers from the material's tendency to wrinkle. This is due to the lowered amount of radial tension (a tension denominated by the actual amount of deformation), which leaves a space for the rise of tangential tension.

### 9-10-4. Air vents

No drawing operation is successful if appropriate air vents are missing from the body of a punch, for a drawn part has a tendency to stick to the punch around which it is wrapped, retained by the force of vacuum thus created.

Air vents serve a dual purpose in the drawing operation (Figs. 9-67 and 9-68). During the drawing cycle, they eliminate the air entrapped between the face of the punch and the bottom of the drawn shell. At the end of drawing, they permit the air to reenter the space between the punch and the part, to aid in the stripping of the latter. In the absence of vents, the drawn shell would either collapse or be impossible to strip off the punch.

### 9-10-5. Drawing inserts

Drawing dies for large production runs may be combined from two parts instead of a single block, complex in shape. Instead of finishing the drawing opening in a block, a press-in ring fabricated to the exact shape may be used (Figs. 9-69 and 9-70). Such rings can be replaced when necessary, with no impairment to the remainder of the block and with minimum downtime.

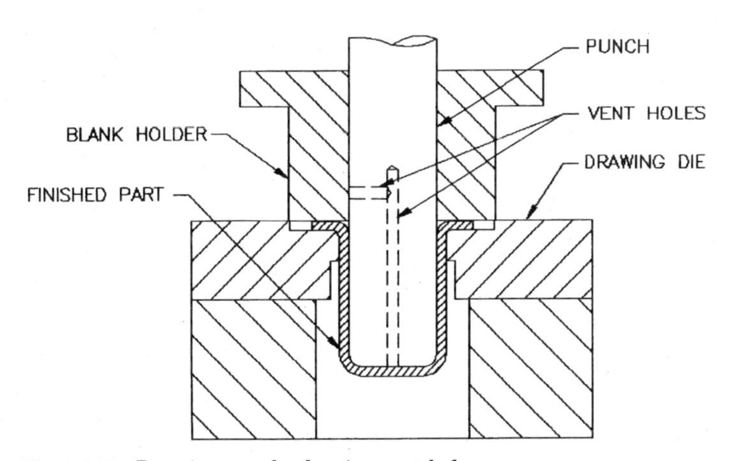


**Figure 9-67** Drawing punch, showing vent holes.

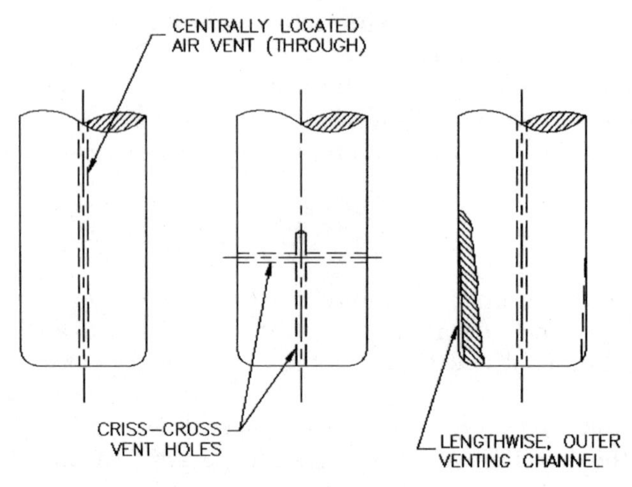


**Figure 9-68** Air vents in tooling.

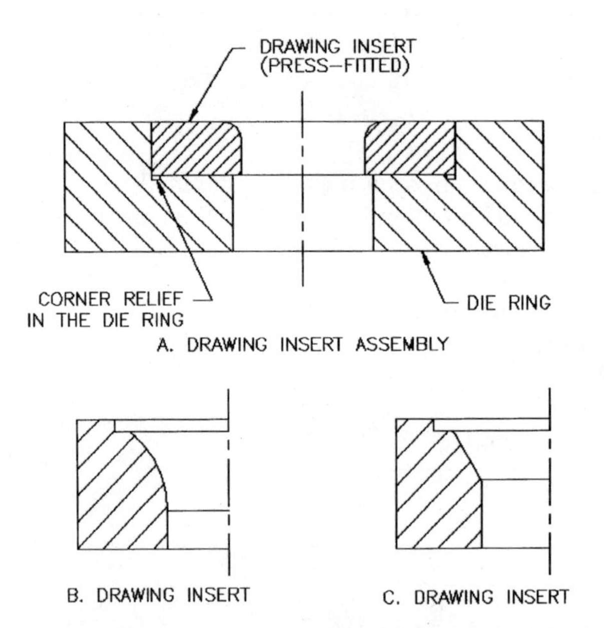


**Figure 9-69** Drawing inserts and their application.

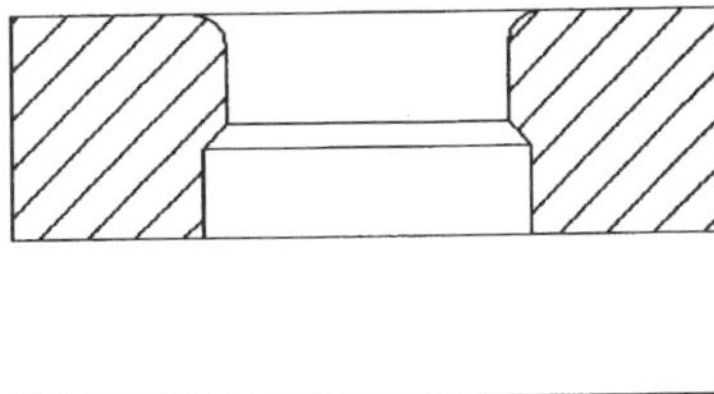

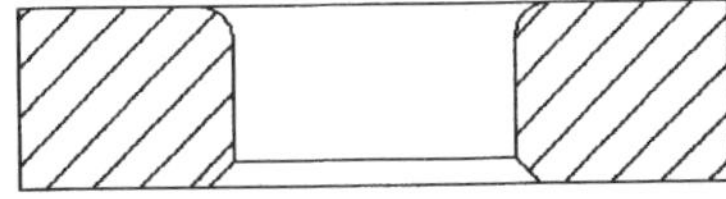

**Figure 9-70** Drawing inserts.

## 9-11. Blank Holders

Blank holders are employed to retain blanks firmly under the punch. Their pressure on the material secures it from being pulled into the die and ruined there, while controlling the amount of metal redistribution from the flange into the height of a shell. With inadequate pressure of the blank holder, the material may not be sufficiently retained, which will result in its greater than planned volume flowing into the body of the shell. Often such action will admit the formation of wrinkles on the surface of the flange or collapsing of the flange and other defects.

As the metal passes over the edge of a die, its flow is no longer controlled by the pressure of a blank holder and it is free to form wrinkles within the body of a shell.

To avoid the formation of wrinkles or buckling, chamfered dies are useful. However, when drawing stock thinner than 0.062 in radiused corners should be used instead. Experiments have proved that buckling does not occur where the ratio of the flange width to metal thickness is less than 3 to 4.

According to their construction, there are several types of blank holders, as shown in Figs. 9-71 and 9-72.

Blank holders, shown in Fig. 9-71*A* and *B,* are of simple, basic construction, permitting an adjustment of their pressure. For this advantage, they are preferred throughout the industry, especially where a thin stock, which is always prone to wrinkling, is drawn.

A blank holder (Fig. 9-72*A*) is used where thickening of the flange during the drawing process is required. The height of the nest is crucial to the proper outcome, as with too large a gap, wrinkles are obtained. The proper space between the flange and the bottom of the blank holder should be 25 to 50 percent smaller than the difference in thickness between the flange before drawing and its anticipated thickness at the end of drawing.

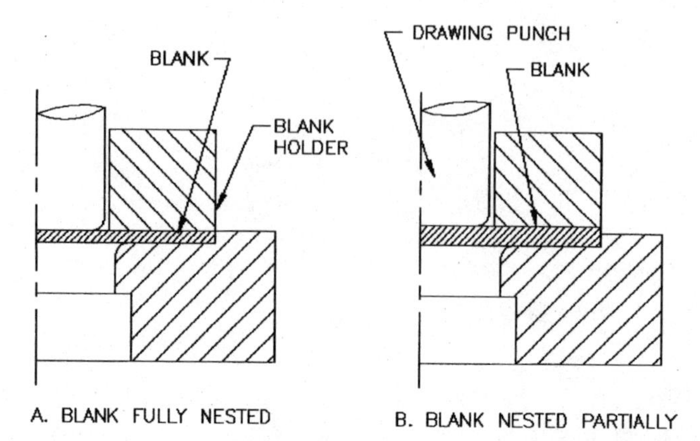


**Figure 9-71** Adjustable-pressure blank holders.

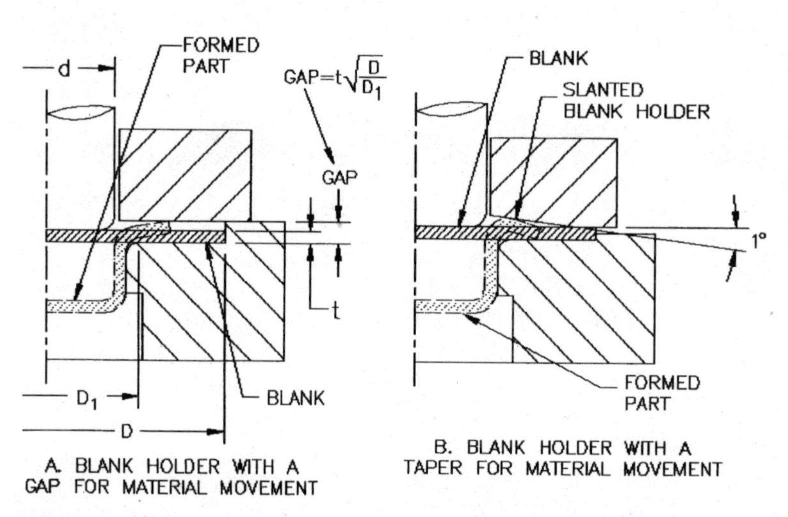


**Figure 9-72** Blank holders.

The height of the thickened flange may be obtained from the equation

$$\frac{t_1}{t} = \sqrt{\frac{D}{D_1}} \tag{9-34}$$

where $t_1$ = thickness of drawn part's flange
$t$ = thickness of the blank
$D$ = diameter of blank
$D_1$ = diameter of flange, or mean diameter of shell without flange

The height of the nest may be assessed as

$$t_{\text{die}} = t\sqrt{\frac{D}{D_1}} \tag{9-35}$$

The blank holder shown in Fig. 9-72*B* is an alternative of the blank holder in Fig. 9-72*A*. Its pressure is confined within the circumferential area of the flange, which allows for gradual thickening of its profile. The angular inclination may be around 1°, as it is not overly critical. This type of blank holder is said to decrease the necessary drawing force.

### 9-11-1. Drawing without a blank holder

Some drawing operations can be performed without blank holders. Evaluation of such a possibility can be done on the basis of several methods of assessment. Some theories consider a thick stock eligible for drawing without a blank holder; others judge the possibility of eliminating the blank holder by altering the geometry of the drawing die.

According to some other opinions, shells may be drawn without a blank holder if the opening in the die is 5 to 6 times the metal thickness per diameter smaller than the blank. The exact amount of this difference in a given case is based on the rate of reduction during the drawing stage. It may be expressed mathematically as a percentage of reduction $R$ percent:

$$R\% = \frac{100t(5 \text{ to } 6)}{D} \tag{9-36}$$

Requirements regarding the geometry of the position of a blank with regard to the drawing edge are shown in Fig. 9-73. These arrangements allow for drawing of a shell without using a blank holder.

For eligibility to draw without the blank holder, the radius or chamfer $R$ of the drawing die must be smaller than $20t$, as shown in Fig. 9-73.

According to Freidling's theory, evaluation of the possibility of drawing without a blank holder depends on the percentage value of the ratio of stock thickness to the diameter of the blank. The formula to use for such an evaluation is described in Sec. 9-8, "Radius of Drawing Punch and Die," Eq. (9-31).

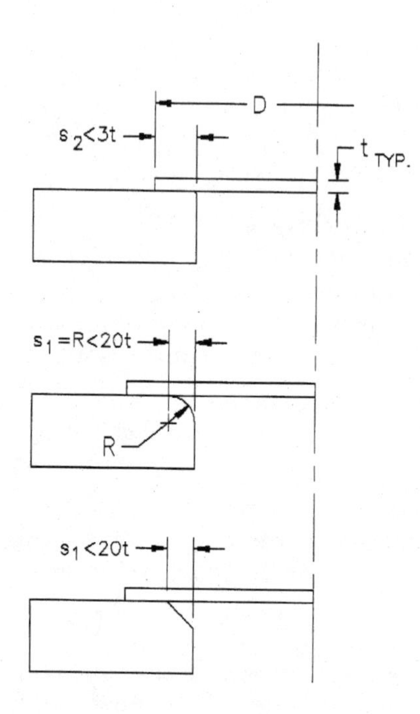


**Figure 9-73** Dimensional requirements for drawing shells without blank holders.

CSN 22 7301 evaluates the need of a blank holder by the formula

$$A = 50\left(z - \frac{\sqrt{t_0}}{\sqrt[3]{D_0}}\right) \tag{9-37}$$

where $z$ = constant, related to the drawn material as follows:
for deep-drawing steel strip, $z = 1.9$
for brass strip, $z = 1.95$
for aluminum and zinc strip, $z = 2.0$

$A \geq \frac{100d}{D_0}$ blank holder is necessary

$A < \frac{100d}{D_0}$ blank holder is not necessary

Deep-drawing materials thinner than 0.020 in must always be drawn with a blank holder.

### 9-11-2. Blank holder in conjunction with draw beads

The lifting force of draw beads against the blank holder is enhanced by the flow of material along their profile. This occurrence may be increased by an inappropriate geometry of the bead or by the inclination of its walls, as well as by the number of successive beads. In order to promote the correct functionability of draw beads, which—in return—means a resistance to drawing, the blank holder's pressure should be as close to its lower limits as possible (Fig. 9-74).

However, with lowering of the pressure of the blank holder, if the bead design has any imperfection, the lifting force of the flange material sliding past may totally disqualify its function. Therefore, the lower limits of the blank holder's pressure, in ideal conditions, should be the same as forces of the sheet-metal flange, causing its lifting.

Calculation of such pressure should consider that deformation of the flange does not alter its thickness. The yield stress should be taken for uniform. Such a calculation should further neglect any frictional resistance along the beads' shape, considering the bead too small in comparison with its diametral value and that of the flange.

The need for a blank holder's pressure is considerably increased where the walls of a bead are made steeper or by an addition of supplementary bead lines. These conditions also decrease the blank holder's sensitivity, which is the ratio of the drawing resistance of the flange to the increase in blank-holding load.

Draw beads on the die surface, with corresponding shapes on the blank holder, affect the material tension, reaching up to the wall of parts, and have a considerable influence on the amount of blank-holding pressure, which is decreased in this case.

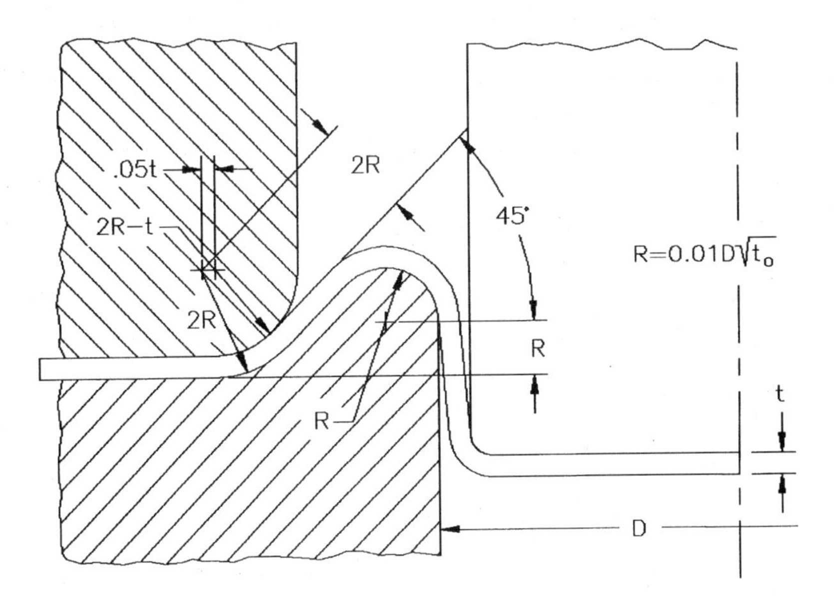


**Figure 9-74** Draw bead design.

### 9-11-3. Blank-holding pressure

The pressure exerted by the blank holder on the flange was found to be somewhere between 0.005 and 0.067 of $S_\Sigma$, where $S_\Sigma$ is the sum of yield strength and tensile strength of the drawn material. The blank-holding pressure was found to vary, even though very slightly, along with the variation of metal thickness.

To prevent wrinkling, the pressure needed for various materials is approximately:

| | |
|---|---|
| Deep-drawing steel | 300–450 lb/in$^2$ |
| Low-carbon steel | 500 lb/in$^2$ |
| Aluminum and aluminum alloys | 120–200 lb/in$^2$ |
| Aluminum alloys, special | 500 lb/in$^2$ |
| Stainless steel, general | 300–750 lb/in$^2$ |
| Stainless steel, austenitic | 1000 lb/in$^2$ |
| Copper | 175–250 lb/in$^2$ |
| Brass | 200–300 lb/in$^2$ |

The blank holder's pressure may be adjusted either by application of springs or by means of hydraulic or pneumatic devices. As far as the spring-loaded support is concerned, urethane pressure pads, strippers, and shedders were found to provide greater pressure per unit area than wound springs.

The numerical value of the blank holder pressure is important mostly for calculations of the total drawing pressure. To actually fine-adjust the blank holder pressure for the given value may prove difficult, as presses generally do not contain such a splendid capacity. In production, the blank holder pressure is adjusted with respect to the part drawn. The object is observed for appearance of wrinkles, buckling, or tearing, and the pressure is corrected accordingly.

## 9-12. Defects Caused by Drawing Process

The drawing process places a considerable strain on the material, which sometimes responds by failing during the manufacturing process. Many defects are mainly visual, such as wrinkles, but there are others, much more insidious, which affect the part's functionality.

### 9-12-1. Breakage of shells

If tearing or breakage occurs in shells drawn from a thin stock, it may be caused by too small radii of the drawing punch or die. Shells also break around their bottom when the ratio of the blank diameter to the shell diameter is too great. Such a drawing process exceeds the strength of a material and a fracture will appear. As a rule, shells should not be drawn deeper (in a single pass) than the amount of their blank diameter. Drawn parts may also break because of various other reasons, such as

- Insufficient clearance between the punch and die
- Excessive pressure of the blank holder
- Excessive friction between the material and tooling

### 9-12-2. Alligator's skin

Alligator's skin may be observed on deep-drawn shells which are made from a material with a hardened surface. This defect appears where the movement of drawn metal is considerable, forcing the inner, annealed layers to flow along more readily and into greater depths than the upper crust, which is hardened. The latter, when stretched beyond its elastic endurance, tends to pull apart and crack.

Hardening of the strip steel surface may be caused by strain hardening of the material during the rolling process. In order to diminish brittleness of the rolled material, the cold-rolled strip is annealed prior to its arrival at the last rolling pass. This last stage of its production is used to bring the material thickness to the correct size. However, should the reduction in thickness be excessive, strain hardening of the upper layer will result.

### 9-12-3. Grain growth

Grain growth is a rather peculiar condition of material that has been cold worked and annealed afterward. It consists of a considerable enlargement of the material grain, which almost explodes in size. The sign of this occurrence is the roughening of the surface, either in spots or in the whole sheet.

Grain growth occurs most often in cold-rolled, low-carbon steels, below 0.2 percent of carbon, and in some aluminum alloys. The annealing process, when performed at 1250 to 1650°F, causes the grain to grow in size, even though the intensity of such growth is closely related to the amount of cold work the material was exposed to previously. Grain growth appears in critically strained areas and impairs the plastic flow of the material, making it highly irregular.

Products affected by grain growth will fail in service when exposed to comparatively minor operational parameters. In deep-drawn shells, grain growth appears at their bottoms, the area most affected by cold working.

One method of preventing grain growth within the material is to choose lower annealing temperatures, which should not bring the material into its recrystallization stage. With steels of a higher carbon content, higher annealing or normalizing temperatures, reaching beyond the critical temperatures of 1450 to 1750°F, may alleviate the problem.

### 9-12-4. Galling

Galling is a result of localized adhesion of the part to its tooling. It is related to the plastic deformation caused by ironing of areas with rough surface finish. The occurrence of galling is considered as caused partially by the breakdown of lubricant, which may have been wiped off during the drawing process, and exposed the drawn material to contact with its tooling. Such sudden contact produces scars on the part, called galling.

A majority of galling occurs in parts where draw beads have been used in the process. Perhaps the increased pressure of draw beads on the material causes the breakdown of a lubricant, and subsequent damage to the part. The smooth surface of a die, turning into a large bearing section, may expose the material to galling.

Galling is expected to greatly impair the drawing capacities of recently designed low- and high-yield-strength steels, such as low-alloy steels containing columbium and titanium, in their decarburized condition. Experimental testings of these alloys found the formation of debris that caused the galling to occur.

Where galling occurs in a drawing process, it is always preceded by the presence of loose metal particles, deposited on the drawing tool and even contained within the lubricant. Some of these particles remain attached to the tooling, creating a buildup of material, especially in the vicinity of high-stressed zones, such as the die radius, blank holder functional surfaces, the area of beads, etc. The final stage occurs when these obstacles, already too bulky, tear off the surfaces they have been adhering to and, placing themselves in the path of the drawn material, produce deep scratches or outright breakage of its surface.

The inclusion of lubricant containing high-pressure additives was previously considered a remedial treatment for galling. Such a solution in itself gave rise to wrinkling of the drawn shell or to the formation of waves within the shell. It made the parts stick to their tooling, caused difficulties in the cleaning process, etc.

However, not much attention was ever given to the fact that galling is mainly caused by the roughness of the surface of the sheet, which—when consisting of too many deep gaps and other irregularities—promotes the occurrence of galling in the drawing process. Experiments (by Japanese researchers, Takahashi, Okada, and Yoshida) have proved that the roughness of the surface is directly proportional to the occurrence of galling.

During the drawing process, various small debris may get entrapped within the rough surface's valleys and gaps, creating minute obstacles around which more debris may accumulate as the process progresses. The highest of irregularities are then subjected to reduction in height, with the removed material being added to the mass already entrapped within the surficial roughness. Such leveling is called "rasp action" by mechanics.

Frictional forces between the drawn material and its tooling, especially during the repeated movements along the same surface, act further upon its already distorted structure. If the valleys in the material are not deep enough to contain all the material sheared by the movement, which in time will always happen, a growth of the deposited material will result, with subsequent tearing and damaging of the drawn product or even its tooling.

## 9-13. Friction and Drawing Lubricants

Lubricants for a drawing process must have an adequate film strength in order to withstand the high pressures and high temperatures associated with drawing. The lubricant's wetting ability should be considerable as well, to spread readily over the ever-expanding surface area of the part. Oiliness control is of essence, since a certain amount of it is necessary, while its excess may prove harmful.

Drawing lubricants, as any other lubricants, must be nonaggressive toward the part, toward the tooling, the machinery, and the operator. Their application and removal must be uncomplicated and economical.

There is a variation in a lubricant's qualities with each material drawn. For example, aluminum alloys are more plastic and may be drawn into greater depths without tearing or fracturing. However, they also possess a high coefficient of friction, which—on the die side—must be decreased with proper lubrication. Such a decrease should be controlled, as an excessive oiliness of the lubricant will certainly produce slippage of the material, with resulting damage to the part and perhaps even to the die.

When selecting the lubricant for a given drawing operation, all subsequent manufacturing processes must be taken into account, such as heat treatment and surface treatment. Some lubricants, usually those of high lubricating qualities, are highly adherent and for that reason difficult to remove off the part.

## 9-14. Drawing Tonnage and Other Calculations

A quick evaluation method, assessing the maximum drawing tonnage based on the thickness of material, shell diameter, ultimate tensile strength of the drawn stock, is as follows:

$$P_{\text{draw}} = 0.00157tdS \tag{9-38}$$

where $S$ = tensile strength of the material

However, a more complex formula used for the assessment of the drawing pressure $P_{DR}$ is

$$P_{DR} = A\,S_y\,\eta_c \ln E_c \tag{9-39}$$

where $A$ = area of cross section of a shell, $A = \pi dt$
$S_y$ = yield strength of the material
$\eta_c$ = deformation efficiency of drawing process (see Fig. 9-75 )
$E_c$ = cupping strain factor, in this case expressed in the form of its natural logarithm ln

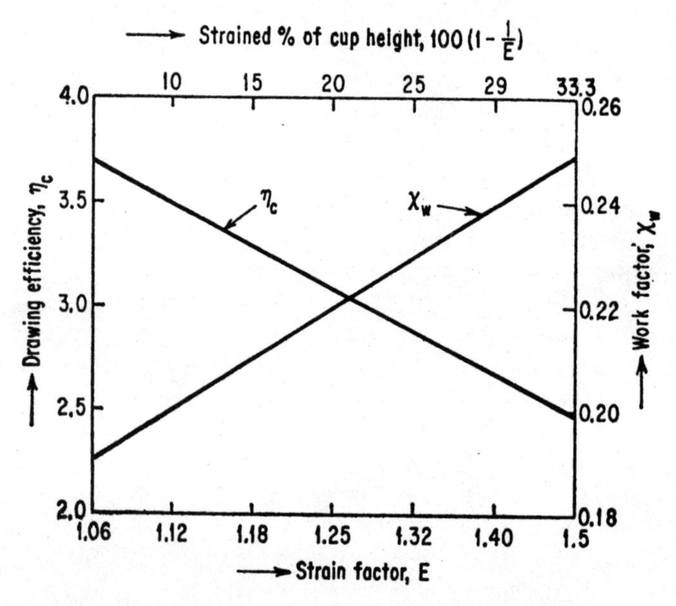


**Figure 9-75** Relationship of the strain factor, work factor, and drawing efficiency. (*From Frank W. Wilson,* "Die Design Handbook," *New York, 1965. Reprinted with permission from The McGraw-Hill Companies.*)

Deformation efficiency factor $\eta_c$ is a friction-related component, depending on the amount of strain factor. The product ($\eta_c$ ln $E_c$) from Eq. (9-39) may be replaced by an expression $[(D/d) - C]$, where $C$ is a constant ranging from 0.6 to 0.7. Equation (9-39) becomes

$$P_{DR} = A\,S_y\left(\frac{D}{d} - C\right) = \pi dt\,S_y\left(\frac{D}{d} - C\right) \tag{9-40}$$

A total drawing force for dies using a blank holder consists of the combined drawing force and the pressure of the blank holder, or

$$P_{\Sigma-DR} = P_{DR} + P_{BH}$$

where $P_{BH}$ is the pressure of the blank holder.

Another method of drawing force evaluation takes into account tensions within the drawn material. The formula to use is

$$P_{DR} = \pi t_0 S d_m \geq \pi t_0 \sigma_\Sigma d_m \tag{9-41}$$

where $d_m$ = mean diameter of shell
$\sigma_\Sigma$ = total of all tension within the material, which may be calculated as follows:

$$\sigma_\Sigma = (\sigma_r + \sigma_f)\,(1 + 1.6\mu) + \sigma_b \tag{9-42}$$

where $\sigma_r$ = radial tension
$\sigma_f$ = tension caused by friction
$\sigma_b$ = tension due to bending
$\mu$ = coefficient of friction

**Ironing.** In the ironing process, the vital information is the percentage of reduction of the wall thickness. This may be assessed by calculating

$$R_{i\%} = \frac{100(t_0 - t_1)}{t_0} \tag{9-43}$$

The reduction percentage, however, does not evaluate the strain of the material $E_i$, which can be expressed as the ratio of thickness variation, or

$$E_i = \frac{t_0}{t_1} \tag{9-44}$$

The result should be equal to

2.0 (50 percent reduction) for a single ironing pass

2.5 (60 percent reduction) for ductile and annealed materials

Ironing strain factors are shown in Table 9-14.

**TABLE 9-14 Ironing Strain-Factor Values and Relative Ability of Ironing of Various Materials**

| $E_{it,max}$ | Relative ironing | Reduction $R_i$, %* | Material (example) |
|---|---|---|---|
| 1.25 | Extremely little | 20.0 | (For third draws) |
| 1.4 | Very little | 28.6 | |
| 1.6 | Little | 37.5 | (For second draws) |
| 1.8 | Medium | 44.4 | Steel |
| 2.0 | Medium-good | 50.0 | Stainless |
| 2.24 | Good | 55.4 | Copper |
| 2.5 | Very good | 60.0 | Aluminum |
| 2.8 | Excellent | 64.3 | |
| 3.15 | Extreme | 68.3 | |

*$R_i = 100(1-t_1/t_0)$.

SOURCE: Frank W. Wilson, *Die Design Handbook,* New York, 1965. Reprinted with permission from The McGraw-Hill Companies.

With multiple ironing operations, the action of strain hardening may decrease the ratio of strain factor of the material in every ironing pass.

The final strain, or a total strain of all combined ironing operations, is calculated as

$$E_{\Sigma i} = E_{i-1} \times E_{i-2} \times E_{i-3} \times \cdots \times E_{i-n} \tag{9-45}$$

The approximate height of the ironed shell $h_i$ may be assessed by using the formula

$$h_i = \frac{h_0 t_0 (D - d_1)}{t_1 (D - d_0)} \tag{9-46}$$

The pressure required for ironing increases with decrease of the die face angle, which is most often between 10° and 45°. An approximate calculation is given in Eq. (9-47).

$$P_i = \pi d_1 t_1 S_{avg} \ln \frac{t_0}{t_1} \qquad (9\text{-}47)$$

where $d_1$ = mean diameter of shell after ironing
$t_0$ = shell thickness before ironing
$t_i$ = shell thickness after ironing
$S_{avg}$ = average value of tensile strengths before and after ironing, or $(S_{before} + S_{after})/2$.
ln = natural logarithm

Chapter

# 10

# Evaluation of a New Die Design

## 10-1. Basic Approach to Die Design

With every new part produced, a complete evaluation of the stamping method and parameters must be performed. Based on the part's flat layout, the sequence of tooling must be designed, which in turn dictates the size of the die. The economies of the strip must be assessed before the rest of the design is finalized. Seemingly small details such as the availability of strip material, the predetermined width, and its thickness and tolerance ranges may turn out to be of tremendous importance when it comes to production.

For selection of the proper press, tonnage requirements must be calculated. Further, the amount of stroke, shut height, mounting arrangement, and other press- and production-related data must be compared to the capacities of the selected press equipment.

Only then may the actual design be started, which always begins with the strip layout and its projection into the cross section of a die. Such a sequence of work process is intentional, as the cross-sectional view provides control of the placement of punches within the assembly. Where punch bodies or heads may be too large to fit the predetermined sequence of operations, one of the stations must be skipped, with subsequent enlargement of the die. This can be readily assessed by comparing the cross-sectional view with the layout of the strip, whereas by looking only at the strip this may pass undetected.

Both strip layout and cross-sectional view should be drawn to size or scaled. With accurately drawn punches and dies, the need for further detailing may often be eliminated. In questionable areas, some dimensions may be added instead of separate sketching or verbal explanations.

## 10-2. Strip Layout and Selection of Tooling

One of the main determinants in an assessment of appropriate strip layout is the production rate expected from the die. As already addressed in the first chapter, a situation where 100,000 pieces are to be delivered within a month is completely different from that where the same number of parts must be produced within a week or perhaps even a day.

To evaluate the problem of production rate properly, a rough estimate of the tonnage and die size must be made for selection of a suitable press. These are preliminary assessments and need not be based on elaborate calculations or sketches. A hand sketch of the strip will often suffice, showing only the sequence of tooling and its location.

Once a press of appropriate tonnage, bed size, stroke, and shut height is selected, scheduling of this machine has to be consulted to find out its availability. It is important to know what other jobs may be running at the time the new production is to begin, if such runs may be interrupted, or if a rigid schedule, denominated by firm deliveries, is to be observed. These aspects must be clarified before the new assignment is committed to the specific pressroom equipment, especially where the proposed die and its size, shut height, and other parameters may not fit any other machine. For an evaluation of the press availability in conjunction with press-room scheduling and other requirements, consult Sec. 1-2, "What Constitutes Suitability for Die Production?"

With flat parts, the stroke of the press should be of no concern, as their height is almost negligible. However, where bent-up parts, drawn shells, and other three-dimensional profiles are to be produced, the stroke must be adequate to clear their height, which—in some cases—may be excessive for the selected pressroom equipment.

With regard to the number of parts to be produced, an accurate strip layout should be drawn next. It will help to evaluate the correctness of the first rough assumption, and it will also establish the exact location of blanks within the strip. Parts may be positioned horizontally, vertically, or at an angle. They may be placed beside each other or intertwined.

Taking as an example the part shown earlier in Fig. 7-3, its vertically oriented placement within the sheet will be as shown later in Fig. 10-1. The amount of feed will be equal to the width of the part, plus a distance $S_p$ between the blanks. Rules governing the distance between parts and their distances off the edge are presented in Table 6-4, Scrap Allowance in Multiple Parts.

## 10-3. Economies of the Strip

With the width of a blank at 1.50 in (Fig. 10-1), the feed, or the amount of progression of the strip during the operation of the die will be 1.562 in. Evaluation of economical aspects of such an arrangement consists of comparing an area of the blank to the total area of the strip needed for production of such a part.

The area of the blank need not be calculated too exactly; its rough outline as indicated in Fig. 10-1 will suffice, provided the same method is used in all subsequent evaluations of differently positioned blanks.

The area of the blank being 9.237 $in^2$ and a corresponding area of the strip being 9.812 $in^2$, can be used in calculation of the percentage of strip taken up by the blank, as

$$100\,\frac{\text{blank area}}{\text{strip area}} = 100\,\frac{9.237\ i^2}{9.812\ i^2} = 94.14\%$$

This means that the part, positioned on the strip per the arrangement shown in Fig. 10-1, uses up 94.14 percent of the area of the strip for its production. Such a high percentage rate is quite impressive, yet a different positioning of parts will often be evaluated for comparison.

The arrangement shown in Fig. 10-2 places the blank on the strip lengthwise. In such a situation, the feed is equal to the entire length of the part plus the spacing in between. The economic utilization of the strip, or the square footage, is calculated the same way:

$$100\,\frac{9.237\ i^2}{10.101\ i^2} = 91.45\%$$

The second strip arrangement uses up 91.45 percent of the strip area for its production. We may immediately observe that the second method of blank orientation is less economical. However, perhaps the result may be improved by sinking the *B* flange into the rectangular cutout of the next piece, should the flange be narrow enough to fit as shown in Fig. 10-3.

The economic evaluation of the square footage in such a case will be

$$100\,\frac{8.673\ i^2}{9.491\ i^2} = 91.38\%$$

The percentage of sheet usage is almost the same as previously, meaning we would not gain much by such an arrangement. Naturally, decreasing the amount of material along the shorter side of the part will never produce results equal to those obtained by decreasing the material over the entire length, which in Fig. 10-1 is the case.

Just by looking at the angular arrangement of the same blanks shown in Fig. 10-4 we may conclude that such a strip layout is the most unreasonable of all. If calculation of a square footage is attempted at all, it should be done as a practice only.

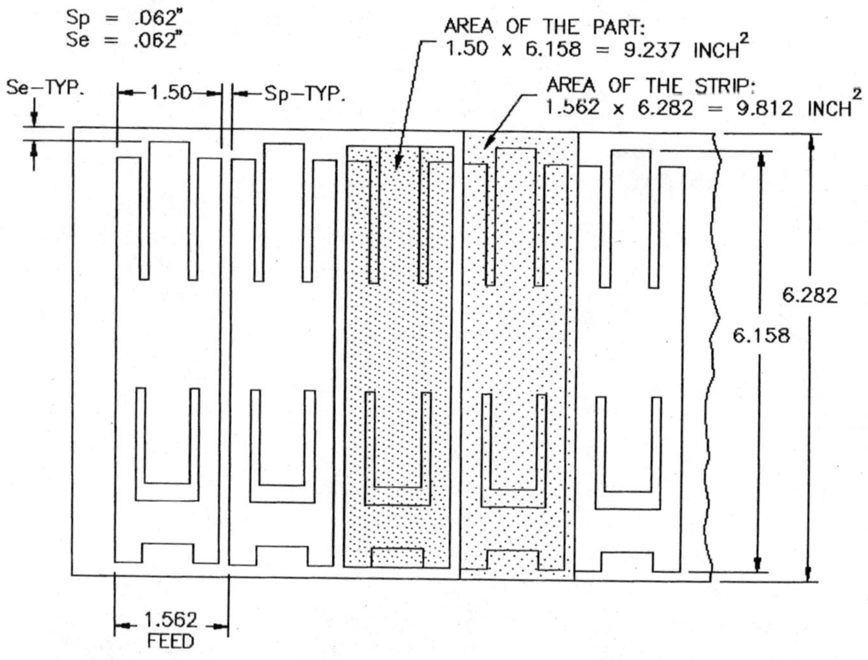


**Figure 10-1** Vertical strip layout of the bracket.

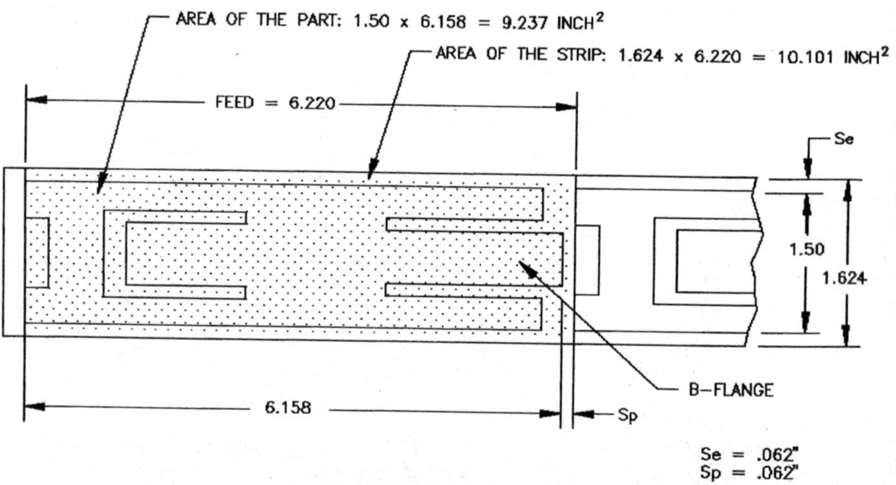


**Figure 10-2** Horizontal strip layout of the bracket.

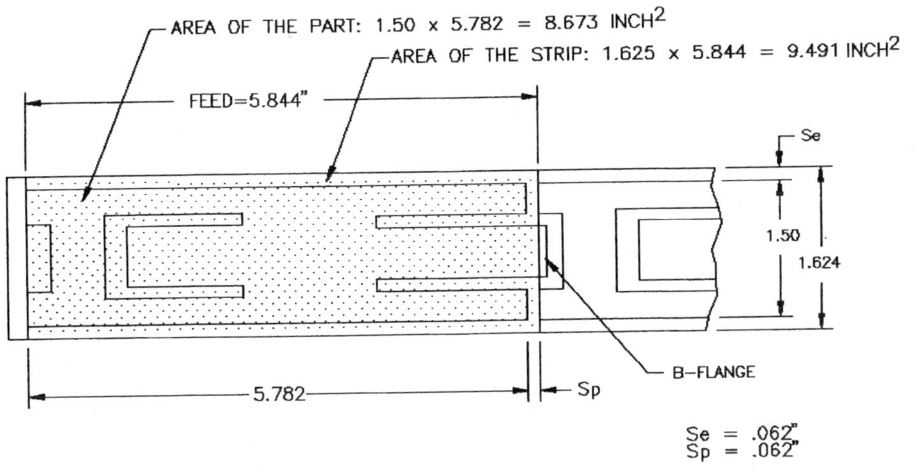


**Figure 10-3** Altered horizontal strip layout of the bracket.

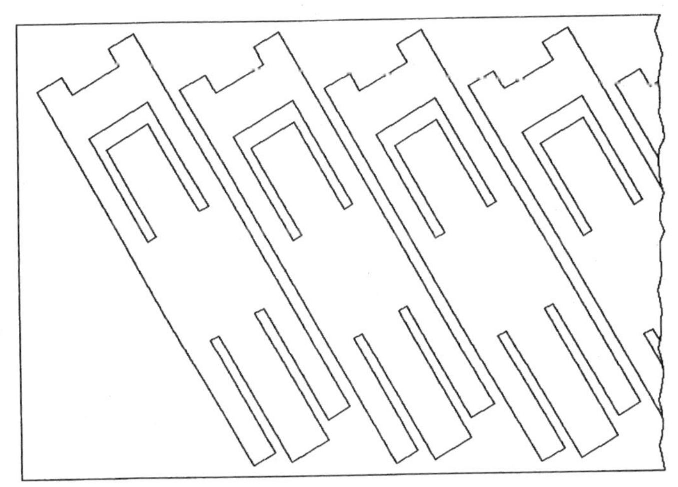

**Figure 10-4** Angular strip layout of the bracket.

This assessment leaves the strip layout presented in Fig. 10-1 as the best economical scenario of all the alternatives presented here.

## 10-4. Tonnage Calculation and Selection of the Press

The next step is to verify the calculation of tonnage needed to produce the part. The total of all cut outlines (per Fig. 10-5) should be obtained by adding up linear distances of the part's periphery, the linear length of round or otherwise shaped cuts, notches, openings, and

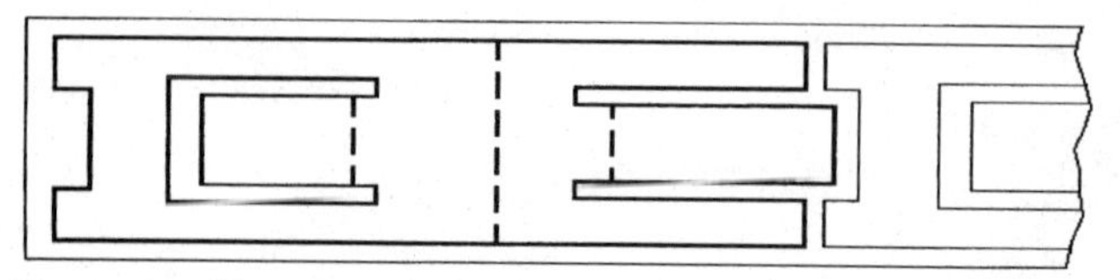

**Figure 10-5** Tonnage calculation.

pilot holes. The total length of all cuts should be multiplied by the thickness of the stock and by the shear strength of the material in tons, presented in Table 6-3, Shear Strength of the Material.

The tonnage needed to produce bends should be included in the calculation to procure the total of all tonnage needed to produce a complete part during each hit of the press. It should not be overlooked that even though piercing and bending is done in stages, it must all be accomplished within a single stroke of the press.

Another influential determinant in the choice of a press is the size of its bed area compared to the size of the die shoe.

## 10-5. Die Shoe Size

A thorough evaluation of the tooling sequence in conjunction with finalization of the strip layout is the next step to take. An accurate sketch of the strip provides overall dimensions of the die shoe not only dimensionally but through a visual verification as well. The die shoe must be ordered with dependence on the suitability of its delivery, most often placing the order as soon as possible.

The evaluation of tooling sequence may often be accomplished by first establishing the method of bending the part. Of importance will be the location of a blank in the die, the direction of bending, and whether the part will be positioned up side down, or down side up. With many parts, such a decision will influence all the preceding die work, affecting the location of tools within either the upper or the lower portion of the die cross section.

The part from Fig. 7-3 as shown in Fig. 10-6 has two somewhat controversial tabs *A* and *B*. Their advantage is that both of them point in the same direction, once the remainder of the part is flat, with the *C* bend flattened out. Therefore, with regard to the bending sequence, we may opt for producing these two flanges first, regardless of whether they are oriented pointing up or down within the die. The sequence of all additional operations should then be arranged accordingly, to first produce these two flanges, with the final bending of corner bends *C* and *D* to be done in the last station.

The first manufacturing method, corresponding with the above requirements, is shown in Fig. 10-7. Both *A* and *B* tabs are here

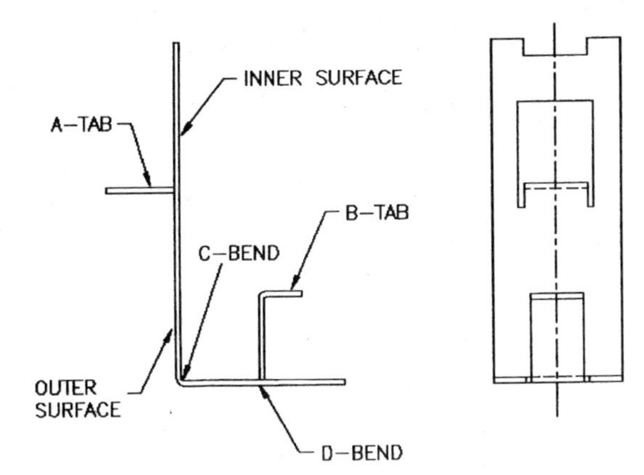


**Figure 10-6** Bent-up bracket.

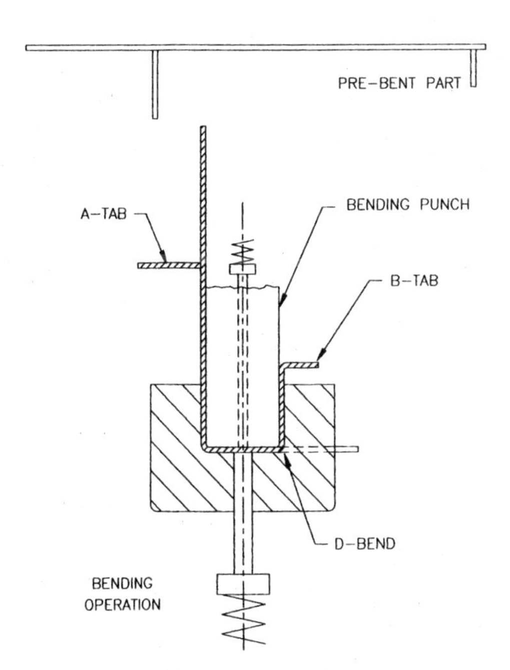


**Figure 10-7** Bending of the bracket.

prebent down. The final bending takes place in the last operation, where the bending punch will force the part into the recess in a die. Bends *C* and *D* are formed simultaneously, and the finished product is ejected upward.

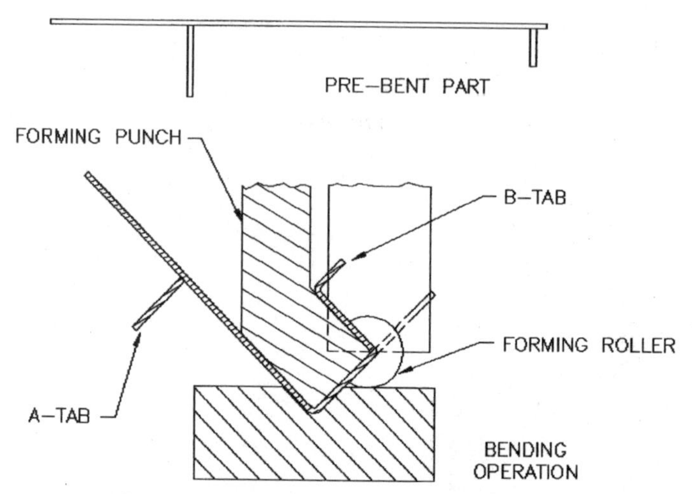


**Figure 10-8** Bending of the bracket, a different approach.

An alternative technique, shown in Fig. 10-8, would be to move the prebent blank (with tabs *A* and *B* pointing downward) toward a V die, perform bending of corner *C,* and obtain corner bend *D* by utilizing a roller.

In Fig. 10-9 the part has two tabs *A* and *B* bent upward. The forming punch wraps bends *C* and *D* around its width, while forcing the blank into a recess in the punch plate. Such a bending method is actually a reversal of that shown in Fig. 10-7.

The fourth alternative, depicted in Fig. 10-10, uses a side action toward bend *D,* whereas bend *C* is provided by the movement of the bending punch.

Regardless of the method chosen, the bending should preferably be done in the last station, and the rest of the strip should be arranged around such a requirement. The choice of method is denominated mainly by the location of burrs. Where tabs *A* and *B* are bent down, burrs will usually be oriented around the outer surface of the part, whereas with tabs bent up, burrs may be found on the inside (refer to Fig. 10-6).

The strip layout in Fig. 10-11 shows the cutting and bending sequence, where all relief slots are provided in the first station, with the removal of the material of both edges in the second and third stations. On arrival of the part in the fourth station, *A* and *B* tabs are produced. The cutoff takes place between the fifth and sixth stations, and in the latter location the final bending takes place. Finished parts are forced back—either up or down, depending on the bending method—to the die block surface level, from where they may be removed by air or allowed to drop down if the die or press is inclined.

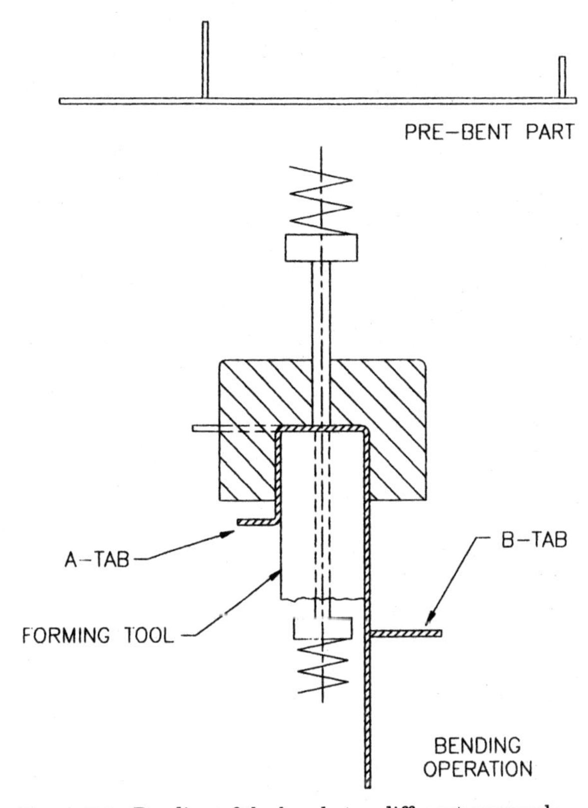


**Figure 10-9** Bending of the bracket, a different approach.

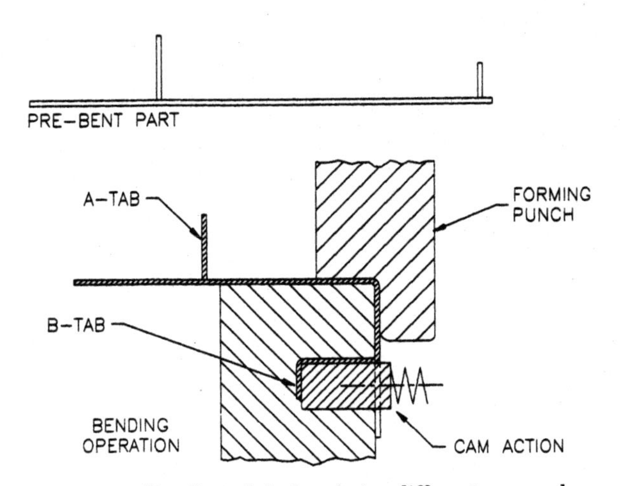


**Figure 10-10** Bending of the bracket, a different approach.

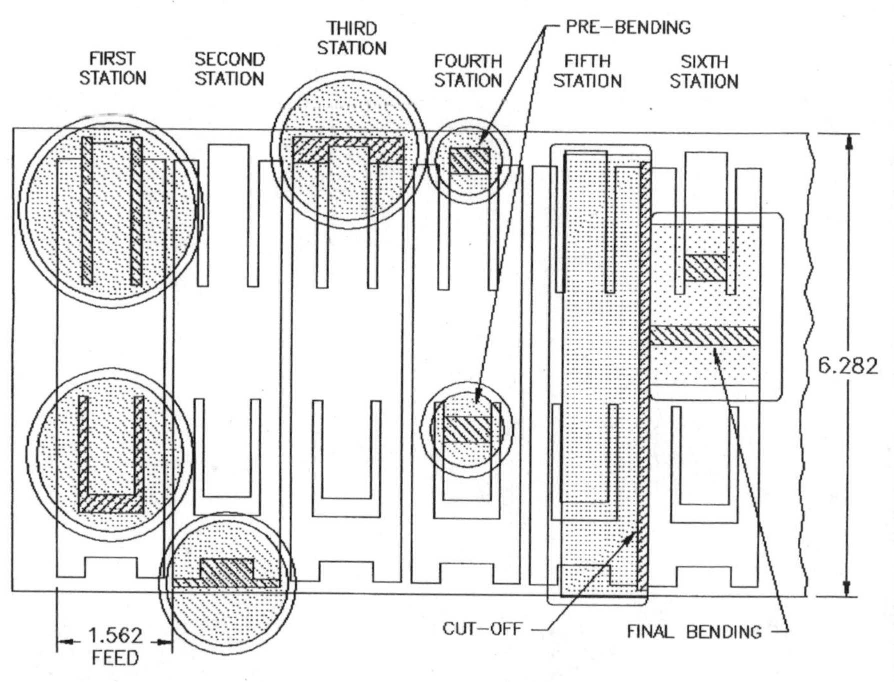


**Figure 10-11** Strip layout and tooling. Cut shapes are shown as hatched; tooling bodies are dotted.

According to the strip layout in Fig. 10-11, we have six stations within this die, spaced 1.562 in from each other, resulting in the total length of

$$6 \times 1.562 = 9.372 \text{ in}$$

Adding material to the sides and assuming 2 in per side is adequate, the size of the die block (and punch plate) is

$$9.372 + 2(2) = 13.372 \doteq 13.5 \text{ in long}$$

and

$$6.282 + 2(2) = 10.282 \doteq 10.5 \text{ in wide}$$

The size of the die block dictates the corresponding size of the die shoe. Here we refer to the manufacturers' tables and charts and find the appropriate die shoe combination. As already noted, the overall size of the die is of considerable importance, since it must fit the bed area of the selected press. Mounting of the die with regard to the mounting arrangement of the chosen press, along with all additional parameters, will have a definitive impact on selection of the proper die shoe.

## 10-6. Method of Parts Ejection

The method of ejection of the part should be addressed, with checking into the ejecting system available to the chosen press. If the table is inclined, parts may be allowed to slide down off the die surface. Where parts are dropped through the bottom die shoe, the opening in the bolster plate of the press must be checked for size to evaluate if the parts will pass through. Where parts may fit when coming down straight, sometimes their passage may be impaired when turning haphazardly sideways, which may often be the case.

Where air ejecting devices are available, we should evaluate their usage with respect to the given shape of the part and see that the die design will incorporate the appropriate mounting openings to contain them.

The method of ejection, the arrangement of tooling sequence, and all previously discussed aspects should still be considered preliminary, as only the final strip layout, along with the accurate drawing of the cross section of the die, will show if parts can be produced as shown. These final details are addressed in the next chapter, which is reserved for such topics.

Chapter

# 11

# Progressive Die—Strip Layout

## 11-1. Washers and Other Round Blanks

As already noted, the overall production requirements are the main factors in considering the size of a die and the number of finished products per stroke of the press. Let us consider a requirement for 1,000,000 washers per week, to be produced in a press capable of delivering 150 strokes per minute. We may easily calculate that 150 hits per minute equals 9,000 strokes per hour and 72,000 blows per workday. This means that if we run a strip through the die, producing one part with each blow, at the end of the week we will have 360,000 pieces.

However, we need a whole million of parts, which is 277.78 percent of whatever we are getting now. For that reason we should increase the output of a press by producing more than one part at a single stroke. To shorten the amount of production, we may want to build a die, producing four parts at a time, or four-ups. The output from such a tool will certainly satisfy the basic demand, and there will still be some time left.

With washers and other round blanks, the strip layout for a four-up die will look as shown in Figs. 11-1 and 11-2. The punches are not mounted along a straight line, as their shanks and heads will never fit the distance in between, aside from the fact that such a crowded arrangement will considerably weaken the punch plate and the die shoe.

Instead, the tooling is spaced along an angular axis, the angle of inclination off the horizontal being either 19.5° or 30°, which is an industry standard.

The first row of washers will be produced as indicated by their hatched shapes, with subsequent hits filling the area of the strip with

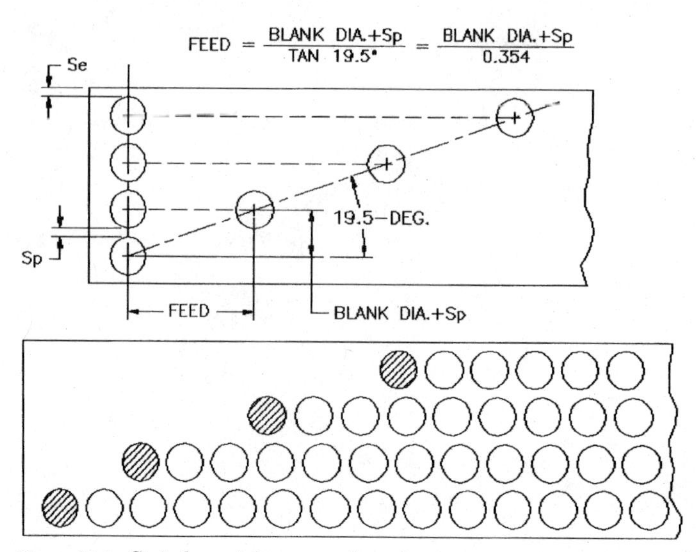


**Figure 11-1** Strip layout for a round washer.

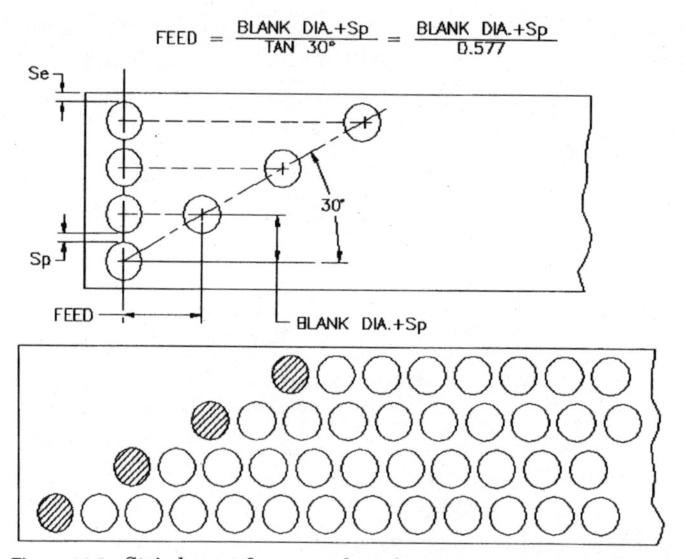


**Figure 11-2** Strip layout for a round washer.

an even and orderly pattern. The feed can easily be calculated by using the appropriate angle and figuring the vertical distance between the centers of blanks to be equal to the blank diameter, plus the required spacing $S_p$ in between. These two values will give us all the additional data for the assessment of feed and strip width.

## 11-2. Pilots and Pilot Holes

Every strip design should begin with an assessment of the location of pilot holes. These openings must be pierced at the first station, serving afterward as a guide and a locating arrangement for the strip on its way through the die.

Pilot holes may be extra openings placed beside the parts, or they may be holes included within the part itself and serving at the same time as the strip guidance. The location of pilot holes should always be at the far opposite sides of the strip, with the greatest possible gap in between. This is to secure the best fixation and positioning of the strip, once the pilots engage in their respective openings.

Pilot holes may not be necessary where producing parts from the strip in a single station of the die, as with production of washers, shown in Figs. 11-1 and 11-2. However, where any amount of subsequent work is to be done to the part, piloting of the strip must be included in the design of such a die.

The strip layout discussed in the preceding chapter would actually be difficult to produce, as shown in Fig. 10-11, for no pilot holes are included within the strip layout and neither is a space left for their allocation. Therefore, the strip width will have to be slightly increased, with some stations shifted around to suit the requirements.

The result of such a rearrangement is shown in Fig. 11-3, with a sequence of material removal illustrated in Fig. 11-4. Pilot holes are produced first on the new strip design, with pilots fitting into these openings right in the second station. At the first station, relief openings are produced in a similar fashion to that shown on the strip layout from the preceding chapter. However, the head of the upper punch had to be altered in shape; otherwise it would not fit between the two pilot punches.

The punch, shaped to remove all material at the upper narrow edge of the shape and used in the second station of Fig. 10-11, is now dismantled, consisting of two separate punches; the first is positioned within the second station, with the second part of the tool being moved all the way toward the end of the strip, between the fourth and the fifth stations. At such a location, the rectangular punch removes the material surrounding the second prong of the preceding part, plus that of the first prong of the following piece. The tool, equipped with a pilot tip for accurate locating prior to piercing, is the last cutting tool of the die, severing the upper edge of the part off the strip.

Lengthwise separation of parts is produced between the second and third stations, with bending of the *B* tab to follow and bending of the *A* tab to be achieved in the fourth station.

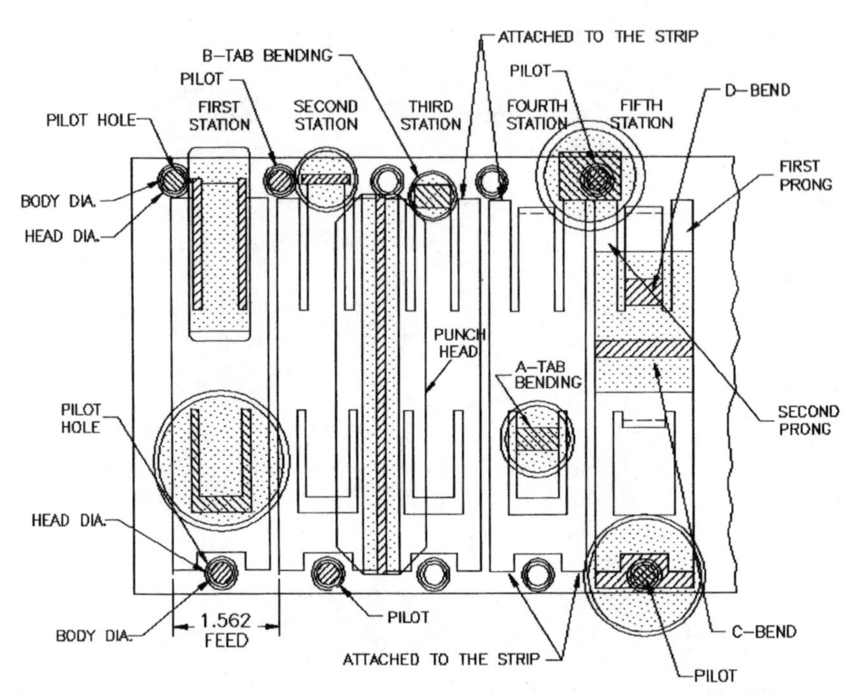


**Figure 11-3** Strip layout, adjusted. Cuts and bends are shown as hatched; tooling bodies are dotted.

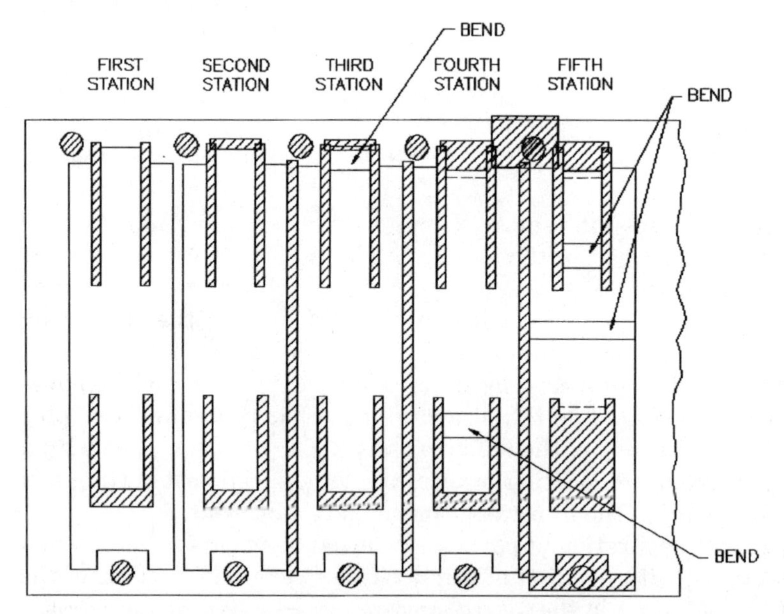


**Figure 11-4** Cutting of the strip. Hatched surfaces are cutouts.

Within the final section, the part's total separation from the strip is produced, and bending of the *C* and *D* flanges occurs. Removal of the lower edge section is achieved by using a punch of the same shape, equipped with a pilot tip for locating purposes.

## 11-3. Skipping of Stations

Stations may be skipped or left nonoperative (empty) for various reasons. Where punch bodies or their heads may become excessive in size, resulting in interference between them, the station in between must be skipped.

In the first and second stations of Fig. 11-5 is a situation where such interference of the pilot head and that of the piercing punch occurs. Since skipping of the station will not help in either case, for the pilot will always be found in exactly the same spot, altering of its head will be the preferable solution, as shown in Figs. 11-6 and 11-7.

Tooling details *C* and *D* (Figs. 11-8 and 11-9) have pilots installed within their bodies to provide for locating of the strip prior to any cutting. The pilot point must always exceed the length of all cutters, so that a proper location is attained prior to any alteration of the strip.

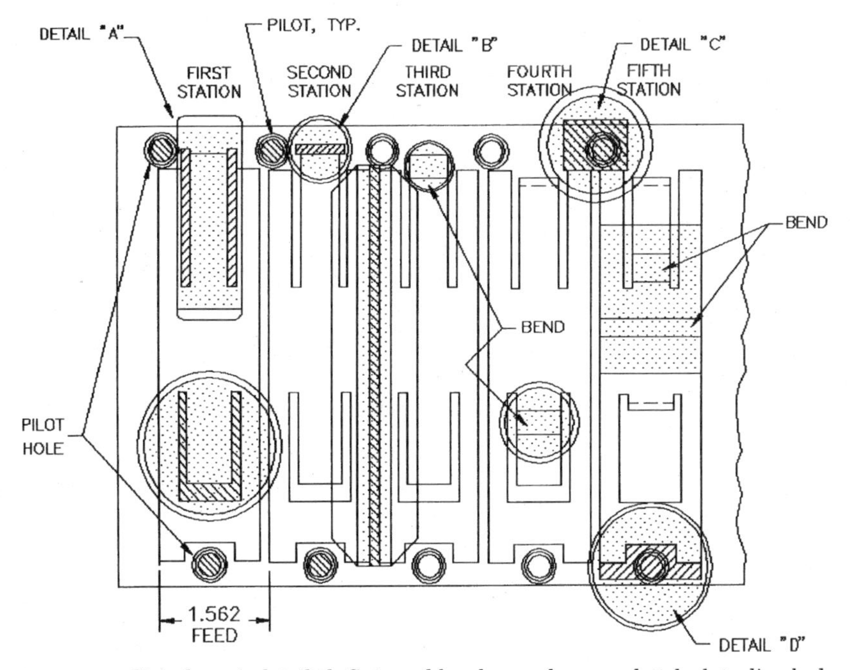


**Figure 11-5** Strip layout, detailed. Cuts and bends are shown as hatched; tooling bodies are dotted.

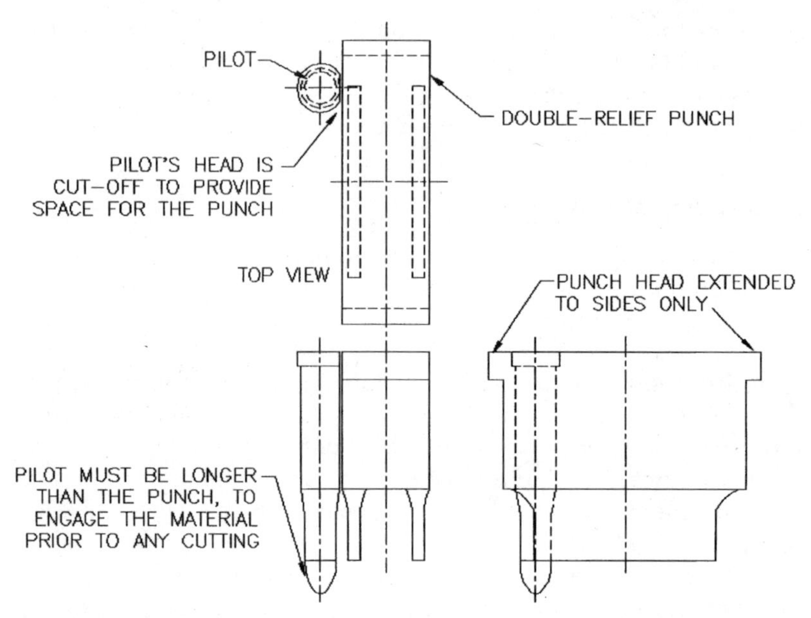


**Figure 11-6** Detail *A*.

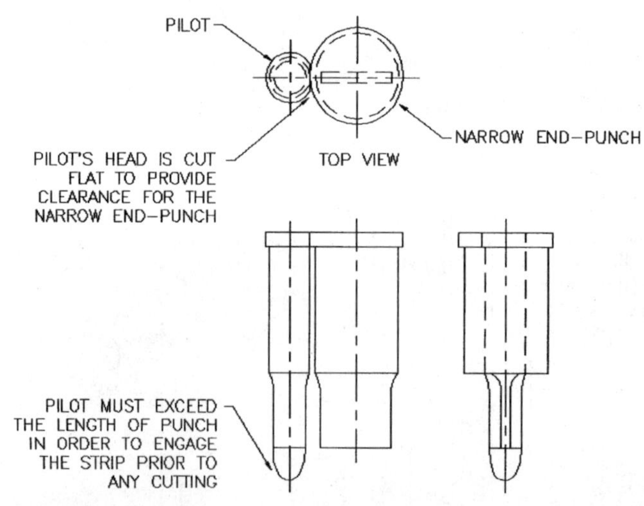


**Figure 11-7** Detail *B*.

Another reason for skipping stations is a situation where the punch (or die) block is too densely perforated by too closely spaced tooling. An extensive drilling of these blocks will impair their strength, which should be prevented by spacing the tooling farther apart.

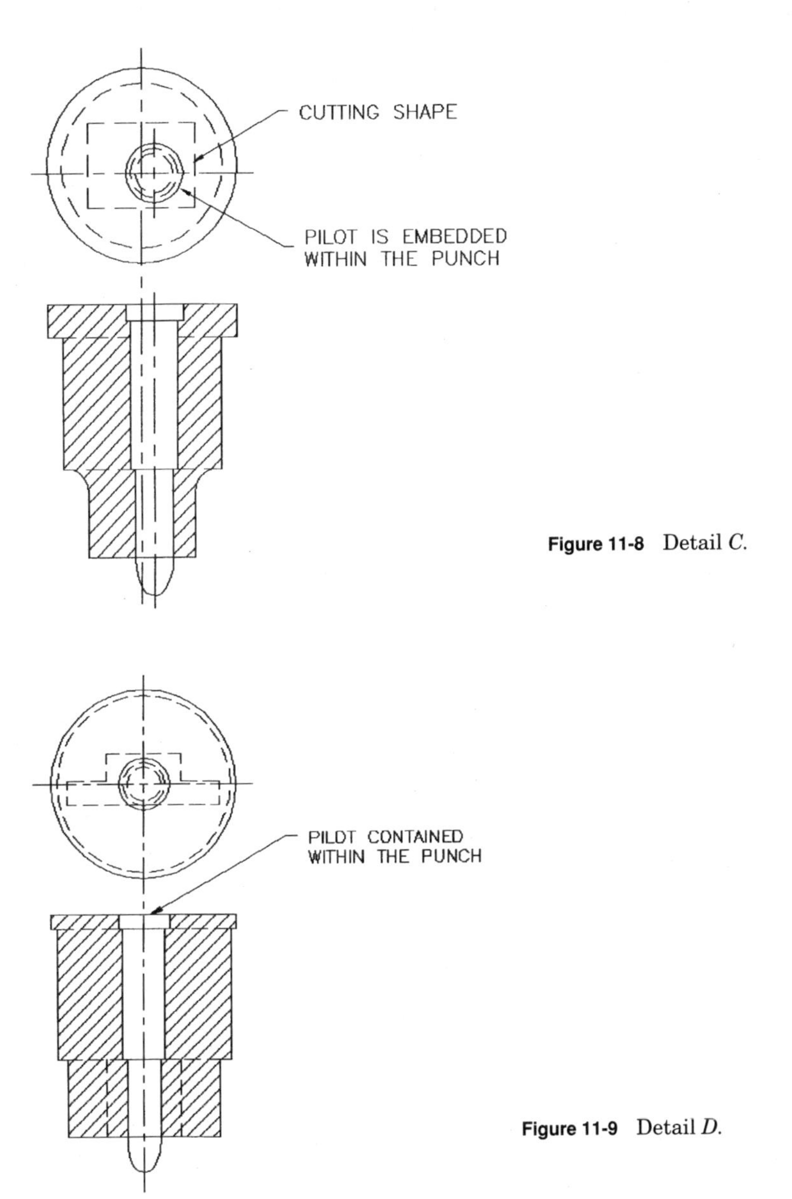


**Figure 11-8** Detail *C*.

**Figure 11-9** Detail *D*.

## 11-4. Nesting and Locating

Nesting of parts is required where the part shape is other than flat. With flanges bent down in one station, all subsequent stations must be provided with a relief opening to contain them not only on arrival of the strip into that particular station but also throughout their passage through the rest of the die.

Nesting of parts is further necessary where some highly accurate work is done to the part, for example, shaving. In such a case, the nest often contains the whole outer shape of the part, serving as a retaining and positioning device at the same time, for the ensuing die operation.

Some nests may consist of pins within the block, in between which the part fits, when deposited by the movement of a strip. Other nests are obtained by cutting the outline of the part into the solid block, where also the depth of the placement is controlled. Nests may also be made of segments (Fig. 11-10).

A method of carrying a strip of sheet metal from station to station and performing various alterations to it is sometimes called the "cut and carry" method. It is a standard procedure of strip movement through the progressive die.

## 11-5. Strip Sample Number Two

A part, shown earlier in Fig. 7-10, was altered for the die production, with its phantom flange removed. The adjusted flat layout is shown in Fig. 11-11.

Positioning of such blanks on the strip may be as shown in Fig. 11-12. Here the pilot holes and 0.343-in-diameter openings are produced in the first station, along with the two long relief slots. In the second station, the middle flange is severed off the strip, and the side flanges are separated between this and the next station. The third and fourth stations share the long rectangular back-line cut, where the second half of the material of the third blank is removed at the same time as the first half of the material of the fourth blank.

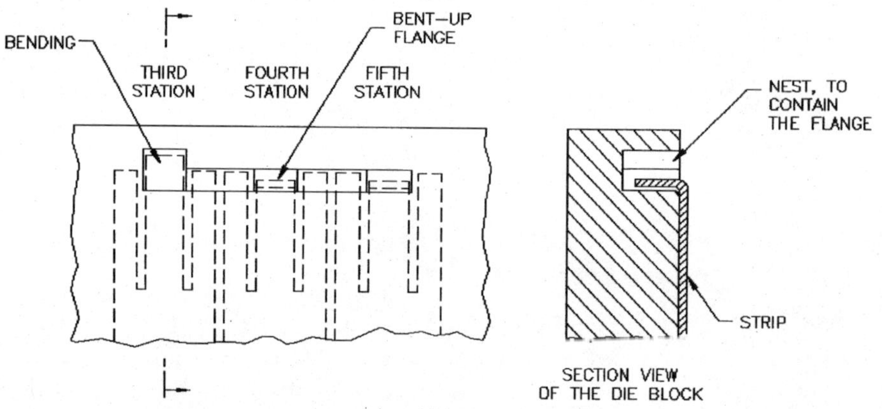


**Figure 11-10** Die block with a nest.

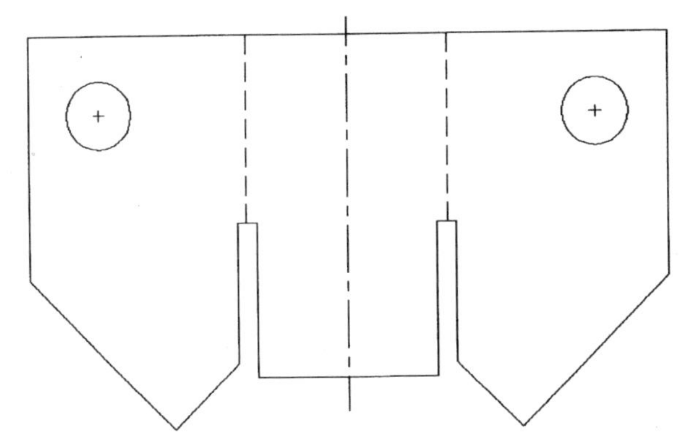

**Figure 11-11** Flat layout, adjusted.

Sp = .062"
Se = .062
BLANK AREA = 3.100 x 2.000 = 6.200 INCH$^2$
STRIP AREA = (3.162 x 4.187):2 = 6.620 INCH$^2$

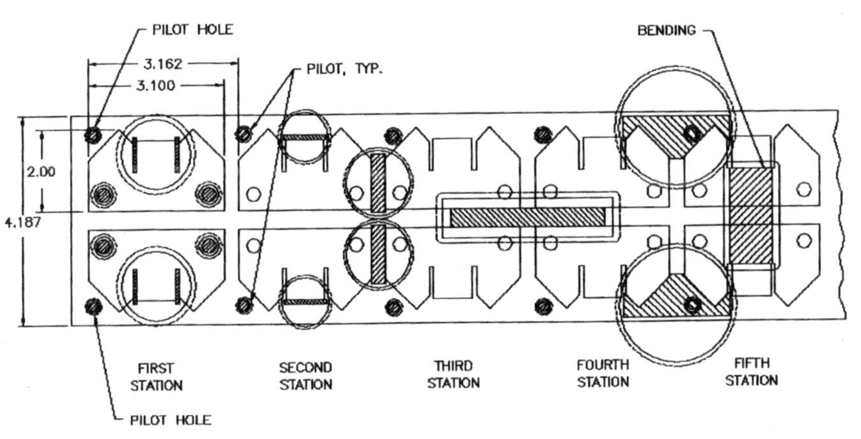


**Figure 11-12** Strip layout.

Between the fourth and fifth station, all angular sections of side flanges are cut off, and bending is then performed in the last, fifth station.

When assessing the strip layout in Fig. 11-12 for the economy of such positioning, only a rough outline of the part may be calculated, provided that the same approach is used in all subsequent evaluations. In order not to break through the edge of the strip, the usually recommended $S_e$ distance will probably have to be increased, yet for the sake of practice, the originally recommended 0.062 in will be kept here. With the area of the part being

$$3.100 \times 2.00 = 6.200 \text{ in}^2$$

and the area of the strip

$$(3.162 \times 4.187):2 = 6.620 \text{ in}^2$$

the square foot percentage will be

$$100 \frac{6.200}{6.620} = 93.65\%$$

However, such a strip layout presents a definite disadvantage in the fact that all tooling will have to be provided twice. This is caused by the mirror-image reversal of the part, which prevents almost any conjunction in tooling arrangement. Where a rather simplified method of tooling order is required, the parts should be as positioned on the strip, shown in Fig. 11-13.

The economical aspects of this strip layout are identical to Fig. 11-12. But where the tooling previously had to be different for each row of parts, here it may be combined, which may produce savings.

In the first station, the two pilot holes and two round openings are pierced simultaneously with the two long relief slots. Between the second and third stations, the angular material of the side flange is cut off, freeing one edge of each part with a single hit of a square tooling.

The third station provides for the removal of the long back strip of material; in the fourth station, the middle section is eliminated.

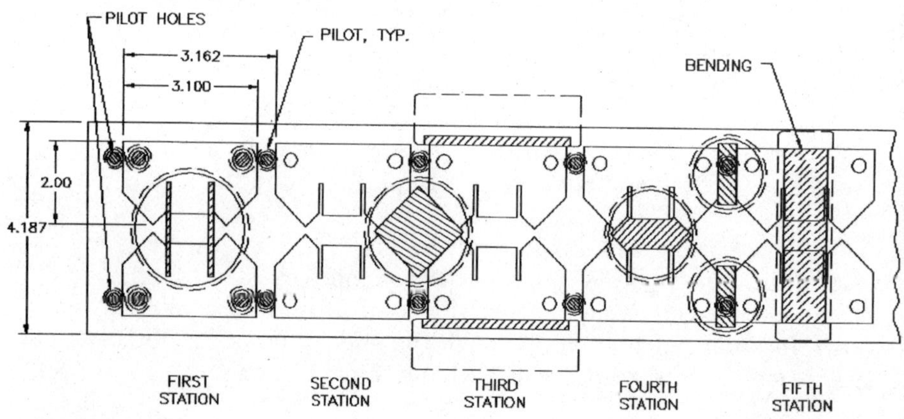


**Figure 11-13** Strip layout.

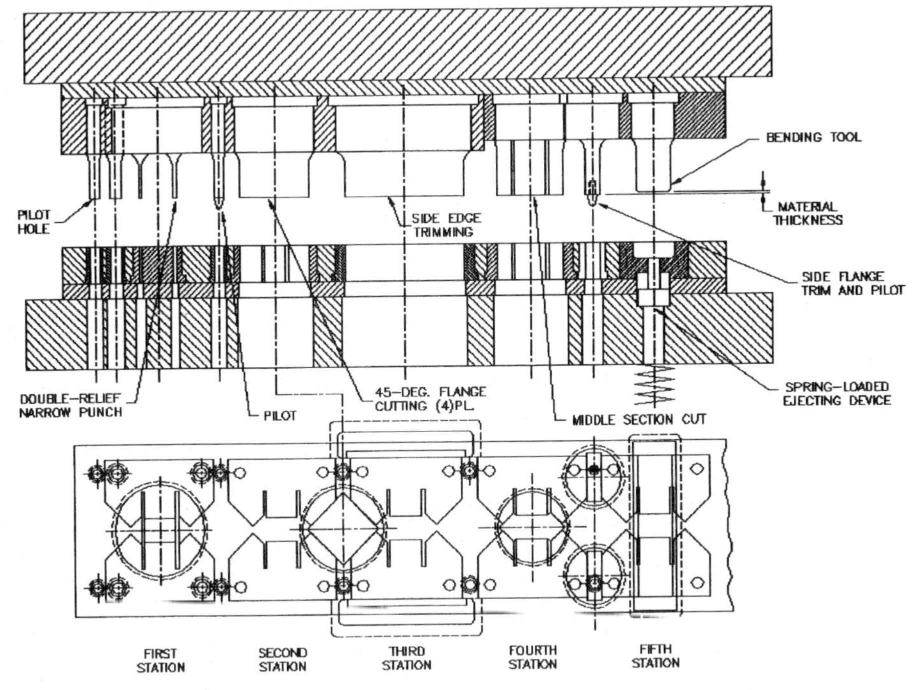


**Figure 11-14** Die layout.

The web between the two parts goes at the transition between the fourth and the fifth stations. The tool used for its removal is basically a rectangle, with a pilot tip attached to its face by a threaded shank.

By cutting the web between, the final blank is separated from the strip, and all forming is provided in the last, fifth station. It should be noted that the forming punch is slightly shorter in order to finish cutting of the side web before any forming takes place.

The strip layout, as positioned on the die block, plus the cross section of the die are shown in Fig. 11-14.

## 11-6. Strip Sample Number Three

Shown in Fig. 11-15 is another blank, already discussed in the preceding chapter. Its positioning within the strip is rather awkward, leaving large areas unused and the economical evaluation of the square footage showing the blank as occupying only 49.4 percent of the strip area.

Utilization of the strip area is improved in the subsequent illustration. Shown in Fig. 11-16, the sets of interlocking parts show the area usage to be at 61.15 percent. This ratio is further enhanced by the strip layout in Fig. 11-17, where the square footage is 72.25 percent.

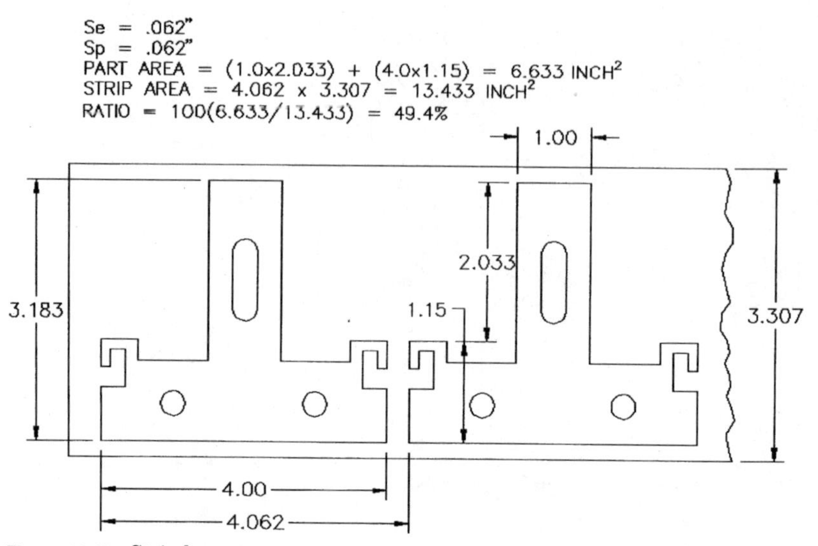


**Figure 11-15** Strip layout.

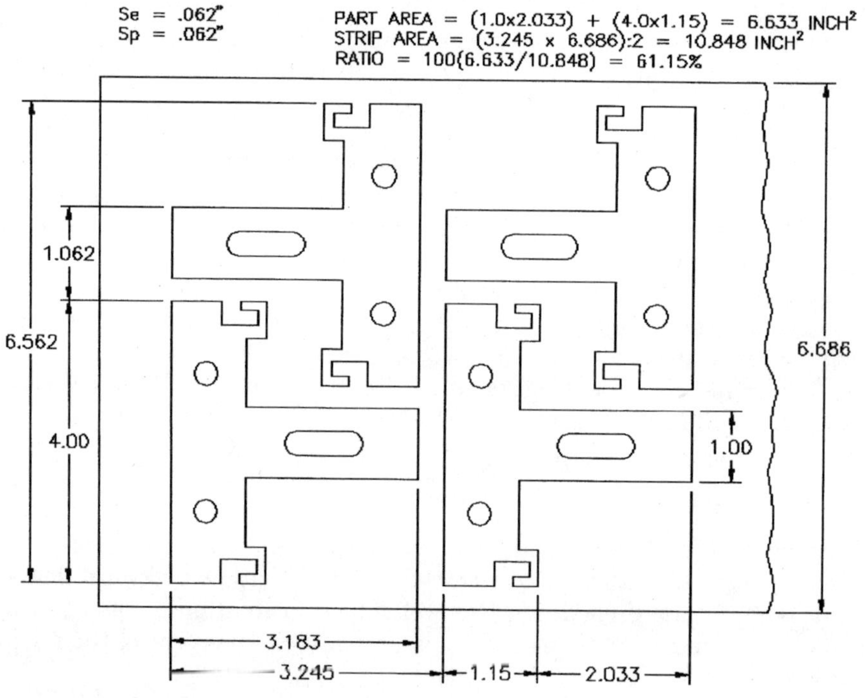


**Figure 11-16** Strip layout variation.

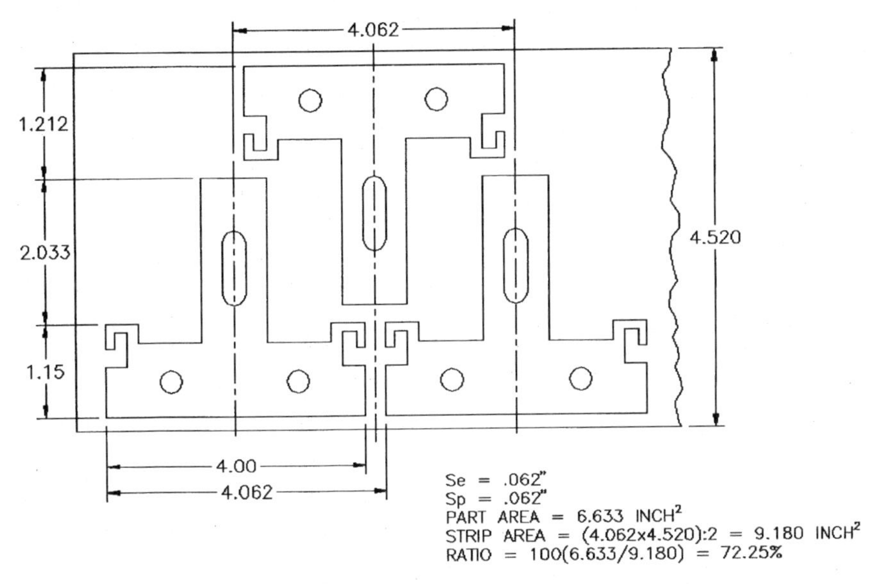


**Figuro 11-17** Strip layout variation.

But the final solution, presented in Fig. 11-18, with the square footage of the blank at 78.3 percent, seems to be the ultimate method of strip layout.

However, there are several problems with the winning strip layout, not obvious from the first fleeting glance. Most of the tooling to pro-

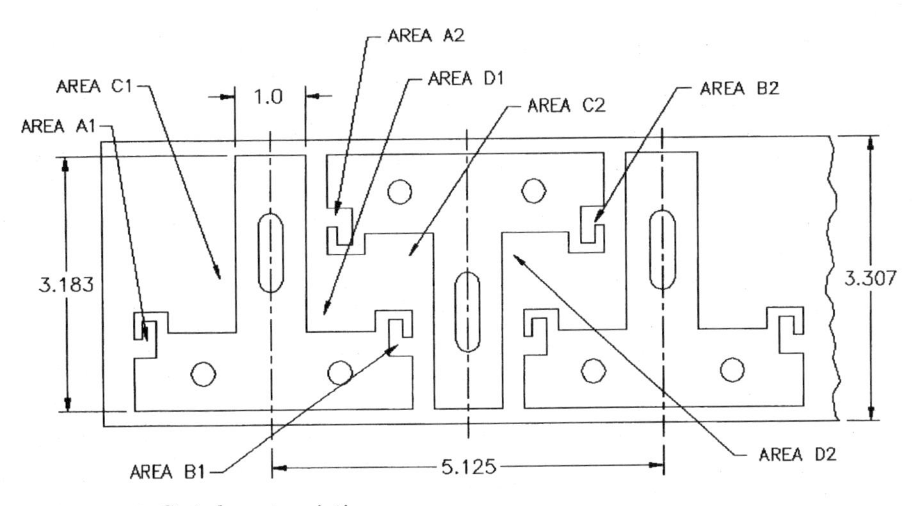


**Figure 11-18** Strip layout variation.

duce this part will have to be doubled, or rather quadrupled, because of the mirror-image reversal of the part. Therefore, we will have specialized tooling for removal of the material in section *A1,* with the same shape of tool, rotated at 180° to be used for removal of section *B1.* The same shape further rotated is used to cut *A2,* and an additional rotation will have to be produced for the removal of section *B2.* Yet, there is not a chance of using a single tool twice or combining it with another shape.

The same applies to the removal of material within areas *C1, D2,* and *C2, D2.* All these shapes are identical; they could be punched with the same tooling, but their mirror-image reversal prevents such a practice, and each shape will have to utilize its own tooling at a different location, as well as rotated around its axis.

The strip layout of the blank placement per Fig. 11-18, with its respective tooling, is not provided. However, if such a sketch is attempted, it will certainly consist of at least 11 stations, which will make die design of such a size quite prohibitive. Just imagine, with the feed of 5.125 in, the linear span of 11 stations will amount to some 56.375 in. This is quite a size for a die, and such a tool will probably never be made just because of that.

Nevertheless, some tooling attempts are contained in the following illustrations, beginning with Fig. 11-19. An avid reader may try to sketch such a strip layout at his or her leisure.

Two tooling variations are presented in Fig. 11-19, each applied to one half of the blanks' combination. Method *A* uses two squares of the same size and two identical rectangles to remove the material between the parts. Naturally, the areas marked *A1, B1, A2,* and *B2*

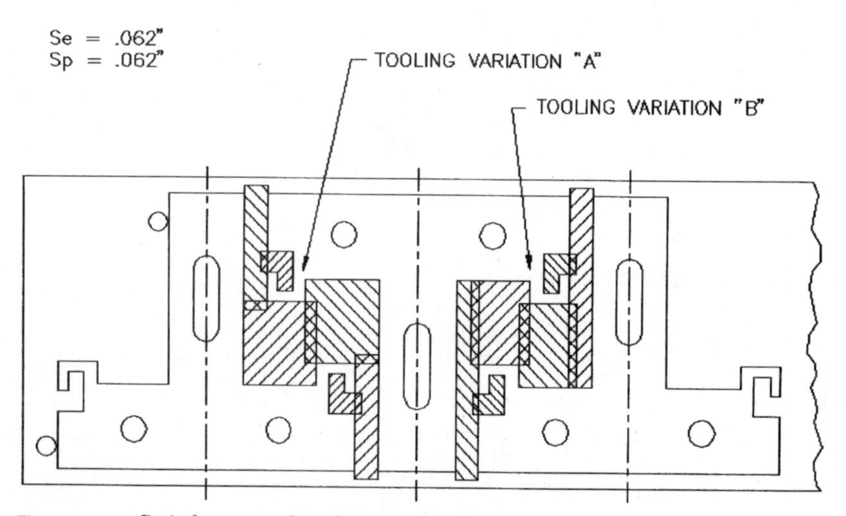


**Figure 11-19** Strip layout with tooling.

previously will have to be cut by the original tooling, rotated with each application, which will give a total of four additional specialized punches.

Method *B* uses rectangles instead of squares, and in its basis this solution is similar to method *A*.

Tooling variation *C*, depicted in Fig. 11-20, replaces the two basic-shaped square or rectangular punches with a single specialized tool, while keeping the two long side rectangles for the removal of the web in between. Such tooling is lesser in quantity, but its special contour will be much more complex to produce than simple basic shapes.

Tooling variation *D*, shown in the same illustration, replaces the three hits of method *C* with a single stroke of a complex-shaped punch. Such a tool will certainly remove all the material in between, and only two of such shapes will be needed for the entire strip, but the procurement of a tool of this complexity and size will be much more costly.

And so, at the end, we may even come back to the original, economically least effective strip layout, presented in Fig. 11-15. Its advantage lies in the fact that the parts are positioned the same way, which means that all tooling will be only doubled instead of quadrupled.

Further, such a layout may even be of advantage in cases where an additional, different part may be produced by utilizing the empty area, unused by the bracket pictured. If this is the case, the additional blanks may either be produced within the same die, or the strip may be run twice: once as shown and a second time with a different strip layout and in a different die, utilizing the remaining material.

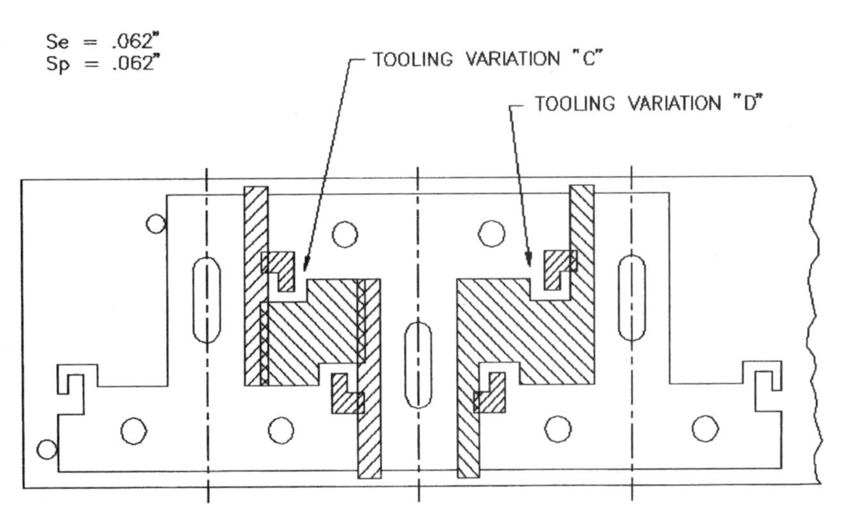


**Figure 11-20** Strip layout with tooling.

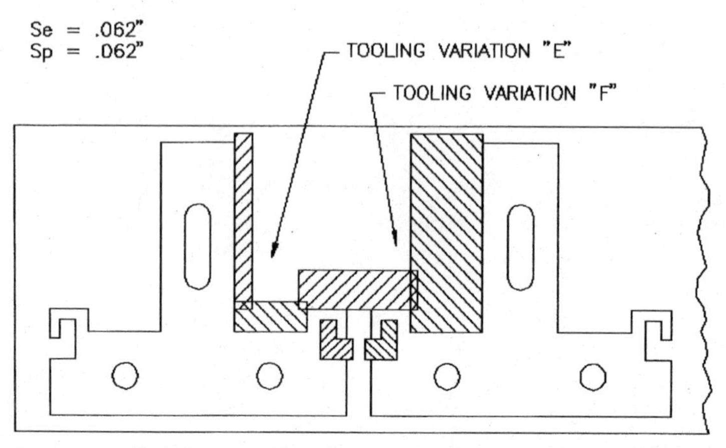


**Figure 11-21** Strip layout with tooling.

Often, the reversal of the strip and running it through the same die may achieve the economies of the strip layouts shown in Figs. 11-18 or 11-19, while using a single set of tooling to produce the parts.

However, coming back to the original layout shown in Fig. 11-15, tooling variation *E* presented in Fig. 11-21 severs the part's outline off the strip with 2.5 hits per part. Tooling variation *F* needs only 1.5 hits, but such a method leaves no material in between to be utilized for production of different parts, unless the rectangular blank itself is the part.

Tooling variation *G* in Fig. 11-22 needs two hits to achieve the same, whereas variation *H,* using a specialized punch, produces the same result with a single blow. The ultimate solution of material

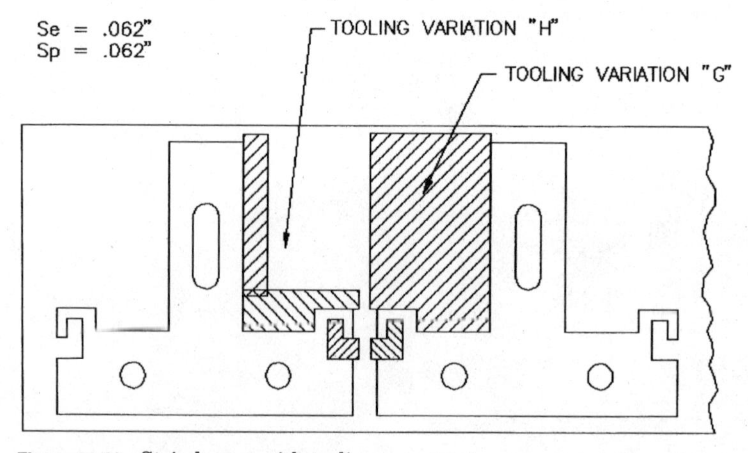


**Figure 11-22** Strip layout with tooling.

Se = .062"
Sp = .062"

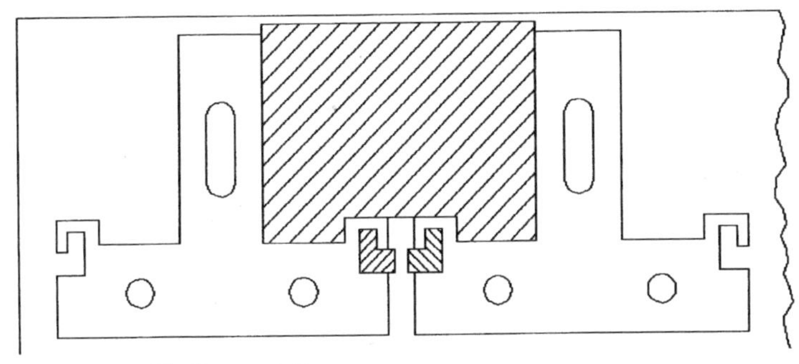

**Figure 11-23** Tooling variation.

removal is the huge tool shown in Fig. 11-23, which removes all the material with a single hit of a punch.

## 11-7. Miscellaneous Strip Arrangements

Several additional strip arrangements are presented in Figs. 11-24 through 11-31. The parts shown are positioned beside each other, opposite each other, or interlocking. With interlocking parts, the tooling is usually very complex, and large runs are often the only justification for attempting such strip layout solutions.

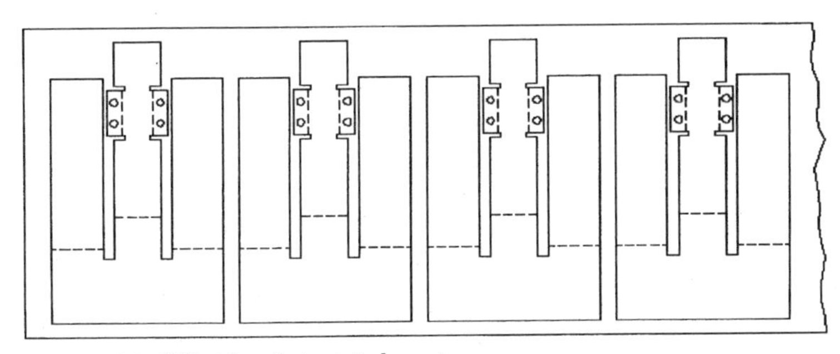

**Figure 11-24** Offset bracket, strip layout.

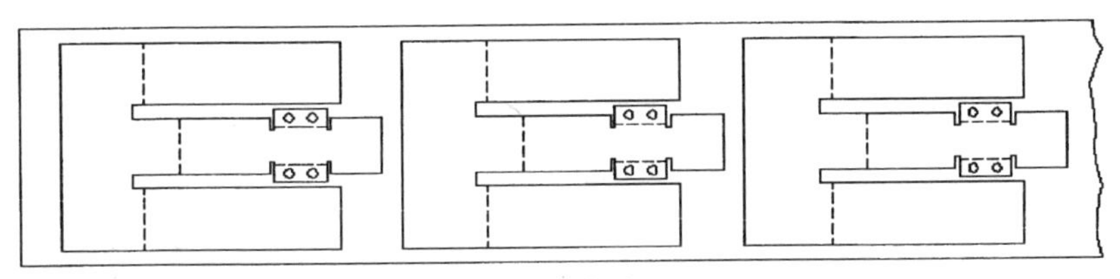

**Figure 11-25** Offset bracket, strip layout variation.

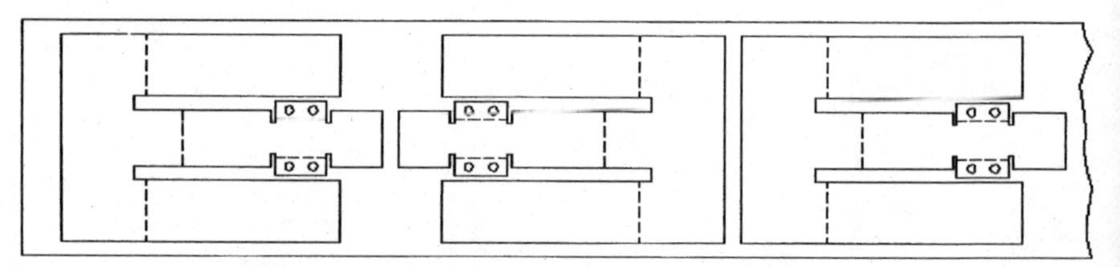

**Figure 11-26** Offset bracket, strip layout variation.

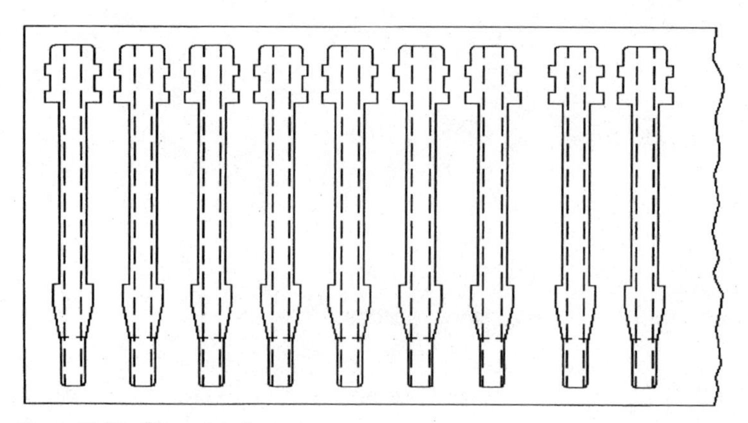

**Figure 11-27** Clip, strip layout.

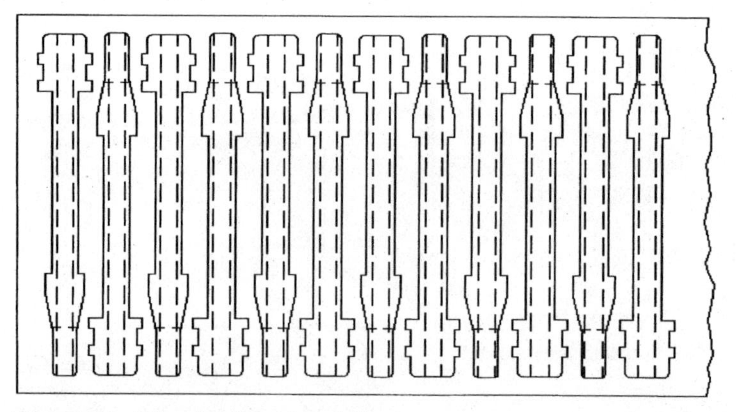

**Figure 11-28** Clip, strip layout variation.

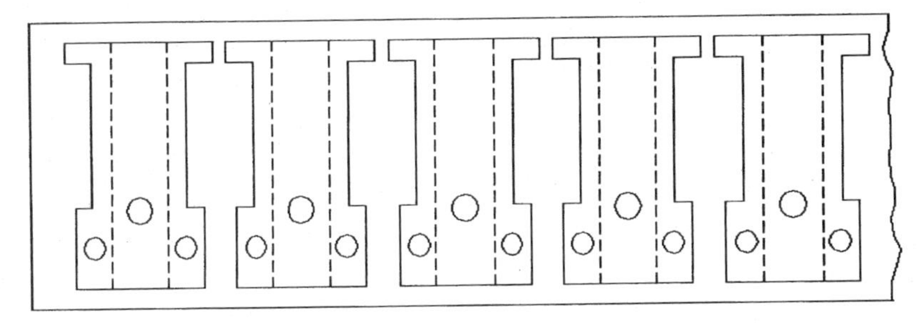

**Figure 11-29** Support, strip layout.

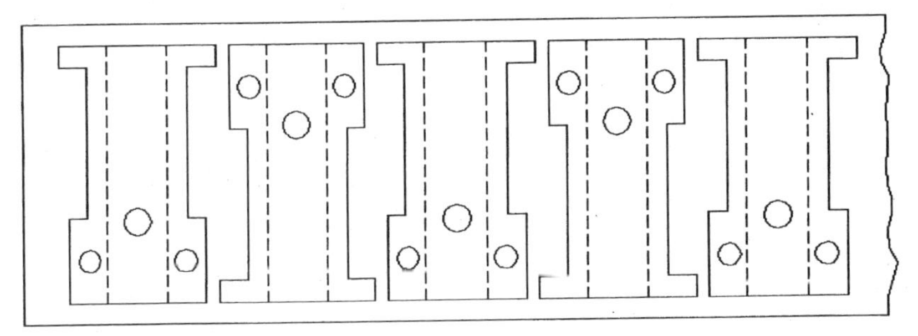

**Figure 11-30** Support, strip layout variation.

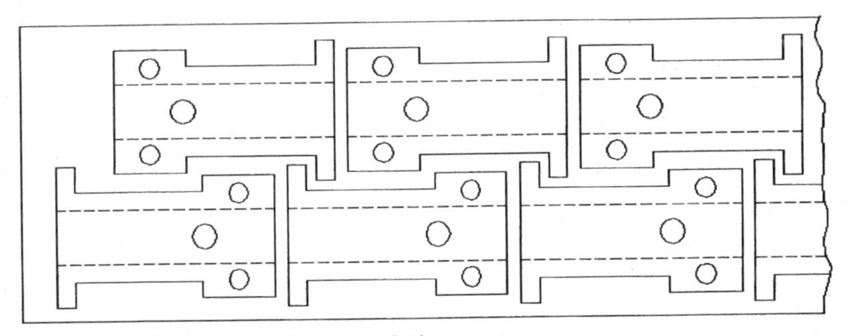

**Figure 11-31** Support, strip layout variation.

Chapter

# 12

# Springs, Their Design and Calculations

## 12-1. Springs and Their Properties

Well-functioning springs are one of the most important prerequisites of a good die function. After all, what good is the drawing operation if the part cannot be stripped off the punch because there is not enough spring power behind the pressure pad? Or—what kind of parts will emerge from a die where the spring stripper is not spring-loaded adequately?

If ample pressure is the absolute basic of a good die operation, then springs are the most vital parts of every die.

### 12-1-1. Spring materials

Springs are elements designed to withstand great amounts of deflection and return to their original shape and size on its release. To be capable of such cyclical loading, spring materials must possess very high elastic limits.

Often materials not specifically made for the spring application are utilized for the purpose because their elastic limits are within the above requirements. Steels of medium-carbon and high-carbon content are considered good spring materials. Where a copper-base alloy is required, beryllium copper and phosphor bronze are utilized.

The surface quality of the spring material has a considerable influence on the function of a spring, namely, on its strength and fatigue. Where possible, the surface finish has to be of the highest grade, preferably polished. This is especially important with closely wound springs, where friction between single coils may create minute defects in the surface, which subsequently will cause the spring to crack. Music wire, the highest-quality spring material, is polished, and its surface is almost defect-free.

Of course, the higher quality the material, the more expensive it is. The designer should strive to find the best combination of price versus quality for each particular job.

A brief description of basic spring materials is included in Table 12-1, which provides a rough comparison of properties, usefulness, and some specific aspects.

**12-1-1-1. High-carbon spring steel wire.** This group of spring materials is lowest in cost, which may account for its widespread use. It does not take impact loading or shock treatment well. Also it should not be used in extreme temperatures, high or low. Main representatives of this group are listed, with the percent of carbon (C) given.

*Music wire, ASTM A228* (0.80 to 0.95 percent C). Good for high stresses caused by cyclic repeated loading. A high-tensile-strength material, available as (cadmium or tin) preplated.

*Oil-tempered MB grade, ASTM A229* (0.60 to 0.70 percent C). A general-purpose spring steel, frequently used in coiled form. It is not good with shock or impact loading. Can be formed in annealed condition and hardened by heat treatment. Forms a scale, which must be removed if the material is plated.

*Hard-drawn MB grade, ATM A227* (0.60 to 0.70 percent C). Used where cost is essential. Not to be used where long life and accuracy of loads and deflections are important. Can be readily plated.

*Oil-tempered HB grade, SAE 1080* (0.75 to 0.85 percent C). With the exception of a higher carbon content and higher tensile strength, this spring steel is almost the same as the previously described MB grade. It is used for more precise work, where a long fatigue life and high endurance properties are needed. If such aspects are not required, an alloy spring steel should be used in replacement.

**12-1-1-2. High-carbon spring steel strip.** The two main types of spring steel in this group are an absolute majority of all flat spring use. However, both are susceptible to hydrogen embrittlement even when plated and baked afterward.

*Cold-rolled blue-tempered spring steel, SAE 1074, plus 1064 and 1070* (0.60 to 0.80 percent C). This steel can be obtained in its annealed or tempered condition: Its hardness should be within 42 to 46 Rockwell hardness Scale C.

*Cold-rolled, blue-tempered spring steel, SAE 1095* (0.90 to 1.05 percent C). It is not advisable to purchase annealed, as this type of steel does not always harden properly and spring properties obtained after

**TABLE 12-1 Typical Properties of Common Spring Materials**

| Common name | Young's modulus $E$[a] | | Modulus of rigidity $G$[a] | | Density[a] | | Electrical conductivity,[a] % IACS | Sizes normally available[b] | | | | Typical surface quality[c] | Maximum service temperature[d] | |
|---|---|---|---|---|---|---|---|---|---|---|---|---|---|---|
| | MPa $10^3$ | lb/in² $10^6$ | MPa $10^3$ | lb/in² $10^6$ | g/cm³ | (lb/in³) | | Min. mm | Min. (in) | Max. mm | Max. (in) | | °C | °F |
| Carbon-steel wires: | | | | | | | | | | | | | | |
| Music[e] | 207 | (30) | 79.3 | (11.5) | 7.86 | (0.284) | 7 | 0.10 | (0.004) | 6.35 | (0.250) | 1 | 120 | 250 |
| Hard drawn[e] | 207 | (30) | 79.3 | (11.5) | 7.86 | (0.284) | 7 | 0.13 | (0.005) | 16 | (0.625) | 3 | 150 | 250 |
| Oil tempered | 207 | (30) | 79.3 | (11.5) | 7.86 | (0.284) | 7 | 0.50 | (0.020) | 16 | (0.625) | 3 | 150 | 300 |
| Valve spring | 207 | (30) | 79.3 | (11.5) | 7.86 | (0.284) | 7 | 1.3 | (0.050) | 6.35 | (0.250) | 1 | 150 | 300 |
| Alloy-steel wires: | | | | | | | | | | | | | | |
| Chrome vanadium | 207 | (30) | 79.3 | (11.5) | 7.86 | (0.284) | 7 | 0.50 | (0.020) | 11 | (0.435) | 1,2 | 220 | 425 |
| Chrome silicon | 207 | (30) | 79.3 | (11.5) | 7.86 | (0.284) | 5 | 0.50 | (0.020) | 9.5 | (0.375) | 1,2 | 245 | 475 |
| Stainless-steel wires: | | | | | | | | | | | | | | |
| Austenitic type 302 | 193 | (28) | 69.0 | (10) | 7.92 | (0.286) | 2 | 0.13 | (0.005) | 9.5 | (0.375) | 2 | 260 | 500 |
| Precipitation hardening 17-7 PH | 203 | (29.5) | 75.8 | (11) | 7.81 | (0.282) | 2 | 0.08 | (0.002) | 12.5 | (0.500) | 2 | 315 | 600 |
| NiCr A286 | 200 | (29) | 71.7 | (10.4) | 8.03 | (0.290) | 2 | 0.40 | (0.016) | 5 | (0.200) | 2 | 510 | 950 |
| Copper-base alloy wires: | | | | | | | | | | | | | | |
| Phosphor bronze (A) | 103 | (15) | 43.4 | (6.3) | 8.86 | (0.320) | 15 | 0.10 | (0.004) | 12.5 | (0.500) | 2 | 95 | 200 |
| Silicon bronze (A) | 103 | (15) | 38.6 | (5.6) | 8.53 | (0.308) | 7 | 0.10 | (0.004) | 12.5 | (0.500) | 2 | 95 | 200 |
| Silicon bronze (B) | 117 | (17) | 44.1 | (6.4) | 8.75 | (0.316) | 12 | 0.10 | (0.004) | 12.5 | (0.500) | 2 | 95 | 200 |
| Beryllium copper | 128 | (18.5) | 48.3 | (7.0) | 8.26 | (0.298) | 21 | 0.08 | (0.003) | 12.5 | (0.500) | 2 | 205 | 400 |
| Spring brass, CA260 | 110 | (16) | 42.0 | (6.0) | 8.53 | (0.308) | 17 | 0.10 | (0.004) | 12.5 | (0.500) | 2 | 95 | 200 |
| Nickel-base alloys: | | | | | | | | | | | | | | |
| Inconel alloy 600 | 214 | (31) | 75.8 | (11) | 8.43 | (0.304) | 1.5 | 0.10 | (0.004) | 12.5 | (0.500) | 2 | 320 | 700 |
| Inconel alloy X750 | 214 | (31) | 79.3 | (11.5) | 8.25 | (0.298) | 1 | 0.10 | (0.004) | 12.5 | (0.500) | 2 | 595 | 1100 |
| Ni-Span-C | 186 | (27) | 62.9 | (9.7) | 8.14 | (0.294) | 1.6 | 0.10 | (0.004) | 12.5 | (0.500) | 2 | 95 | 200 |
| Monel alloy 400 | 179 | (26) | 66.2 | (9.6) | 8.83 | (0.319) | 3.5 | 0.05 | (0.002) | 9.5 | (0.375) | 2 | 230 | 450 |
| Monel alloy K500 | 179 | (26) | 66.2 | (9.6) | 8.46 | (0.306) | 3 | 0.05 | (0.002) | 9.5 | (0.375) | 2 | 260 | 500 |

**TABLE 12-1 Typical Properties of Common Spring Materials (*Cont.*)**

| Common name | Young's modulus $E^a$ MPa $10^3$ | Young's modulus $E^a$ lb/in² $10^6$ | Modulus of rigidity $G^a$ MPa $10^3$ | Modulus of rigidity $G^a$ lb/in² $10^6$ | Density[a] g/cm³ | Density[a] (lb/in³) | Electrical conductivity,[a] % IACS | Sizes normally available[b] Min. mm | Min. (in) | Max. mm | Max. (in) | Typical surface quality[c] | Maximum service temperature[d] °C | °F |
|---|---|---|---|---|---|---|---|---|---|---|---|---|---|---|
| Carbon-steel strip: | | | | | | | | | | | | | | |
| AISI 1050 | 207 | (30) | 79.3 | (11.5) | 7.86 | (0.284) | 7 | 0.25 | (0.010) | 3 | (0.125) | 2 | 95 | 200 |
| AISI 1065 | 207 | (30) | 79.3 | (11.5) | 7.86 | (0.284) | 7 | 0.08 | (0.003) | 3 | (0.125) | 2 | 95 | 200 |
| AISI 1074, 1075 | 207 | (30) | 79.3 | (11.5) | 7.86 | (0.284) | 7 | 0.08 | (0.003) | 3 | (0.125) | 2 | 120 | 250 |
| AISI 1095 | 207 | (30) | 79.3 | (11.5) | 7.86 | (0.284) | 7 | 0.08 | (0.003) | 3 | (0.125) | 2 | 120 | 250 |
| Bartex | 207 | (30) | 79.3 | (11.5) | 7.86 | (0.284) | 7 | 0.10 | (0.004) | 1 | (0.040) | 1 | 95 | 200 |
| Stainless-steel strip: | | | | | | | | | | | | | | |
| Austenitic types 301, 302 | 193 | (28) | 69.0 | (10) | 7.92 | (0.286) | 2 | 0.08 | (0.003) | 1.5 | (0.063) | 2 | 315 | 600 |
| Precipitation hardening 17-7 PH | 203 | (29.5) | 75.8 | (11) | 7.81 | (0.282) | 2 | 0.08 | (0.003) | 3 | (0.125) | 2 | 370 | 700 |
| Copper-base alloy strip: | | | | | | | | | | | | | | |
| Phosphor bronze (A) | 103 | (15) | 43 | (6.3) | 8.86 | (0.320) | 15 | 0.08 | (0.003) | 5 | (0.188) | 2 | 95 | 200 |
| Beryllium copper | 128 | (18.5) | 48 | (7.0) | 8.26 | (0.298) | 21 | 0.08 | (0.003) | 9.5 | (0.375) | 2 | 205 | 400 |

[a]Elastic moduli, density, and electrical conductivity can vary with cold work, heat treatment, and operating stress. These variations are usually minor but should be considered if one or more of these properties is critical.

[b]Diameters for wire; thicknesses for strip.

[c]Typical surface quality ratings. (For most materials, special processes can be specified to upgrade typical values.)
1. Maximum defect depth: 0 to 0.5% of $d$ or $t$.
2. Maximum defect depth: 1.0% of $d$ or $t$.
3. Defect depth: less than 3.5% of $d$ or $t$.

[d]Maximum service temperatures are guidelines and may vary owing to operating stress and allowable relaxation.

[e]Music and hard drawn are commercial terms for patented and cold-drawn carbon-steel spring wire.

Inconel, Monel, and Ni-Span-C are registered trademarks of International Nickel Company, Inc. BARTEX is a registered trademark of Theis of America, Inc.

SOURCE: *Design Handbook*, 1987. Reprinted with permission from Associated Spring, Barnes Group, Inc., Dallas, TX.

forming may be marginal. Its hardness range is 47 to 51 Rockwell hardness Scale C.

**12-1-1-3. Alloy spring steel.** A good spring steel for a high-stress application, with impact loading and shock application involved.

*Chromium vanadium steel, ASTM A231* takes higher stresses than high-carbon steel. It also has a good fatigue strength and endurance.

*Chromium silicon steel, ASTM A401.* This material can be groomed to high tensile stress through heat treatment. Applicable where long life is required in combination with shock loading.

**12-1-1-4. Stainless spring steel.** A corrosion-resistant material. With the exception of the 18-8 type, none of these steels should be used for lower-than-zero temperature applications. High-temperature tolerance is up to 550°F.

*Stainless spring steel 302, ASTM A313* (18 percent Cr, 8 percent Ni). This material has quite uniform properties and the highest tensile strength of the group. It can be obtained as cold drawn, since it cannot be hardened by heat treatment. The slight magnetic properties are due to cold working, as in annealed form it is nonmagnetic.

*Stainless spring steel 304, ASTM A313* (18 percent Cr, 8 percent Ni). Because of its slightly lower carbon content, this material is easier to draw. Its tensile strength is somewhat lower than that of type 302, even though their other properties coincide.

*Stainless spring steel 316, ASTM A313* (18 percent Cr, 12 percent Ni, 2 percent Mo). Less corrosion-prone than the 302 type stainless, with its tensile strength about 12 percent lower. Otherwise it is quite similar to the 302 type.

*Stainless spring steel 17-7 PH, ASTM A313* (17 percent Cr, 7 percent Ni, with trace amounts of aluminum and titanium). The tensile strength of this material is almost as high as that of music wire. This is achieved through forming in a medium hard condition and precipitation hardening at low temperatures.

*Stainless spring steel 414, SAE 51414* (12 percent Cr, 2 percent Ni). Its tensile strength is approximately the same as that of type 316 (above), and it may be hardened through heat treatment. In a high-polished condition this material resists corrosion quite well.

*Stainless spring steel 420, SAE 51420* (13 percent Cr). May be obtained in the annealed state, hardened and tempered. Scales in heat treatment. Its corrosion-resistant properties emerge only after hardening. Clear bright surface finish enhances its corrosion resistance.

*Stainless spring steel 431, SAE 51431* (16 percent Cr, 2 percent Ni). This material has very high tensile properties, almost on a par with music wire. Such a characteristic is achieved through a combination of heat treatment, followed by cold working.

**12-1-1-5. Copper-base spring alloys.** This group of spring materials is more expensive than alloy steels or high-carbon materials. They are, however, very useful for their good corrosion resistance and superb electrical properties. An additional advantage is their usefulness in lower-than-zero temperatures.

*Spring brass, ASTM B134* (70 percent Cu, 30 percent Zn) cannot be hardened by heat treatment and has generally quite poor spring qualities. Even though it does not tolerate temperatures higher than 150°F, it performs well at subzero. It is the least expensive copper-base spring material, with the highest electrical conductivity, outweighed by its low tensile strength.

*Phosphor bronze, ASTM B159* (95 percent Cu, 5 percent Sn). This is the most popular copper-based spring material. Its popularity is due to its favorable combination of electrical conductivity, corrosion resistance, good tensile strength, hardness, and low cost.

*Beryllium copper, ASTM B197* (98 percent Cu, 2 percent Be) is the most expensive material of this group. It is better formed in its annealed condition and precipitation hardened afterward. The hardened material turns brittle and does not take additional forming. The material has a high hardness and tensile strength. It is used where electrical conductivity is of importance.

**12-1-1-6. Nickel-base spring alloys.** These alloys take both extremes in temperature, extremely hot and extremely cold, while being corrosion-resistant. For their high resistance to electricity the materials should not be used with electric current. Their field of application lies with precise measuring instruments such as gyroscopes.

*Monel* (67 percent Ni, 30 percent Cu) cannot be hardened by heat treatment. Its high tensile strength and hardness is obtained through cold drawing and cold rolling. It is almost nonmagnetic and withstands stresses comparable to those beryllium copper can handle. It is the least expensive material of this group.

*K-Monel* (66 percent Ni, 29 percent Cu, 6 percent Al). The material is nonmagnetic, and the small amount of aluminum makes it a precipitation-hardening applicant. Otherwise it is very similar to previously described monel. It can be formed soft and hardened afterward by application of an age-hardening heat treatment.

*Inconel* (78 percent Ni, 14 percent Cr, 7 percent Fe) has higher tensile strength and hardness than K-monel, both of these properties being attributable to cold drawing and cold rolling, as it cannot be hardened by heat treatment. It can be used at temperatures of up to 700°F. It is a very popular alloy because of its corrosion resistance, even though its cost is higher than that of the stainless-steel group, yet not so costly as beryllium copper.

*Inconel-X* (70 percent Ni, 16 percent Cr, 7 percent Fe, with small amounts of titanium, columbium, and aluminum). This nonmagnetic material should be precipitation hardened at high temperatures. It is operable up to 850°F.

*Duranickel* (98 percent Ni) takes slightly lower temperatures than inconel. It is nonmagnetic, resistant to corrosion, and has a high tensile strength. It can be precipitation hardened.

### 12-1-2. Heat treatment of springs

Heat treatment of finished springs is done in two stages. First, following the forming process, a low-temperature heat treatment of 350 to 950°F (175 to 510°C) is applied. Such a treatment causes the material to stabilize dimensionally, while removing some residual stresses developed during the forming operation. Residual stresses come in two groups: Some of them are beneficial to the part's functionality; others are detrimental to it.

A second heat treatment is done at higher temperatures, ranging between 1480 and 1650°F (760 and 900°C). This heat treatment strengthens the material, which is still annealed after forming. Typical heat-treatment temperatures for specific materials are shown in Table 12-2. Usually, a 20 to 30-min-long exposure to these temperatures is considered adequate.

Hardened high-carbon steel parts, when electroplated, are prone to cracking. This is caused by the action of hydrogen atoms, which intermingle with the material's metallic lattice and affect its structure. Such an occurrence is called hydrogen embrittlement. To prevent hydrogen embrittlement in plated springs, heat treatment at low temperatures is used prior to plating, with a baking operation added after forming.

Beryllium copper is strengthened after forming by the application of an age-hardening process; with other materials, tempering may sometimes be utilized.

### 12-1-3. Corrosion resistance

Coatings (zinc, cadmium, and their alloys) are frequently utilized to prevent corrosion damage to springs. These coatings not only act as a blockade between the material and the outside environment. They also protect the part cathodically, often even when scratched or otherwise topically damaged.

Electroplating is another method of protection used with application of metallic coatings. This type of surface finish, however, causes hydrogen embrittlement to appear, and care should be taken to mini-

**TABLE 12-2 Typical Heat Treatment for Springs after Forming**

| Materials | Heat treatment | |
|---|---|---|
| | °C | °F |
| Patented and cold-drawn steel wire | 190–230 | 375–450 |
| Tempered steel wire: | | |
| Carbon | 260–400 | 500–750 |
| Alloy | 315–425 | 600–800 |
| Austenitic stainless-steel wire | 230–510 | 450–950 |
| Precipitation-hardening stainless wire (17-7 PH): | | |
| Condition C | 480/1 h | 900/1 h |
| Condition A to TH 1050 | 760/1 h, cool to 15° C, followed by 565/1 h | 1400/1 h, cool to 60°F, followed by 1050/1 h |
| Monel: | | |
| Alloy 400 | 300–315 | 575–600 |
| Alloy K500, spring temper | 525/4 h | 980/4 h |
| Inconel: | | |
| Alloy 600 | 400–510 | 750–950 |
| Alloy X-750: | | |
| No. 1 temper | 730/16 h | 1350/16 h |
| Spring temper | 650/4 h | 1200/4 h |
| Copper-base, cold-worked (brass, phosphor bronze, etc.) | 175–205 | 350–400 |
| Beryllium copper: | | |
| Pretempered (mill hardened) solution | 205 | 400 |
| annealed, temper rolled or drawn | 315/2–3 h | 600/2–3 h |
| Annealed steels: | | |
| Carbon (AISI 1050 to 1095) | 800–830* | 1475–1525* |
| Alloy (AISI 5160H 6150, 9254) | 830–885* | 1525–1625* |

*Time depends on heating equipment and section size. Parts are austenitized, then quenched and tempered to the desired hardness.

SOURCE: *Design Handbook,* 1987. Reprinted with permission from Associated Spring, Barnes Group, Inc., Dallas, TX.

mize the part's susceptibility. As a means of protection, there should be no stress points in the part, such as sharp corners, sharp bends, or sharp-cornered cuts. Hardness should be at the minimum allowable level, and residual stresses within the material should be relieved by application of the highest possible heat-treating temperatures. After plating, parts should be baked at low temperatures for approximately 2 to 3 h.

Mechanical plating offers an adequate amount of protection against corrosion and hydrogen embrittlement as well. Such surface treatment should be used for parts suffering from high residual stresses after the forming operation. Its drawback lies in the difficulties with plating of tight or inaccessible areas—all part surfaces must be well exposed and clean.

### 12-1-4. Fatigue and reliability

Fatigue is a process that develops slowly and insidiously over the span of three stages: (1) crack induction, (2) crack increase, and (3) failure of the material. It is obvious that fatigue is an irreversible process, detrimental to the functionability of the part. Its development is caused by the emergence of cyclic stresses, accompanied by plastic strains, so common in springs. It may also be caused by the quenching process.

Residual stresses, as found in the spring material after bending, may either increase or diminish its fatigue resistance. This variation in their influence is due to the fact that there actually are two types of residual stresses within the material.

Stresses which counterbalance those accompanying the spring operation are beneficial to the part's longevity. For example, in a compressed coil spring, where a residual tension is encountered at its core, some residual stresses of the compressive type should ideally be near its surface. A condition like this may create an environment within the material of the spring, allowing for increased loads and improving the spring's resistance to fatigue.

However, if the residual stresses are in another (opposite) direction, their contribution to the load-carrying capacity and fatigue resistance of the spring will be negative.

Favorable residual stresses are often introduced to the spring material by the spring manufacturer. After the first stress-relieving heat treatment, a slight plastic deformation is purposely caused to the parts, following the direction of the spring's own elastic deformation later in service. Unfortunately, such prestressing cannot be preformed with all springs, as its subsequent increase in production costs cannot always be justified.

Plated steel springs emerge from the plating operation free from residual stresses, which cannot be reintroduced afterward.

For removal of various residual stresses located near the surface, shot peening is utilized. This procedure, however, decreases the load-carrying capacity of the spring, as it lowers the material's yield strength.

*Reliability* is a fatigue-dependent value, where the decrease in the spring's reliability is always caused by defects produced by fatigue.

Reliability of springs operating at higher temperatures is negatively influenced by so-called *stress relaxation.* It is the decrease in the load-carrying capacity and deflecting capacity of a spring held or cycled under a load. Higher temperatures also affect the tensile strength, fatigue, and modulus of the material.

Stresses and high operating temperatures will in time produce stress relaxation in springs. In opposition to such an influence is the type of alloy: More alloyed materials were found less susceptible to the damage caused by temperature increases.

In static applications, the load-carrying ability of a spring may be impaired by its yield strength and resistance to stress relaxation. To increase the static load-carrying capacity, a longer than necessary spring length should be selected and precompressed to solid in assembly. This process is called *set removal* or *presetting* of the spring, and it may increase the load-carrying ability by 45 to 65 percent. By presetting the spring, favorable residual stresses are introduced into the material. Their type and direction correspond with the spring's own natural (elastic) deformation, attributable to its function.

## 12-2. Springs in Die Design

Types of springs most often used in die and fixture design are coil springs of the compression type. Marginally, extension coil springs and flat springs are utilized.

Compression springs are wound as an open helix (Fig. 12-1) with an open pitch to resist the compressive force applied against it. Overall shapes of these springs are most often straight and cylindrical. But variations in the outline and winding, such as barrel-shaped, conical, hourglass, and variable-pitch springs can be encountered (Fig. 12-2).

Extension springs form a tight helix, and their pitch is limited to the wire thickness (Fig. 12-3). Flat springs may come in many types and shapes (Fig. 12-4).

## 12-3. Helical Compression Springs

These are abundant in die and fixture design, being used to support spring pads, spring strippers, and other spring-loaded arrangements.

### 12-3-1. Spring-related terminology

A certain terminology has been developed over the years, describing various spring attributes, which is used throughout the industry. Terms like *spring diameter, mean diameter, pitch, squareness,* and *parallelism,* among others, are explained further in the text.

*Spring diameter* can be either the outside diameter (OD) or inside diameter (ID) or mean diameter ($D$) of the spring. Mean diameter is equal to the value of OD plus ID divided by 2. It is used for calculations of stress and deflection.

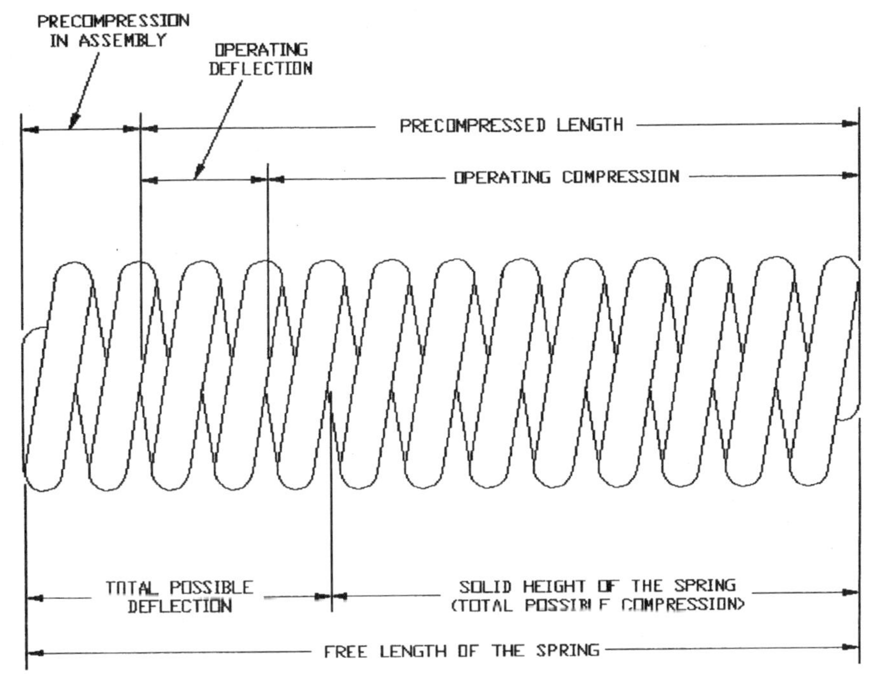


**Figure 12-1** Compression spring and its properties.

Where the OD is specified, the number is given with regard to the spring's working environment, in this case the cavity, where the spring would be retained. With specification of ID, the size of the coil-supporting pin, which is to fit inside the coil, is important.

Minimum clearances between the spring and its cavity or between the spring and the supporting pin (per diameter) are

$$0.10D \text{ where } D_{\text{cavity}} \text{ is less than } 0.512 \text{ in } (13 \text{ mm})$$

$$0.05D \text{ where } D_{\text{cavity}} \text{ is greater than } 0.512 \text{ in } (13 \text{ mm})$$

This is to allow for the increase in diametral size which occurs with the load application on the spring. This increase, seen as a bulging of the spring, is usually quite small, yet it must be taken into account if the function of the spring is not to be impaired. To calculate the increase in size, the following formula is provided:

$$\text{OD}_{\text{solid}} = \sqrt{D^2 + \frac{p^2 - d^2}{\pi^2}} + d \qquad (12\text{-}1)$$

where the values are as shown in Fig. 12-5.

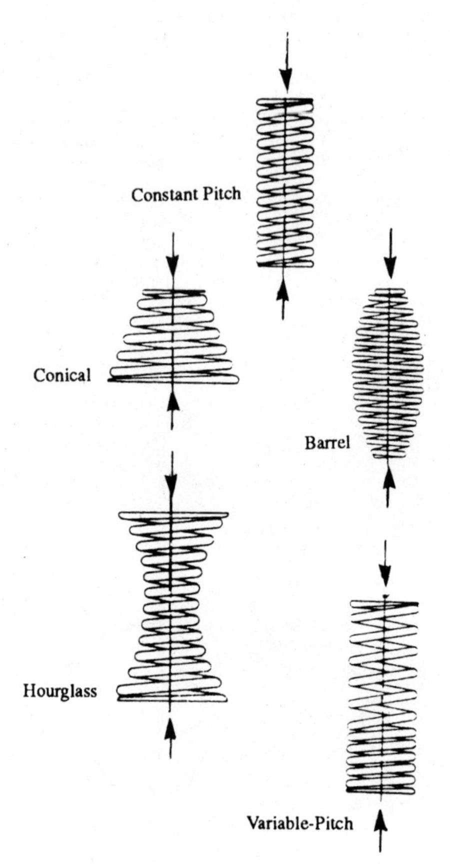


**Figure 12-2** Helical compression springs, round and rectangular wire. (*From* "Design Handbook," *1987. Reprinted with permission from Associated Spring, Barnes Group, Inc., Dallas, TX.*)

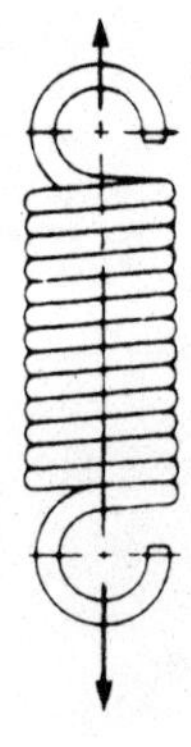

**Figure 12-3** Helical extension spring. (*From* "Design Handbook," *1987. Reprinted with permission from Associated Spring, Barnes Group, Inc., Dallas, TX.*)

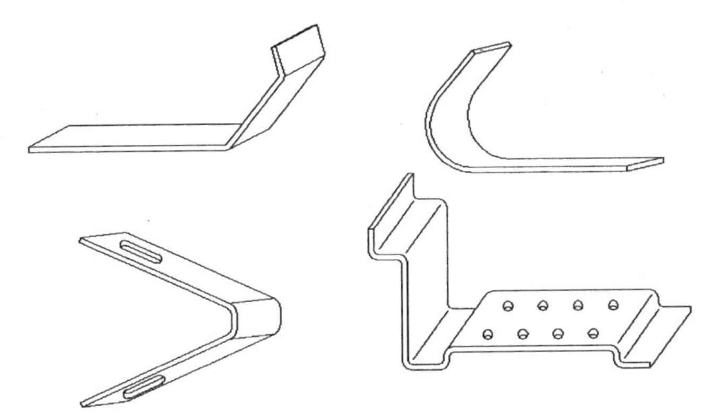

**Figure 12-4** Flat spring samples.

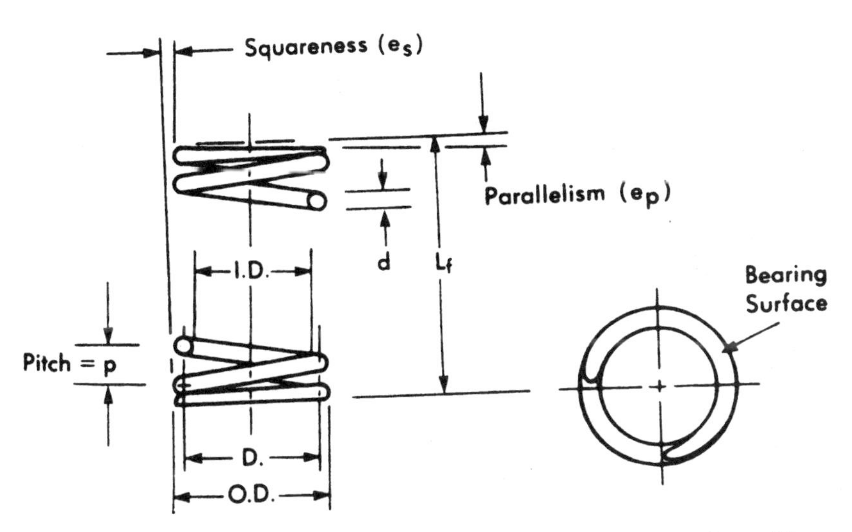


**Figure 12-5** Dimensional terminology for helical compression springs. (*From* "Design Handbook," *1987. Reprinted with permission from Associated Spring, Barnes Group, Inc., Dallas, TX.*)

**Buckling of compression springs.** Long springs may buckle unless they are supported by a pin coming through their center. Buckling may occur where the length of a spring unsupported by any pin exceeds the value of *four times its diameter.* Critical buckling conditions are given in Fig. 12-6. Critical buckling will occur with values to the right of each line.

Curve *A* depicts those springs whose one end is positioned against a flat plate while its other end is free to tip, as shown in Fig. 12-7. This

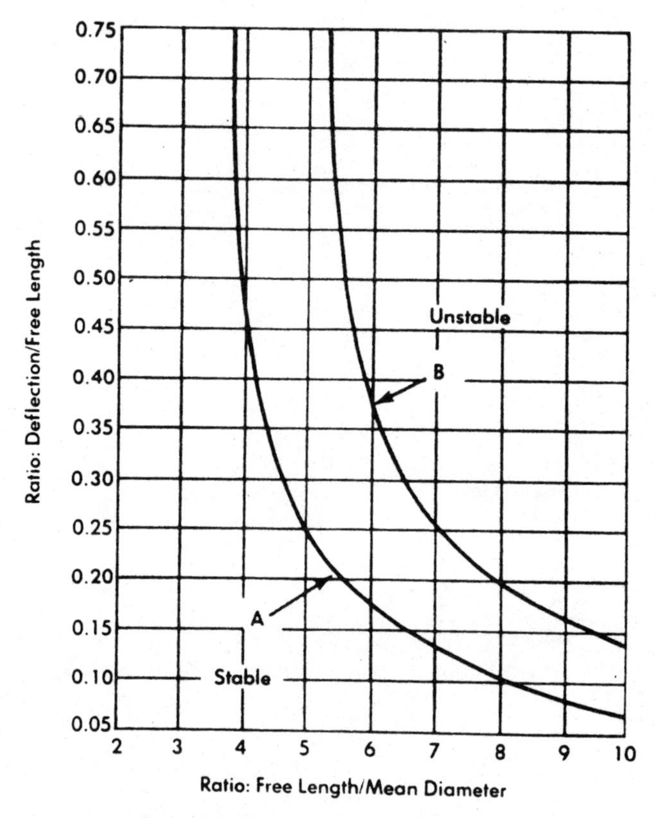


**Figure 12-6** Critical buckling condition curves. (*From* "Design Handbook," *1987. Reprinted with permission from Associated Spring, Barnes Group, Inc., Dallas, TX.*)

curve limits the occurrence of buckling conditions to the right and above its location.

Buckling occurrence is lower in springs retained between two parallel plates, as shown in section $B$ of Fig. 12-7. $B$-line buckling, as observed in the graph in Fig. 12-6, is lessened accordingly.

In cases where large deflections are required, several springs supported by an inner core consisting of rods or shoulder screws may be utilized.

*Spring index C* is the ratio of mean diameter to the wire diameter, or

$$C = \frac{D}{d} \tag{12-2}$$

With spring cross sections other than round, this formula is altered as shown in Fig. 12-8.

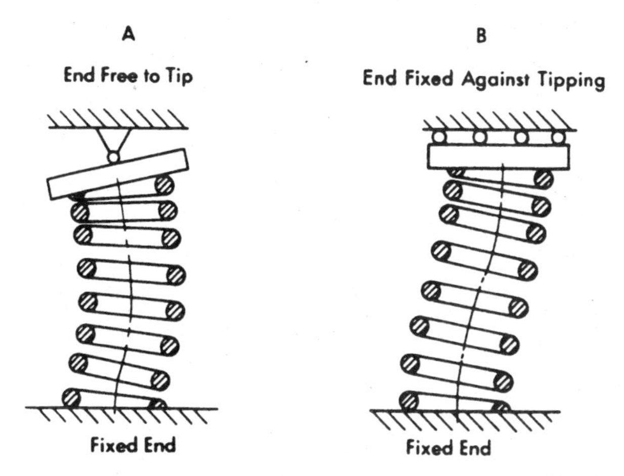


**Figure 12-7** End conditions used to determine critical buckling. (*From* "Design Handbook," *1987. Reprinted with permission from Associated Spring, Barnes Group, Inc., Dallas, TX.*)

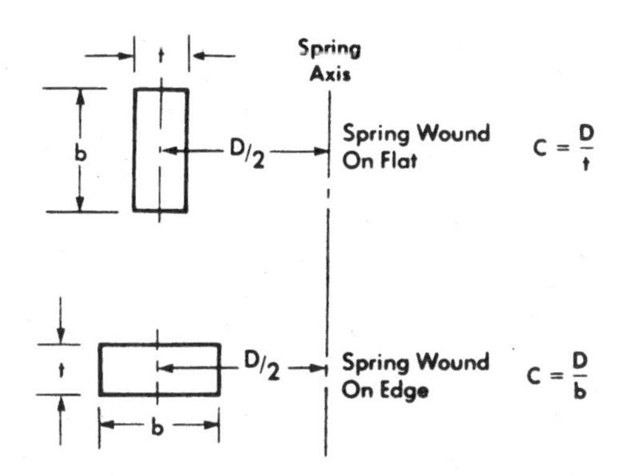


**Figure 12-8** Rectangular wire compression spring wound on flat or edge. (*From* "Design Handbook," *1987. Reprinted with permission from Associated Spring, Barnes Group, Inc., Dallas, TX.*)

The preferable spring index value is 4 to 12. Springs with high indexes may become tangled, requiring individual packaging for shipment, especially where their ends are not squared. Springs with indexes lower than 4 are difficult to form.

**Types of spring ends.** A wide variety of spring ends may be selected, such as plain ends, plain ends ground, square ends, and square ends ground (Fig. 12-9).

A bearing surface of at least 270° serves to reduce buckling. Squared and ground spring ends have a bearing surface of 270 to 330°. Additional grinding of these ends is undesirable, as it may result in further thinning of these sections.

Springs with squared ends only, where no grinding is involved, are naturally cheaper. This type of end should be reserved to springs with

- Wire diameters less than 0.020 in (0.5 mm)
- Index numbers greater than 12
- Low spring rates

**Number of active coils $N_a$.** Springs with squared ends have the number of active coils approximately equal to the total number of coils minus 2. Springs with plain ends usually have more of their coils active, the exact number being dependent on their seating method.

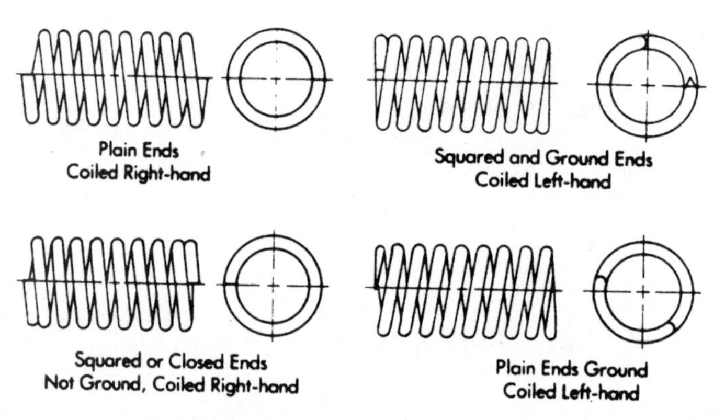


**Figure 12-9** Types of ends for helical compression springs. (*From* "Design Handbook," *1987. Reprinted with permission from Associated Spring, Barnes Group, Inc., Dallas, TX.*)

**TABLE 12-3 Guidelines for Dimensional Characteristics of Compression Springs**

| | Type of ends | | | |
|---|---|---|---|---|
| Dimensional characteristics | Open or plain (not ground) | Open or plain (ground) | Squared only | Squared and ground |
| Solid height $L_s$ | $(N_t + 1)d$ | $N_t d$ | $(N_t + 1)d$ | $N_t d$* |
| Pitch $p$ | $\frac{L_f - d}{N_a}$ | $\frac{L_f}{N_t}$ | $\frac{L_f - 3d}{N_a}$ | $\frac{L_f - 2d}{N_a}$ |
| Active coils $N_a$ | $\frac{L_f - d}{p}$ | $\frac{L_f}{p} - 1$ | $\frac{L_f - 3d}{p}$ | $\frac{L_f - 2d}{p}$ |
| Total coils $N_t$ | $N_a$ | $N_a + 1$ | $N_a + 2$ | $N_a + 2$ |
| Free length $L_f$ | $p N_t + d$ | $p N_t$ | $p N_a + 3d$ | $p N_a + 2d$ |

*For small index springs lower solid heights are possible.

SOURCE: *Design Handbook,* 1987. Reprinted with permission from Associated Spring, Barnes Group, Inc., Dallas, TX.

Guidelines for selection of end types for a particular spring application are given in Table 12-3.

**Solid height.** Solid height of a spring is the length of all coils, pressed together. Solid height of a ground spring can be obtained by multiplying the coil diameter by the number of coils. Nonground springs have solid height equal to the number of coils plus 1, multiplied by the wire diameter.

Plating or other coating will increase the solid height of a spring. The safe amount to add for such an increase is approximately one-half of the wire diameter.

**Direction of coiling.** Helical compression springs are either right- or left-hand wound (Fig. 12-10).

To assess the direction of coiling, the index finger of the right hand should be bent to resemble the shape of a coiled spring, with its tip ending in approximately the same location as the end of the coil. Such a spring, if matching the finger's arrangement, is right-hand-wound. An opposite-side arrangement is a left-hand-wound spring.

**Squareness and parallelism.** Squared and ground springs are usually square within 3° when measured in their free form. However, squareness of free springs may differ from those under a load.

Parallelism has a considerable effect on the function of an unsupported spring. For illustration of squareness and parallelism, refer to Fig. 12-5.

**Hysteresis.** Hysteresis means the loss of mechanical energy in a spring, which is exposed to cyclic loading and unloading within its elastic range. The known reason for such behavior is the friction

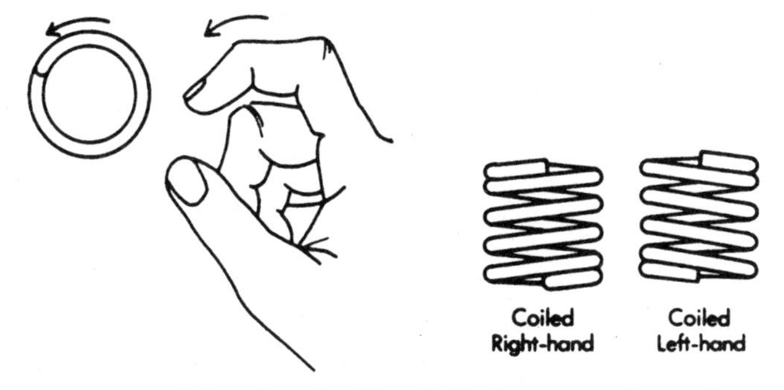


**Figure 12-10** Direction of coiling helical compression springs. (*From* "Design Handbook," *1987. Reprinted with permission from Associated Spring, Barnes Group, Inc., Dallas, TX.*)

between coils, or the friction between the spring and its support during compression.

### 12-3-2. Variable-diameter springs

These springs, as shown in Fig. 12-2, respectively conical, hourglass, and barrel-shaped springs, are utilized where the solid height of the spring must be low or where greater lateral stability and resistance to surging are needed.

The conical spring's solid height may be as low as one coil diameter, for these springs can be designed in such a fashion as to allow each coil to nest in the preceding coil. The spring rate can be made uniform by varying the pitch along the spring length.

When calculating the highest amount of stress at a predetermined load, the mean diameter of the largest active coil should be used.

Solid height of a spring $L_S$, made from a round wire, tapered in shape but not telescoping, with its ends squared and ground, can be estimated by using the formula

$$L_S = N_a \sqrt{d^2 - u^2} + 2d \tag{12-3}$$

where $u$ is equal to

$$u = \frac{\text{OD large end} - \text{OD small end}}{2N_a}$$

where $N_a$ is the number of active coils. To approach this calculation properly, each spring has to be considered to amount to several springs in series. Equation (12-11) may be used for such a purpose.

Barrel-shaped and hourglass springs can be calculated the same way, considering them to be two conical springs, which they incidentally are.

**Variable-pitch springs.** Variable-pitch springs are utilized where the natural spring frequency is near or corresponds with that of the cyclic rate of the load application. As coils of lesser pitch become inactive during the spring's function, the natural frequency of the spring will change. This will result in minimizing of surging and spring resonance. Spring resonance is addressed in Sec. 12-4-6.

### 12-3-3. Commercial tolerances

Standard commercial tolerances for free length of a spring, diameter, and load are presented in Tables 12-4, 12-5, and 12-6. These toler-

**TABLE 12-4 Free Length Tolerances of Squared and Ground Helical Compression Springs**

| Number of active coils per mm (in) | Tolerances: ± in/in of free length — Spring index ($D/d$) 4 | 6 | 8 | 10 | 12 | 14 | 16 |
|---|---|---|---|---|---|---|---|
| 0.02 (0.5) | 0.010 | 0.011 | 0.012 | 0.013 | 0.015 | 0.016 | 0.016 |
| 0.04 (1) | 0.011 | 0.013 | 0.015 | 0.016 | 0.017 | 0.018 | 0.019 |
| 0.08 (2) | 0.013 | 0.015 | 0.017 | 0.019 | 0.020 | 0.022 | 0.023 |
| 0.2 (4) | 0.016 | 0.018 | 0.021 | 0.023 | 0.024 | 0.026 | 0.027 |
| 0.3 (8) | 0.019 | 0.022 | 0.024 | 0.026 | 0.028 | 0.030 | 0.032 |
| 0.5 (12) | 0.021 | 0.024 | 0.027 | 0.030 | 0.032 | 0.034 | 0.036 |
| 0.6 (16) | 0.022 | 0.026 | 0.029 | 0.032 | 0.034 | 0.036 | 0.038 |
| 0.8 (20) | 0.023 | 0.027 | 0.031 | 0.034 | 0.036 | 0.038 | 0.040 |

For springs less than 12.7 mm (0.500 in) long, use the tolerances for 12.7 mm (0.500 in). For closed ends not ground, multiply above values by 1.7.

SOURCE: *Design Handbook,* 1987. Reprinted with permission from Associated Spring, Barnes Group, Inc., Dallas, TX.

**TABLE 12-5 Coil Diameter Tolerances of Helical Compression and Extension Springs**

| Wire dia, mm (in) | Tolerances: ±mm (in) — Spring index ($D/d$) 4 | 6 | 8 | 10 | 12 | 14 | 16 |
|---|---|---|---|---|---|---|---|
| 0.38 (0.015) | 0.05 (0.002) | 0.05 (0.002) | 0.08 (0.003) | 0.10 (0.004) | 0.13 (0.005) | 0.15 (0.006) | 0.18 (0.007) |
| 0.58 (0.023) | 0.05 (0.002) | 0.08 (0.003) | 0.10 (0.004) | 0.15 (0.006) | 0.18 (0.007) | 0.20 (0.008) | 0.25 (0.010) |
| 0.89 (0.035) | 0.05 (0.002) | 0.10 (0.004) | 0.15 (0.006) | 0.18 (0.007) | 0.23 (0.009) | 0.28 (0.011) | 0.33 (0.013) |
| 1.30 (0.051) | 0.08 (0.003) | 0.13 (0.005) | 0.18 (0.007) | 0.25 (0.010) | 0.30 (0.012) | 0.38 (0.015) | 0.43 (0.017) |
| 1.93 (0.076) | 0.10 (0.004) | 0.18 (0.007) | 0.25 (0.010) | 0.33 (0.013) | 0.41 (0.016) | 0.48 (0.019) | 0.53 (0.021) |
| 2.90 (0.114) | 0.15 (0.006) | 0.23 (0.009) | 0.33 (0.013) | 0.46 (0.018) | 0.53 (0.021) | 0.64 (0.025) | 0.74 (0.029) |
| 4.34 (0.171) | 0.20 (0.008) | 0.30 (0.012) | 0.43 (0.017) | 0.58 (0.023) | 0.71 (0.028) | 0.84 (0.033) | 0.97 (0.038) |
| 6.35 (0.250) | 0.28 (0.011) | 0.38 (0.015) | 0.53 (0.021) | 0.71 (0.028) | 0.90 (0.035) | 1.07 (0.042) | 1.24 (0.049) |
| 9.53 (0.375) | 0.41 (0.016) | 0.51 (0.020) | 0.66 (0.026) | 0.94 (0.037) | 1.17 (0.046) | 1.37 (0.054) | 1.63 (0.064) |
| 12.70 (0.500) | 0.53 (0.021) | 0.76 (0.030) | 1.02 (0.040) | 1.57 (0.062) | 2.03 (0.080) | 2.54 (0.100) | 3.18 (0.125) |

SOURCE: *Design Handbook,* 1987. Reprinted with permission from Associated Spring, Barnes Group, Inc., Dallas, TX.

**TABLE 12-6 Load Tolerances of Helical Compression Springs**

| Length tolerance, ±mm (in) | Tolerances: ±% of load. Start with tolerance from Table 12-4, multiplied by $L_F$ | | | | | | | | | | | | | | |
|---|---|---|---|---|---|---|---|---|---|---|---|---|---|---|---|
| | Deflection from free length to load, mm (in) | | | | | | | | | | | | | | |
| | 1.27 (0.050) | 2.54 (0.100) | 3.81 (0.150) | 5.08 (0.200) | 6.35 (0.250) | 7.62 (0.300) | 10.2 (0.400) | 12.7 (0.500) | 19.1 (0.750) | 25.4 (1.00) | 38.1 (1.50) | 50.8 (2.00) | 76.2 (3.00) | 102 (4.00) | 152 (6.00) |
| 0.13 (0.005) | 12 | 7 | 6 | 5 | | | | | | | | | | | |
| 0.25 (0.010) | | 12 | 8.5 | 7 | 6.5 | 5.5 | 5 | | | | | | | | |
| 0.51 (0.020) | | 22 | 15.5 | 12 | 10 | 8.5 | 7 | 6 | 5 | | | | | | |
| 0.76 (0.030) | | | 22 | 17 | 14 | 12 | 9.5 | 8 | 6 | 5 | | | | | |
| 1.0 (0.040) | | | | 22 | 18 | 15.5 | 12 | 10 | 7.5 | 6 | 5 | | | | |
| 1.3 (0.050) | | | | | 22 | 19 | 14.5 | 12 | 9 | 7 | 5.5 | | | | |
| 1.5 (0.060) | | | | | 25 | 22 | 17 | 14 | 10 | 8 | 6 | 5 | | | |
| 1.8 (0.070) | | | | | | 25 | 19.5 | 16 | 11 | 9 | 6.5 | 5.5 | | | |
| 2.0 (0.080) | | | | | | | 22 | 18 | 12.5 | 10 | 7.5 | 6 | 5 | | |
| 2.3 (0.090) | | | | | | | 25 | 20 | 14 | 11 | 8 | 6 | 5 | | |
| 2.5 (0.100) | | | | | | | | 22 | 15.5 | 12 | 8.5 | 7 | 5.5 | | |
| 5.1 (0.200) | | | | | | | | | | 22 | 15.5 | 12 | 8.5 | 7 | 5.5 |
| 7.6 (0.300) | | | | | | | | | | | 22 | 17 | 12 | 9.5 | 7 |
| 10.2 (0.400) | | | | | | | | | | | | 21 | 15 | 12 | 8.5 |
| 12.7 (0.500) | | | | | | | | | | | | 25 | 18.5 | 14.5 | 10.5 |

First load test at not less than 15% of available deflection.
Final load test at not more than 85% of available deflection.
SOURCE: *Design Handbook,* 1987. Reprinted with permission from Associated Spring, Barnes Group, Inc., Dallas, TX.

ances are a good combination of the manufacturing costs and the spring's quality and performance.

The squareness tolerances, as noted, is 3°. Spring life is presented in lieu of fatigue values.

## 12-4. Calculation of Compression Springs

All spring design begins with the application of Hooke's law. This law states that any force acting upon the material is directly proportional to the material's deflection, provided such deflection is within the range of that material's elastic limit.

### 12-4-1. Stress calculation

Compression springs made of round wire subject this wire to a stress classified as a torsional stress. The basic formula to calculate such a stress $S$ is, according to Bernoulli-Euler,

$$S = \frac{Mc}{J} \qquad (12\text{-}4)$$

where $c$ = distance from neutral axis at center of section to outside of material, or one-half of material thickness for a round wire
$J$ = polar moment of inertia

Polar moment of inertia for a round section is

$$J = \frac{\pi d^4}{32}$$

where $d$ = wire diameter
$M$ = torsional moment, calculated as follows:

$$M = PR = \frac{PD}{2}$$

where $P$ = load on spring, lb
$D$ = mean diameter

Adding a stress-correcting factor $K_W$ changes this formula to

$$S = \frac{8PD}{\pi d^3} K_W = \frac{2.546PD}{d^3} K_W \qquad (12\text{-}5)$$

The sudden emergence of the stress-correcting factor $K_W$ is due to the nonuniform distribution of torsional stress across the cross section of the wire. This is caused by the curvature of the coil and a direct shear load. Maximum torsional stress can be found at the inner surface of the spring, and its value is assessed with the aid of stress-correcting factor $K_{W1}$ or $K_{W2}$ (see Table 12-7), attributable to Dr. A. M. Wahl of Westinghouse Electric Co. The formula to calculate this correction factor is as follows:

$$K_{W1} = \frac{4C - 1}{4C - 4} + \frac{0.615}{C} \qquad (12\text{-}6)$$

In some conditions, where resultant stresses are distributed more uniformly around the cross section, the stress-correcting factor $K_{W2}$ can be used as

$$K_{W2} = 1 + \frac{0.5}{C} \qquad (12\text{-}7)$$

Where elevated temperatures are encountered in the spring-operating environment, the stress distribution is more uniform around the cross section and can therefore be estimated, referring to Fig. 12-11.

Maximum allowable torsional stresses for helical compression springs in static applications are as listed in Table 12-7.

### 12-4-2. Diameter of wire *d*

To choose the proper wire diameter for a given load, at an assumed stress, Eq. (12-8*a*) may be used.

$$\text{If } S = \frac{2.546PD}{d^3} \qquad \text{then} \qquad d = \sqrt[3]{\frac{2.546PD}{S}} \qquad (12\text{-}8a)$$

Knowing other pertinent values, we may calculate the spring diameter by using the formula for a round wire:

$$d = \frac{\pi D^2 S\, N_a}{Gf\, K_W} \qquad (12\text{-}8b)$$

where $N_a$ = number of active springs
$G$ = modulus of rigidity
$f$ = deflection, in
$K_W$ = Wahl's correction factor

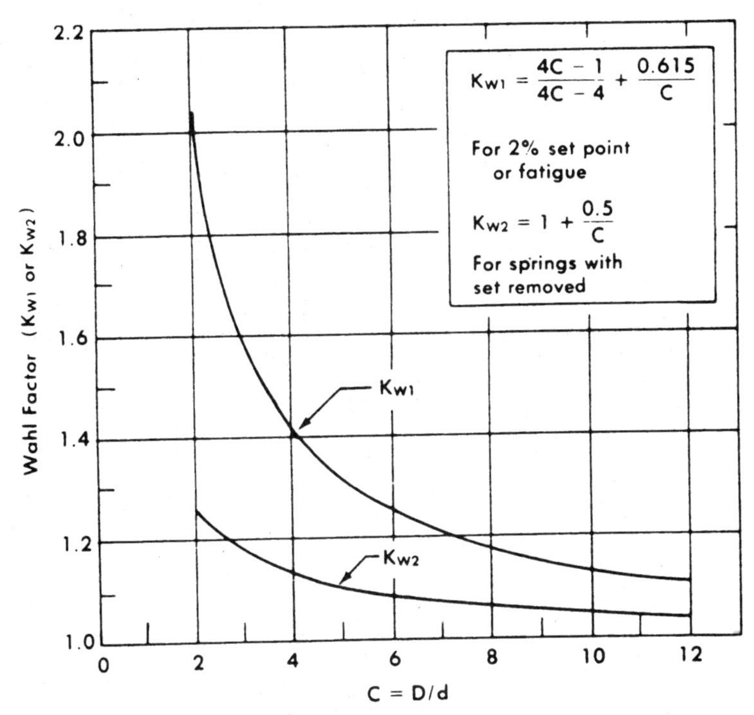


**Figure 12-11** Wahl stress-correction factors for round wire helical compression and extension springs. (*From* "Design Handbook," *1987. Reprinted with permission from Associated Spring, Barnes Group, Inc., Dallas, TX.*)

**TABLE 12-7 Maximum Allowable Torsional Stresses for Helical Compression Springs in Static Applications**

| Materials | Max % of tensile strength | |
|---|---|---|
| | Before set removed ($K_{W1}$) | After set removed ($K_{W2}$) |
| Patented and cold-drawn carbon steel | 45 | 60–70 |
| Hardened and tempered carbon and low-alloy steel | 50 | 65–75 |
| Austenitic stainless steels | 35 | 55–65 |
| Nonferrous alloys | 35 | 55–65 |

Bending or buckling stresses are not included.
SOURCE: *Design Handbook,* 1987. Reprinted with permission from Associated Spring, Barnes Group, Inc., Dallas, TX.

whereas a wire of square cross section can be figured out as

$$d = \frac{2.32\, SD^2 N_a}{Gf K_W} \qquad (12\text{-}8c)$$

where $d$ = length of square side of coil

### 12-4-3. Deflection *f*

Accordingly, the deflection can be assessed on the basis of previous information by using the stress formula described previously:

$$f = \frac{8PD^3N_a}{Gd^4} \tag{12-9}$$

where $N_a$ = number of active coils
$G$ = modulus of rigidity, usually around 11,500,000 lb/in$^2$ for steel wire

The modulus of rigidity, also called the modulus of shear, differs from the modulus of elasticity in that it produces an angular shift in the material's atomic structure. The modulus of rigidity and modulus of elasticity are related:

$$E = 2G(1 + \mu) \tag{12-10}$$

where $\mu$ = Poisson's ratio

### 12-4-4. Spring rate *k*

For helical compression springs, the spring rate is the change in load per unit of deflection:

$$k = \frac{P}{f} = \frac{Gd^4}{8D^3N_a} \tag{12-11}$$

where $P$ = load on spring, lb

Where compression springs are used in parallel, the total rate is equal to the sum of the rates of individual springs. The sum of the rates of compression springs in series is calculable as

$$k = \frac{1}{1/k_1 + 1/k_2 + 1/k_3 + \cdots + 1/k_n}$$

### 12-4-5. Dynamic loading, suddenly applied load

Often, not only the influence of slowly applied loads should be figured out with springs. Suddenly applied loads can have a tremendous impact on the life and performance of a spring. Since the load velocity is usually not exactly known, bodies may contain an unknown amount of kinetic energy.

**TABLE 12-8 Typical Properties of Spring Temper Alloy Steel**

| Material | Tensile strength, MPa ($10^3$ lb/in$^2$) | Rockwell hardness | Elongation,* % | Bend factor* ($2r/t$ transverse bends) | Modulus of elasticity, $10^4$ MPa ($10^6$ lb/in$^2$) | Poisson's ratio |
|---|---|---|---|---|---|---|
| Steel, spring temper | 1700 (246) | C50 | 2 | 5 | 20.7 (30) | 0.30 |
| Stainless 301 | 1300 (189) | C40 | 3 | 3 | 19.3 (28) | 0.31 |
| Stainless 302 | 1300 (189) | C40 | 5 | 4 | 19.3 (28) | 0.31 |
| Monel 400 | 690 (100) | B95 | 2 | 5 | 17.9 (26) | 0.32 |
| Monel K500 | 1200 (174) | C34 | 40 | 5 | 17.9 (26) | 0.29 |
| Inconel 600 | 1040 (151) | C30 | 2 | 2 | 21.4 (31) | 0.29 |
| Inconel X-750 | 1050 (152) | C35 | 20 | 3 | 21.4 (31) | 0.29 |
| Copper-beryllium | 1300 (189) | C40 | 2 | 5 | 12.8 (18.5) | 0.33 |
| Ni-span-C | 1400 (203) | C42 | 6 | 2 | 18.6 (27) | |
| Brass CA 260 | 620 (90) | B90 | 3 | 3 | 11.0 (16) | 0.33 |
| Phosphor bronze | 690 (100) | B90 | 3 | 2.5 | 10.3 (15) | 0.20 |
| 17-7 PH RH950 | 1450 (210) | C44 | 6 | Flat | 20.3 (29.5) | 0.34 |
| 17-7 PH condition C | 1650 (239) | C46 | 1 | 2.5 | 20.3 (29.5) | 0.34 |

*Before heat treatment.

SOURCE: *Design Handbook*, 1987. Reprinted with permission from Associated Spring, Barnes Group, Inc., Dallas, TX.

Considering that work done on springs ( = force × space) equals the energy absorbed by the spring when neglecting the hysteresis, the solution is as follows:

For loads applied very slowly:

$$f = \frac{P}{k} \tag{12-12}$$

For loads applied suddenly:

$$f = \frac{2P}{k} \tag{12-13}$$

For loads dropped from a certain height:

$$f = \sqrt{\frac{2P(s + f)}{k}} \tag{12-14}$$

where $f$ = deflection, in
$k$ = spring rate
$P$ = load on the spring, lb
$s$ = height from which the load was dropped, in

**Dynamic loading, impact.** With a spring being cyclically loaded and unloaded, an emergence of surge wave provides for the transmission of torsional stress from the point of loading application to the point of restraint. This surge wave travels at a velocity one-tenth the velocity of a normal torsion stress wave. Velocity of the torsion stress wave $V_T$ can be calculated:

$$V_T = \sqrt{\frac{Gg}{\rho}} \quad \text{in/s} \tag{12-15a}$$

which in metric version becomes

$$V_T = 10.1\sqrt{\frac{Gg}{\rho}} \quad \text{m/s} \tag{12-15b}$$

where $\rho$ = density, 1/1365 for steel
$g$ = acceleration due to gravity, 32 ft/s$^2$, or 9.8 m/s$^2$

This surge wave limits the springs' absorption and release of energy by restricting its impact velocity $V$, which is a function of stress and material constants, applied in parallel with the spring axis.

Impact velocity may be calculated as follows:

$$V \simeq S \sqrt{\frac{g}{2\rho G}} \qquad \text{in/s} \tag{12-16a}$$

and in metric:

$$V \simeq 10.1S \sqrt{\frac{g}{2\rho G}} \qquad \text{m/s} \tag{12-16b}$$

Impact velocity and stress are actually independent of the configuration of the spring. For steel materials, impact velocity should be in the range of

$$V = \frac{S}{131} \quad \text{in/s} \qquad \text{or} \qquad V = \frac{S}{35.5} \quad \text{m/s} \tag{12-17}$$

### 12-4-6. Dynamic loading, resonance

Springs have a natural inclination to vibration, creating a resonance within their mass. Resonance occurs where the cyclic loading approaches the natural frequency of the spring or its multiples. Resonance may increase the coil deflection and stress level, exceeding all assumed amounts. It can cause the spring to shiver and bounce, with subsequent alteration of its load-carrying capacity and other values.

A natural spring's frequency must be at least 13 times greater than its operating frequency to prevent the emergence of resonance.

The compression spring's natural frequency is inversely proportional to the amount of time needed for a surge wave to traverse the spring. For a spring which has no damping and has both ends fixed, this amounts to

$$n = \frac{d}{9D^2N_a} \sqrt{\frac{Gg}{\rho}} \tag{12-18a}$$

where the value of $n$ for steel is

$$n = \frac{14{,}000d}{N_a D^2} \tag{12-19a}$$

In metric translation, this calculation becomes

$$n = \frac{1.12 \times 10^3 d}{D^2 N_a} \sqrt{\frac{Gg}{\rho}} \tag{12-18b}$$

where the value of $n$ for steel is

$$n = \frac{3.5 \times 10^5 d}{D^2 N_a} \qquad (12\text{-}19b)$$

where $n$ = natural frequency, Hz
$\rho$ = density, 1/1365 for steel
$g$ = acceleration due to gravity, 32 ft/s², or 9.8 m/s²

## 12-5. Special Cross Sections of the Wire

Springs whose cross section is rectangular in shape and oriented with the width of the rectangle perpendicular to the spring axis have a capacity to absorb more work energy in smaller space than equivalent round wire. This is true despite the fact that the distribution of stress around the rectangular section may not be quite as uniform as that of the round wire. Rectangular-shaped wire is also more costly than round wire, with the keystoned wire being the most expensive of the three (Fig. 12-12).

The coiling operation, when applied to a rectangular wire, alters its shape, slanting the rectangle against one of the sides. Keystone wire is manufactured to come out from coiling rectangular.

Rectangular-shaped wire springs can be calculated with slightly altered round-wire formulas. The rate for such a compression spring is

$$k = \frac{P}{f} = \frac{Gbh^3}{N_a D^3} K_2 \qquad (12\text{-}20)$$

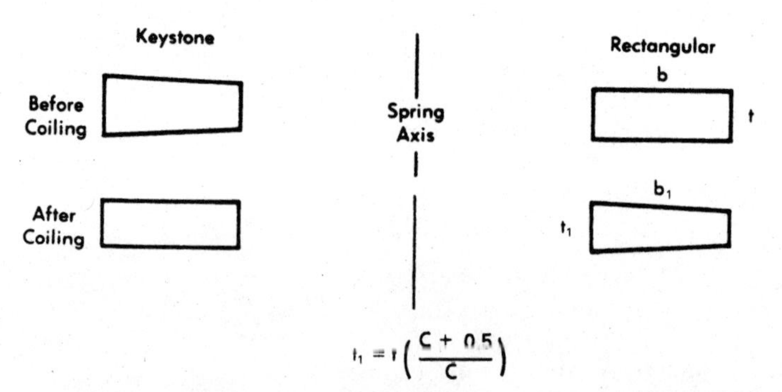


**Figure 12-12** Wire cross section before and after coiling. (*From* "Design Handbook," *1987. Reprinted with permission from Associated Spring, Barnes Group, Inc., Dallas, TX.*)

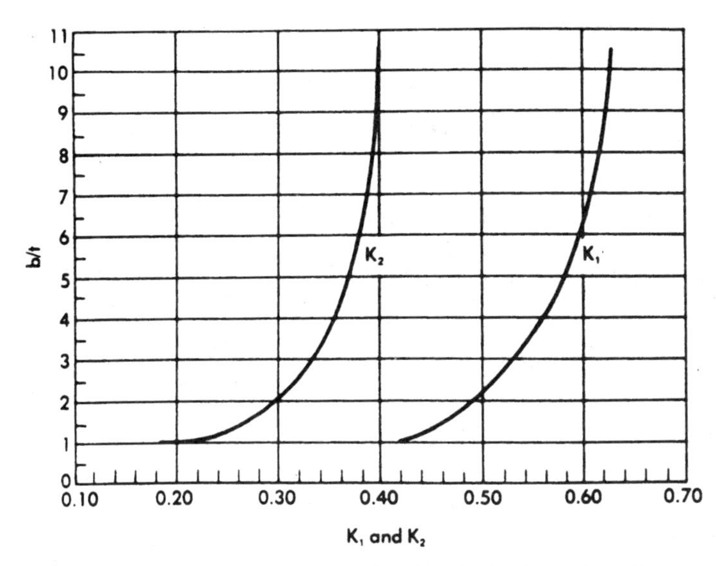


**Figure 12-13** Constants for rectangular wire in torsion. (*From* "Design Handbook," *1987. Reprinted with permission from Associated Spring, Barnes Group, Inc., Dallas, TX.*)

Since the wire is torsionally loaded, the rate is equal if the wire is wound on flat or on edge (see Fig. 12-8). Values of constants $K_1$ and $K_2$ are as shown in Fig. 12-13.

The stress in such a spring can be obtained by using the formula given in Eq. (12-21).

$$S = \frac{PD}{K_1 bh^2} K_E \quad \text{or} \quad \frac{PD}{K_1 bh^2} K_F \tag{12-21}$$

where $K_E$ = stress-correcting factor for springs wound on edge, shown in Fig. 12-14

$K_F$ = stress-correcting factor for springs wound on flat, shown in Fig. 12-15

Where an attempt is made to produce a rectangular-shaped wire by rolling a round material, or where the cross-sectional shape of the wire is not quite round, a correction factor $h'$ should be added to the stress formulas above, to replace $h$:

$$h' = \frac{2d}{1 + b/h} \tag{12-22}$$

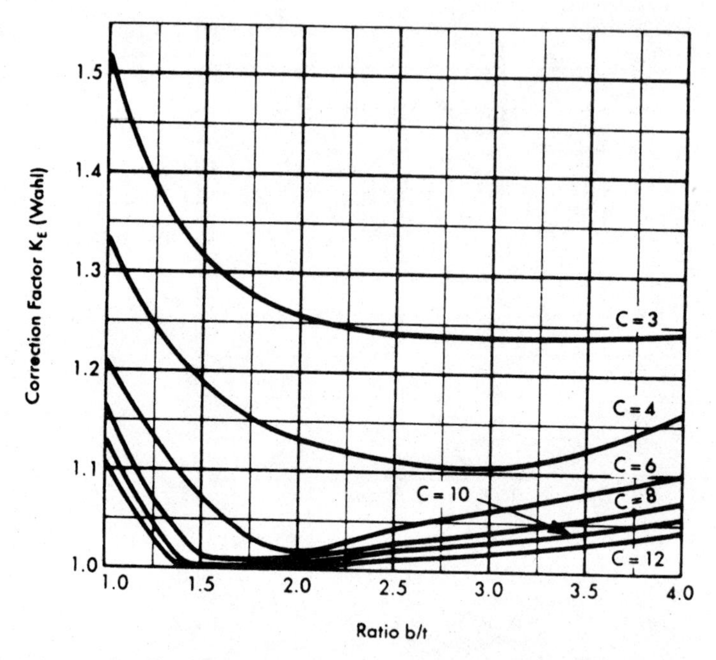


**Figure 12-14** Stress-correction factors for rectangular wire compression springs wound on edge. (*From* "Design Handbook," *1987. Reprinted with permission from Associated Spring, Barnes Group, Inc., Dallas, TX.*)

To figure out the amount of stress and deflection, a triangular cross section of the wire would utilize the formulas

$$S = \frac{20PR}{l^3} \tag{12-23}$$

where $l$ = length of each side of the triangle (see Fig. 12-16)
$R$ = mean radius of the coil

$$f = \frac{290.3PN_aR^3}{Gl^4} \tag{12-24}$$

Hexagonal cross sections, where the inscribed circle's diameter (as shown in Fig. 12-16) is $v$ and the area of the cross section is $A$, can be calculated by using the following formulas:

$$S = \frac{PR}{0.217Av} \tag{12-25}$$

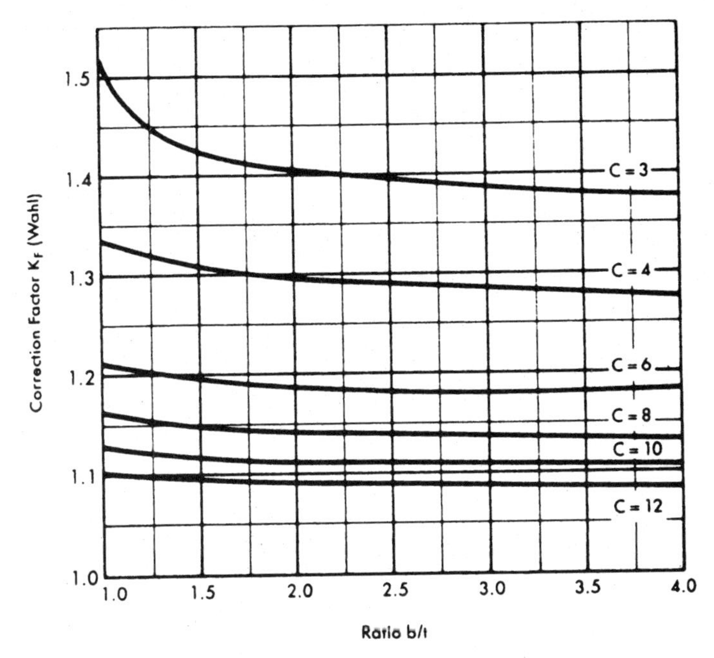


**Figure 12-15** Stress-correction factors for rectangular wire compression springs wound on flat. (*From* "Design Handbook," *1987. Reprinted with permission from Associated Spring, Barnes Group, Inc., Dallas, TX.*)

and

$$f = \frac{47.24PN_aR^3}{Gv^2A} \qquad (12\text{-}26)$$

For octagonal sections (Fig. 12-16), where the inscribed circle's diameter is $v$ and the area of the cross section is $A$, the formula is

$$S = \frac{PR}{0.223Av} \qquad (12\text{-}27)$$

and

$$f = \frac{48.33PN_aR^3}{Gv^2A} \qquad (12\text{-}28)$$

Figure 12-16 shows triangular, hexagonal, and octagonal cross sections. For a regular elliptical section,

$$S = \frac{16PR}{\pi x^2 y} \qquad (12\text{-}29)$$

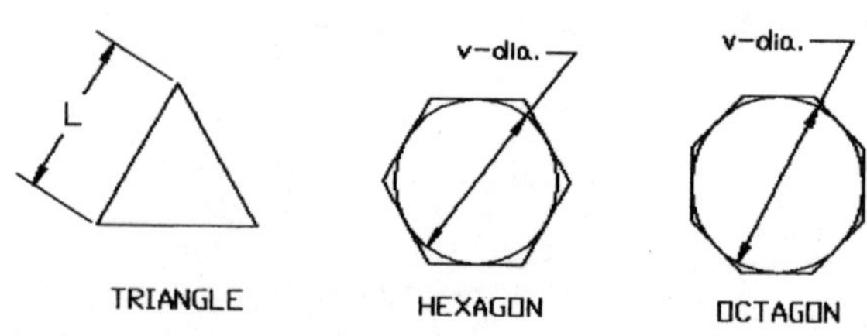


**Figure 12-16** Cross-sectional shapes.

and

$$f = \frac{248.1\, P N_a R^3 J}{G A^4} \tag{12-30}$$

where $x$ = minor axis of ellipse
$y$ = major axis of ellipse
$A$ = cross-sectional area of ellipse
$J$ = polar moment of inertia of the section, which can be calculated as

$$J = \frac{\pi(xy^3 + x^3y)}{64} \tag{12-31}$$

and subsequently,

$$A = \frac{\pi xy}{4}$$

## 12-6. Hot-Wound Springs

Most often, springs are cold-formed up to 3/8 in (10 mm) diameter of the wire or bar size. After this dimension, cold forming becomes difficult, and hot winding of springs is used instead. This type of spring manufacture involves heating of the steel up to the austenitic range, winding it, quenching down to martensitic structure, and tempering to arrive at required properties.

The most often used type of hot-wound spring is the compression spring, utilized as a part of an automobile suspension system or springs used in rail cars.

Marginally some extension, torsion, and volute springs may be hot-wound as well.

### 12-6-1. Design and calculations

Design parameters and calculations for this type of spring are the same as those of other springs. The only exception is that of the spring rate calculation, which here includes an empirical factor $K_H$, providing for the adjustment due to scaling-caused complications.

$$k = \frac{P}{f} = \frac{Gd^4K_H}{8D^3N_a} \tag{12-32}$$

where $k$ = spring rate
$P$ = load, lb
$f$ = deflection, in
$G$ = modulus of rigidity
$N_a$ = number of active threads
$K_H$ = 0.91 for hot-rolled carbon or low-alloy steel materials, which are not centerless ground
= 0.96 for hot-rolled carbon or low-alloy steel materials, centerless ground
= 0.95 for carbon or low-alloy steel material on torsion springs

### 12-6-2. Types of spring ends

The ends of hot-wound springs may be ground, tapered, tangent, or pigtailed, as shown in Fig. 12-17.

Ground ends provide the spring with a good bearing surface and unsurpassed squareness.

Tapered ends are produced by rolling a taper on the bar. During hot winding, these ends must be guided to provide for their proper orientation. Additional grinding improves the spring's squareness and bearing surface quality.

Tangent ends are standard, with no secondary manufacturing procedures involved. Because of the hot-winding process, springs with tangent ends have a straight portion approximately two wire diameters in size at each end. Their bearing surface must be designed in accordance with the requirements of such a shape, as it tends to exceed the outline of the spring.

Pigtailed ends are formed along with the hot-winding process. These ends are popular in situations where the spring must be clamped or bolted to its seat.

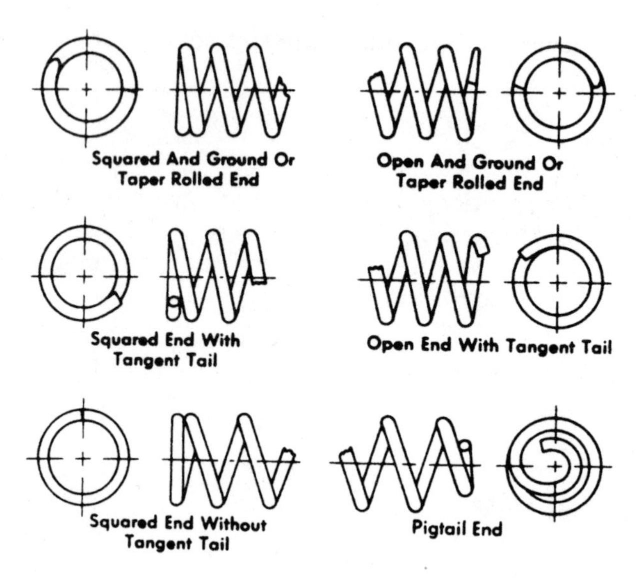


**Figure 12-17** Typical ends of hot-wound compression springs. (*From* "Design Handbook," *1987. Reprinted with permission from Associated Spring, Barnes Group, Inc., Dallas, TX.*)

### 12-6-3. Hot-wound, noncompression springs

Extension and torsion springs, utilizing loops and legs, have these additions to their shapes formed at the same time the spring is wound. For that reason, such shapes should be kept as simple as possible. All elaborate designs that are difficult to achieve involve reaustenitizing of the spring, with subsequent increase in manufacturing costs.

## 12-7. Helical Extension Springs

Helical extension springs are used where a pulling force is needed. They are most often made from round wire, closely wound, with initial tension obtained through stressing them in torsion.

### 12-7-1. Design of extension springs

Design procedures are the same as those of compression springs. It should be remembered, though, that extension springs, when compared to compression-type springs, are slightly different in that they are

- Coiled with initial tension, equal to the minimal amount of force needed to separate adjacent coils

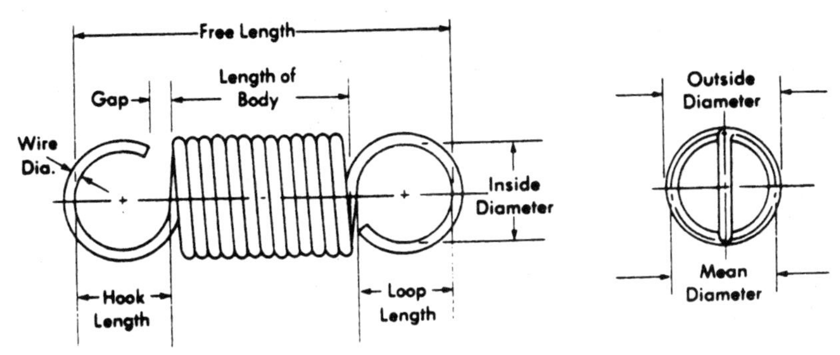


**Figure 12-18** Typical extension spring dimensions. (*From* "Design Handbook," *1987. Reprinted with permission from Associated Spring, Barnes Group, Inc., Dallas, TX.*)

- Coiled (usually) without their set being removed
- Equipped with no fixed stop to prevent their overloading

Figure 12-18 shows typical extension spring dimensions.

**Spring rate of extension springs *k*.** Same as for helical compression springs, the spring rate of the extension spring is the change in load per unit of deflection. The formula to be used with extension springs is, however, slightly different:

$$k = \frac{P - P_I}{f} = \frac{Gd^4}{8D^3N_a} \tag{12-33}$$

where $P_I$ is the initial tension. The applicable stress is given by the formula

$$S = \frac{8PD}{\pi d^3} K_W \tag{12-34}$$

### 12-7-2. Types of spring ends

A wide variety of ends is utilized with this type of spring as a provision for their attachment to other parts of the assembly. There are twist loops, side loops, cross-center loops, and extended hooks. Loops differ from hooks in that their shape will form small gaps, whereas hooks are actually loops with large gaps.

Special types of ends are formed from straight sections of wire tangent to the spring body shape.

Naturally, common loops of standard lengths are most economical to obtain. Figure 12-19 shows common end configurations for helical extension springs.

| Type | Configurations | Recommended Length* Min.- Max. |
|---|---|---|
| Twist Loop or Hook | 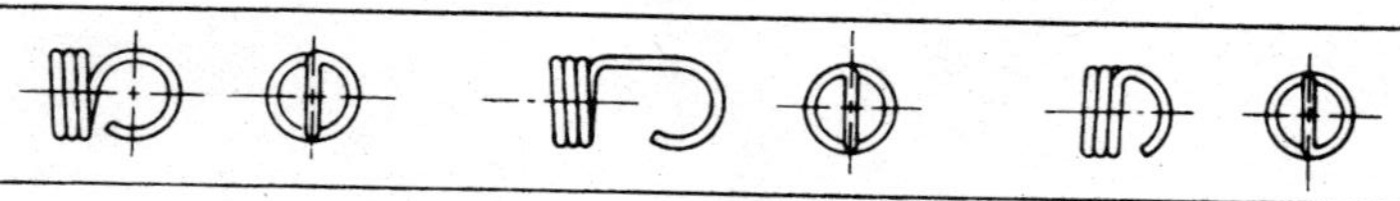 | 0.5-1.7 I.D. |
| Cross Center Loop or Hook | 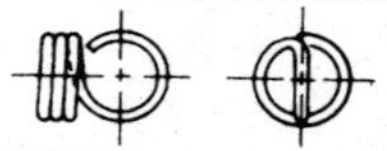 | I.D. |
| Side Loop or Hook | 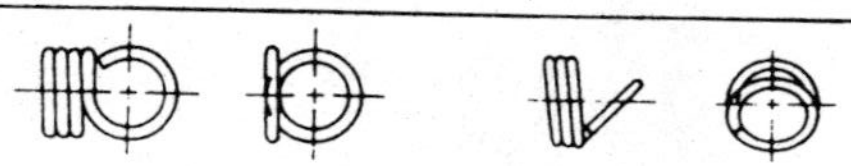 | 0.9-1.0 I.D. |
| Extended Hook | 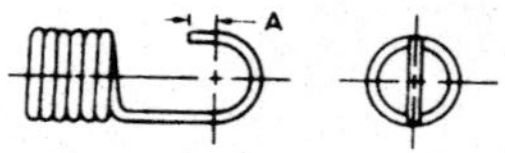 A | 1.1 I.D. and up, as required by design |
| Special Ends | 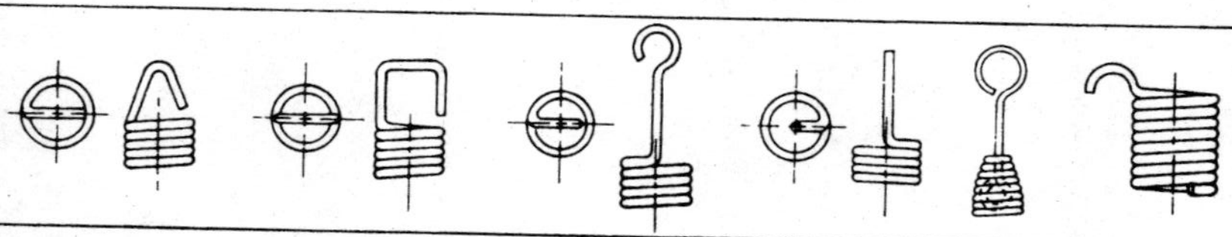 | As required by design |

* Length is distance from last body coil to inside of end. I.D. is inside diameter of adjacent coil in spring body.

**Figure 12-19** Common end configurations for helical extension springs. (*From* "Design Handbook," *1987. Reprinted with permission from Associated Spring, Barnes Group, Inc., Dallas, TX.*)

Stresses in loops or hooks are often higher than those within the spring wound-up body itself. For that reason, liberal bend radii in loops, combined with reduced end coil diameters, should be used to alleviate this problem.

The stress encountered in a full twist loop may reach its maximal value in bending at point $A$ (shown in Fig. 12-20), while the value of the maximum stress in torsion is at its highest at point $B$.

To assess the actual amount of stress at these two locations, the following formulas should be used:

$$S_A = \frac{16DP}{\pi d^3} K_1 + \frac{4P}{\pi d^2} \qquad \text{bending} \tag{12-35a}$$

where

$$K_1 = \frac{4C_1^{\,2} - C_1 - 1}{4C_1 (C_1 - 1)} \tag{12-36}$$

and

$$C_1 = \frac{2R_1}{d} \tag{12-37a}$$

and

$$S_B = \frac{8DP}{\pi d^3} \frac{4C_2 - 1}{4C_2 - 4} \tag{12-35b}$$

and

$$C_2 = \frac{2R_2}{d} \qquad \text{torsion} \tag{12-37b}$$

It is recommended that the value of $C_2$ be greater than 4.

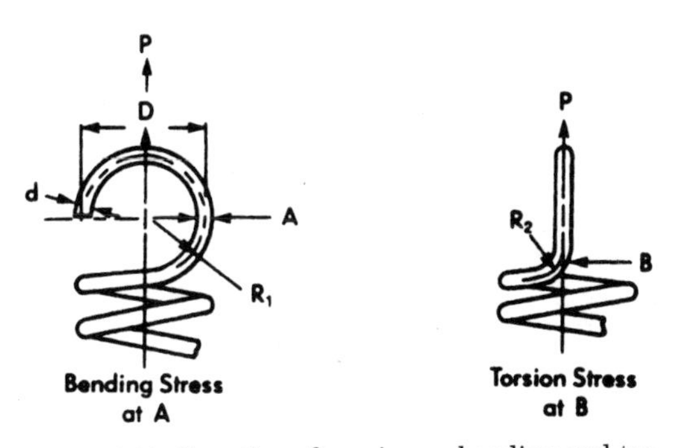


**Figure 12-20** Location of maximum bending and torsion stresses in twist loops. (*From* "Design Handbook," *1987. Reprinted with permission from Associated Spring, Barnes Group, Inc., Dallas, TX.*)

### 12-7-3. Specific recommended dimensions

With the free length of an extension spring being measured between the inner surfaces of both ends (per Fig. 12-18), this dimension should be made equal to the length of the spring body, plus its ends. The spring body can be calculated as

$$L_{body} = d(N + 1)$$

where $N$ = number of coils

The gap in the loop (or hook) opening can vary with the manufacturer. Generally, this gap should not be specified smaller than one-half the wire diameter. If a gap smaller than that is desired, the designer should consult the spring manufacturer about its feasibility.

The number of active extension spring coils is equal to the number of coils it contains. In springs used with threaded inserts and swivel hooks, the number of active coils is lower.

Loops and hooks account for approximately $0.1N_a$ of active coils with one-half twist loops. Up to $0.5N_a$ can be used with some cross-center, full-twist loops, or extended loops.

## 12-8. Flat Springs

Flat springs are those made from strip or sheet material. They may contain bends and their shapes may be quite complex. Often, they perform additional functions in an assembly, such as locating the opening or a part, retaining other parts in an assembly, or banking on them, acting as a latch, or conducting electricity.

The most commonly used flat springs are of a cantilever type (Fig. 12-21). When calculating such springs, all cantilever and simple-beam equations may be used. However, to obtain more accurate results, calculations based on curved beam theory are recommended by Associated Spring's experts. Where the amount of elastic deflection is of importance, Castigliano's method is their additional choice.

### 12-8-1. Materials

Carbon steel materials, most often used in flat spring manufacture, usually belong in one of the following groups:

- 0.70 to 0.80 percent carbon content, which is a slightly less expensive material, more tolerant of sharper bends
- 0.90 to 1.05 percent carbon content with a higher elastic limit

Materials are used either in their annealed form or pretempered. Annealed materials must be heat-treated after forming.

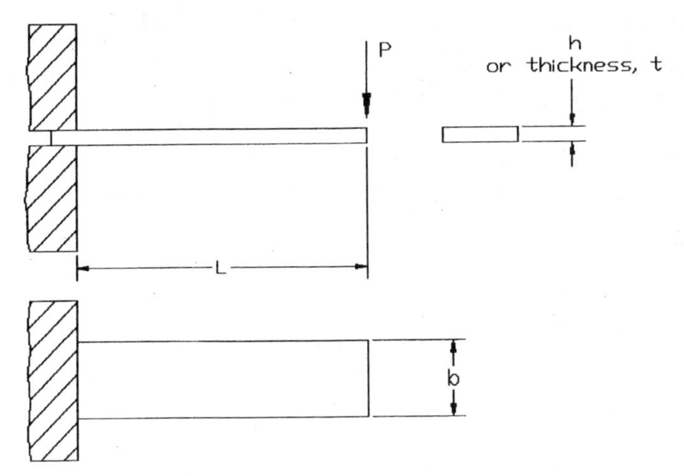


**Figure 12-21** Cantilever spring.

The amount of distortion caused by a heat-treatment is difficult to assess or calculate. Rather, the designer should depend on a spring manufacturer's experience, while avoiding too precise tolerances, sharp corners, and edges on the spring. Parts with thin and wide cross sections will tend to be more distorted, sometimes requiring restriking or other adjusting operation.

Pretempered materials must be hard enough to possess a sufficient elastic limit for their function under desired loads. At the same time, this material should not be too hard, as it may fracture during forming or cause breakage and excessive wear to the tooling. The springback of pretempered material is greater, and an allowance for it must be made in the tool designing stage. Again, an experienced spring manufacturer may be the one to evaluate the amount of necessary alterations.

### 12-8-2. Design and calculations

Mostly all flat springs are preloaded during the bending operation. The surface condition of the material must be smooth, possibly polished, with no dents or nicks. Sharp edges and burrs should be eliminated by design or by abrasive means. Bend radii should be liberal in size, since sharp radii become stress points, exerting damaging influence on the part. Naturally, all bends should be oriented across the material's grain line.

Based on Bernoulli-Euler beam theory for bending of beams, the maximum stress can be calculated as

$$S = \frac{Mc}{I} \tag{12-38}$$

where $c$ = distance from neutral axis to outside or one-half material thickness
$M$ = moment, amounting to distance from support times the load, or $M = PL$
$I$ = moment of inertia. For a rectangular section, moment of inertia can be calculated:

$$I = \frac{bh^3}{12}$$

Combining the above values into the single equation, we get the maximum stress expressed as

$$S = \frac{6PL}{bh^2} \tag{12-39}$$

where $P$ = load on the spring, lb
$L$ = length of lever arm, in

The load value may be calculated by using a formula

$$P = \frac{fEbh^3}{4L^3} \tag{12-40}$$

where $E$ = modulus of elasticity
$f$ = deflection, in

For flat springs, where the width to thickness ratio is relatively small, the maximum stress and deflection formulas are reasonably accurate. Higher width to thickness ratio increases the flexural rigidity of the spring, resulting in the modulus of elasticity $E$ being replaced by $E'$ as follows:

$$E' = \frac{E}{1 - \mu^2} \tag{12-41}$$

where $\mu$ = Poisson's ratio
Then the deflection can be calculated using a standard formula

$$f = \frac{PL^3}{3EI} \tag{12-42}$$

where $E$ = modulus of elasticity
$I$ = moment of inertia; for a rectangular section its value is

$$I = \frac{1}{12} bh^3$$

This formula, when applied to a rectangular section, changes to

$$f_{RT} = \frac{4PL^3}{Ebh^3} \quad \text{or} \quad \frac{2SL^2}{3Eh} \tag{12-43}$$

where $S$ = stress

Since $L$ and $h$ values are raised to the third power, accurate measurements are vital.

All these equations were proved satisfactory where the ratio of deflection to cantilever length $F/L$ was less than 0.3. For larger deflections, $E$ should be replaced by $E'$, as given by Eq. (12-40).

Chapter

# 13

# Spring Washers

Spring washers, even though small in size, may sometimes outperform much larger springs. They are used in offset situations, to provide tension in bolted assemblies, or to furnish the recoil action of springs. Even though their span of deflection is limited because of their size and especially because of their height, a somewhat improved performance may be expected where coupling several washers together. Their stackability, along with their compactibility and versatility, makes spring washers quite advantageous where used in confined spaces or where a stabilizing function is needed.

In bolted assemblies, spring washers are capable of keeping all parts under tension, preventing threaded items from rotating and loosening up. Spring washers can negate the effects of vibration, or they may diminish the side-acting force and control the pressure in vibration mounts, aside from many other applications, where their usage is often taken for granted.

Basically, there are three types of spring washers:

- Cylindrically curved washers
- Wave washers
- Conical disks, or Belleville washers

These three basic variations are capable of covering a wide range of loading applications. Where cylindrically curved washers will sustain a loading of several ounces, the sturdiness of Belleville washers allows for loads ranging within tons (Table 13-1).

The effect of loading force is localized in spring washers, causing a stress response within a small area surrounding the inside diameter of the part. The subsequent deformation tends to increase the affected area in size, which nevertheless is never large enough to affect and alter the height of the washer.

**TABLE 13-1 Characteristics of Three Basic Types of Spring Washers**

| Type | Load capacity | Spring characteristics | Nature of spring contact | Expansion under load | Maximum deflection |
|---|---|---|---|---|---|
| Cylindrically curved washers | Light loads<br>Ounces to a hundred pounds | Spring rate approximately linear over entire deflection range | 4 Contact points (2 top, in line — 2 bottom) | Has most expansion of three basic types | 1/2 of outside diameter |
| Wave washers | Light to medium loads<br>Pounds to hundreds of pounds | Spring rate approximately linear, except near flat position | Number of contact points equals twice the number of waves | Has less expansion than cylindrically curved washers | Approximately 1/4 of outside diameter |
| Belleville washers | Medium to heavy loads<br>Tens of pounds to tons | Can have:<br>(1) Approximately linear spring rate<br>(2) Increasing spring rate<br>(3) Decreasing spring rate<br>(4) Zero spring rate<br>Load capacity is erratic near the flat position | Contact around inner and outer circumferences | Has least expansion of three basic types | (Belleville Criterion) 1/10 of rim width |

SOURCE: Reprinted with permission from H.K. Metalcraft Co. Lodi, N.J.

Conical washers can take up to 200,000 lb/in$^2$ loading, which is the value of their maximum stress in load cycles of 500,000. With more cycles, loading limits must be reassessed on the basis of fatigue testing of actual washers. However, some conical washers will tolerate stresses in the range of two or three times the maximum permissible value.

All spring washers are usually made of spring steel, with some marginal use of spring brass, beryllium copper, phosphor bronze, and other materials. Hardness of the material does not influence the spring rate of the washer in any way as some may have believed. Corrosion resistance is ensured by application of coatings, which may include electrogalvanizing, cadmium plating, black oxide, nickel and chromium plating, etc.

The actual usefulness of each washer varies along with its shape, making each of them restricted in application to specific situations, with their interchangeability outright impossible.

## 13-1. Cylindrically Curved Washers

This type of washer demonstrates a considerable uniformity of spring constant over a wide range of deflections. Cylindrically curved washers are suitable for application of light loads or where repeated cycling with varied range of motion is involved. The recommended range of their maximum height is limited to less than one-half of their outer diameter. Cylindrically curved washers (Fig. 13-1) may be

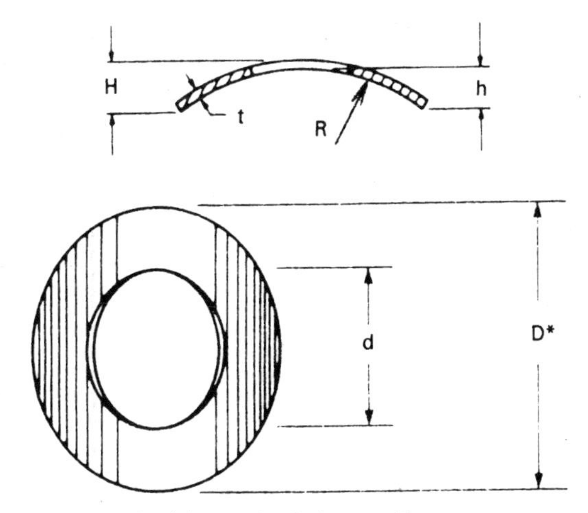


**Figure 13-1** Typical curved spring washer. (*From* "Design Handbook," *1987. Reprinted with permission from Associated Spring, Barnes Group, Inc., Dallas, TX.*)

used where tightening of assemblies is needed, to protect them from looseness and lack of stability.

The functionability of these washers should not be hampered by installing them in restricted or confined spaces, as they need room for diametral expansion under loads. The condition and hardness of their bearing surface is of importance as well, since it must allow for easy sliding of edges during expansion, with no subsequent digging into the material.

To calculate the values of cylindrical washers, the following formulas may be used:

$$P = \frac{4Efbt^3}{D^3K} \tag{13-1}$$

where $P$ = applied load, lb
$E$ = modulus of elasticity, lb/in$^2$
$K$ = empirical correction factor per Fig. 13-2
$D$ = outside diameter, in
$f$ = deflection, in
$b$ = radial width of the material, or $(D - d)/2$
$t$ = material thickness

The maximum induced stress $S$ will be

$$S = \frac{6Eft}{D^2} \tag{13-2a}$$

or

$$S = \frac{1.5P}{t^2}K \tag{13-2b}$$

where $K$ = empirical correction factor per Fig. 13-2

These equations are valid for deflections of up to 80 percent of the washer's height $h$, where the actual amount of deflection $f$ is smaller than $1/3D$. Beyond these ranges, the spring rate, which so far is linear, begins to rise in value, becoming higher than calculated.

The radius of curvature $R$ may be figured as follows:

$$R = \frac{(D/2)^2 + h^2}{2h} \tag{13-3}$$

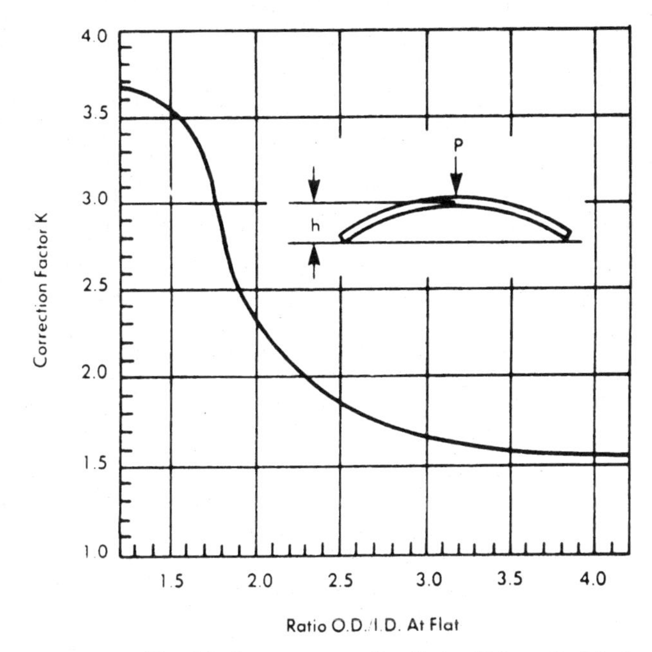


**Figure 13-2** Empirical stress-correction factor $K$ for cylindrically curved spring washers. (*From* "Design Handbook," *1987. Reprinted with permission from Associated Spring, Barnes Group, Inc., Dallas, TX.*)

Equation (13-3) may be used to evaluate the height of the washer $h$,

$$h = R - \sqrt{R^2 - (D/2)^2} \tag{13-4a}$$

or the outside diameter

$$D = 2\sqrt{2hR - h^2} \tag{13-4b}$$

Spring rate $k$ may be obtained from the formula

$$k = \frac{P}{f} = \frac{41.75Ebt^3}{\left(\frac{D+d}{2}\right)^3} \tag{13-5a}$$

where $b$ = radial width of the material, or $(D - d)/2$

$k$ = spring rate, lb/in

## 13-2. Wave Washers

Wave washers may be used for small to moderate static loads, ranging from a few pounds up to hundreds. They are an excellent choice for mounting within tight or restrained areas, as their outside diameter increases in size only very slightly under a load (Fig. 13-3).

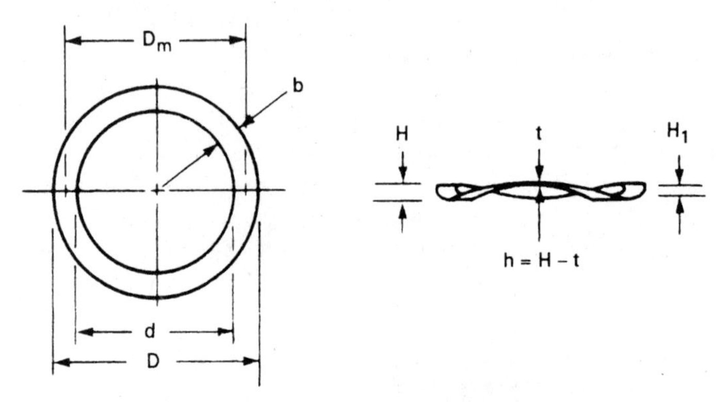


**Figure 13-3** Typical wave spring washer. (*From* "Design Handbook," *1987. Reprinted with permission from Associated Spring, Barnes Group, Inc., Dallas, TX.*)

Wave washers are often utilized in situations where some amount of cushioning is required, to offset components of shaft assemblies, or to prevent loosening of parts due to vibration. These washers may be obtained in a wide range of sizes, but for the best balance between their flexibility and load-carrying capacity, the ratio of mean diameter $D_m$ to the radial width of the washer material $b$ should be kept at the numerical value of 8, or

$$\frac{D_m}{b} = 8$$

Ratios smaller than 8 generate a discrepancy between the calculated and actual values of load and stress. Such an impediment to the washer's performance is caused by the inability of waves to assume their previous shape after deflection. Where the $D_m/b$ ratio should fall considerably below 8, a replacement with a Belleville washer is recommended.

The number of waves must be three or more, with the most commonly used washers having three, four, or six waves. By increasing the number of waves, the washer's thickness may need to be reduced for a required load, with a subsequent decrease in allowable deflection. The number of waves is based on the desirable spring rate and may be calculated by using the formula

$$k = \frac{P}{f} = \frac{Ebt^3 N_a^4 D}{2.40 D_m^3 d} \tag{13-5b}$$

Dimensional uniformity of waves is important, as the actual load deflection rate takes effect only after all waves are equally loaded. For this reason, the load deflection rate should always be verified against the initial preload. With evenly loaded waves, the spring constant is expressed by a linear segment, especially within the range of 20 to 80 percent of the total available deflection of the washer. The point at which the spring rate begins to deviate from its linear representation differs with various types of washers. Its occurrence should be prevented by making the washer height equal to twice the amount of deflection.

**TABLE 13-2 Maximum Recommended Operating Stress Levels for Cylindrically Curved and Wave Washers Made of Steel in Cyclic Applications**

| | Percent of tensile strength |
|---|---|
| Life (cycles) | Maximum stress |
| $10^4$ | 80 |
| $10^5$ | 53 |
| $10^6$ | 50 |

This information is based on the following conditions: ambient environment, free from sharp bends, burrs, and other stress concentrations, AISI 1075.

SOURCE: *Design Handbook,* 1987. Reprinted with permission from Associated Spring, Barnes Group, Inc., Dallas, TX.

Stress range of wave washers (Table 13-2) may be calculated as follows:

$$S = \frac{0.75\pi P (D + d)}{N_a^2 t^2 b} \tag{13-6a}$$

or

$$S = \frac{\left(\frac{48E}{\pi^2}\right) t f N_a^2}{(D + d)^2} \tag{13-6b}$$

where $N_a$ = number of waves

Material thickness $t$ may be related to the following values:

$$t = \frac{0.635\,(D + d)\,(P/bf)^{1/3}}{E^{1/3}\,N_a^{4/3}} \tag{13-7}$$

where $E$ = modulus of elasticity, lb/in$^2$
$P$ = applied load, lb
$b$ = radial width of material, or $(D - d)/2$
$f$ = deflection, in

and the deflection $f$ is tied to a formula

$$f = \frac{\pi^2\,S\,(D + d)^2}{48EtN_a^2} \tag{13-8}$$

**Finger washers.** Finger washers are used in static load cases such as those exerted by ball-bearing races. In such an application finger washers are utilized to reduce vibration and noise. Cyclic loading is not recommended for this type of spring washer (Fig. 13-4).

Finger washers combine the equality of distribution of loading of wave washers with the flexibility of curved washers. Stresses, deflection, and other calculations for this type of washer should be calculated considering fingers to be cantilever springs. Actual samples of the calculated values should be produced and tested prior to use.

Finger washers usually do not retain any favorable residual stresses, and they are usually supplied in their stress-relieved condition. Operating stresses from Tables 13-2 and 13-3 may be used for finger washers as well.

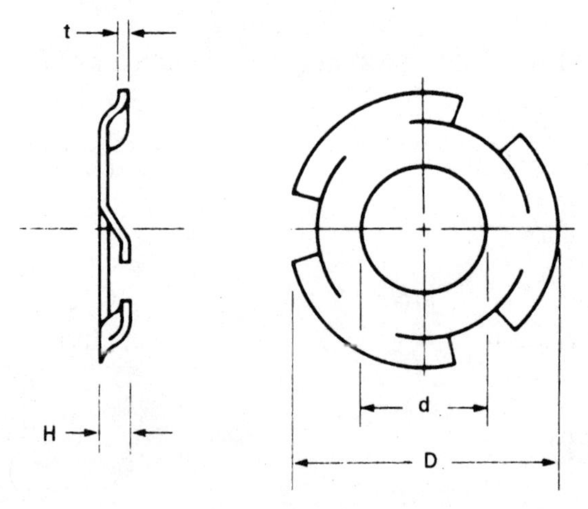


**Figure 13-4** Typical finger spring washer. (*From* "Design Handbook," *1987. Reprinted with permission from Associated Spring, Barnes Group, Inc., Dallas, TX.*)

**TABLE 13-3 Maximum Recommended Operating Stress Levels for Special Spring Washers in Static Applications**

| Material | Percent of tensile strength | |
|---|---|---|
| | Stress-relieved | With favorable residual stresses |
| Steels, alloy steels | 80 | 100 |
| Nonferrous alloys and austenitic steel | | 80 |

Finger washers are not generally supplied with favorable residual stresses.

SOURCE: *Design Handbook,* 1987. Reprinted with permission from Associated Spring, Barnes Group, Inc., Dallas, TX.

**TABLE 13-4 Load Tolerances for Special Spring Washers**

| Washer type | Stock thickness | | ± Load percentage |
|---|---|---|---|
| | in | mm | |
| Curved | 0.004–0.039 | 0.1–1.0 | 15 |
| Wave | 0.004–0.010 | 0.1–0.25 | 33 |
| | 0.010–0.012 | 0.25–0.30 | 25 |
| | 0.012–0.020 | 0.3–0.5 | 20 |
| | 0.020–0.039 | 0.5–1.0 | 15 |
| | 0.039–0.079 | 1.0–2.0 | 12 |

SOURCE: *Design Handbook,* 1987. Reprinted with permission from Associated Spring, Barnes Group, Inc., Dallas, TX.

## 13-3. Belleville Spring Washers

Belleville washers were patented in France by Julien F. Belleville of Paris in 1867. Their use throughout the industry is widespread until this day, for they possess certain qualities not obtainable with any other spring washers.

Being basically coned disk spring washers, the uses of Belleville washers in packing seals, lathes, and clutches are justified by their ability to maintain a constant force regardless of their dimensional variation caused by wear (Fig. 13-5).

Belleville washers' performance is height-sensitive, and with higher cones, there is no increase in spring action, as some may have thought. An increase in height may actually cause an incomplete recovery after deflection, leading to a permanent set of the washer, which may cause the development of internal strains, with a subsequent failure of the part.

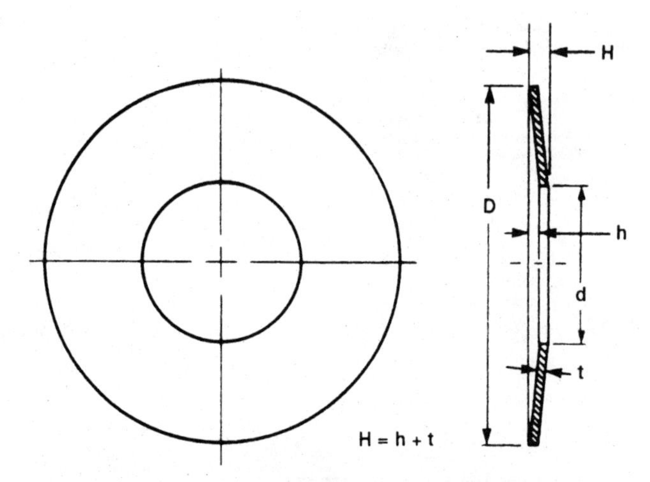


**Figure 13-5** Belleville spring washer. (*From* "Design Handbook," *1987. Reprinted with permission from Associated Spring, Barnes Group, Inc., Dallas, TX.*)

Belleville washers may be produced teethed around both their upper and lower edge. Such teeth may prevent the washer from succumbing to any lateral shifting while not really altering its characteristics.

### 13-3-1. Design guidelines

The rules governing the design of Belleville spring washers amount to several basic requirements, established originally by Julien Belleville, which are as listed below:

- The washer's height to the width of its rim should not exceed 1:10 ratio. A relationship exceeding 1:10 ratio results in a maximum recommended angle of cone equivalent to 5.5°.
- Material thickness should be related to the width of a rim at the ratio of 1:5. This ratio should never exceed 1:10.
- Where the outside diameter is equivalent to roughly twice the inside diameter value, the maximum flexibility is attained. Such a relationship also provides the best ratio of the washer's spring properties to its weight.
- With the ratio of OD/ID = 1.5 to 1.7, both the load and stiffness capacities are at their highest regardless of the value of $t/h$, or thickness to height ratio.
- Formulas used for a washer which is deflected beyond 90 percent of its initial height become inaccurate, with the actual load greater

than the calculated value. For the best results, the washer should be designed to arrive at the predetermined load sooner than becoming totally flattened.

- In screw or bolt assemblies, the load supported with a practically flattened washer should be equal to 50 percent of the screw's or bolt's tensile strength.

### 13-3-2. Height to thickness ratio

Load-deflection curves of Belleville washers for various height to thickness ratios are as shown in the graph in Fig. 13-6. It may be observed that a curve for low $h/t$ values turns into almost a straight line. At $h/t = 1.41$, this curve becomes fixed in its value, dependent on the amount of loading. Such a range lasts for approximately 50 percent of deflection preceding the flat stage, with up to 50 percent of deflection past the flat stage.

With $h/t$ values exceeding the 1.41 range, the load-carrying capacity rapidly diminishes after reaching its peak value, whereas with $h/t$ greater than 2.83, which is exceeding its flattened condition, the washer will snap backward and a force will have to be applied to return it to its original shape.

Belleville spring washers are quite unique in their function, as their stiffness may change over the range of deflection, while the load

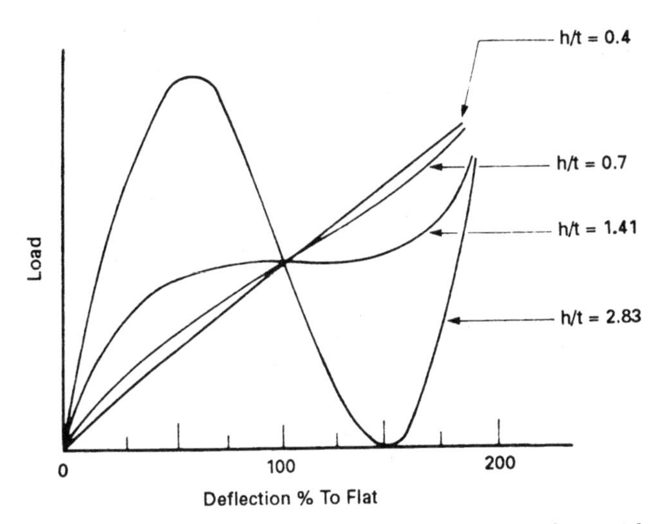


**Figure 13-6** Load-deflection curves for Belleville washers with various $h/t$ ratios. (*From* "Design Handbook," *1987. Reprinted with permission from Associated Spring, Barnes Group, Inc., Dallas, TX.*)

deflection itself may be altered by changing the ratio of free height to thickness, or $h/t$. A comparison of $h/t$ ratios and their characteristics and applications is given in Table 13-5.

A conical washer of $h/t$ greater than 0.4, when exposed to the maximum loading condition, will show an almost uniform spring constant, which should be maintained up to $h/t = 0.8$. Where the $h/t$ ratio exceeds the 1.41 value, the washer's capacity to support a maximum load reaches its peak, decreases with additional deflection, and increases again once the flattened position is passed. With cessation of the pressure the washer snaps back into its original position.

Where the $h/t$ ratio exceeds the value of 2.8, the load becomes negative if the deflection is pursued beyond the washer's flattened position. Under a continuous application of load the washer becomes inverted, turning inside out, in which position it remains until a load applied from the opposite direction returns it to its original shape and position. This is an effect similar to that of a damaged oilcan, which

**TABLE 13-5 Applications and Load-Deflection Characteristics of Conical Spring Washers Having Various *h/t Ratios***

| Characteristics | $h/t$ ratios | Applications |
|---|---|---|
| Constant spring rate | $h/t<0.4$ | Where constant spring rate is required or to obtain very high loads with small deflections |
| Approximately constant spring rate (Belleville's rules apply) | $h/t<0.8$ | Fasteners and in stacks where approximate linear spring rate is acceptable |
| Positive, decreasing spring rate | $A<h/t>1.41$ | Fasteners and in stacks. Higher $h/t$ values may cause snapping if deflection can proceed past flat position ($A = D/d$, see Fig. 13-23) |
| Zero-rate condition over large deflection | $h/t = 1.5$ | To take up wear while establishing constant load. To apply gasket pressure and to provide constant pressure in special brakes |
| Positive, decreasing spring rate which becomes zero, then negative before the flat position is reached | $1.4<h/t>2.83$ | Special purpose: for fasteners not generally loaded to the flat position, and in devices where an "oilcan" characteristic is desired |
| Same as above, except the spring washer can remain stable on either side of the flat position | $h/t>2.83$ | Special purpose: in devices where load is required in opposite direction to restore to working position |

SOURCE: Reprinted with permission from H. K. Metalcraft Co., Lodi, NJ.

may be forced to snap out but, unless a force is applied from another direction, never snaps back by itself.

Where $h/t = 1.5$, such a condition is called zero rate, or constant-load condition, and it spans over a large range of deflection possibilities. Such washers are useful in certain types of disk brakes, where they apply the braking pressure to a disk. The conical washer in such an arrangement maintains constant pressure on the disk even as the brake lining wears out because of friction.

Not all applications of Belleville washers require their total flattening under a load. However, these washers should always be designed to withstand an accidental flattening without succumbing to permanent deformation. Where greater loads and deflections are required, several conical washers should be used in such applications, stacked in various arrangements, as shown later in Fig. 13-16.

### 13-3-3. Mounting of Belleville washers

The method of mounting of Belleville washers will certainly influence the amount of their deflection and therefore influence their performance in service. Where a washer should be mounted on a flat plate, its useful range of deflection will amount to 15 to 85 percent of its height. All deflection rates exceeding 85 percent of height result in higher than calculated loads.

In situations where a Belleville washer is expected to deliver precise loads at the moment its shape is nearing the flattened position, such a washer should be prevented from reversing by banking on a positive stop. Loading in such a case should be applied uniformly over the whole circumference of the washer. The spring rate increases with the deflection beyond the washer's flat position. An applicable method of mounting is shown in Fig. 13-7.

### 13-3-4. Stress and cyclic loading

Belleville washers should be designed for low stress ranges to prevent their setting when accidentally compressed flat. However, various stresses are not uniformly dispersed over the washers' surface, as they accumulate at different points of their cross section (Figs. 13-8 and 13-9). The highest compressive stress can be found around the top inner edge of the washer. Tensile stresses are the highest around the bottom corners. Stress $S_{T2}$ usually exceeds that of $S_{T1}$ in situations where the $h/t$ ratio exceeds the value of 0.6.

Where cyclic loading is expected, both types of these stresses should be calculated. The washer should be designed with stresses of such a low value as to prevent its setting when accidentally compressed flat. Applicable calculations for such an assessment are presented next.

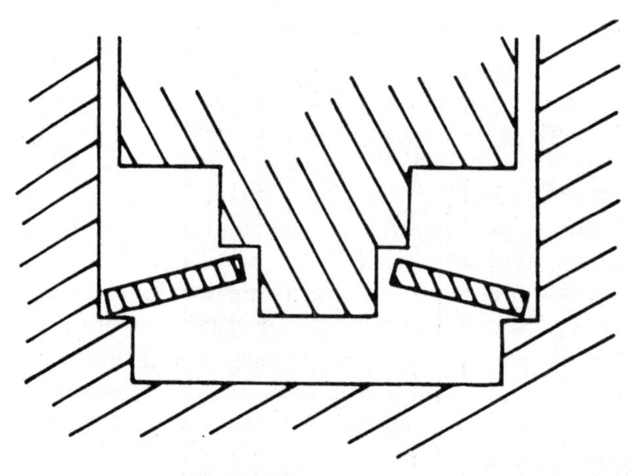

**Figure 13-7** Mounting of a Belleville washer for deflection past the flat position. (*From* "Design Handbook," *1987. Reprinted with permission from Associated Spring, Barnes Group, Inc., Dallas, TX.*)

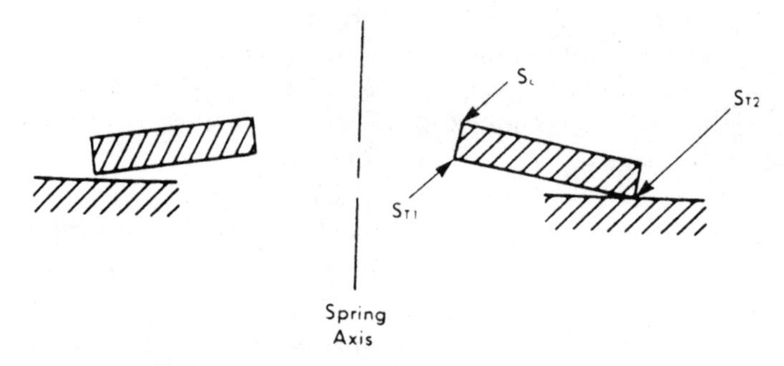


**Figure 13-8** Highest stressed regions in Belleville washers. (*From* "Design Handbook," *1987. Reprinted with permission from Associated Spring, Barnes Group, Inc., Dallas, TX.*)

$$P = \frac{Ef}{MR^2(1-\mu^2)}\left[t^3 + t\left(h - \frac{f}{2}\right)(h - f)\right] \tag{13-9}$$

and

$$S = \frac{Ef}{MR^2(1-\mu^2)}\left[C_1\left(h - \frac{f}{2}\right) + C_2 t\right] \tag{13-10}$$

Equation (13-10) may be adjusted to give the height $h$ as

$$h = \frac{SMR^2(1-\mu^2)}{C_1 Ef} + \frac{f}{2} - \frac{C_2 t}{C_1} \tag{13-11}$$

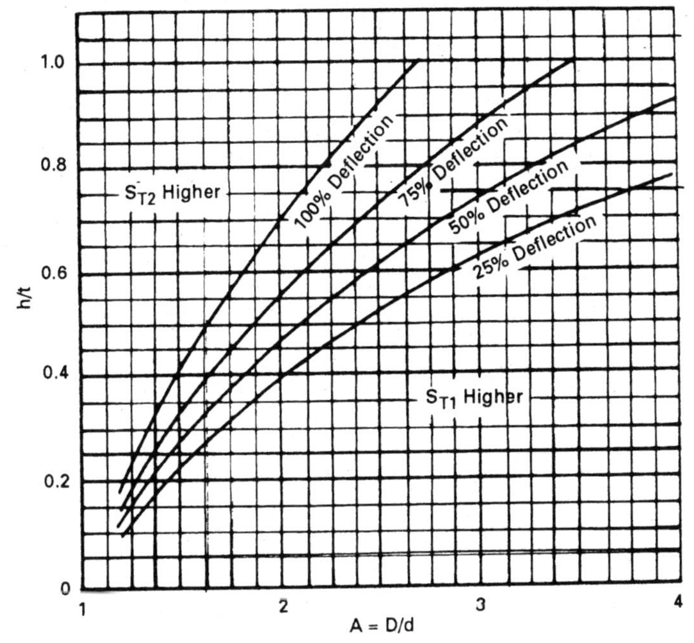


**Figure 13-9** Comparison of $S_{T1}$ and $S_{T2}$ for various deflections, $h/t$ ratios, and diameter ratios ($A = D/d$) of Belleville washers. (*From* "Design Handbook," *1987. Reprinted with permission from Associated Spring, Barnes Group, Inc., Dallas, TX.*)

where $M$ = constant [see Fig. 13-10 or Eq. (13-12)]

$C_1$ and $C_2$ = compressive stress constants from Fig. 13-10 or Eqs. (13-13) and (13-14)

$R = D/2$

$S$ = maximum compressive stress at the circumference, ID convex corner

$P$ = applicable load, lb

$f$ = deflection, in

$h$ = inside height, or $H - t$

$H$ = height, overall

$t$ = material thickness

$\mu$ = Poisson's ratio

$$M = \frac{6}{\pi \ln A} \frac{(A - 1)^2}{A^2} \tag{13-12}$$

$$C_1 = \frac{6}{\pi \ln A} \left( \frac{A - 1}{\ln A} - 1 \right) \tag{13-13}$$

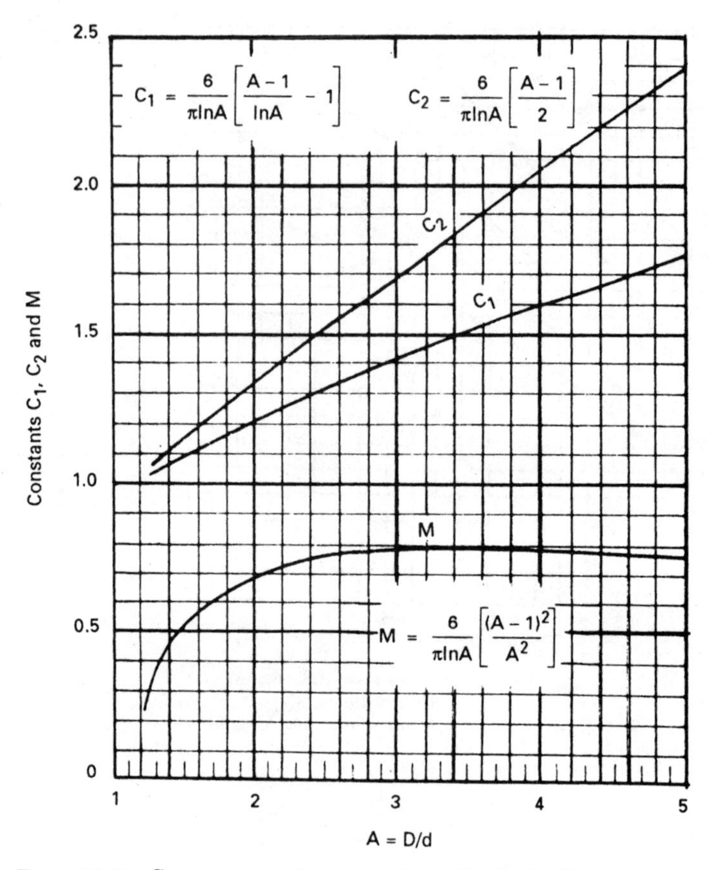


**Figure 13-10** Compressive stress constants for Belleville washers. (*From* "Design Handbook," *1987. Reprinted with permission from Associated Spring, Barnes Group, Inc., Dallas, TX.*)

$$C_2 = \frac{6}{\pi \ln A} \frac{A - 1}{2} \tag{13-14}$$

Other applicable values may be obtained by using the formulas

$$S_c = \frac{-Ef}{MR^2(1 - \mu^2)} \left[ C_1 \left( h - \frac{f}{2} \right) + C_2 t \right] \tag{13-15}$$

$$S_{T1} = \frac{Ef}{MR^2(1 - \mu^2)} \left[ C_1 \left( h - \frac{f}{2} \right) - C_2 t \right] \tag{13-16}$$

$$S_{T2} = \frac{Ef}{R^2(1 - \mu^2)} \left[ T_1 \left( h - \frac{f}{2} \right) + T_2 t \right] \tag{13-17}$$

where $S_c$ = compressive stress at the convex ID corner
$S_{T1}$ = tensile stress at the concave ID corner
$S_{T2}$ = tensile stress at the concave OD corner
$T_1$ = tensile stress constant from Fig. 13-11
$T_2$ = tensile stress constant from Fig. 13-11
$M$ = constant from Fig. 13-10
$R$ = ratio $D/2$
$A$ = ratio $D/d$

With conical spring washers deflected to their flat position where $h = f$, the formulas become

$$P_F = \frac{Eht^3}{MR^2(1 - \mu^2)} \tag{13-18}$$

where $P_F$ = load at flat position

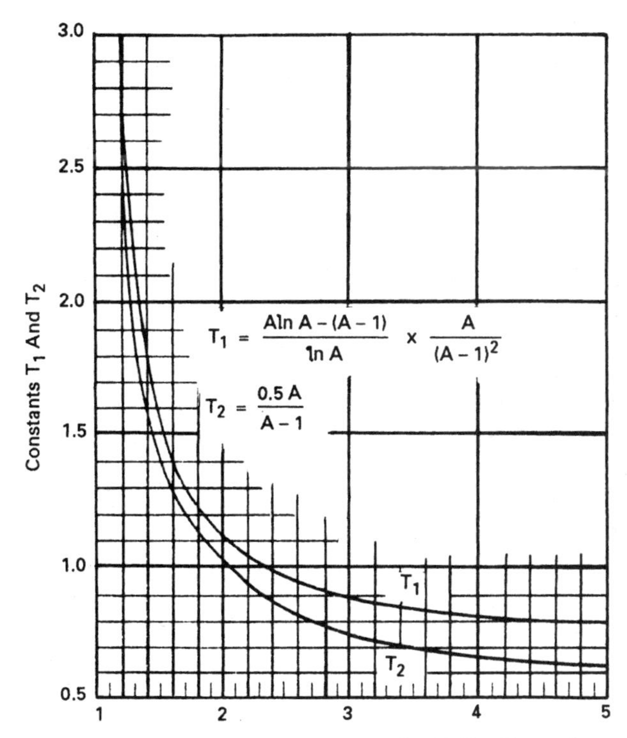


**Figure 13-11** Tensile stress constants for Belleville washers. *(From* "Design Handbook," *1987. Reprinted with permission from Associated Spring, Barnes Group, Inc., Dallas, TX.)*

and
$$S = \frac{Eh}{MR^2(1 - \mu^2)}\left(C_1 \frac{h}{2} + C_2 t\right) \tag{13-19}$$

All the above scenarios assume a uniform distribution of the load, with the angular deflection of the cross section of the washer almost negligible.

With static loading, the compressive stress $S_c$ acting against the inner convex corner is the controlling element of the set point of a spring. Belleville washers made of carbon steel set when such a stress level reaches 120 percent of their tensile strength. Nonferrous and austenitic stainless steel sets at 95 percent of its tensile strength. With set removed, which is the case with the majority of Belleville washers, compressive stress may reach up to 270 percent of tensile strength in carbon steel washers and up to 160 percent in washers made from nonferrous materials or stainless steel. Stresses can be reduced by increasing the outer diameter or by decreasing the $h/t$ ratio, in which case the graph presented later in Fig. 13-23 may be used. Loads and compressive stresses and load deflection characteristics of Belleville washers are given in Figs. 13-12 and 13-13.

Where cyclic loading is required, both minimum and maximum stress levels and their range at both concave corners $S_{T1}$ and $S_{T2}$ should be evaluated by using a modified Goodman diagram, shown in Fig. 13-14.

The amount of maximum stress delineates the success of all stress formulas. At 500,000 cycles, the typical maximum stress for carbon steel Belleville washers is in the vicinity of 200,000 lb/in². However, such a value will change with the type of usage.

### 13-3-5. Spherically curved spring washers

Instead of cylindrical shapes of Belleville spring washers, segments of spheres may be used. Their characteristics are almost the same as those of conical spring washers, with the exception of the load capacity, which is quite different when nearing the flat position. The shape of spherically curved washers exerts a distinct stiffening influence, resulting in higher spring rates (Fig. 13-15).

The spherical forming radius $R_S$ may be determined by using an equation

$$R_S = R^2 + \sqrt{\frac{(R^2 - r^2 - h^2)^2}{4h^2}} \tag{13-20}$$

where $R = D/2$
$r = d/2$

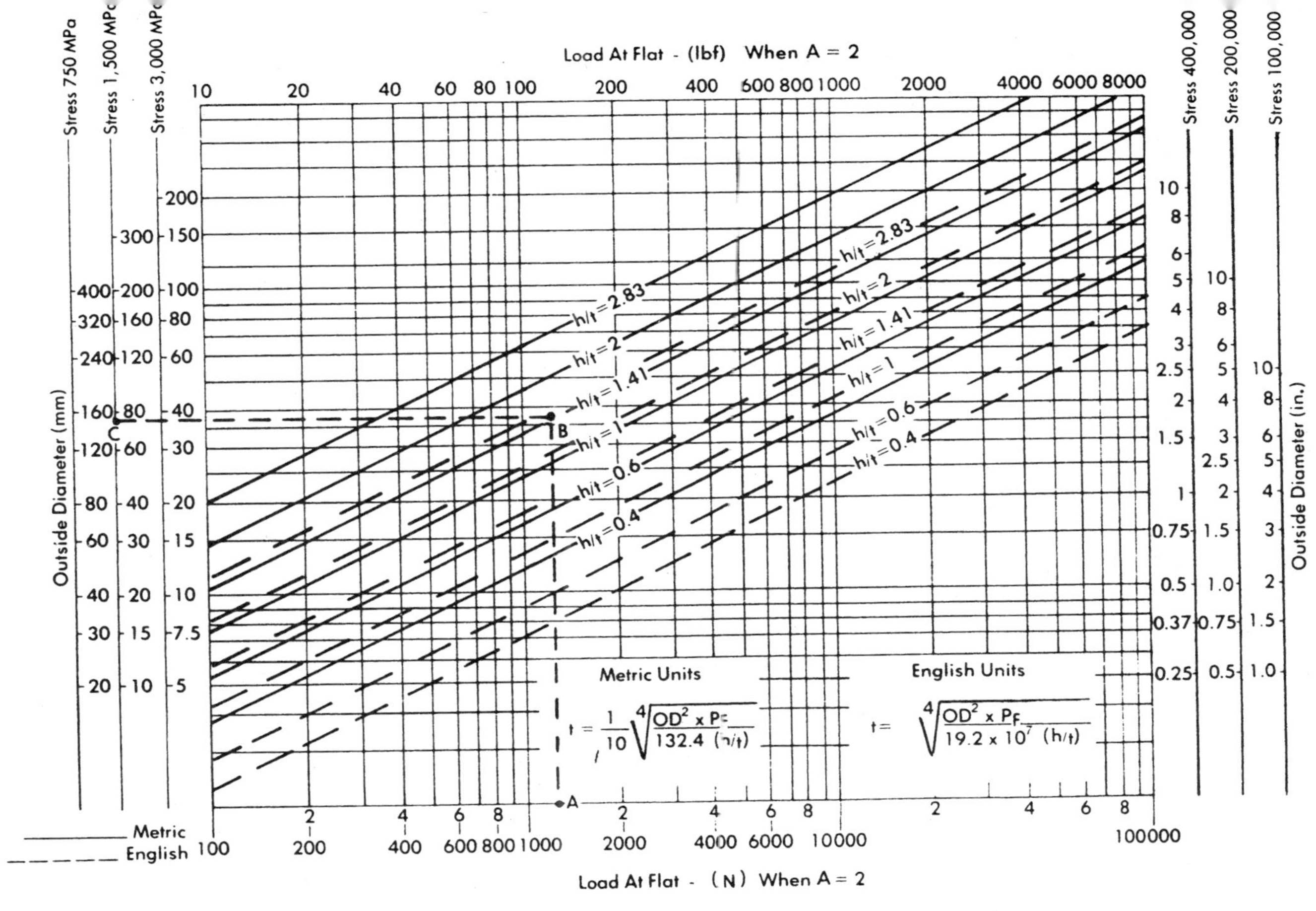


**Figure 13-12** Loads and compressive stresses $S_c$ for Belleville washers with various outside diameters and $h/t$ ratios. (*From* "Design Handbook," *1987. Reprinted with permission from Associated Spring, Barnes Group, Inc., Dallas, TX.*)

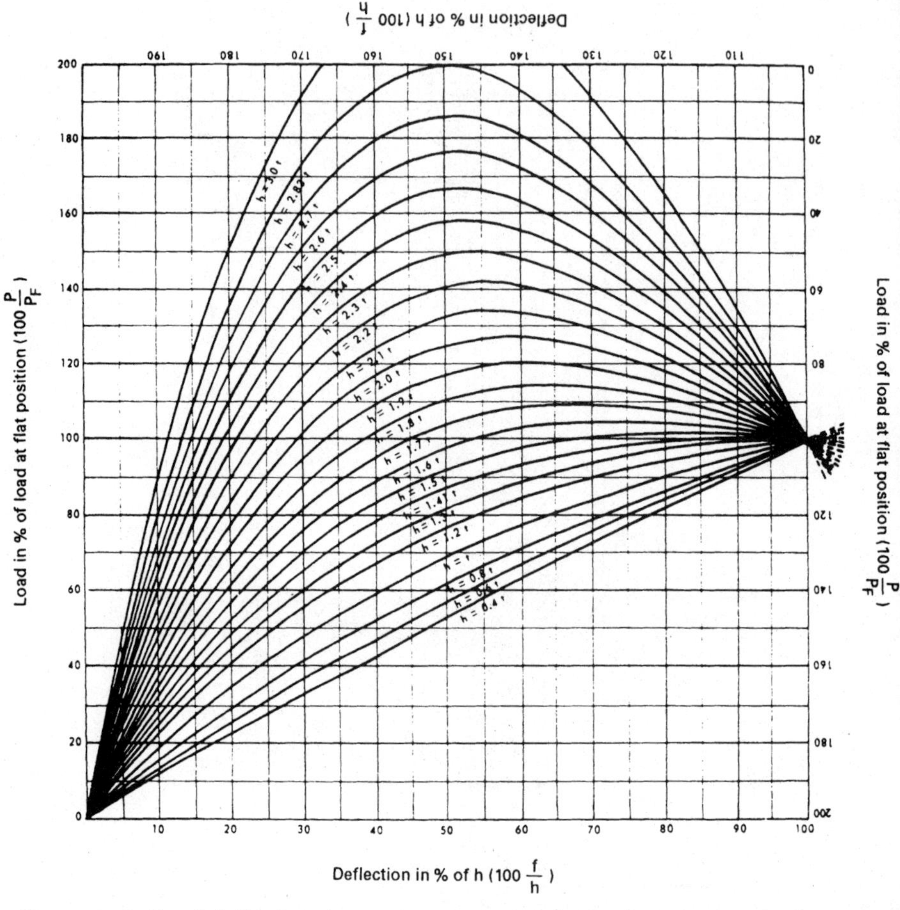


**Figure 13-13** Load deflection characteristics for Belleville washers. (*From* "Design Handbook," *1987. Reprinted with permission from Associated Spring, Barnes Group, Inc., Dallas, TX.*)

The same formula may be used for Belleville spring washers whose height equals one-tenth the width of their rim, or

$$h = \frac{R - r}{10}$$

in which case the condition below applies:

$$R_S = R^2 + \sqrt{\frac{(0.99R^2 - 1.01\,r^2 + 0.02\,Rr)^2}{0.04\,(R - r)^2}} \tag{13-21}$$

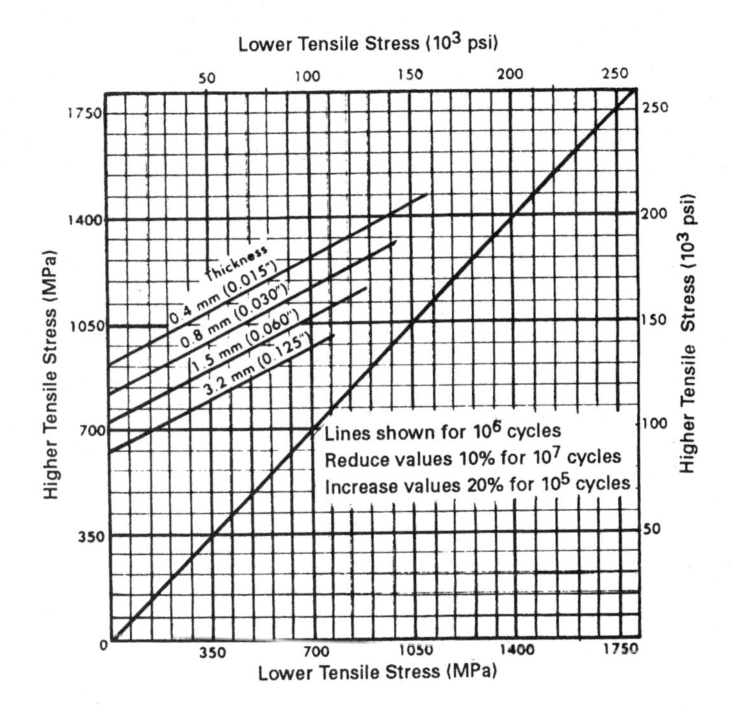


Figure 13-14 may be read as follows:
A belleville washer 0.8 mm (0.030") thick may be expected to have a life of approximately $10^6$ cycles when stressed between either

0–820 MPa (0–117,000 psi)
or 350–990 MPa (50,000–141,000 psi)
or 700–1170 MPa (100,000–167,000 psi)

and may be expected to have a life of approximately $10^7$ cycles when stressed between either

0–740 MPa (0–105,000 psi)
or 315–890 MPa (45,000–127,000 psi)
or 630–1050 MPa (90,000–150,000 psi)

**Figure 13-14** Modified Goodman diagram for fatigue strength of Belleville washers. (Carbon and alloy steel at Rockwell hardness 47 to 49 scale C, with set removed but not shot-peened.) (*From* "Design Handbook," *1987. Reprinted with permission from Associated Spring, Barnes Group, Inc., Dallas, TX.*)

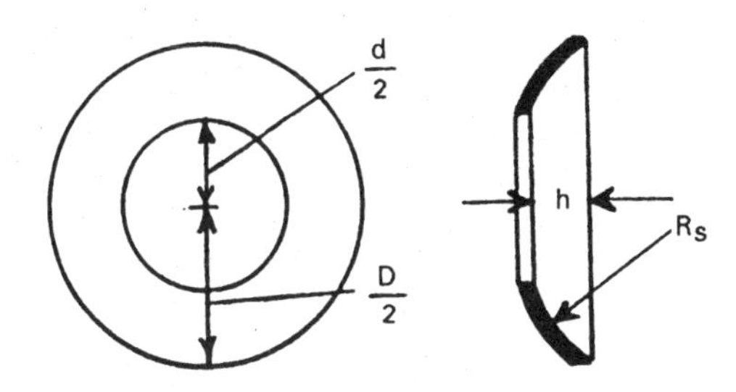


**Figure 13-15** Spherically curved washer. (*Reprinted with permission from H. K. Metalcraft Co., Lodi, NJ.*)

### 13-3-6. Multiple conical spring washer assemblies

Conical washers, when stacked, provide greater deflections, along with larger loading possibilities. Where an indefinite $n$ amount of washers is assembled in series, each washer provides $1/n$-th of the total deflection, while the maximum obtainable load is that of a single washer. Where the $h/t$ ratio exceeds 1.35, such a series of washers will tend to snap back when forced past the flat position.

Loading of a parallel stack of washers is equal to the $1/n$-th load per single washer, multiplied by the number of washers within the assembly. The deflection of the whole stack is equal to the amount of deflection of a single washer (Fig. 13-16).

Hysteresis is greater with parallel-arranged than with series-oriented washers. Its effects may be diminished by proper lubrication (Fig. 13-17).

Where stacks of washers are used, their guidance should be provided by a centrally located pin or tube. Clearance values between these segments should be approximately 1.5 percent of the spring washer inner diameter. Recommended dimensional tolerance values are as shown in Table 13-6.

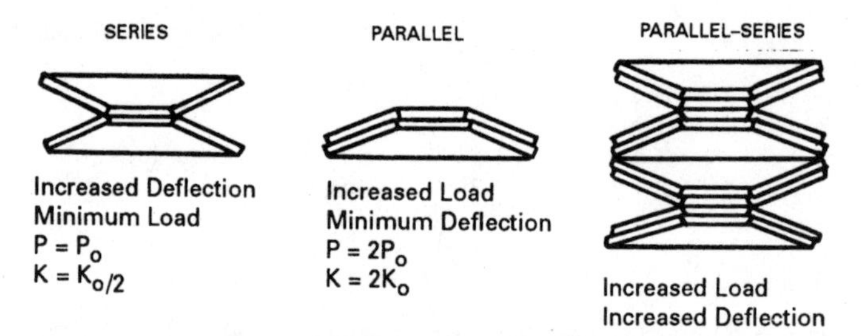


**Figure 13-16** Load and deflection relationships for conical spring washers stacked in series, parallel, and parallel-series combinations. (*Reprinted with permission from H. K. Metalcraft Co., Lodi, NJ.*)

**TABLE 13-6 Belleville Washers' Diameter Tolerance**

| Diameter, mm (in) | OD mm (in) +0.00 | ID mm (in) −0.00 |
|---|---|---|
| Up to 5 (0.197) | −0.20 (−0.008) | +0.20 (+0.008) |
| 5–10 (0.197–0.394) | −0.25 (−0.010) | +0.25 (+0.010) |
| 10–25 (0.394–0.984) | −0.30 (−0.012) | +0.30 (+0.012) |
| 25–50 (0.984–1.969) | −0.40 (−0.016) | +0.40 (+0.016) |
| 50–100 (1.969–3.937) | −0.50 (−0.020) | +0.50 (+0.020) |

Based on $A = 2$; increased tolerances are required for lower $A$ ratios.

SOURCE: *Design Handbook,* 1987. Reprinted with permission from Associated Spring, Barnes Group, Inc., Dallas, TX.

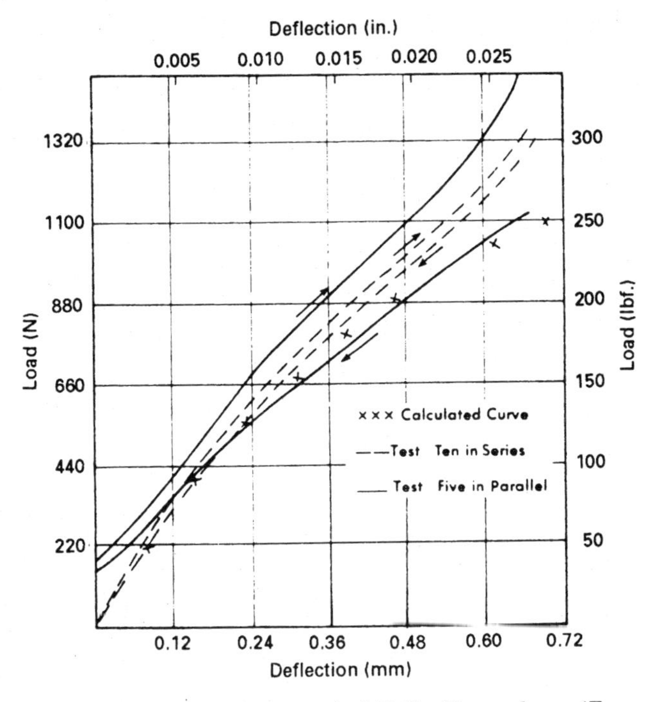


**Figure 13-17** Hysteresis in stacked Belleville washers. (*From* "Design Handbook," *1987. Reprinted with permission from Associated Spring, Barnes Group, Inc., Dallas, TX.*)

Stacking of combined parallel and series arrangements provides greater deflections and higher load capacities. Parameters governing assemblies of parallel sets of spring washers arranged in series may be calculated with the aid of the following formulas.

$$P = wP_F \tag{13-22}$$

$$Y_F = mh \tag{13-23}$$

$$L = m(wt + h) \tag{13-24}$$

$$S_T = \frac{w\,S_{T-F}}{m} \tag{13-25}$$

where
$P$ = applied load, lb
$P_F$ = load necessary to flatten the washer, lb
$w$ = number of spring washers in each parallel set
$m$ = number of parallel sets in a series
$Y_F$ = total deflection at flat height, in
$S_T$ = stiffness at deflection $f$, lb/in
$S_{T-F}$ = stiffness at flat height, lb/in

### 13-3-7. Thickness of washers and load tolerances

It may be established from the observation of deflection formulas that the load capacity of spring washers depends on the cubed thickness of the material used. With any alteration of thickness, a whole array of changes in washer loading may be observed.

To determine the correct material thickness, the following formulas may be used, with their respective values taken off the graph shown in Fig. 13-18.

$$\pm T_t = t\left(\sqrt[3]{1 \pm \frac{P_t}{100}} - 1\right) \tag{13-26}$$

and

$$\pm P_t = 100\left[\left(\pm\frac{T_t}{t} + 1\right)^3 - 1\right] \tag{13-27}$$

where $t$ = material thickness, nominal, in
$P_t$ = load tolerance, percentage of the total load
$T_t$ = thickness tolerance, in

By using the tolerance range in terms of thickness percentage, the graph in Fig. 13-18 may help to assess the variation in load capacity. With demands for tighter load specifications, materials with more accurate thickness tolerances should be used.

Material thickness may also be calculated as follows:

$$t = \sqrt[4]{\frac{D^2 P_F}{19.2 \times 10^7\,(h/t)}} \tag{13-28a}$$

The same formula adapted to the metric system will be

$$t_{mm} = \frac{1}{10}\sqrt[4]{\frac{D^2 P_F}{132.4\,(h/t)}} \tag{13-28b}$$

where $P_F$ = load necessary to flatten the washer, lb

Basic designs of typical spring washers are shown in Fig. 13-19. The most common cylindrically curved spring washer is shown at *A*. It is capable of providing the greatest quantity of spring action for its size and thickness. The washer shown at *B* is a modification of this design, with an increased load capacity, but at reduced springiness. *A* and *B* styles offer the greatest deflection possibilities, but their load-bearing surfaces are the smallest.

Spring washer *C* banks on two flat portions. It is more rigid than the design shown at *A* and is capable of carrying greater loads at the same material thickness. The conical washer *D* and spherical washer

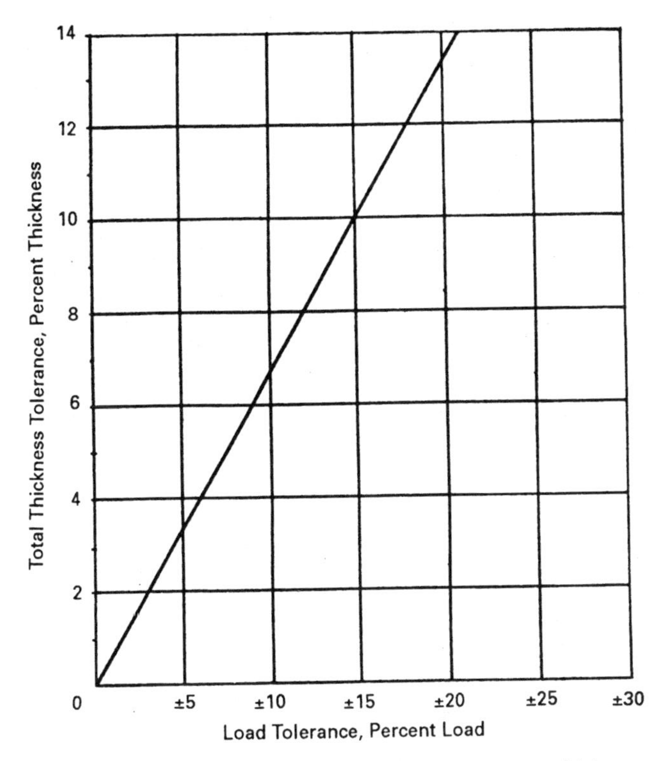


**Figure 13-18** Graph for use in determining the proper thickness and load tolerances for spring washers. (*Reprinted with permission from H. K. Metalcraft Co., Lodi, NJ.*)

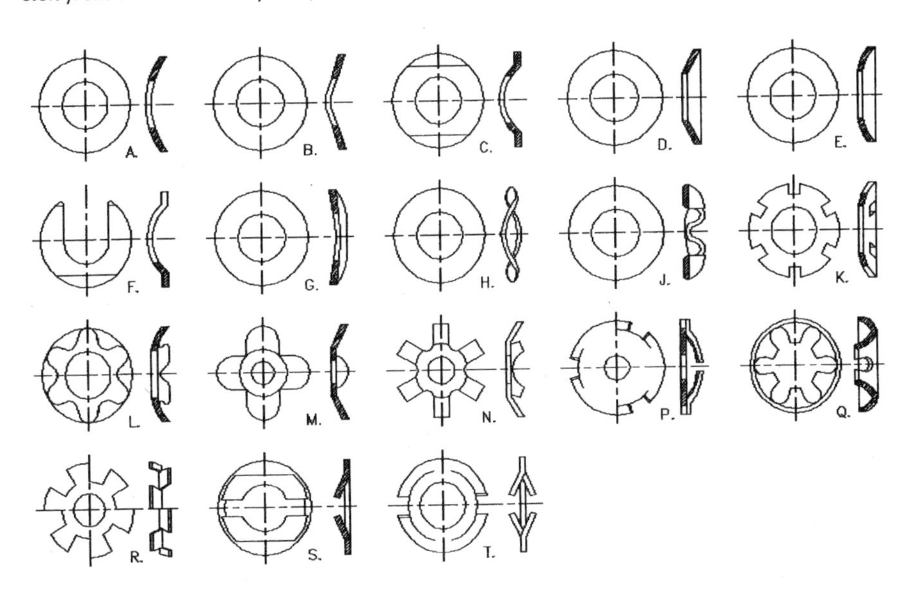


**Figure 13-19** Basic designs of spring washers.

*E* have the best load-carrying capacities when compared to any spring washers of the same size. Their amount of deflection is somewhat smaller, but their springiness is very powerful. A modification of this washer is shown at *F*.

The washer shown at *G* has two waves and two contact surfaces, at which it banks in assembly. The *H* washer is a three-wave type, with three contact points per side, and its load-carrying capacity exceeds that of the *A* and *G* types. Multiple waves shown in washer type *J* increase its load-carrying capacity in comparison to the *H* washer. These two washers offer a high spring force, combined with minimum spring movement.

*K, L, M, N* washers are modified versions of washers *D* and *E*. Decoratively appealing, the scalloping of their circumferences enhances their spring movement, while achieving more uniform distribution of pressure.

Elasticity of fingers, shown at variation *P*, provides for balancing of the applied pressure even at the greater distances off the washer's center. This type of spring washer is often used as a ball-bearing retainer. Washers *Q* and *R* show a difference in arrangement of their teeth, where in designs *S* and *T* these are replaced by fingers.

## 13-4. Lock Washers

Toothed lock washers display an impressive load-supporting capacity at quite a small overall height. Their teeth, which are always in contact with the material, may enhance their action by digging into the material, as shown in Fig. 13-20, type *B*. The type *A* washer in the

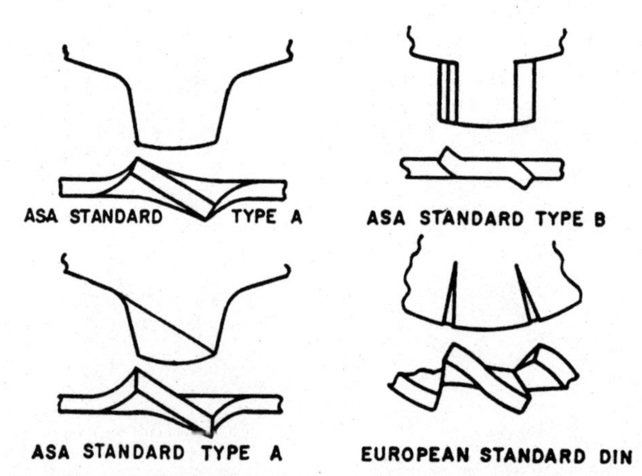


**Figure 13-20** Toothed lock washers are designed to hold either by spring action or by "digging-in" or by a combination of both. Pictured are four commonly used types. (*Reprinted with permission from H. K. Metalcraft Co., Lodi, NJ.*)

same illustration may sometimes have its teeth twisted and provided with a stiffening rib, as shown in the lower left-hand view.

The effectiveness of the lock washer's teeth embedding within the material depends on the amount of the washer's coverage, or its blockage by the screw head. Another influence is exerted in the form of the bearing material hardness, which may or may not allow for the washer's penetrating its surface.

Helical lock washers, shown in Fig. 13-21, deflect in installation, keeping the retaining part, usually a screw or a bolt, under a constant tension. Surfaces of these washers are hardened and tempered, providing for greater applicable tightening torques, combined with excellent thrust characteristics.

A cross section of a helical spring washer, shown in Fig. 13-22, shows slanting of the edge surface, with the *A* dimension emerging smaller than *B*. This alteration in thickness is purposely provided for protection against washers' spreading out and exceeding their diametral size during tightening.

Helical spring washers may be wound on wire-forming machines, in which case their cross section is of trapezoidal shape. However, stamped sheet-metal products are often used as well, with their cross sections either square or rectangular. The free height of the latter parts is usually provided at the range of twice the washer thickness.

Special shapes of helical spring washers, shown in Fig. 13-21, have different usages within the industry. Type *A* washer is recommended for soft bearing surfaces, such as wood, plastic, etc. Its unrestrained

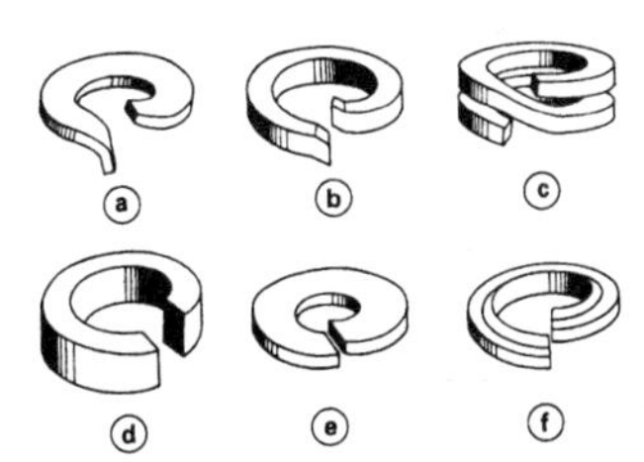


**Figure 13-21** Helical spring lock washers for special applications: (*A*) wood spring lock washer; (*B*) positive-type washer; (*C*) double-coil type; (*D*) Hy-Collar type; (*E*) wide bearing lock washer; (*F*) ribbed type. (*Reprinted with permission from H. K. Metalcraft Co., Lodi, NJ.*)

**Figure 13-22** To prevent worming out, or spreading during tightening, many helical spring lock washers are made so that the peripheral thickness *A* is less than that at the inner surface *B*. (*Reprinted with permission from H. K. Metalcraft Co., Lodi, NJ.*)

end becomes embedded in the material of bearing surface, preventing any rotation of the washer during tightening. The other end, oriented toward the opposite member of the assembly, is driven into that material, to prevent loosening in service.

The *B* type washer provides a similar service without denting the bearing surface. For greater range of functions, a double-coil washer of the *C* type is recommended. It is used in assemblies, where a combination of wood and metal is utilized.

Washer type *D,* designed for socket head cap screws, is utilized in confined spaces or with recessed screw heads. The height of this type of collar washer is much greater than that of the next type *E,* the latter being used instead of regular washers in conditions where an additional spring tension is needed. This type of washer is of advantage where an uneven supporting surface is encountered or with larger than necessary clearance holes.

The *F* type of washer contains a rib surrounding its inner periphery. Such a shape, when compressed, engages the thread of the bolt or screw, locking the whole assembly.

Figure 13-23 provides a graph for determining the constants $M$, $C_1$, and $C_2$ used in the design equations for conical disk washers.

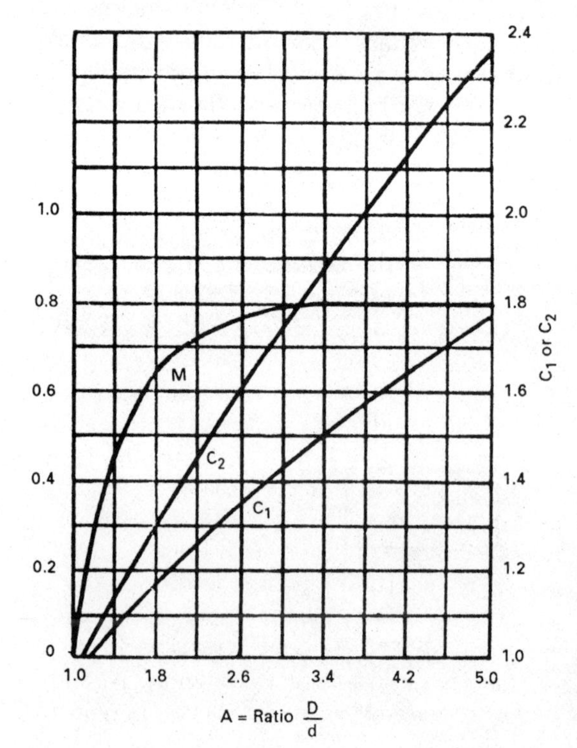


**Figure 13-23** Graph for determining constants $M$, $C_1$, and $C_2$ employed in the design equations for conical disk washers. (*Reprinted with permission from H. K. Metalcraft Co., Lodi, NJ.*)

Chapter

# 14

# Materials and Surface Finish

## 14-1. Metal Materials and Their Properties

Many designers and engineers wonder on a daily basis if they are using the correct material for a given application or if a better choice could be made, because in the field of material selection the choices are vast and are continuously expanding. New materials are developed whenever the old combination does not perform on a par with the hundred percent goal. Material developers don't waste their time: Instead of blaming a failure on design or a faulty manufacturing method, they immediately set to work and begin improving their end of the problem.

Over the years the industry assembled an impressive inventory of various steels, alloys, and other materials, causing many to fear that they may not see the forest for the trees. Therefore, the decision to use a particular material for the given application is often based on hours and hours of laborious research and evaluation. And yet an idea may often lurk somewhere, perhaps in our subconscious, that a better, cheaper, more advantageous, and more appropriate selection could have been made.

The already complex problem of material selection is further impaired by the fact that even though so many materials are now available many designers may not be acquainted with them in detail. There is just too much material to be studied, with too many facts to be assessed. One has to have a mind of a computer to be able to store data in bulk and yet exercise some sort of selective control and evaluating properties, which is quite an unlikely combination. Nobody expects the computer to think and nobody should expect a person to store data the way a computer does.

To select a proper material for an existing problem, the material characteristics must be evaluated first. Limiting properties should be surveyed for a proportion of their disadvantage, compared to a practical usefulness. Beneficial aspects should be looked into, if they're not overly beneficial, to actually impair the process they are supposed to enhance. This may happen surprisingly often. For example, a material perfect for its frictional qualities may fail in operations, where only a certain amount of friction is necessary. Or a material of high elastic limits will be detrimental to the cutting process, where it may behave like chewing gum.

But outright harmful attributes should be surveyed as well, to compare their spectrum of influence to the degree of actual usefulness. Some materials are too brittle, but if used where brittleness is needed, they may prove to be an excellent choice. Another material may suffer from excessive spring-back, or lack of it, and its application should be selected to suit such capabilities. As already said, choices are vast and proper selections can become complex and intricate.

### 14-1-1. Metallurgy of metals

The first aspect to be evaluated when searching through the material jungle is the metallurgical process the particular stock has been subjected to, along with the amount and influence of additives it contains.

Metals, when in their annealed form, consist of a mixture of carbon particles, deposited within a base of low-alloyed iron. In such a state, metals are easily machinable and relatively soft. During heat treatment, when subjected to high temperatures in the vicinity of 1400 to 2300°F, the metal is austenitized. Some of its carbon content melts and dissolves in the iron matrix. On cooling down, or quenching in either water, oils, air, or molten salts, a martensitic transformation occurs, producing a hard and brittle substance. The base iron alloy, even though now considerably more alloyed by the addition of dissolved carbon, suffers from residual stresses and is far from being tough enough to function as an element of tooling.

The metal is subjected to another heating cycle, which is called tempering, at temperatures between 300 and 1200°F. Tempering relieves some residual stresses, while precipitating a portion of alloyed elements. Toughness of the material is enhanced, protecting the finished product from the shock of an impact in cutting and similar environments.

**Rimmed and killed steels.** Rimmed steel is always low-carbon or medium-carbon steel. When poured into the ingot mold, it solidifies quickly around its periphery, which leaves the remaining steel in the middle "shut off" the access to the surface. Without a connection with the outside, gases become entrapped within the mass of an ingot, dispersed in the form of small bubbles. In a solidified ingot, these small bubbles form tiny cavities within the material, disturbing its structure.

The surface of a rimmed product may be perfect, smooth, and of a high finish. But the inner material may be damage-prone, owing to little inner pockets affecting the unity of the material.

Killed steel does not suffer from the emergence of inner gas bubbles, as the solidification process is regulated not to produce them. On solidification, the whole ingot begins to form an indent in the center of its mass alongside its axis, starting from the top. This impression is not of a great depth, but the material from that portion has to be disposed of. Killed steels are usually those containing either higher carbon content or those that are more alloyed. Their surface finish is mostly inferior.

Where rimmed steels do not have a significant amount of silicon (sometimes they have none), the silicon content of killed steels is over 0.15 percent. Silicon aids in deoxidation and degasification of the material.

### 14-1-2. Fe—C phase diagram

The Fe—C or Fe—$Fe_3C$ phase diagram represents the relationship between the iron and carbon within the molten steel material, recording changes in the solidifying mass. These processes, assigned to various temperature ranges and further influenced by the carbon content, influence the structure of material and its reaction to heat treatment during each particular phase (Fig. 14-1).

On solidification, the excess carbon may be segregated in the form of either graphite or iron carbide $Fe_3C$, also known as cementite (containing 6.67 percent C). Iron carbide, when compared to other types of carbon, is very hard, nonductile, and readily affected by stresses within the material.

Ferrite, the $\alpha$ iron, is a ferromagnetic material, quite ductile, with tensile strength under 45,000 lb/in$^2$. Carbon's solubility within ferrite is very limited, because the cubic structure of ferrite, with oblate spaces among atoms, cannot be modified enough to include even a very small atom of carbon.

Ferrite is similar to austenite, or $\gamma$ iron, even though the latter's atoms are more densely spaced. That's why during the heating cycle, when the transfer from the ferrite phase to austenite takes place, contractions within the material amounting to some 0.29 percent may be encountered. This has a considerable importance in heat treatment,

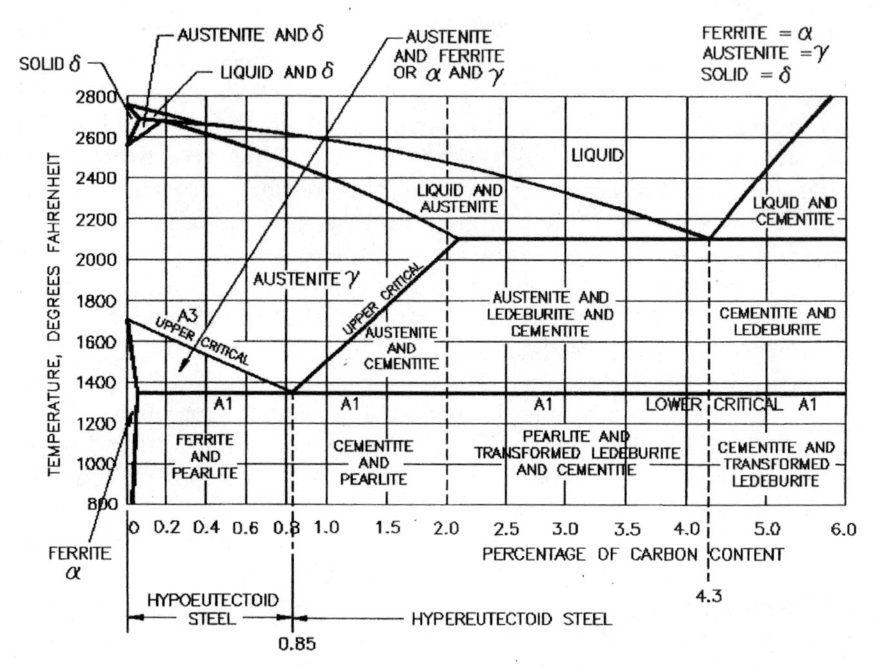


**Figure 14-1** Fe—C phase diagram.

with regard to the development of residual stresses. Austenite is not ferromagnetic.

δ iron, sometimes called δ ferrite, is quite similar to ferrite. It is a solid solution, which means a homogeneous solid mass of a binary or more complex alloy, the chemical composition of which may be altered at a certain range without subsequent modification of its properties. The crystalline lattice of such a solution is the same as that of one of its constituents.

The phase diagram is the basic tool of evaluation of various heat-treating processes for the majority of steels. Their ranges are shown in Fig. 14-2, where their dependence on temperature and carbon content is obvious.

### 14-1-3. Additive elements within the metal material

Various metallic and nonmetallic elements, when dissolved within the iron matrix, have the capacity of affecting and altering the qualities of the finished material. Some additives make the metal more brittle, others make it more ductile, and some are still controversial in their effect.

According to the influence of additives on the metal material, they may be divided into three basic groups:

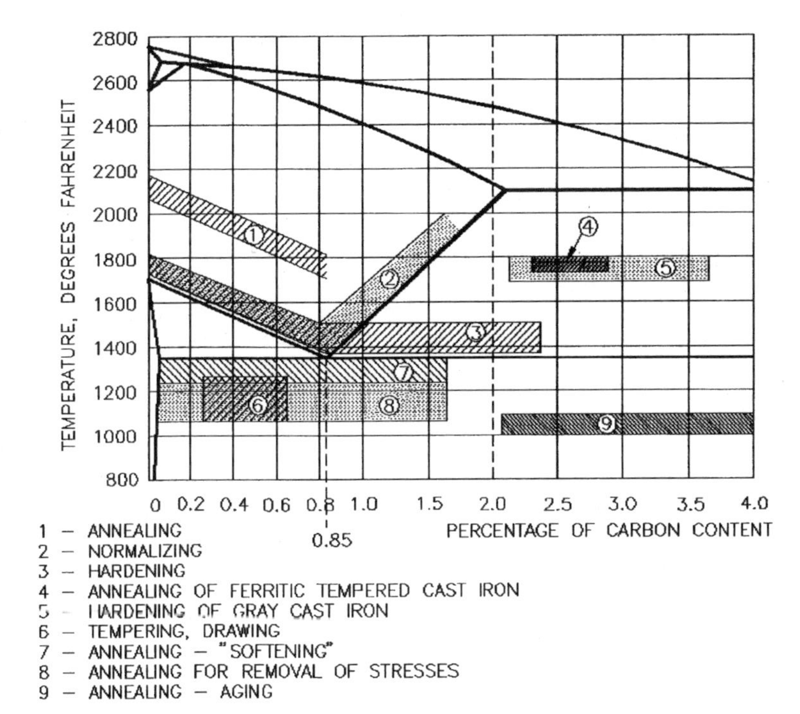


**Figure 14-2** Fe—C phase diagram. Heat-treatment data.

1. Additives detrimental to the material quality
2. Additives beneficial to the material quality
3. Alloying additives

**14-1-3-1. Additives detrimental to the material.** Small quantities of by-products left from the deoxidating process remain within the metal content in the form of inclusions. These are most often oxides, silicates, aluminates, and sulfates, and they may be of

- Endogenous type, or those produced from within
- Exogenous type, produced from without, due to external causes, such as the melting environment and machinery

Where they are too numerous or unequally distributed, these inclusions affect the mechanical properties of the material, making it less fatigue-resistant and less tough. Oxides and nitrides of aluminum are usually distributed evenly; they improve the grain structure and the drawability of material. However, other inclusions, mostly detrimental to the quality and mechanical properties of steel, are undesirable, and an attempt to remove them or minimize their effect is prevalent.

These harmful elements are hydrogen, nitrogen, oxygen, phosphorus, and sulfur.

**Hydrogen (H).** This element may become entrapped within the molten mass in the manufacturing process. During the solidification phase, most of the hydrogen evaporates, until only about one-third of the original amount is left within the material. This hydrogen does not remain intact but strives to separate itself from the melt, being attracted by areas of submicroscopic defects, interstitial impurities, vacancies or microvoids, interfaces, or grain boundaries. There, locked in and compressed by the solidifying mass of metal, it is retained under pressure, until the material cracks later under the force. This type of destructive behavior is called hydrogen embrittlement. Low-alloy steel with some nickel content is especially susceptible to such a defect.

Hydrogen may be removed from steel by an annealing process, where the material is heated for an adequate time and held at such a temperature until the hydrogen content is diminished. A steel manufacturing method of hydrogen prevention rather than removal is vacuum casting.

It is believed, but not yet completely proved, that hydrogen is capable of active interaction with material defects, such as microvoids, grain boundaries or interfaces, dislocations, vacancies, or atoms of impurities of substitutional and interstitial types. A large number of experiments have shown that impurities actually act as hydrogen traps, to which this element is attracted, to form stable, diatomic complexes.

**Nitrogen (N).** This element is readily dissolved in the molten steel, and its amount depends on the type of manufacturing method. Excessive nitrogen is expelled from the solidifying material in the form of nitride $Fe_4N$, only if the metal solidification is adequately slow. At speeding up of this process, nitrogen forms a volatile solution within the $\alpha$ steel.

With greater content of nitrogen, the metal is exposed to various changes from within, which produce a decrease in its notch impact strength, ductility, drawability, and formability. Overall, these changes as grouped together are called *aging* of metal. Aging continues to progress at regular temperatures and may be accelerated by heating the metal to some 730 to 900°F. Metals susceptible to aging include "softer" carbon steels, which may fail in service due to this tendency if used in higher-temperature environments, such as welded steel objects or boiler parts.

The lower the amount of nitrogen within the steel, the less affected by aging it becomes. Therefore, the nitrogen should be forced to form temperature-stable nitrides wherever possible.

In some materials, nitrogen is purposely added to the austenite, since it improves and refines the grain structure. In high-chromium steels, nitrogen is an indispensable element in case hardening.

**Oxygen (O).** Oxygen may become entrapped within the metal mass during the period of oxidation, which depends on a certain amount of it being present within the material. The excess oxygen not utilized by this mechanism cannot simply dissolve itself and evaporate from the metal content, being restricted by the presence of carbon. In the solidified steel, the amount of oxygen may be found at some 0.05 percent.

Oxygen lowers the notch impact strength of the material. It is usually tied to various inclusions, such as oxides (MnO, FeO, $Al_2O_3$) or silicates ($SiO_2$). Other deoxidants are aluminum, silicon, and somewhat titanium, zirconium, and calcium.

**Phosphorus (P).** Dissolvable within the melt $\gamma$, phosphorus raises the temperature of the transformation point $A3$ while lowering that of $A4$. The amount of phosphorus within the steel rarely exceeds 0.1 percent. Like sulfur, it separates from the melt during the process of solidification, impairing the solubility of carbon. Its effect on the mechanical properties of steel consists of negatively affecting the notch impact strength, promoting brittleness, and impairing weldability.

**Sulfur (S).** This element forms a sulfide FeS with iron, which remains insoluble within the solidifying molten mass. Repelled by the solidification mechanism, sulfides migrate to the areas of slowest solidification, which are located at the top of the ingot, concentrated around its central axis.

Sulfides' action consists of their enwrapping the material's austenitic grain structure and impairing its cohesion, which results in an intergranular cracking. They increase the brittleness at higher working temperatures and decrease the toughness, strength, ductility, and machinability of the material.

Sulfur readily combines with manganese, if the latter is present, in which condition it forms MnS of a high melting temperature of approximately 3000°F. Most of this sulfide is expelled from the material along with the slag.

**14-1-3-2. Additives beneficial to the material.** Copper, manganese, and silicon are elements that influence steel positively, enhancing properties that are advantageous in a wide range of applications.

**Copper (Cu).** This element slows down the recrystallization rate and slightly improves the toughness of the finished material. It usually gets into the metal in the form of an ingredient of various ores or from the metal scrap added to the process. Quantitatively, copper hardly ever exceeds 0.2 percent. Amounts ranging at about 0.1 percent improve the resistance to corrosion, weathering effects, and humidity. Larger amounts of copper are not beneficial at all, as they enhance the tendency of the material surface to crack during heat working.

**Manganese (Mn).** A deoxidant and sulfur repellant, manganese is used most often in quantities of 0.1 to 0.8 percent within the steel makeup. When dissolved within the ferritic substance of the material, it somewhat increases the toughness and strength of it, while decreasing its brittleness and improving forgeability. A small portion of manganese dissolved in cementite enhances its stability. Eutectoid concentration displays a marked dependency of carbon on manganese content; with increased percentages, the amount of carbon decreases and vice versa. Manganese further lowers the recrystallization speed while lowering the temperature ranges at which this process takes place as well.

However, manganese alone is not an adequate additive for deoxidation of the material, as it is not capable of preventing the reaction of carbon with the solidifying alloy. An unrestricted carbon action produces a material that is not completely killed, which may be detrimental to parts made out of it later under the assumption that it is killed. To aid the process, an inclusion of silicon, a deoxidant, or a combination of silicon and aluminum is necessary. Other deoxidants are titanium, zirconium, and calcium.

**Silicon (Si).** This is usually added to serve as a deoxidant, in quantities of up to 0.5 percent. It increases the ferritic resistance but lowers the material's formability and machinability, while improving hot-forming properties. All deep-drawing steels must have controlled, low amounts of silicon; otherwise the drawing process will be impaired.

Like manganese, silicon controls the amount of carbon in eutectoid and austenitic steel, making their quantities dependent on its percentage. Silicon further controls the proper dissolving of carbon within the base material, which makes this additive especially useful in the production of cast iron.

Special steels contain up to 1.5 to 2.5 percent silicon, in which case their hardenability, strength, and toughness are enhanced. Silicon, when added in such a large percentage, improves the electrical properties of the material, for which these steels are sometimes called electrical steels.

**14-1-3-3. Alloying additives in steel metallurgy.** Another group of materials to consider in steelmaking practice are alloying elements.

**Aluminum (Al).** This element is an excellent deoxidizer, along with other elements such as titanium, zirconium, and calcium. It is also utilized for controlling the grain size.

**Cobalt (Co).** Cobalt in small amounts supports the hot hardness of alloy. However, in large amounts cobalt is not beneficial, as it reduces the toughness of the material, increases its decarburization tendency, and raises the critical quenching temperature.

**Chromium (Cr).** This element increases the depth of hardness penetration and the material's response to heat treatment. The usual content is 0.5 to 1.5 percent Cr, with the exception of stainless steels, which contain this element in the amounts of 12 to 25 percent. In stainless steel, chromium is usually paired with nickel, providing the alloy with resistance to corrosion and oxidation. A high chromium content lowers the grindability, while raising the hardening temperature of the material, which may lead to deformation of heat-treated parts.

**Columbium (Cb) or niobium (Nb).** Columbium is similar in its effect to titanium, preventing the harmful carbide precipitation which causes an intergranular corrosion.

**Lead (Pb).** Lead improves machinability, being added to the alloy in quantities of 0.15 to 0.35 percent. It must be finely dispersed within the material.

**Molybdenum (Mo).** This element aids the penetration of hardness while increasing the toughness of the steel. It ranges usually from 0.1 to 0.4 percent. In small amounts, molybdenum aids the toughness and deep-hardening properties of the material. In higher concentrations, such as those used for high-speed steels, it replaces tungsten in some cases. Molybdenum will protect the material from effects of creep and improve its hot hardness.

**Nickel (Ni).** Nickel is found in the steel in quantities of 1 to 4 percent, although alloy steels containing up to 35 percent of Ni may be found. Some stainless steels may have up to 20 percent of Ni. Nickel increases the toughness and strength of the material, wear, and impact resistance at low temperatures.

**Tellurium (Te).** Used in the amount of approximately 0.05 percent, tellurium improves machinability.

**Titanium (Ti).** With similar effects to columbium, titanium provides the steel material with resistance to the harmful effects of carbon precipitation. When added to low-carbon steels, titanium makes them more suitable for porcelain enameling.

**Tungsten (W).** This element is often used in large quantities of 17 to 20 percent, usually in combination with chromium and other alloying elements. It is the basic ingredient of high-speed steel, which permits it to retain its hardness even at high temperatures on account of tungsten's good hot-hardness qualities. Tungsten is irreplaceable where a high-heat environment is encountered or where considerable wear resistance is required. In lesser percentages, tungsten is used to produce a fine, dense grain within the material.

**Vanadium (V).** When added to steel, most often in quantities of 0.15 to 0.20 percent, vanadium retards grain growth and refines the carbide structure, which results in improved forgeability. Vanadium also enhances the material's shock resistance and improves its hardness and resistance to wear and abrasion. Too much vanadium content lowers the grindability of the material.

## 14-2. Mechanical Properties of Metal Materials

Mechanical properties influence the behavior of materials and subsequently the behavior of products which they may form already during the manufacturing process and later in service as well. Various mechanical properties should be combined to suit the appropriate application, as some properties may be harmful where others are beneficial, and vice versa.

Mechanical properties may be roughly listed as shown in Table 14-1.

**TABLE 14-1 Mechanical Properties of Materials**

| Property | Symbol | Description | Units of measure | |
|---|---|---|---|---|
| | | | English | SI |
| Stress | $S$ | Force/unit area | lb/in$^2$ | MPa |
| Strain | $e$ | Deformation, $\Delta L/L$ | | |
| Modulus, elasticity | $E$ | Stress/elastic strain | lb/in$^2$ | MPa |
| Modulus, shear | $G$ | Unit stress/unit strain | lb/in$^2$ | MPa |
| Strength, yield | $S_y$ | Resistance to initial plastic deformation | lb/in$^2$ | MPa |
| Strength, tensile | $S_T$ | Maximum strength | lb/in$^2$ | MPa |
| Strength, ultimate | $S_u$ | See Strength, tensile | lb/in$^2$ | MPa |
| Ductility | | Plastic strain at failure | | |
| Elongation | | $(L_1 - L_0)/L_0$ | % | % |
| Reduction of area | | $(A_0 - A_1)/A_0$ | % | % |
| Toughness | | Energy for failure by fracture | ft-lb | joules |
| Hardness | | Resistance to indentation | Per method used | |

### 14-2-1. Strength of materials

The tensile strength of the material is a ratio of the force $P$ to the original cross-section area $A$, against which the force is applied, or

$$S_s = \frac{P}{A} \tag{14-1}$$

The same formula may be used when assessing the real, or true, stress. In such a case, the area of cross section will be that obtained after the cessation of the applied force.

Strength of the material may be described as the amount of force needed to produce its failure. The ability of the material not to become a victim to plastic deformation is expressed by the amount of its *yield strength* $S_y$. It is the ratio of the force within the yield strength range and the cross-sectional area of the test piece. Sometimes a *yield point* is given instead, especially where the material hardness is low.

### 14-2-2. Hardness of materials

Hardness of materials may be tested by several methods, all of them assessing resistance to penetration of the surface. Naturally, hardness and strength are therefore quite similar. Some most often used methods of measurement follow.

The *Rockwell hardness test* (*HR*) is performed by assessing the depth of penetration of a steel ball or a diamond spheroconical penetrating tool into the material's surface. The hardness evaluation is proportional to the depth of the indentation produced, with higher numbers assigned to harder materials.

There are several Rockwell hardness scales, with their field of application reserved for certain areas. The most common scales and their descriptions are shown in Table 14-2.

The *Brinell hardness test* (*HB*) for assessing the hardness of metallic materials is performed by applying a certain specific load on a steel ball of predetermined diameter, forcing it into a tested material. The diameter of resulting indentation is measured and the appropriate Brinell hardness number is calculated with the aid of a formula

$$BHN = \frac{2P}{\pi D(D - \sqrt{D^2 - d^2})} \tag{14-2}$$

where $P$ = load applied, kg
$D$ = diameter of ball, mm
$d$ = diameter of impression, mm

**TABLE 14-2 Rockwell Hardness Scales**

| Scale | Tester | Force, kg | Usage |
|---|---|---|---|
| A | Diamond point | 60 | For extremely hard materials |
| B | $\phi$0.062-in ball | 100 | Medium-hard materials, such as annealed low-carbon and medium-carbon steels |
| C | Diamond point | 150 | Continuation of B scale |
| D | Diamond point | 100 | Same as C-scale testing with a lighter testing load |
| E | $\phi$0.125-in ball | 100 | For very soft materials |
| F | $\phi$0.062-in ball | 60 | Continuation of E scale |
| G | $\phi$0.062-in ball | 48 | Testing of metals harder than those using B scale |
| H | $\phi$0.125-in ball | 60 | For softer materials |
| K | $\phi$0.125-in ball | 150 | For softer materials |

*Vickers hardness testing* (*HV*) is somewhat similar to the Brinell method. The penetrator is a square-based diamond pyramid, angled toward its point at 136°. The numerical value of the hardness is calculated from the ratio of the load to the area of the impression as

$$H_v = \frac{2P\sin(\alpha/2)}{d^2} = \frac{1.8544P}{d^2} \tag{14-3}$$

where $P$ = load applied, kg
$d$ = diagonal of indentation, mm
$\alpha$ = face angle of pyramid, 136°

The *Knoop hardness test* is used for evaluation of extremely thin materials, plated surfaces, very brittle or very hard surfaces, or where the load must be kept below 3.6 kg. The hardness number is obtained by dividing the applied load by the area of indentation. The penetrator in this case is a special rhombic-based pyramid indenter, made of industrial diamond, with one of the angles between the intersections of its four-faceted shape ground to an angle of 172.5° while the other angle is 130°. The calculation of the hardness value is

$$H_K = \frac{P}{0.07028d^2} \tag{14-4}$$

where $P$ = load applied, kg
$d$ = length of a long diagonal, mm
0.07028 = constant, corresponding to the standard angles of the pyramidal shape

*Shore's scleroscope testing* evaluates hardness in terms of elasticity. A diamond-pointed tester is dropped from a predetermined height onto the tested material. On falling upon its surface, the tester rebounds, the height of its rebound indicating the hardness of the material. With harder materials, this distance is greater; with softer materials it is shorter. Shore's hardness is often used to specify the toughness of rubber or urethane.

### 14-2-3. Toughness of materials

The degree of toughness actually assesses the amount of energy required to produce a breakage within the material. This is not the same property as the material's strength, which shows its reaction to the stresses, producing a deformation within its structure.

Toughness is composed of two components: an elastic component (strength) and a plastic component (ductility). The elastic portion of toughness increases in congruence with the increasing hardness, while at the same instant the plastic component decreases.

The toughness of a material is given in terms of energy needed to break a sample piece. This energy is specified in foot-pounds or joules and is measured by using the Izod (or Charpy) testing arrangement, where a test specimen is broken by an impact of a swinging pendulum. Also tensile testing and torsion testing has been used in assessment of material toughness.

### 14-2-4. Deformation and ductility

Deformation of the material occurs where the stress (force) is applied to its mass, coercing its shape to follow the direction of its application. There are two types of deformation: elastic and plastic. In *elastic deformation,* the material is altered within its elastic limits, which means that on release of force it returns to its original size and shape.

*Plastic deformation* occurs where deviation in shape due to the application of force exceeds the material's elastic limit, in which case the deformation is permanent, and on release of the force the material will not return to its original form and shape. Examples of plastic deformation are bending, forming, and drawing operations. Examples of elastic deformation can be found in spring materials.

The initial strain of the material is proportional to the stress applied against it. The modulus of elasticity, or Young's modulus $E$, depicts the stress and strain relationship as

$$E = \frac{S}{e} \qquad \text{lb/in}^2 \text{ or MPa} \tag{14-5}$$

where $S$ = stress

$e$ = strain

**TABLE 14-3 Comparison of Hardness Scales for Hardened Steel**

| Rockwell scale C | Brinell *HB* | Vickers *HV* | Shore scleroscope | Rockwell Scale A | Rockwell Scale D | Rockwell 15-N | Rockwell 30-N | Rockwell 45-N | Rockwell Scale G | Knoop *HK* |
|---|---|---|---|---|---|---|---|---|---|---|
| 20 | 230 | 236 | 34 | 60.5 | 40.0 | 69.5 | 41.5 | 19.5 | 81.0 | 251 |
| 21 | 235 | 241 | 35 | 61.0 | 41.0 | 70.0 | 42.5 | 20.5 | 82.5 | 256 |
| 22 | 240 | 246 | 35 | 61.5 | 41.5 | 70.5 | 43.0 | 22.0 | 83.5 | 261 |
| 23 | 245 | 251 | 36 | 62.0 | 42.5 | 71.0 | 44.0 | 23.0 | 84.5 | 266 |
| 24 | 250 | 257 | 37 | 62.5 | 43.0 | 71.5 | 45.0 | 24.0 | 86.0 | 272 |
| 25 | 255 | 264 | 38 | 63.0 | 44.0 | 72.0 | 46.0 | 25.5 | 87.0 | 278 |
| 26 | 260 | 271 | 38 | 63.5 | 44.5 | 72.5 | 47.0 | 26.5 | 88.0 | 284 |
| 27 | 265 | 278 | 40 | 64.0 | 45.5 | 73.5 | 47.5 | 28.0 | 89.0 | 290 |
| 28 | 270 | 285 | 41 | 64.5 | 46.0 | 74.0 | 48.5 | 29.0 | 90.0 | 297 |
| 29 | 276 | 293 | 41 | 65.0 | 47.0 | 74.5 | 49.5 | 30.0 | 91.0 | 304 |
| 30 | 283 | 301 | 42 | 65.5 | 47.5 | 75.0 | 50.5 | 31.5 | 92.0 | 311 |
| 31 | 290 | 309 | 43 | 66.0 | 48.5 | 75.5 | 51.5 | 32.5 | | 318 |
| 32 | 297 | 317 | 44 | 66.5 | 49.0 | 76.0 | 52.0 | 33.5 | | 326 |
| 33 | 305 | 325 | 46 | 67.0 | 50.0 | 76.5 | 53.0 | 35.0 | | 334 |
| 34 | 313 | 334 | 47 | 67.5 | 50.5 | 77.0 | 54.0 | 36.0 | | 342 |
| 35 | 322 | 343 | 48 | 68.0 | 51.5 | 78.0 | 55.0 | 37.0 | | 351 |
| 36 | 332 | 353 | 49 | 68.5 | 52.5 | 78.5 | 56.0 | 38.5 | | 360 |
| 37 | 342 | 363 | 50 | 69.0 | 53.0 | 79.0 | 56.5 | 39.5 | | 370 |
| 38 | 352 | 373 | 51 | 69.5 | 54.0 | 79.5 | 57.5 | 41.0 | | 380 |
| 39 | 362 | 383 | 52 | 70.0 | 54.5 | 80.0 | 58.5 | 42.0 | | 391 |
| 40 | 372 | 393 | 54 | 70.5 | 55.5 | 80.5 | 59.5 | 43.0 | | 402 |
| 41 | 382 | 403 | 55 | 71.0 | 56.0 | 81.0 | 60.5 | 44.5 | | 414 |
| 42 | 393 | 413 | 56 | 71.5 | 57.0 | 81.5 | 61.5 | 45.5 | | 426 |
| 43 | 404 | 424 | 57 | 72.0 | 57.5 | 82.0 | 62.0 | 46.5 | | 438 |
| 44 | 415 | 435 | 58 | 72.5 | 58.5 | 82.5 | 63.0 | 48.0 | | 452 |
| 45 | 426 | 446 | 60 | 73.0 | 59.0 | 83.0 | 64.0 | 49.0 | | 466 |
| 46 | 437 | 458 | 62 | 73.5 | 60.0 | 83.5 | 65.0 | 50.0 | | 480 |
| 47 | 448 | 471 | 63 | 74.0 | 60.5 | 84.0 | 66.0 | 51.5 | | 495 |
| 48 | 460 | 485 | 64 | 74.5 | 61.5 | 84.5 | 66.5 | 52.5 | | 510 |
| 49 | 472 | 498 | 66 | 75.5 | 62.0 | 85.0 | 67.5 | 54.0 | | 526 |
| 50 | 484 | 513 | 67 | 76.0 | 63.0 | 85.5 | 68.5 | 55.0 | | 542 |
| 51 | 496 | 528 | 68 | 76.5 | 64.0 | 86.0 | 69.5 | 56.0 | | 558 |
| 52 | 509 | 545 | 69 | 77.0 | 64.5 | 86.5 | 70.5 | 57.5 | | 576 |
| 53 | 522 | 562 | 71 | 77.5 | 65.5 | 87.0 | 71.0 | 58.8 | | 594 |
| 54 | 534 | 580 | 72 | 78.0 | 66.0 | 87.5 | 72.0 | 59.5 | | 612 |
| 55 | 547 | 598 | 74 | 78.5 | 67.0 | 88.0 | 73.0 | 61.0 | | 630 |
| 56 | 560 | 617 | 75 | 79.0 | 67.5 | 88.5 | 74.0 | 62.0 | | 650 |
| 57 | 573 | 636 | 76 | 79.5 | 68.5 | 89.0 | 75.0 | 63.0 | | 670 |
| 58 | 587 | 655 | 78 | 80.0 | 69.0 | 89.3 | 75.5 | 64.0 | | 690 |
| 59 | 600 | 675 | 80 | 80.5 | 70.0 | 89.5 | 76.5 | 65.5 | | 710 |
| 60 | 614 | 695 | 81 | 81.0 | 71.0 | 90.0 | 77.5 | 66.5 | | 732 |
| 61 | | 716 | 83 | 81.5 | 71.5 | 90.5 | 78.5 | 67.5 | | 754 |
| 62 | | 739 | 85 | 82.5 | 72.5 | 91.0 | 79.0 | 69.0 | | 776 |
| 63 | | 763 | 87 | 83.0 | 73.0 | 91.5 | 80.0 | 70.0 | | 799 |
| 64 | | 789 | 88 | 83.5 | 74.0 | 91.8 | 81.0 | 71.0 | | 822 |
| 65 | | 820 | 91 | 84.0 | 74.5 | 92.0 | 82.0 | 72.0 | | 846 |
| 66 | | 854 | 92 | 84.5 | 75.5 | 92.5 | 83.0 | 73.0 | | 870 |
| 67 | | 894 | 95 | 85.0 | 76.0 | 93.0 | 83.5 | 74.5 | | 895 |
| 68 | | 942 | 97 | 85.5 | 77.0 | 93.2 | 84.5 | 75.5 | | 920 |
| 69 | | 1004 | | 86.0 | 78.0 | 93.5 | 85.0 | 76.5 | | 946 |
| 70 | | 1076 | | 86.5 | 78.5 | 94.0 | 86.0 | 77.5 | | 972 |

**TABLE 14-4 Comparison of Hardness Scales for Nonhardened Steel, Iron, and Nonferrous Metals**

| Rockwell | | | | Rockwell | | | | | | | |
|---|---|---|---|---|---|---|---|---|---|---|---|
| Scale B | Scale F | Brinell *HB* | Vickers *HV* | Scale G | 15-T | 30-T | 45-T | Scale E | Scale K | Scale A | Knoop *HK* |
| 0 | 57.0 | 53 | | | | 15.0 | | 57.0 | 21.0 | | 67 |
| 1 | 57.5 | | | | 61.0 | 16.0 | | 57.5 | 22.0 | | |
| 2 | 58.0 | 54 | | | 61.5 | 16.5 | | 58.0 | 23.0 | | 68 |
| 3 | 59.0 | | | | | 17.0 | | 58.5 | 23.5 | | |
| 4 | 59.5 | | | | 62.0 | 18.0 | | 59.0 | 24.5 | | |
| 5 | 60.0 | 55 | | | 62.5 | 18.5 | | 60.0 | 25.5 | | 69 |
| 6 | 60.5 | | | | | 19.5 | | 60.5 | 26.0 | | |
| 7 | 61.0 | 56 | | | 63.0 | 20.0 | | 61.0 | 27.0 | | |
| 8 | 61.5 | | | | 63.5 | 20.5 | | 61.5 | 28.0 | | 71 |
| 9 | 62.0 | | | | | 21.5 | | 62.0 | 29.0 | | |
| 10 | 63.0 | 57 | | | 64.0 | 22.0 | | 62.5 | 29.5 | | |
| 11 | 63.5 | | | | | 23.0 | | 63.5 | 30.5 | | 73 |
| 12 | 64.0 | | | | 64.5 | 23.5 | | 64.0 | 31.5 | | |
| 13 | 64.5 | 58 | | | 65.0 | 24.0 | | 64.5 | 32.0 | | |
| 14 | 65.0 | | | | | 25.0 | | 65.0 | 33.0 | | |
| 15 | 65.5 | 59 | | | 65.5 | 25.5 | | 65.5 | 34.0 | 20.0 | 76 |
| 16 | 66.0 | | | | 66.0 | 26.0 | | 66.5 | 35.0 | 20.5 | |
| 17 | 66.5 | 60 | | | | 27.0 | | 67.0 | 35.5 | 21.0 | |
| 18 | 67.0 | | | | 66.5 | 27.5 | | 67.5 | 36.5 | | |
| 19 | 68.0 | 61 | | | 67.0 | 28.5 | | 68.0 | 37.5 | 21.5 | 79 |
| 20 | 68.5 | | | | | 29.0 | | 68.5 | 38.0 | 22.0 | |
| 21 | 69.0 | 62 | | | 67.5 | 29.5 | | 69.5 | 39.0 | 22.5 | |
| 22 | 69.5 | | | | | 30.5 | | 70.0 | 40.0 | 23.0 | |
| 23 | 70.0 | 63 | | | 68.0 | 31.0 | | 70.5 | 41.0 | 23.5 | 82 |
| 24 | 70.5 | | | | 68.5 | 32.0 | | 71.0 | 41.5 | 24.0 | |
| 25 | 71.0 | 64 | | | | 32.5 | | 72.0 | 42.5 | | |
| 26 | 72.0 | 65 | | | 69.0 | 33.0 | | 72.5 | 43.5 | 24.5 | |
| 27 | 72.5 | | | | 69.5 | 34.0 | | 73.0 | 44.5 | 25.0 | 85 |
| 28 | 73.0 | 66 | | | | 34.5 | | 73.5 | 45.0 | 25.5 | |
| 29 | 73.5 | | | | 70.0 | 35.5 | 1.0 | 74.0 | 46.0 | 26.0 | |
| 30 | 74.0 | 67 | | | 70.5 | 36.0 | 2.0 | 75.0 | 47.0 | 26.5 | |
| 31 | 74.5 | 68 | | | | 36.5 | 3.0 | 75.5 | 48.0 | 27.0 | 88 |
| 32 | 75.0 | | | | 71.0 | 37.5 | 4.0 | 76.0 | 48.5 | 27.5 | 89 |
| 33 | 75.5 | 69 | | | | 38.0 | 5.0 | 76.5 | 49.5 | | 90 |
| 34 | 76.5 | 70 | | | 71.5 | 38.5 | 6.0 | 77.0 | 50.5 | 28.0 | 91 |
| 35 | 77.0 | 71 | | | 72.0 | 39.5 | 7.0 | 78.0 | 51.5 | 28.5 | 92 |
| 36 | 77.5 | | | | | 40.0 | 8.0 | 78.5 | 52.0 | 29.0 | 93 |
| 37 | 78.0 | 72 | | | 72.5 | 40.5 | 9.0 | 79.0 | 53.0 | 29.5 | 94 |
| 38 | 78.5 | 76 | | | 73.0 | 41.5 | 10.0 | 79.5 | 54.0 | 30.0 | 95 |
| 39 | 79.0 | 74 | | | | 42.0 | 11.0 | 80.0 | 54.5 | 30.5 | 96 |
| 40 | 79.5 | | | | 73.5 | 43.0 | 12.5 | 81.0 | 55.5 | | 97 |
| 41 | 80.5 | 75 | | | 74.0 | 43.5 | 13.5 | 81.5 | 56.5 | 31.0 | 98 |
| 42 | 81.0 | 76 | | | | 44.0 | 14.5 | 82.0 | 57.5 | 31.5 | 99 |
| 43 | 81.5 | 77 | | | 74.5 | 45.0 | 15.5 | 82.5 | 58.0 | 32.0 | 100 |
| 44 | 82.0 | 78 | | | 75.0 | 45.5 | 16.5 | 83.5 | 59.0 | 32.5 | 101 |
| 45 | 82.5 | 79 | | | | 46.0 | 17.5 | 84.0 | 60.0 | 33.0 | 102 |
| 46 | 83.0 | | | | 75.5 | 47.0 | 18.5 | 84.5 | 61.0 | 33.5 | 103 |
| 47 | 84.0 | 80 | | | 76.0 | 47.5 | 19.5 | 85.0 | 61.5 | 34.0 | 104 |

**TABLE 14-4 Comparison of Hardness Scales for Nonhardened Steel, Iron, and Nonferrous Metals (*Continued*)**

| Rockwell | | | | Rockwell | | | | | | | |
|---|---|---|---|---|---|---|---|---|---|---|---|
| Scale B | Scale F | Brinell *HB* | Vickers *HV* | Scale G | 15-T | 30-T | 45-T | Scale E | Scale K | Scale A | Knoop *HK* |
| 48 | 84.5 | 81 | | | | 48.5 | 20.5 | 85.5 | 62.5 | 34.5 | 105 |
| 49 | 85.0 | 82 | | | 76.5 | 49.0 | 22.0 | 86.5 | 63.5 | | 106 |
| 50 | 85.5 | 83 | | 2.5 | 77.0 | 49.5 | 23.0 | 87.0 | 64.5 | 35.0 | 107 |
| 51 | 86.0 | 84 | | 4.0 | | 50.5 | 24.0 | 87.5 | 65.0 | 35.5 | 108 |
| 52 | 86.5 | 85 | | 5.5 | 77.5 | 51.0 | 25.0 | 88.0 | 66.0 | 36.0 | 109 |
| 53 | 87.0 | 86 | | 7.0 | 78.0 | 51.5 | 26.0 | 89.0 | 67.0 | 36.5 | 110 |
| 54 | 87.5 | 87 | | 8.5 | | 52.5 | 27.0 | 89.5 | 68.0 | 37.0 | 111 |
| 55 | 88.0 | 89 | 100 | 10.0 | 78.5 | 53.0 | 28.0 | 90.0 | 68.5 | 37.5 | 112 |
| 56 | 89.0 | 90 | 101 | 11.5 | 79.0 | 54.0 | 29.0 | 90.5 | 69.5 | | 114 |
| 57 | 89.5 | 91 | 103 | 13.0 | | 54.5 | 30.0 | 91.0 | 70.5 | 38.0 | 115 |
| 58 | 90.0 | 92 | 104 | 14.5 | 79.5 | 55.0 | 31.0 | 92.0 | 71.0 | 38.5 | 117 |
| 59 | 90.5 | 94 | 106 | 16.0 | 80.0 | 56.0 | 32.0 | 92.5 | 72.0 | 39.0 | 118 |
| 60 | 91.0 | 95 | 107 | 17.5 | | 56.5 | 33.5 | 93.0 | 73.0 | 39.5 | 120 |
| 61 | 91.5 | 96 | 108 | 19.0 | 80.5 | 57.0 | 34.5 | 93.5 | 74.0 | 40.0 | 122 |
| 62 | 92.0 | 98 | 110 | 20.5 | | 58.0 | 35.5 | 94.5 | 74.5 | 40.5 | 124 |
| 63 | 93.0 | 99 | 112 | 22.0 | 81.0 | 58.5 | 36.5 | 95.0 | 75.5 | 41.0 | 125 |
| 64 | 93.5 | 101 | 114 | 23.5 | 81.5 | 59.5 | 37.5 | 95.5 | 76.5 | 41.5 | 127 |
| 65 | 94.0 | 102 | 116 | 25.0 | | 60.0 | 38.5 | 96.0 | 77.5 | | 129 |
| 66 | 94.5 | 104 | 117 | 26.5 | 82.0 | 60.5 | 39.5 | 97.0 | 78.0 | 42.0 | 131 |
| 67 | 95.0 | 106 | 119 | 28.0 | 82.5 | 61.5 | 40.5 | 97.5 | 79.0 | 42.5 | 133 |
| 68 | 95.5 | 107 | 121 | 29.5 | | 62.0 | 41.5 | 98.0 | 80.0 | 43.0 | 135 |
| 69 | 96.0 | 109 | 123 | 31.0 | 83.0 | 62.5 | 42.5 | 99.0 | 81.0 | 43.5 | 137 |
| 70 | 97.0 | 110 | 125 | 32.5 | 83.5 | 63.5 | 43.5 | 99.5 | 81.5 | 44.0 | 139 |
| 71 | 97.5 | 112 | 127 | 34.5 | | 64.0 | 44.5 | 100.0 | 82.5 | 44.5 | 141 |
| 72 | 98.0 | 114 | 130 | 36.0 | 84.0 | 65.0 | 45.5 | | 83.5 | 45.0 | 143 |
| 73 | 98.5 | 116 | 132 | 37.5 | 84.5 | 65.5 | 46.5 | | 84.5 | 45.5 | 145 |
| 74 | 99.0 | 118 | 135 | 39.0 | | 66.0 | 47.5 | | 85.0 | 46.0 | 147 |
| 75 | 99.5 | 120 | 137 | 41.0 | 85.0 | 67.0 | 48.5 | | 86.0 | 46.5 | 150 |
| 76 | | 122 | 139 | 42.5 | | 67.5 | 49.0 | | 87.0 | 47.0 | 152 |
| 77 | | 124 | 141 | 44.0 | 85.5 | 68.0 | 50.0 | | 88.0 | 48.0 | 155 |
| 78 | | 126 | 144 | 46.0 | 86.0 | 69.0 | 51.0 | | 88.5 | 48.5 | 158 |
| 79 | | 128 | 147 | 47.5 | | 69.5 | 52.0 | | 89.5 | 49.0 | 161 |
| 80 | | 130 | 150 | 49.0 | 86.5 | 70.0 | 53.0 | | 90.5 | 49.5 | 164 |
| 81 | | 133 | 153 | 51.0 | 87.0 | 71.0 | 54.0 | | 91.0 | 50.0 | 167 |
| 82 | | 135 | 156 | 52.5 | | 71.5 | 55.0 | | 92.0 | 50.5 | 170 |
| 83 | | 137 | 159 | 54.0 | 87.5 | 72.0 | 56.0 | | 93.0 | 51.0 | 173 |
| 84 | | 140 | 162 | 56.0 | 88.0 | 73.0 | 57.0 | | 94.0 | 52.0 | 176 |
| 85 | | 142 | 165 | 57.5 | | 73.5 | 58.0 | | 94.5 | 52.5 | 180 |
| 86 | | 145 | 169 | 59.0 | 88.5 | 74.0 | 58.5 | | 95.5 | 53.0 | 184 |
| 87 | | 148 | 172 | 61.0 | 89.0 | 74.5 | 59.5 | | 96.5 | 53.5 | 188 |
| 88 | | 151 | 176 | 62.5 | | 75.0 | 60.5 | | 97.0 | 54.0 | 192 |
| 89 | | 154 | 180 | 64.0 | 89.5 | 75.5 | 61.5 | | 98.0 | 55.0 | 196 |
| 90 | | 157 | 185 | 66.0 | 90.0 | 76.0 | 62.5 | | 98.5 | 55.5 | 201 |
| 91 | | 160 | 190 | 67.5 | | 77.0 | 63.5 | | 99.5 | 56.0 | 206 |
| 92 | | 163 | 195 | 69.0 | 90.5 | 77.5 | 64.5 | | 100.0 | 56.5 | 211 |
| 93 | | 167 | 200 | 71.0 | 91.0 | 78.0 | 65.5 | | | 57.0 | 216 |
| 94 | | 171 | 205 | 72.5 | | 78.5 | 66.0 | | | 57.5 | 221 |
| 95 | | 175 | 210 | 74.0 | 91.5 | 79.0 | 67.0 | | | 58.0 | 226 |
| 96 | | 179 | 216 | 76.0 | | 80.0 | 68.0 | | | 59.0 | 231 |

**TABLE 14-4 Comparison of Hardness Scales for Nonhardened Steel, Iron, and Nonferrous Metals (*Continued*)**

| Rockwell | | | | Rockwell | | | | | | | |
|---|---|---|---|---|---|---|---|---|---|---|---|
| Scale B | Scale F | Brinell *HB* | Vickers *HV* | Scale G | 15-T | 30-T | 45-T | Scale E | Scale K | Scale A | Knoop *HK* |
| 97 | | 184 | 222 | 77.5 | 92.0 | 80.5 | 69.0 | | | 59.5 | 236 |
| 98 | | 189 | 228 | 79.0 | | 81.0 | 70.0 | | | 60.0 | 241 |
| 99 | | 195 | 234 | 81.0 | 92.5 | 81.5 | 71.0 | | | 61.0 | 246 |
| 100 | | 201 | 240 | 82.5 | 93.0 | 82.0 | 72.0 | | | 61.5 | 251 |

A major characteristic of the material, influencing its behavior under the application of stress (force), is its ductility. *Ductility* may be expressed as the percentage of elongation, in which case it depends on the result of the relationship

$$\frac{L_1 - L_0}{L_0} = \frac{\Delta L}{L_0} \tag{14-6}$$

where $L_0$ = original length of part
$L_1$ = final length of part

Ductility assessed through this method is based on the amount of elongation of the sample specimen. Since the greater amount of plastic deformation occurs in the necked area of the tested piece, the percentage of elongation must be specified as dependent on the length of the gauge (see Fig. 14-3).

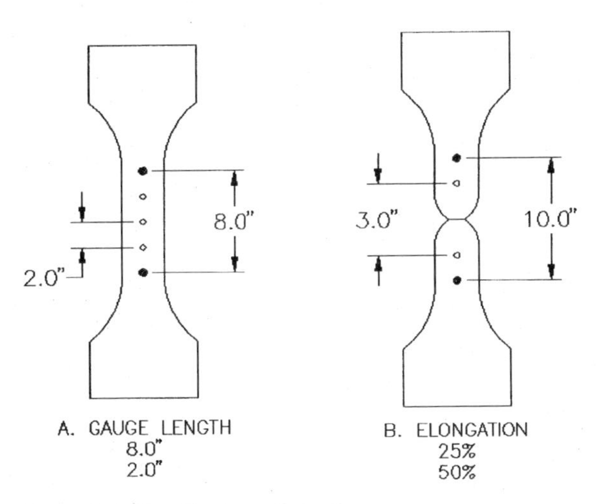


**Figure 14-3** Tensile strength testing.

Ductility may also be evaluated on the grounds of the reduction in area at the point of fracture as

$$\frac{A_0 - A_1}{A_0} = \frac{\Delta A}{A_0} \qquad (14\text{-}7)$$

High ductility means the neck of the fracture will be quite small in cross section before it breaks off.

*Deformation* tendency of a steel may be caused by a combination of its properties, as brought together by different manufacturing methods. For example, a previously cold-worked steel material will display a marked tendency to distortion after machining. It should never be finished to size without first being stress-relieved. The heat-treatment process may produce additional dimensional changes within such material, or a deviation from straight, round, etc. Therefore, only after stress relieving should such parts be finished and their critical dimensions tightened. Areas that are most susceptible to distortions are those where the greatest amount of machining was performed and subsequently where the greatest amount of stress was produced.

Even a previously cold-rolled material, when cold-worked afterward, will experience changes within its structure, affecting it into its very core. The amount of additional cold working may not be extensive; a 5 percent reduction due to rolling will suffice to bring about considerable changes.

For one thing, the cold working produces an increase in the material's hardness, or so-called strain hardening. With approximately 15 percent reduction in thickness, the effect of hardening on the layers of material closest to the surface is guaranteed. However, with a reduction in size amounting to up to 25 percent, the whole thickness of the stock will be hardened, or strain-hardened.

Hot-rolled steels are not excessively influenced by the cold work instituted afterward. Even though such treatment increases their hardness, the amounts are small in comparison with cold-rolled stock. Where in hot-rolled steel the surface hardness may be increased some 30 percent, the same amount of cold work will produce roughly a 60 percent increase of hardness in cold-rolled material.

### 14-2-5. Thermal properties

Where the temperature is used to measure the amount of thermal activity, heat content depicts the amount of thermal energy.

*Heat capacity* is the change in heat content of the test specimen, given in either °F or °C or other units. The *specific heat* is the ratio of the heat capacity of that particular material to the heat capacity of water.

*Thermal conductivity* $k$ is a measure of heat transfer, utilized with the aid of a given material. The amount of thermal conductivity is a constant, relating the heat flux $Q$ to the thermal gradient $\Delta T/\Delta x$ or

$$Q = k\frac{\Delta T}{\Delta x} = k\frac{T_1 - T_0}{x_1 - x_0} \tag{14-8}$$

where $T$ = temperature value
$x$ = thickness of material

The value of the thermal conductivity itself $Q$ can be expressed as the ratio of energy to area $A$ by time $t$, or

$$Q = \frac{\text{energy}}{At}$$

*Thermal expansion* measures the expansion in length of a heated specimen. The increase in length $\Delta L$ is considered proportional to the change in temperature $\Delta T$, or

$$\frac{\Delta L}{L} = \alpha_L \Delta T \tag{14-9}$$

where $\alpha_L$ = coefficient of linear expansion

Where assessing the volumnar expansion, the coefficient $\alpha_v$ depicts the volumnar change $\Delta V/V$. $\alpha_v = 3\alpha_L$, and the formula becomes

$$\alpha_v \Delta T = 3(\alpha_L)\Delta T = \frac{\Delta V}{V} \tag{14-10}$$

### 14-2-6. Electrical properties

Metals, like semiconductors, conduct an electrical charge when positioned within an electrical field. The conductivity depends on the charge carried, on the mobility of the carrier, and on the number of carriers. The conductivity is the opposite, or the reciprocal of the resistivity $\rho$. It can be expressed as

$$\sigma = 1/\rho = nq\mu \tag{14-11}$$

where $\sigma$ = conductivity
$\rho$ = resistivity
$n$ = number of carriers
$\mu$ = carriers' mobility
$q$ = charge

Resistivity is considered to be an electrical property of material. In parts of uniform geometry, resistivity may be converted to the material's resistance $R$, as

$$R = \frac{\rho L}{A} \tag{14-12}$$

where $A$ = cross-sectional area
$L$ = length of the piece

### 14-2-7. Endurance and fatigue

Fatigue will impair the performance of parts or assemblies over a period of time and after going through a certain number of work cycles. The effects of fatigue may be seen on a round metal bar, with both ends subjected to loading of the same type as that of a beam. A compressive force is acting against its upper surface, creating a tension within the bottom layers. Such an arrangement, if allowed to rotate, will definitely produce a dimensional alteration of the specimen. After the completion of a few hundred thousand cycles, even a test piece made of good grade steel may begin to show a fatigue-dependent distortion, such as the one in Fig. 14-4.

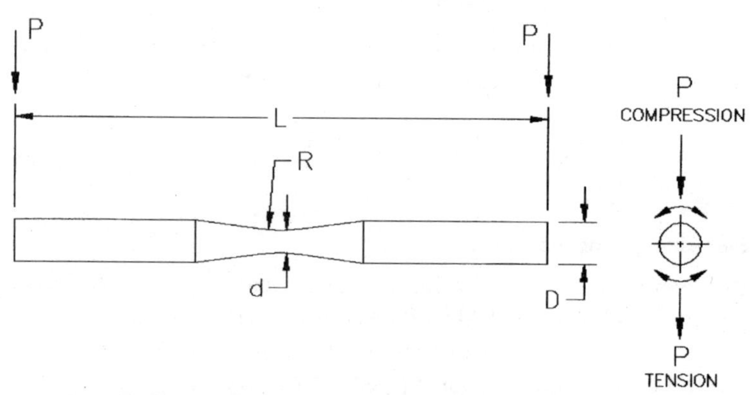


**Figure 14-4** Endurance limit testing.

For comparison, a similar specimen exposed to a loading of approximately 60,000 lb/in$^2$ broke apart under the effect of fatigue on completion of 100,000 revolutions.

The endurance limit is greater where the specimen's surface is polished. Subsequently, corrosion lowers this limit, as well as any sharp corners, notches, and similar sharp indentations of the part's surface.

Even the width of a groove on a shaft has a definite effect on the endurance limit of such a part. If such a shaft has an endurance limit of 35,000 lb/in$^2$, a circular notch may lower this value to some 85 percent, which is 30,000 lb/in$^2$.

A comparison of the performance of notched parts is offered here, as based on the actual testing, where a specimen of 1-in-diameter heat-treated steel bar was subjected to a surface alteration by notching and rotated in the testing machine described earlier. The endurance limit of the material was 75,000 lb/in$^2$. A nice, rounded fillet, as shown in Fig. 14-5*A*, made the part sustain 80,000 revolutions before it broke apart. A bit sharper fillet, shown in Fig. 14-5*B*, produced 75,000 revolutions. A sharp, square groove (Fig. 14-5*C*) lasted 20,000 revolutions. A very sharp V notch (Fig. 14-5*D*) broke the test piece after 14,000 revolutions.

From the above, the influence of the sharpness of grooving on the life expectancy of the part is more than obvious.

Another detrimental effect on the part's endurance limit may be attributed to the decarburization process applied to it. Where an annealed steel has a strength of approximately 60,000 lb/in$^2$, its high-strength counterpart boasts 200,000 lb/in$^2$, with a fatigue strength of some 110,000 lb/in$^2$. However, if a part made of such high-strength material is decarburized, its endurance limit drops to 60,000 lb/in$^2$. There is no size change, no length alteration, no difference in appearance. Only the endurance limit drops to a meager 60,000 lb/in$^2$, which is equivalent to the strength of an annealed steel.

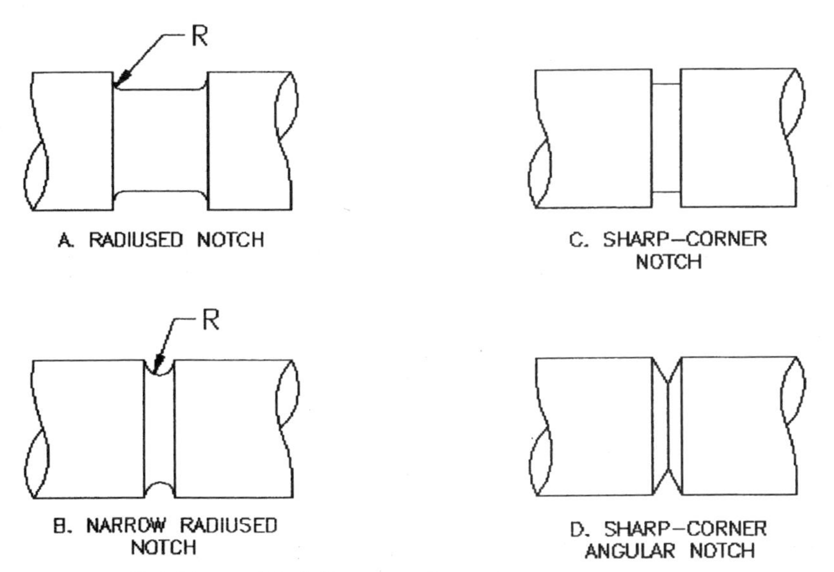


**Figure 14-5** Notching and its influence on the part.

Where a grooved shaft was subjected to decarburization, it broke off within the area of the groove. Leaf springs and valve springs are greatly affected by decarburization, which usually results in their failure later in service.

In Fig. 14-6, depicting the drop in the endurance limit for various materials, the lines shown are for the specific material's surface conditions: (*a*) polished, (*b*) ground, (*c*) roughed, (*d*) with a sharp circumferential indentation, or grooved, (*e*) with hardened skin after cold rolling, (*f*) surface corroded by water, (*g*) surface corroded by salt water.

Notching is not the only means of producing fatigue and lowering the endurance limit of parts. The same may be achieved with any sharp edge, any thickness inequality, and many other deviations from

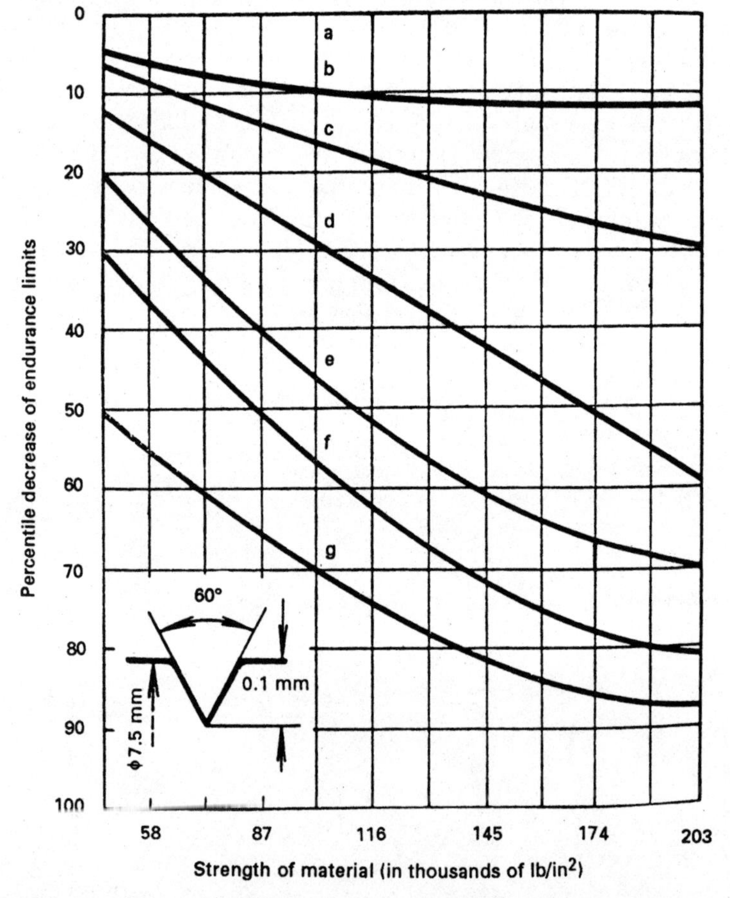


**Figure 14-6** Lowered endurance limit due to variation of the surface finish. (*From Svatopluk Černoch,* "Strojně technická příručka," *1977. Reprinted with permission from SNTL Publishers, Prague, CZ.*)

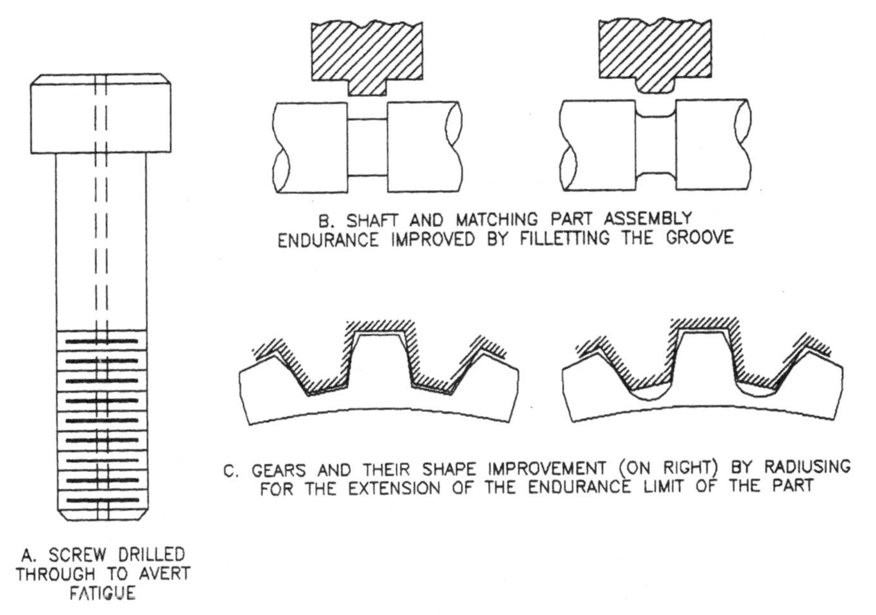


**Figure 14-7** Alterations to improve the endurance limit of parts.

a sound design. Finite-element analysis reveals clearly and in colors, the accumulation of stresses within a body of a part. The most crowded areas are exactly those mentioned earlier.

A screw, shown in Fig. 14-7, will gain in strength if a hole is drilled through its central portion. Naturally, it must not be in any way located near the roots of threads. A small opening proportionally equal to the one shown will relieve some inner stresses imposed on the screw by the process of tightening it, which actually is a process of applying pressure toward some portions of its thread. With such an empty space to accommodate for any irregularities due to the loading, the thread grows stronger and the functionability of the screw is enhanced.

Another contributing factor is a shape of the central opening, which is that of a continuous supporting arch, not unlike those used in medieval times for support of heavy stone structures. The support of such a continuous arch is flexible enough, yet firm and unyielding.

### 14-2-8. Wear

Wear may be classified as removal of surficial layers of material, caused by the part's performance in the work process. Most wear is attributable to adhesion and diffusion of material, chipped off the tooling or workpiece, but additional causes of wear mechanisms are given in the following list:

- **Abrasion wear** is usually caused by impurities such as dirt, fragments of metal, and similar objects within the working areas of parts. By passing over the tool surface, they may mechanically attack it by scratching its outer layer.
- **Adhesion wear** is caused by the friction between the tooling and the workpiece. The heat of friction tears off small pieces of the tool material, depositing it either with chips or with the workpiece.
- **Diffusion wear** occurs where a localized high heat, produced by the manufacturing process, forces the atoms within the material's metallic crystal to move from one point of the lattice to another point. Such a movement creates a shift within the material, during which the elements transfer in the direction of a different rate of their concentration. The surface of the part becomes depleted and may disintegrate and dust away or be removed otherwise.
- **Electrochemical wear** is found where an interaction between the material of tooling and that of the part is achieved in an environment created by a cutting fluid. Such a wear mechanism forces the ions to pass freely between the tool and the workpiece, resulting in oxidation of the tool surface.
- **Pitting** may be classified as a result of compression-related surface failure. It is caused by a repeated action against the material surface, during which stresses are applied against the material outer layers. Pitting is frequently found to be a common defect in gear teeth.
- **Galling** is a metal-to-metal contact during the movement of tooling in the operation. It is described in Sec. 9-12-4, "Galling."
- **Creep** occurs in a stressed metal material, with neither the stress force nor the temperature being too high. The material, exposed to regular stresses of the work cycle, responds to their application by deforming elastically, which produces slight plastic changes along the grain boundaries or at possible internal flaws. With the continuation of stress application, the material begins to very slowly flow under its force, until finally, when a sufficient strain is created, it extends in shape, with subsequent reduction of the area closest to the stress application. If such a process continues, it leads to total fracture of that material.

### 14-2-9. Corrosion

The most frequent corrosion is caused by oxidation of metals and alloys, in which case the reaction itself produces a deposition within the area of affliction. The speed and the spread of corrosion depend on the characteristics of this growth and on the possibility of its penetration enhanced by additional influences. Corrosion is often connected with the transfer of electrical particles on the atomic level, and for that reason the electrical properties of the corrosive growth are pertinent.

Once a deposit produced by the corrosive action is formed, its kinetics is given by the character of its particles. The main influence is the diffusion of oxygen ions and ions of metal through its barrier.

However, not all materials form layers of residue of their corrosive action. Some materials, such as potassium, lithium, barium, and strontium, actually don't. Also the reaction of chloride with iron at high temperatures does not produce the corrosive residue.

With electrochemical corrosion, the mechanism is the same as that of galvanic cell: With greater difference between the potentials of two metals, the initial corrosive reaction is more pronounced. For example, zinc and aluminum, when immersed in a solution of sodium chloride, have a difference of potential of 300 mV, whereas zinc with copper has a 700 mV difference. This means that when paired with copper, zinc will corrode faster.

Every corrosion contains two opposite sides of its reaction: anodic and cathodic. Cathodic reaction, being depolarizating in its origin, reduces the amount of anodic influence, while anodic reaction, which is oxidation, promotes corrosion. The two combined interactions are generally summed under the title of corrosion, which is an oxidating and reductive process.

Where a protective coating is created within the metal material surface, it is called a passive coating. Passivity may be defined as a state in which metals or alloys are found to be resistant to corrosion, even though being exposed to a corrosion-inducing environment. Some consider passivity to be due to a retardation of either the anodic or the cathodic reaction. Others promote the *adsorption theory,* which claims the passivity to be a phenomenon depending on the adsorption of some elements, mainly oxygen, from the solution off the metal's surface. Still another theory considers passivity as attributable to a thin film of a compound, such as a third phase on the metal surface.

Metals that can be passivated are aluminum, chromium, manganese, titanium, vanadium, iron, nickel, cobalt, molybdenum, niobium, tungsten, and others. For industrial purposes several of their alloys are important, such as carbon and low-carbon steel, stainless steel, with alloying elements such as chromium, titanium, and nickel. The composition of the metal is the main factor constituting its suitability for passivation.

## 14-3. Testing of Mechanical Properties

Mechanical properties of materials are assessed through actual testing, with results of these tests recorded for future reference. The most

**TABLE 14-5 Materials with Corrosive Influence**

| No. | Material, basic | Influence-exerting material | | | | | | | | | | |
|---|---|---|---|---|---|---|---|---|---|---|---|---|
| | | 1* | 2* | 3* | 4* | 5* | 6* | 7* | 8* | 9* | 10* | 11* |
| 1. | Carbon and low-carbon steel | - | B | B | B | B | B | A | A | A | B | B |
| 2. | Stainless steel | A | - | A | A | A | A | A | A | A | A | A–B |
| 3. | Nickel and its alloys | A | A | - | A | B | A | A | A | A | A | A–B |
| 4. | Chromium | A | A | A | - | A | A | A | A | A | A | A |
| 5. | Copper and its alloys | A | B | B | B | - | A | A | A | A | A | B |
| 6. | Aluminum and alloys | B | B | B | B | C | - | A | A | B | A | C |
| 7. | Zinc and its alloys | C | C | B | B | C | B | - | B | B | B | C |
| 8. | Cadmium | B | B | B | B | B | B | A | - | A | A | B |
| 9. | Magnesium and alloys | C | C | C | C | C | B | B | B | - | C | C |
| 10. | Lead, tin, and alloys | A | B | A | B | B | A | A | A | A | - | B |
| 11. | Ag, Au, Pd, Pt, Ta, Ti, Zr | A | A | A | A | A | A | A | A | A | A | - |

*Numbers with an asterisk correspond with numbers of particular materials in the far left column. The reaction between any two materials may be read by finding the intersection of their respective lines and rows.

The explanation of listed values is as follows:

A: Corrosion within atmospheric environment is not possible.

B: A mild corrosive reaction may be observed between the two materials.

C: Corrosive reaction will occur.

NOTE: Ag = silver; Au = gold; Pd = palladium; Pt = platinum; Ta = tantalum; Ti = titanium; Zr = zirconium.

SOURCE: Svatcpluk Černoch, *Strojně technická příručka,* 1977. Reprinted with permission from SNTL Publishers, Prague, CZ.

common test is the hardness test, which is performed using several testing methods. The process is described in Sec. 14-2-2, "Hardness of Materials." Additional testing of mechanical properties is as follows.

### 14-3-1. Static tensile testing

Static tensile testing is performed by applying a static load to the tested material. The testing process is free from sudden impacts, load cycling, or abrupt changes. The amount of force, if increased, is regulated slowly and gradually.

Static load testing is performed mainly to assess the elongation of the material, and the testing records should include the amount of elongation with reference to the applied force. Marginally, torsion testing and compression testing is performed similarly.

### 14-3-2. Izod impact testing

Izod impact testing (or Charpy) is used to assess the toughness of a material. A notched bar of steel is attached into the vise and a pendulum type of hammer is allowed to drop onto the tested piece to break it (Fig. 14-8). A ductile material will need a greater amount of energy or more blows to be split than its nonductile counterpart, even if they are both of the same strength. The ductile material is therefore tougher.

The orientation of applied force with reference to the material grain is important, as every material displays different results along its grain from those against it. Testing records must therefore include this information as well.

The impact testing method is very sensitive, assessing the properties of materials with regard to their manufacturing process and heat treatment. This type of testing is an irreplaceable method of evalua-

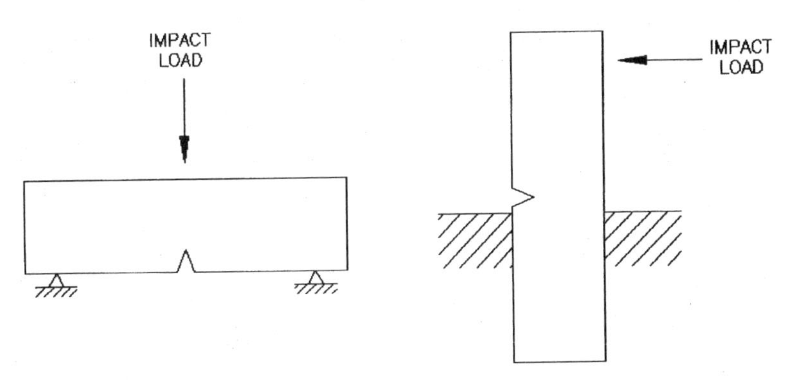


**Figure 14-8** Impact load testing.

tion of the quality of weldments. It is used for selection of proper materials for low-temperature applications, for testing of materials exposed to cyclic loading, and for evaluation of suitability of materials for steel constructions where the threat of fractures due to brittleness may be encountered.

The impact strength of a material is given in foot-pounds, indicating the amount of energy required for breaking the specimen in a single blow.

#### 14-3-3. Fatigue strength testing

Testing of materials under cyclic loading is concerned with the fatigue testing of such materials. Fatigue-related failures are known for starting at much lower tensions than those of the material's tensile strength. However, some materials may sustain almost unlimited cyclic loading without succumbing to any changes. For that reason, cyclic loading, used in the fatigue assessment, is limited to a certain number of cycles. For steel, 10,000,000 cycles is often chosen, while for nonferrous materials up to 100,000,000 cycles may be performed.

Fatigue testing is often based on the evaluation of bending fatigue and is classified according to the direction of the load as

- Axial testing, or simple bending
- Bending during rotation
- Torsional testing

#### 14-3-4. Testing of breakability

Testing of a material's breakability is used for evaluation of its deformation properties in cold working and for assessment of its brittleness. This testing method is frequently used where a welded assembly's breakage is of concern or where a comparison of different welding methods is needed (Fig. 14-9).

The amount of breakability is given by the size of angle $\alpha$, and the width of the pin is specified in relation to the thickness of the tested material.

#### 14-3-5. Testing of drawability

Erichsen's test of drawability shown in Fig. 14-10 uses a $\phi$20-mm drawing punch to press into a firmly retained sheet of metal until a breakage occurs. The result is given in the form of the height of a cup $h$, while a radial breakage spells the unsuitability of material for drawing. The condition of the drawn surface is observed; a rough surface coincides with a coarse grain.

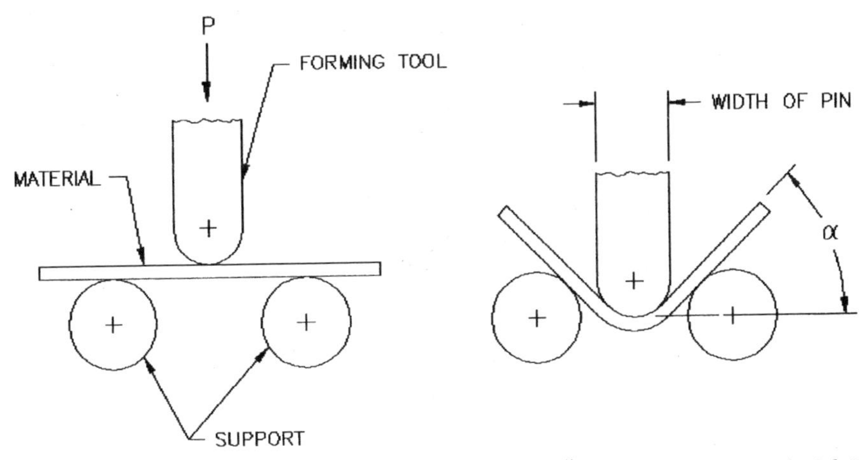


**Figure 14-9** Test of breakability. (*From Svatopluk Černoch,* "Strojně technická příručka," *1977. Reprinted with permission from SNTL Publishers, Prague, CZ.*)

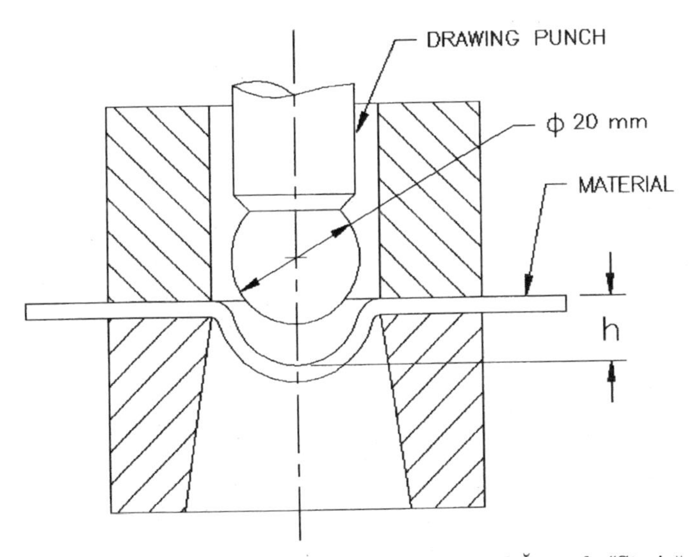


**Figure 14-10** Drawability testing. (*From Svatopluk Černoch,* "Strojně technická příručka," *1977. Reprinted with permission from SNTL Publishers, Prague, CZ.*)

Evaluation of a material's anisotropy $r$ is obtained through a cupping procedure, shown in Fig. 14-11.

The amount of anisotropy is given by the difference of measurements $V_1$ and $V_2$. Lately, the amount of normal anisotropy is often calculated from the results of regular tensile testing as in Eq. (14-13).

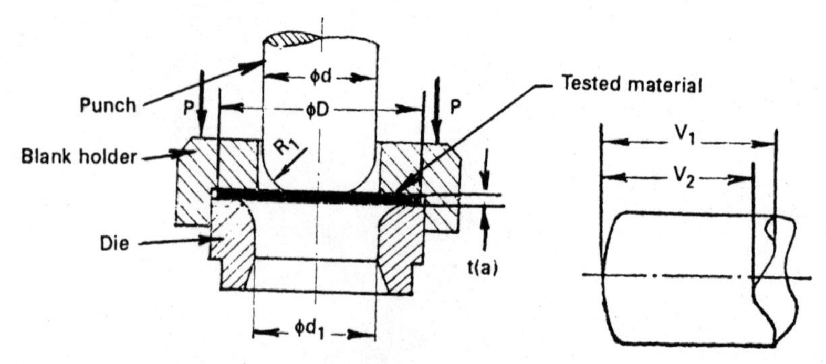


**Figure 14-11** Cupping test for drawing purposes. (1) Punch. (2) Die. (3) Blank holder. (4) Tested strip. ($v_1$, $v_2$) Heights of cup sides, affected by anisotropy. (*From Svatopluk Černoch,* "Strojně technická příručka," *1977. Reprinted with permission from SNTL Publishers, Prague, CZ.*)

$$r = \frac{\log b_0/b}{\log a_0/a} \tag{14-13}$$

where $b_0, a_0$ = width or thickness, original
$b, a$ = width or thickness after tensile testing

## 14-4. Materials Used for Tooling Applications

The appropriate tool material must be carefully selected to be in accordance with requirements of the machining process and the workpiece. The most important points to consider are the necessary strength of the tooling and its toughness, hot hardness, thermal shock resistance, and chemical stability.

Since all tool steel materials are heat-treated for attainment of best properties, the behavior of steel material during the process of heat treatment is important. Any distortion produced by heat treatment should call for attention toward the selection of material, or part design, or manufacturing flawlessness, in this order. Some materials are more prone to distortion caused by heat treatment than others. Less stable steels need more conservative design methods in order to sustain the stress placed upon them by heat treatment. Long thin sections should be avoided or properly supported. Sharp corners, holes placed too close to the edges, or mixing of thin and thick sections should definitely be avoided. Some ideas are presented in Fig. 14-12.

The depth of surface hardening and the material's resistance to decarburization are important aspects of proper tool material selection. Naturally, the machinability of the material is of essence, as a

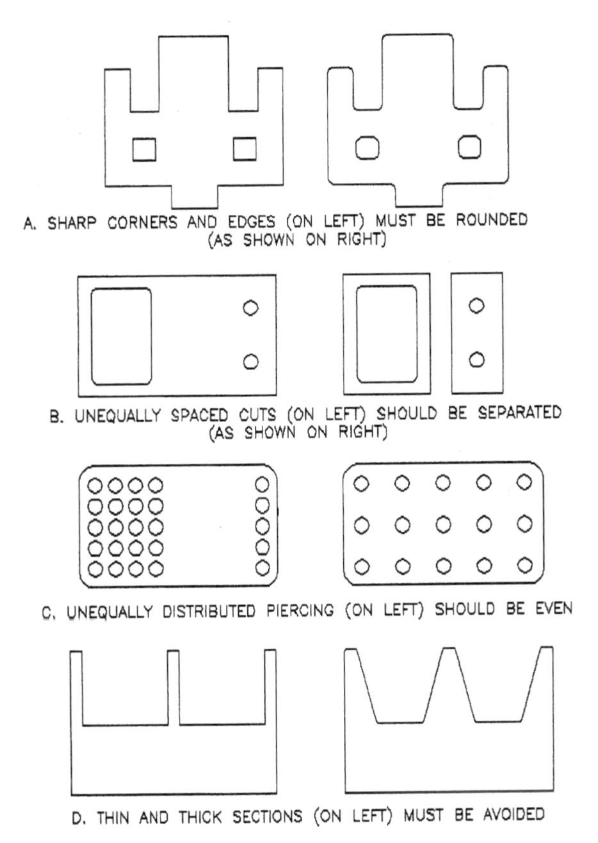


**Figure 14-12** Design changes for less distortion in heat treatment.

tool made of material which is difficult to machine will cost much more to produce than its counterpart made of easily machinable material.

Wear resistance and hot hardness are other important segments of tooling material suitability. These two properties are somewhat dependent on each other, as tool wear is enhanced with lower hot hardness. Hot hardness of a material is its resistance to softening produced by high operating temperatures. Hot hardness should be greater where the high-speed work and subsequently high-temperature working environment is expected.

Toughness of materials is the demarcation of their ability to withstand shocks, interruptions of their work path, sudden application of loading, and all abrupt changes without deformation or breakage. The material's elastic limit is important in this evaluation, as such parts should be rigid, not succumbing to deformation, be it elastic or plastic, and yet not be brittle or unyielding to break easily in service.

**TABLE 14-6 Commonly Used Tooling Materials and Their Qualities**

| Material | Hardness, Vickers scale | Transverse rupture strength, lb/in$^2$ |
|---|---|---|
| High-speed steel | 100–1,100 | 440,000–750,000 |
| Cemented carbides | 900–1,800 | 120,000–400,000 |
| Ceramics | 1,700–2,100 | 73,000–160,000 |
| Cubic boron nitride | 4,000–6,000 | 15,000–29,000 |
| Diamond | 8,000–10,000 | 7,000–20,000 |

The most common tool materials used today are as shown in Table 14-6. They are grouped according to their hardness and transverse rupture strengths. The two properties are dependent on each other.

The values shown in Table 14-6 are approximate only, presented for comparison. The Vickers hardness denomination shows quite high ranges, considering that 1700 $HV = 78\ HRc$.

### 14-4-1. Tool steel materials

A lot is now expected from tool steel materials. With costs being cut everywhere, prices of finished goods being slashed to the bare minimum, and production running on diminishing margins, tooling just should not break or need to be sharpened too often.

Demands are abundant, and so are the choices. Tool steels, which may often accommodate situations almost controversial, are here for the asking. The continuous research of material manufacturers produced results in the form of an emergence of more versatile tool steel materials of either general or narrowly specialized qualities.

Under the sponsorship of AISI and SAE, a classification system for tool steel has been developed, grouping all materials into several categories, shown in Table 14-7.

The quick reference data presented in Table 14-8 should serve as a rough identifier of available tooling materials for a given application. This is a starting point in the evaluation of materials, with a finer distinction to be based on the ratings of their properties, on their availability, cost, economical aspects, and all additional applicable factors. Vital mechanical properties of tooling materials were described at the beginning of this section.

The most often used steel types in the die-building practice are W1, W2, O1, A2, D2, D4, M2, S1, and S5. They represent a majority of selections design engineers make when deciding on the proper tooling material. All these steels may be categorized according to the additives constituting their main properties, according to their applica-

**TABLE 14-7 Classification of Tool Steels**

| Type of tool steel | AISI letter symbol | Main distinction |
|---|---|---|
| High-speed steel | M | Molybdenum content |
| | T | Tungsten content |
| Hot-work steel | H1–H19 | Chromium content |
| | H20 and up | Tungsten content |
| | H40 and up | Molybdenum content |
| Cold-work steel | D | High carbon and high chromium |
| | O | Oil-hardening steel |
| | A | Medium alloy, air hardening |
| Shock-resisting steel | S | Low carbon content, alloying elements vary with type |
| Mold steel | P | Very low carbon content |
| Special-purpose steel | L | Low-alloy |
| | F | Carbon-tungsten, low-alloy |
| Water-hardening steel | W | Minimum of alloying elements |

**TABLE 14-8 Tool Selection Guide**

*High-speed steel application:* Cold and hot dies, roller bearing
M category For high abrasion areas
T category Where high hot hardness is needed

*Cold-work steel application:* Cutting tools for medium speeds, and where heat-treatment stability is required
D category Cutting, coining, drawing tools, thread rolling dies. Tooling for long runs
O category Coining tools for medium runs
A category Bushings, cutting, trimming, forming, and bending tools. Tooling for medium runs

*Shock-resisting steel application:* Hot work (punching, shearing), cold work. For hobbing, hot swaging, compression molding applications

*Special-purpose steel application:* Cutting tools and knives, blanking and trimming sets. Used where exceptional toughness is required

*Water-hardening steel application:* Where high abrasion resistance and hot hardness are needed. Cold-work tooling, such as cutting tools, cold heading dies. Hot-work tooling application, such as drop forging dies. Tooling for short runs

tions, according to their method of hardening, etc. (Table 14-9 through 14-11).

The properties with the greatest impact on the tooling material application are

- Resistance to softening at high temperatures, or hot hardness
- Depth of hardness penetration during the heat-treating process
- Abrasion and wear resistance

**TABLE 14-9 Tool and Die Steels**

| AISI steel type* | Nominal composition, % | | | | | | | |
|---|---|---|---|---|---|---|---|---|
| | C | Mn | Si | W | Cr | Mo | V | Other |
| W1 | 1.00 | | | | | | | |
| W2 | 1.00 | | | | | | 0.25 | |
| O1 | 0.90 | 1.00 | | 0.50 | 0.50 | | | |
| O2 | 0.90 | 1.60 | | | | | | |
| O6 | 1.45 | | 1.00 | | | 0.25 | | |
| A2 | 1.00 | | | | 5.00 | 1.00 | | |
| A4 | 1.00 | 2.00 | | | 1.00 | 1.00 | | |
| A6 | 0.70 | 2.00 | | | 1.00 | 1.00 | | |
| A7 | 2.25 | | | 1.00 | 5.25 | 1.00 | 4.75 | |
| A8 | 0.55 | | | 1.25 | 5.00 | 1.25 | | |
| A10 | 1.35 | 1.80 | 1.25 | | | 1.50 | | 1.80 Ni |
| D2 | 1.50 | | | | 12.00 | 1.00 | | |
| D3 | 2.25 | | | | 12.00 | | | |
| D4 | 2.25 | | | | 12.00 | 1.00 | | |
| D5 | 1.50 | | | | 12.00 | 1.00 | | 3.50 Co |
| D7 | 2.35 | | | | 12.00 | 1.00 | 4.00 | |
| S1 | 0.50 | | | 2.50 | 1.50 | | | |
| S2 | 0.50 | | 1.00 | | | 0.50 | | |
| S4 | 0.55 | 0.80 | 2.00 | | | | | |
| S5 | 0.55 | 0.80 | 2.00 | | | 0.40 | | |
| S7 | 0.50 | | | | 3.25 | 1.40 | | |
| T1 | 0.70 | | | 18.00 | 4.00 | | 1.00 | |
| T15 | 1.50 | | | 12.00 | 4.00 | | 5.00 | 5.00 Co |
| M1 | 0.80 | | | 1.50 | 4.00 | 8.00 | 1.00 | |
| M2 | 0.85 | | | 6.00 | 4.00 | 5.00 | 2.00 | |
| M4 | 1.30 | | | 5.50 | 4.00 | 4.50 | 4.00 | |
| L3 | 1.00 | | | | 1.50 | | 0.20 | |
| F2 | 1.25 | | | 3.50 | | | | |

*W, water-hardening; O, oil-hardening, cold work; A, air-hardening, medium alloy; D, high-carbon high-chromium, cold work; S, shock-resisting; T, tungsten-base, high-speed; M, molybdenum-base, high-speed; L, special-purpose, low-alloy; F, carbon-tungsten, special-purpose.

SOURCE: Frank W. Wilson, *Die Design Handbook,* New York, 1965. Reprinted with permission from The McGraw-Hill Companies.

**14-4-1-1. High-speed tool steels.** The two groups of high-speed tool steels are based on the content of their main alloying element, which may be either tungsten or molybdenum.

**Molybdenum high-speed tool steels.** Molybdenum is the main constituent of this group, even though in some combinations certain percentages of other significant elements, such as tungsten and cobalt, may also be present. Steels of a higher than usual carbon content and steels with an additional content of vanadium for increased resistance to abrasion may be found within this group.

Properties of various molybdenum-based steel types from Table 14-12 and their application are as follows:

*AISI M1* steel is a cheaper substitute for the well-known T1 steel group, with molybdenum used in place of tungsten. Properties of both

**TABLE 14-10 Recommended Tool Steels for Press Tooling**

| Application | AISI steel type* | Hardness, Rockwell C |
|---|---|---|
| Blanking dies and punches (short runs) | W2 | 57–65 |
| | O1 | 58–62 |
| | A2 | 58–62 |
| Blanking dies and punches (long runs) | A2 | 58–62 |
| | D2 | 58–62 |
| | M4 | 58–62 |
| Bending dies | O1 | 58–62 |
| | A2 | 58–62 |
| | D2 | 58–62 |
| Coining dies | S1 | 52–55 |
| | W1 | 58–62 |
| | A2 | 58–62 |
| | D2 | 58–62 |
| | D4 | 58–62 |
| Drawing dies | W1 or W2 | 58–64 |
| | O1 | 58–62 |
| | O6 | 58–62 |
| | A2 | 58–62 |
| | D2 | 58–62 |
| | D4 | 58–62 |
| Dies (cold extrusion) | D2 | 60–64 |
| | M4 | 63–65 |
| Dies (embossing) | O1 | 59–61 |
| | O2 | 59–61 |
| | A2 | 59–61 |
| | D2 | 59–61 |
| Dies (lamination) | M2 | 58–62 |
| | D2 | 58–62 |
| | D4 | 58–62 |
| | M4 | 60–62 |
| | T15 | 60–62 |
| | D7 | 60–62 |
| | A7 | 60–62 |
| Dies (sizing) | W2 | 61–64 |
| | M2 | 61–64 |
| | D2 | 61–64 |
| | M4 | 61–64 |
| Dies and punches (trimming) | W2 | 57–60 |
| | A2 | 57–60 |
| | D2 | 57–60 |
| | D4 | 57–60 |
| | M4 | 57–60 |
| Punches (embossing) | S1 | 59–61 |
| | S5 | 59–61 |
| Punches (trimming) | W2 | 57–60 |
| | O1 | 57–60 |
| Punches (notching) | W2 | 57–60 |
| | M2 | 57–60 |

*Several of the other tool steels listed in Table 14-9 are also used for these applications, depending on the preferences, special applications, and heat-treating facilities found in certain plants.

SOURCE: Frank W. Wilson, *Die Design Handbook,* New York, 1965. Reprinted with permission from The McGraw-Hill Companies.

**TABLE 14-11 Comparison of Basic Characteristics of Tool and Die Steels**

| AISI steel type | Non-deforming properties | Safety in hardening | Toughness | Resistance to softening effect of heat | Wear resistance | Machinability |
|---|---|---|---|---|---|---|
| W1 | Low | Fair | Good | Low | Fair | Best |
| W2 | Low | Fair | Good | Low | Fair | Best |
| O1 | Good | Good | Good | Low | Fair | Good |
| O2 | Good | Good | Good | Low | Fair | Good |
| O6 | Good | Good | Good | Low | Fair | Best |
| A2 | Best | Best | Fair | Fair | Good | Fair |
| A4 | Best | Best | Fair | Fair | Good | Fair |
| A6 | Best | Best | Fair | Fair | Good | Fair |
| A7 | Best | Best | Low | Fair | Best | Very low |
| A8 | Good | Best | Best | Good | Good | Fair |
| A10 | Best | Best | Fair | Fair | Good | Good |
| D2 | Best | Best | Low | Fair | Very good | Low |
| D3 | Good | Best | Low | Fair | Very good | Low |
| D4 | Best | Best | Low | Fair | Very good | Low |
| D5 | Best | Best | Low | Fair | Very good | Low |
| D7 | Best | Best | Low | Fair | Best | Very low |
| S1 | Fair | Good | Best | Fair | Fair | Fair |
| S2 | Fair | Good | Best | Fair | Low | Fair |
| S4 | Fair | Good | Best | Fair | Low | Fair |
| S5 | Fair | Good | Best | Fair | Low | Fair |
| S7 | Fair | Good | Best | Fair | Good | Fair |
| T1 | Good | Good | Good | Best | Very good | Fair |
| T15 | Good | Good | Low | Best | Best | Very low |
| M1 | Good | Good | Good | Best | Very good | Fair |
| M2 | Good | Good | Good | Best | Very good | Fair |
| M4 | Good | Good | Good | Best | Very good | Fair |
| L3 | Fair | Good | Good | Low | Fair | Good |
| F2 | Fair | Fair | Low | Low | Good | Fair |

SOURCE: Frank W. Wilson, *Die Design Handbook,* New York, 1965. Reprinted with permission from The McGraw-Hill Companies.

**TABLE 14-12 Molybdenum High-Speed Tool Steels**

| | M1 through M3 | M4 | M7 through M10 | M30 and up |
|---|---|---|---|---|
| Depth of hardening | 5 | 5 | 5 | 5 |
| Decarburization resistance | 3–4 | 4 | 3 | 3 |
| Safety in hardening | 2 | 2 | 2 | 2 |
| Shape stability in heat treatment (air) | 3 | 3 | 3 | 3 |
| Shape stability in heat treatment (oil) | 2 | 2 | 2 | 2 |
| Machinability | 1–2 | 2 | 2 | 2 |
| Wear resistance | 4 | 5 | 4 | 4 |
| Toughness | 1 | 1 | 1 | 1 |
| Hot hardness | 4 | 4 | 4 | 5 |

Relative evaluation, where 1 is the lowest rating and 5 is the greatest.

steel types are quite similar, the only difference being greater sensitivity of molybdenum steel to heat treatment. M1 steel's application includes drills, reamers, milling cutters, and lathe cutters for light-duty work.

*AISI M2* is similar to M1, with a portion of its molybdenum content replaced by tungsten. It is suitable for general work, displaying improved toughness and wear resistance, while being less problem-prone in hardening and economically more advantageous. It is a good choice of steel for all cutting applications, be it drills, mills, lathe tooling, etc.

*AISI M3* with vanadium added for improving its wear resistance is preferred especially where the property is required. Otherwise it is widely used for broaches, milling cutters, reamers, etc.

*AISI M7* has a higher carbon and vanadium content, which readily raises its cutting efficiency but without lowering its toughness. Otherwise M7 steel is similar to M1. Greater care should be exercised in heat treatment, where a salt bath or a controlled atmosphere is recommended because of the material's proneness to decarburization. Applications include blanking and trimming dies, shear blades, thread rolling dies, and lathe tooling.

*AISI M10* has excellent wear resistance owing to its higher vanadium content. Various cutters and lathe tooling are made from this type of steel, with usage for blanking dies and punches, shear blades, etc.

*AISI M42* has good hot hardness as well as regular hardness produced by the addition of cobalt. Forming tools, fly cutters, thread rolling dies, tool bits, shaving tools, and similar tools may be made from this type of material. All M40 group members are also utilized for tooling on materials with a low machinability index.

**Tungsten high-speed tool steels.** These materials are irreplaceable where a high-temperature working environment is encountered. They are also more permissible in heat treatment, having a distinct resistance to decarburization while not being excessively prone to distortion.

Properties of tungsten-based steels from Table 14-13 and their applications are as follows:

*AISI T1* is also known as the 18-4-1 type of steel; its designation refers to the percentages of the primary alloying elements: W-Cr-V. A high toughness, combined with better machinability and permissibility in heat treatment, makes this type of steel an excellent general-purpose material. It is used in all multiflute cutters, threading taps and dies, machine knives, lathe tooling, and punches and dies.

*AISI T2* is a very agreeable material with regard to machining and heat treatment, where it is comparable to T1 material, in spite of its higher carbon content and twice the percentage of vanadium. Because

TABLE 14-13 Tungsten High-Speed Tool Steels

| | T1, T2 | T4 | T5, T6 | T8, T15 |
|---|---|---|---|---|
| Depth of hardening | 5 | 5 | 5 | 5 |
| Decarburization resistance | 5 | 4 | 3 | 4 |
| Safety in hardening | 3 | 2 | 2 | 2 |
| Shape stability in heat treatment (air) | 3 | 3 | 3 | 3 |
| Shape stability in heat treatment (oil) | 2 | 2 | 2 | 2 |
| Machinability | 2 | 2 | 1–2 | 1–2 |
| Wear resistance | 4 | 4 | 4 | 4–5 |
| Toughness | 1 | 1 | 1 | 1 |
| Hot hardness | 4 | 5 | 5 | 5 |

Relative evaluation, where 1 is the lowest rating and 5 is the greatest.

of its improved wear resistance, T2 is chosen where finer and more precise cutting is needed. It is an excellent choice for all finishing tools and forming inserts. Generally it may be used in all situations where T1 steel would be selected.

*AISI T5* has a higher cobalt and vanadium content, the combination of which produces excellent wear resistance combined with a very high hot hardness. Its tendency to decarburization should be controlled in heat treatment, utilizing a slightly reducing atmosphere. This steel may be recommended for all single-fluted tooling, for high speeds and feeds, for cutting of materials with lower machinability index, and for removal of larger chunks of material.

*AISI T15* boasts excellent wear resistance and unsurpassed cutting ability, along with increased hot hardness. The higher vanadium content, aided by a very high percentage of carbon and some cobalt, produces a steel of the best working qualities within the high-speed steel group. As a permissible material in heat treatment, T15 steel can withstand quite high heat-treating temperatures, followed usually by more than one tempering pass. The machinability index of this material is very high, even though its grindability suffers. Because of its lower toughness, this steel material may not be used where greater shock resistance is required. Otherwise its applications include cold-work tooling such as dies and punches, either blanking or forming.

**14-4-1-2. Hot- or cold-work steels.** Another distinction within tool steel materials divides them into hot- and cold-working materials.

**Hot-work tool steels,** including all H-type materials, are not described here since their application range covers mostly the mold-building branch of industry, even though some marginal uses within the field of stamping dies may certainly be found.

**Cold-work tool steels** can be grouped into three major categories: (1) high-carbon high-chromium type steels, (2) oil-hardening cold-

**TABLE 14-14 Cold-Work Tool Steels**

| | D2 | D3 | D4, D5 | O1, O2 | O6 | A2, A3 | A6 |
|---|---|---|---|---|---|---|---|
| Depth of hardening | 5 | 5 | 5 | 4 | 4 | 5 | 5 |
| Decarburization resistance | 4 | 4 | 4 | 5 | 5 | 4 | 5 |
| Safety in hardening | 5 | 3 | 5 | 4 | 4 | 5 | 5 |
| Shape stability in heat treatment | 5 | 4 | 5 | 4 | 4 | 5 | 5 |
| Machinability | 1 | 1 | 1 | 3 | 4 | 2 | 1 |
| Wear resistance | 4 | 4 | 3–4 | 2 | 2 | 3–4 | 2 |
| Toughness | 1 | 1 | 1 | 2 | 2 | 2 | 2 |
| Hot hardness | 3 | 3 | 3 | 1 | 1 | 3 | 2 |

Relative evaluation, where 1 is the lowest rating and 5 is the greatest.

work tool steels, and (3) medium-alloy, air-hardening, cold-work tool steels.

**High-carbon, high-chromium, cold-work tool steels** were mainly intended for die work, even though other general applications may be found.

*AISI D2* is an air-hardening die steel, high in hardness, abrasion resistance, and resistance to deformation. Its machinability is quite good and may be improved by a slightly greater amount of sulfur within the material makeup. Well-dispersed particles of sulfide considerably improve the material's machinability and surface finish. Heat treating to a lower hardness positively affects the material's toughness. D2 steel is frequently used for making all types of dies, be it cutting dies, forging dies, or other die-related tooling.

*AISI D3* oil-hardened steel offers excellent resistance to abrasion and wear. It is immune to deformation and displays a superior compressive strength under a gradually increasing load. Its deep-hardening properties make it an excellent choice where frequent regrinding of tools is necessary. Otherwise it is utilized for blanking and other cutting punches and dies intended for long production runs, and it is recommended wherever a high resistance to wear is required.

**Oil-hardening, cold-work tool steels** contain a significant amount of manganese, but the percentage of other alloying elements is quite low. This inexpensive tool steel does not resist wear and deformation, but the depth of its hardness penetration is impressive. A good machinability index makes this type of steel a good choice where short runs of parts are produced.

*AISI 01* steel has a minimal tendency to warpage and shrinking during the heat-treating process, but its resistance to high heat is not commendable. It is often used for slow-running cutting tools such as taps, drills, and reamers, and for all cutting and forming die tooling for short or medium production runs.

*AISI 02* also displays nondeforming properties of the previous type, coupled with an ease of machining, good wear resistance, and safety in hardening at low temperatures. It is mainly used for cutting tools of low- and medium-speed ranges, for making of thread rolling dies, forming tools, bushings, and gauges.

*AISI 06* is a graphitic type of steel, owing to small particles of graphitic carbon equally dispersed throughout its content. Approximately one-third of the total percentage of carbon is present in the form of free nodular graphites which supplement ease of machining. Parts made from this type of steel often utilize these free graphites as a sort of lubricant, with subsequent reduction in wear and galling. O6 steel presents no problems in hardening, utilizing quite low quenching temperatures while attaining a deep-hardness penetration with almost no dimensional change. This steel is most often used for bushings, for holders of tool cutter's inserts, for shanks of cutting tools, for arbors, jigs, gauges, cutting and forming punches and dies, and where stability of the material is of greater importance than its wear resistance.

**Medium-alloy, air-hardening, cold-work tool steels** have a high machinability index accompanied by high toughness and nondeformity in heat treatment. Even though the alloy content is low, hardening by air quenching is acceptable. However, wear resistance of these materials is not impressive.

*AISI A2* alloy has a low chromium content that places it within a competitively priced range of steels, with its resistance to deformation equal to that of high-chromium materials. An inclusion of sulfur improves the machinability, while a reduced wear resistance is balanced by an increased toughness. This type of steel is most often used for punches and dies, either cutting or forming, for cold and hot trimming dies, and for thread rolling dies.

*AISI A6* may be air hardened from a low temperature, which makes this steel comparable to oil-hardening types, with an additional advantage of improved stability in the heat-treating process. Its main usage is within the die-building field as forming tools, gauges, and tooling that does not need a high degree of wear resistance.

**14-4-1-3. Shock-resisting, mold, and special-purpose tool steels.** Each of these steel types was originally developed for quite a specific use, but gradually they all advanced into additional fields of application.

**Shock-resisting steels.** These are steels of increased toughness, even though suffering from lowered wear resistance (Table 14-15). Their carbon content is quite low, and their alloying elements vary with each steel's composition. The diversification in alloying additives produces different properties, such as hot hardness, machinability, and abrasion resistance.

*AISI S1* is a chromium-tungsten alloy, demonstrating in its hardened condition a great toughness, with considerable strength and

**TABLE 14-15 Shock-Resistant and Special-Purpose Tool Steels**

| | S1 | S2 | S5 | L2 | L6 | F2 |
|---|---|---|---|---|---|---|
| Depth of hardening | 4 | 4 | 4 | 4 | 4 | 3 |
| Decarburization resistance | 4 | 3 | 3 | 5 | 5 | 5 |
| Safety in hardening | 3 | 1 | 3 | 2 | 3 | 1 |
| Shape stability in heat treatment | 2 | 1 | 2 | 1–2 | 3 | 1 |
| Machinability | 2 | 2 | 2 | 3 | 2 | 2 |
| Wear resistance | 1–2 | 1–2 | 1–2 | 1 | 2 | 3–4 |
| Toughness | 4 | 5 | 5 | 4 | 4 | 1 |
| Hot hardness | 2 | 1 | 1 | 1 | 1 | 1 |

Relative evaluation, where 1 is the lowest rating and 5 is the greatest.

hardness. The influence of low carbon content on the wear resistance of this material may be alleviated by carburizing, which will not diminish its generous shock-resisting qualities. Main uses are for piercing and forming tools, drop forging die inserts, heavy shear blades, tooling for shock loads, etc.

*AISI S2,* with its moderate wear resistance, presents good resistance to rupturing. It is mainly used in the manufacture of pneumatic tools, hand chisels, and a limited amount of hot work. For this last application it must be heat-treated in a neutral atmosphere in order to avoid either carburization or decarburization of its surface.

*AISI S5* is a silicon-manganese steel with some chromium, vanadium, and molybdenum added to obtain an improvement in the depth of hardness penetration and a refinement of the inner grain structure. These steels are quenched in oil, although a water quench may also be used where the shape of a part does not contain sharp corners, drastic deviations in thickness, and other designing faults. High elastic limit and good ductility are among the most sought after properties of this type. The material is used for heavy-duty punches, bending rolls, shear blades, and also for shanks of carbide tools and machine parts that are subject to shock.

**Mold steels.** These steels have a quite low carbon content, in which they differ from other types of tool steels. They therefore require carburizing, while their high resistance to decarburization protects them from the reversal of this process. Their dimensional stability is very good, accompanied by an excellent surface finish and the possibility of altering their shape by hobbing instead of regular machining.

**Special-purpose tool steels.** These low-alloyed steels have various properties, often even conflicting ones. These materials are cheaper than higher-alloyed tool steels.

*AISI L6* is resistant to deformation, and its toughness is quite good in comparison with that of other oil-hardening types of steel. The alloy content and abrasion resistance is low. This tool steel material is recommended for use where moderate shock and wear resistance are required, as in forming and trimming dies, knuckle pins, clutch members, etc.

*AISI F2* is a carbon-tungsten type of material, sensitive to thermal variations, even to heat treatment, during which it may succumb to distortions. Such a tendency, along with its low resistance to cracking during the hardening process, makes the application limited to tools of simple shapes. Hardening must be very shallow, yet it produces a tough core. Regardless of all drawbacks, this type of material is quite resistant to abrasion.

**14-4-1-4. Water-hardening tool steels.** These steels contain a minimum of alloying elements, with their carbon content varying from group to group. Quenching should be harsh, utilizing water or brine. Where the carbon content is higher, such steel is more sensitive to emergence of defects due to heat treatment. Steels of this type are then used for more demanding applications. Table 14-16 rates their characteristics.

**Water-hardening steel type W-1 (plain carbon).** *Group I,* with carbon content between 0.7 and 0.9 percent, is tough and may be used where shock or harsh treatment is expected. Applications include cold punches, fixture elements, anvil faces, chuck jaws, screwdriver blades, chisels, etc.

*Group II,* of carbon content 0.9 to 1.1 percent, has a moderate ability to withstand shock, which is accompanied by greater hardness and quite satisfactory toughness. It is used for manufacture of hand tools, threading dies, wood augers, die parts for drawing and heading dies, drill bushing, lathe centers, collets, and fixturing elements.

*Group III* (1.05 to 1.20 percent C) has improved hardness penetration and a good resistance to shock application. Its toughness is

**TABLE 14-16 Water-Hardening Tool Steels**

| | All grades |
|---|---|
| Depth of hardening | 3 |
| Decarburization resistance | 5 |
| Safety in hardening | 2 |
| Shape stability in heat treatment | 1 |
| Machinability | 5 |
| Wear resistance | 1 |
| Toughness | 2–3 |
| Hot hardness | 1 |

Relative evaluation, where 1 is the lowest rating and 5 is the greatest.

reduced, but the wear resistance and cutting ability is high. It is used for hand tools, such as knives, chisels, slow-running cutting tools, center drills, and parts of blanking, bending, and coining dies.

*Group IV* (1.2 to 1.3 percent C) may be case hardened to a greater depth. This steel has improved wear and abrasion resistance, offset by a sensitivity to shock loads and to application of concentrated stresses. Often this type of material is used for finishing cutters and forming and burnishing tools.

### 14-4-2. Cemented carbides

Prior to the practical and widespread use of cemented carbides, which began in the thirties, cutting tool materials depended mainly on heat treatment to obtain properties needed for machining. At higher speeds, accompanied by higher temperatures, these tools were subjected to the effects of yet another heat treatment, the one that subsequently ruined them.

Not too many materials even today can take the heat produced by a work cycle without being affected by its detrimental influence. Aside from ceramics, cemented carbides are probably the only additional tooling material. Made from a mixture of carbide and an iron filler with cobalt, they are compacted and sintered. They display superior hardness even at higher temperatures, for which property they can be used where higher machining speeds are necessary or where machining of harder materials has to be done. Cemented carbides come in several forms. The most common types are tungsten carbides and titanium carbides.

**Tungsten carbides.** These may further be subdivided into two groups: (1) Two-phase-type tungsten carbide with a cobalt binder (WC-Co) is the most commonly used type. It displays an extreme hardness and resistance to abrasion, which predisposes it to be used where wear by abrasion is expected. (2) Alloyed tungsten carbide, alloyed with either titanium carbide (TiC) or tantalum carbide (TaC) or both, should be used where wear of tooling by cratering is possible.

Cobalt in tungsten carbide tooling increases its strength; however, in greater quantities it lowers the wear resistance, hot hardness, cratering resistance and resistance to deformation due to higher temperatures. The addition of titanium carbide works oppositely: It increases the wear resistance, hot hardness, cratering resistance, and tendency to deformation at higher temperatures of the material, but it decreases its strength. Tantalum carbide in tungsten carbide material decreases the strength and wear resistance while increasing the resistance to deformation caused by the material's exposure to high temperatures.

Depending on the cobalt content, tungsten carbide's Rockwell A hardness is very high and may range between A89 and A94.

**Titanium carbides.** These are usually made of a titanium carbide combination with additional nickel and molybdenum (TiC-Ni-Mo). The Rockwell A hardness ranges of commercial grades are between 92 and 93, and they are recommended only for high-speed precision machining or finishing operations at higher speeds and lower cutting feeds and depths. Titanium carbides display a superior strength and can operate at high-speed rates, of up to 1500 surface ft/min. They produce a superior surface finish, for they have no tendency to adhere to the workpiece material. An exception to this is their use with aluminum-based materials, which produce no satisfactory results. Because of their excellent flank wear resistance, they can be used to attain high-tolerance-range work.

### 14-4-3. Ceramics

Ceramic tools are brittle, but their hardness and durability are admirable. They retain their high range of hardness even at extreme temperatures of 1400°F, at which the hot hardness of all other tooling fails. Ceramics are therefore excellent for all machining work which requires a good surface finish or for machining of already hardened parts (up to Rockwell C 65). Ceramic tooling does not produce a welding tendency between the workpiece and the tool, as it generates less heat during the machining process than, for example, carbide tooling. Their main source of wear is oxidation.

Ceramic materials are most often based on aluminum oxide ($Al_2O_3$) but with some marginal utilization of zirconium oxide ($ZrO_2$) or magnesium oxide (MgO) and even nonoxides such as silicon nitride ($Si_3N_4$) or silicon carbide (SiC). The internal structure of ceramics is very dense yet porous. Because of their low transverse rupture strength, they should not be used where interrupted cuts are to be produced. They are also quite sensitive to notches, which become stress-concentrating points, fracturing the material under considerable low tension.

The low ductility of the material, resulting in its brittleness, often renders ceramics unreasonable to use in some applications. Also their low tensile strength, which makes them susceptible to ruptures and failures due to minute chipping, may be quite restrictive. To enhance their tensile qualities, ceramic materials must be kept under compression. Further, they should never be exposed to bending, twisting, or pulling influences. Heavy cuts are also not recommended for this type of tooling. But the advantage of their excellent wear resistance combined with low thermal conductivity (half of that of carbide tooling) often outweighs their drawbacks. They are irreplaceable where

**TABLE 14-17 Recommended Cutting Speeds for Ceramic Cutting Tools by Material Classification**

| Material to be machined | Material condition or type | Roughing (over 1/16 in depth, 0.015 to 0.030 in feed) | Finishing (under 1/16 in depth, under 0.010 in feed) | Recommended tool geometry (rake angles) | Coolant |
|---|---|---|---|---|---|
| Carbon and tool steels | Annealed | 300–1,500 | 600–2,000 | Negative | |
| | Heat-treated | 300–1,000 | 500–1,200 | Negative | N.R.* |
| | Scale | 300–800 | | Negative with edge hone | |
| Alloy steels | Annealed | 300–800 | 400–1,400 | Negative | |
| | Heat-treated | 300–800 | 300–1,000 | Negative with edge hone | N.R. |
| | Scale | 300–600 | | Negative with edge hone | |
| High-speed steel | Annealed | 100–800 | 100–1,000 | Negative | |
| | Heat-treated | 100–600 | 100–600 | Negative with edge hone | N.R. |
| | Scale | 100–600 | | Negative with edge hone | |
| Stainless steel | 300 series | 300–1,000 | 400–1,200 | Positive and negative | Sulfur-base oil |
| | 400 series | 300–1,000 | 400–1,200 | Negative | |
| Cast iron | Gray iron | 200–800 | 200–2,000 | Positive and negative | |
| | Pearlitic | 200–800 | 200–2,000 | Negative | |
| | Ductile | 200–600 | 200–1,400 | Negative | N.R. |
| | Chilled | 100–600 | 200–1,400 | Negative with edge hone | |
| Copper and alloys | Pure | 400–800 | 600–1,400 | Positive and negative | Mist coolant |
| | Brass | 400–800 | 600–1,200 | Positive and negative | Mist coolant |
| | Bronze | 150–800 | 150–1,000 | Positive and negative | Mist coolant |
| Aluminum alloys† | | 400–20,000 | 600–3,000 | Positive | N.R. |
| Magnesium alloys | | 800–10,000 | 800–10,000 | Positive | N.R. |
| Nonmetallics | Green ceramics | 300–600 | 500–1,000 | Positive | |
| | Rubber | 300–1,000 | 400–1,200 | Positive | N.R. |
| | Carbon | 400–1,000 | 600–2,000 | Positive | N.R. |
| Plastics | | 300–1,000 | 400–2,000 | Positive | N.R. |

*N.R., coolant is not required. If a coolant must be used, it is recommended that the tool be flooded to eliminate the possibility of heat checking.

†On certain aluminum alloys, ceramic tools have the tendency to develop a built-up edge.

SOURCE: Haldon J. Swinehart, *Cutting Tool Material Selection*. Reprinted with permission from the Society of Manufacturing Engineers, formerly ASTME, Dearborn, MI.

higher cutting speeds and their variations are needed, for wide, facing cuts, or for cutting of highly abrasive materials. Tables 14-17 through 14-19 give cutting speeds, a troubleshooting guide, and selected properties of ceramic materials.

The tool life of ceramics can be lowered by an increase in cutting feed or speed. This is caused by an increase in temperature, which will—in turn—negatively affect the wear mechanism of the tooling. Wear resistance of ceramics is improved with the formation of interfacial residual compounds such as aluminate formation on alumina tooling material or silicate layers on SiAlON. These protective layers begin to form on the surface of tooling during work cycles conducted at higher cutting speeds, and such coating produces more protection from wear than the basic material from which it evolved.

It is speculated that the formation of these protective layers is facilitated by a low thermal conductivity of ceramics in general, which at high cutting speeds suffers from the localized overheating of the cutting edge. The formation of a protective layer is therefore a subsequent reaction to such thermal abuse of the material.

### 14-4-4. Diamond tooling

Diamond is the hardest natural material, with a resistance to scratching five times greater than that of carbides. Properly utilized,

**TABLE 14-18 Troubleshooting Guide for Ceramic Tools**

| Problem | Possible causes |
|---|---|
| Chipping | Lack of rigidity. "Sawtoothed" or too keen a cutting edge. Chip breaker too narrow or too deep. Chatter. Scale or inclusions. Improper grinding. Too much relief. Defective tool holder |
| Cracking or breaking | Insert surfaces not flat. Insert not seated solidly. Stopping workpiece while tool is engaged. Worn or chipped cutting edges. Feed is too heavy. Improperly applied coolant. Too much rake or relief. Too much overhang or tool too small. Lack of rigidity in setup. Speed too slow. Too much variation in cut for size of the tool. Chatter. Grinding cracks |
| Chatter | Tool not on center. Insufficient relief and/or clearance. Too much rake. Too much overhang or tool too small. Nose radius too large. Feed too heavy. Lack of rigidity. Insufficient horsepower or slipping clutch |
| Torn finish | Lack of rigidity. Dull tool. Speed too slow. Chip breaker too narrow or too deep. Improper grinding |
| Wear | Speed too high or feed too light. Nose radius too large. Improper grinding |
| Cratering | Chip breaker set too tight. Nose radius too large. Side cutting edge angle too great |

SOURCE: Haldon J. Swinehart, *Cutting Tool Material Selection.* Reprinted with permission from the Society of Manufacturing Engineers, formerly ASTME, Dearborn, MI.

**TABLE 14-19 Selected Properties of Ceramic Materials**

| Material | Chemical symbol | Density, $g/cm^3$ | Hardness, Rockwell A scale | Hardness, Vickers | Modulus of elasticity $\times 10^6$ $lb/in^2$ (in tension) | Thermal conductivity W/m/°K at 25°C | Coefficient of linear thermal expansion, $\times 10^{-6}$/°K at 25°C |
|---|---|---|---|---|---|---|---|
| Alumina | $Al_2O_3$ | 3.9–4.0 | 93–95 | 1000–2400 | | 30.0 | 6.5–7.0 |
| Aluminum nitride | AlN | 3.26 | | 200 | | 2.7 | |
| Silicon nitride | $Si_3N_4$ | 3.1–3.3 | 86–95 | 1800–2000 | 20–32 | 15–25 | 2.7–3.5 |
| Silicon carbide | SiC | 2.6–3.3 | | | 35–58 | 20–95 | 3.5–4.5 |
| Tungsten carbide | WC | 15.7 | | 1400–1800 | 105 | 22 | 5.2 |
| Titanium carbide | TiC | 5.0 | | 1700 | 47 | 21 | 7.4 |
| Titanium nitride | TiN | 5.4 | 92 | | 38 | 30 | 9.35 |
| Tantalum carbide | TaC | 14.5 | | 1800 | 42 | 22 | 6.3 |
| Magnesium oxide | MgO | 3.6 | | | | 53 | |
| Zirconia | $ZrO_2$ | 5.7–6.1 | 91–94 | 1600–2200 | 29 | 1.8–2.2 | 9.0–11.0 |
| Diamonite | $Cr_2O_3$ | 4.15 | | | | 22 | 8.0 |

diamond tooling can produce 30 to 300 percent more parts than carbides are capable of turning out between resharpenings.

Diamond crystal is the crystalline form of carbon, very efficient at cutting of nonferrous materials, including aluminum, phenolics and other plastics, graphite, sintered carbides, or fiberglass-filled materials. Because of the smoothness of its surface, chips do not adhere to such tooling even when machining nonferrous materials. Diamond tooling is not recommended for cutting of alloy steels, ferrous metals, and similar harder materials or where a large amount of material is to be removed at once.

Diamond tooling is brittle and should be protected from all shock, be it in the form of machine vibration or that created by a sudden contact with the machined part. Diamonds are also sensitive to higher heat conditions, which may cause them to crack. They operate well at the highest speeds and feeds, with up to 6000 surface ft/min and 0.001 in per revolution.

## 14-5. Types of Steel and Alloys: Properties and Classifications

Various types of standard steels, as used in metal-stamping practice and other applications, have been grouped according to SAE steel specifications. A numeral descriptive system designates values to specific places within the material code, which provides information about the material's manufacture, type, and carbon and major alloy content.

### 14-5-1. Standard grades of steel

The basic code system for these materials is as follows:

_10XX

A prefix A, B, C, D, E may be added in front of the coded description of the material. The meanings of different AISI prefix letters are:

A - Basic open-hearth steel

B - Acid bessemer carbon steel

C - Basic open-hearth carbon steel

D - Acid open-hearth carbon steel

E - Electric furnace steel

The first numeral, shown above as 1, describes the type of steel or its major alloying element. The numerals and their meanings area listed as follows:

1 - Carbon steel
2 - Nickel steel
3 - Nickel-chromium steel
4 - Molybdenum steel
5 - Chromium steel
6 - Chromium-vanadium steel
8 - Triple-alloy steel
9 - Silicon-manganese steel

The last two (sometimes three) numbers of the material code give the content of carbon within the material. For example, 1050 steel has a carbon content of 0.5 percent.

Typical mechanical properties of selected carbon and alloy steels are given in Table 14-20. A brief description and comparison of various steel types according to their common qualities and drawbacks is included below.

*Carbon steels, SAE 1006 through 1015,* are steels with the lowest carbon content, as their numbers indicate. They have low tensile strengths and may be obtained in variations of (1) rimmed steels of deep drawing qualities and (2) killed steels of nonaging tendencies. Steels within this range are susceptible to grain growth due to the amount of cold working, which produces brittleness of the material. Welding of these materials presents no difficulties, while their free machining qualities are impaired. A cold-drawing process may improve the machinability. Yet these steels should not be used for parts where broaching or turning is required and better surface finish is to be produced. The best application of this type of material includes car body covers, oil pans and other deep-drawn parts (rimmed steel), and intricate stampings (killed steel).

*SAE 1017 through 1027 steels* are the carburizing or case-hardening grades, with an increase in their hardness and strength and lowered cold-forming properties. A greater carbon content and an inclusion of manganese raise their depth of hardness penetration while improving the hardness of the core. An increased manganese content aids the machinability, whereas lowered manganese aids the formability. For a greater uniformity of heat treatment, killed steel should be used. Brazing and welding of these materials poses no problems. Their uses include frames, fan blades, welded tubing (1020), and bolts of low strength.

*SAE 1030 through 1052 steels* are medium-carbon materials, heat-treatable, with a possibility of selective hardening (flame or induction hardening), useful for cold work and where their higher mechanical properties are required. They are used as forgings for parts like truck

**TABLE 14-20 Properties of Selected Steel Materials**

| Material | Mfg. process | Tensile strength, KSI | Yield strength, KSI | Elongation, % in 2 in | Reduction of area, % | Hardness, Brinell |
|---|---|---|---|---|---|---|
| 1018 | Hot rolled | 58 | 32 | 25 | 50 | 115 |
| | Cold drawn | 64 | 54 | 15 | 40 | 125 |
| 1020 | Hot rolled | 55 | 30 | 25 | 50 | 110 |
| 1030 | Hot rolled | 80 | 50 | 32 | 57 | 180 |
| | Normalized | 75 | 50 | 32 | 60 | 150 |
| 1035 | Hot rolled | 72 | 40 | 18 | 40 | 144 |
| | Tempered | 103 | 72 | 23 | 59 | 200 |
| 1040 | Hot rolled | 90 | 60 | 25 | 50 | 200 |
| | Normalized | 85–86 | 54 | 28 | 55 | 170 |
| 1045 | Hot rolled | 82 | 45 | 16 | 40 | 162 |
| | Cold drawn | 91 | 77 | 12 | 35 | 178 |
| 1095 | Hot rolled | 120 | 66 | 10 | 25 | 245 |
| | Tempered | 200 | 137 | 12 | 37 | 385 |
| 1137 | Hot rolled | 91 | 55 | 28 | 61 | 190 |
| | Normalized | 97 | 57–58 | 22 | 48 | 196 |
| 1141 | Hot rolled | 94 | 50–51 | 15 | 35 | 186 |
| | Cold drawn | 105 | 88 | 10 | 30 | 210 |
| 1340 | Annealed | 102 | 63 | 25 | 57 | 207 |
| 4130 | Hot rolled | 90 | 52 | 28 | | 178 |
| 4150 | Annealed | 106 | 55 | 20 | 40 | 196 |
| | Normalized | 167 | 106 | 12 | 31 | 320 |
| 4340 | Annealed | 100 | 58 | 21 | 45 | 230 |
| | Normalized | 185 | 125 | 12 | 36 | 360 |
| 5150 | Annealed | 98 | 50 | 22 | 43 | 195 |
| | Normalized | 125 | 75 | 20 | 59 | 254 |
| 6150 | Annealed | 97 | 55 | 23 | 45 | 195 |
| | Normalized | 135 | 88 | 22 | 61 | 270 |
| 8630 | Annealed | 81 | 54 | 29 | 59 | 155 |

front axles and tractor and automobile components. A proper heat treatment of these products is necessary for their machinability. Another use is bar stock for machining, with or without heat treatment. Where utilized for stampings, flat parts or parts with only simple bends are recommended.

*SAE 1055 through 1095 steels* boast a high carbon content, greater than necessary for attainment of maximum hardness. The higher carbon content makes these materials suitable for manufacture of coiled springs or cutting tools, with the necessity of heat treatment after machining. Their cold-forming properties are extremely limited, and metal-stamping applications are reserved for flat parts, such as washers. Most use is for farming machinery.

*SAE 1111, 1112, 1113 steels, or free-cutting steels,* are easily machinable, in return for their poor cold forming, welding, and forging characteristics. Their machinability improves with a higher sulfur content. They are also known as bessemer screw stock.

*SAE 1109 through 1126 steels* combine a good response to heat treatment with good machining properties. Grades with lower carbon content are used for parts to be cyanided or carbonitrided. Steels with greater manganese content have better hardenability tendencies.

*SAE 1132 through 1151 steels* are specified for various machining purposes, as they are frequently used for manufacture of threaded nuts, bolts, and studs. Greater manganese content of some steels aids their hardenability.

### 14-5-2. Stainless steels

Stainless steels are iron-base alloys containing 10.5 percent or more chromium in their composition. Such a high chromium content provides resistance to corrosion by forming a thin and transparent film of chromium oxide over their surface. Properties of this protective coating are quite amazing: If such a layer is disrupted or ruined, a new surface layer emerges in its stead. The functionability of this surface protection is applicable to all normal atmospheric conditions, including exposure to humidity and weathering. It further expands in the form of resistance to various chemical influences, which are usually limited to a certain group of specific steel type. Such applications are listed in every steelmaker's catalog.

A further improvement in surface protection may be achieved by including additional amounts of chromium, as combined with nickel, molybdenum, and similarly acting alloying elements.

In secluded areas, if the protective layer of chromium oxide is disrupted, pitting of the material surface may occur. Where such a possibility is expected, the choice of stainless-steel grade should be 316 or 317 where a higher molybdenum content prevents such defects.

However, corrosion may at times affect stainless-steel materials in its intergranular form. This may happen where the material is heated to 800 to 1650°F and then cooled. During this stage, the chromium content of the layers closest to the surface may combine with carbon to form chromium carbides. This reaction, called carbide precipitation, or sensitization, directly influences the appearance of intergranular corrosion and its magnitude.

To prevent such corrosion, stainless steels specified for high-heat applications like welding should be limited to materials with carbon levels under 0.03 percent. Also a selection of stabilized types, such as type 321, which is stabilized with titanium, or type 347, stabilized with columbium, is recommended.

Where intergranular corrosion cannot be averted, parts may be annealed to dissolve the carbides tied up in the chromium carbide formations.

**14-5-2-1. Types of stainless steel.** The AISI numbering system is used to categorize all stainless-steel materials within three categories:

**Austenitic stainless steels.** These are AISI types 201, 202, 205, 301 through 305, 308 through 310, 314, 316, 317, 321, 329, 330, 347, 348, and 384. The austenitic stainless steel may be strengthened by cold working, while retaining a good ductility and toughness, but they cannot be hardened through heat treatment. A sample hardness of 201 and 301 type of stainless steels is as follows:

| | Minimum tensile strength, lb/in² | Minimum yield strength, lb/in² |
|---|---|---|
| ¼ hard | 125,000 | 75,000 |
| ½ hard | 150,000 | 110,000 |
| ¾ hard | 175,000 | 135,000 |
| Full hard | 185,000 | 140,000 |

Physical properties of austenitic stainless steels are comparable to those of ferritic and martensitic grades.

**Ferritic stainless steels.** These contain AISI types 405, 409, 429, 430, 434, 436, 442, and 446. In the annealed condition, their toughness decreases simultaneously with the increase of their chromium content. Ferritic stainless steels have 12 percent or more of chromium in their composition. Their ductility tends to be elevated by the inclusion of additional molybdenum, while greater carbon content tends to decrease this property. Exposure to heat, such as that produced by welding, causes enlargement of the grain size, which in turn decreases the material's mechanical properties.

Ferritic stainless steels are used in the automotive industry, for kitchen utensils and equipment, and as appliance parts, fasteners, and various other machined components.

**Martensitic stainless steels.** These include AISI types 403, 410, 414, 416, 420, 422, 431, and 440. They are called martensitic steels because when they are heated above the critical temperature of 1600°F and cooled rather rapidly, their metallurgical structure turns into martensite. A stress-relieving treatment or tempering is required for these materials to attain the necessary ductility, impact strength, and corrosion resistance.

The hardness range of martensitic stainless steels in their annealed state is approximately 24 Rockwell C. When hardened, they can be divided into two groups: (1) low-carbon steels, with maximum hardness of 45 Rockwell C; and (2) high-carbon steels, with maximum hardness of 60 Rockwell C. Their hardness, however, increases along with their toughness, for which reason nickel is used to enhance the toughness. Nickel also improves their notch impact strength and corrosion and wear resistance, for it permits a higher chromium content within the material.

Martensitic steels have considerable abrasion and wear resistance and lower modulus of elasticity, but they tend to suffer from temper brittleness if heat-treated within the range of 800 to 1050°F. Properties of stainless steels are listed in Table 14-21.

**14-5-2-2. Surface finish of stainless steel.** The standard mechanical finish may be divided into two groups: rolled finish (not polished) and polished finish. *Rolled finish* contains categories 1, 2D, and 2B. Groups 1 and 2D are dull in appearance, while the third group, 2B, is bright, as this annealed and descaled material is given a final run through the polished rolls.

*Polished finish* contains groups 3, 4, 6, 7, and 8. Groups 3 and 4 are obtained by finishing the stock with 100-grit or 150-grit abrasives. No. 6 finish is a satin finish, with a lowered reflectivity. No. 7 finish is highly reflective, obtained by buffing. But the most reflective surface finish is no. 8, which is obtained through polishing with fine abrasive elements and buffing as well.

### 14-5-3. Aluminum

Grading of aluminum alloys uses a methodology similar to that used for other groups of materials. The first digit identifies the type of alloy. The second digit shows its modification, and the third and fourth digits are reserved for demarcation of aluminum's purity within the given alloy.

The first digit of the alloy designation distinguishes between following types of aluminum alloys, grouped according to their second main constituent to the basic aluminum:

1 - Aluminum, 99 percent minimum and up

2 - Copper

3 - Manganese

4 - Silicon

5 - Magnesium

6 - Magnesium and silicon

7 - Zinc

8 - Other elements

9 - Not used

**TABLE 14-21 Properties of Stainless Steel**

| Material type | 201 | 302 | 303 | 304 | 304L | 309 |
|---|---|---|---|---|---|---|
| Structure of material | Austenitic | Austenitic | Austenitic | Austenitic | Austenitic | Austenitic |
| Hardening properties | Nonhardening | Nonhardening | Nonhardening | Nonhardening | Nonhardening | Nonhardening |
| Tensile strength, lb/in$^2$ × $10^3$ | 95–115 | 90 | 90 | 85 | 80 | 90–100 |
| Yield strength, lb/in$^2$ × $10^3$ | 45–55 | 40 | 35–40 | 35 | 30 | 35–45 |
| Elongation, %, 2 in | 40–55 | 50–55 | 50 | 55 | 55 | 45 |
| Reduction of area, % | 40 | 60–70 | 55 | 70 | 70 | 65 |
| Modulus of elasticity, in tension, lb/in$^2$ × $10^6$ | | 28 | 28–29 | 29 | 29 | 29 |
| Hardness, Brinell | 179 | 150 | 160 min | 180 max | 180 max | 200 max |
| Hardness, Rockwell | B90 | B85 | B80 min | B80–90 | B90 max | B95 max |
| Magnetic permeability at 200 H, annealed | 1.02 | 1.02 | 1.02 | 1.02 | 1.02 | 1.02 |
| Electrical resistivity, μΩ, cm at 68°F | 69.0 | 72.0 | 72.0 | 72.0 | 72.0 | 78.0 |
| Heat resistance, max. °F, intermittent service | 1500 | 1450 | 1450 | 1600 | 1600 | 1800 |
| Machinability index, % | 45 | 50 | 70 | 50 | 50 | 40–45 |
| Drawability, stamping | Good | Good | Fair | Good | Good | Good |
| Weldability | Excellent | Excellent | Limited | Excellent | Excellent | Good |
| Approximate price range, % | 45 | 25 | 50 | 50 | 55 | 70–75 |
| Carbon (C), % | 0.15 | 0.15 | 0.15 | 0.08 | 0.03 | 0.20 |
| Manganese (Mn), % | 5.5–7.5 | 2.0 | 2.0 | 2.0 | 2.0 | 2.0 |
| Silicon (Si), % | 1.0 | 1.0 | 1.0 | 1.0 | 1.0 | 1.0 |
| Chromium (Cr), % | 16–18 | 17–19 | 17–19 | 18–20 | 18–20 | 22–24 |
| Nickel (Ni), % | 3.5–5.5 | 8–10 | 8–10 | 8–10.5 | 8–12 | 12–25 |
| Other elements, % | 0.25 N | | | | | |

| Material type | 310 | 316 | 316L | 321 | 347 | 405 |
|---|---|---|---|---|---|---|
| Structure of material | Austenitic | Austenitic | Austenitic | Austenitic | Austenitic | Ferritic |
| Hardening properties | Nonhardening | Nonhardening | Nonhardening | Nonhardening | Nonhardening | Hardenable |
| Tensile strength, $lb/in^2 \times 10^3$ | 90 | 85 | 75 | 85–90 | 95 | 65–70 |
| Yield strength, $lb/in^2 \times 10^3$ | 45 | 35 | 30 | 35 | 40 | 40 |
| Elongation, %, 2 in | 40–50 | 50–60 | 50–60 | 55 | 45–50 | 25–30 |
| Reduction of area, % | 50–60 | 70 | 70 | 65 | 60–65 | 60 |
| Modulus of elasticity, in tension, $lb/in^2 \times 10^6$ | 20–30 | 29 | 29–30 | 29 | 28 | 29 |
| Hardness, Brinell | 180 max | 200 max | 180 max | 200 max | 185 max | 150 |
| Hardness, Rockwell | B90 max | B95 max | B90 max | B95 max | B90 max | B75 |
| Magnetic permeability at 200 H, annealed | 1.01 | 1.02 | 1.02 | 1.02 | 1.02 | Magnetic |
| Electrical resistivity, $\mu\Omega$, cm at 68°F | 78.0 | 74.0 | 72.0 | 72.0 | 72.0 | 57.0 |
| Heat resistance, max. °F, intermittent service | 1900 | 1600 | 1600 | 1600 | 1450 | 1250 |
| Machinability index, % | 40–45 | 45 | 45 | 50 | 50 | 60 |
| Drawability, stamping | Good | Good | Good | Good | Good | Fair |
| Weldability | Good | Excellent | Excellent | Excellent | Excellent | Good |
| Approximate price range, % | 83–110 | 60–80 | 60–90 | 60 | 60–85 | 40–55 |
| Carbon (C), % | 0.25 | 0.08 | 0.03 | 0.08 | 0.08 | 0.08 |
| Manganese (Mn), % | 2.0 | 2.0 | 2.0 | 2.0 | 2.0 | 1.0 |
| Silicon, (Si), % | 1.5 | 1.0 | 1.0 | 1.0 | 1.0 | 1.0 |
| Chromium (Cr), % | 24–26 | 16–18 | 16–18 | 17–19 | 17–19 | 11.5–14.5 |
| Nickel (Ni), % | 19–22 | 10–14 | 10–14 | 9–12 | 9–13 | |
| Other elements, % | | 2–3 Mo | 2–3 Mo | | | |

**TABLE 14-21 Properties of Stainless Steel (*Continued*)**

| Material type | 410 | 416 | 420 | 430 | 440C |
|---|---|---|---|---|---|
| Structure of material | Martensitic | Martensitic | Martensitic | Ferritic | Martensitic |
| Hardening properties | Hardenable | Hardenable | Hardenable | Hardenable | Hardenable |
| Tensile strength, lb/in$^2 \times 10^3$ | 75 | 75 | 95 | 75 | 110 |
| Yield strength, lb/in$^2 \times 10^3$ | 40 | 40–45 | 55 | 45–50 | 65 |
| Elongation, %, 2 in | 35 | 30 | 25 | 30 | 14 |
| Reduction of area, % | 70 | 65 | 55 | 65 | 25 |
| Modulus of elasticity, in tension, lb/in$^2 \times 10^6$ | 29 | 29 | 29 | 29 | 30 |
| Hardness, Brinell | 200 max | 180 max | 190 max | 200 max | 250 max |
| Hardness, Rockwell | B95 max | B90 max | B95 | B95 max | B105 max |
| Magnetic permeability at 200 H, annealed | Magnetic | Magnetic | Magnetic | Magnetic | Magnetic |
| Electrical resistivity, $\mu\Omega$, cm at 68°F | 57.0 | 57.0 | 55.0 | 60.0 | 60.0 |
| Heat resistance, max. °F, intermittent service | 1250 | 1250 | N/A | 1500 | 1400 |
| Machinability index, % | 60 | 93 | 54 | 55 | 40 |
| Drawability, stamping | Fair | Fair | Fair | Fair | Fair |
| Weldability | Fair | Limited | N/A | Fair | Limited |
| Approximate price range, % | 40–45 | 40 | 45 | 40 | 51 |
| Carbon (C), % | 0.15 | 0.15 | 0.15 min | 0.12 | 0.95–1.2 |
| Manganese (Mn), % | 1.0 | 1.25 | 1.0 | 1.0 | 1.0 |
| Silicon (Si), % | 1.0 | 1.0 | 1.0 | 1.0 max | 1.0 |
| Chromium (Cr), % | 11–14 | 12–14 | 12–14 | 16–18 | 16–18 |
| Nickel (Ni), % | 0.5 | 0.5 | | 0.5 max | |
| Other elements, % | | | | | 0.75 Mo |

Where X precedes the four digits' code, for example, X8280, the aluminum alloy is experimental.

Temper designation follows the basic coded description, separated by a dash, as 6061-T4. The additional coded information may be explained as follows:

- -F As fabricated
- -0 Annealed and recrystallized, wrought materials only
- -H Strain-hardened, without heat treatment, wrought materials only. This designation is always followed by a number, indicating the particular strain-hardening process.
  - -H1 Strain-hardened only
  - -H2 Strain-hardened and partially annealed
  - -H3 Strain-hardened and stabilized
- -T Thermally treated material, obtaining other than previously described qualities. This designation is always followed by another digit as shown:
  - -T1 Partially solution heat-treated and naturally aged
  - -T2 Annealed, cast materials only
  - -T3 Solution heat-treated and cold-worked
  - -T4 Solution heat-treated and naturally aged
  - -T5 Artificially aged
  - -T6 Solution heat-treated and artificially aged
  - -T7 Solution heat-treated and stabilized
  - -T8 Solution heat-treated, cold-worked, and afterward artificially aged
  - -T9 Solution heat-treated, artificially aged, and afterward cold-worked
  - -T10 Artificially aged and afterward cold-worked
  - -T42 Solution heat-treated to attain different mechanical properties from those of -T4 temper
  - -T62 Solution heat-treated, artificially aged, to obtain different mechanical properties from those of -T6 temper
  - -TX52 Stress relieved by compressing
  - -TX53 Stress relieved by thermal treatment
- -W Solution heat-treated

The final amount of strain hardening is indicated by a number following the designation -H1, -H2, -H3, as:

H111 Strain-hardened less than the controlled H11 temper

H112 Temper acquired from shaping processes without being particularly controlled

H311 Strain-hardened less than the controlled H31 temper

**Types of aluminum alloys.** Aluminum alloys, as grouped into the above series, display various properties particular to their designation.

*1000 series* are almost pure aluminum, with mechanical properties at the low level but with a high workability, ductility, corrosion resistance, chemical and weathering effect resistance, and conductivity, thermal and electrical. It is frequently used for reflectors, fan blades, dials, nameplates, and ductwork.

*2000 series aluminum alloys* are stronger than the previous group, with the same ductility but lower corrosion resistance, as it may become affected by intergranular corrosion. Sheet material is usually coated with a high-purity alloy, to provide for galvanic protection of the core. The material may be solution heat-treated for optimal properties. In its 2024 form, this alloy is used mainly for aircraft applications.

*3000 series aluminum alloys*; in this series, 3003 is the most popular grade. It displays good ductility, which designates it for drawing applications, and is widely used in the manufacture of chemical equipment, kitchenware, truck and van panels, shelves, and refrigerator liners.

*4000 series aluminum alloys* can have their melting point lowered without subsequent brittleness of the material. For these properties such materials are used as welding rods, brazing rings, and similar applications. Most of 4000 series alloys cannot be heat-treated.

*5000 series aluminum alloys* are widely utilized for automotive parts, such as body trims, as well as gas tanks. Applications further include aircraft parts, pressure vessels, marine products, welded constructions, refrigerator components, etc. This type of material offers good welding characteristics, along with a resistance to corrosion even in marine applications. Alloys with greater than 3.5 percent of magnesium content should not be exposed to either increased amounts of cold work or heat. Higher operating temperatures, such as those exceeding 150°F, may bring about a risk of stress-corrosion appearance within these materials.

*6000 series aluminum alloys* are heat-treatable and combine good formability, along with good corrosion resistance. Their strength is somewhat smaller than that of 2000 series alloys. The most often used alloy of this group of materials is 6061, which is valued for its versatility. This material is utilized for manufacture of tubes, pipes, tanks, furniture, chemical equipment, sailboats, fittings, couplings, apparatus components, etc.

*7000 series aluminum alloys* display considerable strength, for which reason these materials are often used for manufacture of highly stressed parts. Materials of this group may be heat-treated. Their additional uses are within aircraft parts manufacturing as a material for keys, and where higher strengths than those exhibited by 2024 alloy are needed. Tables 14-22 through 14-24 show various properties and characteristics of aluminum alloys.

### 14-5-4. Nickel and nickel alloys

Nickel is known for its high heat resistance, corrosion resistance, and electricity-conductive properties. The usage of this material includes various chemical and electrical-related applications, with heavy involvement within the food and drug industry. Table 14-25 lists mechanical properties of nickel.

### 14-5-5. Copper alloys

Copper is usually alloyed with other ingredients such as silicon or beryllium and cobalt. Copper-silicon alloys are materials of high strength, resistant to corrosion, with free machining qualities.

Beryllium copper alloys may be separated into two basic groups: those with beryllium content exceeding 1 percent, which are alloys of a considerable hardness and strength, and those with a beryllium content of less than 1 percent, which are valued for good thermal and electrical qualities, such as good conductivity and nonmagnetism. Physical properties are given in Table 14-26.

**TABLE 14-22 Mean Values of Materials (at Room Temperature)**

| Material | Modulus of elasticity, KSI | Shear strength, KSI | Poisson's ratio |
|---|---|---|---|
| Steel | 30.5 | 11.75 | 0.30 |
| Gray iron | 16 | 6.25 | 0.25 |
| Copper | 17.4 | 6.37 | 0.35 |
| Bronze | 16 | 6.1 | 0.35 |
| Brass | 13.75 | 5 | 0.35 |
| Aluminum alloys | 10.16 | 3.9 | 0.33 |
| Magnesium alloys | 4.9 | 2 | 0.30 |
| Zinc, drawn | 12 | 4.79 | 0.27 |
| Lead | 2.5 | 0.87 | 0.45 |
| Glass | 8.7 | 3.48 | 0.23 |

**TABLE 14-23 Properties of Selected Aluminum Alloys**

| Material type | Tensile strength, KSI | Yield strength KSI | Shear strength, KSI | Modulus of elasticity, KSI | Elongation, % in 2 in | | Endurance limit, KSI | Hardness, Brinell, 500 kg, ϕ10 mm |
|---|---|---|---|---|---|---|---|---|
| | | | | | 1/16 in thick | 1/2 in thick | | |
| 1100-0 | 13 | 5 | 9–9.5 | 10.0 | 35 | 45 | 5 | 23 |
| 1100-H12 | 15.5–16 | 14–15 | 10 | 10.0 | 12 | 25 | 6 | 28 |
| 1100-H14 | 17.5–18 | 16–17 | 11 | 10.0 | 9 | 20 | 7 | 32 |
| 1100-H16 | 20–21 | 18–20 | 12 | 10.0 | 6 | 17 | 8.5–9 | 38 |
| 1100-H18 | 24 | 22 | 13 | 10.0 | 5 | 15 | 8.5–9 | 44 |
| 2011-T3 | 55 | 43 | 32 | 10.2 | | 15 | 18 | 95 |
| 2011-T6 | 57 | 39 | 34 | 10.2 | | 17 | 18 | 97 |
| 2011-T8 | 59 | 45 | 35 | 10.2 | | 12 | 18 | 100 |
| 2014-0 | 27 | 14 | 18 | 10.6 | | 18 | 13 | 45 |
| 2014-T4 | 62 | 40–42 | 38 | 10.6 | | 20 | 20 | 105 |
| 2014-T6 | 70 | 60 | 42 | 10.6 | | 13 | 18 | 135 |
| 2017-0 | 26 | 10 | 18 | 10.5 | | 22 | 13 | 45 |
| 2017-T4 | 62 | 40 | 38 | 10.5 | | 22 | 18 | 105 |
| 2018-T61 | 61 | 46 | 39 | 10.8 | | 12 | 17 | 120 |
| 2024-0 | 17 | 11 | 18 | 10.6 | 19–20 | 22 | 13 | 47 |
| 2024-T3 | 70 | 50 | 41 | 10.6 | 18 | | 20 | 120 |
| 2025-T6 | 58 | 37 | 35 | 10.4 | | 19 | 18 | 110 |
| 2219-0 | 25 | 10 | | 10.6 | 20 | | | |
| 2219-T31 | 52–54 | 36.5 | 33 | 10.6 | 17 | | | |
| 3003-0 | 16 | 6 | 11 | 10.0 | 30 | 40 | 7 | 28 |
| 3003-H12 | 19 | 18 | 12 | 10.0 | 10 | 20 | 8 | 35 |
| 3003-H14 | 22 | 21 | 14 | 10.0 | 8 | 16 | 9 | 40 |
| 3003-H16 | 26 | 25 | 15 | 10.0 | 5 | 14 | 10 | 47 |
| 3003-H18 | 29 | 27 | 16 | 10.0 | 4 | 10 | 10 | 55 |
| 3004-0 | 26 | 10 | 16 | 10.0 | 20 | 25 | 14 | 45 |
| 3004-H32 | 31 | 25 | 17 | 10.0 | 10 | 17 | 15 | 52 |
| 3004-H34 | 35 | 29 | 18 | 10.0 | 9 | 12 | 15 | 63 |
| 3004-H36 | 38 | 32 | 20 | 10.0 | 5 | 9 | 15.5 | 70 |

| | | | | | | | | |
|---|---|---|---|---|---|---|---|---|
| 3004-H38 | 41 | 35 | 21 | 10.0 | 5 | 6 | 16 | 77 |
| 3105-0 | 17 | 8 | 12 | 10.0 | 24 | | | |
| 3105-H12 | 22 | 19 | 14 | 10.0 | 7 | | | |
| 3105-H14 | 25 | 22 | 15 | 10.0 | 5 | | | |
| 3105-H16 | 28 | 25 | 16 | 10.0 | 4 | | | |
| 4032-T6 | 55 | 46 | 38 | 11.4 | | 9 | 16 | 120 |
| 5005-0 | 18 | 6 | 11 | 10.0 | 30 | | | 28 |
| 5005-H12 | 20 | 19 | 14 | 10.0 | 10 | | | |
| 5005-H14 | 23 | 22 | 14 | 10.0 | 6 | | | |
| 5005-H16 | 26 | 25 | 15 | 10.0 | 5 | | | |
| 5005-H18 | 29 | 28 | 16 | 10.0 | 4 | | | |
| 5005-H32 | 20 | 17 | 14 | 10.0 | 11 | | | 36 |
| 5005-H34 | 23 | 20 | 14 | 10.0 | 8 | | | 41 |
| 5005-H36 | 26 | 24 | 15 | 10.0 | 6 | | | 46 |
| 5050-0 | 21 | 8 | 14.5 | 10.0 | 24 | | 12 | 36 |
| 5050-H32 | 25 | 21 | 16.5 | 10.0 | 9 | | 13 | 46 |
| 5050-H34 | 28 | 24 | 17.5 | 10.0 | 8 | | 13 | 53 |
| 5050-H36 | 30 | 26 | 18.5 | 10.0 | 7 | | 14 | 58 |
| 5052-0 | 27.5 | 12.5 | 18 | 10.2 | 25 | 30 | 16 | 47 |
| 5052-H32 | 33.5 | 27.5 | 20 | 10.2 | 12 | 18 | 17 | 60 |
| 5052-H34 | 37.5 | 31 | 21 | 10.2 | 10 | 15 | 18 | 67.5 |
| 5052-H36 | 39.5 | 34.5 | 23 | 10.2 | 8 | 10 | 19 | 73.5 |
| 5056-0 | 42 | 22 | 26 | 10.3 | | 35 | 20 | 65 |
| 5056-H18 | 63 | 59 | 34 | 10.3 | | 10 | 22 | 105 |
| 5086-0 | 38 | 17 | 23 | 10.3 | 22 | 30 | | |
| 5086-H32 | 42 | 30 | 25 | 10.3 | 12 | | | |
| 5086-H34 | 47 | 37 | 27.5 | 10.3 | 10 | 14 | | |
| 5154-0 | 35 | 17 | 22 | 10.2 | 27 | 30 | 17 | 58 |
| 5154-H32 | 39 | 30 | 22.5 | 10.2 | 15 | 18 | 18 | 67 |
| 5154-H34 | 42 | 33 | 24 | 10.2 | 13 | 16 | 19 | 73 |
| 5154-H36 | 45 | 36 | 26 | 10.2 | 12 | 14 | 20 | 78 |
| 5252-H25 | 34 | 25 | 21 | 10.0 | 11 | | | 68 |
| 5252-H38 | 41 | 35 | 23 | 10.0 | 5 | | | 75 |
| 5254-0 | 35 | 17 | 22 | 10.2 | 27 | 30 | 17 | 58 |

**TABLE 14-23 Properties of Selected Aluminum Alloys (*Continued*)**

| Material type | Tensile strength, KSI | Yield strength KSI | Shear strength, KSI | Modulus of elasticity, KSI | Elongation, % in 2 in | | Endurance limit, KSI | Hardness, Brinell, 500 kg, $\phi$10 mm |
|---|---|---|---|---|---|---|---|---|
| | | | | | 1/16 in thick | 1/2 in thick | | |
| 5254-H32 | 39 | 30 | 22.5 | 10.2 | 15 | | 18 | 67 |
| 5254-H34 | 42 | 33 | 24 | 10.2 | 13 | | 19 | 73 |
| 5254-H36 | 45 | 36 | 26 | 10.2 | 12 | 14 | 20 | 78 |
| 5357-0 | 19 | 7 | 12 | 10.2 | 25 | | | |
| 5357-H25 | 27 | 23 | 16 | 10.2 | 10 | | | |
| 5456-0 | 45 | 23 | 27 | 10.3 | 24 | 20 | | |
| 5456-H24 | 54 | 41 | 31 | 10.3 | 12 | | | |
| 5557-0 | 16 | 6 | 10 | 10.2 | 25 | | | |
| 5557-H25 | 25 | 21 | 15 | 10.2 | 10 | | | |
| 5652-0 | 28 | 13 | 18 | 10.2 | 25 | 30 | 16 | 47 |
| 5652-H32 | 33 | 28 | 20 | 10.2 | 12 | 18 | 17 | 60 |
| 5652-H34 | 38 | 31 | 21 | 10.2 | 10 | 14 | 18 | 68 |
| 5652-H36 | 40 | 35 | 23 | 10.2 | 8 | 10 | 19 | 73 |
| 6061-0 | 18 | 8 | 12 | 10.0 | 25 | 30 | 9 | 30 |
| 6061-T4 | 35 | 21 | 24 | 10.0 | 22 | 25 | 14 | 65 |
| 6061-T6 | 45 | 40 | 30 | 10.0 | 12 | 17 | 14 | 95 |
| 6062-0 | 17.5 | 7–8 | 12 | 10.0 | | 30 | 8.5 | 28 |
| 6062-T4 | 35 | 21 | 24 | 10.0 | | 25 | 13.5 | 65 |
| 6062-T6 | 45 | 40 | 30 | 10.0 | | 17 | 13.5 | 95 |
| 6063-0 | 13 | 7 | 10 | 10.0 | | | 8 | 25 |
| 6063-T1 | 22 | 13 | 14 | 10.0 | 20 | | 9 | 42 |
| 6063-T4 | 25 | 13 | 16 | 10.0 | 22 | | | |
| 6063-T5 | 27 | 21 | 17 | 10.0 | 12 | | 10 | 60 |
| 6063-T6 | 35 | 31 | 22 | 10.0 | 12 | 18 | 10 | 73 |

| | | | | | | | | |
|---|---|---|---|---|---|---|---|---|
| 7001-0 | 37 | 22 | | 10.3 | | 14 | | 60 |
| 7001-T6 | 98 | 91 | | 10.3 | | 9 | 22 | 160 |
| 7075-0 | 33 | 15 | 22 | 10.4 | 17 | 16 | | 60 |
| 7075-T6 | 82.5 | 72.5 | 48.5 | 10.4 | 11 | 11 | 23.5 | 150 |
| 7178-0 | 33 | 15 | 22 | 10.4 | 15 | 16 | | |
| 7178-T6 | 88 | 78 | 52 | 10.4 | 10 | 11 | | |

***Note:*** Values presented in this table are to be considered average, with no variation owing to shape and size of parts or the method of their production taken into account.

**TABLE 14-24 Aluminum—Working Characteristics**

| Type | Condition | Corrosion resistance | Cold-working suitability | Machinability |
|---|---|---|---|---|
| 1100-0 | Annealed | 5 | 5 | 4 |
| 1100-H18 | Hard | 5 | 4 | 4 |
| 3003-0 | Annealed | 5 | 5 | 4 |
| 3003-H18 | Hard | 5 | 3 | 4 |
| 3004-0 | Annealed | 5 | 5 | 4 |
| 3004-H38 | Hard | 5 | 3 | 4 |
| 2011-T3 | Heat-treated | 2 | 3 | 5 |
| 2014-T4 | Heat-treated | 3 | 3 | 5 |
| 2014-T6 | Heat-treated and aged | 3 | 3 | 5 |
| 2017-T4 | Heat-treated | 3 | 4 | 5 |
| 2024-T3 | Heat-treated | 3 | 3 | 5 |
| 5050-0 | Annealed | 5 | 5 | 4 |
| 5050-H38 | Hard | 5 | 3 | 4 |
| 5052-0 | Annealed | 5 | 5 | 4 |
| 5052-H38 | Hard | 5 | 3 | 4 |
| 6053-T4 | Heat-treated | 5 | 4 | 4 |
| 6053-T6 | Heat-treated and aged | 5 | 3 | 4 |
| 5056-0 | Annealed | 5 | 5 | 4 |
| 5056-H38 | Hard | 4 | 4 | 4 |
| 5154-0 | Annealed | 5 | 4 | 3 |
| 5154-H34 | Hard | 5 | 4 | 3 |
| 5154-H38 | Hard | 5 | 4 | 3 |
| 6061-T4 | Heat-treated | 5 | 4 | 4 |
| 6061-T6 | Heat-treated and aged | 5 | 4 | 4 |
| 7075-T6 | Heat-treated and aged | 3 | 2 | 5 |

Relative evaluation, where 1 is the lowest rating and 5 is the greatest.

**TABLE 14-25 Mechanical Properties of Nickel**

| Material type | Material condition | Tensile strength, KSI | Yield strength, 0.2% offset, KSI | Elongation, % in 2 in | Modulus of elasticity, KSI | Hardness, Rockwell |
|---|---|---|---|---|---|---|
| Monel, 400 | Annealed strip | 70 | 25–30 | 35 | 26 | 68B |
| Monel, K-500 | Annealed strip | 90 | 40 | 20 | 26 | 75B |
| Monel, R405 | Annealed rods | 75 | 35 | 35 | | |
| Inconel, 600 | Annealed strip | 80 | 30 | 30 | 31 | 84B |
| Inconel, 800 | Annealed strip | 75 | 30 | 30 | 31 | 84B max |
| Nickel, 200 | Annealed strip | 55 | 15 | | 30 | 64B |
| Duranickel, 301 | Annealed strip | 90 | 35 | | 30 | 90B |

## 14-6. Comparison of Materials Worldwide

A comparison of designations for various material types worldwide is included in Table 14-27. It contains the material denominations used in England, Germany, France, Italy, Japan, Sweden, and the Czech Republic. Properties and composition of materials are not described,

**TABLE 14-26 Physical Properties of Copper Alloys**

| Material type | Material condition | Tensile strength, KSI | Yield strength, KSI | Elongation, % in 2 in | Hardness, Rockwell |
|---|---|---|---|---|---|
| Copper, 110 | Annealed sheet | 33 | 10 | 35 | F35 |
| Bronze, 210 | Annealed sheet | 35 | 11 | 38 | F45 |
| Bronze, 220 | Annealed sheet | 37 | 12 | 40 | F45 |
| Brass, yellow, 268 | Annealed sheet | 45 | 17 | 60 | B15 |
| Brass, yellow, 274 | Annealed sheet | 54 | 20 | 45 | B45 |
| Aluminum bronze | Annealed sheet | 55 | 22 | 65 | B35 |
| Everdur | Annealed sheet | 58 | 22 | 60 | B35 |
| Cupronickel, 30% | Annealed sheet | 55 | 22 | 40 | B35 |
| Nickel-silver, 10% | Annealed sheet | 55 | 20 | 42 | B30 |
| Nickel-silver, 18% | Annealed sheet | 60 | 22 | 45 | B45 |
| Phosphor bronze, 50 | Annealed sheet | 40 | 14 | 48 | F60 |
| Phosphor bronze, 51 | Annealed sheet | 48 | 20 | 50 | B28 |
| Phosphor bronze, 52 | Annealed sheet | 60 | 24 | 65 | B50 |

the list being limited to their equivalent names or codes within that particular nation's system.

## 14-7. Heat Treatment

One of the most important factors to be considered when evaluating the possibility of heat treatment for any particular material is the effect it may exert on the size of its grain. Since all fine-grained structures display much better toughness and less warpage inclination at higher attained hardnesses, materials of grain structure not affected by heat treatment are definitely preferable for this procedure.

The greatest danger of the emergence of grain-related irregularities may be encountered at temperatures above the critical range, in parts previously cold-worked, when an effect known as grain growth may occur. This condition was discussed in Sec. 9-12-3, "Grain Growth."

**Suitability of materials for heat treatment.** The suitability of materials for heat treatment is given by the ease of the hardening process or by the depth of hardness penetration achievable within that material. Such suitability, otherwise called hardenability of the material, is closely tied to the carbon content of the particular stock: With greater carbon content the hardenability is improved. Low-carbon steels sometimes must first be saturated with carbon elements, or carburized, in order to attain the necessary hardness range.

Hardenability is further affected by the cooling rate of steel, or the speed at which the material must be cooled in order to harden. The depth of hardness penetration with regard to the length of exposure to hardening influences is often assessed as an indicator of hardenability. The depth is always more pronounced in materials with higher carbon

**TABLE 14-27 Worldwide Steel and Alloy Comparison Chart**

| U.S. AISI-ASTM | Germany No. | Germany DIN | Belgium, NBN | France, AFNOR | Great Britain, B.S. |
|---|---|---|---|---|---|
| 1010 | 1.0301 | C10 | | AF34C10 | 045M10, 040A10 |
| | 1.1121 | Ck10 | C10-2 | XC10 | 1449 10CS |
| 1015 | 1.0401 | C15 | | AF37C12, XC18 | 040A15, 080M15 |
| | 1.1141 | Ck15 | C16-2 | XC15 | 1449 17CS |
| 1020 | 1.0402 | C22 | C25-1 | AF42C20, XC25 | 050A20, 055M15 |
| 1020, 1023 | 1.1151 | Ck22 | C25-2 | XC25, XC18 | 050A20 (055M15) |
| 1022 | 1.1133 | 20Mn5 | | 20M5 | 120M19 |
| 1023, 1020 | 1.1151 | Ck22 | C25-2 | XC25, XC18 | 050A20 (055M15) |
| 1025 | 1.0406 | C25 | C25-1 | AF50C30 | 070M26 |
| | 1.1158 | Ck25 | C25-2 | XC25 | |
| 1035 | 1.0501 | C35 | C35-1 | AF55C35, XC38 | 060A35, 080M36 |
| | 1.1180 | Cm35 | C35-2 | XC32 | 1449 40CS |
| | 1.1181 | Ck35 | C35-3 | XC38H1, XC32 | |
| | 1.1183 | Cf35 | C36 | XC38H1TS | 060A35 |
| 1039 | 1.1157 | 40Mn4 | — | 35M5 | 150M36 |
| 1040 | 1.0511 | C40 | C40-1 | AF60C40 | 060A40 |
| | 1.1186 | Ck40 | C40-2 | XC42H1 | 080A40, 080M40 |
| 1045 | 1.0503 | C45 | C45-1 | AF65C45 | 080M46 |
| | 1.1191 | Cf45 | C45-2 | XC42H1, XC45 | 080M46, 060A47 |
| | 1.1193 | Ck45 | C45-3 | XC48 | |
| | 1.1201 | Cm45 | C46 | XC42H1TS | |
| 1050 | 1.1206 | Cf53 | C53 | XC48H1 | 060A52, 070M55 |
| | 1.1213 | Ck50 | | XC48H1TS | 080M50 |
| 1055 | 1.0535 | C55 | C55-1 | | 070M55 |
| | 1.1203 | Ck55 | C55-2 | XC55H1 | 070M55, 060A57 |
| | 1.1209 | Cm55 | C55-3 | XC55H1 | 070M55 |
| 1060 | 1.0601 | C60 | C60-1 | AF70C55 | 080A62 |
| | 1.1221 | Ck60 | C60-2 | XC60 | 080A62, 060A62 |
| 1070 | 1.1231 | Ck67 | | XC68 | 060A67 |
| 1078, 1080 | 1.1248 | Ck75 | | XC75 | 060A78 |
| 1086 | 1.1269 | Ck85 | | XC90 | |
| 1095 | 1.1274 | Ck101 | | XC100 | 060A96 |
| 1108 | 1.0721 | 10S20 | | 10F1 | (210M15) |
| 1139 | 1651-70 | 35S20 | | 38C4 | 212M36 |
| 1140 | 1.0726 | 35S20 | | 35MF6 | 212M36 |
| 1146 | 1.0727 | 45S20 | | 45MF4 | 212M44 |
| 1212 | 1.0711 | 9S20 | | | 220M07 |
| 1213 | 1.0715 | 9SMn28 | | S250 | 230M07 |
| 1215 | 1.0736 | 9SMn36 | | S300 | 240M07 |
| 1330 | 1.1165 | 30Mn5 | 28Mn6 | 20M5 | 120M36, 150M28 |
| | 1.1170 | 28Mn6 | | 35M5 | |
| 1335 | 1.1167 | 36Mn5 | | 40M5, 35M5 | 150M36 |
| 1518 | 1.1133 | 20Mn5 | | 20M5 | 120M19 |
| 1536 | 1.1166 | 34Mn5 | | | |
| 2515 | 1.5680 | 12Ni19 | 12Ni20 | Z18N5 | |
| 3115 | 1.5713 | 13NiCr6 | | 10NC6 | |
| 3135 | 1.5710 | 36NiCr6 | | 35NC6 | 640A35 |
| 3140 | 1.5711 | 40NiCr6 | | | 640M40 |
| 3310, 3415 | 1.5752 | 14NiCr14 | 13NiCr12 | 12NC15 | 655M13, 655A12 |
| 3415 | 1.5752 | 14NiCr14 | 13NiCr12 | 12NC15 | 655M13, 655A12 |
| 3415, 3310 | 1.5732 | 14NiCr10 | | 14NC11 | |
| 3435 | 1.5736 | 36NiCr10 | | 30NC11 | |
| 4130 | 1.7218 | 25CrMo4 | 25CrMo4 | 25CD4 | 1717CDS110 |
| 4135 | 1.2330 | 35CrMo4 | | 34CD4 | 708A37 |

| Italy, UNI | Japan, JIS | Sweden, SS | Czech, CSN | Spain, UNE |
|---|---|---|---|---|
| C10 | S10C | | | F.151, F.151.A |
| | S10C, S9CK | 1265 | 12 010 | F.1510-C10k |
| C15, C16 | | 1350 | 12 023 | F.111, F1110-C15k |
| | S15C, S15CK | 1370 | | F.1511-C16k |
| C20, C21 | | 1450 | 11 353, 12 021 | F.112 |
| C20, C25 | S20C, S20CK, S22C | | | F.1120-C25k |
| C22Mn3 | SMnC420 | | | F.1515-20Mn6 |
| C20, C25 | S20C, S20CK, S22C | | | F.1120-C25k |
| C25 | S25C | | 12 024 | F.1120-C25k |
| C35 | | 1550 | 11 453 | F.113 |
| | | 1572 | 12 040 | F.1135-C35k-1 |
| C35 | S35C | | | F.1130-C35k |
| C36, C38 | S35C | | | |
| C40 | S40C | | 12 041 | F.114A |
| C42, C43, C45 | S45C, S50C | 1650 | 12 050 | F.114 |
| C46 | | 1660 | | F.1140-C45k |
| | | 1672 | | F.1145-C45k-1 |
| | S50C | 1674 | 12 051 | |
| C55 | | 1655 | 12 060 | |
| C55 | S55C | | | F.1150-C55k |
| | | | | F.1155-C55k-1 |
| C60 | | | 12 061 | |
| C60 | S58C | 1665, 1678 | | |
| C70 | | 1770 | 13 180 | |
| C75 | | 1774, 1778 | | |
| C90 | | | | |
| C100 | SUP4 | 1870 | | |
| CF10S20 | | | 11 110 | F.2121-10S20 |
| | | 195703 | 11 140 | |
| | | 1957 | | F.210.G |
| | | 1973 | | |
| CF9S22 | SUM21 | | | |
| CF9SMn28 | SUM22 | 1912 | 11 109 | F.2111-11SMn28 |
| CF9SMn36 | | | | F.2113-12SMn35 |
| C28Mn | SMn433H, SCMn2 | | 13 141 | F.8211-30Mn5 |
| | SCMn1 | | | F.8311-AM30Mn5 |
| | SMn438 (H), SCMn3 | 2120 | | F.1203-36Mn6 |
| | | | | F.8212-36Mn5 |
| G22Mn3 | SMnC420 | | | F.1515-20Mn6 |
| | SMn433 | | | TO.B |
| | | | 16 520 | |
| 16CrNi4 | | | 16 220 | |
| SNC236 | | | 12 042, 16 240 | |
| | SNC815(H) | | | |
| | SNC815(H) | | | |
| 16NiCr11 | SNC415(H) | | | |
| 35NiCr9 | SNC631(H) | | | |
| 25CrMo4(KB) | SCM420, SCM430 | 2225 | 15 130, 15 131 | F.8372-AM26CrMo4 |
| | | | | F.8330-AM25CrMo4 |
| | | | | F.1256-30CrMo4-1 |
| 35CrMo4 | | 2234 | 15 141 | F.8331-AM34CrMo4 |

**TABLE 14-27 Worldwide Steel and Alloy Comparison Chart (*Continued*)**

| U.S. AISI-ASTM | Germany | | Belgium, NBN | France, AFNOR | Great Britain, B.S. |
|---|---|---|---|---|---|
| | No. | DIN | | | |
| 4135, 4137 | 1.7220 | 34CrMo4 | 34CrMo4 | 35CD4 | 708A37 |
| 4137 | 1.7225 | 42CrMo4 | | 38CD4 | 708A37 |
| 4140 | 1.7225 | 42CrMo4 | 42CrMo4 | 42CD4 | 708M40 |
| 4140, 4142 | 1.7223 | 41CrMo4 | 41CrMo4 | 42CD4 TS | 708M40 |
| 4142 | 1.2332 | 42CrMo4 | 42CrMo4 | 42CD4 | |
| | 1.7225 | 47CrMo4 | | | |
| 4142, 4140 | 1.7223 | 41CrMo4 | 41CrMo4 | 42CD4 TS | 708M40 |
| 4145 | 1.7228 | 50CrMo4 | | | 708H45 |
| 4147 | | | | | 708A47 |
| 4150 | 1.7228 | 50CrMo4 | | | 708A47 |
| 4340 | 1.6562 | 34CrNiMo6 | 35CrNiMo6 | 35NCD6 | 311-Type 6 |
| | 1.6565 | 40NiCrMo6 | | | 817M40 |
| | 1.6582 | 40NiCrMo73 | | | |
| 4419 | 1.5419 | 22Mo4 | | | |
| 4520 | 1.5423 | 16Mo5 | 16Mo5 | | 1503-245-420 |
| 5015 | 1.7015 | 15Cr3 | 15Cr2 | 12C3 | 523M15 |
| 5045, 5046 | 1.7006 | 46Cr2 | 46Cr2 | 42C2 | |
| 5115 | 1.7131 | 16MnCr5 | 16MnCr5 | 16MC5 | 527M17 |
| 5120 | 1.7147 | 20MnCr5 | | 20MC5 | |
| 5130 | 1.7030 | 28Cr4 | | | |
| | 1.7033 | 34Cr4 | | 32C4 | 530A30 |
| 5132 | 1.7037 | 34CrS4 | 34Cr4 | 32C4 | 530A32 |
| 5135 | 1.7034 | 37Cr4 | 37Cr4 | 38C4 | 530H36 |
| 5140 | 1.7035 | 41Cr4 | 41Cr4 | 42C4 | 530M40, 530A40 |
| | 1.7045 | 42Cr4 | | 42C4 TS | 530A40 |
| 5155 | 1.7176 | 55Cr3 | 55Cr3 | 55C3 | 527A60 |
| 6150 | 1.8159 | 50CrV4 | 50CrV4 | 50CV4 | 735A50 |
| 8620 | 1.6523 | 21NiCrMo2 | | 20NCD2 | 805M20 |
| 8720 | 1.6543 | 21NiCrMo22 | | | 805A20 |
| 8740 | 1.6546 | 40NiCrMo22 | 40NiCrMo2 | 40NCD2 | 311-Type 7 |
| 9255 | 1.0903 | 51Si7 | 50Si7 | 51S7 | 250A53 |
| | 1.0904 | 55Si7 | 55Si7 | 55S7 | |
| 9260 | 1.0909 | 60Si7 | 60Si7 | 60S7 | 250A58 |
| 9262 | 1.0961 | 60SiCr7 | 60SiCr8 | 60SC7 | 250A61 |
| 9314 | 1.5752 | 14NiCr14 | 13NiCr12 | 12NC15 | 655M13, 655A12 |
| 9840 | 1.6511 | 36CrNiMo4 | | 40NCD3 | 816M40 |
| 301 | 1.4310 | X12CrNi177 | | Z12CN17.07 | 301S21 |
| | | | | Z12CN18.07 | |
| 303 | 1.4305 | X10CrNiS189 | | Z10CNF18.09 | 303S21 |
| 304, 304H | 1.4301 | X5CrNi1810 | | Z6CN18.09 | 304S15, 304S16 |
| | | | | | 304S31 |
| 308, 305 | 1.4303 | X5CrNi1812 | | Z8CN18.12 | 305S19 |
| 304L | 1.4306 | X2CrNi1911 | | Z2CN18.09 | |
| | | G-X2CrNi189 | | Z2CN18.10 | |
| | | | | Z3CN19.10M | |
| 304LN | 1.4311 | X2CrNiN1810 | | Z2CN18.10Az | 304S62 |
| 308, 305 | 1.4303 | X5CrNi1812 | | Z8CN18.12 | 305S19 |
| 309 | 1.4828 | X15CrNiSi2012 | | Z15CNS20.12 | (309S24) |
| 309S | 1.4833 | X7CrNi2314 | | Z15CN24.13 | 309S24 |

| taly, JNI | Japan, JIS | Sweden, SS | Czech, CSN | Spain, UNE |
|---|---|---|---|---|
| 35CrMo4 | SCM432, SCCrM3 | 2234 | | F.8231-34CrMo4 |
| | SCM435H | | | F.1250-35CrMo4 |
| 42CrMo4 | SCM440 | | 15 142 | |
| 42CrMo4 | SCM440 (H) | 2244 | 15 142 | F.8332-AM42CrMo4 |
| | | | | F.8232-42CrMo4 |
| | | | | F1252-40CrMo4 |
| 41CrMo4 | SCM440 | 2244 | | |
| 40CrMo4 | SCM440(H) | 2244 | | F.8232-42CrMo4 |
| 42CrMo4 | | | | F.8332-AM42CrMo4 |
| 41CrMo4 | SCM440 | 2244 | | F.1252-40CrMo4 |
| | SCM445 | | | |
| | SCM445 | | | |
| | SCM445 (H) | | 15 261 | |
| 35NiCrMo6 KB | SNB24-1-5 | | | |
| | | 2541 | 16 243 | F.1272-40NiCrMo7 |
| 40NiCrMo7(KB) | SNCM439 | | 16 341 | |
| | SNCM447 | | 16 342 | |
| G22Mo5 | SCPH11 | | | |
| 16Mo5 | | | | F.2602-16Mo5 |
| | SCr415(H) | | 14 120 | |
| 45Cr2 | | 2511 | | |
| 16MnCr5 | | | | F.1516-16MnCr5 |
| | | | | F.1517 |
| 20MnCr5 | SMnC420H | | | F.150D |
| | | | 15 231 | |
| | SCr435 | | 14 141 | |
| 34Cr4(KB) | SCr430(H) | | 14 141 | F.8221-35Cr4 |
| 36CrMn4, 38Cr4 | SCr435H | | 14 140 | F.1201-38Cr4 |
| 41Cr4 | SCr440 | 2245 | 14 140 | F.1202-42Cr4 |
| | SCr440(H) | | | |
| 55Cr3 | SUP9 (A) | 2253 | | F.1431-55Cr3 |
| 50CrV4 | SUP 10 | 2230 | 15 260 | F.1430-51CrV4 |
| 20NiCrMo2 | SNCM220(H) | 2506 | 16 125, 15 124 | F.1522-20NiCrMo2 |
| | | | | F.1534-20NiCrMo31 |
| | | | | F.1524-20NiCrMo3 |
| 40NiCrMo2(KB) | SNCM240 | | | F.1204-40NiCrMo2 |
| 48Si7, 50Si7 | | 2085, 2090 | 13 261 | F.1440-56Si7 |
| 55Si8 | | | | F.1450-50Si7 |
| | | | 13 270 | F.1441-60Si7 |
| 60SiCr8 | SUP7 | | | F.1442-60SiCr8 |
| | SNC815(H) | | | |
| 38NiCrMo4(KB) | | | 16 243 | F.1280-35NiCrMo4 |
| X12CrNi1707 | SUS301 | | 17 241; 17 242 | F.3517-X12CrNi1707 |
| X10CrNiS1809 | SUS303 | 2346 | 17 243 | F.3508-X10CrNiS18-09 |
| X5CrNi1810 | SUS304 | 2332, 2333 | | F.3504-X6CrNi1910 |
| | | | | F.3541-X5CrNi18-10 |
| | | | | F.3551-X5CrNi1811 |
| X8CrNi1910 | SUS305 | | | F.3513-X8CrNi18-12 |
| X2CrNi1811 | SCS19, SUS304L | 2352, 2333 | | F.3505-X2CrNi19-10 |
| GX2CrNi1910 | | | | |
| X2CrNiN1811 | SUS304LN | 2371 | | |
| X8CrNi1910 | SUS305 | | | F.3513-X8CrNi18-12 |
| | SUH309 | | 17 251 | F.3312-X15CrNiSi20-12 |
| X6CrNi2314 | SUS309S | | | |

**TABLE 14-27 Worldwide Steel and Alloy Comparison Chart (*Continued*)**

| U.S. AISI-ASTM | Germany No. | Germany DIN | Belgium, NBN | France, AFNOR | Great Britain, B.S. |
|---|---|---|---|---|---|
| 310, 314 | 1.4841 | X15CrNiSi2520 | | Z12CNS25.20<br>Z15CNS25.20 | |
| 310S | 1.4845 | X12CrNi2521 | | Z12CN25.20 | 310S24 |
| 314, 310 | 1.4841 | X15CrNiSi2520 | | Z12CNS25.20<br>Z15CNS25.20 | |
| 316 | 1.4401 | X5CrNiMo17122 | | Z6CND17.11 | 316S16, 316S31 |
| | 1.4436 | X5CrNiMo17133 | | Z6CND17.12 | |
| 316L | 1.4404 | X2CrNiMo17132 | | Z2CND17.12 | 316S11, 316S12 |
| | 1.4435 | X2CrNiMo18143 | | Z2CND17.13 | |
| | | G-X2CrNiMo1810 | | Z2CND18.13<br>Z3CND19.10M | |
| 316LN | 1.4406 | X2CrNiMoN17122 | | Z2CND17.12Az | 316S61 |
| | 1.4429 | X2CrNiMoN17133 | | Z2CND17.13Az | 316S62 |
| 316Ti | 1.4571 | X6CrNiMoTi17122 | | Z6CNDT17.12 | 320S17, 320S31 |
| | 1.4573 | X10CrNiMoTi1812 | | | 320S33 |
| 317 | 1.4449 | X5CrNiMo1713 | | | 317S16 |
| 317L | 1.4438 | X2CrNiMo18164 | | Z2CND19.15 | 317S12 |
| 318 | 1.4583 | X10CrNiMoNb1812 | | | |
| 321 | 1.4541 | X6CrNiTi1810 | | Z6CNT18.10 | 321S12, 321S20 |
| | 1.4878 | X12CrNiTi189 | | Z6CNT18.12(B) | 321S31 |
| 329 | 1.4460 | X8CrNiMo275 | | | |
| 330 | 1.4864 | X12NiCrSi3616 | | Z12NCS35.16<br>Z12NC37.18<br>Z12NCS37.18 | NA17 |
| 347 | 1.4550 | X6CrNiNb1810 | | Z6CNNb18.10 | 347S17, 347S31 |
| 348 | 1.4546 | X5CrNiNb1810 | | | 347S17, 347S18 |
| 403, 410S | 1.4001 | X7Cr14 | | Z3C14 | |
| 405 | 1.4002 | X6CrAl13 | | Z6CA13 | 405S17 |
| 409 | 1.4512 | X5CrTi12 | | Z6CT12 | 409S19 |
| 410 | 1.4006(G-) | X10Cr13 | | Z12C13 | 410S21, 410C21 |
| 410S, 403 | 1.4000 | X6Cr13 | | Z6C13 | 403S17 |
| 416 | 1.4005 | X12CrS13 | | Z12CF13 | 416S21 |
| 420 | 1.4021 | X20Cr13 | | Z20C13 | 420S37 |
| 430 | 1.4016 | X6Cr17 | | Z8C17 | 430S15 |
| | 1.4742 | X10CrAl18 | | Z10CAS18 | |
| 430Ti, XM8 | 1.4510 | X6CrTi17 | | Z8CT17 | |
| 431 | 1.4057 | X20CrNi17 2 | | Z15CN16.02 | 431S29 |
| 433 | 1.4113 | X6CrMo17 | | Z8CD17.01 | 434S17 |
| 446 | 1.4762 | X10CrAl24 | | Z10CAS24 | |
| 430F | 1.4104 | X12CrMoS17 | | Z10CF17 | |
| 440C | 1.4125 | X105CrMo17 | | Z100CD17 | |
| A2 | 1.2363 | X100CrMoV51 | | Z100CDV5 | BA2 |
| D2 | 1.2379 | X155CrVMo121 | | Z160CDV12 | BD2 |
| D3 | 1.2080 | X210Cr12 | | Z200 C12 | BD3 |
| L2 | 1.2210 | 115CrV3 | | | |
| L3 | 1.2067 | 100Cr6 | | Y100 C6 | BL3 |
| L6 | 1.2713 | 55NiCrMoV6 | | 55NCDV7 | |
| M2 | 1.3343 | S6-5-2 | | Z85WDCV 06-<br>-05-04-02 | BM2 |
| M3 | 1.3342 | SC6-5-2 | | Z90WDCV 06-<br>-05-04-02 | |
| M7 | 1.3348 | S2-9-2 | | Z100DCWV 09-04-02-02 | |
| XM8, 430Ti | 1.4510 | X6CrTi17 | | Z8CT17 | |
| M33 | 1.3249 | S2-9-2-8 | | | BM34 |

| Italy, UNI | Japan, JIS | Sweden, SS | Czech, CSN | Spain, UNE |
|---|---|---|---|---|
| X16CrNiSi2520 | SUH310 | | | F.3310-X15CrNiSi25-20 |
| (X6CrNi2520) | SUH310, SUS310S | 2361 | | F.331 |
| X16CrNiSi2520 | SUH310 | | | F.3310-X15CrNiSi25-20 |
| X5CrNiMo1712 | SUS316 | 2343 | 17 352 | F.3534-X6CrNiMo17-12-03 |
| X5CrNiMo1713 | | 2347 | | F.3543-X5CrNiMo17-12 |
| X2CrNiMo1712 | SUS316 | 2348 | | F.3533-X2CrNiMo17-12-03 |
| X2CrNiMo1713 | SUS316L | 2353 | | |
| GX2CrNiMo1911 | | | | |
| | SUS316L | 2348 | | |
| X2CrNiMoN1712 | SUS316LN | 2375 | | |
| X2CrNiMoN1713 | | | | |
| X6CrNiMoTi1712 | | 2350 | | F.3535-X6CrNiMoTi17-12-03 |
| X6CrNiMoTi1713 | | | | |
| (X5CrNiMo1815) | SUS317 | | | |
| X2CrNiMo18.15 | SUS317L | 2367 | | |
| X6CrNiMoNb1713 | | | | |
| X6CrNiTi1811 | SUS321 | 2337 | 17 246, 17 247 | F.3523-X6CrNiTi1811 |
| | | | 17 248 | F.3553-X7CrNiTi18-11 |
| | SUS329 J1 | 2324 | | F.3309-X8CrNiMo27-05 |
| | SCH11, SCS11 | | | |
| | SUH330 | | | F.3313-X12CrNiSi36-16 |
| X6CrNiNb1811 | SUS347 | 2338 | 17 245 | F.3524-X6CrNiNb18-11 |
| | | | | F.3552-X7CrNiNb18-11 |
| X6CrNiNb1811 | | | | |
| | SUS410S | | | F.8401-AM-X12Cr13 |
| X6CrAl13 | SUS405 | 2302 | 17 125 | F.3111-X6CrAl13 |
| X6CrTi12 | SUH409 | | | |
| X12Cr13, X10Cr13 | SUS410 | 2302 | 17 021 | F.3401-X12Cr13 |
| X6Cr13 | SUS403 | 2301 | | F.3110-X6Cr13 |
| X12CrS13 | SUS414 | 2380 | | F.3411-X12CrS13 |
| X20Cr13 | SUS420 J1 | 2303 | 17 024, 17 029 | F.3402-X20Cr13 |
| X18Cr17 | SUH21, SUS430 | 2320 | 17 041 | F.3113-X8Cr17 |
| | | | | F.3153-X10CrAl18 |
| X6CrTi17 | SUS430LX | | | F.3114-X8CrTi17 |
| X16CrNi16 | SUS431 | 2321 | 17 145 | F.3427-X15CrNi16 |
| X8CrMo17 | SUS434 | 2325 | | — |
| X16Cr26 | | | | F.3154-X10CrAl24 |
| X10CrS17 | SUS430F | 2383 | | F.3117-X10CrS17 |
| | SUS440C | | | |
| X100CrMoV51 KU | SKD12 | 2260 | | F.5227, X100CrMoV5 |
| X155CrVMo121KU | | | | |
| X205Cr12 KU | SKD1 | | 19 436, 19 437 | F.5212, X210Cr12 |
| 107CrV3 KU | | | 19 423 | |
| | | | | F.5230, 100Cr6 |
| | SKT4 | | | |
| HS 6-5-2 | SKH51 | 2722 | 19 829, 19 830 | F.5603, 6-5-2 |
| HSC 6-5-3 | | | | |
| HS 2-9-2 | | 2782 | | F.5607, 2-9-2 |
| X6CrTi17 | SUS430LX | | | F.3114-X8CrTi17 |
| | | | | F.5611, 2-9-2-8 |

**TABLE 14-27 Worldwide Steel and Alloy Comparison Chart (*Continued*)**

| U.S. AISI-ASTM | Germany No. | Germany DIN | Belgium, NBN | France, AFNOR | Great Britain, B.S. |
|---|---|---|---|---|---|
| M34 | 1.3249 | S2-9-2-8 | | | BM34 |
| M41 | 1.3246 | S7-4-2-5 | | Z110WKCDV 07-05-04-04-02 | |
| M42 | 1.3247 | S2-10-1-8 | | Z110DKCWV 09-08-04-02-01 | BM42 |
| O1 | 1.2510 | 100MnCrW4 | | | BO1 |
| O2 | 1.2842 | 90MnCrV8 | | 90MV8 | BO2 |
| S1 | 1.2542 | 45WCrV7 | | | BS1 |
| T1 | 1.3355 | S18-0-1 | | Z80WCV18-04-01 | BT1 |
| T4 | 1.3255 | S18-1-2-5 | | Z80WKCV18-05-04-01 | BT4 |
| T5 | 1.3265 | S18-1-2-10 | | | BT5 |
| T15 | 1.3202 | S12-1-4-5 | | | BT15 |
| W1 | 1.1625<br>1.1750 | C75W, C80W2 | | | BW1A, BW1B |
| W108 | 1.1525 | C80W1 | | Y1 90, Y1 80 | |
| W110 | 1.1545 | C105W1 | | Y1 105 | |
| W112 | 1.1663 | C125W | | Y2 120 | |
| W210 | 1.2833 | 100V1 | | Y1 105V | BW2 |

***Note:*** Exchangeability of presented material specifications is possible only after a thorough examination of each material's composition and manufacturing methods.

content, where the difference between the hardened case and softer core is more apparent. According to Grossmann, a part is heat-treated when its core contains less than 50 percent of martensite.

Some applications rely on low hardenability of steel, however. These are instances where the material is subjected to welding and other temperature-dependent treatments.

Hardenability of a material may be evaluated by heating and quenching a round bar, which is then cut across and the depth of its hardness with reference to the outer circumferential surface is measured on the cross section.

The Jominy test of hardenability uses a testing bar heated to a specified temperature and held there for 30 min. The bar must be previously normalized and free of decarburization, the removal of which may be achieved through machining away the upper surface. One end of the heated bar, as held in a vertical position, is then quenched in water. Its hardness is measured along the length, distancing the measurements in 0.062-in intervals off the quenched end, and the differences of these values are evaluated.

**Heat-treating process.** Heat-treating furnaces can be heated by gas, oil, or electricity. Their atmosphere may be either composed of air, or controlled, in which case it is selectively affected by residues of various burning gases or by removal of carbon dioxide or by devaporizing of the furnace area. In salt bath furnaces, parts are heated by means of electrodes surrounding the salt bath. Their location and design together create an electromagnetic influence within the bath, which by stirring the content aids the distribution of temperature. Salt

| Italy, UNI | Japan, JIS | Sweden, SS | Czech, CSN | Spain, UNE |
|---|---|---|---|---|
| | | | | F.5611, 2-9-2-8 |
| HS7-4-2-5 | | | 19 851 | F.5615, 7-4-2-5 |
| HS 2-9-1-8 | | | | F.5617, 2-10-1-8 |
| 95MnWCr5 KU | | 2140 | 19 314 | F.5220, 95MnCrW5 |
| 90MnVCr8 KU | | | 19 312, 19 313 | |
| 45WCrV8 KU | | 2710 | | F.5241, 45WCrSi8 |
| HS 18-0-1 | SKH2 | | 19 824 | F.5520, 18-0-1 |
| HS 18-1-1-5 | SKH3 | | 19 855 | F.5530, 18-1-1-5 |
| HS 18-0-1-10 | SKH4A | | | F.5540, 18-0-2-10 |
| | | | 19 857 | F.5563, 12-1-5-5 |
| | SKC3, SK5, SK6 | | 19 132, 19 133 | F.5107, C80 |
| | | | 19 191, 19 192 | |
| C80KU | | | 19 152 | |
| C100KU | | 1880 | 19 221, 19 222 | |
| C120KU | SK2 | | 19 255 | F.5123, C120 |
| 102V2 KU | SKS43 | | 19 356 | |

baths, however, may cause a decarburization of parts if the solution content is not properly controlled.

Cooling of heat-treated parts, otherwise called quenching, is achieved through their immersion in liquids or through their exposure to air, gases, or solids. The liquid cooling media may be water, oil, salt bath, soap bath, lead bath, or a brine, consisting of either sodium carbonate or sodium hydroxide or even sulfuric acid.

The differentiation between various quenching media is based on the speed of the cooling process they can provide, even though some additional aspects are attributable to the outcome as well. With quenching in oil, the hardness of the core comes out smaller yet tougher than that quenched in water. Oil-quenched materials are also less prone to distortion when compared to water-quenched parts. Where water quench may be considered mild for a 5-in-diameter round part, it may certainly be too drastic for a part a quarter of this size. Still smaller parts are best when hardened in air, while thicker products may benefit by oil quenching.

Generally, steels with lower thermal conductivity and greater thermal expansion coefficient will suffer from small depth of heat-treatment penetration, of coarser grain and greater distortions during the heat-treating process. Another cause contributing to the emergence of inner stresses within the heat-treated material may be found in

- Unequal distribution of the heat
- Uncompleted austenization
- Decarburization of the parts' surface

Naturally, a proper selection of the heat-treating method, its temperature range, and the quenching media is vital to assess the successful outcome of this operation.

Another considerable influence is exerted in the form of the shape of heat-treated parts and their size. The inner tension is always greater within the larger parts, where the difference between the temperature of the core and that of the surface may be increased. Sharp edges, sharp corners, thin walls, and notches are all detrimental to the results of the heat-treating process (Fig. 14-13). Variations in outcome of the heat-treatment process are included in Table 14-28.

Certain areas which need to be protected from the effect of heat treatment are filled with a physical barrier of insulating type for the duration of the process. Such sections are especially the transitions between different thicknesses, or excessively thin walls. Notches and sharp corners may be protected by an encirclement of several strands of wire. Tables 14-29 through 14-32 cover various aspects of hardening, tempering, and heat treatment.

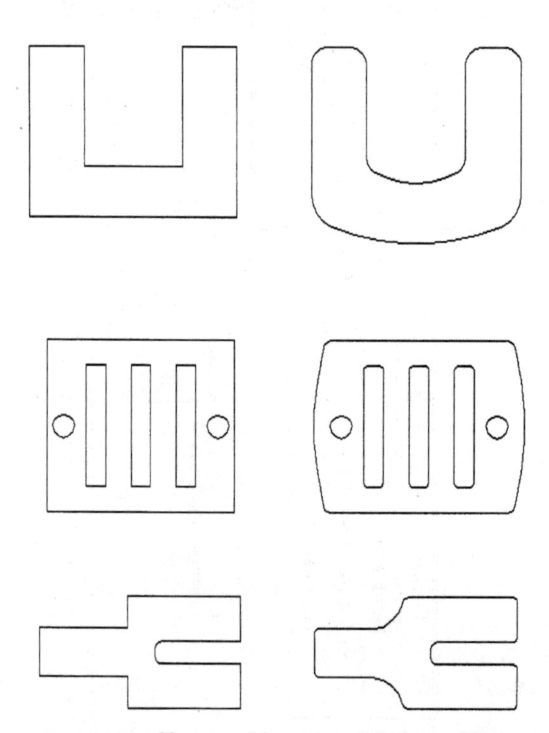

**Figure 14-13** Shapes of heat-treated parts. Wrong shapes are on left, correct shapes on right.

**TABLE 14-28 Variation of the Heat-Treatment Process**

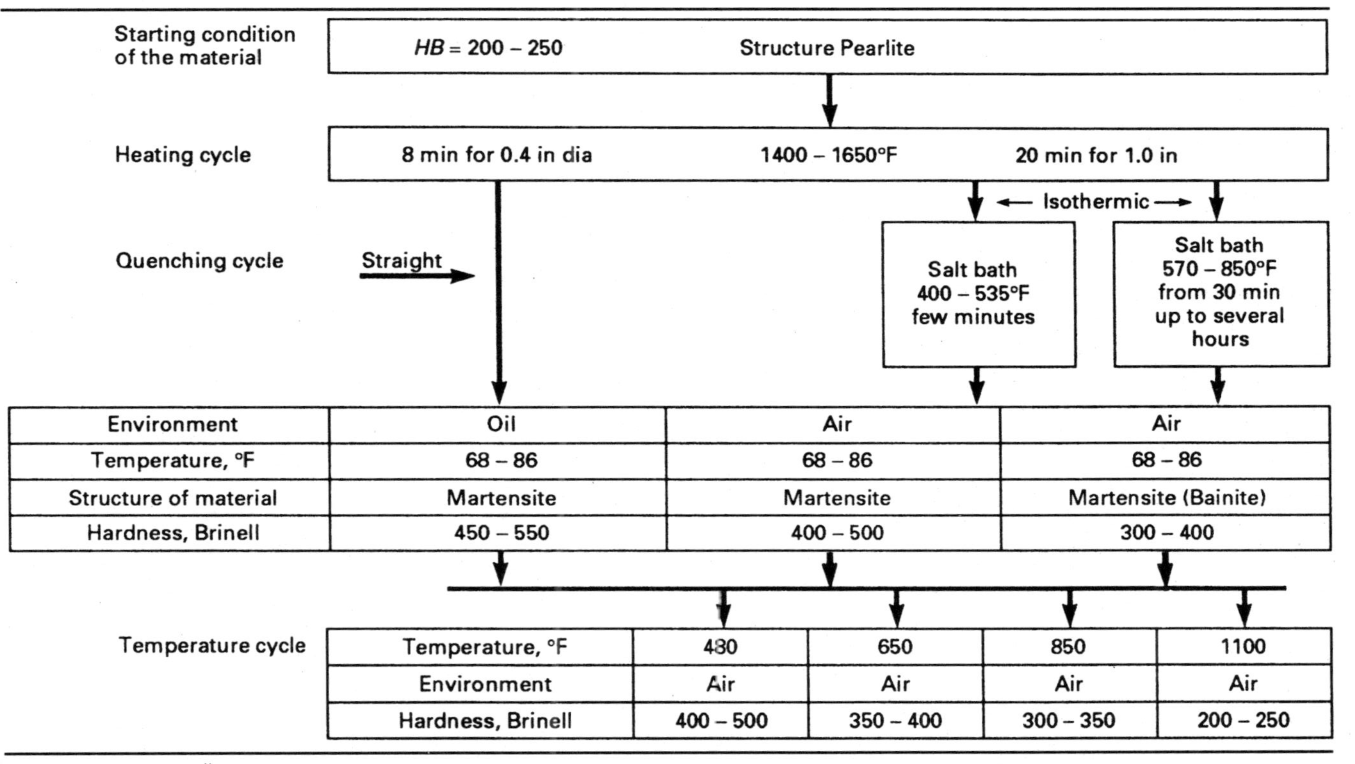


| Environment | Oil | Air | Air |
|---|---|---|---|
| Temperature, °F | 68 – 86 | 68 – 86 | 68 – 86 |
| Structure of material | Martensite | Martensite | Martensite (Bainite) |
| Hardness, Brinell | 450 – 550 | 400 – 500 | 300 – 400 |

Temperature cycle

| Temperature, °F | 430 | 650 | 850 | 1100 |
|---|---|---|---|---|
| Environment | Air | Air | Air | Air |
| Hardness, Brinell | 400 – 500 | 350 – 400 | 300 – 350 | 200 – 250 |

SOURCE: Svatopluk Černoch, *Strojně technická příručka,* 1977. Reprinted with permission from SNTL Publishers, Prague, CZ.

**TABLE 14-29 Comparison of Properties, Case-Hardening and Hard-Tempering Steels**

| | Case-hardening steel | Hard-tempering steel |
|---|---|---|
| Cost of machining | Lesser | Greater |
| Cost of forging | Lesser | Greater |
| Cost of annealing (forgings) | Lesser | Greater |
| Cost of heat treatment | Greater | Lesser |
| Distortion due to heat treatment | Greater | Lesser |
| Straightening difficulties | Lesser | Greater |
| Hardness of surface | Greater | Lesser |
| Surface toughness | Lesser | Greater |
| Wear resistance | Greater | Lesser |
| Overall strength | Lesser | Greater |
| Tendency to upset under loading | Greater | Lesser |
| Tendency to pit under pressure | Lesser | Greater |

Initial cost of material is considered equal.

### 14-7-1. Carburizing

Often heat treatment of metal parts is used to either increase or decrease their hardness, relieve inner stresses, or aid the even distribution of their properties. In the carburizing process, the expected outcome is somewhat different. Here the solid iron-base alloy is heated to a temperature below its melting point and left in the oven to absorb carbon from purposely added carbonaceous materials, be it solids or gases. Carburizing is often followed by quenching, which produces a hardened skin on the part, otherwise called a hardened case.

The reverse of this process is *decarburization,* or loss of carbon content from the material surface, which occurs when it is heated in an environment reacting with its carbon content.

Carburizing is a diffusion process of adding extra carbon particles into the surficial layers of parts. It does not affect the parts' core, as its influence seeps into the body of metal very, very slowly through the surface, affecting only the immediate layers. A thin layer, or a case, becomes austenitized by the increased content of carbon.

Carbonaceous material utilized for the purpose of carburization can be powdered, liquid, solid, or gas. The latter has the advantage of producing cleaner parts in less time, with more balanced carbon distribution.

Temperatures of the carburizing process are in the vicinity of 1650 to 1700°F for carbon steels. The speed of a process may be increased by raising the temperature, but the austenitic grain will emerge coarser, with too sudden a transition between the core and the surface, the latter being prone to peeling off during the heat treatment.

**TABLE 14-30 Hardening and Tempering Treatments for Tool and Die Steels***

| AISI steel type | Preheat temp., °F | Rate of heating for hardening | Hardening temp., °F | Minimum time at temp., min | Quenching medium | Tempering temp., °F | Depth of hardening | Resistance to decarburization | Maximum tempered hardness, Rockwell C |
|---|---|---|---|---|---|---|---|---|---|
| W1 | | Slow | 1425–1500 | 15 | Brine or water | 325–550 | Shallow | Best | 64 |
| W2 | | Slow | 1425–1500 | 15 | Brine or water | 325–550 | Shallow | Best | 65 |
| O1 | 1200 or none† | Slow | 1450–1500 | 15 | Oil | 325–500 | Medium | Very good | 62 |
| O2 | 1200 or none† | Slow | 1400–1475 | 15 | Oil | 325–500 | Medium | Very good | 62 |
| O6 | 1200 or none† | Slow | 1450–1500 | 30 | Oil | 300–600 | Medium | Very good | 63 |
| A2 | 1250 | Slow | 1725–1800 | 30 | Air | 350–950 | Deep | Fair | 62 |
| A4 | 1250 | Slow | 1500–1600 | 20 | Air | 300–800 | Deep | Good | 62 |
| A6 | 1250 | Slow | 1500–1600 | 20 | Air | 300–800 | Deep | Good | 60 |
| A7 | 1250 | Slow | 1700–1800 | 30 | Air | 300–1000 | Deep | Fair | 67 |
| A8 | 1250 | Slow | 1800–1850 | 30 | Air | 300–950 | Deep | Low | 60 |
| A10 | | Slow | 1450–1500 | 30 | Air | 300–600 | Deep | Good | 63 |
| D2 | 1450 | Very slow | 1800–1850 | 45 | Air | 300–950 | Deep | Fair | 62 |
| D3 | 1450 | Very slow | 1750–1825 | 45 | Oil | 300–950 | Deep | Fair | 63 |
| D4 | 1450 | Very slow | 1775–1850 | 45 | Air | 300–950 | Deep | Fair | 62 |
| D5 | 1450 | Very slow | 1800–1850 | 45 | Air | 300–950 | Deep | Fair | 63 |
| D7 | 1450 | Very slow | 1875–2000 | 45 | Air | 300–1000 | Deep | Fair | 67 |
| S1 | | Slow | 1650–1800 | 20 | Oil | 300–650 | Medium | Fair to good | 58 |
| S2 | | Slow | 1525–1575 | 10 | Brine or water | 300–800 | Medium | Fair to good | 60 |
| S4 | | Slow | 1600–1700 | 10 | Brine or water | 300–800 | Medium | Low | 60 |
| S5 | | Slow | 1600–1700 | 10 | Oil | 300–800 | Medium | Low | 60 |
| S7 | | Slow | 1700–1750 | 20 | Air or oil | 400–1000 | Medium | Fair | 59 |
| T1 | 1500–1600 | Rapid from preheat | 2150–2300‡ | 2 | Air, oil, or salt | 1050–1150 | Deep | Good | 64 |
| T15 | 1500–1600 | Rapid from preheat | 2125–2270‡ | 2 | Air, oil, or salt | 1000–1150 | Deep | Fair | 67 |
| M1 | 1500 | Rapid from preheat | 2125–2175‡ | 2 | Air, oil, or salt | 1000–1150 | Deep | Low | 65 |
| M2 | 1500 | Rapid from preheat | 2125–2225‡ | 2 | Air, oil, or salt | 1025–1150 | Deep | Fair | 65 |
| M4 | 1500 | Rapid from preheat | 2125–2225‡ | 2 | Air, oil, or salt | 1025–1150 | Deep | Fair | 66 |
| L3 | | Slow | 1425–1600 | 20 | Brine or oil | 300–800 | Medium | Good | 62 |
| F2 | 1200 | Slow | 1450–1600 | 20 | Brine or water | 300–500 | Shallow | Good | 66 |

*Based on AISI *Tool Steels Manual,* 1955, and modified by committee.
†Preheating necessary for complicated shapes and shapes of widely differing cross section.
‡Salt-bath basis.
SOURCE: Frank W. Wilson, *Die Design Handbook,* New York, 1965. Reprinted with permission from the McGraw-Hill Companies.

**TABLE 14-31 Typical Heat-Treatment Comparison Chart**

| SAE | Normalizing temperature, °F | Annealing temperature, °F | Carburizing temperature, °F | Cooling method | Reheated, °F | Cooling method | Tempering temperature, °F |
|---|---|---|---|---|---|---|---|
| 1010–1022 | | | 1650–1700 | A | | | 250–400 |
| | | | 1650–1700 | B | 1400–1450 | A | 250–400 |
| | | | 1650–1700 | C | 1400–1450 | A | 250–400 |
| | | | 1650–1700 | C | 1650–1700 | B | 250–400 |
| | | | 1350–1575 | D | | | Open |
| 1024 | 1650–1750 | | 1650–1700 | E | | | 250–400 |
| | | | 1350–1575 | D | | | Open |
| 1109–1120 | | | 1650–1700 | A | | | 250–400 |
| | | | 1650–1700 | B | 1400–1450 | A | 250–400 |
| | | | 1650–1700 | C | 1400–1450 | A | 250–400 |
| 1111–1113 | | | 1500–1650 | B | | | Open |
| | | | 1350–1575 | D | | | Open |
| 1320 | 1700–1750* | | 1650–1700 | C | 1400–1450 | E | 250–350 |
| | 1700–1750* | | 1650–1700 | E | 1475–1525 | E | 250–350 |
| | 1700–1750* | | 1650–1700 | E | | | 250–350 |
| | 1700–1750* | Yes† | 1650–1700 | E | 1375–1425 | E | 250–350 |
| 2317 | 1700–1750* | Yes† | 1650–1700 | C | 1475–1525 | E | 250–350 |
| | 1700–1750* | Yes† | 1650–1700 | E | | | 250–350 |
| 3115–3120 | 1700–1750* | | 1650–1700 | C | 1400–1450 | E | 250–350 |
| | 1700–1750* | | 1650–1700 | C | 1500–1550 | E | 250–350 |
| | 1700–1750* | | 1650–1700 | C | 1400–1450 | E | 250–350 |
| 4317–4621 | 1700–1750* | Yes† | 1650–1700 | E | 1475–1525 | E | 250–350 |
| | 1700–1750* | Yes† | 1650–1700 | C | 1425–1475 | E | 250–350 |
| | 1700–1750* | Yes† | 1650–1700 | C | 1475–1525 | E | 250–350 |
| | 1550–1700* | | 1500–1650 | E | | | 250–350 |
| | 1700–1750* | Yes† | 1650–1700 | E | 1375–1425 | E | 250–350 |
| 4812–4820 | 1700–1750* | Yes† | 1650–1700 | C | 1450–1500 | E | 250–350 |
| | 1700–1750* | Yes† | 1650–1700 | E | 1375–1425 | E | 250–350 |
| | | | 1650–1700 | E | | | 250–350 |

| | | | | | | | |
|---|---|---|---|---|---|---|---|
| 5115–5120 | 1700–1750* | | 1650–1700 | E | 1500–1550 | E | 250–350 |
| | 1700–1750* | | 1650–1700 | E | 1425–1475 | E | 250–350 |
| | 1700–1750* | | 1650–1700 | C | 1425–1475 | E | 250–350 |
| 8615–8720 | 1700–1750* | Yes† | 1650–1700 | E | 1475–1525 | E | 250–350 |
| | 1700–1750* | Yes† | 1650–1700 | C | 1475–1525 | E | 250–350 |
| | 1700–1750* | Yes† | 1650–1700 | E | | | 250–350 |
| | 1700–1750* | Yes† | 1650–1700 | E | 1525–1575 | E | 250–350 |
| | 1700–1750* | Yes† | 1500–1650 | E | | | 250–350 |
| 9310–9317 | 1700–1750* | | 1650–1700 | C | 1500–1525 | E | 250–350 |
| | 1700–1750* | | 1650–1700 | E | 1400–1450 | E | 250–350 |

A = cooling in water or brine, B = cooling in water or oil, C = slow cooling, D = cooling in air or oil, E = cooling in oil only.

*Air cooling to follow.

† Cycle anneal. Temperature to be identical with normalizing temperture. Rapid cooling to 1000–1250°F followed by air cooling 3 hours later.

Heat treatment of carburized parts may be initiated during the carburizing process. The simplest procedure is to continue with the heat-treating cycle right at the termination of the carburizing process. The temperature of the inner core is truly suitable; however, a danger of overheating of the outer surface should be looked into. Since deformations may be the by-product of the method, only simple and straightforward parts should be subjected to such a procedure.

### 14-7-2. Hardening

The hardening process consists of heating the material to at least some 100°F above the temperature of the transformation point, during which the inner pearlitic structure turns austenitic. Such a temperature is held for a certain amount of time and followed by a rapid cooling, or quenching.

**Critical points.** The critical or transformation point is also called the *decalescence point.* On reaching such a temperature range, the steel material ceases to increase its own temperature, even though its surroundings are growing hotter. During the cooling sequence, a similar point called the *recalescence point* is encountered. It marks the transformation of austenite back into pearlite. On reaching this temperature range, the steel material, until now continuously releasing its heat and decreasing in temperature, will go through a sudden momentary wave of increased temperature.

These two critical points have considerable importance in the hardening process. If the temperature of the decalescence point is not fully passed, the material will not harden. Subsequently, if the steel is not cooled suddenly before it reaches its recalescence point, no hardening takes place. Usually the recalescence point is anywhere from 85 to 215°F lower than the decalescence point.

### 14-7-3. Case hardening

The case-hardening procedure is used to make the outer surface of a part hard while the core remains softer and keeps its toughness. For case-hardening of low-carbon steel, the surface of parts must first be exposed to the effect of the carburizing process to absorb an adequate amount of carbon. Only afterward these parts may be case hardened and quenched by immersion in water, oil, brine, etc.

**Depth of hardened surface.** The desired depth of the case-hardened surface dictates the duration of the process and all its other parameters. With heavy or thick cases, exposure to case-hardening temperatures is longer, yet the maximum amount of carbon at the surface level is not found equally increased because the carbon's continuous yet slow seepage into deeper layers of the material is promoted by longer exposures to the case-hardening environment.

However, the actual linear depth of the hardened surface is a controversial subject, as the most influential factor in this sense is the size of hardened parts: Where 0.015 in may be a heavy case on one part, 0.035 in may be considered a light case on another.

When deciding on the depth of hardened surface, various aspects should be evaluated. These are

- Maximum permissible surficial wear of the part
- Amount of balance between properties of the core and those of the case
- Overall strength of the part after hardening

A heavy case must withstand quite heavy loading without succumbing to breakage and total collapse, since, if the case-hardened surface breaks, the inner core underneath it will not have sufficient strength to sustain undue loading. The resistance to wear as well as to crushing may be attained by two basic procedures: Either a heavy case should be produced on a poorly hardenable steel, or a light case should be given to steel that will harden well even without any case. (See Fig. 14-14 and Table 14-32.)

Excessive grinding after the case hardening should be avoided, as it diminishes the hardened surface's thickness, wasting the expense of such treatment.

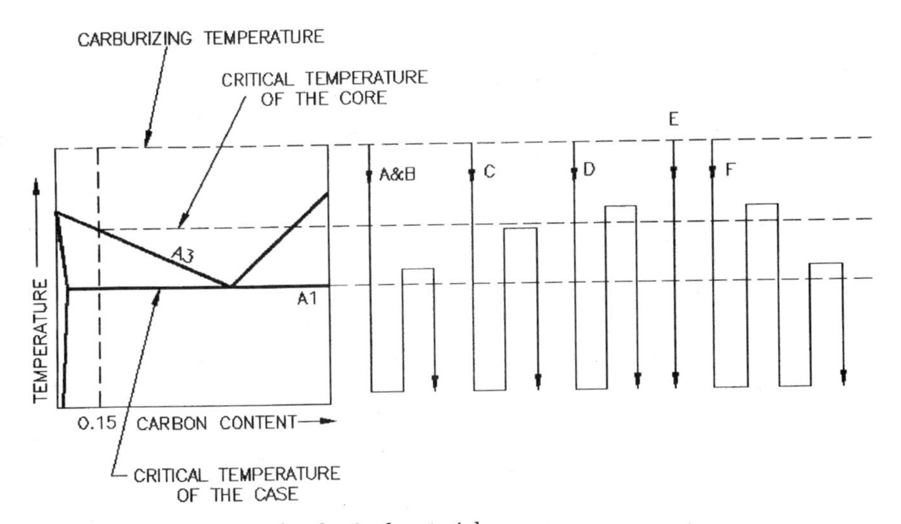


**Figure 14-14** Hardening of carburized material.

**TABLE 14-32 Heat Treatment (Hardening) for Carburized Steel (Refer to Fig. 14-14)**

| Treatment, steel condition | Condition of case | Condition of core |
|---|---|---|
| A, fine-grain | Refined, excess carbide not dissolved | Not refined, soft and machinable |
| B, fine-grain | Refined, excess carbide dissolved, austenite retention minimized | Unrefined, reasonably tough |
| C, fine-grain | Coarsened, excess carbide partially dissolved | Partially refined stronger and tougher than A |
| D, fine-grain | Coarsened, excess carbide partially dissolved. Austenite retention in high-alloy steel | Refined, max strength and hardness. Better combination of strength and ductility than C |
| E, fine-grain | Unrefined, excess carbide dissolved, austenite retained, minimal distortion | Unrefined, hardened |
| F, coarse-grain | Refined, excess carbide dissolved, austenite retention minimized | Refined, soft and machinable, maximum toughness and impact resistance |

**Pack hardening.** To prevent breakage of sensitive parts and to protect them from scaling while minimizing the danger of cracking or warpage, pack hardening is utilized. It consists of packaging the parts to be hardened into a carbonaceous material, which serves not only as their protection but also as a supply of carbon. For pack hardening, only the lowest temperatures should be used, between 1400 and 1450°F.

**Surface hardening in liquid baths.** These may be cyanide baths, which are used where a very hard and thin case is needed on a low-carbon steel without producing shock-resisting qualities of the material. Other carburizing baths are of the sodium cyanide type. The advantage of surface hardening in liquids lies in rapid progress of a process, which also provides a uniform dispersion of carbon with a minimum of distortion and less nitrogen absorption. Portions that are not to be carburized may be protected by being copper plated.

**Localized hardening.** This is used where the parts are too large to fit the furnace or the bath, or where only certain portions of the product should be case hardened. Usually an oxyacetylene torch is used to heat the surface quickly; it is quenched afterward. Often tempering or drawing of the treated surface is recommended.

**Induction hardening.** The induction-hardening cycle is short, often lasting only several seconds. The depth of a case and its area or amount of hardness are controllable, with no additional influences in the form of decarburization or oxidation. The best-suited parts for induction hardening are those of complicated shapes or those requiring only localized hardening. Also benefiting are parts such as cams which should be protected from distortion of their shape.

### 14-7-4. Annealing

Annealing is the opposite of hardening of the surface. It is used where stresses induced by previous cold-working operations are to be removed or where the crystal structure has to be refined, grain orientation reassessed, the hardness of the material lowered for subsequent machining, or mechanical and physical properties altered. Annealing is further used for removal of gases from the material and for changes in the material microstructure.

During the process of annealing, material is heated slightly above the

- Lower critical point for hypereutectoid steels, which are materials of greater than 0.85 percent carbon content
- Upper critical point for hypoeutectoid steels, which are materials of less than 0.85 percent carbon content

This temperature range is held for as long as the material pearlite structure needs to thoroughly dissolve and transform into austenite. Maintaining such temperatures also dissolves all available ferrite or cementite, turning them into austenite as well. The temperature range at which only an austenitic structure remains within the material is called the upper critical point (see 14-7-2 "Hardening"). After reaching such a range, the material is slowly cooled down.

*Normalizing* is an alternative of the annealing process, used to obtain uniformity within the part, free its structure of stresses, and restore proper grain size. Normalizing temperatures are somewhat higher than those used for annealing, the difference being usually 100°F. Parts are allowed to cool in still air at room temperature. The most often normalized objects are usually forgings, where a proper response to a subsequent heat treatment has to be ensured.

When normalizing precedes annealing for machinability, the process is called a *double annealing.*

*Spheroidizing* is used to obtain a spherically shaped form of carbide within the material. When spheroidized, high-carbon steels have their machinability improved, while the strength of low-carbon steel may be altered for subsequent heat treatment. Such a process also increases resistance to abrasion.

*Stress relieving,* or *aging,* is concerned with removal of the instability of material after quenching or with removal of strains imposed by cold working. Aging is performed at slow rates and parts are cooled at room temperatures. The change in material structure is followed by a change of its physical properties.

### 14-7-5. Tempering

Tempering, sometimes called *drawing,* is a method used for removal of both brittleness and internal strains from the hardened material. Tempering consists of heating of the material to the temperature of 300 to 750°F, at which its martensitic structure changes into slightly softer and tougher troostite. Additional heating to 750 to 1300°F produces another alteration within the material structure, turning it into sorbite, with much greater ductility, even though less strength.

The correct temperature range is judged on the basis of the color change of an oxide layer which develops on the material surface during the process of tempering. Such a layer forms on the surface of steel heated in an oxidizing atmosphere. Some basic colors and their corresponding temperature ranges are:

- Light yellow at 440°F
- Yellow-brown at 490°F
- Brown-purple at 520°F
- Dark blue at 570°F
- Light blue at 640°F

Tempering is performed in oil, salt baths, lead baths, or sand. Oil tempering is used for the treatment of tools and the tempering temperature is limited to the range of 500 to 600°F. Salt baths run hotter, between 300 and 1100°F, and the possibility of temperature range control is greater, with much faster heating and greater uniformity in heat dispersion.

Tempering in a lead bath is used where both tempering and hardening are performed at the same time. To lower the melting temperature of lead, which is 620°F, tin is usually added to the bath.

## 14-8. Surface Cleaning

Metal parts, as manufactured, may contain residues of lubricants, shop dirt and dust, abrasives, burrs and nicks of materials, and a host of other impurities or contaminants. Often these parts have to be cleaned in order to prepare the surface for some other finishing process, such as painting or other coating application.

The proper cleaning method of such parts must be well chosen, with many factors in mind. First, the type of soil or contaminant to be removed has to be identified, since a different method of surface cleaning is needed for removal of grease than for metal chips. The surface requirements of the finished part must be taken into account in order not to use a method which may become detrimental to some special feature of the product. As an example, openings for certain sheet-metal hardware should *not* be deburred, as the roughness of one side is important for its installation.

Further, the problem has to be assessed with regard to the subsequent finishing processes, while bearing in mind the cleaning capacities of the particular company or plant.

There are several methods of parts cleaning, each using a different principle and each being applicable to a different range of cleaning applications. Some attack the elements to be removed by mechanical means; others use chemical compounds or steam or electrolytes or ultrasound, salt baths, and other variations. Main categories of these cleaning processes are listed below.

### 14-8-1. Mechanical cleaning

Mechanical cleaning utilizes a mechanical action of abrasives and other objects, which are used in processes such as those of grinding, polishing, buffing, blast cleaning, or shot peening. Abrasive particles may be either dry or as contained in a liquid and applied against the surface of the part. Other objects used in mechanical cleaning may be anything from rags up to glass beads or buffing compounds.

This type of cleaning method may be used for removal of dirt, rust, flash, for deburring of parts, or just for roughing of the surface for subsequent finishing. The actual procedure depends on the particular part and the expected outcome.

*Vibration cleaning* is frequently used for small metal-stamped parts, where these are mixed with abrasives in the form of small stones or similar materials and placed in large drums, which are either vibrating or rotating. The simultaneous movement of parts and abrasive elements is capable of removing burrs, smoothing the surface, and to some degree finishing the edges and removing their sharpness. Larger-sized parts are deburred and surface-cleaned by an abrasive method of running them through an equipment which scrubs their surface by contact with an abrasive belt.

*Blast cleaning* uses abrasive particles, propelling them against the part to be cleaned. It is a cleaning method used with ferrous and nonferrous forgings and castings or to clean weldments, etc.

*Shot peening* differs from blast peening in that its cleaning action is merely an addition to its actual purpose of improving the fatigue strength of the material. This type of finishing is also capable of relieving tensile stresses that would otherwise produce stress-corrosion cracking. In shot peening, the objects propelled against the part are not of abrasive origin. They attack the surface by creating a multitude of shallow indents, which makes the process easily comparable to cold working of the material surface.

*Cleaning of the surface with glass beads* is used for parts of all sizes. As a cleaning method, it surpasses that using an abrasive slurry within a liquid. Glass bead cleaning may be utilized in preparation for painting, brazing, welding, and other manufacturing processes. It produces a matte finish, for which reason it may also be utilized for decorative purposes. A definitive advantage of cleaning with glass beads is that while the surface is being cleaned, no measurable amount is removed.

### 14-8-2. Alkaline cleaning

The most often used industrial cleaning method is alkaline cleaning, the action of which is basically physical as well as chemical, aided by combinations of surfactants, emulsifiers, separating agents, saponifiers, and wetting agents all attacking the part to be cleaned. The solution may be heated or agitated in motion by stirring.

Dissolvable particles of dirt are washed away. Solid particles are separated from the part and allowed to either settle in the form of sludge to the bottom or be floated away and removed from the solution by means of filtering and similar devices.

Alkaline cleaning may be used for removal of wax-type solids, metallic particles, oil, grease, dust, and other contaminants. The application of the process is by immersion in liquid or by spraying or emulsification. Such a cleaning process is often followed by a water rinse and a drying cycle.

### 14-8-3. Electrolytic cleaning

This process is a specialized type of immersion cleaning, with the inclusion of electrodes within the process. A direct current is conducted through the solution, where the part to be cleaned serves as the anode while the electrode acts as the cathode. Some processes alternate the cathode-anode designation. The cleansing action of oxygen, which develops at the anode during the cleaning cycle, may further aid the operation.

This type of cleaning may be used for removal of rust, in preparation for phosphating, chromating, painting, and especially for electroplating, the latter demanding a higher degree of cleanliness.

### 14-8-4. Emulsion cleaning

This process uses two basic materials, insoluble within each other, such as water and oil, combined with an emulsifying agent capable of forcing them to emulsify. This type of cleaning is used with heavily soiled parts, and the cycle is usually followed by alkaline cleaning for final removal of very minute contaminants.

Emulsifiers are of two types: (1) emulsifiers that aid the formation of emulsion which consists of a solvent in water, and (2) emulsifiers that aid the formation of emulsion which consists of water in solvent.

Frequently used emulsifiers are nonionic polyethers, hydrocarbon sulfonates, amine soaps, amine salts, glycerols, or polyalcohols. Solvents usually are of petroleum origin, such as naphthenic hydrocarbons (kerosene).

### 14-8-5. Solvent cleaning

This consists of an application of solvents to the organic contaminants such as oils or grease, in an attempt to remove them from the surface of parts. Sometimes this cleaning method has to be followed by an alkaline wash, in order to remove the solvent itself from the part surface. This type of cleaning may also be used for removal of water from electroplated parts.

Solvents may be either petroleum-based (such as naphtha, mineral spirits, or kerosene) or chlorinated hydrocarbons (trichloroethane, trichloroethylene, methylene chloride) or alcohols (isopropanol, methanol, ethanol). Other solvents include but are not restricted to benzol, acetone, and toluene.

The mechanism of cleaning is applicable mainly to contaminants of organic origin, such as grease or oils. These impurities may be easily solubilized and removed, or washed off the part's surface.

**Vapor degreasing.** Vapor degreasing with solvents is a specialized branch of solvent cleaning. It uses chlorinated or fluorinated solvents for removal of soils such as grease, waxes, or oil. This degreasing process functions by placing the objects to be degreased within a tank, where a heated solvent is boiled. Objects are degreased by the action of vapors, which—being heavier than air—displace the latter from the volume of the tank. On reaching the upper cooler zones, these heated vapors condense and drip back into the area where they are reheated.

### 14-8-6. Acid cleaning

Acid cleaning uses a solution containing organic acids, mineral acids, and acid salts, combined with a wetting agent and detergent for cleaning of iron and steel. Such a cleaning method may be used to remove oil, grease, oxide, and other contaminants without additional application of heat.

Acid cleaning and acid pickling are quite similar processes, with acid pickling being much more aggressive treatment, used for removal of scale from forgings or castings and from various half-finished mill products.

Mineral acids and salts are numerous, forming either inorganic (mineral) acid solutions or solutions of acid salts or acid-solvent mixtures. Organic components of these cleaning solutions may be oxalic, tartaric, citric, acetic, and other acids, with acid salts such as sodium acid sulfate, bifluoride salts, or sodium phosphates. Solvents used within this process may be ethylene glycol or monobutyl (and other) ethers.

### 14-8-7. Pickling

Pickling of metal materials removes the oxides, or scale, off the surface of parts. It may be used for removal of other contaminants as well, by immersing the parts in a liquid solution of acid. Such a solution may vary in its composition, temperature, and selection of ingredients, the most common pickling bath being sulfuric acid. Hydrochloric acid is utilized where etching prior to galvanizing is needed. For pickling of stainless steel, nitric-hydrofluoric acid is used.

The mechanism of pickling functions by penetration of the scale through the cracks and chemical reaction of the pickling solution with the metal underneath. In order for the pickling solution not to attack the metal, inhibitors in the form of gelatin, flour, glue, petroleum sludge, and other substances are added. Inhibitors can minimize the loss of iron surface and reduce the range of hydrogen embrittlement while protecting the metal from pitting, which may occur where pickling becomes excessive.

### 14-8-8. Salt bath descaling

The salt bath descaling process is used for removal of scale and it must—for a complete removal—be followed by acid pickling or acid cleaning. Salt bath descaling may be divided into three groups: oxidizing type, reducing type, and an electrolytic method. The latter may be used even in conjunction with the previous two processes.

Oxidizing type of salt bath descaling is the most often used method of scale removal because of its simplicity, even though the electrolytic method offers greater scale-removing capabilities. The reducing method's advantage is lower temperatures of the salt bath.

The removed scale, along with the descaling salts, forms an insoluble sludge, which must be removed mechanically. For that reason such impurities are allowed to settle into a pan placed there for their collection.

### 14-8-9. Ultrasonic cleaning

Ultrasonic energy, when applied to the solution of chlorinated hydrocarbon solvents or to water and surfactants or to any other type of cleaning solution, will boost the cleaning process, removing various types of contaminants. It may be used for removal of fine particles embedded within the material, or for cleaning of complex parts, precious metals, or hermetically sealed units, and also for cleaning where extreme cleanliness is required.

The disadvantage of the ultrasonic process is its high cost, which is due to the much higher initial cost of the equipment and its maintenance. However, this type of cleaning has been found beneficial where previously only hand-cleaning methods worked.

## 14-9. Surface Coating

Surface coating should be chosen with regard to the application it has to serve, along with a consideration for the basic metal it has to cover. Some coatings are used as a protection against abrasion, corrosion, oxidation, and for a host of other reasons. Surface coating creates a barrier between the basic metal itself and the environment, which may be detrimental to its stability. There are coatings to alter the frictional properties and to enhance the esthetic appeal of the part. Various coatings may be used for various applications but are most often chosen to protect the basic metal, the basic product, from outer influences.

Even two metallic parts within an assembly are capable of attacking each other by forming a galvanic cell, the same way a basic material may react adversely to its coating if chosen improperly. Evaluation of the possibility of a galvanic couple formation must therefore be considered when choosing the type and amount of protection a coating should offer. This involves a survey of whether the coating is in its nature cathodic or anodic toward the metal underneath it.

For example, a steel may be protected from other influences by nickel or zinc, even though nickel is cathodic to iron and zinc is anodic. Nickel protects the steel by successfully blocking the influence of the outer corrosive environment on the material, and for that reason the nickel coating should be free of pores. Zinc provides protection by corroding more readily than steel, and a by-product of the corrosive reaction, zinc oxide, being quite sizable, impairs the corrosive process and protects the coated material.

Many metals are capable of forming oxide films, which—when stabilized—act as a protective coating for that particular material. Aluminum oxides thrive in acidic atmospheres, where they form thick protective layers, but once the basic alloy is anodized, the coating shrinks, turning thin, hard, and stable. Some oxides, such as those of tin, zinc, titanium, etc., could be stabilized by an additional chemical or electrochemical treatment, which turns them into protective layers for the basic metal material.

The success of such protection depends on proper analysis of the galvanic-cell process, during which an anodically dissolvable metal must be protected by an equal and opposite cathodic reaction.

### 14-9-1. Electroplating

The electroplating process should actually be called galvanizing, since it uses the principle of a galvanic couple between the plated part and plating material to transfer particles of material to the surface of the part. In this process, a direct electric current is applied to a solution of metal salts in which the parts to be coated are deposited. These parts assume the role of the cathode, or negative pole, by being connected to the negative end of the source of energy. Large parts are left hanging off a copper bar attached to the negative pole of the source, and small items, such as washers or bolts, are placed in wire baskets. The coating metal itself acts as an anode, and it is added to the bath in the form of plates, bars, or extruded shapes.

When affected by the electric current, the anodic metal material slowly ionizes, its particles entering the solution of the bath. These little ions travel toward the cathodic-polarized part, on whose surface they become deposited in the form of metal crystals. Some types of metal-coating processes require coating baths to be heated and sometimes a liquid-stirring action is added to enhance the uniformity of the film.

The speed of the development of coating depends on the intensity of electric current and temperature of the bath. With a warmer bath or with higher amperage of the current, the coating process becomes faster. However, with too high an intensity or with too warm a solution, the coating emerges coarse and inadequate. The electric current has to be low in *voltage* (often few volts will suffice), but the *intensity* must be quite high, with 0.1 to 2 amperes or more per each square foot of the coated surface.

Organic compounds are sometimes added to the bath and their minute quantities alter the properties of the coating film to a considerable extent. Their influence is oriented mostly toward the esthetic appearance, with subsequent smoothing of the coated surface and providing it with a sheen. These strictly optical enhancements are outweighed by a diminished protection against corrosion they offer.

Almost all metals can be galvanically applied as coatings by using modern methods and modern technology. However, with some, the process is so costly that it remains only a technical curiosity.

The four most common processes of galvanized coating are

- Acidic galvanic coating, in which the metal is present as a cation in a simple salt solution, such as that of sulfates, sulfamates, fluoborates, or chlorides. This process is used for the application of nickel, copper, zinc, and tin coating.
- Complex alkaline cyanide baths, with the metal particle in the form of an anion, connected to the cyanide portion of the solution. This type of bath is utilized for application of copper, cadmium, zinc, silver, and gold coating.
- Complex acid baths, where the cathodic deposition is achieved through an intermediate stage, or as a cathodic film. An example is chromic acid, which forms mono- and dichromate ions.
- Alkaline baths for metals, forming amphoteric oxides, such as alkaline stannate bath, which contains sodium or potassium stannate, which is stabilized by ions of hydroxyl.

Parts to be finished by electrodeposition must be deburred, cleaned, and their previous coating—if any—completely removed. For better adherence, pickling or acid dip may be used. The cleaning process is vital to the success of the plating operation, because a maximum adhesion of the coating to the basic metal is necessary.

*Copper electroplating* is usually used as a bottom layer for additional plating. Rarely is copper used alone as a coating material, since it scratches and stains readily and tarnishes from weathering. If a bright copper surface is required, it must be protected by at least a coat of clear lacquer. Copper is usually plated in cyanide baths or in acid plating baths.

*Chromium electroplating,* or industrial chromium plating, is corrosion-resistant and extremely hard. It differs from decorative chromium plating in that industrial-type deposits are applied to the basic metal without intermediate undercoats. Industrial chromium plating is intended only for the protection of parts, for extending their life in service by shielding them from wear, corrosion, or heat effects. This type of deposit, as opposed to decorative chromium plating, is also thicker, ranging from 0.1 to 20 mils, whereas decorative chromium plating uses thicknesses of 0.005 to 0.05 mil.

*Hard chromium plating* is being widely applied to various types of tooling, in which case the coating extends the life of the tool, improves its performance, or even repairs worn out surfaces. By the application of hard chromium coating to plastic molds, these tools are protected from the destructive effect some plastic materials (i.e., vinyl) may have on the metal material. Cutting tools, deep drawing tools, various machine parts, and other products may greatly benefit from the chromium plating. However, with parts exposed to high heat and pressure, chromium coating will not perform well, as it may crack in service.

*Nickel electroplating* may be produced in Watts baths or in sulfamate or fluoborate baths. Nickel plating is one of the oldest surface-protecting metallic coatings of steel because of its good appearance, combined with resistance to corrosion. Today, these coatings are used for protection of iron-, copper-, or zinc-based alloys against corrosion.

*Cadmium electroplating* serves as a protection against corrosion. Cadmium, being anodic to iron, conserves the basic ferrous material even when scratched or otherwise damaged. Aside from acting as an anticorrosive layer, cadmium coating has lubricating properties, considerable electrical conductivity, and low contact resistance.

*Tin electroplating* allows for thin layers of tin to be used as a protective barrier against tarnishing and corrosion, while enhancing the solderability of coated material. The bulk of tin plating applications is within the mass-packaging of food, where it is used as a liner for steel cans. In the absence of oxygen, tin protects the food within the cans from coming into contact with the material of the can. Another usage of tin plating is within the electronic, agricultural, and transportation industry.

*Zinc electroplating* offers a suitable surface protection for materials with low melting point, such as iron or steel. Zinc is a nontoxic metal and the plating process is relatively inexpensive, offering an excellent protection from weathering and corrosive influences, in which way it surpasses nickel coating. In heavily corrosive environments, such as marine applications, zinc is outperformed by cadmium. However, with the worldwide progressing ban on cadmium plating, alternative resources, such as alloyed zinc coatings, are being investigated.

*Miscellaneous plating materials* include silver, gold, brass, and bronze among others. Plating with combinations of metals, or alloy-plating, is a modern attempt at the greater control of the result, aiming at the reduction of the progress of corrosion. These alloyed coatings may be zinc-iron, zinc-cobalt, zinc-nickel, zinc-lead, and other. Zinc-nickel coating is found widespread in the modern fastener, automotive, and communication industry, where it is utilized in some heavily corrosive areas. The protective coat of zinc-nickel plating was found capable of preventing the basic metal of a fastener from forming a galvanic corrosive cell with aluminum, which possibly opens a new field of application within the airline industry. Zinc-cobalt coating displays a tremendous resistance to atmospheric influences, even to those enhanced by a greater content of sulfur.

### 14-9-2. Electroless plating

Electroless plating uses no electric current. The plating procedure of electroless zinc plating, for example, consists of depositing the plating material by means of an autocatalytic chemical reduction of zinc ions by hypophosphite, aminoborane, or borohydride compounds. There are two types of baths for electroless plating: (1) hot acid baths, to plate steel and other metals, and (2) alkaline baths, for plating of plastics and other nonmetallic materials.

*Nickel plating* provides the basic material with excellent protection against corrosion. Where applied to aluminum, nickel plating provides a solderable surface.

*Zinc plating* is attained in cyanide baths, alkaline noncyanide baths, or acid chloride baths, the latter being the fastest-growing method of plating.

### 14-9-3. Hot dip coating

The hot dip method of coating uses a bath of molten metal material to dip the objects to be coated in. Often, this method is called "hot dip galvanized coating" in the literature, where the word "galvanized" is not correct. Galvanizing always refers to the process, implementing a galvanic cell within the principle of the operation. With the galvanic cell, there is always an electric current involved as well, which—in hot dip coating—is not present.

The temperature of the bath, combined with the length of immersion, govern the speed of coating application. The visibility of the crystallike grain of the solidified coating, which often "decorates" its entire surface, may sometimes be optically disturbing.

A thorough cleaning of objects to be dipped is required. All products must be cleaned to be free from grit, oils, and grease, drawing lubricants, and other contaminants, to ensure a proper adherence of the coating to the basic metal.

*Hot dip zinc coatings* are readily attacked by sulfur dioxide and other industrial pollutants, and for that reason their longest life expectancy is in rural areas where industrialization is not yet widespread.

*Hot dip tin coatings* when applied to the cast iron or steel provide the basic material with a nontoxic coating, often used in the food-processing or electronics industry. Decorative coatings of the hot dip type are also common. The tin coating improves solderability of the basic metal and can be used as an adhesion promoting agent with subsequent coatings.

*Hot dip lead coatings* usually employ lead-tin or antimony combinations to coat washers, bolts, and other mounting hardware, and small metal-stamped parts, such as brackets, plates, and various fixturing elements. Lead alone is not capable of combining with iron into a coating, as it either separates into free lead crystals, or turns into a fungus-resembling layer.

Lead-tin alloys form a layer of excellent adhesion, which acts also as a lubricant, an agent to improve solderability, or just a protective barrier against corrosion. This type of layer is called a *terne coating.* Two kinds of such surface protection are used widely: *Short terne,* which is very light in thickness (0.01 in), and its thicker equivalent, *long terne,* ranging between 0.01 to 0.08 in.

### 14-9-4. Chemical coating processes

These types of coatings provide the part with the surficial layer of mostly nonsoluble salts, a result of the chemical reaction between the material of the coated part and the chemical components of the bath. The acidic bath attacks the metal surface, dissolving the most outer layers and turning them into ions, which readily combine with the chemical content of the bath. Inorganic and marginally also organic salts constitute the basis of the bath, their action being supplemented by activators or oxidating agents.

Parts to be coated are allowed to remain immersed for a certain period of time. They must be thoroughly cleaned, often pickled, free from grease, oils, and other contaminants. Complexity of the shape presents no restriction, as the chemical reaction takes place simultaneously over the whole surface, whenever the bath solution gets into contact with it.

Chemical coatings do not provide the parts with the best corrosion or abrasion protection. The coating is limited in thickness and it is useful mainly with small parts, for those with very accurate dimensions, or where the equal distribution of coating over a complex surface is important, such as with objects containing an inner thread or other types of complex crevices. The coating can be easily damaged by mechanical means. For these reasons, chemical coatings are the best used in conjunction with other coating processes, as adhesion promoting layers with subsequent coatings, especially with paints.

According to the process used, chemical coatings can be divided into several categories: chromating (chromate conversion coating or passivating), phosphating, and oxidating.

*Chromate conversion coating.* Parts, as immersed in an aqueous solution of chromic acid or chromium salts (sodium or potassium chromate or dichromate) or in various other acids combined with activators and modifiers in the form of chlorides, fluorides, sulfates, complex cyanides, and phosphates, form their own protective coatings within their surface. Such a coating is actually the material's response to the chemical attack of the bath. It is composed of nonsoluble metal salts, obtained by a partial dissolution of the material surface and their combination with the chromate ions of the bath.

Metals such as zinc, magnesium, tin, and aluminum may be coated this way for protection against rust or corrosion. There are two forms of chromate conversion treatments: (1) those providing a film of their own on the material's surface, and (2) those supplementing or securing another type of nonmetallic protective coating, such as that of oxide or phosphate.

Chromate conversion coatings may be colored or clear, the colors being influenced by the type of modifiers and accelerators in combination with the basic metal material. For example, fluorides and sulfates will produce a bright blue film on an electroplated zinc material, fluorides and ferricyanides will result in a gold film on aluminum.

*Passivating.* The possibility of forming a protective layer of its own is a sign of passivity of a material's surface, or its ability to remain unaltered in appearance, even though being subjected to corrosive attacks of its surrounding. Because most conversion coatings dissolve very slowly in water, passivation serves as a simple means of protection against corrosion in milder or indoor environments.

*Phosphate coating* is a method of surface protection for steel or iron consisting of an application of diluted phosphoric acid and its salts, combined with metals (zinc or iron or manganese) and other chemicals, to the surface of material. The reaction between these elements and the base metal produces a layer of insoluble crystalline phosphate at the interface, capable of protecting the material from abrasion by mechanical means or atmospheric corrosion.

This type of coating is also used as a base coating for further application of paint or additional corrosion-resisting material. Phosphate coating also provides the surface with lubricity and with protection against wear and galling.

The three principal types of phosphate coatings are distinguished according to their additives as those using zinc, iron, and manganese. Zinc phosphate coating varies in colors and their intensity: The higher the carbon content of the material, the darker the hue. Iron phosphate coatings have excellent adherence to the basic material, which they protect from flexing when painted or from flaking under an impact. Manganese phosphate coatings are used for bearings, gears, etc., to prevent galling.

After the treatment, the remaining phosphate is rinsed off in water, which must be free from chlorine or sulfates to avoid an attack on the fresh protective layer.

Surface preparation for phosphate coating should be very thorough, as the chemical reaction between the basic material and the phosphating solution depends on the amount of contact between them.

*Chemical oxidizing* is a process similar to chromating and it is sometimes quite difficult to distinguish one from the other. In oxidizing baths, a surface coating, consisting of oxides of the coated metal in conjunction with other ingredients of the bath, is formed. Chemical oxidizing sufficiently controls the stability of the esthetic appearance of parts exposed to the indoor pollution. Combined with an additional paint, oxidation is a valid barrier against the general atmospheric corrosion. It is reserved for smaller objects made of aluminum and aluminum alloys or for objects with a long shelf life, for those with very close tolerances, optical devices, and firearms.

### 14-9-5. Anodizing

Anodizing may be defined as an electrolytic process which produces a thickening and stabilizing of oxide layers within the surface of the base material. Anodizing provides the coated part with wear, corrosion, heat, and abrasion resistance. A film created by anodizing serves also as an electric insulator.

The method of coating consists of immersing the part to be anodized in an electrolyte (a 15 percent solution of either sulfuric acid or various organic acids), to which an increasing voltage is applied. The current is usually direct, but may also be alternating. It converts the immediate surface of the anode, which is the part to be coated, to an oxide. This oxide is electrically nonconductive and being almost nonsoluble in the electrolyte, it remains attached to the part, forming a continuous, solid coating. The bulk of such a layer automatically slows down and finally stops the additional electrolytic process, for which reason the coating of anodized objects can be produced up to a specific thickness. The thickness of anodized coating is uniform throughout the part, regardless of the complexity of its shape. It is being developed from the outside toward the core of the part and its thickness is always greater than that of the original layer of material, utilized for its development.

The anodized surface can further be altered in appearance prior to sealing, since the oxide pores are still open and able to absorb various colloidal substances, such as coloring agents or hydroxides of metals. Dyes, when combined with specialized anodizing procedures allow for attainment of various colors, or imitation-look of pewter, copper, bronze, and other special finishes. Unfortunately, corrosive influences may attack such a part; therefore the oxide pores must be sealed for the protection of the coating.

Various techniques can be used for sealing the anodized surface. For a clear finish, boiling in deionized water converts the amorphous form of aluminum oxide to a more stable form of crystalline hydrate. To improve corrosion resistance, a dichromate sealing method is used; a color-stained anodized surface must be sealed in nickel acetate to prevent bleeding.

The anodizing process is not restricted to aluminum; it can be applied to other light metals such as magnesium, titanium, and their alloys. The hardness of an anodized coating equals that of a diamond, making this type of surface finish an excellent barrier to corrosion, providing the parts with wear and abrasion resistance, and enhancing the esthetic appearance of the parts.

### 14-9-6. Thermodiffusion process

Coatings produced by thermodiffusion are formed at high temperatures and in controlled atmospheres of specific content. Diffusing materials of gaseous, solid (powder), or molten form are placed in contact with the part to be coated and allowed to enter its surface. The coated material, usually steel or iron, forms an alloy with the diffusing components within the upper layers of the coated surface. The coating emerges uniform in thickness throughout the part.

The temperature of the process is somewhere near the melting point of the diffused metal and the heating procedure is conducted in an oven. Various processes use different temperature settings: These are either below or above the melting point preferences, according to the diffusion substance used. The temperature of the process influences not only the speed of the coating operation, but also the character and texture of the finish as well.

The most common thermodiffusion processes are cementing and nitriding, but other applications utilizing chromium, aluminum, sulfur, and zinc are being widely used. With zinc, the process is called *sheradizing,* and the metal material is added in the form of powder. With the melting point of zinc at 786°F, sheradizing is usually performed at temperatures ranging from 600 to 700°F. Thermodiffusing of sulfur is performed along with nitrogen, and the process is almost the same as that of nitriding.

Newer diffusing processes, utilizing boron and silicon, were developed for attainment of an extra high surface hardness, abrasion and wear resistance, and resistance to high temperatures. Another new technique involves a combination of the thermodiffusion process with electrolysis of the salt melt.

Thermodiffusion is preferred as the surface treatment of small parts, since a distortion of products and their dimensional alterations may occur with larger objects. A considerable variation in wall thickness or sharp corners on the part will magnify these complications.

### 14-9-7. Thermal spray coatings

Thermal spray coatings are applied at high temperatures, using a high-velocity stream of compressed air or gas in combination with an electric arc, plasma arc, or arc flame. The coating material is melted by the temperature of the heat source and propelled against the coated part. This process is often called *metallizing*.

The coating material is supplied in the form of rods, wires, or powder. After its melt is force-deposited on the coated surface, it is retained by either becoming embedded in the material, or by bonding with it through the process of either diffusion or alloying. A possibility of a combination of all three retaining methods within a single coating process is probable.

The thermal spray method is not restricted to coating with metals; ceramic materials, or those of combined metal-ceramic content can be utilized. Mechanical properties of the coating material change on thermo-deposition, since it turns into a hard, brittle, and nonhomogeneous layer of metals and their oxides, with only marginal tensile strength and a great resistance to pressure. Such a coating offers less protection against corrosion since the thermal spray coating process produces layers perforated with open pores.

The thermal spray method is employed where electrical resistance, or electrical conductivity, or electromagnetic and thermal shielding of the part are required. The attainment of either of these properties depends on the coating material and the coating process used.

### 14-9-8. Vacuum coating

Vacuum coating is applied using three basic techniques of the coating material disintegration: evaporation, ion implantation, and sputtering. The appliable condition of metal is achieved with the aid of electron beam or ion beam gun, resistance or induction heating, plasma discharge method, or electron-emitting arrangement.

*Evaporation using vacuum coating process* is conducted in the vacuum of $10^{-2}$ Pa or greater, where the coating material is heated by either resistance or induction heating process or laser beam application, up to its melting point. The object to be coated is purposely kept distanced from the source of heat to remain colder. Evaporating gases condense on the colder surface of the coated part, covering only those portions exposed to their influence. The thickness of coating can be well regulated, and for the best adherence of the film, parts must be thoroughly cleaned and sometimes even pretreated, especially where thicker deposits are desired.

Almost all metals can be used as vacuum-applied coatings, even though practically the process is restricted to aluminum and chromium and some of their alloys, selenium, germanium, selected oxides, and fluorides. Coated materials can be almost anything as well: metal, glass, aluminum, paper, and other. Vacuum coating with aluminum is selected for its high gloss, used for production of reflective surfaces in reflectors. Vacuum coating is irreplaceable as a costume jewelry coating and in other decorative applications, along with a heavy involvement in electronics industry.

The *ion implantation method* is employed for complex shaped parts, and it uses a bombardment with high-energy ions, produced in a glow discharge of the gas. The part to be plated is conductively attached to a high-voltage electrode, insulated from its surroundings. A negative current of 3 to 5 kV is used in this process, as applied across the electrode, with the ground connected to the system.

At the beginning of plating process, the plating chamber is pumped down to $10^{-3}$ or $10^{-4}$ Pa and then partially refilled with a controlled amount of argon, to a vacuum of approximately $10^{-2}$ Pa. At the application of electric current, the part is first bombarded by ions of argon, which clean the surface for further processing. By adding the vaporized coating material into the glow discharge, it is propelled against the coated part. The coating film produced by this method emerges uniform in thickness, no matter how intricate a shape the coated part possesses.

*Sputtering* also uses a heavy inert gas, most often argon, in a glow discharge for the bombardment of the coated surface. As with ion planting, the chamber is evacuated and refilled with argon until reaching a desirable vacuum. The coating material is considered a cathode, receiving a negative bias from the high-voltage source of energy, which is supplying the process with 1 to 5 kV. Positive plasma ions are accelerated by the high electric power source and sputtered against the cathode, striking and ejecting it against the coated part.

With such a method of coating, almost any material can be used to produce the film, as its turning into a vapor phase is achieved by mechanical (exchange of momentum), rather than electrical or other means.

### 14-9-9. Painting

Painting is application of a thin layer of organic coating (liquefied or as a paste) to the treated surface. Paints may be separated into the groups listed on the following page:

- Waterborne paints, or those which are dilutable with water. These may be solution coatings, based on water-soluble binders (alkyds, acrylics, epoxies) or colloidal dispersions (small elements of binder material, dispersed in water) or emulsions (latex).
- Enamels, forming a smooth, high-gloss surface. They may be dried in air or cured in an oven.
- Lacquers, or synthetic thermoplastic materials, capable of creating a film soluble in an organic solvent.

Painting has certain advantages over other coating processes. Paints are easy to apply, the necessary equipment is less costly, and a wide variety of pigments allows for easier matching of a hue, while the coating itself protects the basic material against different types of corrosive influences. Paints can prevent an emergence of galvanic corrosion between dissimilar materials. Many pigments contained in paints are conductive enough to offer protection against static electricity.

### 14-9-10. Porcelain enameling

This process is applied to steel or cast-iron material. It is essentially a glass coating, which must be matured in the oven at higher temperatures of approximately 800°C. Porcelain enameling of aluminum or copper is rare. If enameled, aluminum cannot be heated to such a high temperature, with 580°C considered sufficient.

The process of porcelain enameling starts with an application of frits to the surface. Frits are smelted complex glass or ceramic materials in an aqueous solution. Some types of frits are applied in their powdered form, in which case an electrostatic spraying is the method of their application. In spite of considerable content of metallic oxides, these materials behave similarly to glass; they are brittle, display a great resistance to chemicals and higher temperatures, and have a limited resistance to thermal or mechanical shocks.

After being treated with frits, parts are placed in an oven and heated to the desired temperature. Most often, a single coat, preceded by a base coat of limited spectrum of colors, is produced. The top coat may bc opaque, in which case it is mostly white. But pigmenting for a wide array of colors is possible and clear or semiopaque coats can be attained as well.

The design of enameled parts must take into consideration the coating process and its demands: All corners must be rounded, with small-sized radii totally excluded, and a too diverse combination of surfaces must be avoided. With thicker enamel coatings, the requirements are still more demanding.

Porcelain enameling offers an excellent protection against abrasion and corrosion, coupled with greater than normal weather and chemical resistance. Enamels can resist an attack of acids even at higher temperatures. However, they can be affected by phosphoric acid or by fluorides.

### 14-9-11. Miscellaneous coating techniques

These include, but are not limited to, various methods listed.

*The chemical vapor deposition method* is a process similar to carburizing. In this method of coating, a reactant gaseous material is introduced into a heating chamber, where a part to be coated is deposited. The gas is then allowed to settle and decompose upon the part's surface. The coating material, admitted in a gaseous form as well, becomes absorbed by the surface of the coated part.

Reactant atmosphere consists of fluorides, chlorides, bromides, hydrocarbons, and other compounds. The coating materials used in chemical vapor deposition coating are

- Chromium, which is used for steel and its alloys, and the coating process resembles that of pack cementation.
- Tungsten, used most often with ferrous alloys and iron and requiring a nickel undercoat.
- Nickel, a coating used for plastic molds or for specific inaccessible areas.
- Titanium in the form of titanium carbide or titanium nitride is useful for coating of cemented carbide inserts, threaded parts, different types of tools, and other small items.

These coating are used mainly for preventing wear and corrosion of the base material.

*Babbitting* consists of attaching a layer of softer metal (usually a tin-lead composition) to a part of much sturdier composition which acts as a supporting element. The soft layer, or the babbitt, has excellent antifrictional properties. In shafts, the babbitt averts galling and scoring of the surface, while the inner, stiff core acts as its support in torsion, when rotating.

Babbitting is used with bearing shells, hardware elements, automotive connecting rods, jewelry, and numerous other applications. The babbitt is attached to the supporting metal by either of two methods:

- Mechanical bonding of babbitting is performed by using fasteners, dovetails, and other grooves.
- Heating of the babbitting material along with its supporting part and allowing the assembly to first cool at the area of contact between the babbitt and its support. This method is useful with shells, where the babbitt is introduced in the form of a mandrel.

*Electropolishing* acts almost like etching, and it can smooth the metal surface by anodic means, using a concentrated acid or alkaline solution for removal of burrs produced by conventional machinery, or for improving the appearance of the parts and enhancing their resistance to corrosion. The process can be further expanded to prepare the surface for subsequent coating, to improve reflective properties of parts, and to remove stressed or distorted surficial layers. Electropolishing is used for surface treatment of turbine blades, surgical instruments, nameplates, reflectors, jewelry, watch cases, piston rings, valves, ornaments and trims, and many other items.

*Coating with plastics.* Plastics can be applied in the form of foils, paints, bonded and baked-on layers, or shrink-wrapping elements. Plastic coats are usually quite bulky and their use is reserved for specific situations only.

*Cladding,* otherwise called a solid-to-solid diffusion, is based upon diffusion of various solid materials such as metal laminates or composites. Aluminum cladding is used for mild steel and aluminum alloy. Nickel cladding on steel is advantageous, because the ductility and thermal properties of both materials are similar. Stainless-steel cladding to carbon steel is aided by electrowelding or by hot pressing or casting. Copper cladding of steel may be attained through casting, and it is particularly applicable to cladding of wire and tubing.

Chapter

# 15

# Die Cost Estimating

## 15-1. Trends in Sheet-Metal Manufacturing

Metal stamping and sheet-metal production are constantly growing technologies. True, there have been tendencies to replace many metal parts with plastics, since the cost of dies is sometimes considered too high. (As if the cost of molds were lower!) However, these tendencies probably will be somewhat balanced in the future, when manufacturers will finally realize—by adding up numbers on paper—that plastics cannot always replace sheet metal nor may sheet metal always replace plastics.

Everything has its own sphere of usage and application. Plastics are great materials, yielding, permissive but easily deteriorating, changing colors, crumbling away under the effect of either sun and weather or time. Sheet metal, once a part is made of it, remains stable without being affected by time, environment, an operator, or anything at all, with the exception of hammers and corrosives and extreme heat, of course.

A great advance in sheet-metal work was accomplished by implementing numerical control machinery like automatic drill presses, automatic punch presses, and lathes. These manufacturing tools are real workhorses, hurling out products of superior quality and continuity. Naturally, this is true only if the programs they run on are of an equivalent quality.

The discovery of EDM machinery was another step forward, as these machines, quite unheard of several years ago, quietly and efficiently produce parts previously considered impossible to make. Additional new technologies emerge on the market continuously, many of them quite beneficial to the manufacturing field, many still controversial. Lasers, plasma technology, electronic rays, electroerosion, chemical erosion, electrospark coating systems, anodomechanic machining, electroresistant machining, vibratory machining, ultrasound machining, pressure-waves machining (explosives)—all these new manufacturing processes are here to make our lives easier and more productive.

The number of innovations within the manufacturing field has been tremendous, considering that a host of the new methods mentioned did not take long to arrive in our shops. They are quite numerous and their technical advancement superior. Where before we could have a simple little press with limited controls and a manually lowered safety grating, we now show off programmable, fully controllable, and fully cooperating machines.

It almost seems that if we go ahead at this rate, we shall soon live in a utopia where machines do our bidding and we are finally able to devote more time to areas so far neglected because, even though important, they were not considered profitable.

## 15-2. Basic Approach to Cost Estimating

In order to fit comfortably within the new world of technical modernization, we must create a completely new approach to various supportive factions of the industry. Many aspects to consider may not be strictly technical, yet they are essential to the overall outcome of the manufacturing process. These portions of production are planning, inventory control, advertising, selling, and naturally—cost estimating.

Fortunately we have some excellent manufacturing-related computer programs which allow for production planning and scheduling of quite complex operations. They keep track of numerous bills of materials, consisting of myriads of components stored within a dense forest of shelves. The mind of a computer, if it has any, can easily sort its way out of such chaos and instantaneously find the correct procedure to follow, correct papers to issue, or the most pertinent parts to send to the production line. On the basis of existing production rate, it warns us when the stock is getting low or when deliveries are exceeding the rate of usage. It shows the numerous locations where a particular item is used. It alerts us to a change of design, a change in manufacturing method, a change in delivery, and any other change at all. It actually does all the work of technical aide to the designer/engineer team.

Computers are capable of incessantly aiding the manufacturing process, the administrative process, bookkeeping, planning, inventory control, and a host of other pertinent segments within an industry. They achieve this complex task by aligning various data, matching results, sorting information, and grouping it, listing it, and referencing it.

But how would even the most advanced computers aid a process where no data is stored, no records of previous operations accumulated, and no procedures to follow outlined? This is basically the existing condition of the cost estimating field within many current industrial enterprises. This portion of manufacturing did not always advance as readily as other sections; it took upon itself the role of "sleeping beauty" who is still waiting for someone to awaken her.

Over the years, people within the manufacturing field came up with many ideas on how to estimate the cost of a new work, but their methods were either too complex or too tedious or too inaccurate. Where some estimating methods worked well, they were probably kept secret, since a well-functioning estimating procedure is equal to real money in the bank.

The method of estimating the product's cost on the basis of projected sales outcome may work, and it may fail at the same time. What is true today may not be true tomorrow, and we can hardly predict sales revenues for next year when we don't know if the product will sell at all. Today, if everyone's buying tractors, who knows if they will keep on buying them next year? Perhaps they will, perhaps they won't. It all depends on the saturation of the market with that particular product. It also depends on the accessibility of funds, on the exchange rates, on the company's solvency, on the amount of enthusiasm of its leadership, and on many other subtle or obvious aspects, which govern every financial decision of every firm, person, or government.

A basic lesson of economy maintains that every product has several stages of its useful life. First, there is the phase of development, followed by advertising and the product's introduction into the market. Then, ideally, sales picks up and the plotted line goes straight up at 45° or steeper if possible, until a large curve bends it down again. The top of such a curve, or the peak of that part's performance as a salable item, marks the saturation point within the segment of applicable buyers. From then on, there will be nothing but a decline, usually gradual but quite sudden if a new and more appealing object of the same category emerges, replacing the old one.

Some speculate, taking off the peak of the line with another 45° (or steeper) line, representing a market, within a different part of the world. Where a market is declining in one corner of our little global plantation, the other corner of the world may have a genuine interest in such a market's implementation. This is true, naturally, if the world is staged as found today: Some nations are better off, some are more developed, other nations are not so much advanced, and some are outright backward. Naturally, all backward nations will gladly purchase a useful item even though it has completed its life cycle within the more advanced world of developed countries. This means that—provided these people have means of payment—and provided we have the means of supplying them with that particular item, we have a deal.

But what happens once all nations are on the same level? Where will we find additional dump grounds for our overexploited or unutilized stock? Who will buy an item which does not sell at home anymore, when "home" is the worldwide term?

Some are preparing for this situation by outpricing their competitors, meaning lowering their own prices. There are now various means of lowering prices. Products may be manufactured elsewhere, where the cost of labor is low. Products may be manufactured from inferior materials, the design may be cheapened, or a company's profit may be slashed. Yes, there are means of lowering prices, but who will guarantee that such means are not lowering the salability of the particular product as well?

In its extreme, such an effort may result in dumping of products upon a foreign country's market, paying all the importing costs and duties, and still charging a lower price than a local manufacturer. This is not a fantasy; it's an often encountered reality.

However, it may be stipulated that lowering a price is a relatively feasible approach to capturing the market. But with respect to what are we lowering it? What are the criteria for any existing price range anyway? The manufacturing cost plus the profit margin plus overhead? Certainly not! At least not always and not everywhere. Sometimes a tiny pill two times a day may cost several times its weight equivalent in gold.

But markets have been created and prices have been set, and actually we don't even know if the price ranges are correct. For, should we—for example—show a brand new product to some people, explain its function along with its advantages, and ask how much they will pay for such an item? Without a doubt, we will get a whole range of prices, and perhaps none of them will be in the vicinity of the actual cost. A $12 item may be assessed at $80, and a $1200 article may be valued at $20. The sky is the limit; prices may fluctuate incessantly and—what's worse—people will be willing to pay them.

Because of prices' fluctuation and also because of their frequent nonrelatedness, it often helps to assess costs rather by percentages instead of assigning them actual monetary values. Especially when using a computerized approach, formulas may be implemented and stored in the computer, and once an actual numerical input is added, the system will come up with a whole range of variations.

### 15-2-1. Pricing history

For obvious reasons, the most important commodity in cost estimating is the price history of each particular item. Usually, when figuring the cost of a part or product, estimators tend to send out numerous requests for quotation, eagerly awaiting the answers, which they add up to obtain the cost of the item. This method, even though it works in the short run, is actually disaster-prone, because when one or more manufacturers do not submit their proposals, what numbers will the estimator use in the absence of the object's pricing history? Or—what if only one manufacturer who always comes up with the highest cost responds? It will be risky to use that quote, because the price will be too high then. It will be equally risky to lower the price, because what if the other respondents didn't quote because they were not able to price the item within their usual cost range or because they were not able to manufacture it at all?

Therefore, every estimator should strive to accumulate his or her own pricing history of each quoted product. For example, Table 15-1 shows estimates five vendors submitted over a 5-year span. This is valid information, which may be assessed in several ways.

Rates comparing various estimates to the original amount submitted by the particular vendor should be calculated to evaluate the degree of increase for each manufacturer (Table 15-2). Knowing such a probable rate of increase may prove useful when pricing other items from the same source: An older estimate may be used with a proper increase added to it.

The highest increase in price is 12 percent (year to year), which is a rather reasonable raise. Usually, unless something is wrong with the economy, or unless the business field is quite unique, with metal man-

**TABLE 15-1 Cost-Estimating History**

| Vendor | Jan 92 | Jan 93 | Jan 94 | Jan 95 | Jan 96 |
|---|---|---|---|---|---|
| First | $3.25 | $3.65 | $3.65 | $4.80* | $3.95 |
| Second | $3.75 | $4.20 | $4.25 | $4.15 | $4.15 |
| Third | $2.60† | $2.75 | $2.85 | $2.85 | $2.90 |
| Fourth | $3.35 | $3.50 | $3.55 | $3.65 | $3.80 |
| Fifth | $3.25 | $3.50 | $3.80 | $3.80 | $3.90 |

*A sudden increase in price may have several reasons, such as (1) a mistake, (2) a new, inexperienced estimator, (3) a sudden surge in costs within that particular area, (4) the company does not want this business.

†Too low cost in comparison with other vendors may be caused by one (or more) factors: (1) cheap, but poor quality workmanship, (2) new vendor, making an attempt to pay entry into a field, (3) the company has a large stock of some material it is trying to dispose of, (4) an estimate copied over and over, each time with a slight increase, but overall wrong.

**TABLE 15-2 Percentages of the Cost History**

Percentages of increase or decrease were calculated on the difference between successive years.

| Vendor | Jan 92 | Jan 93 | Jan 94 | Jan 95 | Jan 96 |
|---|---|---|---|---|---|
| First | Base | +12 | ±0 | +31.5 | −21.5 |
| Second | Base | +12 | +1 | −2.5 | ±0 |
| Third | Base | +6 | +4 | ±0 | +2 |
| Fourth | Base | +4.5 | +1.5 | +3 | +4 |
| Fifth | Base | +8 | +8.5 | ±0 | +2.5 |

ufacturing, an increase of 6 to 12 percent on the average may be expected.

These percentages of change may allow us to estimate subsequent costs on the basis of a few old quotes. But beware, in several instances this may not be true:

- When ownership changes, companies may suddenly use completely different rates, with estimates coming out within a totally different range.
- Vendor ownership may change or a supplier may be replaced. Beware, as you may not know the new vendor's quality range.
- A mistake could be made and copied over a span of time, increasing slightly in value each year. True, the company still must perform the job for the estimated cost, but they may cut corners in order to fit within the estimated amount.
- An estimate may be made of another estimate. Someone looked at the object and came up with the price offhand. Such costs usually don't make it, and if the production is forced upon the vendor, the quality suffers.

Percentage rates may be calculated on additional information and will certainly prove to be very useful. For example, if one vendor is always slightly ahead of the other vendor in cost, but the quality of the first source is superior, the ratio of the usual difference between these two vendors may be established, and if a quote from one of them is received, the second cost may be proportioned. Sometimes it pays, if the product to be made is difficult to produce, to know such ratios of difference and opt for a vendor who is perhaps expensive but very thorough. In such a case, proportioned costs may often be used, since we still have the lower-pricing vendor to fall back on.

In order not to err in such an assumption, an additional cost adjustment may be appropriate. For this reason, the history of all costs over a period of time should be continuously compared to the estimator's own percentages of difference, to be able to assess a ratio of error.

If—for example—the cost was $3.35 in 1993 and $4.40 in 1995, there is a 31 percent increase in 2 years. However, if a new cost estimate arrives, pricing the job at $4.50, there is a difference of 3.5 percent, which is the deviation between the assumption and the actual estimate. Therefore, an estimate based on a percentage prorated cost variation should include an additional 3.5 percent for the protection of correctness.

Ratios between costs of various vendors may supply additional information. By assessing the difference between their quotes, we may find that one vendor is always 10 percent higher than the other, but the quality is always better, even though deliveries suffer. Or the quality may stink and deliveries are still worse.

Variations are endless, and each assessment should rate vendors accordingly, since only this way may the actual condition of the order be predicted. Perhaps sometimes it would be appropriate if the delivery is made slightly later, as long as the quality is there, and sometimes a lower quality suffices as long as parts are delivered on time, and so on.

### 15-2-2. Work intensity history

It surely pays to know how much work was done for the price charged. Sometimes drilling a hole in a part may produce a whole scale of various prices for such a task. A good estimator will always note such differences not only by judging the price difference from vendor to vendor but by assessing the amount of work performed.

Let's say we have a part as shown in Fig. 15-1. It is a punch, into the flange of which four holes have to be drilled. Their size is quite precise; therefore, we may immediately assume three passes of the tool with a single setup (1—spot drill, 2—roughing drill, 3—finishing

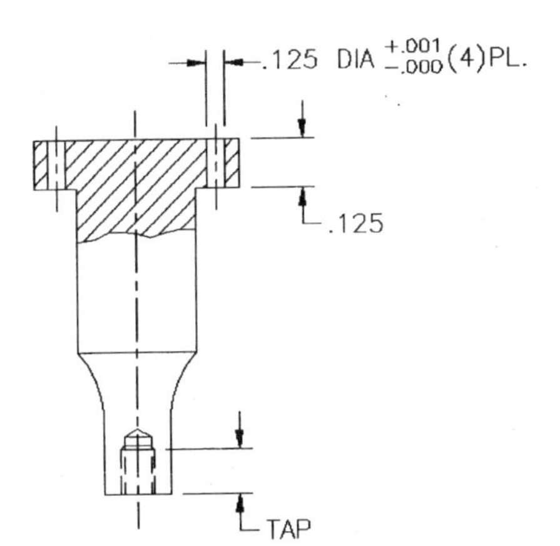


**Figure 15-1** Workpiece.

reamer). Off the other side, a tapped hole has to be added, which means that the part has to be removed from a fixture, reversed, and fixtured again. The hole has to be spotted, drilled, and tapped. This means two drilling passes, one tapping, and an additional setup cost. We may sum up the above information as follows:

| | |
|---|---|
| 2 setups | 12 units of work |
| 2 spot drills | 2 units of work |
| 2 drills | 3 units of work |
| 1 finishing tool (reamer) | 2 units of work |
| 1 tap | 3 units of work |

We assume that to drill a 1/8-in-thick punch flange with a 0.093-in-diameter drill may be done in a single pass. But with the 3/8-in depth of the tapped hole, the scenario may be different: We may need to drill such an opening at no less than two passes in order to accomplish a decent cut. Therefore, this hole will count as two drilling passes, as the list above indicates.

Three operational units for the tap drill are given in order to equalize its work rate with that of drills and a reamer. The tap drill will certainly run at a slower speed; therefore, its work cycle must be made proportional to the other tooling.

Assigning values to the above, we may figure that one unit of work is worth (in this case) 10 s of time, with each setup being equivalent to a minute. To assess the total cost of such a job, we simply add up all the units and multiply the number by a shop rate for the particular machine, adding a prorated overhead, or:

$$22 \text{ units} \times 10 \text{ s (each)} = 3.67 \text{ min}$$

Considering the shop rate for the particular machine to be $25 per hour, the amount of this cost per minute will be $0.4167.

Therefore,

$$3.67 \text{ min} \times 0.4167 = \$1.53$$

The cost of drilling one punch (shown in Fig. 15-1) will be $1.53. This cost, however, does not include the setup time of tooling, which has to be divided into the whole order or into applicable segments of it. To retain a single tool within a tool holder, along with the time to locate such a tool, will take approximately 3 to 5 min, depending on the order of a shop. We have five different tools (1—spot, 2—rough drills, 1—reamer, 1—tap), which utilize 6 tool-changing charges (Fig. 15-2).

Considering each tool change $TC$ at a rate of 5 min, the total cost is

$$TC = 0.4167(6\ S/U \times 5 \text{ min}) = \$4.40$$

where 0.4167 = tool shop rate per minute, or \$25/60

The cost of \$4.40 applies to the total order of pieces unless this amount is separated by partial delivery requirements. Assume we need 200 pieces of these parts, which finalizes the calculation

$$\begin{aligned} 200 \times \$1.53 \text{ each} &= \$306.00 \\ TC \text{ charge, all} &= \quad \$4.40 \\ \text{Total cost} &= \$310.40 \end{aligned}$$

To get the actual cost of each drilled punch, the total amount must be divided into the number of parts, to arrive at the adjusted price, which will include the tool change charge, as

$$\$310.40{:}200 = \$1.552 = \$1.56 \text{ each}$$

Obviously, the more pieces produced on the same setup, the more advantageous costwise such an arrangement is.

Where evaluating a cost submitted by another company, all the above items must be assessed the same way. The difference between the bare cost (\$1.56/piece) and their estimate is attributable to the

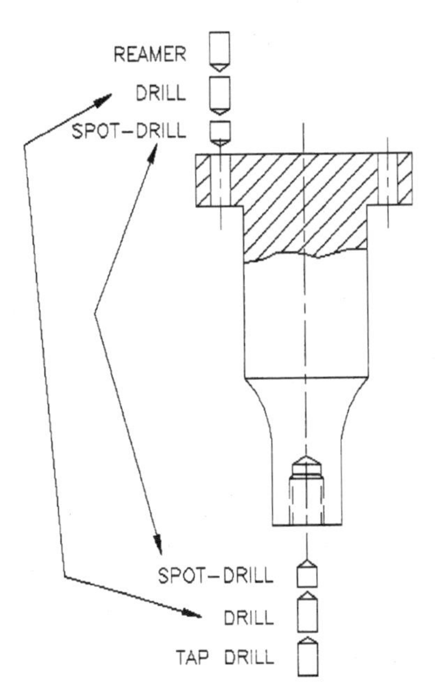


**Figure 15-2** Work sequence.

cost of overhead and company's profit. Overhead, on the average, may run anywhere from 150 to 350 percent and even up, with dependence on the type of work, type of product, range of processes and services it covers, and many other aspects.

Companies sometimes operate on a 60 percent margin, which means that the total manufacturing cost of the product makes 40 percent of the price. Some companies, however, may operate on a 50 percent or 40 percent margin or any other fractional value of the total cost. By a careful scrutiny of their estimates, this information may be extracted and used for our own purpose.

### 15-2-3. Additional costs

The above cost estimate is bare, without any expenses for inspection, overhead, company profit, and other costs and charges, whatever they may be. These naturally have to be added on in order to complete the calculation. Since these charges may vary from place to place, they are not included here.

Inspection costs are based on the accuracy of dimensions to be checked and the frequency of checking procedure (Table 15-3). Some production runs may need every 100th piece inspected, others every 25th piece, and some operations have to be inspected step by step, every piece and every addition to it. Dimensions with greater tolerance take less time to inspect than those that are tight in their tolerance ranges, because often a different set of checking instruments must be used. An approximate time comparison for checking dimensions is included in Table 15-4.

With a series of products to manufacture, usually a first-piece inspection has to be performed. This means that the first article or

**TABLE 15-3 Checking Frequency**

| Tolerance, in | Ratio—1 part to |
|---|---|
| Less than 0.0005 | ½ |
| 0.0005–0.001 | 1 |
| 0.002 | 5 |
| 0.003 | 10 |
| 0.004 | 15 |
| 0.005 | 20 |
| ±0.005 | 25 |
| ±0.010 | 50 |
| ±0.031 | 100 |

SOURCE: W. A. Nordhoff, *Machine Shop Estimating*, McGraw-Hill, New York, 1960. Reprinted with permission from The McGraw-Hill Companies.

**TABLE 15-4 Time in Minutes to Check One Dimension**

| Instrument | Min |
|---|---|
| With rule | 0.10 |
| Outside micrometer | 0.15 |
| Inside micrometer | 0.30 |
| Depth micrometer | 0.20 |
| Dial micrometer | 0.30 |
| Outside calipers | 0.05 |
| Inside calipers | 0.10 |
| Plug gauge | 0.20 |
| Snap gauge | 0.10 |
| Surface gauge | 0.20 |
| Thread snap gauge | 0.15 |
| Thread gauge (male, female) | 0.30 |
| Thread micrometer | 0.25 |
| Vernier calipers | 0.50 |

SOURCE: W. A. Nordhoff, *Machine Shop Estimating,* McGraw-Hill, New York, 1960. Reprinted with permission from The McGraw-Hill Companies.

first few pieces are scrutinized for any discrepancies so that the rest of the run will not come out of production faulty.

On machining of the first piece, the operator must submit it to inspection and cannot continue producing other parts until the first product is approved. Such inspection may be time-consuming, or the quality control department may be jammed up with additional work. In either case, the operator must stand by and wait, unable to change the setup in the machine because that would render the first-piece inspection useless. And unless another job can run on the same setup or another machine can be used to work on something else, it's just stand by and wait.

Such waiting time may add up to 25 percent of the total cost of that part. Naturally, in some instances the amount of standby time may be lower, but depending on the situation, it may also be higher.

Another set of work-influencing factors are personal adjustments, which usually run within 5 to 10 percent and perhaps even higher. This is the time an operator needs for personal survival of the workday, including trips to the lavatory, water drinking, etc.

Personal fatigue caused by the monotony of work is another factor slowing down the production. Where the work-learning curve is raising production rate with dependence on the time spent on a job, at a certain moment it is replaced by the boredom caused by a too well known set of movements.

Additional factors enhancing the fatigue and boredom may include:

- Semiautomatic work processes, where the handling time is much lower than the actual part-producing time. These losses may amount to 5 to 10 percent, depending on the ratio of handling to the work-producing time.
- Close-tolerance work of ±0.0003 to ±0.005 in and lathe operating cycles of 30 or fewer per hour. Such losses may run in the vicinity of 15 percent.
- Cycles, starting abruptly or fast, regardless of their total length. Extreme close-toleranced work, short cycles. Also blind hole drilling and tapping, filing, and bench work. The time loss may add up to 20 percent.
- Physically demanding work, plus deep hole drilling, high-speed milling, or operations with handling time much greater than actual machining time. Additionally, all hazardous procedures, such as sandblasting, torch cutting, and buffing. Up to 25 percent should be figured for such losses.

Some companies may include many such losses within their overhead allowance, along with the scrap rate, cost of inspection, cost of adjustments, etc. Such possibilities should be investigated prior to committing the basic cost estimate to further adjustments.

### 15-2-4. Machinability of materials

Machinability of materials presents another variable in the cost estimate. Not all steels may be machined equally fast, and the percentage of impairment their machinability index may cause should be included in every estimate as well (Table 15-5).

The main factors influencing machinability are the hardness of the material and the amount of some of its alloying elements. Materials of hardness approximately $HB = 180$ may be considered easily machinable, especially if their ductility is lower. High-speed steel tooling may use 25 percent higher cutting speeds than those generally given in charts for cutting of annealed materials. Machinability of cold-rolled material is approximately 20 percent better than that of hot-rolled stock.

The machinability of steel goes down with increase of carbon content. Lead, manganese, and sulfur improve the machinability, but greater amounts of sulfur may be detrimental to it. This applies especially to sulfur in the form of sulfides, which are found distributed throughout the material in the form of small but quite hard particles. These small pieces tend to dull the tooling, aside from the fact that sulfur enhanced the brittleness of materials.

**TABLE 15-5 Materials Grouped by Machinability***

| Group | Group name | Materials | | | Cutting speed, ft/min.† |
|---|---|---|---|---|---|
| I | Magnesium | Magnesium alloys:<br>Sand castings<br>Permanent-mold castings<br>Extruded bars, rods, and shapes<br>Die castings<br>Forgings | | | 2000 |
| II | Aluminum | Aluminum alloys:<br>Sand castings<br>Permanent-mold castings<br>Extruded bars, rods, and shapes<br>Die castings<br>Forgings<br>Phenolic | | | 1000 |
| III | Brass | Copper alloys:<br>Free-machining yellow brass<br>Phosphor bronze (free-cutting)<br>Bearing brass<br>Hardware bronze<br>Red bronze (80%)<br>Commercial bronze<br>Copper (leaded)<br>Brass forging<br>High-strength commercial bronze<br>Naval brass (leaded)<br>Government babbitt<br>Plastics<br>Formica<br>Micarta<br>Zinc alloys<br>Kirksite | | | 250 |
| IV | Screw stock | Carbon steels:<br>B-1006<br>B-1010<br>C-1005<br>C-1006<br>C-1008<br>C-1010<br>C-1012<br>C-1013<br>C-1015<br>C-1016<br>C-1017<br>C-1018<br>C-1019<br>C-1020<br>C-1021 | Free-cutting steels:<br>C-1106<br>C-1108<br>C-1109<br>C-1110<br>C-1111<br>B-1111<br>B-1112<br>B-1113<br>C-1113<br>C-1114<br>C-1115<br>C-1116<br>C-1117<br>C-1118<br>C-1119 | Miscellaneous:<br>Cast iron (soft)<br>Malleable iron (soft)<br>Leaded phosphor bronze (5%)<br>Yellow brass<br>Muntz metal<br>Red brass 85%<br>Red brass 80%<br>Naval brass<br>Tobin bronze<br>Manganese bronze<br>Hard rubber<br>Bakelite | 125 |

**TABLE 15-5 Materials Grouped by Machinability* (*Continued*)**

| Group | Group name | Materials | | | Cutting speed, ft/min.† |
|---|---|---|---|---|---|
| IV | Screw stock (*Cont.*) | Carbon steels:<br>C-1022<br>C-1023<br>C-1090<br>C-1095 | Free-cutting steels:<br>C-1120 | | 125 |
| V | Mild steels | Carbon steels:<br>C-1024<br>C-1025<br>C-1026<br>C-1027<br>C-1029<br>C-1030<br>C-1033<br>C-1034<br>C-1035<br>C-1036<br>C-1038<br>C-1039<br>C-1040<br>C-1041<br>C-1042<br>C-1043<br>C-1045<br>C-1046<br>C-1049<br>C-1050<br>C-1051<br>C-1052<br>C-1054<br>C-1055<br>C-1057<br>C-1060<br>C-1061<br>C-1062<br>C-1064<br>C-1065<br>C-1066<br>C-1069<br>C-1070<br>C-1071<br>C-1074<br>C-1075<br>C-1078<br>C-1080<br>C-1084<br>C-1085 | Free-cutting steels:<br>C-1125<br>C-1126<br>C-1132<br>C-1137<br>C-1138<br>C-1140<br>C-1141<br>C-1146<br>C-1151<br>Manganese steels:<br>1315<br>1320<br>1321<br>1330<br>1335<br>1340<br>Nickel steels:<br>2135(CR)‡<br>2320(A)§<br>2340(A)<br>2512<br>2515<br>2517<br>Nickel-chromium steels:<br>3120(CR)<br>3130(A)<br>3140(A)<br>3220(A)<br>3230(A)<br>3312(A) | Molybdenum steels:<br>4130(A)<br>4140(A)<br>4150(A)<br>4340(A)<br>4615(A)<br>4615(CR)<br>4620(A)<br>4620(CR)<br>4640(A)<br>4815(A)<br>Chromium-vanadium steels:<br>6120(A)<br>6130(A)<br>6135(A)<br>6140(A)<br>6150(A)<br>Miscellaneous:<br>Cast iron (medium)<br>Malleable iron (medium)<br>Meehanite (soft)<br>Copper (electrolitic)<br>Bronze (commercial 90%)<br>Phosphor bronze<br>Aluminum bronze<br>Spinning brass<br>Copper tubes<br>Copper wire | 75 |

**TABLE 15-5 Materials Grouped by Machinability* (*Continued*)**

| Group | Group name | Materials | | | | | Cutting speed, ft/min.† |
|---|---|---|---|---|---|---|---|
| VI | Medium steels | Nickel steels: | Molybdenum steels: | Chromium steels: | Nickel-chromium steels: | Stainless steels: | 50 |
| | | 2317 | 4017 | 5045 | 8615 | 303 | |
| | | 2320 | 4023 | 5046 | 8617 | 403 | |
| | | 2330 | 4024 | 5120 | 8620 | 405 | |
| | | 2335 | 4027 | 5130 | 8622 | 410 | |
| | | 2340 | 4028 | 5132 | 8625 | 412 | |
| | | 2345 | 4032 | 5135 | | 414 | |
| | | 2350 | 4037 | 5140 | 8627 | 416 | |
| | | 2515 | 4042 | 5145 | 8630 | 430 | |
| | | Nickel- | 4047 | 5147 | 8632 | 430F | |
| | | chromium | 4053 | 5150 | 8635 | 440F | |
| | | steels: | 4063 | 5152 | 8637 | | |
| | | 3115 | 4068 | E50100 | 8640 | | |
| | | 3120 | 4130 | E51100 | 8641 | | |
| | | 3130 | E4132 | E52100 | 8642 | | |
| | | 3135 | 4135 | Chromium- | 8645 | | |
| | | 3140 | 4137 | vanadium | 8647 | | |
| | | 3141 | E4137 | steels: | 8650 | | |
| | | 3145 | 4140 | 6120 | 8653 | | |
| | | 3150 | 4145 | 6130 | 8655 | | |
| | | | 4147 | 6135 | 8660 | | |
| | | 3220(CR) | 4150 | 6140 | 8719 | | |
| | | 3230 | 4317 | 6145 | 8720 | | |
| | | 3240 | 4320 | 6150 | 8735 | | |
| | | | E4337 | 6152 | 8740 | | |
| | | 3240(CR) | 4340 | | 8742 | | |
| | | 3250 | E4340 | | 8743 | | |
| | | E3310 | 4608 | | 8747 | | |
| | | 3312 | E4617 | | 8750 | | |
| | | E3316 | 4620 | | Miscellaneous: | | |
| | | 3340 | X4620 | | Cast iron (hard) | | |
| | | 3450 | E4620 | | Malleable iron (hard) | | |
| | | | 4621 | | Meehanite (medium) | | |
| | | | 4640 | | Cast steel (soft) | | |
| | | | E4640 | | Beryllium copper | | |
| | | | 4812 | | Drill rod | | |
| | | | 4815 | | Titanium: RC-70 | | |
| | | | 4817 | | | | |
| | | | 4820 | | | | |
| VII | Hard steel | Silicon-manganese steels: | Stainless steels: | Miscellaneous: | | | 35 |
| | | | 301 | 18-8 stainless (alloys 302, 304) | | | |
| | | 9255 | 302 | K-monel | | | |
| | | 9260 | 304 | Inconel | | | |
| | | 9261 | 308 | Heat-treated steels | | | |
| | | 9262 | 309 | (160.000–180,000 lb/in$^2$) | | | |

**TABLE 15-5 Materials Grouped by Machinability* (*Continued*)**

| Group | Group name | Materials | | | Cutting speed, ft/min.† |
|---|---|---|---|---|---|
| VII | Hard steels (*Cont.*) | Silicon-manganese steels: E9310, E9315, E9317, 9437, 9440, 9442, 9445, 9747, 9763, 9840, 9845, 9850 | Stainless steels: 310, 316, 317, 321, 329, 330, 347, 431, 440, 442, 443, 446, 501, 502 | Miscellaneous: Meehanite (hard), Cast steel (medium), Beryllium copper (heat-treated), Titanium: RC-130B, TI-150A, Timken, Konal | 35 |
| VIII | High-temperature alloys | Stainless steels: 420, 420F, 440A, 440B, 440C | | Miscellaneous: Cast iron (chilled), Heat-treated steels (180,000–200,000 lb/in$^2$), Cast steel (hard), Titanium: MST, Discalloy, Hastalloy B, Hastalloy C | 25 |

*Based upon high-speed-steel single-point turning tools.
†Cutting speeds are based on depth of cuts of 0.125, feed rates of 0.020 in/rev., and a tool life of 2.0 h.
‡CR = cold-rolled or cold-drawn.
§A = annealed.
SOURCE: W. A. Nordhoff, *Machine Shop Estimating,* McGraw-Hill, New York, 1960. Reprinted with permission from The McGraw-Hill Companies.

Generally, gray iron is easily machinable but may be hard and brittle and difficult to work with if rapidly cooled during its manufacturing process. With this type of material, the surface displays worse machinability than the core.

Machinability of materials is further influenced by the type of tooling used and its quality. Various tooling materials work at different rates of metal removal, with variation in the obtained surface finish as well. Some numerical differences between tooling materials are given in Table 15-6. The multiplying factor in this table clearly indicates the difference between tooling material qualities, and it should be applied to the given tool rates as listed in various charts.

**TABLE 15-6 Cutting Tool Material Multiplying Factor**

| Quality of tool material | Multiplying factor |
|---|---|
| Carbon tool steel | 0.50 |
| High-speed tool steel | 1.00 |
| Super-high-speed tool steel (Stellite, etc.) | 1.50 |
| Carbide (Carboloy, etc.) | 3.50 |
| Aluminum oxide and other ceramics | 7.00 |

Machinability of material is further influenced by the type of cutting tool used. Single-point tooling has the best stock-removing capacity, while drills and reamers are quite slow at cutting, using approximately half of the single-point tooling's cutting speeds. An additional factor applicable to all cutting speed charts compares different machining operations by showing percentage values of applicable speed, as shown in Table 15-7. This multiplying factor should be used with Table 15-5 to fine-tune the given values for the appropriate machining operation.

However, even with all these adjustments there may be situations when we cannot evaluate the actual process of producing a given part, as sometimes the surface finish of the cut or some other precision-related demands may lower the speed and feed of the tooling still further, throwing off all the expectations of the estimator.

The depth of the cut is another factor influencing the machinability of materials. Block 1/4 in thick will certainly be drilled differently than the same opening in 1-in block. Going through such a thickness, the tool cannot accomplish the task at a single pass. It must come down, return to dispose of chips, and come down again, sometimes repeating this procedure several times.

Spindle speeds and cutting speeds are given by a ratio:

$$C = \frac{\pi DN}{12} \qquad \text{or} \qquad N = \frac{12C}{\pi D}$$

where $C$ = cutting speed, ft/min
$D$ = diameter of work cutter, in (Fig. 15-3)
$N$ = number of revolutions of spindle, rev/min

Setup, fixturing, and support of the machined part are of extreme importance for its machinability. If the part is loose, machining feeds and speeds will be impaired by the amount of its chatter. Naturally, the size of the cutting tool is of importance as well, since small drills (and first-use drills in general) cannot operate at such cutting para-

**TABLE 15-7 Cutting Tool Machining Factors**

| Operation | | | | Factor | Max. spindle speeds |
|---|---|---|---|---|---|
| Turning | | | | 1.00 | No limit |
| Boring | | | | 0.75 | No limit |
| Broaching | | | | 0.25 | |
| Counterboring: | | | | | |
| Solid counterbores (piloted) | | | | 0.50 | No limit |
| Spot facer (piloted) | | | | 0.50 | No limit |
| Inverted spot facer (piloted) | | | | 0.50 | No limit |
| Back spot facer (piloted) | | | | 0.50 | No limit |
| Center reaming (solid center reamer) (for internal chamber or countersink) | | | | 0.50 | No limit |
| Countersinking (combination drill and countersink) | | | | 0.50 | No limit |
| Cutting off | | | | 1.00 | No limit |
| Drilling | | | | 0.50 | No limit |
| Start drilling | | | | 0.75 | No limit |
| Center drilling | | | | 0.75 | No limit |
| Forming | | | | 1.00 | No limit |
| Gear-shaping cutters | | | | 0.75 | 200–450 strokes per min |
| Gear-generating tools | | | | 0.75 | No limit |
| Gear-shaving cutters | | | | 1.50 | No limit |
| Hobs | | | | 0.50 | No limit |
| Hollow milling | | | | 1.00 | No limit |
| Knurling: | | | | | |
| Aluminum Magnesium Brass (500 ft/min) | Screw stock Steel, mild (150 ft/min) | Steel, medium Steel, hard (100 ft/min) | | | |
| Milling (general) | | | | 1.00 | No limit |
| Metal-slitting saws | | | | 0.50 | No limit |
| Pointing and facing tools | | | | 1.00 | No limit |
| Reaming: | | | | | |
| Ordinary reaming; reaming for size | | | | 0.75 | No limit |
| Reaming for high degree of finish | | | | 0.25 | No limit |
| Recessing tools: | | | | | |
| End cut | | | | 1.00 | No limit |
| Inside cut | | | | 0.75 | No limit |
| Threading (without lead screw) | | | | | |
| Dies | Aluminum Magnesium Brass | Screw stock steel, mild | Steel, medium steel, hard | | |
| (Self-opening) | 30 ft/min | 20 ft/min | 10 ft/min | | 250 |
| Dies (button) | 30 ft/min | 20 ft/min | 10 ft/min | | 250 |
| Taps (solid) | 30 ft/min | 20 ft/min | 10 ft/min | | 250 |
| Threading (with leadscrew) | | | | | |
| Dies | | | | | |
| (Self-opening) | 30 ft/min | 20 ft/min | 10 ft/min | | No limit |
| Dies (button) | 30 ft/min | 20 ft/min | 10 ft/min | | No limit |
| Taps (solid) | 30 ft/min | 20 ft/min | 10 ft/min | | No limit |

**TABLE 15-7 Cutting Tool Machining Factors (*Continued*)**

| Operation | Factor | Max. spindle speeds |
|---|---|---|
| Threading (single-point high-speed-steel tool) | 0.75 | 150 |
| Thread milling | 1.50 | No limit |
| Thread rolling | 1.00 | No limit |

SOURCE: W. A. Nordhoff, *Machine Shop Estimating,* McGraw-Hill, New York, 1960. Reprinted with permission from the McGraw-Hill Companies.

meters as heavy, sturdy, large-diameter tools. For more on cutting of materials, see Sec. 4-3 "Machining of Blocks."

Generally accepted working feeds for carbide single-point tools for turning, boring, and facing are presented in Table 15-8.

Feeds and speeds of cutting tools influence not only the condition of a cut, such as its surface finish, straightness, roundness, and concentricity, but also the tooling itself, causing it to wear and tear under unreasonable operating parameters. (See Figs. 15-4 through 15-6.) Tool life, if shortened unnecessarily, drives the cost of production high not only because of its breakage but also because of the greater than usual setup charges, tool-installation charges, number of rejects before the dulling or breakage is registered, or outright a number of ruined pieces and cost of their material.

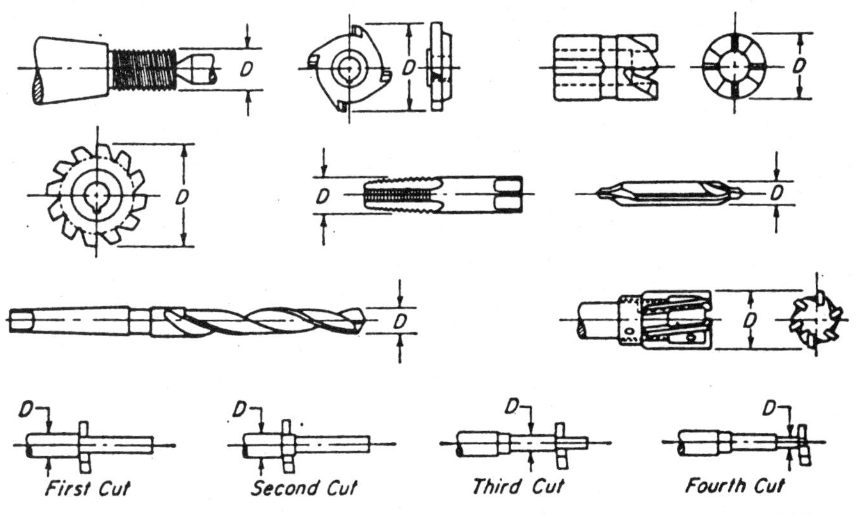


**Figure 15-3** Examples showing the value of $D$ in the formula $C = \pi DN/12$ (in). (*From W. A. Nordhoff,* "Machine Shop Estimating," *McGraw-Hill, New York, 1960.*)

**TABLE 15-8 Feeds for Single-Point Carbide Tools**

<table>
<tr><th rowspan="2">Materials</th><th colspan="2">Feeds</th></tr>
<tr><th>Rough</th><th>Finish</th></tr>
<tr><td>Carbon steels (10xx)</td><td colspan="2">0.015–0.020</td></tr>
<tr><td>Free-machining (11xx)</td><td colspan="2">0.010–0.020</td></tr>
<tr><td>Manganese (13xx)</td><td colspan="2">0.015–0.025</td></tr>
<tr><td>Nickel steels (23xx) (25xx)</td><td colspan="2">0.012–0.022</td></tr>
<tr><td>Nickel-chrome (31xx) (33xx)</td><td colspan="2">0.010–0.020</td></tr>
<tr><td>Molybdenum (40xx) (41xx)</td><td colspan="2">0.010–0.020</td></tr>
<tr><td>(46xx)</td><td colspan="2">0.015–0.030</td></tr>
<tr><td>Chromium (50xx)</td><td colspan="2">0.010–0.020</td></tr>
<tr><td>(86xx)</td><td colspan="2">0.010–0.020</td></tr>
<tr><td>Stainless steels</td><td>0.005–0.015</td><td>0.003–0.007</td></tr>
<tr><td>Titanium</td><td colspan="2">0.008–0.015</td></tr>
<tr><td>Heat-resistant alloys</td><td colspan="2">0.015 min</td></tr>
<tr><td>Cast iron</td><td>0.015–0.025</td><td>0.010–0.015</td></tr>
<tr><td>Malleable iron</td><td>0.015–0.020</td><td>0.010–0.015</td></tr>
<tr><td>Nickel alloys (monel, K-monel)</td><td>0.010–0.020</td><td>0.003–0.010</td></tr>
<tr><td>Copper alloys:</td><td></td><td></td></tr>
<tr><td>Free-cutting</td><td>0.007–0.020</td><td>0.005–0.009</td></tr>
<tr><td>Average machinability</td><td>0.007–0.018</td><td>0.003–0.008</td></tr>
<tr><td>Difficult to machine</td><td>0.003–0.015</td><td>0.003–0.005</td></tr>
<tr><td>Aluminum alloys</td><td>0.007–0.012</td><td>0.003–0.008</td></tr>
<tr><td>Magnesium alloys</td><td>0.010–0.040</td><td>0.005–0.010</td></tr>
<tr><td>Plastics</td><td colspan="2">0.003–0.008</td></tr>
</table>

SOURCE: W. A. Nordhoff, *Machine Shop Estimating,* McGraw-Hill, New York, 1960. Reprinted with permission from The McGraw-Hill Companies.

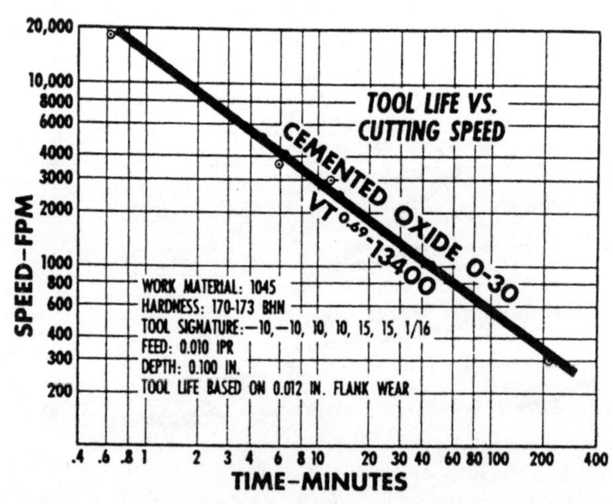


**Figure 15-4** Cemented oxide tools show a straight tool-life over a broad range of speeds. (*From R. G. Brierley and H. J. Siekmann,* "Machining Principles and Cost Control," *McGraw-Hill, New York, 1964. Reprinted with permission.*)

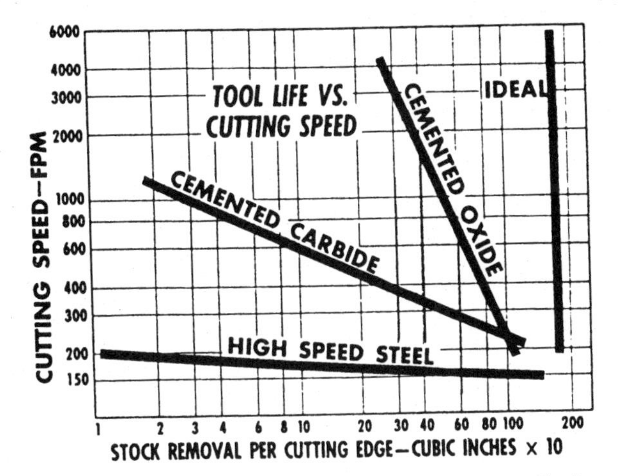


**Figure 15-5** Modern cutting-tool materials approach the ideal concept of a constant amount of stock removal per cutting edge at any speed. (*From R. G. Brierley and H. J. Siekmann,* "Machining Principles and Cost Control," *McGraw-Hill, New York, 1964. Reprinted with permission.*)

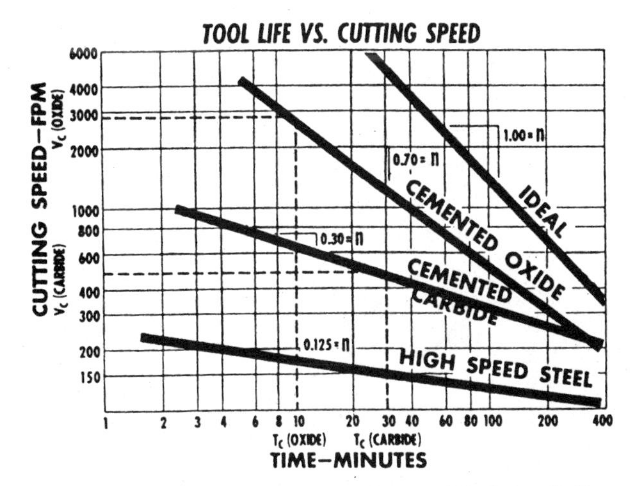


**Figure 15-6** Different classes of tool materials have distinctively different tool-life line slopes. (*From R. G. Brierley and H. J. Siekmann,* "Machining Principles and Cost Control," *McGraw-Hill, New York, 1964. Reprinted with permission.*)

**TABLE 15-9 Normal Tool Life, High-Speed Tool Steel**

| Type of cutting tool | Aluminum brass | Mild steel | Hard steel |
|---|---|---|---|
| Drills: | | | |
| Under 1/16 in diameter | 30.0 | 20.0 | 10.0 |
| 1/16–1/2 in | 45.0 | 30.0 | 15.0 |
| 1/2 in and over | 60.0 | 45.0 | 30.0 |
| Taps: | | | |
| Under 1/16 in | 60.0 | 40.0 | 20.0 |
| 1/16–1/2 in | 90.0 | 60.0 | 30.0 |
| 1/2 in and over | 120.0 | 90.0 | 60.0 |
| Milling cutters: | | | |
| Metal-slitting saws<br>End mills under 1 in diameter<br>Slotting cutters 3/32 in and under<br>Woodruff key seaters | 60.0 | 45.0 | 30.0 |
| Slotting cutters over 3/32 in<br>Plain-milling cutters<br>End mills over 1 in diameter<br>Angle-milling cutters | 120.0 | 90.0 | 60.0 |
| Hobs: | | | |
| Gear<br>Spline<br>Serration<br>Thread | 360.0 | 270.0 | 180.0 |
| Turning and boring tools<br>Shaper tool<br>Thread chasers | 120.0 | 90.0 | 60.0 |

SOURCE: W. A. Nordhoff, *Machine Shop Estimating,* McGraw-Hill, New York, 1960. Reprinted with permission from The McGraw-Hill Companies.

A comparison of tool life is given in Table 15-9 in an attempt to relate the actual situation to that written on paper or, worse, estimated. The amount of tool life is given in arbitrary units.

### 15-2-5. Cost of material

The cost of material was not included in the calculation in Sec. 15-2-2 "Work Intensity History," since the basic part, the punch, was being supplied by the vendor. However, where the material has to be obtained, its pricing must be included in the cost.

Material cost should be handled in the same fashion as any other cost. Prices of different suppliers may be prorated, to assess the percentage difference among various sources. In order to calculate a whole range of steels from a single material estimate, the financial difference between materials has to be assessed as well. After all, alloying elements such as nickel, chromium, vanadium, titanium, and others increase the cost of material in approximately the same way. Also their pricing is rather consistent, unless a disaster strikes the place on

**TABLE 15-10 Comparison of Cost of Steel Materials**

| Material | Alloying elements | Percentage of price |
|---|---|---|
| 1300 | 0.28–0.33 C, 1.6–1.9 Mn | 100 |
| 2517 | 0.15–0.2 C, 0.45–0.6 Mn, 4.75–5.25 Ni | 189 |
| 3135 | 0.33–0.38 C, 0.6–0.8 Mn, 1.1–1.4 Ni, 0.55–0.75 Cr | 122 |
| 4130 | 0.28–0.33 C, 0.4–0.6 Mn, 0.8–1.1 Cr, 0.15–0.25 Mo | 118 |
| 4615 | 0.13–0.18 C, 0.45–0.65 Mn, 1.65–2.00 Ni, 0.2–0.3 Mo | 140 |
| 5120 | 0.17–0.22 C, 0.70–0.90 Mn, 0.70–0.90 Cr | 110 |
| 6120 | 0.17–0.22 C, 0.70–0.90 Mn, 0.70–0.90 Cr | 140 |
| 9260 | 0.55–0.65 C, 0.70–1.00 Mn | 111 |

SOURCE: W. A. Nordhoff, *Machine Shop Estimating,* McGraw-Hill, New York, 1960. Reprinted with permission from The McGraw-Hill Companies.

earth where such and such an element is mined or otherwise obtained.

For example, should we consider SAE No. 1300 alloy to be a basis for the new comparison chart, we may use such a base for development of other materials' prices, as shown in Table 15-10.

Cold-rolled stock is usually higher in cost than hot-rolled material. The difference may be in the neighborhood of 20 percent, depending on the actual mill process. It is worthwhile to question the sales personnel of a particular mill and find out exactly what's involved in each manufacturing method.

The cost ratios presented in Table 15-10 should be considered arbitrary, as every mill may have a slightly different set of prices, which will vary with dependence on sources of ingredients, the manufactured amounts, the demand for the particular material, and many additional factors tied to its production and sale. The percentages and monetary values given throughout this chapter are merely guidelines for a person who is eager to do not only die design, die manufacturing, and die-related production, but the costing as well. These numbers have to be adjusted or fine-tuned for each applicant's scenario and sources.

Prorating of materials according to their ingredients and manufacturing processes may serve as yet another set of guidelines for all subsequent quoting work. Alloying elements and their quantity should be of special interest, as inclusion of some will considerably influence the final cost. We may observe that wherever a higher percentage of nickel is added, the cost of such steel jumps up. The same applies to molybdenum.

Nickel is obtained mostly from nickeliferous and cupriferous pyrrhotite, by smelting in blast furnaces. Molybdenum is a metallic element of the chromium group, occurring in a combination of molybdenite, wulfenite, and molybdite. Molybdenum is of American origin (Arizona, Montana, Nevada, Utah), but nickel is produced by Myanmar (formerly Burma), China, Albania, Botswana, Indonesia,

Zimbabwe, New Caledonia, and similar countries, aside from some Canadian and Norway production. Naturally, following the world news should prove helpful to the eager estimator.

In the evaluation of any particular material quantity needed for production, the so-called scrap rate should not be overlooked. Scrap rate is the percentage amount of ruined and otherwise destroyed material, which is usually used for testing, for trial runs, or that which is accidentally crushed, ruined, etc. Scrap rate may also contain material supplied in larger than necessary sizes, where a certain percentage must be added to a rough cut, to be later removed by the machining process. Various types of materials and their shapes may use the stock-removing allowances shown in Table 15-11.

Scrap rate may run in percentages of 5 percent and up, with dependence on the type of production, sensitivity of handled material, training of the personnel, and numerous other aspects.

### 15-2-6. Evaluation

In spite of all guidelines and tables presented here, and elsewhere, the estimator's job is never an easy one. First of all, every evaluation is based solely on assumptions. Existing time studies don't help, because they always pertain to jobs already done, in which case they were studied after a number of successive repeats, when the operator had already acquired considerable skill doing the same job over and over. The estimator cannot depend on such time and monetary values. Unknown obstacles the new production presents and little surprises that come out of hiding in the woodwork once the job hits the shop floor remain unpredictable.

Naturally, this applies to production runs of multitudes of pieces. With an individual work, or research and development projects, which die design and die building often are, every new assignment may be completely different from the previous one. An approach that worked well once may be useless the second time around, and the

**TABLE 15-11 Stock Allowances for Rough Cutting, to Be Removed in Machining**

| Material type and shape | Amount to be added for machining (per cut), in |
|---|---|
| Hot-rolled stock, general | 0.125–0.375 |
| Cold-rolled stock, general | 0.015–0.187 |
| Cold-rolled flat stock, thickness | 0.015–0.187 |
| Cold-rolled flat stock, width | 0.015–0.093 |
| Round stock, length | 0.062–0.187 |
| Any stock used for chucking | 1.25–2.00 |

speed of a work process is determined more by a lack of difficulties encountered than by the preventive treatment of their expected appearance.

Further, all textbook types of calculations and evaluations don't work the same way for each particular machine shop. In some places the order of the shop floor may be exemplary, with properly marked tools placed within well accessible cabinets. Other places may have chaos instead of an organizational strategy, with toolmakers borrowing the tools from each other, with many tool sizes missing, or with chipped or dull tools that nobody sharpens. Factors such as these may never be included in the form of any ratios or as tabulated. These influences are unaccountable, as they change with the day, with the job, with the number of people who decided to appear for work.

Some estimators like to work by the book, and they cannot be blamed, because it is much easier. However, how much will these nicely added up costs of all necessary operations fare when compared to actual shop floor values? Does the company really support the formulas they're using? Is the leadership willing and able to provide a workplace where things can be done such a way? And do the workers appreciate such efforts? Are they really conscientiously striving to produce their best? Or is it all just one large utopia depicting how we all would like to work and operate?

If the latter is true, a company is heading for ruin, because estimates, no matter how much estimated they are, must be in congruence with actual numbers. How else can a firm make a profit if the real time to make an item is twice its estimated value?

For these obvious reasons, sometimes it pays to further evaluate each job visually as well, imagining operations as they are performed, trying to foresee obstacles and assess parameters and guidelines. If the formula says it should be done in 5 minutes—can it really be done at such a time? Well, perhaps Mike would do it if he gets the job, but Tony will never fit into such numbers. Yet Tony delivers high-quality workmanship and Mike tends to be sloppy at times. Comparisons such as these are additional factors influencing the work flow and its quality.

When it comes to such influences, the work history and history of the work intensity may prove invaluable for every estimator, designer, engineer, industrial engineer, and everybody else up to the general manager of the plant. These records, when continuously updated and compared with real outcomes of previously estimated jobs, will give the historically correct information about all work already done. In such a case it will be much easier to estimate a job which looks like something else, something already known and familiar. In this way the history may guide and focus the estimating process, perfecting the outcome up to a total flawlessness.

Needless to add, all records should be written, preferably stored in a computer. To depend on memories of those who were then involved in such and such a process is a highly risky and inaccurate manufacturing method. It may suffer from impairment caused by an occasional memory loss, or become inappropriately influenced by seasonal activities such as vacations, or be severely crippled by the particular key person not showing up for work.

Where the actual records are found different from the basic estimate, a correction factor should immediately be established and used in subsequent quotations. Such a correction factor may be calculated by using these formulas:

$$CF = 100\,\frac{A}{H} \qquad \text{or} \qquad CF = 100\,\frac{A}{E}$$

where $CF$ = correction factor, percentage
$A$ = actual data
$H$ = historic data
$E$ = estimated data

Some may argue that such a procedure is very time-consuming and costly, which may or may not be true. It all depends on the way such a method is set up, and it all depends on the equipment available for the purpose. Where the estimator may use a computer, which is now mostly the case, any basic spreadsheet program such as Lotus 1-2-3 may do the job. If properly formatted, the software will calculate the appropriate percentages, ratios, additions or subtractions, and similar mathematical tasks by itself, as soon as the basic numerical input is typed in.

Sometimes such a procedure is much cheaper than sending out desperate faxes begging for a quote, and wasting precious time and stomach lining over the cost that is just not coming in.

More advanced computerized help may be obtained by employing any programming language, such as C-programming or C++. Such assistance is immeasurable, especially where whole charts of information are created, evaluating an array of outcomes from a limited numerical input. By giving the basic solid information to the computer, the compiler spits a complex charted answer in a matter of seconds, provided, of course, that the initial set of directing commands is correct.

**Quality control input.** The quality control department should have a definite say in the estimating procedure, even though not many estimators realize it. We should not forget that all the incoming parts, materials, assemblies, subassemblies, and other objects to be included in the manufacturing process first land in the quality control department. There they are checked out, compared to drawings or specs, and on the basis of their compliance are either released into production or rejected.

An experienced quality control inspector knows each particular vendor's strength and weakness, who to give a deep-hole drilling job, and who will mess it up. The inspector can pinpoint the shop capable of producing high-precision parts and avoid outfits that they just slap things together and hope for the best. Quality control people also know which company is responsive and which company's representative is rude or uncooperative or just never there.

An input of these people should be emphasized, as their experience with various suppliers may be worth a lot, if not exactly in monetary terms, then certainly in measures of avoidance of problems.

**The lowest-cost syndrome.** Some manufacturers tend to think that the vendor who submits the lowest cost is the best candidate for the job. This, however, may not always be true. Some vendors are low in cost and high in quality of their output, and other vendors are low in cost and still lower in quality. There can be jobs which two different vendors may accomplish at a price difference of hundreds of dollars. But there is a chance that thousands may be spent on bringing the produced objects within the drawing's requirements.

A manufacturer who goes to faraway countries to produce dies may save some money on such expensive tooling, but in the end, may be lucky to show a profit. After all tariffs, duties, and taxes, plus shipping costs and currency-converting fees are paid, the expected difference in price may easily disappear. The die itself may be another disappointment, for the faraway shop knows well that the overseas customer will not be returning it for repair, and some shops may even count on it. They may use inferior materials or inferior manufacturing methods, knowing that these are extremely difficult to assess and to prove. The life of such tooling may be impaired just there, at the beginning. Materials may chip off, peel off, become fractured, rusty, and who knows what, and the die that was to produce parts for many years to come may not survive two winters.

For example, in plastic injection molds' manufacture, some faraway masterminds use epoxy to "repair" cracked or faulty steel blocks so that visually the surface arrives perfect. At any rate, not much is to be seen on an assembled mold, and to take it apart is too costly and time-consuming. Well, the epoxy soon melts away, when exposed to a hot plastic material, and suddenly products emerge with craters, gaps, and valleys and all kinds of distortions, and nobody's the wiser as to where they came from.

Therefore, it may perhaps be advisable to keep a local vendor who produces parts of a good quality, even though the prices are not always what we may want them to be. Maybe the vendor's workload is prohibitive. However, it pays to investigate these possibilities and to work with such a source on an improvement. After all, a really good vendor will meet the other party halfway, provided an honest effort is produced.

## 15-3. Die Building Estimates

Included below are some personal observations which may be used as guidelines in die building and die design estimating. However, some adjustments for the particular application may be required, since not all people work the same way and therefore not all work is ever produced at the same speed.

We may begin with an assumption that every article produced in a die will have some alterations done to its outer shape as well as to its inner surface area. Some parts are pierced, some are formed, and others are drawn. On the average, to produce a single punch and corresponding die button and stripper guide for a piercing station always takes a certain number of hours, with dependence on the size of tooling. According to my experience, every linear distance of 10 mm, or 0.394 in (roughly 0.4 in), as produced by a single continuous cut, takes the average of some 2 h for a set of tooling, described previously. The cut may be straight or curved, as long as it is continuous. With round tooling, its circumference should be calculated to obtain its linear distance. With square or rectangular cuts, each side of the square or rectangle will count as a separate cut.

On the basis of such a rough assessment, time values for different sizes of tooling may be calculated as shown in Table 15-12. Here the time intervals in hours are given with respect to various sizes of the punch and with consideration of quantities of the same tooling to be produced. A time interval to machine a 0.375-in-diameter punch and die and guide bushing is figured as follows:

$$\text{Circumference} = \pi D = \pi(0.375\text{ in}) = 1.178\text{ in}$$

$$1.178\text{ in} \div 0.394\text{ in} \doteq 3$$

$$3 \times 2\text{ h} = 6\text{ h}$$

We need a total of 6 h to produce this size of tooling; with two punches of the same size it will be 12 h; three tooling sets will be 18 h, etc.

To machine punch plate, die block, stationary stripper, and two backup plates for the above tooling, the values are given in the chart with dependence on sizes as well. The basic amount of time is 20 h for the 1/4-in-diameter punch. This interval increases 1 h with every 1/8-in enlargement of the punch diameter size. A set of plates for a 0.5-in-diameter station will take 23 h, which consists of 20 basic hours plus 1 h for each 1/8-in increase in diametral size.

With multiple tooling of the same shape and size, each additional tool within the same set of blocks not only will increase their proportions but will demand more machining as well. Such an increase usually amounts to +5 h for each additional tool set.

From Table 15-12, to build a die consisting of three 0.5-in-diameter punches will amount to

$$3 \times 8 \text{ h} = 24 \text{ h to machine punches, dies, and stripper guides}$$

$$23 \text{ h} + (2 \times 5 \text{ h}) = 33 \text{ h to make all necessary plates}$$

The total will come to 57 h.

Another example: A die with five 1-in-diameter tools is calculated similarly:

$$5 \times 16 \text{ h} = 80 \text{ h for punches, dies, stripper guides}$$

$$27 \text{ h} + (4 \times 5 \text{ h}) = 47 \text{ h for all plates}$$

The total time to make such die: 127 h.

Small fragile punches are more difficult to handle, which impairs their machining time. For that reason their time values were

**TABLE 15-12 Die-Building Cost Estimate**

| Hole diameter, in | Linear dimensions or circumference, in | To build punch and die and guide, h | Punch plate, die block, backup plates, h for a number of tools below | | | | | |
|---|---|---|---|---|---|---|---|---|
| | | | 1st | 2nd | 3rd | 4th | 5th | 6th |
| 0.125 | 0.393 | 1.50(2) = 3 | +20 | +7.5 | +7.5 | +7.5 | +7.5 | +7.5 |
| 0.25 | 0.785 | 1.25(4) = 5 | +21 | +6.25 | +6.25 | +6.25 | +6.25 | +6.25 |
| 0.375 | 1.178 | 6 | +22 | +5 | +5 | +5 | +5 | +5 |
| 0.5 | 1.571 | 8 | +23 | +5 | +5 | +5 | +5 | +5 |
| 0.625 | 1.963 | 10 | +24 | +5 | +5 | +5 | +5 | +5 |
| 0.75 | 2.356 | 12 | +25 | +5 | +5 | +5 | +5 | +5 |
| 0.875 | 2.75 | 14 | +26 | +5 | +5 | +5 | +5 | +5 |
| 1 | 3.142 | 16 | +27 | +5 | +5 | +5 | +5 | +5 |
| 1.125 | 3.534 | 18 | +28 | +5 | +5 | +5 | +5 | +5 |
| 1.25 | 3.927 | 20 | +29 | +5 | +5 | +5 | +5 | +5 |
| 1.375 | 4.32 | 22 | +30 | +5 | +5 | +5 | +5 | +5 |
| 1.5 | 4.712 | 24 | +31 | +5 | +5 | +5 | +5 | +5 |
| 1.625 | 5.105 | 26 | +32 | +5 | +5 | +5 | +5 | +5 |
| 1.75 | 5.498 | 28 | +33 | +5 | +5 | +5 | +5 | +5 |
| 1.875 | 5.89 | 30 | +34 | +5 | +5 | +5 | +5 | +5 |
| 2 | 6.283 | 32 | +35 | +5 | +5 | +5 | +5 | +5 |

increased accordingly, with 150 percent increase across the board for 0.125-in-diameter tool and plates set and 125 percent increase for 0.250-in-diameter tooling.

These guidelines need not be restricted to cutting tools. Actually, all punches and dies may be calculated using the same approach, with few exceptions, listed in Table 15-13.

The cost of material and the cost of any additional expenses such as heat treatment, cost of the die shoe, springs, mounting hardware, switches, cams, and all other single parts or their subassemblies must be added to the above assessment. Also not included is the cost of any special alterations of the blocks, such as nests, lift pins, and openings for ejection devices.

But—since not every toolroom produces the tooling at the same rate as shown above, a good amount of personal observation and comparison of the above values to the actual situation should be attempted first. A subsequent modification or fine tuning of this data should follow if necessary. Blank charts are provided in Appendixes A and B for such use.

## 15-4. Design and Development Costs

The cost of research, design, and development is often impossible to estimate. For that reason many research activities are government-subsidized, either in the form of straightforward allocation of funds to universities and colleges or as grants and other funding for a variety of privately conducted research. Such a solution is outright necessary, because as much as research and development are necessary, they are almost equally unprofitable.

**TABLE 15-13 Time Values for Building of Various Die Elements**

| Tooling to be produced | Time needed |
|---|---|
| Forming, shaving | 100–120% of HPS* |
| Drawing | 120–150% of HPS* |
| Guiding inserts | 110–120% of HPS* |
| Embossing, extruding, swaging | HPS* |
| Strip stop, manual and automatic | Each: 5–10 h |
| Spring stripper | Add: 15–40 h |
| Knock-outs, nests, grip stripper | Each: 15–25 h |
| Horn, cam, side-acting tool | 30–60 h |
| Pinch trim, curling tool | 150–200% of HPS* |

*HPS = hours per size, as in Table 15-12.

Yet, every manufacturer and every producer must conduct some sort of research and new product development, or his production standards will lag behind until finally run over by the competition. For today's industry, the design of new parts, products, or systems, and their development is an absolute must; it is a matter of survival.

Various companies comply with these demands differently. Some charge the cost of research and development (further R&D) toward their overhead, embedding it within the piece price. Others keep these expenditures separately, hoping that the price will still look good enough to be deemed acceptable, and yet they would be able to fit—miracle provided—such activities within its range.

Design and R&D is an ongoing process, lasting a whole year, and a year after a year. Designers and engineers should always be on the lookout for possible improvements and innovations, and the company's leadership should always be patient enough to lend them an ear and sincerely evaluate their proposals and ideas. Because anyone who stands still and does not progress is actually moving backward. All the others will precede him soon, no matter how slowly they move.

Perhaps the best way to account for the design and R&D, if not proportioning its cost within the piece price, is to charge a flat rate. A simple die design will be a flat rate of—let's say—two thousand; a more complex die, twice the amount; a superdifficult die, a multiple of the basic flat rate.

The development itself is a slightly different story. Here we have more people involved, which brings an equal number of ideas, which are continuously evaluated, discussed, and evaluated and discussed again. Perhaps it would be advisable to use a rule of thumb and make two assessments of a time interval needed to come up with a solution to a given problem. One assessment takes a view of an optimist who sees the world through rosy glasses and considers a rare instance where everything works fine the first time. Then take a view of the pessimist and imagine all obstacles lurking in the background and waiting for an appropriate, or rather for the most inappropriate moment to strike.

These two ranges should give a nice overview of the situation, with the mean between the two extremes, or any other angle of preference, serving as the basis for an estimate. However, we should never forget that where two solutions to a given problem are present, the worst of them usually occurs. A view of a realistic optimist will then be to take the worst situation for valid and be pleasantly surprised if it turns out slightly better.

A problem of the cost of new development should be assessed on the basis of work-hours. How many hours are needed to produce a punch and die assembly is fairly well known. If the punch and die is complex and needs some grooming in between, the number of hours spent on such job should be expected to rise in accordance with its complicity. The total number of estimated hours should be multiplied by 3 to give a range of the time span for the development of an unknown product.

The reason behind multiplying the number of hours by 3 is as follows: one-third will be worker(s) salary, one-third will be overhead, and one-third will go against the company's profits. If one of the thirds falls short, the existence of the other two may compensate. If one of the thirds soars, it may provide some cushioning for the next time.

Another approach to the design and development evaluation is to charge a percentage of the actual die cost. For example, if building a die that is to blank and draw the part, 25 to 45 percent of the hours needed to build the drawing station alone should go against the design and development. With a blanking and forming die, the forming station will be used in such an assumption. As a rule, always the most difficult station made should serve as the basis for design and development cost assessment.

Naturally values may change from company to company. It would be worthwhile to design several methods of evaluation, use them consistently, and compare results with the actual hours spent on the job.

Considering the basic die design, a simple and straightforward die, consisting of 3 to 6 stations (some may be empty) will take approximately 8 to 10 h to come up with the strip layout, as positioned on the die block. This includes all calculations, economical evaluation, and a rough blank calculation. Another 15 to 25 h will be needed to produce detail drawings, write bills of material, and do the necessary paperwork.

With a more complex die, the initial strip layout may well take 15 to 20 h, with an additional 20 to 30 h for details and paperwork.

However, this estimate may be considered valid only where no complications are involved, where all punches and dies are fairly simple, with no tryout runs needed, and where dealing with no new or unpredictable materials or either complex or unknown processes.

## 15-5. Estimate Format and Terminology

Every submitted estimate must reflect not only the price ranges for the services rendered but also other terms of the production and its delivery, the time span needed to produce the job, where to deliver, who will pay for what, the number of pieces delivered, and their packaging. Otherwise such demand may be summed up as:

who-does-what-for-whom

how-why-where-when-at what price

under which conditions and restrictions

provided payment is made, when and how and in which currency

The requirements considering all the above factors may be numerous and if not properly evaluated for their full impact, may greatly impair the meaning of the estimate, with subsequent injury to the actual production of the merchandise and financial standing of the firm. Beware of conditions not specified in the estimate—they may be implied, which does not always work in your favor.

### 15-5-1. Delivery timing

The time element is of extreme importance in every estimating job. Not only is it applicable to the actual length of time needed to produce the parts. The delivery date must further include the time to receive raw materials and the time to inspect products and do adjustments if they are necessary.

The actual span of time before the arrival of parts at the client's door is vital with reference to the time interval promised: if 5 to 6 weeks are estimated for the delivery, which is more valid, 5 weeks or 6? And what does such a number of weeks really mean—is it the time the products leave the gate of the manufacturer, or does it mark the time of their arrival at the client's place?

Is it possible that the receiving party expects their quality control procedures to be included in those 5 to 6 weeks? In such a case the 5 to 6 weeks' interval will mean the time of the physical placement of parts or tooling in service on the production line. In such a scenario, the conditions of sale may be very disagreeable if—for example—the client's quality control finds some problems and wants them repaired prior to the end of the 5 to 6 weeks' period.

Even though such delivery arrangements are not common, they may sometimes be demanded, especially where a delivery to the heavily scheduled production line is being made. Estimators must therefore be aware of all such little ambiguities and possible misunderstandings, which they may not be familiar with but which nevertheless exist. They must carefully guard the wording of the quotation, specifying each term in clear and understandable language. After all, an estimate becomes a binding contract once accepted.

The form of submission of the quotation should be somewhat standardized and well thought over. It may utilize generally known statements, such as:

- COD, cash-on-delivery or collect-on-delivery, means the merchandise is to be deposited with the receiving party only after all costs are settled with either the shipper or the vendor's trucker. Such costs, payable in cash, will certainly consist of the shipping charges, usually combined with the total cost of the merchandise. In such a scenario, the buyer carries the risk of losing the shipment and finances as well should an accident happen on the way and the goods be destroyed.

- FOB, or freight-on-board or free-on-board, means that the cost and shipping terms of the estimate are valid up to the point when the merchandise is placed with the shipping forwarder. At that moment, a copy of the bill of lading must be submitted to the buyer, who takes full responsibility for the shipment and its safety from that point on. FOB New York means that the merchandise is brought to the shipyard (or a train station or a trucking terminal) in New York, and all subsequent expenses and problems are the buyer's responsibility.
- FAS, or free-alongside-ship, has a similar meaning to FOB.
- EX, or ex factory, ex warehouse, ex city, or ex origin are additional clarifications of FOB.
- CIF, or cost-insurance-freight. CIF New York means that the merchandise is delivered to the New York port (or railroad station or trucking company) with its passage and insurance fully paid by the seller up to that point. If such merchandise is coming from Africa, the cost of shipping includes the cost of its ocean passage. The buyer must receive from the seller the bill of lading, an invoice, an insurance agreement, and a receipt, confirming the payment of shipping costs.
- C and F, a variation of CIF, meaning cost-and-freight, without insurance.
- ARO—on receipt of an order, or after receipt of an order.
- Bill of lading is an agreement between the seller and the shipper specifying the terms of the shipping contract, while at the same time serving as the receipt for the goods accepted by the shipper. A copy of the bill of lading must be forwarded to the buyer, which indicates ownership. Anyone presenting a bill of lading to the shipper may claim the merchandise.
- Lead time is the length of the production period. There are various types of lead times. There may be a lead time on design, lead time on the first piece inspection, or lead time on a production run. Each of these time intervals covers a different span of time, specified in detail within the estimate, as shown in Fig. 15-7.
- Terms: Net 30 days. This phrase means that if the seller receives full payment within 30 days after the delivery of goods, the price will be as shown. Any additional time span will accumulate an interest per-

(Today's date)

1. Die design, per Dwg. No. AK-386-C, Rev. C — $3,500.00
   Lead time: 4 weeks ARO
2. Die manufacture, per design shown above — $40,000.00
3. Twelve (12) preproduction samples submitted for inspection
   Lead time: 6 weeks after design approval
4. Production run:

| | |
|---|---|
| 1,000 Pcs. | $3.95 Ea. |
| 5,000 Pcs. | $3.50 Ea. |
| 10,000 Pcs. | $3.15 Ea. |
| 50,000 Pcs. | $3.10 Ea. |
| 100,000 Pcs. | $3.05 Ea. |

***Delivery schedule:***

| | |
|---|---|
| First 1,000 Pcs. | 1 week after approval of preproduction of samples |
| Each 10,000 Pcs. | 2–4 weeks production lead time |

***Method of payment:***

One-third (⅓) of the total amount (or $14,500.00) to accompany the order

One-third (⅓) of the total amount (or $14,500.00) payable on receipt of preproduction samples

One-third (⅓) of the total amount (or $14,500.00) payable on the approval of preproduction samples

***Terms:***

Net 30 days.
All prices are quoted in US dollars
Delivery terms: FOB origin
Expiration of the quote: 2 months from the date shown above

**Figure 15-7** Cost estimating format—lead time values.

centage. Sometimes the amount of such a percentage may be specified alongside, for example, Net 30 days 1½ percent. This means that every additional month or fraction of it that the payment is overdue will cause the original amount to increase 1½ percent.

An offer shown in Fig. 15-7, when translated into everyday language, means that we are willing to design a die, to produce part no. AK-386-C Revision C, for the amount of US$3,500 and subsequently build such a tool for an additional US$40,000. We start working on the design only when we receive a written and signed order (ARO), accompanied by a check for the first third of the total sum, or $14,500. The design stage will be completed in 4 weeks from the day the order and the check arrive.

After such a span of time we must submit the finished design for approval. The client may express approval at leisure, but only after we receive written consent with the submitted design shall we proceed with actual manufacture of the die. It is of no consequence how long it will take us to build the die, but we must submit the first 12 pieces of parts produced by that tooling within 6 weeks from the time the client approved the design.

On submission of these preproduction samples, another check must be sent to us, or another bank transfer of funds for an additional $14,500 must take place, regardless of whether the samples are approved by the client or not.

Within a week of approval of preproduction samples, accompanied by a check for the third portion of total cost, 1000 production pieces must be sent to the client. Naturally, these production pieces will be charged separately, at $3.95 each, depending on the total number ordered.

Should preproduction samples not be approved, it will be our duty to alter the die in accordance with the client's demands. However, the claim will be valid only if the die would not perform in accordance with drawing number AK-386-C Revision C. If the request for alteration demands the die to produce parts per the next Rev. D or any other previous or subsequent revision, the client must pay for such alteration separately.

The lead time on the first 1000 pieces is shortened on purpose. At the time the die is built and tested, the very first pieces must be submitted for inspection within our own plant. On the basis of the quality control outcome we actually already know whether the die performs properly or not. While running the 12 preproduction samples for the client, it usually pays to include the first 1000 or so, to eliminate the extra setup required otherwise.

The clause "All prices are quoted in U.S. dollars" is important when sending or receiving a quote from a foreign country. Most of us assume the dollar as a currency term to be an American unit only, but this is wrong. There are Canadian dollars, Hong Kong dollars, Singapore dollars, New Zealand and Australian dollars, East Caribbean dollars, Slovenian tolar, Liberian dollar, Solomon Island dollars, and who knows what additional dollars financiers may come up with. The main problem with these variations of currency is that each has a different value and is affected by that currency fluctuation. It is therefore best to specify the amount to be paid in U.S. funds.

Overseas manufacturers do not like this trend at all, because if their currency changes value in comparison with the U.S. dollar, they may actually suffer a loss on the business deal. Evaluation of this problem and its solution exceeds the scope of a die design handbook, and it will not be discussed further.

Naturally, when estimating the delivery of parts, the component that takes the longest time to obtain is used as a guideline. What good is it to promise delivery of 1000 compression springs within 2 weeks if it takes 4 weeks to get the wire to make them and we have none in stock?

### 15-5-2. Packaging and shipping

Products may be perfect when produced, but if the packaging was not designed properly, they may be damaged when they get to the client. We all know that we should not package a set of crystal in a paper box. But few can imagine what happens to sheet-metal parts that are thrown haphazardly into a container and sent out by "Your-Good-Neighbor" forwarder.

When a multitude of relatively cheap parts are packaged, the cost of organizing them in layers, placing protective sheets or barriers in between, and placing each package, not heavier than 25 lb, in a wooden crate, sealed for shipping would be outrageously high in comparison with the cost of parts. Therefore, sheet-metal parts are usually sent out in barrels or in other types of containers that are often returned to the supplier.

How many pieces may be placed in a single container without damaging the bottom layers? This assessment is usually based on the approximate weight the contents of the container will exert on its bottom. Therefore, the height of the container, as well as the method of stacking containers on the truck, is crucial.

Except for flat round shapes, sheet-metal parts tend to get scratched when packaged in drums. Sometimes they must be separated in layers with protective barriers in between. The packaging containers may also be smaller. But all these alterations are very costly, and if that is not included in the original estimate, guess who will foot the bill?

For obvious reasons, it is preferable to specify the problem of packaging right at the beginning, write the terms into the estimate or contract, and include the cost of any extras. A client who demands that each layer of flat, 1-in$^2$ parts of 0.062-in thickness be separated by a neoprene sheeting should have it—but at a price.

When shipping overseas, the rules are somewhat different, and appropriate packaging is much more costly and demanding. Restrictions are placed on what must be written on the shipping container and how. Relative heights of various lettering are important, and the (last) country of origin must be specified according to the customs-issued rules and regulations. Often, a double packaging is used, the first being the original container and the second a shipping crate.

Additional problems may arise when selecting an appropriate shipper, in the absence of in-house trucking. Naturally, the cost of shipping is of essence here, but that's only a small portion of the problem. Loading and unloading methods may be detrimental to the whole order. Shaking of the truck on the road may force parts to vibrate and scratch their surface. When accidentally dropped, damage-prone items will certainly become damaged. And what if the package is loaded wrong side up? Yes, we know that the safety of the merchandise's passage is often the buyer's responsibility. But will buyers continue to order from us if the parts always arrive damaged?

A standard drop test for shipping purposes, used by major shipping contractors, is conducted from the height of 3 feet, and tested parts, in their original packaging, should land on at least three sides, two edges and one corner. This sentence should tell us all about the dangers of shipping.

### 15-5-3. Extra costs and regulations

Extra costs may be incurred everywhere if something is omitted or forgotten. Shipping companies charge a fee for storage if the merchandise is not picked up on time. A vendor charges a fee if the approval of samples is not on time and the setup machine has to stand waiting. There will be a fee for changing money from one currency to another, a charge for a payment by a letter of credit, a charge for late-received payment, and so on. Shipping brokers charge a fee (usually a flat fee, for example, $100 per shipment of any size) to issue the accompanying papers. There may be fees for loading and unloading of the merchandise when sending it out or receiving it. There are fees and payments and payments and fees. They must all be added up somewhere, and it better be within the cost estimate.

Customs fees are of additional concern to those who may think of importing or exporting. Customs fees may run from zero percent of the merchandise's value up to 50 percent. Perhaps some items may run still higher, who knows; these tariffs tend to change quickly and without any warning.

The range of these fees is so wide and so diverse that it really pays to get in touch with the Bureau of Customs and ask specifically about each exported or imported item, specifying its function and usage, because even the product's usage, the advancement of assembly, the finalization of finishing, and other pertinent details matter when it comes to customs fees.

However, sometimes these fees can be altered in our favor, and it is perfectly legal. For example, let's say the tariff for cultured pearls is 2.5 percent and that for cut, nonset stones starts at zero, up to 5 percent. Now, if we import these two types of merchandise and manufacture gilded sheet-metal back scratchers, set with real pearls and precious stones, we will be getting quite a deal on such a business proposal, especially when considering that duties on completed jewelry items run between 12 and 27.5 percent.

# Appendix

## A. Die-Building Cost Estimate Form for Standard Openings and Their Circumferences

See page A-2. This form shows the hole diameter and linear dimension's circumference for standard openings.

## B. Die-Building Cost Estimate Form for Special-Sized Holes

See page A-3. This form is blank, allowing the user to add the hole diameter and circumferences of special-sized openings to fit a particular job.

## Appendix A Die-Building Cost Estimate Form, Standard Openings and Their Circumferences

| Hole diameter, in | Linear dimensions circumference | To build punch and die and guide, h | Punch plate, die block, backup plates, h for a number of tools below | | | | | |
|---|---|---|---|---|---|---|---|---|
| | | | 1st | 2nd | 3rd | 4th | 5th | 6th |
| 0.125 | 0.393 | | | | | | | |
| 0.25 | 0.785 | | | | | | | |
| 0.375 | 1.178 | | | | | | | |
| 0.5 | 1.571 | | | | | | | |
| 0.625 | 1.963 | | | | | | | |
| 0.75 | 2.356 | | | | | | | |
| 0.875 | 2.75 | | | | | | | |
| 1 | 3.142 | | | | | | | |
| 1.125 | 3.534 | | | | | | | |
| 1.25 | 3.927 | | | | | | | |
| 1.375 | 4.32 | | | | | | | |
| 1.5 | 4.712 | | | | | | | |
| 1.625 | 5.105 | | | | | | | |
| 1.75 | 5.498 | | | | | | | |
| 1.875 | 5.89 | | | | | | | |
| 2 | 6.283 | | | | | | | |

## Appendix A Die-Building Cost Estimate Form, Standard Openings and Their Circumferences

| Hole diameter, in | Linear dimensions circumference | To build punch and die and guide, h | Punch plate, die block, backup plates, h for a number of tools below | | | | | |
|---|---|---|---|---|---|---|---|---|
| | | | 1st | 2nd | 3rd | 4th | 5th | 6th |
| | | | | | | | | |
| | | | | | | | | |
| | | | | | | | | |
| | | | | | | | | |
| | | | | | | | | |
| | | | | | | | | |
| | | | | | | | | |
| | | | | | | | | |
| | | | | | | | | |
| | | | | | | | | |
| | | | | | | | | |
| | | | | | | | | |
| | | | | | | | | |
| | | | | | | | | |
| | | | | | | | | |
| | | | | | | | | |

# Bibliography

American Society for Metals, *Metals Handbook,* 9th ed., 1978.

Antonidis, J. E., *Condensed Practical Aids for the Experienced Die Engineer, Die Designer and Die Maker,* "Die Techniques" Publisher, Medinah, Ill., 1972.

Baldwin, Edward N., and Benjamin W. Niebel, *Designing for Production,* Richard D. Irwin, Inc., Homewood, Ill., 1957.

Beach, Sidney C., and Donald P. Lawley, "Nickel Plating," *Products Finishing Directory,* page 120, 1993.

Beneš, Milan, and Bohumil Maroš, *Křivky přetvárných odporů ocelí, 4. díl,* Czech language (*Graphs of Resistance to Forming of Steels*), Poradenská příručka 33/4. Technicko-ekonomický výzkumný ústav hutního průmyslu, Prague, 1986.

Beránek, Jiří, and Miroslav Přibyl, *Tvorba cen montážních prací na základě časového fondu zakázky,* Czech language (*Cost Estimates of Assembly Works*), SNTL, Prague, B.n., 1992.

Betzalel, Avitzur, *Metal Forming: Processes and Analysis,* McGraw-Hill, New York, 1968.

Blazynski, T. Z., *Metal Forming: Tool Profiles and Flow,* Wiley, New York, 1976.

Brierley, Robert G., and H. J. Siekmann, *Machining Principles Cost Control,* McGraw-Hill, New York, 1964.

Brumer, Herbert, "Chromating," *Product Finishing Directory,* page 206, 1993.

Castlfberry, Guy A., "Designing for Contact Stresses," *Machine Design,* page 75, August 8, 1985.

Černoch, *Svatopluk Strojně Technická, příručka, 2-svazky,* Czech language Engineering Handbook, 2 vols.), SNTL, Prague, 1977.

Chase, H., *Handbook of Designing for Quantity Production,* 2d ed., McGraw-Hill, New York, 1950.

Číhal, Vladimír, *Intergranular Corrosion of Steels and Alloys,* Elsevier, Amsterdam, 1984.

Colvin, Fred H., and Frank A. Stanley, *American Machinist's Handbook, Machinery's Handbook,* 20th ed., McGraw-Hill, New York, 1940.

Cookson, William, *Advanced Methods for Sheet Metal Work,* 6th ed., Oxford Technical Press, 1975.

Dallas, Daniel B., *Progressive Dies, Design and Manufacturing,* McGraw-Hill, New York, 1962.

Donaldson, Cyril, and George H. LeCain, *Tool Design,* Harper, New York, 1943.

Dorazil, Eduard, *Strojírenské materiály a povrchové úpravy,* Czech language (Metal-Working Materials and Surface Treatment), Vysoké učení technické, Fakulta strojní, Brno, 1985.

Dowd, Albert A., and Frank W. Curtis, *Tool Engineering: Punches, Dies and Gages,* McGraw-Hill, New York, 1925.

Elfmark, Jiří, et al., *Tváření kovů; technický průvodce,* Czech language (Press-Working, A Technical Guide), SNTL, Prague, 1992.

Fiala, Jaromír, Adolf Bebr, and Zdeněk Matoška, *Strojnické tabulky,* Czech language (Machinery's Standards), SNTL, Prague, 1990.

Freeman, D. B., *Phosphating and Metal Pre-treatment,* Industrial Press, New York, 1986.

Gabe, D. R., *Principles of Metal Surface Treatment and Protection,* Pergamon Press, Oxford, 1972.

Gopinathan, V., *Plasticity Theory and Its Application in Metal Forming,* Wiley, New York, 1982.

Hanišová, Hana, *Nové materiály a technologie jejich výroby a zpracování, 3,* Czech language (New Materials and Technology of Their Production), SIVO, Prague, 1989.

Hecker, J. G., Jr., "Anodizing Aluminum," *Product Finishing Directory,* page 266, 1993.

Hinman, C. W., *Die Engineering Layouts and Formulas,* McGraw-Hill, 7th impression, New York, 1943.

Hlavenka, Bohumil, *Racionalizace technologických procesů,* Czech language (Rationalizing of Technologic Processes), Vysoké učení technické, Brno, 1987.

Hrivňák, Andrej, Michal Podolský, and Vuko Domazetovič, *Teoria tvárnenia a nástroje,* Slovak language (Theory of Metal Forming and Tooling), ALFA, Bratislava, 1992.

Hromková, Ludmila, *Automatizované systémy řízení strojírenských podniků,* Czech language (Automated Systems of Metal Production), SNTL, Prague, 1988.

International Deep Drawing Research Group, *Sheet Metal Forming and Energy Conservation,* American Society for Metals, Ohio, 1976.

Jeffries, William R., *Tool Design,* Prentice-Hall, New York, 1955.

Jones, Franklin D., *Die Design and Diemaking Practice,* 13th ed., The Industrial Press, New York, 1951.

Jones, Franklin, D., *Jig and Fixture Design,* Machinery, New York, 1937.

Kamelander, Ivan, *Tvářecí stroje I,* Czech language (Press Machines), Vysoké učení technické v Brně, 1989.

Kelly, Roger, "Principles of Fastener Pretreatment," *Metal Finishing,* page 15, April 1988.

Kempfer, Lisa, "The Many Faces of Boron Nitride," *Manufacturing Engineering,* page 41, November 1990.

Kempfer, Lisa, "Diamond: A Gem of a Coating," *Manufacturing Engineering,* page 26, May 1990.

Kempfer, Lisa, "Glass Goes Ceramic," *Manufacturing Engineering,* page 21, September 1990.

Kempster, M. H. A., *Principles of Jig and Tool Design,* Hart Publishing, New York, 1968.

Kent, William, *Mechanical Engineer's Handbook,* McGraw-Hill, New York, 1938.

Klofáč, Jozef, *Tváření neželezných kovů,* Czech language (Press Working of Nonferrous Metals), SNTL, Prague, 1982.

Knight, David, "Three Ways to Eliminate Springback," *Modern Machine Shop,* page 82, March 1992.

Kolář, Jiří, and Zdeněk Stoud, *Výběr norem pro dílny kovoprůmyslu,* Czech language (Excerpts from Industrial Standards), Práce, 1985.

Korten, Tristram, "How to Get the Most Out of Metals," *Design News,* page 110, June 25, 1990.

Kotouč, Jiří, *Teorie tváření,* Czech language (Theory of Metal Forming), České vysoké učení technické, Prague, 1991.

Kotouč, Jiří, Jan Šanovec, Jan Čermák, and Luděk Mádle, *Tvářecí nástroje,* Czech language (Press Tools), Vydavatelství ČVUT, 1993.

Kováč, Andrej, and Bedřich Rudolf, *Tvárniace stroje,* Slovak language (Press Machines), Alfa, vydavatelstvo technickej a ekonomickej literatúry, Bratislava, 1989.

Lange, Kurt, *Handbook of Metal Forming,* McGraw-Hill, New York, 1985.

Lascoe, Orville D., *Handbook of Fabricating Processes,* ASM International, Metals Park, Ohio, 1988.

Laughner, Vallory H., and Augustus D. Hargan, *Handbook of Fastening and Joining of Metal Parts,* McGraw-Hill, New York, 1956.

Lewis, Clifford F., "Zirconia: The Tough Contender," *Manufacturing Engineering,* page 43, May 1989.

Macháček, Zdeněk, Karel Novotný, *Speciální technologie 1. Plošné a objemové tváření,* Czech language (Surficial and Volumnar Presswork), Vysoké učení technické, Brno, 1986.

Machek, Václav, *Tenké ocelové pásy a plechy válcované za studena,* Czech language (Thin Sheet Metal Strips, Cold-Rolled), SNTL-Nakladatelství technické literatury, 1987.

*Mathematics for Sheet Metal Fabrication,* Delmar Publishers, Albany, N.Y., 1970.

Meyer, Leo A., *Sheet Metal Shop Practice,* American Technical Publishers, Alsip, Ill., 1975.

Molloy, E., *Press Tools,* Chemical Publishing Co., New York, 1943.

Muraski, Stephanie J., "Understanding the Effect of Bolt Finishes," *Machine Design,* page 95, July 20, 1989.

Nordhoff, W. A., *Machine Shop Estimating,* 2d ed., McGraw-Hill, New York, 1960.

Novák, Pavel, *Anodická protikorozní ochrana,* Czech language (Anodic Anti-Corrosion Prevention), SNTL, 1987.

*Nové poznatky z tváření kovů,* Czech language (New Developments in Press Working, a College Text), Vysoké učení technické, Brno, 1987.

*Nové poznatky z oblasti tváření,* Czech language (New Developments in Metal Forming), Dům techniky ČSVTS, Ústí nad Labem, 1988.

*Ochrana strojírenských výrobků proti vlivům prostředí.* Sborník přednášek z konference, Czech language (Prevention of Metals against Influences of Their Environment), Prague, 1985.

Paquin, J. R., and R. E. Crowley, *Die Design Fundamentals,* The Industrial Press, New York, 1987.

Parker, Daniel W., "Plasma Coats With Many Colors," *Materials Engineering,* page 21, November 1991.

Polák, Karol, and Alexi Höbemägi, *Výroba dutinových nástrojov tvárněním,* Slovak language (Manufacture of Tools by Forming), ALFA, Bratislava, 1988.

*Press Die Design and Construction,* Parts 1, 2, Machinery Publishing, London, 1963.

Raymond, Howard Monroe, *Modern Machine Shop Practice,* American Technical Society, Chicago, 1941.

Rumíšek, Pavel, *Automatizace výrobních procesů II (tváření),* Czech language (Automated Manufacturing Methods), Vysoké učení technické, Brno, 1990.

Sachs, George, *Principles and Methods of Sheet Metal Fabricating,* Reinhold, New York, 1951.

Šafařík, Miloslav, *Nástroje pro tváření kovů a plastů,* Czech language (Tooling for Presswork in Metal and Plastic), Vysoká škola strojní a textilní, Liberec, 1991.

Schmidt, Z.D., and Bohumil Dobrovolný, *Technická Příručka,* Nakladatelství Práce, Prague, 1956.

Shigley, Joseph E., and Charles R. Mischke, *Standard Handbook of Machine Design,* McGraw-Hill, New York, 1986.

Sidorin, Evžen, *Rozvoj pružných výrobních soustav,* Czech language (Development of Adjustable Manufacturing Groups), ÚVTEI, Prague, 1987.

Snyder, Donald L., "Copper Plating," *Product Finishing Directory,* page 132, 1993.

Stanley, Frank A., *Punches and Dies,* McGraw-Hill, New York, 1950.

Straka, Kamil, *Nové materiály a technologie jejich výroby a zpracování, 2,* Czech language (New Materials and Technology of Their Production) SIVO, Prague, 1988.

Švehla, Štefan, Oleg Abramov, and Ivan Chorbenko, *Využitie ultrazvuku v strojárstve a metalurgii,* Slovak language (Ultrasound in Metalworking and Metallurgy), ALFA, Bratislava, 1986.

Swinehart, Haldon J., *Cutting Tool Material Selection,* American Society of Tool and Manufacturing Engineers, Dearborn, Mich., 1968.

Tobias, S. A., *Advances in Machine Tool Design and Research,* Proceedings of the 5th International M.T.D.R. Conference, University of Birmingham, September 1964, Pergamon Press, Oxford, 1965.

"Tooling 'extras' Make Dies More Productive," *Precision Metal,* page 13, April 1986.

Town, H. C., *Technology of the Machine Shop,* Longmans, Green, London, 1951.

Turnbull, Virginia H., "Designing for Performance," *Plastic Design Forum,* page 64, July/August 1985.

Vasilash, Gary S., "A New Approach to Bending," *Manufacturing Engineering,* March 1982.

Vezzani, A. A., *Manual of Instructions for Die Design,* ASTME, Royalle Publishing Company, Detroit, Mich., 1964.

Walker, William Francis, *Guide to Press Tool Design,* Newnes-Butterworths, London, 1970.

Wendes, Herbert, *Sheet Metal Estimating Handbook,* Van Nostrand, New York, 1983.

Williams, D. J., *Selection of Tool Materials* (from: *Design of Tools for Deformation Processes*), Elsevier Applied Science Publishers, London & New York, 1986.

Williams, Gordon T., *What Steel Shall I Use?* American Society for Metals, Cleveland, Ohio, 1941.

Wilson, Frank W., *Die Design Handbook,* McGraw-Hill, New York, 1965.

Wilson, Frank W., *Handbook of Fixture Design,* McGraw-Hill, New York, 1962.

Yoshida, Makoto, and Akira Kokaji, "Firing Up the Future with Ceramic Engine Parts," *Machine Design,* page 58, October 26, 1989.

Zaki, Nabil, "Zinc Alloy Plating," *Product Finishing Directory,* page 199, 1993.

# Index

## ABOUT THE AUTHOR

Ivana Suchy is currently working on a privately subsidized research and development project. She previously worked as a cost estimator for Kreisler Industrial Corporation, where she was responsible for the cost analysis and estimate of subassemblies used in aircraft engines and turbines. She also worked as a project engineer for Sunbeam Precision Instrument and as an associate engineer for the Singer Sewing Machine Comapny.